Urban Stormwater Management

Urban Stormwater Management, Third Edition, covers the design, installation, and maintenance of stormwater management systems, addresses the impact of urban development on runoff and infiltration, and focuses on stormwater management relative to flooding and water pollution. Recognizing that urbanization increases and accelerates runoff, reduces infiltration, and deteriorates water quality, the author proposes stormwater runoff as a resource that can be conserved for reuse. He suggests the reuse of stormwater runoff in general, and rainwater from roofs in particular, as a cost-effective means of achieving long-term sustainability. In addition, the book explores green infrastructure as the future of stormwater management and introduces techniques that can help reduce the thermal impacts of stormwater management practices.

- Provides an updated presentation of urbanization's impact on stormwater.
- Presents further analysis of the universal runoff model and the application of this model to non-uniform rainfalls.
- Offers a more detailed presentation of stormwater management systems, especially bio-filtration basins.
- Includes a comparative analysis of the effectiveness and costs of best management practices (BMPs).
- Contains an in-depth discussion of the means of collecting stormwater, such as roof rain for outdoor and certain indoor uses.

Urban Stormwater Management

3rd Edition

Hormoz Pazwash

CRC Press
Taylor & Francis Group
Boca Raton London New York

CRC Press is an imprint of the
Taylor & Francis Group, an **informa** business

Designed cover image: Shutterstock

Third edition published 2026
by CRC Press
2385 NW Executive Center Drive, Suite 320, Boca Raton FL 33431

and by CRC Press
4 Park Square, Milton Park, Abingdon, Oxon, OX14 4RN

CRC Press is an imprint of Taylor & Francis Group, LLC

First edition published by CRC Press 2011

Second edition published by CRC Press 2016

ISBN: 978-1-032-60672-9 (hbk)
ISBN: 978-1-032-60674-3 (pbk)
ISBN: 978-1-003-46012-1 (ebk)

DOI: 10.1201/9781003460121

Typeset in Times LT Std
by Apex CoVantage, LLC

Contents

Preface

Urbanization has had a drastic impact on the natural process of stormwater runoff. It has increased both the peak and the volume of runoff, has reduced infiltration, and has caused water pollution. Traditionally, runoff used to be conveyed by storm drains directly into streams and lakes. To prevent increased flooding, towns and states adopted regulations that initially mandated maintaining the peak rates of runoff from urban developments. However, since this did not fully address the flooding problem, some regulatory agencies later required certain reductions in the peak rates of runoff.

To address the concerns on substantial pollution from nonpoint sources, the US Environmental Protection Agency (EPA) promulgated the National Pollutant Discharge Elimination System (NPDES), Phase I in 1990, and Phase II, in December 1999 permitting program under the Water Quality Act of 1987. This act applies to all municipal separate storm and sewer systems (MS4s) and construction sites disturbing over 1 acre of land. Many states have since adopted regulations to extend the NPDES to smaller sites. New Jersey, among others, has also adopted regulations that include groundwater recharge. To address the evolving regulations, stormwater management practices have been continually changing during the past 40 years, and they continue to change.

Aside from teaching and practicing in various domains of hydraulic engineering since receiving my PhD in February 1970 under the supervision of the late Dr. Ven Te Chow, Dr. Pazwash has been extensively involved in the fields of drainage design, flood control and stormwater management from 1985 till his retirement in 2021. His involvements have included design of drainage and stormwater management systems for numerous projects and teaching pertinent courses, including Urban Stormwater Management, Drainage Design, Watershed Modeling, and Advanced Hydraulics at Northeastern University in Boston, MA, and the Stevens Institute of Technology in Hoboken, NJ.

In particular, his experience has included performing analysis and design of numerous drainage and stormwater management projects and preparation of calculation reports and plans for same, and review of several hundred drainage and stormwater management plans prepared by consulting engineers and offering corrective recommendations. Based on his expertise, a significant number of civil engineers need more knowledge to accurately perform runoff calculations and properly design stormwater management systems. Thus, there has been and still is a great need for a practical, concise, yet thorough book that will guide practicing engineers and municipal planners in the design of Stormwater management elements. This third edition of Dr Pazwash's book on the subject is intended to meet this objective and serve as a "cookbook" on urban stormwater management, which is one of the most challenging and dynamic fields of engineering.

The book covers all the subject matter that are needed for proper and cost-effective design of drainage and stormwater management systems. It includes numerous examples of hydrologic and hydraulic analyses of relevance to this field. The book also contains ample case studies that exemplify the methods and procedures for the design of drainage networks and structural and non-structural stormwater management systems, such as extended detention basins, infiltration basins, underground retention-infiltration basins, rain gardens, disconnection of pavements, green roofs, pervious pavements, and vegetative buffers.

Since a vast majority of practicing engineers in the United States are yet to become familiar with the System International (SI) system of units, English units are used throughout the book. However, all equations and a large number of examples and case studies are presented in SI as well, so the book can be used internationally.

The book contains 10 chapters. Chapter 1 provides an introduction to the impacts of urban developments on the quality and quantity of stormwater runoff. Chapter 2 presents an overview of pipe and open channel flow equations, supplemented with charts and tables to simplify hydraulic design of runoff conveyance systems. Chapter 3 covers the elements of the rainfall-runoff process and includes several examples that guide practitioners to perform runoff calculations properly. Included in this chapter is Dr. Pazwash's own method of runoff calculations which is a physically based model and which is termed Universal Method. Chapter 4 covers the design of inlets, storm drains, culverts, vegetative swales, and erosion control systems. The examples in this chapter show that roadway inlets have less capacity in intercepting runoff than many engineers perceive. This chapter also includes analysis for design of inlets on roadways at 0% longitudinal slope which has been developed by this author.

Chapter 5 contains an overview of the EPA stormwater rules and excerpts from stormwater management regulations manuals for New Jersey, Maryland and New York which are among the most stringent regulations in the nation. This chapter also discusses the shortcomings of the regulations and suggestions for improvements. Chapter 6 provides a description of various types of manufactured stormwater treatment devices that are increasingly employed to address the applicable water quality requirements either solely or in combinations with structural stormwater systems,

Chapter 7 covers design of various types of s stormwater management structures such as detention/retention, infiltration basins and water treatment structures. It also includes examples and case studies to guide the practitioner for proper design of structural and non-stormwater management systems. Chapter 8 provides a description of various types of source-reduction measures such as porous pavements, rain gardens, green roofs, blue roofs, dry wells and filter strips, among others, some of which are generally more effective than structural systems. These measures, also referred to as green infrastructure, have become increasingly popular in recent decades and are forming the future trend in stormwater management measures. This chapter also provides the author's own procedure for proper analysis of retention capacity and peak discharge of green roofs. It also includes comparative cost analysis of various green infrastructure practices. This chapter also discusses stormwater utility fees which have been introduced by communities as a means of providing fund to cover the cost of maintenance. While these fees have risen during recent decades, the future will tell whether or not they will fully cover the costs of compliance with applicable regulations

Chapter 9 provides an overview of the installation procedures of drainage and stormwater management facilities. It also includes measures for maintenance of stormwater management elements which in spite of being crucial for their proper functioning are at times neglected. Chapter 10 discusses measures to reduce indoor and more importantly outdoor water use. It displays that the stormwater runoff is a vast resource that can be and should be collected and reused. This view of stormwater runoff as a resource is contrary to the traditional view that the runoff is a waste to be disposed of in a regulated manner. The chapter further discusses collection of rainwater from roofs and suggests the proper sizing of rain tanks and rain barrels to harvest it and indicates that though the rainwater harvesting is now becoming popular, the rain barrels used by homeowners are often much smaller than his suggested capacities.

This book is prepared based on extensive experience of the author and is targeted at a large audience. The subject matter is of interest to all professionals involved in the design of drainage and stormwater management systems for urban developments and roadway projects, as well as professionals engaged in urban planning including municipal engineers and planners. In addition, it can be used as a textbook in an upper undergraduate and graduate-level course on stormwater management. Because of numerous examples and case studies and the inclusion of approximately 220 problems/questions together with a solutions manual, it would be an ideal selection for classroom teaching. Due to an ever-increasing demand for stormwater management practitioners, more and more universities around the country are offering courses on this subject.

Because of his busy schedule, it took the author many evenings, weekends, and holidays to prepare the 1st edition of Urban Stormwater Management, published in 2011. He also spent months to update the book, cover additional material, and add more worked examples as well as nearly three times as many problems/questions in the 2nd edition in 2016. In this third addition he has further updated the subject matter and expanded on green stormwater systems. However; he has removed some of the figures and many of the appendices which can be readily downloaded online. This is intended to not only produce a concise book but also to reduce the cost of its production.

Lasty a note on the term "stormwater." Early practices and literature used "storm water." The 1971 EPA's SWMM (Stormwater Management Model) and SWPPP (Stormwater Pollution Prevention Plan) serve as examples SW stood for abbreviation of storm water. In fact, both Oxford English Dictionary and Collins Dictionary show "storm water" as two words. However, in current publications including those by EPA stormwater read as a single word. Therefore, to conform with the current trend, the phrase *Urban Stormwater Management* is chosen as the title of this edition of the book.

Author

Hormoz Pazwash, PhD, earned his BS, CE with highest honor among the entire graduating class of 1963 from Tehran University. He continued his graduate studies under the supervision of the late Dr. Ven Te Chow at the University of Illinois in Urbana-Champaign, earning an MS and a PhD in civil engineering. In 1970 he joined the Faculty of Engineering at Tehran University and, in the next 7 years, held the positions of assistant professor, associate professor, and chairman of the Department of Civil Engineering. His other academic appointments have included visiting professorships at the University of California, Berkeley; the University of Akron in Ohio; and an associate professorship at Northeastern University in Boston. He has also served as an adjunct professor at Stevens Institute of Technology in Hoboken, New Jersey. Dr. Pazwash has received various academic awards, including a fellowship at Tehran University and a Fulbright Scholarship at the University of California, Berkeley.

He is listed in the 1992–1993 premier edition of Marcus Who's Who in Science and Engineering and the International Who's Who of Professionals. From mid1985 when he ventured from full time teaching at Northeastern University to engineering practice until his retirement in 2021, Dr. Pazwash has been engaged in engineering practice. In those 36 years, he has been involved with the design and consultation of drainage systems, flood control and above all stormwater management projects. He has also taught a number of senior- and graduate-level courses in the field of water resources and stormwater management at Stevens Institute of Technology.

Dr. Pazwash is the author of over 50 papers, seven books including 1st and 2nd editions of Urban Stormwater Management and a chapter of the Encyclopedia of Environmental Management, published September 2014. His professional experience covers a broad range of disciplines in water resources and hydrologic/hydraulic engineering. He has been involved in projects that include evaluation of regional water resources; the design of pipelines, channels, and culverts; hydrologic and hydraulic analysis of rivers and streams; flood control projects; reservoir and dam safety studies; and in particular the design of drainage and stormwater management systems. Dr. Pazwash is well respected in the field of engineering due to his professional accomplishments.

Dr. Pazwash had held Professional Engineering licenses in New Jersey and New York. He is a life member and Fellow of the American Society of Civil Engineers (ASCE) and a Diplomat of the American Academy of Water Resources Engineers (AAWRE), D.WRE. He is also a member of the American Water Resources Association (AWRA).

Abbreviations

AASHTO American Association of State Highway Officials

AAWRE American Academy of Water Resources Engineers

ac acre = 43,560 square feet

ACIS Applied Climate Information System

ANSI American National Standards Institute

ASCE American Society of Civil Engineers

ASTM American Society for Testing and Materials

AWRA American Water Resources Association

AWWA American Water Works Association

BDF basin development factor

BMP best management practice

°C degrees Celsius, temperature in metric units

CAFRA Coastal Area Facility Review Act

CCF A unit of tap water use, equal to 100 cubic feet

Cd candle: nit of luminous

Cf cubic feet

Cfs cubic feet per second

CGP the Environmental Protection Agency's Construction General Permit

CLSM controlled low-strength material

cm centimeter

COD chemical oxygen demand

csm cubic feet per second per square mile (in TR-55 method)

CSO combined sewer overflow

CU customary units (English units)

CWA Clean Water Act

DPW Department of Public Works

D.WRE Diplomate, Water Resources Engineer

ECB erosion control blanket

ECOS Environmental Council of States

EGL energy grade line

EISA Energy Independence and Security Act

EPA Environmental Protection Agency

ESD environmental site design

°F degrees Fahrenheit, temperature in English units

FEMA Federal Emergency Management Agency

FHWA Federal Highway Administration

ft foot, feet

GPCD gallons per capita per day

gpd gallons per day

gpf gallons per flush

GPM gallons per minute

GSR-32 New Jersey Geological Survey Report-32 (specific to New Jersey)

h hour

ha hectares

Hg mercury

HGL hydraulic grade line

in inch, inches

kg kilogram
km kilometer
kN Kilonewton
L liter
LEED Leadership in Energy and Environmental Design
LOD limit of disturbance
LPCD liters per capita per day
Lpf Liters per flush
LPM Liters per minute
m meter
mgd million gallons per day
mi mile = 5280 ft
μm micron, micrometer
mm millimeter
mol mole
MOU memorandum of understanding
MS4 municipal separate storm sewer system
MSGP Multi-Sector General Permit
MTD manufactured treatment device
N newton (unit of force in SI)
NAHB National Association of Home Builders
NAS National Academy of Sciences
NEH National Engineering Handbook
NGVD National Geodetic Vertical Datum
NJAC New Jersey Administrative Code
NJCAT New Jersey Corporation for Advanced Technology
NJDEP New Jersey Department of Environmental Protection
NJPDES New Jersey Pollutant Discharge Elimination System
NOAA National Oceanic and Atmospheric Administration
NPDES National Pollutant Discharge Elimination System
NPS nonpoint source
NRC National Research Council
NRCS Natural Resources Conservation Service
NSF National Science Foundation
NSPS nonstructural strategies point system
NURP Nationwide Urban Runoff Program
OSHA Occupational Safety and Health Administration
Pa Pascal (N/m^2)
RCP reinforced concrete pipe
RECP rolled erosion control products
ROW right of way
s second
SCS Soil Conservation Service (now NRCS)
SI System International; metric units
SPDES State of New York PDES
SRI solar reflectance index
SWMM EPA's Stormwater Management Model
SWPPP Stormwater Pollution Prevention Plan
TARP technology acceptance and reciprocity partnership
TCLP toxicity characteristic leach procedure
TN total nitrogen

TP	total phosphorous
TRM	turf reinforcement mat
TSS	total suspended solids
USCS	Unified Soil Classification System
USDA	US Department of Agriculture
USDOT	US Department of Transportation
USGBC	US Green Building Council
USGS	US Geological Survey
USSR	Union of Soviet Socialistic Republics
WEF	Water Environment Federation
WQE	water quality event
WS	water surface
ZLD	zero liquid discharge

Notations

α	alpha: angle
γ	gamma: specific gravity
θ	theta: angle, moisture content
μ	mu: dynamic viscosity
ν	nu: kinematic viscosity
π	pi: 3.1415
ρ	rho: fluid density
σ	sigma: sum
ψ	psi: soil suction
A	area; cross section area of flow in pipe or open channel: Amperes
C	constant; runoff coefficient
CN	Soil curve number in SCS/NRCS runoff calculations
d	depth of flow in channels
D	pipe diameter
g	acceleration of gravity: 9.81 m/s^2, 32.2 ft/s^2
I	rainfall intensity (in./h, mm/h)
Ia	initial abstractions
K	Kelvin (thermodynamic temperature); also a pipe/channel flow factor
n	Manning's roughness coefficient
P	wetted perimeter
q	discharge per unit width
Q	discharge, peak discharge (cfs, m^3/s)
R	runoff depth (in., cm, mm)
R, r	hydraulic radius
Sc	critical slope
Sg	specific gravity
t	time
V	velocity
Vc	critical velocity
y	depth of flow
yc	critical depth

1 Urbanization's Impacts on Runoff

Urban development alters the natural hydrologic process. It increases and accelerates runoff, reduces infiltration, and deteriorates water quality. Neglecting these impacts in the past has aggravated flooding, caused stream pollution, and lowered water tables. These impacts are discussed in this chapter. In addition, urbanization causes thermal pollution from dark impervious surfaces, such as roofs and streets. This latter impact, which is linked to global warming, is, however, beyond the scope of this book.

1.1 IMPACTS ON STORMWATER AMOUNT

Rain falling on virgin lands such as forests, grasslands, and wetlands is partly intercepted by vegetation, partly retained by surface depressions and puddles, and partly infiltrates into the ground; only a small fraction of it flows overland. Urbanization disturbs land and replaces natural vegetation with impervious surfaces such as roads, driveways, parking lots, and building roofs, and it also compacts soils. Thus, it reduces interception, infiltration, and surface retention.

Urban development begins with clearing woods and grading the ground. The trees that intercept rainfall are removed, and natural depressions and puddles that temporarily impound rainfall are flattened to uniform grade. The humus layer of forest and the organic matter that act like a sponge and absorb rainfall are scraped off, and instead of allowing infiltration, the soil is severely compacted. The land can no longer prevent rainfall from being rapidly turned to runoff, and the result is increased flooding and a significant reduction in groundwater recharge. Because of soil compaction by heavy construction equipment, the impacts are more profound in areas where the soil has high permeability or has high organic matter. Healthy soil, containing 5% to 6% organic matter, can hold water up to 40% by its volume (Hudson, 1994).

In addition, drainage systems comprising inlets, pipes, and drainage channels concentrate overland flow into concentrated runoff. The result is reduced rainfall abstraction and a shorter time of concentration (the time it takes for the runoff from the entire watershed area to reach its outlet point). Unnaturally high stormwater discharges overwhelm the capacity of streams and rivers and cause flooding. According to the EPA, approximately 24% of rivers and streams, nearly 458,000 km (271,000 miles), suffer from decreased vegetation cover, which leads to more erosion and water pollution (EPA, 2013, 2013a).

Figure 1.1 visually depicts these impacts of urban development on runoff and infiltration, and Figure 1.2 does so schematically in runoff hydrograph, a graph of runoff discharge with time. These figures indicate that urbanization increases the peak and volume of runoff, shortens the time of concentration, and reduces the infiltration. By lowering the water table and reducing base flow, the hydrograph of a developed watershed has a sharper peak and lower base flow during dry weather. This impact is more severe in areas with soils of high permeability. In deep sandy soils, rainfalls, however intense, infiltrate the ground, and there is little runoff.

Before the eighteenth century, over 95% of the world's population lived in rural areas and farmed in the absence of any machinery. In the United States, the population was approximately four million people in 1800 (less than 1.2% of the 2024 population of approximately 340 million), and 95% of them lived in rural areas. The world's population was ninefold smaller than that

DOI: 10.1201/9781003460121-1

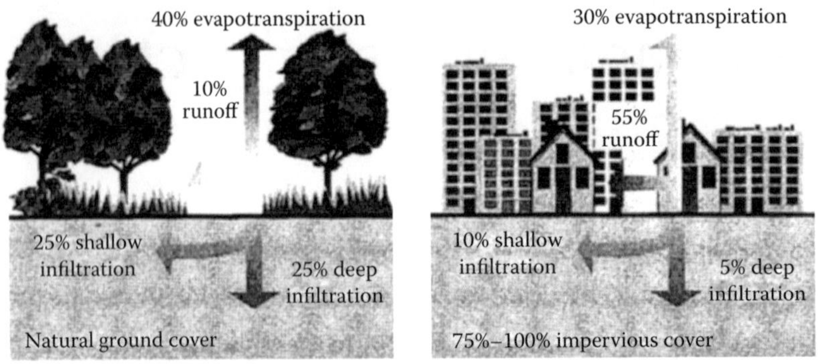

FIGURE 1.1 Visual impacts of urbanization on infiltration and runoff.

Sources: US Environmental Protection Agency, Protecting Water Quality from Urban Runoff, EPA 841-F-03-003, National Research Council, October 2008, Urban Stormwater Management in the United States, National Academies Press, Washington, DC)

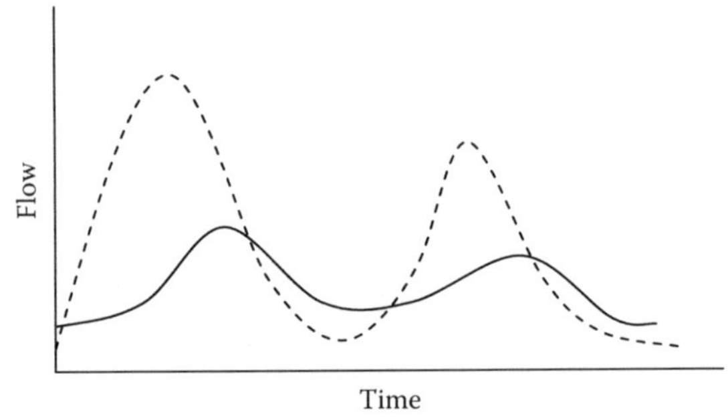

FIGURE 1.2 Effects of urbanization on runoff hydrograph

of today,[1] and there was far less per capita food and goods production and consumption as such, there were insignificant adverse impacts to the environment in general and stormwater runoff in particular. The Industrial Revolution, which developed in England between 1750 and 1850 and rapidly spread to other European countries, changed lifestyles and stimulated the growth of cities. Throughout Europe, the percentage of people who lived in cities rose from 17% in 1801 to 54% in 1891.

In the United States, urbanization was a long and gradual process. Massachusetts and Rhode Island were among the northeastern states that already had an urban majority in 1850, but most of the southern states were still rural until after World War I, which ended in 1918.

Traditional stormwater management practices considered runoff a waste to be removed and disposed of quickly. Thus, the runoff from urban and suburban developments and municipalities used to be collected in drainage systems and directly discharged into lakes and streams. The result, apart from importing impurities to receiving water bodies, was more flash floods, higher flood levels, and expanded flood-prone areas, which adversely impacted properties along and adjacent to streams and lakes.

[1] In 1999, the United Nations estimated that the world population had grown from 980 million in 1800 to 2.52 billion in 1950, reflecting an annual population growth rate of 0.6% during that period. The population grew at a rate of 1.59% from 1950 to 2024, when it reached 8.1 billion.

To offset these adverse impacts on runoff quantity, municipalities and states have adopted stormwater management regulations. Earlier regulations (and practices) were aimed at maintaining the peak rate of runoff for a given storm frequency, commonly 10 or 25 years; however, solving for a single storm frequency could not sufficiently attenuate runoff to maintain predevelopment peaks for other storm frequencies. The regulations were subsequently amended to cover multiple storm events, covering a large range in frequency, such as 1- or 2, 10 and 100-year storms. To address these regulations, detention basins and ponds were employed in practice. These practices, too, were found not to fully address the impact of a development. Because of having a larger volume than predevelopment conditions, the outflow hydrograph from a detention basin/pond was more prolonged than that of pre-development. As a result, the composite discharges from detention basins/ponds in a watershed exceeded the predevelopment runoff in that watershed. To compensate for these effects, as will be discussed in a later chapter, some states, New Jersey included, adopted regulations that require certain reductions in the peak rates of runoff.

1.2 IMPACTS ON WATER QUALITY

Stormwater has long been regarded as a major source of urban flooding; however, its role in degrading the water quality of streams, lakes, and rivers has been given attention only in the past fifty years. Still, many people do not believe that rain falling on pavements and carried by storm drainage systems becomes polluted enough to need treatment.

The porous natural terrain of varied landscapes like forests, grasslands, and wetlands absorbs rainfall and allows it to filter as it seeps into the ground. Urbanization replaces native soil cover with pavements, such as roofs, roads, and driveways, and eliminates both absorption and filtration. Most of the pure rain and snow melt mix with man-made or natural pollutants on the ground, flow into drainage systems, and are quickly transported into streams and lakes. The forceful power of this flow erodes the bed and banks of the receiving water bodies and creates more sediment pollution. It also results in destruction of interstitial spaces within substrate where invertebrates live. Sensitive species, such as stonefish and caddis flies, have begun to disappear.

It is to be noted that mining, septic systems, agriculture, and airborne sediment also contribute to stormwater pollution. Among these, agriculture is the leading source of contamination of streams and lakes in the United States. However, unlike golf courses and landscaping, agriculture is exempt from the Clean Water Act of 1972 and will remain a source of water pollution until it is properly regulated.

According to the US Environmental Protection Agency (EPA), stormwater runoff is the most common source of water pollution (EPA, 2000). The pollutants that are present in urban areas come from various diffuse or nonpoint sources. A nonpoint source pollutant is contrasted with a point source pollutant where discharge to a water body occurs at a single location, such as outfalls from a chemical factory or sewage treatment plant. Because of originating from many different sources and spreading out overland, nonpoint source (NPS) pollution is more difficult to control and treat than point source pollution, such as municipal waste. Also because of the nonlinear relationship between uses and pollutant loading, a conclusive cause–effect relationship cannot be established between the two.

What adds to this difficulty is the varied nature of stormwater runoff. The composition and magnitude of urban runoff is highly time dependent. Unlike municipal sewage, which is continuous and varies little, the flow of stormwater is intermittent; both the flow rate and the pollutant loads can vary by several orders of magnitude during a storm. Consequently, the traditional centralized water quality treatment practices, also referred to as end-of-pipe practices, are far less effective than sewage treatment plants. Most common pollutants present in urban stormwater originate from soil erosion from bare land, an acute problem during construction; road salt and other deicing substances, household products (fertilizers, pesticides, paints, thinners, solvents, cleaning agents, etc.), heavy metals and roof soil (leaks and illicit disposal), atmospheric and automobile (tire and pavement wear), and failing septic systems and illicit sewer connections.

TABLE 1.1
Median Event Mean Concentrations of Pollutants in Urban Land Uses

		Residential		Mixed		Commercial		Open/Nonurban	
Pollutant	**Units**	**Median**	**COV**	**Median**	**COV**	**Median**	**COV**	**Median**	**COV**
BOD	mg/L	10	0.41	7.8	0.52	9.3	0.31	-	-
COD	mg/L	73	0.55	65	0.58	57	0.39	40	0.78
TSS	mg/L	101	0.96	67	1.14	69	0.85	70	2.92
Total lead	µg/L	144	0.75	114	1.35	104	0.68	30	1.52
Total copper	µg/L	33	0.99	27	1.32	29	0.81	-	-
Total zinc	µg/L	135	0.84	154	0.78	226	1.07	195	0.66
Total Kjeldahl nitrogen	µg/L	1900	0.73	1288	0.50	1179	0.43	965	1.00
Nitrate + nitrite	µg/L	736	0.83	558	0.67	572	0.48	543	0.91
Total phosphorus	µg/L	383	0.69	263	0.75	201	0.67	121	1.66
Soluble phosphorus	µg/L	143	0.46	56	0.75	80	0.71	26	2.11

Source: US Environmental Protection Agency, *Results of the Nationwide Urban Runoff Program: Vol. 1 – Final Report*, Water Planning Division, Washington, DC, December 1983, National Technical Information Service (NTIS) Publication No. 83-185552.

Note: COV: coefficient of variation.

In addition, runoff from developed areas can raise temperature in streams. Table 1.1 presents a comparison of pollutants in urban runoff of various land uses.

Causes of urban pollution are readily detectable and include

- Water decolorization
- Excessive plant growth in streams and lakes
- Scum and algae floats near lake shores
- Unpleasant odors
- Fewer fish and wildlife
- Fish kill
- Sediment accumulation in storm drains and ditches

1.3 NPS POLLUTANTS AND THEIR IMPACTS

As indicated, many pollutants are present in urban runoff. The sources and impacts of pollutants have been researched by many organizations including the EPA. The EPA in cooperation with the US Geological Survey (USGS) conducted a comprehensive study of urban stormwater pollution across the United States between 1979 and 1983 The result of this research project, called Nationwide Urban Runoff Program (NURP), was published in *Vol. 1 Final Report* (EPA, 1983) and *Executive Summary* (EPA, 1983a). Among the conclusions of the study are the following: heavy metals (especially copper, lead, and zinc) are by far the most prevalent pollutants found in urban runoff, with copper suggested to be the most significant threat of the three. Coliform bacteria are present at high levels in urban runoff, and nutrients are as well but at generally lower concentrations than those for other possible discharges, but oxygen-demanding substances are present in urban runoff at concentrations approximating those in secondary treatment plant discharges (EPA, 1983, 1983a).

TABLE 1.2
Sources of Pollutants in Urban Runoff

Pollutant	Source
Floatables	Shopping centers, streets, parking lots, parks and recreational areas
Sediment	Construction sites, roads, lawns and gardens
Nitrogen and phosphorus	Lawn fertilizers, detergents, animal waste, automobile deposition
Organic materials	Lawns and gardens, parks, golf courses, leaves, animal waste
Pesticides and herbicides	Lawns and gardens, roadside channels, parks, golf courses
Metals	Atmospheric deposition, automobiles, industrial sites, steel bridge corrosion
Oil and grease	Parking lots, truck stops, roads, driveways, car washes and gas stations, car and truck service stations, and illicit dumping
Coliform bacteria	Lawns, roads, septic systems, leaky sanitary sewers, animal waste

The physical aspects of urban runoff such as erosion and scour can be significant causes of habitat disruption. Detention basins and recharge systems are highly effective at removing pollutants from urban runoff wet basins (basins that maintain a permanent water pool) perform the best, and wetlands are considered promising means of controlling urban runoff quality. Organic priority pollutants in urban runoff do not appear to pose a general threat to freshwater aquatic life (EPA, 1983, 1983a).

In 1987, the EPA amended the Clean Water Act of 1972 (established to regulate the discharge of point source pollutants into the waters of the United States) requiring states, local governments, and industry to address pollution sources indicated in the 1983 report. This amendment mandated any construction activity that disturbs five acres (±2.0ha) or more and all municipal separate sewer stormwater systems (MS4s) to obtain a National Pollutant Discharge Elimination System (NPDES) permit. In addition, it required that same activities prepare Stormwater Pollution Prevention Plans (SWPPPs). As will be discussed in Chapter 5, the NPDES permit has been amended to cover smaller sites.

Major pollutants generated by urban stormwater runoff are sediment, lawn fertilizers and nutrients, heavy metals, hydrocarbons, and coliform. Table 1.2 lists sources of contaminants in urban stormwater runoff, and Table 1.3 shows typical concentrations of stormwater pollutants.

In national water quality inventory 2000 report, the EPA cited NPS pollution as the top reason that some national waterways were too polluted for swimming and fishing. Of the 33% of US water bodies that were assessed in this inventory, 40% of streams, 45% of lakes, and 50% of estuaries were not clean enough for same activities. The 2000 EPA's report identified urban runoff as one of the leading sources of water quality impairment both in surface water and groundwater.

In March 2013, the EPA (2013) released the *National Rivers and Streams* Assessment on findings from a survey conducted in 2008 and 2009 covering a total of 1924 randomly selected sites across the contiguous 48 states; of these, half were large streams and rivers. The report established that 55% of 1.2 million river and stream miles (1.92 million kilometers) were in poor biological condition, 23% in fair condition, and only 21% in good condition. Biological condition, which is the most comprehensive indicator of water body health, is related to total nitrogen, total phosphorous, and acidification, of which phosphorous and nitrogen are by far the most widespread. Of the nation's rivers and streams, 40% have high levels of phosphorous, and 28% have high nitrogen levels. Acidification, although a problem in less than 1% of the surveyed rivers and stream lengths, has a significant impact on biological conditions; poor biological condition is 50% more likely in waters affected by acidification (EPA, 2013a).

TABLE 1.3
Typical Pollutant Concentrations in Urban Stormwater

Pollutant	Typical Concentration	Unit
Total suspended sediment	54.5	mg/L
Total phosphorus[a]	0.26	mg/L
Total nitrogen[b]	2.0	mg/L
Total organic carbon	12.7	mg/L
Fecal coliform bacteria	3600	MPN/100 mL
Escherichia coli bacteria	1450	MPN/100 mL
Oil and grease	3	mg/L
Petroleum hydrocarbons	3.5	mg/L
Cadmium	2	μg/L
Copper	10	μg/L
Lead	18–50.7	μg/L
Zinc	120	μg/L
Chlorides (winter only)	116–230	mg/L
Insecticides	0.1 to 2.0	μg/L
Herbicides	1 to 5.0	μg/L

Sources: Table 1.1 in *Maryland Stormwater Design Manual*, vol. 1, May 2009; Table 2.1 in *New York State Stormwater Management Design Manual*, Draft – May 2022

An EPA (2024) report titled: *National water quality inventory report to Congress 2025* (issued October 29, 2024) includes the following findings on the water quality of water bodies:

River and streams: A total of 1,593,290 rivers and stream miles were assessed in 2018–2019. It was found that only 29% were found to be in good condition, 25% were in fair, and 47% were in poor condition.

Lakes, ponds, and reservoirs: An assessment of 224,916 lakes and reservoirs in 2017 showed that 43% were in good, 29% in fair, and 24% in poor condition.

Wetlands: An assessment of 95,694,241 acres (387,561 square km) of wetlands in 2016 showed that 47% are in good, 19% in fair, and 34% are in poor condition.

Coastal estuaries: 27,478 square miles (711,680 square km) of coastal estuaries were assessed in 2015. It was found that 71% were in good, 15% in fair, and 7% in poor condition. The remaining 7% were not studied.

Great Lakes, nearshore waters: A monitoring of 7,118 square miles (18,436 square km) of near shore waters of Great Lakes showed that 31% in good, 15% in fair and 21% in poor condition. The remaining 33% were not studied.

For more information on the national statistical surveys, visit https://www.epa.gov/national-aquatic-resource-surveys.

Also, The report which represents a snapshot of the state submissions as of July 2016 indicate how:

- Mercury (primarily in in fish tissue) as leading causes of impairment in assessed waters.
- Leading known sources included atmospheric deposit and agricultural activities.

To lessen adverse impacts of the urban runoff pollutants on the environment, proper measures should be taken to control them through source reduction and removal. The primary urban runoff

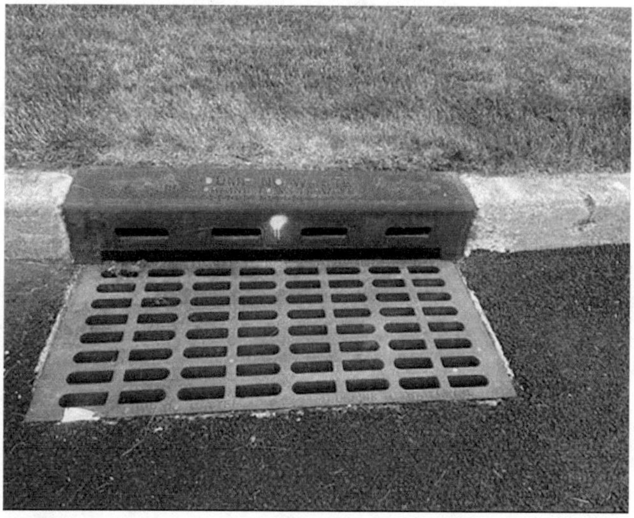

FIGURE 1.3 Combination inlet with curb piece.

Source: Photo by the author (2023).

pollutants and means of their control are briefly discussed in the following sections. More detailed information on the runoff quality can be found in Schueler (1987, 1997), US EPA (1983, 2008), Walker (1987), Terrene Institute (1994), and Caltrans (2010).

1.3.1 Floatable Objects

Floatable objects include cans, bottles, jars, nylon bags, paper, cardboard, leaves, and branches. These materials are encountered in surface water and are no concern for groundwater contamination. Plastic materials are generally non-degradable and may last for centuries, so they build up behind culverts and clog storm drains; see Figure 1.3. The most effective means of controlling these materials is public education. if everyone recycled and no one littered, there would be far less floatable waste. An effective means of avoiding floatable objects from entering drainage systems is to either not using curb inlets or installing a curb piece on the curb opening. Figure 1.3 shows a combination inlet with a curb piece which blocks paper cups, cans and plastic bottles from entering the inlet.

1.3.2 Sediment and TSS

Sediment, which is one of the most prevalent pollutants in urban runoff, is generated due to developments. The largest amount of sediment load is created during the construction phase of a development project. Thus, it is imperative to install and maintain adequate sediment and erosion control measures during construction to avoid discharge of large quantities of sediment in the form of muddy water into downstream drainage systems, waterways, and lakes. Along with sediments, the runoff from urban areas picks up other organic matter and debris collectively termed total suspended solids (TSSs). Suspended solids and sediments are not soluble, and therefore, they can be more readily removed from stormwater than some other pollutants such as nutrients.

The sediment load discharged from development increases significantly with the rainfall intensity (Pazwash, 1982b) a large storm event may deposit more sediment in a siltation basin than the cumulative load during the balance of the construction period. The same phenomenon is evidenced for lakes and reservoirs. A case in point was Sefidrud Dam in northern Iran, which was constructed during the 1954 to 1962 period. Sediment measurements during 1954 to 1976 indicated a range in the annual sediment inflow of 14 million metric tons to 218.3 million metric tons in the 1955 and 1969 water years, respectively.[2] The river discharge during the same period was measured at 3 billion m^3 (106 million ft^3) in 1955 and 14 billion m^3 (494 bilion ft^3) in 1969. A major flood on March 10, 1969, carried 15.55 million tons of sediment into the reservoir, more than the sediment inflow during the entire water year of 1955. The data also revealed that the ratio of the sediment load to the stream discharge in the 1969 water year (a wet year) was over 3.3-fold greater than that in 1955, which was a dry year (Pazwash, 1982a).

A feasible means of removing TSS from stormwater runoff is to route the runoff through a detention basin or pond (wet basin). A detention basin functions like a sediment basin in a water supply treatment plant: It retards the flow and allows the sediment to settle to the bottom. The sediment removal effectiveness (also called trap efficiency) of a pond or detention basin depends on the length and depth of the pond and the residence time, namely the time it takes for the runoff to be discharged from the basin. More importantly, this efficiency depends on the size, shape, and type of sediment material. The falling velocity of a particle varies exponentially with its size and can be calculated from the following equation:

$$C_{\mathrm{D}}\left(\frac{\pi d^2}{4}\right)\left(\frac{\rho V_{\mathrm{f}}^2}{2}\right)=\left(\frac{\pi d^3}{6}\right)\left(\gamma_{\mathrm{s}}-\gamma_{\mathrm{w}}\right) \tag{1.1}$$

The term on the left side of the equation represents drag force on the particle as it falls in water, and the term on the right side is the net weight of the particle in water. The parameters in this equation are

C_D= drag coefficient, function of the Reynolds number ($R_e = Vd/\upsilon$)
d= particle size, diameter for spherical particles
ρ = density of fluid
V_f = falling velocity
γ_s = unit weight of particle
γ_w = unit weight of water

Rewriting the preceding equation in terms of the falling velocity yields

$$V_{\mathrm{f}}=2\left[\frac{g(S-1)}{3}\right]^{1/2}\left[\frac{d}{C_{\mathrm{D}}}\right]^{1/2} \tag{1.2}$$

Where S is the specific weight of the particle. Approximating $S = 2.65$ the preceding equation further simplifies as

$$V_{\mathrm{f}}=0.147\left(\frac{d}{C_{\mathrm{D}}}\right)^{1/2} \quad \mathrm{SI}(\text{metric unit}) \tag{1.3}$$

$$V_{\mathrm{f}}=2.43\left(\frac{d}{C_{\mathrm{D}}}\right)^{1/2} \quad \mathrm{CU}(\text{customary unit}) \tag{1.4}$$

[2] The water year begins October 1 and ends September 30 of the following year. It is designated by the year it ends.

In the preceding equations, d is expressed in millimeters and inches, respectively. For very small particles where $R_e \le 1$ the drag coefficient is given by Stokes's law:

$$C_D = \frac{24}{R_e}; R_e = \frac{Vd}{\upsilon} \tag{1.5}$$

where υ = the kinematic viscosity of water.

Combining Equations 1.2 and 1.5 yields

$$V_f = \frac{g(s-1)d^2}{18\upsilon} \tag{1.6}$$

where V_f = settling velocity m/s (ft/s) = and d = particle size, m (ft)

For larger Reynolds numbers, a number of equations have been proposed. The following simple equation is proposed by Pazwash (2007):

$$C_D = \left(\frac{24}{R_e}\right)\left(1 + R_e^{2/3}/6\right) \tag{1.7}$$

This equation is in good agreement with experimental data for $1 < R_e < 1000$. For large granular particles such as coarse sand and gravel, where $R_e > 1000$, the drag coefficient ranges from 0.4 to 0.45 up to $R_e = 2 \times 10^5$ at which the flow becomes turbulent, and the drag coefficient drops to 0.2. Table 1.4, prepared by Pazwash (2007, 2011), lists settling velocities of spherical particles in water.

Table 1.4, prepared by the Pazwash (2007), reflects that sand particles (0.05 – 2.0mm) settle in a matter of minutes and silt (0.002-0.05mm) in hours in a typical 1.5 m (5 ft) deep pond, but clay particles ($d < 0.002$mm) take days to settle. Therefore, the suspended sediment removal rate of a detention basin/pond is controlled by the type of sediment carried by runoff. That is, contrary to some publications (NJDEP, 2023), TSS removal in a detention basin is not merely a function of detention time but more importantly depends on the size and type of sediment material (fine sand, silt, or clay) as well as its density and therefore is site specific.

TABLE 1.4
Settling Velocities of Spherical Particles in Water[a]

Particle Diameter		Fall Velocity		Particle Size		Fall Velocity[b]	
mm	in	m/h	ft/h	mm	in	m/s	ft/s
0.001	4×10^{-5}	3.2×10^{-3}	1.05×10^{-2}	0.20	0.008	0.02	0.07
0.002	8×10^{-5}	1.3×10^{-2}	4.3×10^{-2}	0.30	0.012	0.04	0.13
0.005	2×10^{-4}	0.08	0.26	0.4	0.016	0.06	0.20
0.01	4×10^{-4}	0.32	1.05	0.5	0.020	0.08	0.26
0.02	8×10^{-4}	1.3	4.27	0.6	0.024	0.10	0.33
0.05	0.002	8.1	26.6	0.8	0.03	0.13	0.43
0.10	0.004	32.4	106.3	1.0	0.04	0.15	0.49

[a] At 20°C (68°F).

[b] Rounded to second decimal place.

1.3.3 Nutrients and Pesticides

A comprehensive national study of nutrients in streams and groundwater was conducted by the US Geological survey from 1992 through 2004. The study results were published in the USGS (2010a) circular 1350, *titled The Quality of Our Nation's Water Nutrients in the Nation's Streams and Groundwater, 1992±2004*, and highlighted in the USGS fact sheet (2010–3078; USGS 2010b). The results indicate that excessive nutrient concentrations were a widespread cause of ecological degradation and that despite major federal, state, and local regulations to control point and nonpoint sources and transport of nutrients, concentrations of nutrients have remained the same or increased in many streams and aquifers nationwide since the early 1990s.

The USGS study finds that Nitrate levels in the Illinois River decreased by 21% between 2000 and 2010; however, the level continued to increase in the Mississippi and Missouri Rivers. The increase between 2000 and 2010 was 29% in the upper Mississippi River and 43% in the Missouri; at the Mississippi River outlet to the Gulf of Mexico, the increase was 12% in the same period. Excessive nitrate and other nutrients from the Mississippi River strongly affect the extent and severity of the hypoxic zone, known as the dead zone, that forms in the northern Gulf of Mexico every summer. The dead zone is characterized by extremely low oxygen levels in the bottom or near-bottom waters, degraded water quality, and impaired marine life. The 2013 Gulf dead zone covered 15,12 km^2 (5840 square miles), an area the size of the state of Connecticut (USGS, 2013).

Phosphorus and nitrogen are the essential nutrients for plants and are also the prime nutrients in urban stormwater runoff. These substances are mostly inorganic, comprising orthophosphates, nitrates, and ammonia. Nitrogen commonly reaches watersheds as organic N or ammonia (NH_3) and stays attached to the soil until oxidation converts it to nitrate; therefore, N is most often transported by water as nitrate (NO_3). In rural and residential areas, significant amounts of nutrients originate from fertilizers, manure, dairy farming and animal waste, detergents, and raw sanitary waste. Urban watersheds typically generate 5 to 20 times as much phosphorus per unit volume of runoff per year than underdeveloped watersheds in a given region (Walker, 1987).

The sources, dispersion, transport, and fate of phosphorus in the environment are highly complex. In urban and suburban stormwater runoff, phosphorus sources include detergents, fertilizers, lubricants, household cleaners, paints, and of course natural soil. Lawns are significant contributors of nutrients in stormwater runoff; four times higher than other land uses such as roads, driveways, and streets. So, fertilizers from lawn and landscaping are major sources of nutrients. Soil tests in New Jersey indicate that most soils have plenty of phosphorus for plant growth. Therefore, the use of fertilizer containing phosphorus has been banned in many municipalities in New Jersey (http://www.nj.gov/dep/watershedmgt).

Phosphates often attach to fine soil particles and remain in the soil until they are either utilized by plants or carried away with the soil as suspended sediment. Nitrates, however, are much more soluble and during late winter and after heavy rainfalls penetrate the water table and contaminate groundwater. High levels of nitrogen and phosphorous increase algae growth, which harms water quality, food resources, and habitats and decreases oxygen, which adversely affects the fish and other aquatic life. at high concentration, nitrates can also cause a public health hazard to drinking water.

According to the EPA's *National Rivers and Streams Assessment Report 2013*, over 27% of the streams and rivers have excessive levels of nitrogen, and 40% have excessive phosphorous. Phosphorus and nitrogen in water bodies increase biological productivity, resulting in eutrophic conditions. Eutrophication mostly occurs in small, semi-stagnant agriculture ponds and urban lakes where the water is retained for over two weeks. In the growing season, these water bodies experience chronic eutrophication, whose symptoms are water discoloration, unpleasant odors, algae scum, low oxygen levels, release of toxins, and ill effects on fish life. Eutrophication is a major environmental problem in all coastal areas in the United States.

Barnegat Bay in New Jersey and Chesapeake Bay in Maryland are the two worst basins for eutrophication in the nation. The watershed boundaries of Barnegat Bay almost fully coincide with the jurisdictional boundaries of Ocean County, which has been one of the fastest-growing development areas in New Jersey. The vast development in this county since 2000 has been a prime reason for degradation of Barnegat Bay. According to a 2010 summit meeting at Rutgers University, Barnegat Bay absorbs an estimated 1.4 million lb. (60,000kg) of nitrogen load annually, over two-thirds of which is estimated to originate from stormwater runoff.

The highly permeable soils in this area aggravate the impacts of development on runoff, because not only impervious cover but also soil compaction, reduce infiltration: Whereas natural soil has a specific gravity of 1.5, the specific gravity of turf in residential developments that were constructed in the 1970s was measured to range from 1.75 to 1.9, a loss of soil porosity from less than 30% to over 40%. To restore soil permeability, the compacted soil should be loosened, and amendments such as lime, organic matter, gypsum, and proper fertilizer should be added to the soil layers. As a part of Chesapeake Bay restoration effort, Maryland has committed to reducing its TSS and nutrient loading by 40% (EPA, 2017).

Total nitrogen (TN) and total phosphorous (TP) are expensive to remove from stormwater; average costs for a number of stormwater management systems are estimated at $530/lb ($1175/kg) and ($2750/lb) ($6100/kg), respectively (England and Listopad, 2012). The concentrations listed in Table 1.1 are significantly lower than those in treated wastewater, but it should be noted that stormwater volumes are far greater than sanitary flows during wet weather conditions.

Pesticides, which include insecticides, herbicides, rodenticides, and fungicides, are routinely used in urban areas and in agricultural lands. These substances contaminate soil, water, and air and have toxic effects on the ecosystem and human life. Pesticides decrease aquatic populations either directly by damaging the food chain or indirectly by reducing the population of phytoplankton, which lowers oxygen levels in the water. Highly carcinogenic pesticides (such as DDT, dieldrin, and chlordane) have been removed from the market; the EPA banned diazinon and chlorpyrifos for urban use in 2004 and 2005, respectively. These pesticides have been replaced by others, especially the pyrethroid pesticides in urban areas. A number of these pesticides, however, are more toxic to fish and zooplankton than the phased-out pesticides. Moreover, many of the pyrethroid pesticides tend to bond strongly to soil particles and accumulate in sediments. Urban waterways in California experience widespread toxicity caused by application of EPA's-registered insecticides. Other commonly used pesticides, such as malathion, are suspected carcinogens through direct contact (Lee and Jones-Lee, 2005).

Upon application, pesticides leave the site by becoming dissolved in stormwater runoff or by binding to suspended sediment carried in runoff. Pesticides can also contaminate groundwater through infiltration. The transport and fate of pesticides depend on their physical and chemical interaction with soil and water. The effects of fertilizer and pesticides on the environment depend on our gardening and lawn-care habits. Our gardening habits have significantly improved since decades ago when homeowners burned autumn leaves and applied harmful pesticides and too much synthetic fertilizer to their lawns and garden plants. But we still overfertilize and have a long way to go to minimize adverse effects on the environment. It is time to realize that lawns, apart from requiring repeated fertilizer applications, use tremendous amounts of water. Thus, we should consider downsizing lawn area and instead landscape with low-maintenance native plants.

1.3.4 Heavy Metals

The main sources of metals in urban stormwater runoff are automobiles and industry. According to the aforementioned NURP study, copper, lead, and zinc are by far the most prevalent heavy metals found in urban runoff. Atmospheric deposition both wet and dry also makes significant contributions in some parts of the country. Metals can also occur naturally in soil and the quantities of metals leaching into water from natural sources elevates as water pH drops (EPA, 1983, 1983a).

Zinc is one of the most prevalent heavy metals found in stormwater runoff here in the United States and abroad. In New Zealand, for example, zinc generated from galvanized roofing is among the primary contaminants in urban runoff (Tveten and Williamson, 2006). Copper, zinc, and mercury can cause health problems, but lead is the prime concern in toxicity and public health; it tends to precipitate in aquatic systems and accumulate in soils and sediment, has cumulative neurological adverse effects, and is particularly harmful to children. Tetraethyl lead in gasoline had been one of the principal sources of lead in stormwater runoff. However, with the use of unleaded gasoline, the pollution from this source has diminished (Lee and Jones-Lee, 2006). The effective control of heavy metals is not simple. Although better gas mileage will reduce the production of heavy metals, much more significant reduction can only be brought up by a change in our lifestyle, discussed later in this chapter.

1.3.5 Pathogens and Fecal Coliform

Pathogens are dangerous microscopic viral or bacterial organisms that cause illness. Fecal waste from animals are prime sources of pathogens in urban stormwater runoff. However, overflow from combined storm and sanitary is by far the largest source of pathogens. Many older, more intensely developed cities and municipalities in the United States do not have separate storm and sanitary sewers. Examples include parts of Fort Lee, North Bergen, and Ridgefield, New Jersey, and the boroughs of Brooklyn and Queens in New York City. Improperly sized or located septic systems contaminate groundwater, especially in areas of very permeable soil and/or high-water tables and where fractured rocks and well casings create entry routes.

Stormwater runoff poses little threat to human health if it does not come in contact with domestic wastewater or fecal waste. However, when inadequately treated combined sewer flows are discharged to beaches and lakes, there is a public health risk due to pathogen contamination. Likewise, the risk exists where runoff comes into contact with shellfish beds and in swimming ponds where large flocks of Canadian geese may roam.

Runoff contaminated with fecal sources transmit a number of human diseases. Some well-known bacteria agents include *Salmonella*, which causes typhoid fever and intestinal fever, and the Shigella genus which causes bacterial dysentery. Other bacterial agents include *E. coli* and *Vibrio cholerae*, the latter of which can cause cholera. Gastroenteritis, which is caused by *E. coli*, is the leading waterborne human infectious disease in the United States.

The EPA (1983, 1983a) studied fecal coliform concentrations at 17 sites for 156 storm events and showed that coliform bacteria are present at high levels in urban runoff and exceed EPA water quality criteria during and immediately after storm events. Coliform counts during warmer periods were 20 times greater than those found during colder months potentially from animal waste in rain runoff, which is far greater in summer than snow melt in winter.

Concentrations of coliform bacteria and *E. coli* bacteria serve as indicators for the presence of pathogenic bacteria. The 2006 bacteria TMDL (total mean daily load) adopted by Ventura County in California included both a summer dry weather single day maximum of 235 MPN (most probable number) per 100 milliliters and a 30-day geomean at 126 MPN. In the city of Thousand Oaks, California, bacterial exceeded the summer dry weather standard of 30% to 40% of the time (Carson and Sercu, 2013). In a study of impaired waters and total maximum daily loads, the EPA found that pathogens are by far the leading cause of impairment in 303(d)-listed[3] waters nationwide with the most common being fecal coliform or *E. coli* (Kaspersen, 2009). The afore mentioned 2013 EPA study finds that about 9% of streams and rivers have high levels of bacteria. A number of attempts have been made to chase away Canadian geese from the pond

[3] Under Section 303(d) of the 1972 Clean Water Act, states, territories, and authorized tribes were required to develop lists of impaired waters (those that do not meet water quality standards).

perimeters. None have been successful. The author suggests taking away the birds' fresh eggs to reduce Canadian geese populations.

1.3.6 Road Salt

Road salt is the most widespread deicing substance used in many states; sodium chloride which is readily available and inexpensive, depresses the freezing point of ice. New Hampshire was the first state in the United States to use salt on an experimental basis to melt snow on roadways in 1938 (Richardson, 2012). Three years later, nearly 5,000 tons of salt were spread on US highways. According to the EPA, more than 13 million tons (11.8×10^6 metric tons) of salt were applied to roads annually throughout the country (*Civil Engineering*, 2005). Today over 18 million tons (16.3 million metric tons) are used for de-icing in the United States each year (Ton = 2000 lb = 907 kg; metric ton = 1000 kg.).

Massachusetts, New Hampshire, and New York had higher than average use, with use highest in Massachusetts at 19.94 tons per lane-mile (11.2 metric tons/lane km) per year. New York City used an average of 500,000 tons of salt (16.6 tons/lane-mile/year or 9.4 metric tons/lane-km/year). Studies by the USGS and private agencies such as the Cary Institute (a nonprofit environmental research and education organization in Millbrook, NY), have shown that salt does not disperse as quickly as was previously perceived and that concentrations of 100 to 200 mg/L, which are common, have considerable adverse effects on aquatic life. The high use of salt has resulted in widespread contamination in the northeast United States. Chloride concentrations have at times exceeded 5 g/L(0.5% by weight) in urban areas. This is nearly one-fourth of the salt concentration in sea water (*Civil Engineering*, 2005).

Salt also enters drainage systems from precipitation falling on salt stockpiles and from salt application to roadways, driveways, and parking areas. Salt in runoff infiltrates into the ground and contaminates groundwater. Because of long residence time, salt concentrations tend to build up in groundwater, reaching their highest level in summer. Excessive concentrations of salt in fresh waters can retard springtime mixing due to density gradient, and also decrease oxygen levels causing high mortality among fish and bottom-living organisms. It can also alter natural saline concentrations in estuaries and bays, disrupting shellfish reproduction. Increased salt concentrations produce a displeasing water taste, which may require expensive treatment for domestic use.

Salt can affect the water chemistry miles downstream of where it is applied. Dr. Stephen Norton, PhD, a professor of geological sciences at the University of Maine, reported that salt causes certain minerals to leach out of soils in the streams. At high concentration, salt increases the acidity of water, with effects similar to acid rain (*Civil Engineering*, 2005). Other studies of road salt have found ecological consequences of salt including adverse impacts on macro-invertebrates. High levels of salt in drinking water can result in high blood pressure and kidney malfunctioning as well as damage to eggs and embryos of wood frogs and salamanders.

Sodium chloride is one of the emerging contaminants of great concern in urban runoff. Although road salt is often overapplied, it is possible to reduce its use, for instance by plowing before applying salt because snow requires a great deal of salt to melt. Additionally, using wet salt before freezing occurs, reduces the salt from 25% for dry salt to only 4%; mixing sand and salt and applying it in parallel strips is also effective. The trucks should also move at moderate speed; at 50 km/h (30 miles/h) speed disperses 30% of salt beyond the roadway. Temperature also as an effect on salt dispersal; less salt should be used at lower temperatures: One kilogram or pound of salt melts 40 kg or lbs of snow at −1°C(30°F) but approximately 5 kg or lbs at −12°C(10°F). Finally, it is wisest to never salt before snow falls. Weather forecasts are not always reliable; snow may not fall or may turn to rain.

A number of deicing materials have been tested as alternatives to road salt. Among these, calcium chloride ($CaCl_2$), calcium magnesium acetate (CMA), and potassium acetate (KA) are the most notable. $CaCl_2$ requires special handling and is more expensive than salt. However, it is very effective and fast acting at temperatures below -18°C (0°F) CMA is slower acting than salt when applied during or after snow and its effectiveness diminishes below –5°C(23°F). In addition, CMA costs $600 to $700 per metric ton compared with $60 to $70 per metric ton for road salt (WisDOT, 1987).

KA is often used as a base for chloride-free liquid deicing material. Advantages of KA include low corrosion, relatively high performance, and, above all, low environmental impact. However, KA costs many times more than salt its cost is comparable with CMA. Studies show that among all deicers, sodium chloride has the most deteriorating effect on concrete surfaces and is significantly more harmful than calcium chloride (Mishra, 2001).

Calcium chloride also has been employed as an anti-icing agent. A deicer agent called Verglimit is an additive containing $CaCl_2$ encapsulated in linseed oil and caustic soda can be used for bituminous or concrete pavements. It is most suited for bridge decks, steep slopes, and shaded areas, which are more prone to icing. Verglimit has been used in Europe since 1974, in North America since 1976, and in Japan since 1978. Tests in New Jersey have found the material to be effective at - 4°C (Clines, 2003).

In airports including Newark International Airport (EWR) in New Jersey, where managing stormwater is critical to their operations, glycol is used as a deicer, but it has negative environmental effects because it depletes. An EPA ruling limiting the discharge of glycol and other pollutants from airports drove changes to the traditional deicing process. Airports such as Buffalo Niagara International installed an underground wetland system to treat glycol (Shane, 2007a). The stormwater management system in each airport will need to be designed based on the existing drainage infrastructure and de-icing operations.

1.3.7 Petroleum Hydrocarbons

Petroleum hydrocarbons include oil and grease, the BTEX compounds (benzene, toluene, ethyl benzene, and xylene), and a variety of polynuclear aromatic hydrocarbons (PAHs). These pollutants enter stormwater runoff from garages, parking lots, roadways, leaky storage tanks, auto emissions, and improper or illicit disposal of waste oil. It may be hard to believe that there are still people who dump their used car oil (and antifreeze) into municipal storm drains. Petroleum hydrocarbons have acute toxicity to humans at low density (Schueler, 1987). They also render water unsuitable for designated uses.

A study involving measurement of petroleum hydrocarbon concentrations in urban runoff from a variety of impervious areas in the District of Columbia and suburban Maryland showed that the amount of car traffic affects the hydrocarbon concentration in runoff, with the medium concentrations ranging from 0.6 to 6.6 mL/L. These concentrations exceed the maximum concentrations recommended for the protection of drinking water supplies and fisheries, in the range of 0.01 to 0.1 mL/L (Shepp, 1996).

Hydrocarbons are harmful to water supply in a liquid state; however, they are absorbed and adsorbed onto sediment and solid particles so rapidly that they are found mainly as particulate in water. Only considerable masses of oil, such as oil spills, will remain in a liquid state. Petroleum hydrocarbons are biodegradable in an aerobic environment, although at a very slow rate. The rate can be significantly hastened by forced aeration.

Building more efficient cars, avoiding unnecessary trips, and, more importantly, carpooling will reduce the hydrocarbon concentration in urban runoff. Carpooling lanes were employed in New Jersey in the mid-1990s. However, the program was abandoned after nearly three years as the high-occupancy car lanes were not effectively utilized. The lack of success was partly due to low gas prices, which kept it convenient to have one's own car. More importantly, the program failed because it required a minimum of three passengers in a car, which limited the number of people with common commutes. In Southern California and other heavy traffic areas, carpool lanes that only require two vehicle occupants have been fairly successful.

Removing hydrocarbons and fuels at airports and seaports has been given special attention in recent years. Albany International Airport has installed a treatment system to address the water quality of runoff from fuel facility expansion discharged to Shaker Creek, which leads to a downstream drinking water intake in Albany County. The system includes trench-drains in refueling areas, conveying the runoff to a stormwater lift station, where the runoff is pumped to a vault with a bank of Smart Sponge absorption filters. The filters supplied by Clear Water Solutions produced superior effluent quality at a much lower cost than traditional activated carbon filter media. These filters do not leach contaminants; rather, they transform them into manageable solid waste. The filters are easy to maintain and replace and can be disposed of as solid waste at a relatively low cost (Shane, October 2007).

A number of other airports in the Northeast use similar filters to treat runoff, such as EWR in New Jersey and Westchester County Airport in New York State (HPN). In 2002, Westchester County Airport installed Ultra Urban filters (AbTech Industries, Scottsdale, AZ, USA) in 54 selected critical inlets at the airport. The infiltration medium, called Smart Sponge, comprises a blend of polymers and effectively absorbs contaminants from water; the polymers are composed of nonleaching molecules that bond to gasoline, oil, and grease, transforming them to a gel-like material and saturating at two- to fivefold their dry weight. The filters come in two standard shapes or modular units, one for curb inlets and a single unit for drop inlets. The filters were good for two years with no spills before they needed to be replaced. The removed filters were tested to ensure that they had no hazardous substances before they were hauled away for recycling (Shane, November–December, 2007).

1.3.8 Atmospheric Dust

Atmospheric dust is defined as minute particles suspended by slight currents and slowly settling in calm air. Generally, less dust exists at high elevations over the ocean and more at low levels over cities. Atmospheric dust is created by activities such as overgrazing, deforestation, and improper agricultural and construction practices. However, atmospheric dust is mostly produced by natural causes, such as climatic conditions and in particular wind. Arid and semiarid regions generate significantly more dust than areas of high rainfall where the ground surface is protected with vegetation. Wind erosion is the largest source of dust pollution; it erodes bare ground and detaches and entrains solid particles in the air. Strong winds can create a thick cloud of dust in the sky. Figure 1.4 shows a dust cloud during a strong wind.

FIGURE 1.4 Dust caused by wind erosion clouding the sky and forming a dark shade.

Source: Photo by the author.

Not only is dust an air pollutant, but airborne dust collects on urban surfaces such as roads, driveways, roofs, and pools and washes away with rain to form water pollution. Large dust particles quickly settle on the ground, but large quantities of fine dust particles remain suspended in the air for a long time. The effects of the August 1883 volcanic eruption of Mt. Krakatau in Indonesia were observed years after its occurrence. Generally, dust particles larger than 1 μm (0.001mm) settle as dry deposition because of their larger settling velocity; smaller particles are primarily removed by wet deposition. Specifically, the small particles adhere to rain and fall with raindrops. This is best observed by dirt spots on cars and pavements due to rain after days of dry weather. In the 2013 EPA study, a little over 1% of streams and rivers had excessively high levels of mercury, primarily from the atmospheric deposition of coal ash.

Studies estimate global dust-emission rates within a range of 1000 to 3000 teragrams (10^{12} g = 10^6 metric tons) per year. About 80% of the dust is from the Northern Hemisphere, in a dust belt extending from North Africa to the Middle East to Central and South Asia and to China. The largest source of dust is the Sahara Desert with an estimated range of 160 to 760 Tg (teragrams) per year (*Encyclopedia of Earth*, 2007). It is worth noting that winds, apart from carrying dust, also transport plant seeds, bacteria, viruses, fungi, and various pollutants, not all of which are harmful; some are food for earth.

1.4 MANAGEMENT OF STORMWATER RUNOFF

National, state, and other jurisdictional stormwater management regulations govern the quantity and quality of runoff from development sites that disturb land beyond certain threshold limits. To address these regulations, a variety of measures have been developed. Because of the diversity of measures, the umbrella term "best management practices" (BMPs) was introduced to the field of stormwater management (see EPA, 1999). BMPs refer to measures or structures that under given conditions control the quantity and improve the quality of stormwater runoff in the most cost-effective manner. BMPs can be either preemptive and aimed at lowering the production of runoff and pollutants from a site or corrective and aimed at reducing the amount of runoff and pollutants generated. These approaches are also referred to as nonstructural or structural (or end-of-pipe) BMPs, respectively.

Structural measures include detention/retention basins, ponds, infiltration basins, and the like. Nonstructural measures, on the other hand, are measures that reduce the runoff volume and pollutant generation; reducing imperviousness, disconnecting impervious areas, and avoiding excessive use of lawn chemicals are examples of nonstructural measures. In recent decades a large number of manufactured devices have been introduced to the market for the treatment of stormwater quality. Structural, nonstructural, and manufactured water-quality devices will be discussed in separate chapters in this book.

Addressing adverse impacts of developments on the environment requires challenging stormwater management actions in the future. A concern with stormwater runoff is that it becomes more polluted as it travels through street gutters and drainage systems to get to a facility to receive treatment. The traditional approach to stormwater management, namely the end-of-pipe methods, will be neither feasible nor cost effective.

A single BMP usually will not address all stormwater problems. Each BMP has its own limitations depending on its intended objectives and specific site conditions. In general, the best solution to control pollution is to reduce its production in the first place, noting that prevention is a more effective solution than cure. The National Research Council (NRC) under a contract with the EPA released a report titled *Urban Stormwater Management in the United States*, on October 15, 2008. This 529- page report, which is the product of a 26-month study by a 15-member committee, includes a description of the history of stormwater management in the United States and an overview of stormwater regulations and the federal regulatory program. The authors of the report concluded that radical changes to the EPA stormwater program are necessary to reverse degradation of freshwater resources. Among the key findings and general recommendations, the authors called for

- changing from the EPA's current piecemeal regulatory system to a new watershed-based permitting system that focuses on all discharges to streams and water bodies.
- focusing more on the increased volume of water and less on chemical pollutants.

In the study authors also determined that areas of disturbance due to urban land were expanding more rapidly than populations. The impact of this trend must be considered in the EPA's stormwater regulatory program.

Following the NRC's recommendation, the EPA has selected a number of projects for pilot study. The EPA has also incorporated certain findings and recommendations of the study in the current Construction General Permit adopted in 2012.

As indicated, a most cost-effective and environmentally friendly solution to manage water quality is to reduce pollution in the first place, for instance by reducing imperviousness, retaining more rainfall and directing runoff from impervious surfaces (roofs and driveways) to lawns/landscaping. Public education will be beneficial regarding reducing fertilizer and pesticide use, properly disposing of household chemicals, and generally avoiding littering by properly disposing of trash and recycling.

Some of these measures will be explored later in this book. It should be noted, however, that most important of all is to redefine the criterion of the impact of a development on the environment. The state-of-the-art practices, such as low-impact development, green infrastructure, and sustainable development, aim at reducing per-lot or per-site impacts. Basing disturbance per unit area of land is a misleading criterion because the land is disturbed to provide people with housing and other amenities. Therefore, the criterion for impact on the environment should be not how much disturbance a project creates but the amount of disturbance per capita (Pazwash, 2011, 2016). On this basis, as will be shown later in this book, compact developments such as condominiums, multifamily residential buildings, and. above all, city living have a far less adverse impact on the environment than single-family homes.

Therefore, to reduce impacts of urbanization on stormwater runoff, our urban planning should be based on compact mixed-use developments, building around existing transport stations, and, above all, building cities having streets with walkways and bikeways. This will minimize the impervious coverage per capita, resulting in a proportionate reduction in the quantitative and qualitative impacts on stormwater. Compact developments and, in particular, city living significantly lessen, if not nearly eliminate, the use of private cars, which in our society form the general means of commuting to work and moving around by the public, and which generate traffic jams and create air and water pollution. This will significantly reduce air pollution and greenhouse gas emission, road rage, and countless hours of nonproductive daily life. In addition, they will reduce energy consumption and overall adverse impacts to the environment. City living, of course, entails a drastic change from our suburban lifestyle and may not be acceptable by many people but is the best solution to reduce our adverse impacts on stormwater runoff and the environment as a whole.

As indicated earlier, land disturbance due to urbanization is growing faster than population growth. Thus, our adverse impacts to the environment in general and stormwater in particular will not diminish; rather, they will grow with time. It is a misperception to believe that sustainable developments will prevent compromising the ability of future generations to meet their needs. True sustainability is an unrealistic goal that cannot be achieved. The best we can hope for is to minimize our ill impacts on the environment.

Man's adverse impacts on the environment are not limited to stormwater runoff and water pollution. We pollute the air, contaminate the earth, and change natural processes. Although technology has reduced the amount of per-car emission per kilometer (mile), there has been over 300% increase in vehicular travel distance since 1970. Also, the number of cars on the road has doubled during the past thirty years.

Ill impacts of our activities on the natural world are enormous and beyond imagination. There should be greater concern regarding the destruction of numerous species of insects and microbial organisms that may be vital to the maintenance of the life cycle on earth. Dr. Edward Wilson, professor emeritus of Harvard University and two-time Pulitzer Prize winner, thinks that it is quite possible that we could destroy the rest of the natural world and with it as many as one-half of plants and animals on earth by the end of the century (Wilson, 2006; Schulte, 2006).

PROBLEMS

1.1 Why does urbanization impact stormwater?
1.2 What are the main impacts of urbanization on stormwater runoff?
1.3 What are the impacts of urbanization on runoff quantity?
1.4 Why does urbanization increase stormwater volume?
1.5 Does urbanization increase the peak rate of runoff? If your answer is yes, what are the reasons?
1.6 Does urbanization impact stormwater quality? If so, what are the changes?
1.7 Does urbanization increase or decrease the sediment load? Explain your answer.
1.8 Why does sediment contribute to stormwater contamination?
1.9 Is the sediment a problem during construction? If so, is it more of a problem than that of post-construction? Explain your answer.
1.10 What does the settling velocity of a particle in water depend on? How is it related to particle size?
1.11 How is sediment commonly removed from stormwater?
1.12 Does urbanization increase nutrients in runoff?
1.13 Name the prime nutrients present in stormwater.
1.14 What are the sources of nutrients in urban stormwater?
1.15 Do nutrients have adverse impacts on water bodies? If so, list the impacts.
1.16 Are there any heavy metals in urban runoff? If so, what are they?
1.17 Where does the heavy metal in urban runoff originate?
1.18 Are pathogens bacterial or viral organisms? What are their sources in stormwater runoff?
1.19 Can pathogens cause human diseases? Name an infective disease associated with pathogen bacteria.
1.20 What is the main source of sodium chloride (salt) in urban runoff?
1.21 Can salt in drinking water affect our health?
1.22 Outline the most effective measures to reduce the use of salt on roads.
1.23 Are petroleum hydrocarbons present in urban runoff? If so, where do they come from?
1.24 Do single-family homes or compact developments have a smaller impact on runoff? Explain your answer.
1.25 To properly compare impacts of developments on runoff, what should the impacts be based on?

REFERENCES

Caltrans, July 2010, *Stormwater quality handbooks project planning and design guide (PPDG)*, State of California, Department of Transportation, CTSW-RT-10-254.03.

Carson, R.A. and Sercu, B., July–August 2013, Efforts to achieve compliance with coliform plan objectives, *Stormwater*, pp. 10–19.

Civil Engineering, May 2005, News brief: Salting road during winter dangers neighboring ecosystems, pp. 36–37.

Clines, K., April 2003, Bid list materials for deicing and anti-icing, *Better Roads Magazine Encyclopedia of Earth*, 2007, Global dust budget.

England, G. and Listopad, C., May 2012, Use of TMDL credits for BMP comparisons, *Stormwater*, pp. 38–43.

EPA (US Environmental Protection Agency), 1983, *Results of the nationwide urban runoff program: Vol. 1 – final report*, Office of Water Planning Division, EPA, Washington, DC, National Technical Information Service (NTIS) Publication No. 83-185552.

EPA (US Environmental Protection Agency), 1983a, *Results of the nationwide urban runoff program, executive summary*, Office of Water Planning Division, EPA, Washington, DC, National Technical Information Service (NTIS), Accession No. PB84-185545.

EPA (US Environmental Protection Agency), August 1999, *Preliminary data summary of urban stormwater best management practices*, Office of Water: EPA, EP-821-R-99-012.

EPA (US Environmental Protection Agency), 2000, *National water quality inventory report to Congress* Office of Water: EPA, Washington, DC, EPA 841-R-00-001.

EPA (US Environmental Protection Agency), October 2008, *Protecting water quality from urban runoff*, National Research Council, National Academies Press, Washington, DC, EPA 841-F-03-003.

EPA (US Environmental Protection Agency), 2009, *Urban stormwater management in the United States*, National Research Council: EPA, p. 460.

EPA (US Environmental Protection Agency), February 28, 2013, *National rivers and streams assessment 2008-2009: a collaborative survey*, Draft, EPA/841/D-13/001.

EPA (US Environmental Protection Agency). March 26, 2013a, Water headlines, http://water.epa.gov/about/owners/waterheadlines/2013.

EPA (US Environmental Protection Agency), August 2017, *National water quality inventory; report to Congress*, EPA 841-R-16-0.11.

EPA (US Environmental Protection Agency), December 2020, *National rivers and streams assessment 2013–2014: A collaborative survey*, EPA 841-R-19-001.

EPA (US Environmental Protection Agency) October 29, 2024, *National water quality inventory report to Congress 2025,* EPA report 841-R-23-001, p. 29.

Hudson, B.E., 1994, Soil organic matter and available water capacity, *Journal of Soil and Water Conservation*, 49 (2): 189–194.

Kaspersen, J., March–April 2009, The great bug hunt, editor's comments, *Stormwater*, p. 6.

Lee, G.F. and Jones-Lee, A., 2005, Urban stormwater runoff quality issues, in *Water encyclopedia: Surface and agricultural water*, Wiley, Hoboken, NJ, pp. 432–437.

Lee, G.F. and Jones-Lee, A., September 2006, Lead as a stormwater runoff pollutant, *Stormwater*, pp. 88–91.

Mishra, S.K., December 2001, *A mixture for snow and ice, roads and bridges*. pp. 18–21, http://www.roads-bridges.com.

New York State Department of Environmental Conservation. *New York State-Stormwater Management Design Manual, draft—New York State-Stormwater Management Design Manual*, May 2022 Draft.

New York State Department of Environmental Conservation. *New York Sate-Sormwater Management Design Manual,* January 2015.

NJDEP (New Jersey Department of Environmental Protection), July 2023, *Stormwater best management practices manual*, NJDEP.

Pazwash, H., August 24–26, 1982a, *Sedimentation in reservoirs, case of Sefidrud dam*, Vol. C, Proceedings of 3rd Congress of the Asian and Pacific Regional Division of the I.A.H.R., Bandung, Indonesia, pp. 215–223.

Pazwash, H., December 5–10, 1982b, *Erosion and sedimentations, effect of reservoirs*, Proceedings of 1982 International Symposium on Surface Mining Hydrology, Sedimentology, and Reclamation, University of Kentucky, Lexington, pp. 457–461.

Pazwash, H., 2007, *Fluid mechanics and hydraulic engineer*, Tehran University Press, Iran, p. 671.

Pazwash, H., August 1, 2012, *Development sustainability, proper basis*, presented at OIDA International Conference on Sustainable Development, Montclair State University, Montclair, NJ

Pazwash, H., 2011, *Urban stormwater management*, 1st, ed., CRC Press, Boca Raton, FL.

Richardson, D.C., January–February 2012, Ice school, melding the science and craft of winter road maintenance, *Stormwater*, pp. 14–21.

Schueler, T., 1987, *Controlling urban runoff: A practical manual for planning and designing urban BMPs*, Metropolitan Washington Council of Governments, Washington, DC.

Schueler, T., 1997, Comparative removal capability of urban BMPs: A reanalysis, *Watershed Protection Techniques*, 1 (2): 515–520.

Schulte, B., September 4, 2006, Q&A: Edward Wilson, *U.S. News & World Report* https://www.usnews.com.

Shane, J.I., October 2007, Airport support Albany international protects local waterways from fuel facility expansion runoff using absorption filter media, *Stormwater Solutions*, pp. 30–33.

Shane, J.I., November–December 2007, Westchester County airport meets tough international standards, *Stormwater Solutions*, pp. 72–82.

Shepp, D.L., 1996, *Petroleum hydrocarbon concentrations observed in runoff form discrete, urbanized automotive-intensive land uses*, Metropolitan Washington Council of Governments, Washington, DC.

Terrene Institute, March 1994, *Urbanization and water quality, a guide to protecting the urban environment, in cooperation with EPA*, Terrance Institute, Washington, DC.

Tveten, R. and Williamson, B., 2006, Zinc found as one of the primary contaminants in New Zealand's urban stormwater, *ASCE EWRI (Environmental and Water Resources Institute)*, 8 (1): 2.

USGS (US Geological Survey), 2010a, *Circular 1350: Nutrients in the nation's streams and groundwater*, USGS.

USGS (US Geological Survey), 2010b, *Nutrients in the nation's streams and groundwater, national findings and implications, fact sheet, 2010–3078*, http://pubs.usgs.gov/fs2010/3078/.

USGS (US Geological Survey), October 30, 2013, *News release,* http://www.usgs.gov/newsroom/article.asp?ID=3715.

Walker, W.W., 1987, Phosphorus removal by urban runoff detention basins, *Lake and Reservoir Management*, III: 314–326.

WisDOT (Wisconsin Department of Transportation), 1987, *Field deicing tests of high-quality calcium magnesium acetate (CMA)*, WDOT.

Wilson, E.O., 2006, *The creation: An appeal to save life on earth*, W. W. Norton, New York.

2 Pipe and Open Channel Flow
A Review

Hydraulic principles including energy equations, specific energy, and critical flow are briefly discussed in this chapter. Also presented in this chapter is flow in pipes and open channels and losses in manholes and junctions.

2.1 FLOW CLASSIFICATIONS

Since the design equations for pipes and channels have been developed for specific flow conditions, it is first necessary to classify the types of flow. Depending on temporal and spatial variations, the flow is classified as steady or unsteady, uniform or nonuniform, gradually varied or rapidly varied. In the design of pipes and channels in urban stormwater management systems, the flow is generally considered steady and uniform, in that the flow velocity, discharge, and depth are assumed to remain unchanged with time or distance along a reach of a conduit. The depth of uniform flow is called the "normal" depth.

Uniform flow can occur as pressure flow, such as full flow in pipes, or gravity (also called free surface) flow, such as flow in channels and pipes when partially full. Nonuniform flow is a flow with changing depth and velocity along the channel or conduit. This type of flow can be gradually varied if the changes in depth and velocity are gradual and occur over a considerable length, or rapidly varied flow where the changes in flow are abrupt and occur over a very short distance. Overland flow on paved surfaces, gutter flow along roadways, and flow in natural streams are examples of gradually varied flow. Flow over emergency spillways, hydraulic jumps, and flow under sluice gates are examples of rapidly varied flow. Uniform flow equations are commonly applied to short-distance intervals in gradually varied flows; however, these equations are not applicable to rapidly varied flows.

2.2 ENERGY EQUATIONS

A flowing fluid has potential, pressure, and kinetic energies at any given point. The sum of these three interchangeable energy components, is called total energy. The principle of conservation of energy states that for an ideal fluid with no external energy sources or sinks, the total energy remains constant along the flow though the contribution from the three components changes. In a real fluid, energy losses including frictional and form losses (also called local or minor losses) should be accounted for in the energy equation. In a system with a pump, the head added by a pump also enters the equation. In a channel flow (see Figure 2.1), the total energy head, defined as the total energy per unit weight at a point, is given by

$$H = z + d + \frac{p}{\gamma} + \frac{V^2}{2g} \tag{2.1}$$

where

H is the total energy head

Z is the elevation of channel bed above an arbitrary, or fixed, datum (such as the National Geodetic Vertical Datum (NGVD) 29

d is the distance of the point above the bed

DOI: 10.1201/9781003460121-2

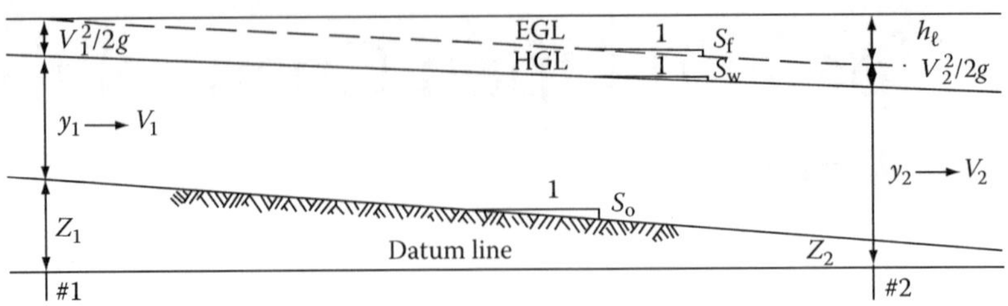

FIGURE 2.1 Energy components in an open channel flow.

p is the fluid pressure at the point
V is the flow velocity
γ is the fluid unit weight

The figure shows that H, which represents energy per unit weight, has units of length and thus is termed the energy head or total head. The pressure head, p/γ, is the pressure per unit weight, and $V^2/2g$ is the kinetic energy per unit weight and is known as the velocity head. The sum of $p/\gamma + d$, which represents the pressure head at the streambed, can be approximated by the flow depth, y, in channels of mild slope. In this case, the energy equation applied between sections 1 and 2 is expressed in the following simple form:

$$H_1 = H_2 + h_l \tag{2.2}$$

where

$$H_1 = z_1 + y_1 + \frac{V_1^2}{2g};\ H_2 = z_2 + y_2 + \frac{V_2^2}{2g}$$

and

$$h_1 = \text{head loss between sections 1 and 2}$$

The line representing the sum of terms $z + y$ is known as the hydraulic grade line (HGL). In open channel flow, the HGL coincides with the water surface profile, and the line lying at a distance equal to the velocity head above the HGL is called the energy grade line or simply energy line (EL). The slope of the EL in a uniform channel represents the energy loss (due to friction) per unit length of channel. In closed conduit flows such as pipes flowing full, the HGL generally does not coincide with the crown of the pipe. In this case, the HGL, which is the sum of $p/\gamma + d$ (d being the pipe diameter), can lie above or below the crown of the pipe depending on whether the pipe is flowing under higher or lower pressure than the atmospheric pressure, respectively. In a storm drain running full and under pressure, the hydraulic grade line lies above the crown of the pipe.

2.3 SPECIFIC ENERGY AND CRITICAL FLOW

If the elevation is taken at the channel bed or invert of a partially full pipe, Equation 2.2 simplifies as

$$E = y + \frac{V^2}{2g} \tag{2.3}$$

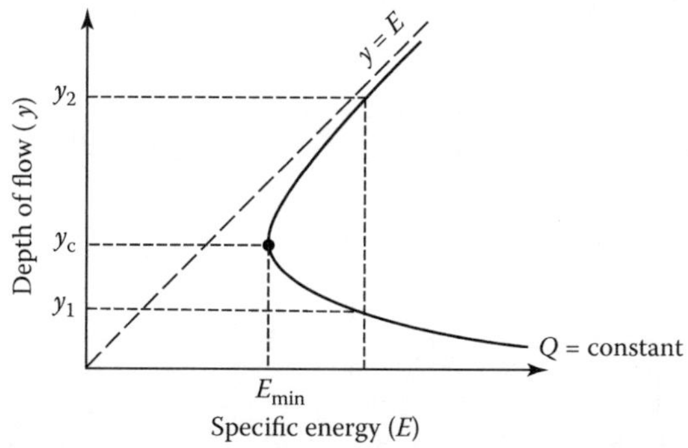

FIGURE 2.2 Specific energy as a function of depth.

Where E is the specific energy. As the depth of fluid increases, $V^2/2g$ decreases, and at a certain depth, E, which is the sum of flow depth and velocity head becomes minimal (see Figure 2.2). The depth at which the energy becomes minimal is known as the critical depth, and the velocity associated with this depth is called the critical velocity

2.3.1 Critical Depth

In a channel of any geometry, the specific energy equation is expressed as

$$E = y + \frac{V^2}{2g} = y + \frac{Q^2}{2gA^2} \tag{2.4}$$

where
A = flow area m² (ft²)
Q = discharge m³/s (cubic feet per second [cfs])

The critical depth, which corresponds with minimum specific energy, satisfies the following equation:

$$\frac{dE}{dy} = 1 - \frac{Q^2}{gA^3} x \frac{dA}{dy} = 0 \tag{2.5}$$

Since dA = Tdy, where T is the top width, Equation 2.5 simplifies as

$$\frac{Q^2 T}{gA^3} = 1 \tag{2.6}$$

or

$$Q = \left(\frac{gA^3}{T}\right)^{1/2} \tag{2.7}$$

Substituting $Q = AV$ and $T = A/D$ in Equation 2.6 results in

$$\frac{V_c^2}{gD} = 1 \tag{2.8}$$

or

$$\frac{V_c^2}{2g} = \frac{D}{2} \tag{2.9}$$

and

$$F_r = \frac{V_c}{\sqrt{gD}} = 1 \tag{2.10}$$

Where $D = A/T$ is called the mean flow depth, and F_r is the Froude number.

Thus, for any channel including partly full pipes, the critical flow occurs when the Froude number, F_r, equals one or velocity head equals one-half of mean flow depth and

$$E = y_c + \frac{D}{2} \tag{2.11}$$

The critical depth and critical discharge for any section can be calculated from Equation 2.6 through an iterative procedure.

2.3.2 Critical Flow in Rectangular Channels

In rectangular channels (and partly full rectangular culverts), the area is related to the depth of flow by

$$A = by \tag{2.12}$$

Where b is the width of the channel and y is the depth of flow. For a rectangular channel, the general Equation 2.6 simplifies as

$$\frac{q^2}{gy_c^3} = 1 \tag{2.13}$$

or

$$y_c = \left(\frac{q^2}{g}\right)^{1/3} \tag{2.14}$$

where $q = Q/b$ represents the discharge per unit width of the channel.

Substituting $q = Vy_c$ in Equation 2.14 and rearranging gives

$$y_c = \frac{V^2}{g} \tag{2.15}$$

and

$$V_c = \sqrt{gy_c} \tag{2.16}$$

Combining Equations 2.3 and 2.15 yields

$$E = 1.5y_c \tag{2.17}$$

or

$$y_c = \left(\frac{2}{3}\right)E \tag{2.18}$$

Since the flow in box culverts (rectangular flumes) with free water surface resembles an open channel flow, Equations 2.12 through 2.18 are equally applicable to box culverts.

Rearranging Equation 2.14, the critical depth can be calculated directly from the following equations:

$$y_c = 0.467\left(\frac{Q}{b}\right)^{2/3} \quad \text{SI} \tag{2.19}$$

$$y_c = 0.314\left(\frac{Q}{b}\right)^{2/3} \quad \text{CU} \tag{2.20}$$

Q = discharge, m^3/s (cfs)
y_c = critical depth, m (ft)
b = bottom width, m (ft)

2.3.3 Critical Flow in Trapezoidal Channels

In a trapezoidal channel, the area, A, and the top width, T, are presented by the following equations:

$$A = by + my^2 \tag{2.21}$$

$$T = b + 2my \tag{2.22}$$

where b, y, and m are the bottom width, depth, and side slope (horizontal to vertical ratio) of the channel, respectively. Combining Equations 2.7, 2.21, and 2.22 and simplifying results in

$$Q_c = \sqrt{g}K_c b y^{3/2} \tag{2.23}$$

Where

$$K_c = \left\{\frac{\left[1 + m\left(\frac{y}{b}\right)\right]^3}{\left[1 + 2m\left(\frac{y}{b}\right)\right]}\right\}^{1/2} \tag{2.24}$$

is a dimensionless number. Table 2.1, prepared by the author, lists values of K_c for a wide range in the y/b, side slope, m. Since K_c is dimensionless, this table is equally applicable to SI and CU.

Example 2.1

Calculate the critical depth for a discharge of 120 cfs in a 4 ft wide box culvert.

TABLE 2.1
K_c Values for Trapezoidal Channels[a]

	m			
y/b	**1**	**1.5**	**2.0**	**3.0**
0.1	1.053	1.082	1.111	1.172
0.2	1.111	1.172	1.235	1.364
0.3	1.172	1.267	1.364	1.565
0.4	1.235	1.364	1.498	1.770
0.5	1.299	1.464	1.633	1.976
0.6	1.364	1.565	1.770	2.185
0.7	1.431	1.667	1.907	2.394
0.8	1.498	1.770	2.046	2.603
0.9	1.565	1.873	2.185	2.813
1.0	1.633	1.976	2.324	3.024
1.2	1.770	2.185	2.603	3.445
1.4	1.907	2.394	2.883	3.868

[a] $\alpha = \cos^{-1}(1 - 2\,y/D)$.

Solution

Using Equation 2.1

$$q = \frac{120}{4} = 30$$

$$y_c = \left(30^2/32.2\right)^{1/3} = 3.035 \approx 3.04 \text{ ft}$$

Using Equation 2.20,

$$y_c = 0.314\left(120/4\right)^{2/3} = 3.032 \text{ ft}$$

Example 2.2

Calculate the critical discharge in a trapezoidal channel of 1.5 m bottom width and 2:1 side slope for a depth of 1 m.

Solution

The solution uses both Equations 2.23 and 2.24 and Table 2.1.

Using Equation 2.24

$$\frac{y}{b} = \frac{1}{1.5} = 0.667$$

$$K_c = \left[(1+2\times 0.667)^3/(1+2\times 2\times 0.667)\right]^{1/2}$$

$$K_c = 1.861$$

$$Q = 1.861\times\sqrt{9.81}\times 1.5\text{x}1^{3/2} = 8.74 \text{ m}^3\text{/s}$$

Using Table 2.1, interpolate K_c between $y/b = 0.6$ and $y/b = 0.7$ under the $m = 2$ column for

$$\frac{y}{b} = 0.6,\ K_c = 1.770$$

$$\frac{y}{b} = 0.7,\ K_c = 1.907$$

Interpolate:

$$K_c = [(1 + 2x0.667)^3 / (1 + 2x2x0.667)]^{1/2}$$

$$K_c = 1.861$$

$$Q = 1.861x\sqrt{}9.81x1.5x1^{3/2} = 8.74\ m^3/s$$

Using Table 2.1, interpolate K_c between $y/b = 0.6$ and $y/b = 0.7$ under the $m = 2$ column for

$$For\frac{y}{b} = 0.6,\ K_c = 1.770$$

$$For\frac{y}{b} = 0.7,\ K_c = 1.907$$

Interpolate:

$$K = 1.77 + \frac{(1.907 - 1.770)}{(0.7 - 0.6)} \times (0.667 - 0.6) = 1.862$$

$$Q = \sqrt{}9.81x1.862x1.5x1^{1.5} = 8.755\ m^3/s$$

Example 2.3

Water flows at a rate of $2\ m^3/s$ in a trapezoidal channel lined with a Reno mattress (see Chapter 9. The channel has 1 m bottom width and 1:1 side slopes. Calculate the critical depth in this channel.

Solution

$$A = y + y^2$$

$$T = 1 + 2y$$

$$f(y) = \frac{Q^2T}{gA^3} = \frac{4(1+2y)}{9.81(y+y^2)^3} = 0.40775\frac{(1+2y)}{(y+y^2)^3}$$

The critical depth can be calculated by solving the equation $f(y) = 1$ by trial and error and these are tabulated values.

y	1.00	0.50	0.600	0.610	0.602
f(y)	0.15	1.93	1.014	0.959	1.002

Thus, $y_c = 0.602$ m.

2.3.4 Critical Flow in Partly Full Circular Pipes

For partly full circular sections (Table 2.2), A and T are expressed by

$$A = \left(\frac{D^2}{8}\right)(2\alpha - \sin 2\alpha) \tag{2.25}$$

and

$$T = D \sin \alpha \tag{2.26}$$

TABLE 2.2
Critical Flow in Partly Full Circular Pipes

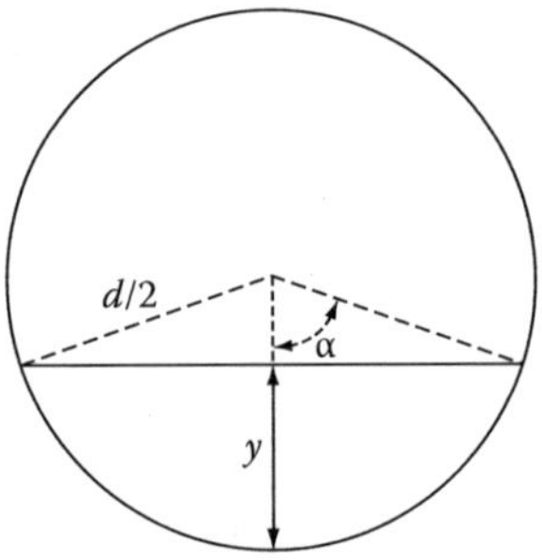

y_c/D	α^a	K_c
0.05	25.84	0.0027
0.10	36.87	0.0107
0.15	45.57	0.0238
0.20	53.13	0.0418
0.25	60.00	0.0647
0.30	66.42	0.0921
0.35	72.54	0.1241
0.40	78.46	0.1605
0.45	84.26	0.2012
0.50	90.00	0.2461
0.55	95.74	0.2952
0.60	101.54	0.3487
0.65	107.46	0.4068
0.70	113.58	0.4700
0.75	120.00	0.5397
0.80	126.87	0.6181
0.85	134.43	0.7102
0.90	143.13	0.8294
0.92	147.14	0.8923
0.94	151.64	0.9731
0.96	156.93	1.0895
0.98	163.74	1.3060

[a] $\alpha = \cos^{-1}(1 - 2\,y/D)$.

Substituting these equations in the general Equation 2.6 gives

$$\frac{\left(64Q^2\sin\alpha\right)}{\left[gD^5\left(\alpha-\frac{1}{2}\sin 2\alpha\right)^3\right]}=1 \tag{2.27}$$

This equation simplifies as

$$Q=K_c g^{1/2}D^{5/2} \tag{2.28}$$

where

$$K_c=\frac{\left(\alpha-0.5\sin 2\alpha\right)^{3/2}}{8\left(\sin\alpha\right)^{1/2}} \tag{2.29}$$

In the preceding equation, the angle α is expressed in radians and is associated with the flow depth by

$$\frac{y_c}{D}=\frac{\left(1-cos\ \alpha\right)}{2} \tag{2.30}$$

or

$$\alpha=\cos^{-1}\left(1-\frac{2y_c}{D}\right) \tag{2.31}$$

The angle α, calculated from Equation 2.31 and substituted into Equation 2.29, can be determined for any given y/D. Table 2.2 prepared by this author lists K_c for a wide range of y_c/D. It should be noted that K_c is dimensionless; therefore, the table applicable to both metric and customary units.

Figure 2.3 (a) plots the critical discharge as a function of flow depth for circular pipes ranging in diameter from 300 to 4500 mm, and Figure 2.3 (b) plots the same for pipes of 1.0 ft (12 in.) to 15 ft (180 in.) in diameter. Figures 2.3 (a) and 2.3 (b), though less accurate than Table 2.2, allow for a direct estimation of critical depth versus discharge or vice versa for various pipe sizes.

Example 2.4

Calculate the critical discharge for a 48-inch reinforced concrete pipe, if y_c = 2.5 ft.

Solution

$$\cos^{-1}(1-1.25)=\cos^{-1}(-0.25)$$

$$\alpha=104.48°=104.48\times\frac{\pi}{180}=1.824\text{ radians}$$

Then calculate K_c using Equation 2.29:

$$\sin\alpha=0.968$$
$$\sin 2\alpha=-0.484$$

$$K_c=0.377$$

$$Q=K_c\times\sqrt{g}\,D^{5/2}=0.377\times(32.2)^{1/2}(4)^{5/2}=68.5\text{ cfs}$$

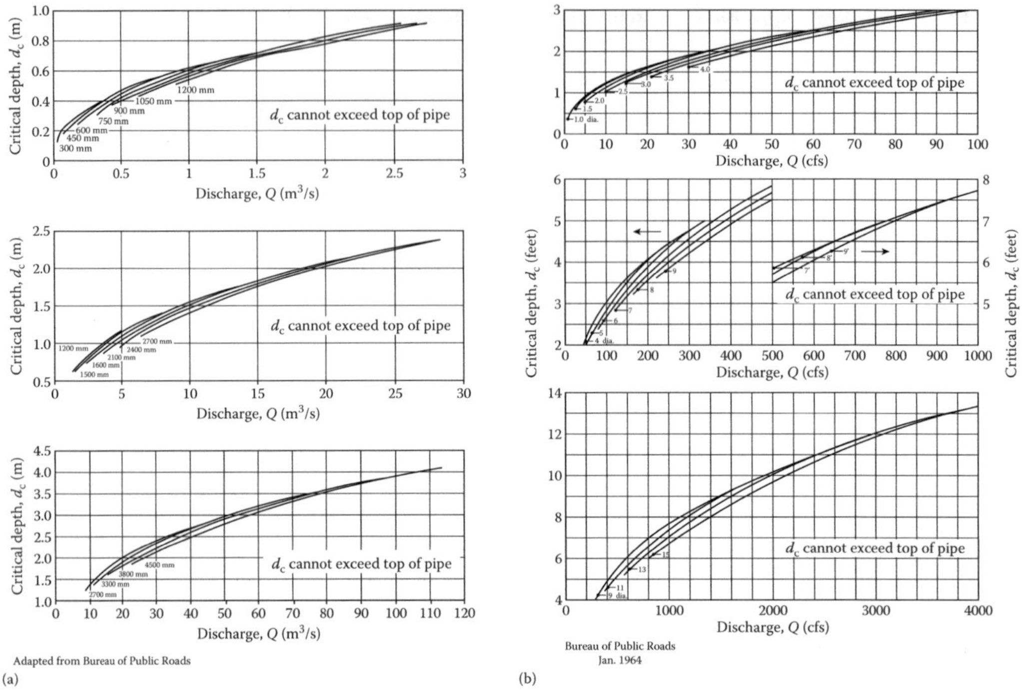

FIGURE 2.3 Critical depth–discharge relationships for circular pipes in (a) SI and (b) CU.

Note: The critical depth, y_c, is designated as d_c in this figure.

Alternatively, interpolate α and K_c between $d_c/D = 0.6$ and 0.65 in Table 2.2:

$$\alpha = \frac{(101.54 + 107.46)}{2} = 104.50°$$

$$K_c = \frac{(0.3487 + 0.4068)}{2} = 0.3778$$

$$Q = 68.6\ cfs$$

Exercise: Calculate the critical discharge in a 1200 mm pipe for $y_c = 750$ mm.

2.4 NORMAL DEPTH

Normal depth is defined as the depth of flow for a constant discharge and under the force of gravity. Specifically, normal depth is associated with a free surface flow and is differentiated from pressure flow, where the pressure at the top of conduit deviates from atmospheric pressure. In a uniform flow, the slope of the channel, the slope of HGL, and the slope of EL are equal, and the HGL and EL are parallel to the channel bed. A number of equations have been introduced for analyzing the normal flow in channels. Two of the more popular equations are discussed next.

2.4.1 Chezy's Formula

In 1775, Chezy presented the following formula for the average flow velocity in a channel:

$$V = C\sqrt{RS} \tag{2.32}$$

where R is the hydraulic radius, defined as the area divided by the wetted perimeter ($R = A/P$), S is the energy gradient, which is the same as channel slope in uniform flow, C is Chezy's coefficient, which has the dimension of $L^{1/2}/T$. A number of formulae have been proposed for C, some of which are:

a. Pavlosky presented the following formula in metric units for C:

$$C = \left(\frac{1}{n}\right) R^{\alpha} \tag{2.33}$$

where $\alpha = 1.5\sqrt{n}$; for $R \le 1$ m and $\alpha = 1.3\sqrt{n}$; for $R > 1$ m.
In this formula, n is the Manning's roughness coefficient, to be defined in the next section.

b. The Prague Hydraulics Institute proposed Chezy's C to be a function of the hydraulic radius, R, and the mean stone diameter, d_{50}, on the channel bed and in SI is expressed by the following formula:

$$C = 18\,log\left(\frac{R}{d_{50}}\right) + 3 \tag{2.34}$$

c. Bazin developed the following relation between C and R:

$$C = \frac{87}{\left(1 + \frac{m}{\sqrt{R}}\right)} \tag{2.35}$$

where m is Bazin's coefficient, which varies from 0.07 for very smooth concrete channels to 1.90 for rock-lined channels. These coefficients correspond to Manning's n values of 0.011 and 0.04, respectively. Table 2.3 provides the relation between Manning's n and Bazin's m for lined channels and natural streams (Pazwash, 2007). However, since these three proposed formulas do not yield consistent results for C, Chezy's formula is not used in the United States.

TABLE 2.3
Typical Manning's *n* and Bazin's *m* (in Chezy's Formula) Values for Lined Channels and Natural Streams

Type of Cover	Manning's *n*	Bazin's *m*
Concrete, very smooth	0.010	0.07
Concrete, normal	0.015	0.25
Brick/mortar stone lined	0.016	0.35
Earth channel, smooth	0.023	0.85
Earth channel, semismooth	0.027	1.30
Rocky channels	0.040	1.90
Streams/rivers, normal	0.030	1.65

2.4.2 Manning's Formula

In 1889, Manning presented the following formula for the normal flow velocity in a channel in English units:

$$V = \left(\frac{1.486}{n} \right) R^{2/3} S^{1/2} \tag{2.36}$$

In SI, Manning's formula, which is known as Gauckler–Manning–Strickler's formula in Europe, reads as

$$V = \left(\frac{1}{n} \right) R^{2/3} S^{1/2} \tag{2.37}$$

where

V = flow velocity, m/s
R = hydraulic radius, m
S = energy slope, m/m
n = Manning's roughness coefficient

The coefficient n varies with the surface roughness of channel. From studies of streams with gravel beds in Switzerland in 1923, Strickler proposed the following equation relating n with the gravel size:

$$n = 0.0417 d_{50}^{1/6} \tag{2.38}$$

where d is the median size of the bed material in meters. In CU, with d expressed in feet, this equation becomes (Henderson, 1966)

$$n = 0.034 d_{50}^{1/6} \tag{2.39}$$

Equating the expressions in Equations 2.32 and 2.39 relates Chezy's coefficient (C) to Manning's coefficient (n) as follows:

$$C = \frac{1.49 R^{1/6}}{n} \tag{2.40}$$

The constant 1.486 in the Manning's formula in CU represents the ratio of meter to foot to 1/3 power. More accurately, this constant would be

$$(3.2808)^{1/3} = 1.4859$$

In practice, however, this constant is commonly rounded to 1.49. Thus, it appears that Manning's formula must have been originally developed in metric units. Hagen first proposed Manning's formula in SI in 1876 (Chow, 1959).

The literature generally ascribes a dimension $L^{1/6}$ to n and $L^{1/2}/T$ to 1.49. However, the same n is used in the CU and SI versions of the Manning formula. For this reason, it is more appropriate to treat n as a dimensionless number and ascribe dimension $L^{1/3}/T$ to the coefficients 1.49 and 1.0 so that the Manning formula is dimensionally homogeneous. This statement is verified by the relation $(1 \text{ m/s})^{1/3} = (3.28 \text{ ft/s})^{1/3} = 1.486 (\text{ft/s})^{1/3}$.

In terms of discharge, the Manning formula is given as

$$Q = \left(\frac{1}{n}\right) AR^{2/3} S^{1/2} \tag{2.41}$$

$$Q = \left(\frac{1.49}{n}\right) AR^{2/3} S^{1/2} \tag{2.42}$$

in SI and CU, respectively.

In uniform channel flow where, S and S_o—the slope of channel bed (or pipe flowing partially full) are equal, Manning's formula is expressed as

$$Q = \left(\frac{1}{n}\right) AR^{2/3} S_0^{1/2} \text{ SI} \tag{2.43}$$

$$Q = \left(\frac{1.49}{n}\right) AR^{2/3} S_0^{1/2} \text{ CU} \tag{2.44}$$

where
Q = discharge, m^3/s (cfs)
A = flow area, m^2 (ft^2)
S_o = channel slope, m/m (ft/ft)

Notably, Manning's formula is also employed for pipe flow calculations and design of storm drains.

Table 2.4 provides equations for areas and hydraulic radius of pipe and channel geometries that are commonly used. The Manning formula is used not only for analyzing uniform flow but also for gradually varied flow in pipes and open channels. Table 2.5 lists typical Manning's n for

TABLE 2.4
Geometric Properties of Channel Sections

Section	Area A	Wetted Perimeter P	Hydraulic Radius R	Top Width T
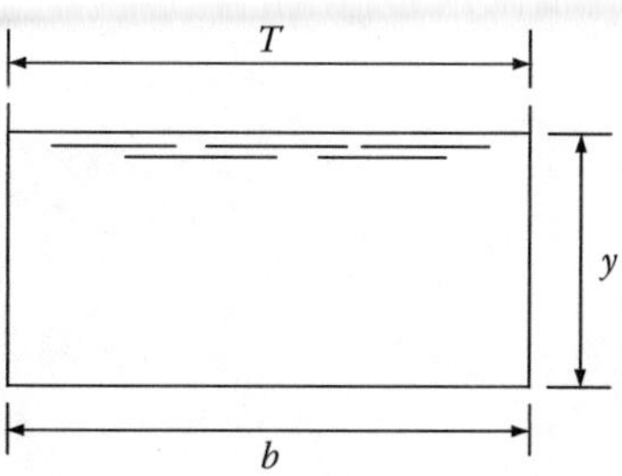 Rectangle	by	$b+2y$	$\frac{by}{b+2y}$	b
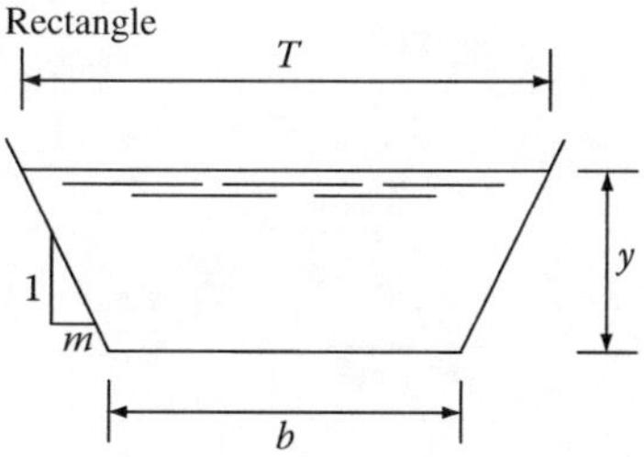				

(Continued)

TABLE 2.4 (*Continued*)
Geometric Properties of Channel Sections

Section	Area A	Wetted Perimeter P	Hydraulic Radius R	Top Width T
Trapezoid 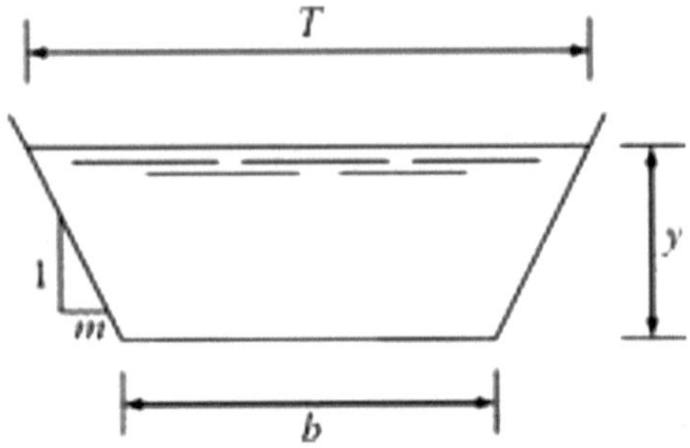		$b+2y\sqrt{1+m^2}$	$\dfrac{(b+my)y}{b+2y\sqrt{1+m^2}}$	$b+2my$
Triangle 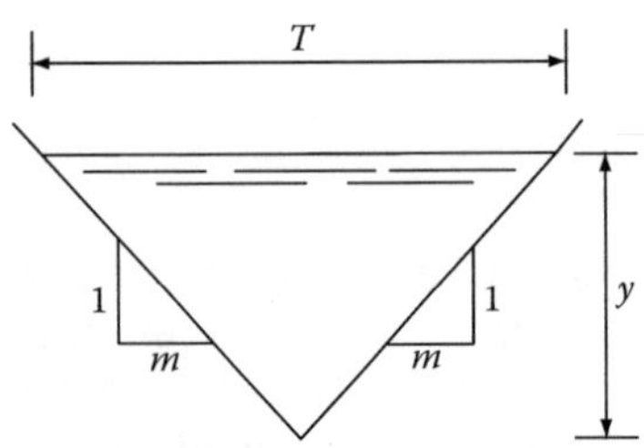	my^2	$2y\sqrt{1+m^2}$	$\dfrac{my}{2\sqrt{1+m^2}}$	$2my$
Circle 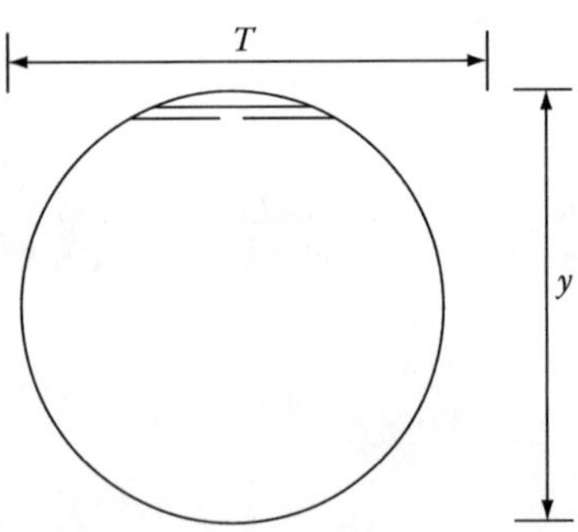	$\dfrac{\pi D^2}{4}$	πD	$\dfrac{D}{4}$	0
Semicircle 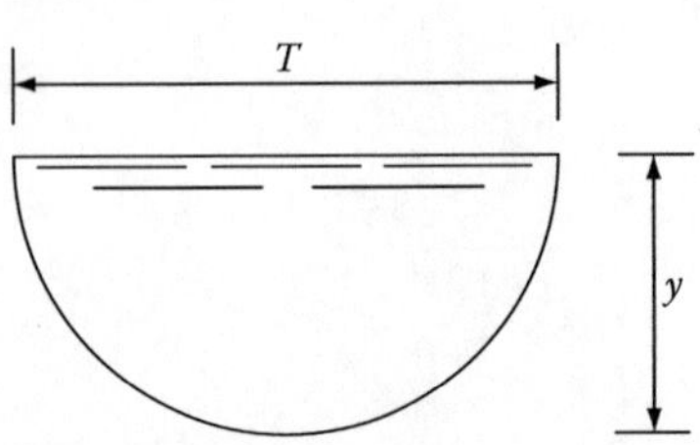	$\dfrac{\pi D^2}{8}$	$\dfrac{\pi D}{2}$	$\dfrac{D}{4}$	D
Parabola 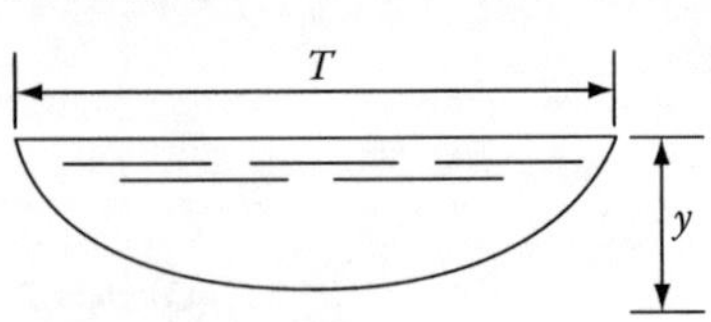	$\dfrac{2}{3}Ty$	$T+\dfrac{8}{3}\dfrac{y^2}{T}$	$\dfrac{2T^2y}{3T^2+8y^2}$	$\dfrac{3}{2}\dfrac{A}{y}$

Note: when $T > 4y$.

TABLE 2.5
Manning's Roughness *n* for Pipes and Channels

Conduit Material	*n*
Concrete pipes	0.012–0.013
Ductile iron pipes	0.011–0.015
PVC pipes	0.010–0.011
HDPE pipes	0.011–0.012[a]
Vitrified clay pipes	0.011–0.015
Corrugated metal pipes (1/2–3 in. [12.5–75 mm] corrugation)	0.022–0.028
Old brick lined conduits	0.013–0.017
Open Channels (Lined)	
Brick	0.012–0.018
Concrete	0.011–0.016
Riprap	0.025–0.040
Grass	0.030–0.30[b]
Dredged Channels	
Earth, straight and uniform	0.020–0.03
Earth, winding	0.025–0.04
Rock, 2–12 in. (5–30 cm)	0.025=0.06
Natural Channels and Streams	
Fairly uniform	0.030– 0.05
Irregular	0.040–0.08
Overbanks	0.080–0.12

[a] ADS, the largest manufacturer of HDPE pipes in the United States, recommends using n = 0.012.

[b] Varies with grass height, depth, and velocity of flow (see Chapter 4).

different makes of pipes and types of linings of manmade or natural channels (also see Barnes, 1987; Brater et al., 1996; Chow, 1959; Henderson, 1966; French 1885; and HEC-22, 2013, 2024).

Figures 2.4a and b present graphical solution of Manning's formula for full flow in pipes in metric and customary units, respectively. Tables 2A.1 to 2A.3 in Appendix 2A provide hydraulic properties of round and elliptical concrete pipes, respectively. These tables facilitate performing design calculations for storm drains.

2.5 CALCULATION OF FLOW DEPTH

Discharge in a conduit depends on both the area and the hydraulic radius, each of which not only is a function of depth but also depends on the conduit geometry. Therefore, a simple relationship cannot be established between discharge and flow depth. In practice, the depth of flow is commonly calculated using an iterative process. Simplified procedures for calculating the flow depths in circular pipes (flowing partly full) and trapezoidal channels are derived in the following sections.

2.5.1 Circular Sections

For circular sections such as full or partly full drainage pipes Manning's formula is written as

$$Q = \left(\frac{1}{n}\right) K D^{8/3} S^{1/2} \text{ SI} \tag{2.45}$$

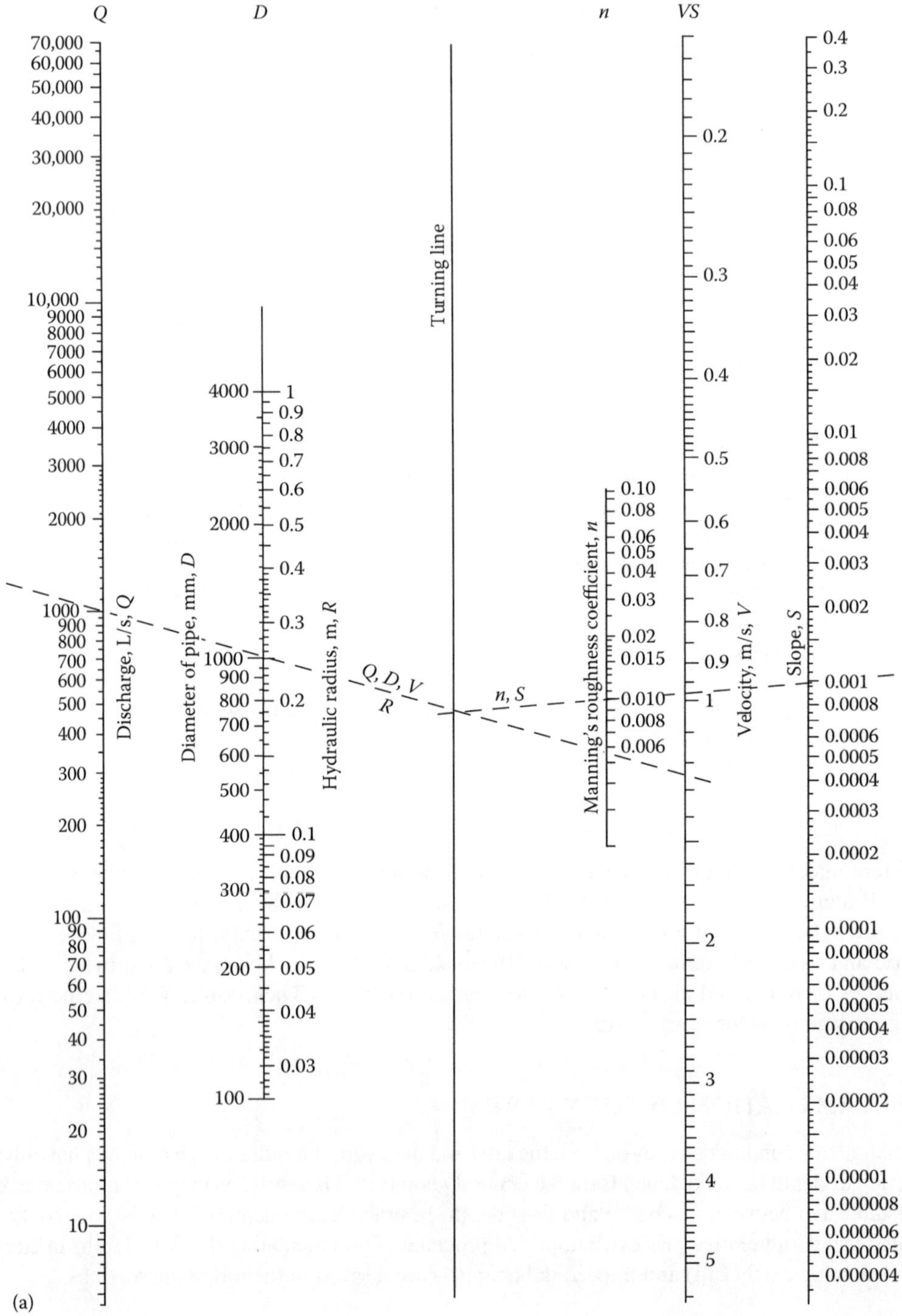

FIGURE 2.4A Manning's formula nomograph (SI).

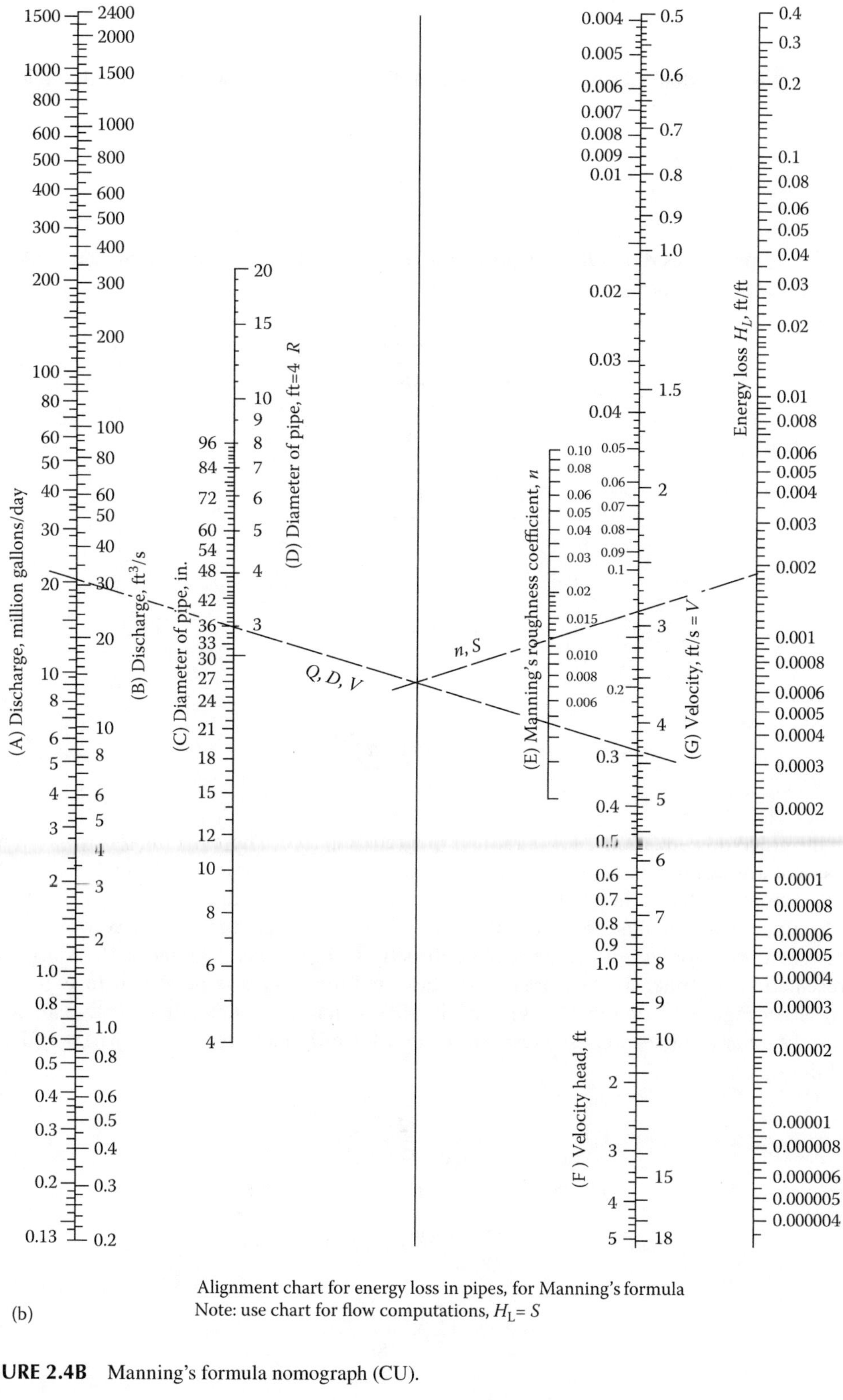

(b) Alignment chart for energy loss in pipes, for Manning's formula
Note: use chart for flow computations, $H_L = S$

FIGURE 2.4B Manning's formula nomograph (CU).

$$Q = \left(\frac{1.49}{n}\right) KD^{8/3}S^{1/2} \text{ CU} \tag{2.46}$$

where D is pipe diameter in m (ft) and K is a dimensionless number given by the following equation

$$K = \frac{AR^{2/3}}{D^{8/3}} \tag{2.47}$$

Hydraulic properties A and R are expressed in terms of pipe geometrical parameters in the following dimensionless forms

$$\frac{A}{D^2} = \frac{(2\alpha - \sin 2\alpha)}{8} \tag{2.48}$$

$$\frac{P}{D} = \alpha \tag{2.49}$$

$$\frac{R}{D} = \frac{1}{4}\left[1 - \frac{(\sin 2\alpha)}{2\alpha}\right] \tag{2.50}$$

where

$$\alpha = \cos^{-1}(1 - 2y/D) \tag{2.51}$$

therefore

$$K = \frac{(2\alpha - \sin 2\alpha)^{5/3}}{32\alpha^{2/3}} \tag{2.52}$$

In the preceding equations, α is in radians (degrees x π/180). These parameters together with the ratio of average depth, D_a, to pipe diameter, D, are listed in Table 2.6. This table shows that the maximum discharge in partly full pipes occurs at approximately $y/D = 0.94$ the highest K. More accurate calculations show that the maximum discharge occurs at $y/D = 0.938$. Since K is dimensionless, it can be applied directly to Equations 2.45 and 2.46. A graphical representation of dimensionless area, velocities, and discharges is presented in Figure 2.5. Using either Figure 2.5 or Table 2.6 greatly simplifies pipe flow calculations (Pazwash, 2007). Figure 2.5 shows dimensionless flow parameters for elliptical pipes under partly full flow conditions.

$$\alpha = \left(\frac{1-2y}{D}\right)$$

$$T = D\sin\alpha = 2\big(y(D-y)\big)^{0.5}\text{, top width}$$

$$A/D^2 = \frac{(2\alpha - \sin 2\alpha)}{8}$$

$$R = A/P = \frac{D(2\alpha - \sin 2\alpha)}{8\alpha}$$

TABLE 2.6
Hydraulic Properties of Partly Full Circular Sections

y/D	*T/D*	*R/D*	*A/D*²	*K*
0.05	0.4359	0.0326	0.0147	0.0015
0.10	0.6000	0.0635	0.0409	0.0065
0.15	0.7141	0.0929	0.0739	0.0152
0.20	0.8000	0.1206	0.1118	0.0273
0.25	0.8660	0.1466	0.1535	0.0427
0.30	0.9165	0.1709	0.1982	0.0610
0.35	0.9539	0.1935	0.2450	0.0820
0.40	0.9798	0.2142	0.2934	0.1050
0.\]\45	0.9950	0.2331	0.3428	0.1298
0.50	1.0000	0.2500	0.3927	0.1558
0.55	0.9950	0.2649	0.4426	0.1826
0.60	0.9798	0.2776	0.4920	0.2094
0.65	0.9539	0.2881	0.5404	0.2358
0.70	0.9165	0.2962	0.5872	0.2610
0.75	0.8660	0.3017	0.6319	0.2842
0.80	0.8000	0.3042	0.6736	0.3047
0.85	0.7141	0.3033	0.7115	0.3212
0.90	0.6000	0.2980	0.7445	0.3322
0.92	0.5426	0.2944	0.7560	0.3345
0.94	0.4750	0.2895	0.7662	0.3353
0.96	0.3919	0.2829	0.7749	0.3339
0.98	0.2800	0.2735	0.7816	0.3294
1.00	0.0000	0.2500	0.7854	0.3117

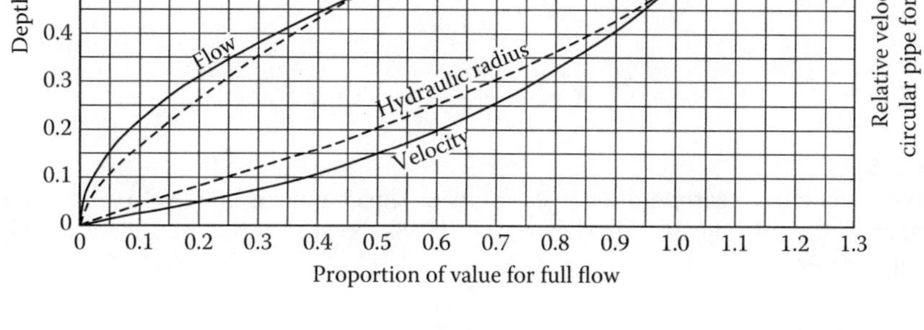

FIGURE 2.5 Relative velocity and flow in a partly full circular pipe.

Using K-value for full flow condition

$$D = 1.55\left(\frac{Q \cdot n}{\sqrt{s}}\right)^{\frac{3}{8}}, \text{ SI}$$

$$D = 13\left(\frac{Q \cdot n}{\sqrt{s}}\right)^{\frac{3}{8}}, \text{ CU}$$

where D is the required pipe diameter.

2.5.2 Trapezoidal Sections

In a trapezoidal section, Manning's formula is written as

$$Q = \left(\frac{K}{n}\right) b^{8/3} S^{1/2} \text{ SI} \tag{2.53}$$

$$Q = 1.49\left(\frac{K}{n}\right) b^{8/3} S^{1/2} \text{ CU} \tag{2.54}$$

where

$$K = \frac{AR^{2/3}}{b^{8/3}} = \frac{\left[\left(\frac{y}{b}\right) + m\left(\frac{y}{b}\right)^2\right]^{5/3}}{\left[1 + 2\left(\frac{y}{b}\right)\sqrt{\left(1 + m^2\right)}\right]^{2/3}} \tag{2.55}$$

b = bottom width
y = flow depth
m = side slope (H/V).

Table 2.7, prepared by the author, lists dimensionless K values for a wide range in the (y/b) ratio and side slope, m. The column $m = 0$ in this table represents a rectangular section such as a box culvert. Since K is dimensionless, this table applies equally to SI and CU. Figure 2.6 graphically depicts K as a function of y/b and m. The use of either Table 2.7 or Figure 2.6 eliminates the need for an iterative solution and greatly simplifies open channel flow calculations.
where

$$K = \frac{Q \cdot n}{b^{8/3} S^{1/2}} \text{ SI}$$

$$K = \frac{Q \cdot n}{1.49 b^{8/3} S^{1/2}} \text{ CU}$$

Example 2.5
A roadside channel is to carry a discharge of 1 m^3/s. The channel is lined with 5 cm stone, has a 1 m bottom width, 2:1 (2H:1V) side slopes, and 0.75% longitudinal slope. Calculate the normal depth of flow in this channel.

Solution

The calculations are performed using both the iterative procedure and Table 2.7.

TABLE 2.7
Hydraulic Parameter K for Rectangular and Trapezoidal Channels

$$K = \frac{AR^{2/3}}{b^{8/3}} = \frac{\left[\left(\dfrac{y}{b}\right) + m\left(\dfrac{y}{b}\right)^2\right]^{5/3}}{\left[1 + 2\left(\dfrac{y}{b}\right)\sqrt{\left(1+m^2\right)}\right]^{2/3}}$$

y/b	$m = 0$	$m = 1/4$	$m = 1/2$	$m = 3/4$	$m = 1.0$	$m = 3/2$	$m = 2.0$	$m = 3.0$
0.00	0.0000	0.0000	0.0000	0.0000	0.0000	0.0000	0.0000	0.0000
0.01	0.0005	0.0005	0.0005	0.0005	0.0005	0.0005	0.0005	0.0005
0.02	0.0014	0.0014	0.0015	0.0015	0.0015	0.0015	0.0015	0.0015
0.03	0.0028	0.0028	0.0028	0.0029	0.0029	0.0029	0.0029	0.0029
0.04	0.0044	0.0045	0.0046	0.0046	0.0047	0.0047	0.0048	0.0049
0.05	0.0064	0.0065	0.0066	0.0067	0.0067	0.0069	0.0070	0.0071
0.06	0.0085	0.0087	0.0089	0.0090	0.0091	0.0093	0.0095	0.0098
0.07	0.0109	0.0112	0.0114	0.0116	0.0118	0.0121	0.0123	0.0128
0.08	0.0135	0.0139	0.0142	0.0145	0.0147	0.0152	0.0155	0.0162
0.09	0.0162	0.0167	0.0172	0.0176	0.0179	0.0185	0.0190	0.0199
0.10	0.0191	0.0198	0.0204	0.0209	0.0214	0.0221	0.0228	0.0241
0.15	0.0356	0.0376	0.0394	0.0409	0.0422	0.0445	0.0466	0.0504
0.20	0.0547	0.0589	0.0627	0.0659	0.0687	0.0737	0.0783	0.0868
0.25	0.0757	0.0832	0.0898	0.0956	0.1007	0.1099	0.1182	0.1340
0.30	0.0983	0.1100	0.1205	0.1298	0.1382	0.1532	0.1669	0.1928
0.35	0.1220	0.1392	0.1547	0.1686	0.1812	0.2038	0.2246	0.2641
0.40	0.1468	0.0167	0.1922	0.2118	0.2297	0.2621	0.2919	0.3486
0.45	0.1723	0.0198	0.2330	0.2596	0.2840	0.3283	0.3693	0.4472
0.50	0.1984	0.2390	0.2770	0.3119	0.3440	0.4027	0.4571	0.5606
0.55	0.2251	0.2761	0.3243	0.3688	0.4100	0.4856	0.5558	0.6896
0.60	0.2523	0.3150	0.3748	0.4304	0.4822	0.5773	0.6660	0.8350
0.65	0.2799	0.3557	0.4286	0.4968	0.5605	0.6781	0.7880	0.9974
0.70	0.3079	0.3980	0.4856	0.5681	0.6453	0.7884	0.9222	1.1777
0.75	0.3361	0.4421	0.5460	0.6442	0.7367	0.9083	1.0692	1.3766
0.80	0.3646	0.4879	0.6096	0.7254	0.8347	1.0383	1.2293	1.5946
0.85	0.3934	0.5354	0.6767	0.8118	0.9297	1.1785	1.4030	1.8325
0.90	0.4223	0.5846	0.7471	0.9033	1.0516	1.3292	1.5907	2.0911
0.95	0.4514	0.6354	0.8210	1.0002	1.1708	1.4908	1.7927	2.3708
1.00	0.4807	0.6879	0.8984	1.1024	1.2973	1.6636	2.0095	2.6725

a. Iterative solution:

Express the Manning's formula as

$$AR^{2/3} = \frac{nQ}{S^{1/2}}$$

$$A = by + my^2 = y + 2y^2$$

$$P = b + 2y\sqrt{1+m^2} = 1 + 2\sqrt{5}y$$

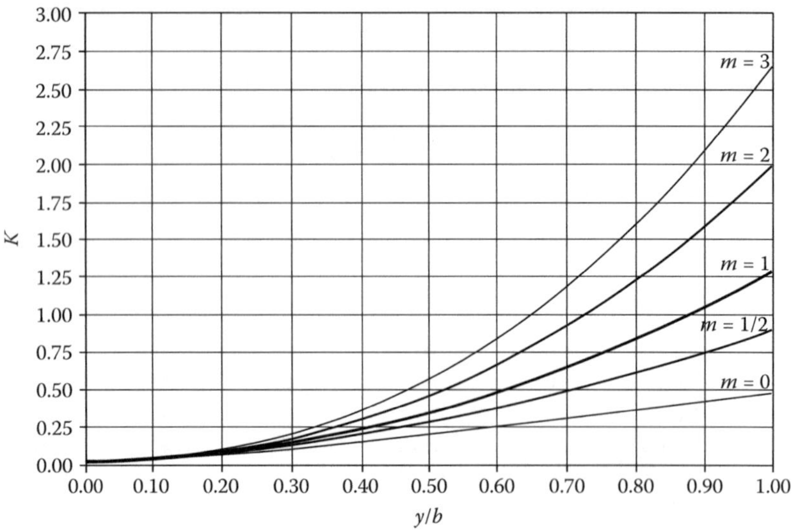

FIGURE 2.6 *K* values with *y*/*b* for rectangular and trapezoidal channels.

$$R = \frac{A}{P} = \frac{(y + 2y^2)}{(1 + 2\sqrt{5}y)}$$

$$AR^{2/3} = \frac{(y + 2y^2)^{5/3}}{(1 + 2\sqrt{5}y)^{2/3}}$$

Using Table 2.5, estimate n value at 0.025

$$\frac{nQ}{\sqrt{S}} = \frac{0.025 \times 1}{(0.0075)^{1/2}} = 0.2887$$

Then

$$AR^{2/3} = \frac{[y(1+2y)]^{5/3}}{[1+2\sqrt{5}y]^{2/3}} = 0.2887$$

Raise both sides of the preceding equation to 3/2 power:

$$\frac{[y(1+2y)]^{5/2}}{(1+4.472y)} = (0.2887)^{1.5} = 0.1551$$

Now solve the equation $f(y) = 0.1551$; where

$$f(y) = \frac{[y(1+2y)]^{5/2}}{(1+4.472y)}$$

First try $y = 0.5$ m

$$f(y) = 0.2975 > 0.1551$$

Second try $y = 0.4$ m

$$f(y) = 0.1577 > 0.1551$$

Third try $y = 0.39$ m

$$f(y) = 0.1413 < 0.1551$$

Fourth try $y = 0.395$ m

$$f(y) = 0.1467 < 0.1551$$

Interpolate y between 0.395 and 0.4 to get 0.398 m.

b. Using Table 2.7:
Under the $m = 2$ column, search for $K = 0.2887$:
y/b lies between 0.35 and 0.4. At

$$\frac{y}{b} = 0.35 \quad K = 0.2248;$$

$$\frac{y}{b} = 0.4 \quad K = 0.2919$$

Calculate y by interpolating y/b between these ratios, as follows:

$$\frac{y}{b} = 0.35 + \frac{0.2887 - 0.2248}{0.2919 - 0.2248} \times (0.4 - 0.35) = 0.35 + 0.952 \times 0.05$$

$$\frac{y}{b} = 0.3976, say, \frac{y}{b} = 0.398$$

$$y = 0.398 \times 1 = 0.398m$$

Example 2.6

To carry stormwater runoff at a rate of 15 cfs, an 18 in. RCP will be used. Calculate the minimum slope required for the pipe to carry this flow under normal flow conditions. What is the depth of flow if the pipe is laid at 3% slope?

Solution

Use $n = 0.012$ for smooth pipe $A = \frac{\pi D^2}{4} = 1.767sf$

$$R = \frac{D}{4} = 0.375 \text{ ft; see Table 2.4}$$

Full flow:

$$Q = \frac{149}{0.012} \text{x} 1.767 \text{x} (0.375)^{2/3} \times S^{1/2}$$

$$Q = 114 \times S^{1/2}$$

Alternatively, we can use Table 2A.2, to read $K = 114.0$.

$$S = \left(\frac{15}{114}\right)^2 = 0.017; 1.7\%$$

Partly full flow:
Instead of performing iterative calculations, use Table 2.6 to facilitate the solution.
Calculate $K = AR^{2/3}/D^{8/3}$.
Using Equation 2.46,

$$K = \frac{Qn}{1.49(D)^{8/3}\,S^{1/2}}$$

$$K = \frac{15 \times 0.012}{1.49(1.5)^{8/3} \times (0.03)^{1/2}} = 0.2366$$

According to Table 2.6, y/D lies between 0.65 and 0.7, though much closer to the former. At

$$\frac{y}{D} = 0.65 \quad K = 0.2358$$

$$\frac{y}{D} = 0.70 \quad K = 0.2610$$

By interpolation:

$$\frac{y}{D} = 0.65 + \frac{0.2366 - 0.2358}{0.2610 - 0.2358} \times (0.7 - 0.65)$$

$$\frac{y}{D} = 0.6516$$

$$y = 0.6516 \times 15 = 9.78\ in., say, 9.8\ in.$$

Alternatively: solve this problem using Figure 2.5.
First, calculate the full flow capacity at 3% slope:

$$Q_F = K\sqrt{S} = 114 \times (0.03)^{\frac{1}{2}} = 19.75\ cfs$$

Then calculate the Q/Q_F ratio:

$$\frac{Q}{Q_F} = \frac{15}{19.75} = 0.76$$

For this flow ratio, Figure 2.5 gives

$$\frac{y}{D} = 0.65$$

$$y = 9.75\ in., say, 9.8\ in$$

Example 2.7

A 38 in. × 60 in. horizontal elliptical RCP is to convey a design discharge of 130 cfs at 70% of full depth. What should the pipe slope be?

Solution

For $y/d = 0.7$, Figure 2A.1 in Appendix 2A shows $Q/Q_F = 0.86$.
Thus $Q_F = 130/0.86 = 151.2$ cfs

For $n = 0.012$ Table 2A.3 gives $K = 1565$

$$Q_F = 1565(S)^{1/2}$$

$$S = \frac{(151.2)^2}{1565} = 0.0093$$

$$S = 0.93\%$$

Example 2.8

During a heavy storm, the depth of water in the pipe of the previous example is measured at 31.5 in. Calculate the discharge.

Solution

$$\frac{y}{D} = \frac{31.5}{38} = 0.83$$

Using Figure 2A.1 with relative depth of 0.83 gives

$$\frac{Q}{Q_F} = 1.05$$

$$Q = 1.05 \times 145.3 = 152.6\ cfs$$

It should be noted that because drainage pipes are commonly designed to carry a discharge less than or equal to their normal capacity, flow in culverts can exceed the normal capacity due to inlet control condition. The inlet control capacity of circular pipes is shown in Figure 2A.2a and b in Appendix 2A. More information on this matter is provided in Chapter 4.

2.6 ENERGY LOSSES IN PIPES AND CULVERTS

Energy losses, also called head losses, occur due to friction along flow and at locations where there is a rapid change either in the cross-section or the flow direction. In design of drainage systems, these losses must be calculated to define the hydraulic grade line and/or energy line along the flow. While frictional losses are gradual and fairly uniform, the losses at junctions are abrupt and localized. These losses are commonly called minor losses, but in a reach of pipe or culvert with many junctions and bends and/or partial obstructions, the latter losses may surpass frictional losses. Thus, these losses are termed local rather than minor in this book. Example 2.9 later in this section exemplifies the matter. The following sections present equations for estimating head losses in drainage systems.

2.6.1 Friction Losses

The lead loss due to friction is calculated as follows:

$$h_f = S_f L \tag{2.56}$$

where

h_f = friction loss, m (ft)
S_f = friction slope, m/m (ft/ft)
L = length of pipe, m (ft)

The frictional slope, S_f, in this equation is the same as the hydraulic grade line. In uniform flow, as previously indicated, this slope is also equal to the slope of the pipe or channel. The friction slope in uniform flow can be calculated using the Manning's formula rearranged as follows:

$$S_f = c\frac{n^2V^2}{R^{1.33}} \tag{2.57}$$

where
R = hydraulic radius
c = 1.0 metric (0.45 CU)

The frictional slope in terms of the velocity head is expressed as follows:

$$S_f = \frac{19.62n^2}{R^{1.33}}\frac{V^2}{2g}\ SI \tag{2.58}$$

$$S_f = \frac{29n^2}{R^{1.33}}\frac{V^2}{2g}\ CU \tag{2.59}$$

Combining Equations 2.56 and 2.58/2.59 gives

$$h_f = \frac{19.62n^2L}{R^{1.33}}\frac{V^2}{2g} \tag{2.60}$$

$$h_f = \frac{29n^2L}{R^{1.33}}\frac{V^2}{2g} \qquad (2.61)^1$$

2.6.2 Local Losses

In practice, local losses are commonly expressed in terms of the velocity head, as follows

$$h_l = k\frac{V^2}{2g} \tag{2.62}$$

where
h_ℓ = local head loss
k = dimensionless loss coefficient
$V^2/2g$ = velocity head

In water pipes, local losses are also expressed in terms of equivalent pipe length producing the equal amount of loss. This length is calculated as follows:

$$L_l = \frac{h_L}{S_f} \tag{2.63}$$

Although not commonly practiced, the same procedure may be used in culverts under full-flow conditions. This matter is illustrated by Example 2.9.

[1] Using 1.486 rather than 1.49 in Manning's formula, the coefficient 29 will read 29.2.

When there is a sudden change in cross-section, such as an abrupt contraction or expansion, Equation 2.62 may be written in terms of the difference in the velocity head across the junction, as follows:

$$h_L = k\Delta\frac{V^2}{2g} \tag{2.64}$$

where $\Delta V^2/2g$ represent the change in the velocity head.

The loss coefficient, k , depends on the type of flow, namely open channel or pressure, and varies from subcritical to supercritical. More information on this matter can be found in Chow (1959), Henderson (1966), French (1985), Brater et al. (1996), and FHWA (2013). Head loss equations for some of the common drainage structures are presented in the following sections.

2.6.2.1 Entrance and Exit Losses

At the inlet face of a pipe or culvert, the flow velocity increases; as a result, the pressure drops. Neglecting the velocity head at the section just before the entrance, Equation 2.64 simplifies as

$$h_l = k_e\frac{V^2}{2g} \tag{2.65}$$

where V is the flow velocity in the pipe. The entrance loss coefficient, k_e, for pipes varies from 0.1 for rounded entrance to 0.7 for mitered to slope connections. Table 2.8 lists entrance loss coefficients for pipes and culverts under various wing wall conditions.

At the exit section from a pipe or culvert to a body of water, the flow undergoes a sudden expansion. The exit loss coefficient at this section depends on the flow condition of the

TABLE 2.8
Entrance Loss Coefficient k_e for Pipes and Culverts

Type of Structure	k_e
Concrete Pipe	
Mitered to conform to slope	0.7
End section	0.5
Projecting, square cut	0.5
Projecting, groove end	0.2
Square-edge at headwall	0.5
Rounded at headwall	0.2
Groove end	0.2
Beveled edges	0.2
Box Culvert	
Wing walls parallel (extending at sides)	0.7
Wing walls at 30°±75° to culvert	0.5
Headwall along embankment (no wing walls)	0.5
Rounded entrance at headwall	0.2
Tapered to slope	0.2
Plastic Pipes	
End section	0.2
Square cut	0.5

receiving water body. For a stagnant body of water, the entire velocity head is lost at the outlet, and therefore,

$$h_o = \frac{V_o^2}{2g} \tag{2.66}$$

However, if the body of water is in motion, such as a channel or stream, the exit loss is

$$h_o = \frac{V_o^2}{2g} - \frac{V_d^2}{2g} \tag{2.67}$$

where

V_o = outlet velocity
V_d = channel velocity in the direction of pipe flow

When the outlet is perpendicular to the receiving channel, $V_d = 0$ and the preceding equations simplify as

$$h_o = \frac{V_o^2}{2g} \tag{2.68}$$

reflecting an exit loss coefficient, $k_o = 1$.

2.6.2.2 Sudden Expansions or Contractions

The loss in an abrupt expansion results from separation of the flow from the wall within a distance from the expansion section (see Figure 2.7). Applying the momentum and energy equations between sections 1 and 2 results in the following equation for head loss:

$$h_e = \frac{(V_1 - V_2)^2}{2g} \tag{2.69}$$

This equation may be expressed as

$$h_e = \frac{kV_1^2}{2g} \tag{2.70}$$

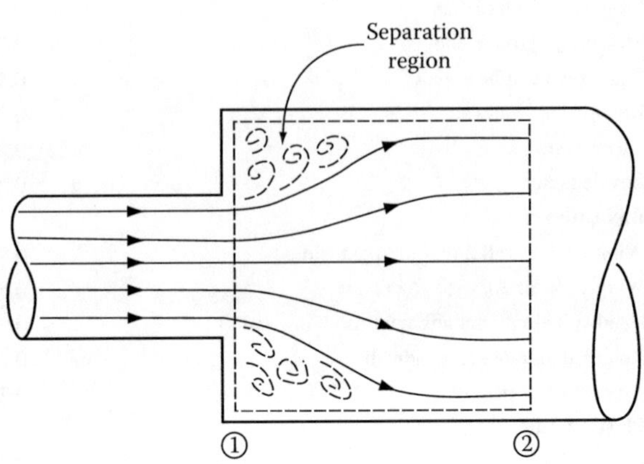

FIGURE 2.7 Sudden expansion.

TABLE 2.9
Sudden Contraction k_c for Pipes

D_2/D_1	0.90	0.80	0.60	0.40	0.20	0.10
k_c	0.05	0.15	0.28	0.36	0.42	0.45

where

$$k = \left[1 - \left(\frac{D_1}{D_2}\right)^2\right]^2 \tag{2.71}$$

A special case is a pipe ending at a reservoir. In this case, velocity V_2 is negligible compared with V_1 and the preceding equation simplifies to $h_l = \frac{V_1^2}{2g}$, which is the same as the exit loss

In a sudden contraction, the equation for head loss is expressed as

$$h_l = \frac{k_c V_2^2}{2g} \tag{2.72}$$

where k depends on the ratio of the areas before and after the contraction section. Typical values of k_c are listed in Table 2.9.

2.6.2.3 Bend Losses

At a pipe bend, the head loss is expressed by the following equation:

$$h_b = k_b \frac{V^2}{2g} \tag{2.73}$$

where

$$k_b = 0.003(\Delta\theta) \tag{2.74}$$

and $\Delta\theta$ = angle of curvature in degrees.

For a 90° bend, the head loss coefficient is approximately 0.3. The head loss in a culvert flowing full may be estimated using Equation 2.73 with a bend factor given by

$$k_b = 0.5\left(\frac{\Delta\theta}{90}\right)^{1/2} \tag{2.75}$$

For a 90° bend, this equation gives $k_b = 0.5$. The preceding equations are for sharp bends. When using a circular bend with a radius four times or greater than the culvert diameter, the head loss can be neglected.

2.6.2.4 Head Loss at Transitions

In gradual expansions and contractions, the head loss depends on the angle of expansion or contraction. It also depends on the ratio of pipe diameters, which reflects the length of the transition. In general, the head loss is given by

$$h_l = k_t\left(\frac{V_1^2}{2g} - \frac{V_2^2}{2g}\right) \tag{2.76}$$

where k_t depends on the angle of expansion/contraction as depicted in Figure 2.8 and the pipe diameters.

Table 2.10 provides a sample of expansion loss coefficients; for more information, the reader is referred to FHWA (2012) and ASCE and Water Environment Foundation (1992). In a gradual contraction, the lead loss is smaller than that of an expansion for a given angle. The head loss coefficient in this case may be approximated as one-half of the expansion coefficients listed in Table 2.10.

2.6.2.5 Junction Losses

At a pipe junction, such as that shown in Figure 2.9, the head loss can be calculated applying the momentum equation as follows:

$$h_j = \frac{\left[(Q_o V_o - Q_i V_i) - (Q_l V_l)\cos\theta_j\right]}{\left[0.5g(A_o + A_i)\right]} + h_i - h_o \quad (2.77)$$

where

Q_o, Q_i, and Q_ℓ = outlet, inlet, and lateral flows, respectively, m^3/s (ft^3/s)
V_o, V_i, and V_ℓ = outlet, inlet, and lateral velocities, respectively, m/s (ft/s)
h_o, h_i = outlet and inlet velocity heads, m (ft)
A_o, A_i = outlet and inlet cross-sectional areas, m^2 (ft^2)
θ_j = angle between the trunk pipe and lateral pipe

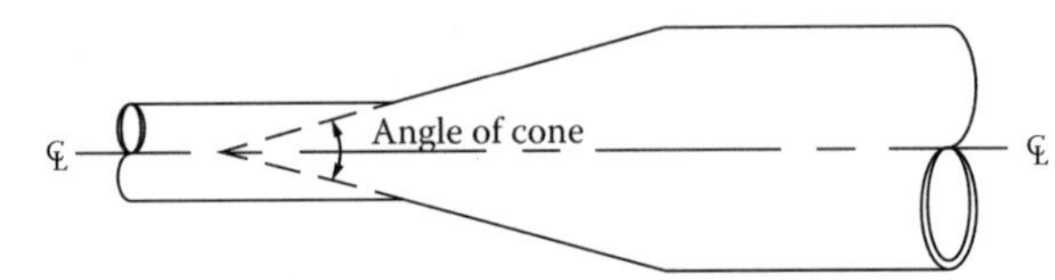

FIGURE 2.8 Sketch of an expansion transition.

TABLE 2.10
Gradual Expansion Loss Coefficients for Angles Ranging from 10° to 180°

	Angle of Cone						
D_2/D_1	10°	20°	45°	60°	90°	120°	180°
1.5	0.17	0.40	1.06	1.21	1.14	1.07	1.00
3	0.17	0.40	0.86	1.02	1.06	1.04	1.00

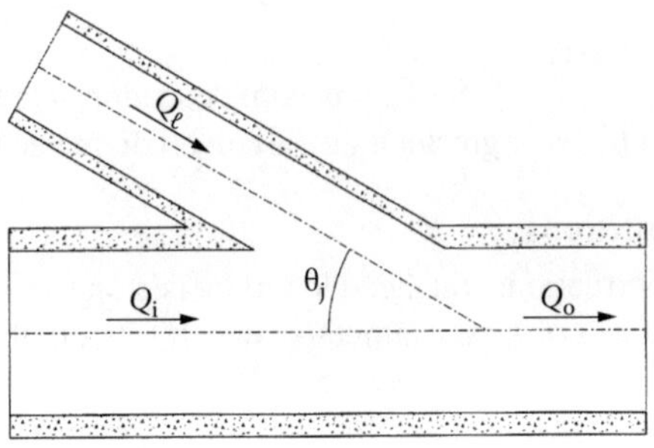

FIGURE 2.9 Pipe junction definition.

2.6.2.6 Losses at Access Holes and Inlets

At an access hole or junction hole (commonly referred to as manhole), there is a loss of head due to velocity change, especially if the incoming and outgoing pipes differ in size. In addition, a loss occurs when the flow direction changes and this loss can be greater than the former. To reduce the loss at a bending access hole, curved vane, or deflector may be installed in the direction of flow. Losses due to a bend at an access hole may be estimated using Figure 2.10. The head loss on this figure is calculated based on the outflow velocity head. The velocity head differential of inflow and outflow pipes may be added to account for the velocity change.

The head loss at an inlet depends largely on the change in the flow direction between the inflow and outflow pipes. The head loss at inlets can be estimated using the following equation:

$$h_i = k_i\left(\frac{V_1^2}{2g}\right) \tag{2.78}$$

where
V_1 = inflow velocity
k_i = head loss coefficient

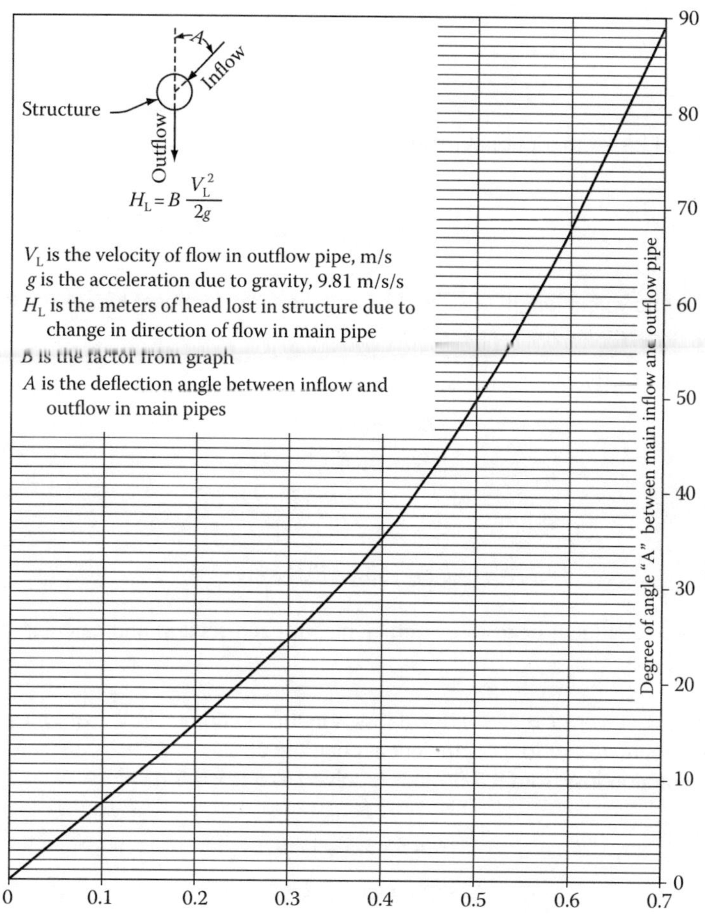

FIGURE 2.10 Bend loss factor for access holes.

The head loss coefficients for a number of common inlet arrangements are listed here:

- Straight run-square edge, $k_i = 0.5$
- Angled through 90°, $k_i = 1.5$

For intermediate angle of deflection in flow direction, k_i may be interpolated between 0.5 and 1.5.

Example 2.9

A drainage system consisting of a series of box culverts is shown in Figure 2.11. This system is aged and in part is located under a building in a town in northern New Jersey. Express local losses in each section in terms of its equivalent length.

Solution

First relate frictional losses in each section to the velocity head using Equation 2.59:

$$S_f = \left(\frac{29.0\,n^2}{R^{1.33}}\right)\frac{V^2}{2g}$$

Estimate $n = 0.015$ for the old culvert:

$$S_f = \left(\frac{6.53\times10^{-3}}{R^{1.33}}\right)\frac{V^2}{2g}$$

Then equate local losses to frictional losses:

$$h_l = \frac{kV^2}{2g} = S_f L_{eq}$$

where L_{eq} is the equivalent length of culvert having the same frictional loss as the local loss.
Combining these equations yields

$$L_{eq} = \frac{k}{\left(\dfrac{6.53\times10^{-3}}{R^{1.33}}\right)}$$

Or

$$L_{eq} = 153.1\times R^{1.33}\times k$$

Now estimate total local head loss coefficients for each section and calculate the equivalent length:

a) 5 ft × 3.5 ft box culvert; $L = 223$ ft; local losses in this section include exit loss at chamber and a bend loss. Estimate exit loss coefficient at chamber at 0.5.
   The contraction loss from 9.5 ft × 4.5 ft culvert to 5 ft × 3.5 ft

$$A = 5\times3.5 = 17.5\ ft^2$$

$$A_2 = 9.5\ ft\times4.5\ ft = 42.75\ ft^2$$

$$\frac{A_2}{A_1} = 0.41$$

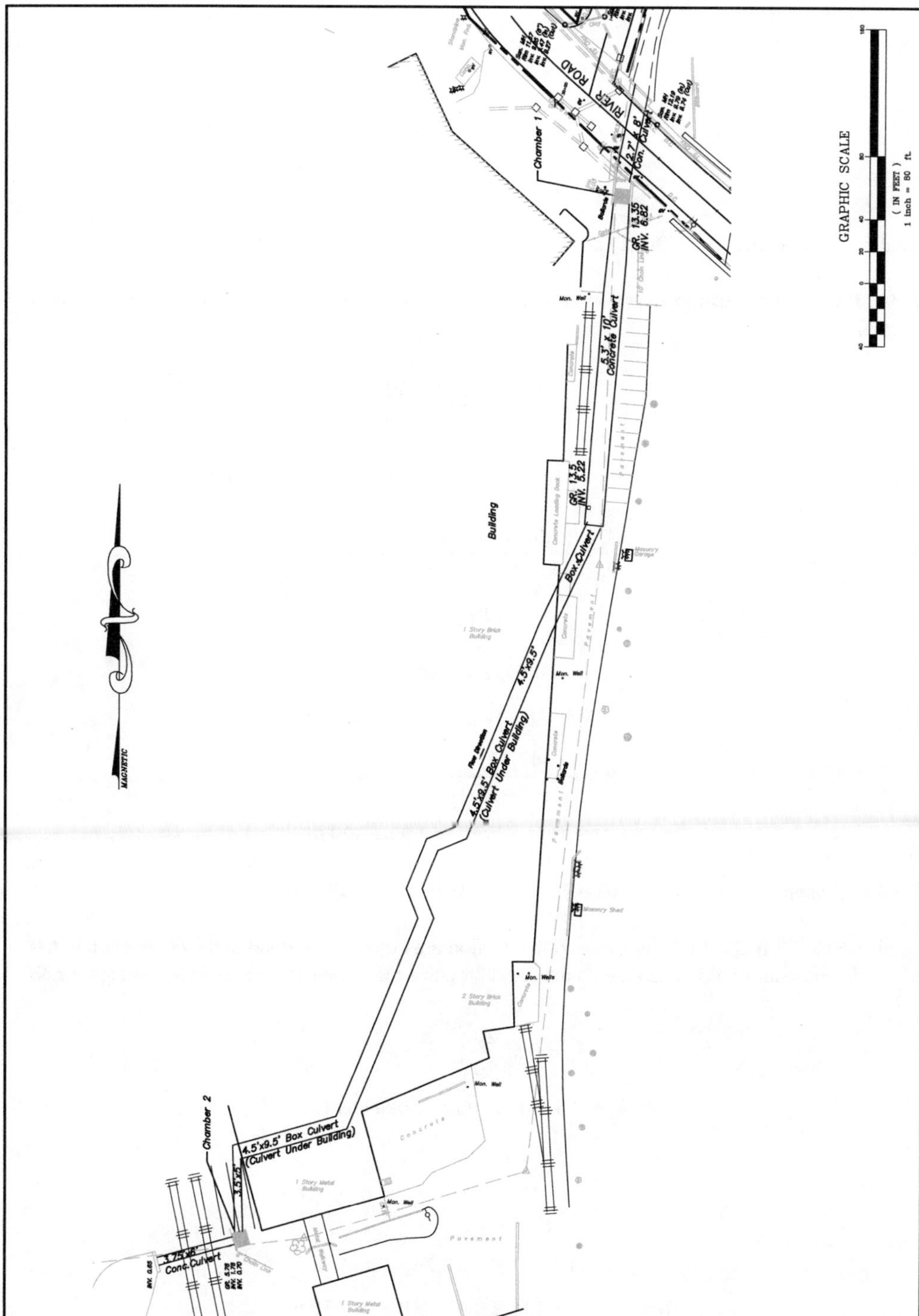

FIGURE 2.11 Drainage system consisting of a series of box culverts.

$$\frac{D_2}{D_1} = (0.41)^{1/2} = 0.64$$

Table 2.9 gives $k = 0.26$

$$\Sigma k = 0.5 + 0.26 = 0.76$$

$$R = \frac{(5\times 3.5)}{\left[2(5+3.5)\right]} = 1.03\ ft$$

$$L_{eq} = 153.1\times(1.03)^{1.33}\times 0.76 = 121.0$$

Total equivalent length = 223 + 121.0 = 344.0 ft.

a) 9.5 ft × 4.5 ft culvert; this section includes five 45° bend and one ±60° bend. Using Equation 2.75:

$$\Sigma k_b = \left[5\times 0.5\times\left(\frac{45}{90}\right)^{1/2} + 0.5\left(\frac{60}{90}\right)^{1/2}\right] = 2.18$$

Also, there is a contraction from the upstream 10 ft × 5.3 ft culvert to this culvert:

$$\frac{A_2}{A_1} = \frac{(9.5\times 4.5)}{(10\times 5.3)} = \frac{42.75}{53} = 0.8$$

$$\frac{D_2}{D_1} = (0.8)^{1/2} = 0.9$$

$$k_c = 0.05;\ see\ Table\ 2.9$$

$$\Sigma k = 2.18 + 0.05 = 2.23$$

$$R = \frac{(42.75)}{\left[2(9.5+4.5)\right]} = 1.527\ ft$$

$$L_{eq} = 153.1\times(1.527)^{1.33}\times 2.23 = 599\ ft$$

Total equivalent length = (70 + 140 + 20 + 30 + 20 + 210) + 599 = 1089 ft.

a) 10 ft × 5.3 ft culvert; local losses in this section are due to a 20° bend at the connection to the downstream section and an entrance loss (this culvert is shown as 5.3′×10′ HxW on Figure 2.4):

$$k_b = 0.5\left(\frac{20}{90}\right)^{0.5} = 0.24$$

$$k_e = 0.5\ entrance\ loss\ coefficient$$

$$\Sigma k = 0.74$$

$$R = \frac{(10\times 5.3)}{\left[2(10+5.3)\right]} = 1.73\ ft$$

$$L_{eq} = 153.1\times 1.73^{1.33}\times 0.74 = 234.9\ ft$$

$$Total = 223 + 234.9 = 457.9\ ft; say,\ 458\ ft$$

$$k_c = 0.05;\ see\ Table 2.9$$

$$\Sigma k = 2.18 + 0.05 = 2.23$$

$$R = \frac{(42.75)}{[2(9.5+4.5)]} = 1.527\ ft$$

$$L_{eq} = 153.1 \times (1.527)^{1.33} \times 2.23 = 599\ ft$$

Total equivalent length = (70 + 140 + 20 + 30 + 20 + 210) + 599 = 1089 ft.

a) 10 ft × 5.3 ft culvert; local losses in this section are due to a 20° bend at the connection to the downstream section and an entrance loss:

$$k_b = 0.5\left(\frac{20}{90}\right)^{0.5} = 0.24$$

$$k_e = 0.5\ entrance\ loss\ coefficient$$

$$\Sigma k = 0.74$$

$$R = \frac{(10 \times 5.3)}{[2(10+5.3)]} = 1.73\ ft$$

$$L_{eq} = 153.1 \times 1.73^{1.33} \times 0.74 = 234.9\ ft$$

$$Total = 223 + 234.9 = 457.9\ ft; say,\ 458\ ft$$

PROBLEMS

2.1 For a wide rectangular channel, assume a velocity distribution of

$$\frac{v}{u^*} = 5.75\ log\ log \frac{d}{k} + 8.5$$

where v is the flow velocity at a distance of d from the bed, k is the surface roughness, and $u^* = \sqrt{\frac{\tau_o}{\rho}}$ is the shear velocity. Show that the velocity at 0.6 depth from the water surface approximates the average velocity in a section.
Also show that the mean of the velocity readings at the relative depths of 0.2 and 0.8 along a vertical line from the water surface approximates the average velocity in the section.

2.2 A riprapped trapezoidal channel with a longitudinal slope of 0.3%, bottom width of 2 m, side slopes of 2:1 (2H,1V) is to carry 10 m^3/s. Calculate the normal depth of water in this channel. Use: $n = 0.03$.

2.3 Solve Problem 2.2 for a channel width of 6 ft and 350 cfs discharge.

2.4 Calculate the normal capacity and full flow velocity for reinforced concrete pipes (RCPs) of 15, 18, 24, 30, and 36 in. diameter. The pipes are at 2% grade. Use Manning's $n = 0.012$.

2.5 Solve the preceding problem in metric units for RCP pipes of 300, 375, 450, 600, 750, and 900 mm in diameter.

2.6 Calculate the normal flow depth in a 24 in. pipe laid at 2% slope for a discharge equal to 65% of the pipe capacity.

2.7 Calculate the normal depth of flow in a 600 mm concrete pipe for a discharge equal to 70% of the pipe capacity. The pipe is at 1.5% slope. Base your calculations on $n = 0.013$.

2.8 The design flow for a 24 in. RCP is calculated at 30 cfs. The pipe is at 2% slope and emanates from a manhole 3.5 ft deep. The invert of the pipe is set at the bottom of the manhole. Is this pipe adequate for the design flow?

2.9 A 500 mm concrete pipe is employed for conveying stormwater at a rate of 400 L/s. The pipe is at 2% slope and emanates from an inlet 1.0 m deep. The invert of the pipe is set at the bottom of the inlet. Perform calculations to determine whether or not this pipe is adequate for the design flow. Base your calculations on $n = 0.013$.

2.10 A semicircular concrete channel 1.5 m in diameter is laid on a 0.5% slope. Calculate the maximum discharge, in m^3/s, that this channel can carry when the brim full. Take $n = 0.013$.

2.11 Solve Problem 2.10 for a semicircular channel of 60 in. diameter.

2.12 Calculate the normal depth of flow in a 120 cm concrete culvert ($n = 0.012$) on a slope of 0.5% for a discharge of 1 m^3/s.

2.13 Solve Problem 2.12 for a 48 in. culvert and 35 cfs flow.

2.14 Select the size of pipe needed to carry a discharge of 10 m^3/s at no more than 65% full. The pipe is at 1% grade. Use $n = 0.013$. Note that large pipe comes in 150 mm diameter intervals.

2.15 A circular concrete pipe culvert is to carry a discharge of 450 cfs on a slope of 0.008 and is to run no more than 60% (by depth) full. Concrete pipes are available in internal diameters that are multiple of 6 in. Choose a suitable culvert diameter. Use $n = 0.012$.

2.16 The critical depth in a storm drain, 60 in. diameter, is 2.2 ft. What is the critical discharge?

2.17 Solve Problem 2.16 for a critical depth of 66 cm in a 150 cm pipe.

2.18 Calculate the critical depth in Example 2.4 in this chapter using Table 2.2.

2.19 Water flows at a rate of 3.0 m^3/s in a 3 m wide rectangular concrete channel at a depth of 0.5 m.
- Calculate the slope for $n = 0.012$.
- What slope is required to produce the critical flow for the given discharge? Also, what is the critical depth?

2.20 Solve Problem 2.19 for a discharge of 100 ft^3/s in a 10 ft wide rectangular channel, and at a depth of 1.5 ft.

2.21 Water discharges at a rate of 8 m^3/s in a trapezoidal channel of bottom width of 2 m and side slopes of 2:1. Calculate the critical depth and the critical velocity for this channel. What will be the critical slope, if the channel is riprap lined ($n = 0.035$)?

2.22 Solve Problem 2.21 for a discharge of 280 cfs in a trapezoidal channel of 6 ft bottom width.

2.23 Calculate the normal depth of flow in a roadside trapezoidal shaped channel for a 1.5 m^3/s design discharge. The channel is at 1% slope, lined with 100 mm stone riprap, and has a bottom width of side slopes of 1.2 m and 2:1, respectively. Estimate $n = 0.04$.

2.24 Solve Problem 2.23 for the following flow parameters:
a) $Q = 50$ cfs
b) $b = 4$ ft
c) Riprap size = 4 in.

2.25 Solve Example 2.6 in this chapter for a discharge of 0.5 m^3/s and 450 mm pipe.

2.26 Flow parameters in a junction similar to that shown in Figure 2.9 are
- $Q_o = 0.8\ m^3/s$
- $Q_\ell = 0.2\ m^3/s$
- $D_i = 0.6$ m; $D_o = 0.675$ m; $D_\ell = 0.3$ m
- $\theta_j = 30°$

Calculate the head loss in this junction.

2.27 Solve Problem 2.26 for the following case:
$Q_o = 30$ cfs
$Q_\ell = 15$ cfs
$D_i = 24$ in., $D_o = 30$ in., $D_\ell = 18$ in.

2.28 Calculate the head loss in the drainage system of Example 2.9 for a discharge of 250 cfs.

2.29 Solve Problem 2.28 for a discharge of 7.5 m^3/s.

APPENDIX 2A: HYDRAULIC PROPERTIES OF ROUND AND ELLIPTICAL PIPES

TABLE 2A.1
Hydraulic Properties of Circular Pipes Flowing Full, in SI Units

Pipe Diameter (mm)	Area (m²)	Hydraulic Radius (m)	$K = A \times R^{2/3}/n$[a] $n = 0.010$	$n = 0.011$	$n = 0.012$	$n = 0.013$
150	0.018	0.0375	0.20	0.18	0.16	0.15
200	0.031	0.0500	0.43	0.39	0.36	0.33
250	0.049	0.0625	0.77	0.70	0.64	0.59
300	0.071	0.0750	1.26		1.05	0.97
375	0.110	0.0938	2.28	2.07	1.90	1.75
450	0.159	0.1125	3.71	3.37	3.90	2.85
525	0.216	0.1313	5.59	5.08	4.66	4.30
600	0.283	0.1500	7.98	7.26	6.65	6.14
675	0.358	0.1688	10.93	9.93	9.11	8.14
750	0.442	0.1875	14.47	13.16	12.06	11.13
825	0.535	0.2063	18.66	16.96	15.55	14.35
900	0.636	0.2250	23.53	21.39	19.61	18.10
1050	0.866	0.2625	35.50	32.27	29.58	27.31
1125	0.994	0.2813	42.67	38.79	35.56	32.82
1200	1.131	0.3000	50.68	46.07	42.23	38.99
1350	1.431	0.3375	69.38	63.08	57.82	53.37
1500	1.767	0.3750	91.89	83.54	76.58	70.69
1800	*2.545*	*0.4500*	*149.4*	*135.8*	*124.5*	*114.9*

Source: American Concrete Pipe Association, *Concrete Pipe Design Manual*, 19th printing, 2005.

[a] *K = conveyance factor; Q = KS*$^{1/2}$.

TABLE 2A.2
Hydraulic Properties of Circular Pipes Flowing Full, in C

D Pipe Diameter (in.)	A Area (sq. ft)	R Hydraulic Radius (ft)	$K = \frac{1.49}{n} \times A \times R^{\frac{2}{3}}$ $n = 0.010$	$n = 0.011$	$n = 0.012$	$n = 0.013$
8	0.349	0.167	15.8	14.3	13.1	12.1
10	0.545	0.208	28.4	25.8	23.6	21.8
12	0.785	0.250	46.4	42.1	38.6	35.7
15	1.227	0.312	84.1	76.5	70.1	64.7
18	1.767	0.375	137.0	124.0	114.0	105.0
21	2.405	0.437	206.0	187.0	172.0	158.0
24	3.142	0.500	294.0	267.0	245.0	226.0
27	3.976	0.562	402.0	366.0	335.0	310.0
30	4.909	0.625	533.0	485.0	444.0	410.0
33	5.940	0.688	686.0	624.0	574.0	530.0
36	7.069	0.750	867.0	788.0	722.0	666.0
42	9.621	0.875	1308.0	1189.0	1090.0	1006.0
48	12.566	1.000	1867.0	1698.0	1556.0	1436.0
54	15.904	1.125	2557.0	2325.0	2131.0	1967.0
60	19.635	1.250	3385.0	3077.0	2821.0	2604.0
66	23.758	1.375	4364.0	3967.0	3636.0	3357.0
72	28.274	1.500	5504.0	5004.0	4587.0	4234.0

Source: American Concrete Pipe Association, *Concrete Pipe Design Manual*, 19th printing, 2007.

[a] K = conveyance factor; $Q = KS^{1/2}$.

TABLE 2A.3
Hydraulic Parameters Elliptical Concrete Pipe under Full-Flow Conditions, in CU

Pipe Size $R \times S$ (HE) $S \times R$ (VEO) (in.)	Approximate Equivalent Circular Diameter (in.)	A Area (sq. ft)	R Hydraulic Radius (ft)	Value of $C_1 = \frac{1.486}{n} \times A \times R^{\frac{2}{3}}$			
				$n = 0.010$	$n = 0.011$	$n = 0.012$	$n = 0.013$
14 × 23	18	1.8	0.367	138	125	116	108
19 × 30	24	3.3	0.490	301	274	252	232
22 × 34	27	4.1	0.546	405	368	339	313
24 × 38	30	5.1	0.613	547	497	456	421
27 × 42	33	6.3	0.686	728	662	607	560
29 × 45	36	7.4	0.736	891	810	746	686
32 × 49	39	8.8	0.812	1140	1036	948	875
34 × 53	42	10.2	0.875	1386	1260	1156	1067
38 × 60	48	12.9	0.969	1878	1707	1565	1445
43 × 68	54	16.6	1.106	2635	2395	2196	2027
48 × 76	60	20.5	1.229	3491	3174	2910	2686
53 × 83	66	24.8	1.352	4503	4094	3753	3464
58 × 91	72	29.5	1.475	5680	5164	4734	4370
63 × 98	78	34.6	1.598	7027	6388	5856	5406
68 × 106	84	40.1	1.721	8560	7790	7140	6590
72 × 113	90	46.1	1.845	10,300	9365	8584	7925
77 × 121	96	52.4	1.967	12,220	11,110	10,190	9403
82 × 128	102	59.2	2.091	14,380	13,070	11,980	11,060
87 × 136	108	66.4	2.215	16,770	15,240	13,970	12,900
92 × 143	114	74.0	2.340	19,380	17,620	16,150	14,910
97 × 151	120	82.0	2.461	22,190	20,180	18,490	17,070
106 × 166	132	99.2	2.707	28,630	26,020	23,860	22,020
116 × 180	144	118.6	2.968	36,400	33,100	30,340	28,000

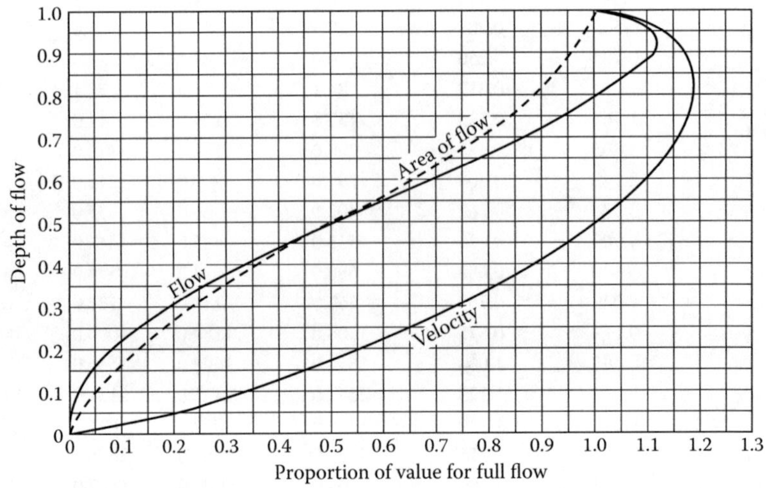

FIGURE 2A.1 Relative flow parameters versus flow depth in horizontal elliptical concrete pipes, in CU and SI

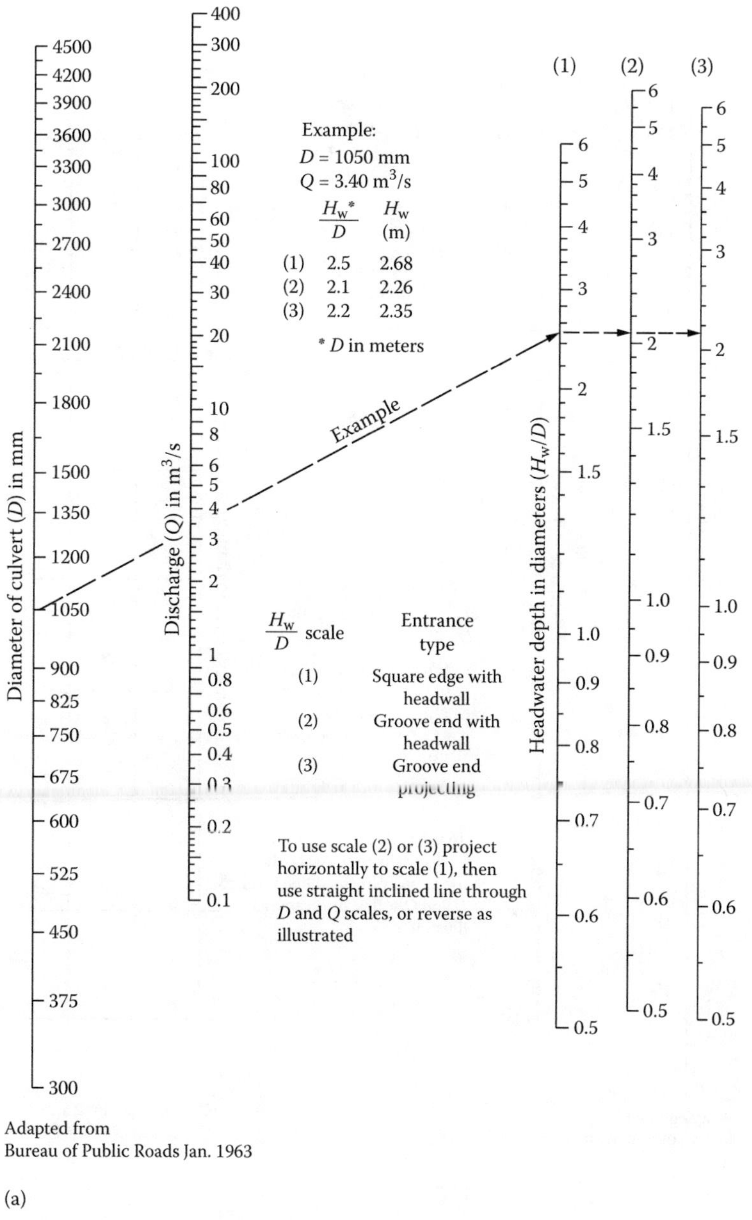

(a)

FIGURE 2A.2A Inlet control headwater depth for round pipes, in SI

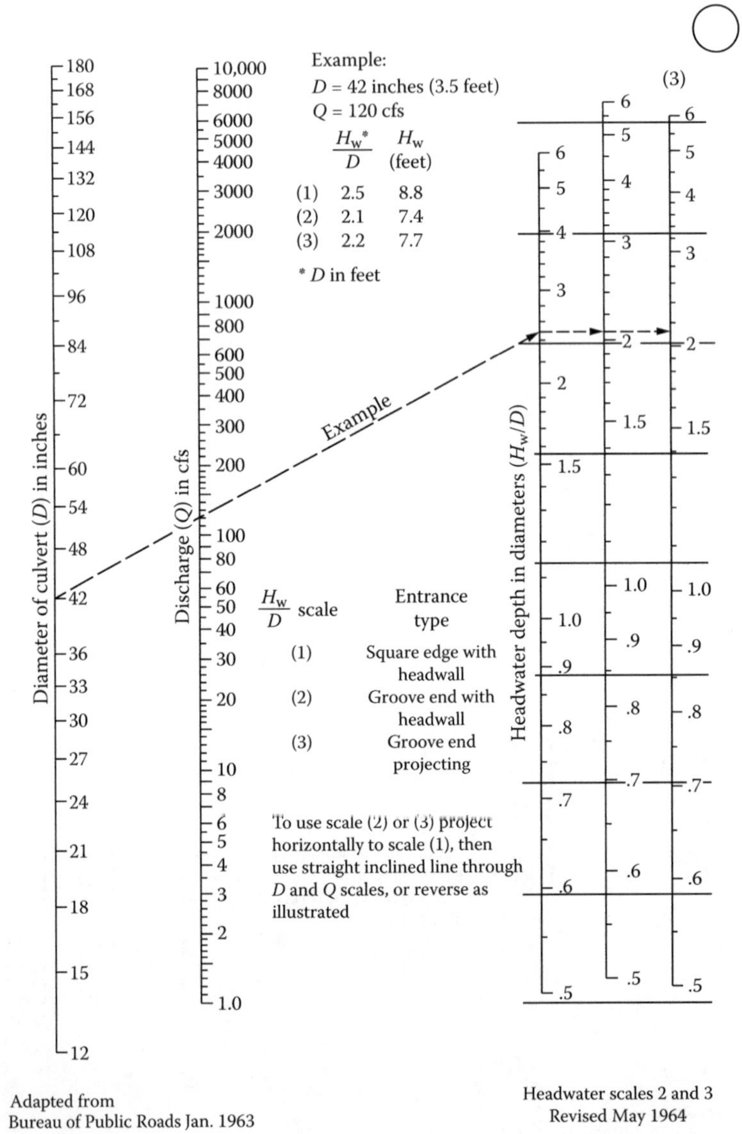

(b)

FIGURE 2A.2B Inlet control headwater depth for round pipes, in CU.

REFERENCES

American Concrete Pipe Association, 2007, *Concrete Pipe Design Manual*, 19th printing.

ASCE and WEF, 1992, *Design and construction of urban stormwater management systems*, American Society of Civil Engineers Manuals and Reports of Engineering Practice No. 77 and Water Environment Federation Manual of Practice FD-20.

Barnes, H.H., 1987, *Roughness characteristics of natural channels, water supply paper 1849*, US Geological Survey.

Brater, E.F., King, H.W., Lindell, J.E. and Wei, C.Y., 1996, *Handbook of hydraulics*, 7th ed., McGraw-Hill, New York.

Chow, V.T., 1959, *Open channel hydraulics*, McGraw-Hill, New York.

FHWA (US Department of Transportation), April 1, 2012, *Hydraulic design of highway culverts*, 3rd ed., FHWA, Hydraulic Design Series No. 5 (HDS-5).

FHWA (US Department of Transportation), August 2013, *Urban drainage design manual, hydraulic engineering circular (HEC-22)*, 3rd ed., FHWA, Archived.

French, R.H., 1985 *Open channel hydraulics*, Chapter 4, McGraw-Hill, New York.

Henderson, F.M., 1966, *Open channel flow*, Chapter 4, Macmillan, New York.

Pazwash, H., 2007, *Fluid mechanics and hydraulic engineering*, in Farsi, Tehran University Press, Iran.

3 Hydrologic Calculations

This chapter covers the principles of hydrologic calculations as they relate to stormwater management practices. Discussed in this chapter are those elements of the hydrologic cycle that affect rainfall – runoff relation. These elements include the rainfall process, vegetal interception, surface retention, and infiltration. Also discussed in this chapter are runoff calculation methods and their limitations. A physically based model developed by the author is also presented in this chapter. Ample examples and case studies are included to illustrate proper application of hydrologic methods.

3.1 RAINFALL

Precipitation is a dynamic process. It has both spatial and temporal variation. Not only varies from one location to another, but at a given location, rainfall also has a varied pattern. Precipitation can fall at a faster rate at the beginning, middle, or end of a storm's duration. A rainstorm can also have more than one maximum or minimum. The falling rate of rain rate of rain with time is called rainfall intensity.

3.1.1 Intensity–Duration–Frequency Curves

To simplify rainfall–runoff dynamics, some methods of calculating the peak rate of runoff assume a constant rainfall intensity equal to its mean during the storm period. This intensity, which represents the rainfall depth divided by the rainfall duration, is expressed in millimeters per hour (mm/h) or inches per hour (in./h).

Rainfall intensity is dependent on the rainfall duration. It also varies with the frequency of the rainfall event; the less frequent the storm is, the larger its intensity will be. Curves representing the rainfall intensity–duration–frequency (IDF) relationship have been developed using long-term rainfall data from precipitation stations in several states and large cities in the United States and many other countries. Figures 3.1a and b present rainfall IDF curves in New Jersey in metric (SI) and customary units (CU), respectively. These figures show that, for example, the intensity of a 10-year, 60-minute storm is approximately 2 in. per h (50 mm/h).

Regional I-D-F curves in the United States are available in the Department of Transportation drainage manuals and/or Department of Environmental Protection/Conservation publications in some states. Based on the analysis of precipitation stations, National Oceanic and Atmospheric Administration (NOAA) has developed point precipitation frequency estimates for the entire United States. These frequency estimates are published in NOAA Atlas 14 and can be downloaded for any observation site free of charge from their website (http://hdsc.nws.noaa.gov). This website allows the user to print the table or the table and color graph of rainfall intensity – duration – frequency or rainfall depth – duration – frequency estimates for any precipitation station in the United States. The website also gives the option of downloading the data in English or metric units.

Tables 3.1 and 3.2 provide rainfall depth-duration-frequency estimates in metric units for Trenton State College in Trenton, New Jersey (one of dozens of NOAA precipitation stations in that state) and for University of Southern California (USC) in Downtown Los Angeles, California, respectively. A comparison of Tables 3.1 and 3.2 reveals a significant difference in rainfall patterns between the East Coast and the West Coast in the United States. For the 10-year, 60-minute storm, for example, the tables indicate a rainfall depth of 52 mm/h in Trenton but 26 mm/h in Los Angeles. Similar data for San Francisco show a 10-year 60-min rainfall intensity of 22 mm/h.

For a long-term comparison, recorded data for Baltimore and Los Angeles are reviewed herein. The city of Baltimore has monthly rainfall data which goes back to 1871. The data for the 1991–2020

DOI: 10.1201/9781003460121-3

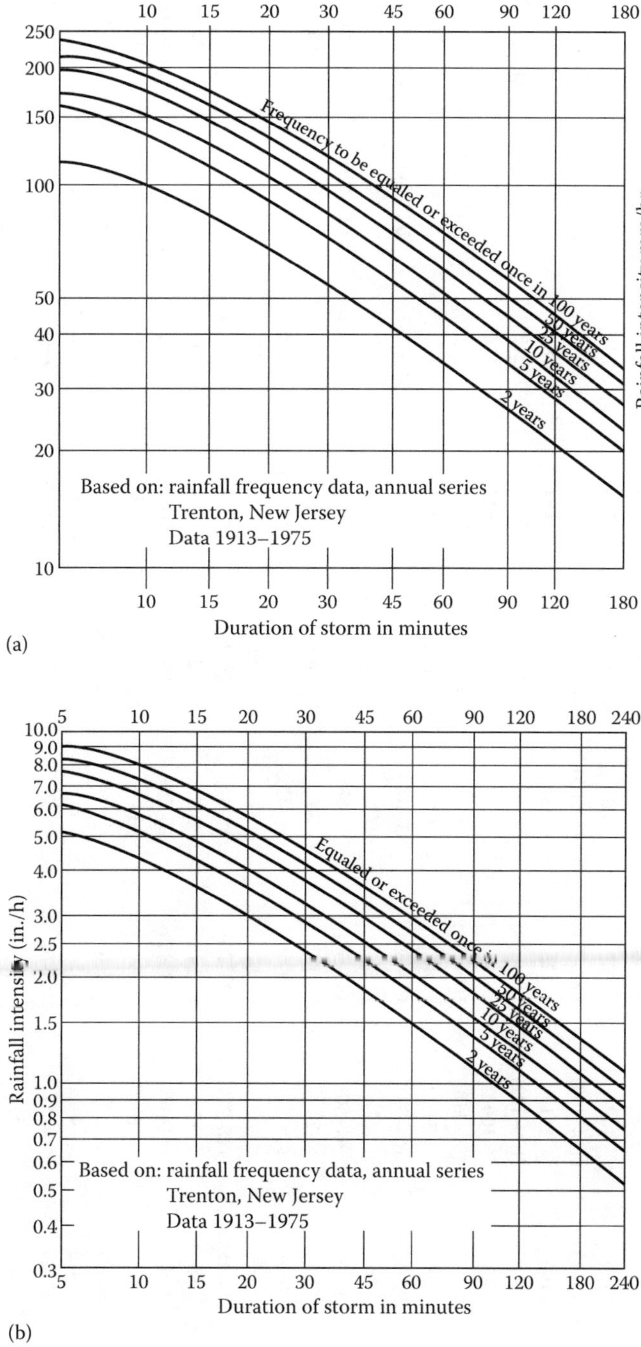

FIGURE 3.1 Rainfall curves in New Jersey: (a) SI, (b) CU.

period shows a monthly average of 2.98 in. in February to 4.40 in. in July. The average annual precipitation during this period was 45.09 in. Los Angeles Almanac shows an average annual rainfall of 11.7 in. (297 mm) at the USC campus in Downtown Los Angeles. The data shows the water year 1984 with 32.18 in. (817 mm) of precipitation as the wettest year and the 2007 water year with 3.21 in. (81.5 mm) as the driest year in the record.

TABLE 3.1
Precipitation–depth–frequency estimates for Trenton, New Jersey

	PDS-based point precipitation frequency estimates with 90% confidence intervals (in millimeters/hour)[1]									
	Average recurrence interval (years)									
Duration	**1**	**2**	**5**	**10**	**25**	**50**	**100**	**200**	**500**	**1000**
5-min	**104** (95–115)	**124** (113–137)	**148** (134–163)	**165** (149–181)	**186** (167–205)	**201** (180–222)	**216** (193–239)	**231** (204–255)	**248** (218–277)	**262** (228–294)
10-min	**83** (76–92)	**100** (90–110)	**111** (100–122)	**125** (112–138)	**135** (121–149)	**145** (129–160)	**154** (136–170)	**183** (162–203)	**197** (172–219)	**207** (179–231)
15-min	**69** (63–76)	**83** (76–92)	**100** (90–110)	**111** (100–122)	**125** (112–138)	**135** (121–149)	**145** (129–160)	**154** (136–170)	**165** (145–184)	**173** (150–194)
30-min	**47** (43–52)	**58** (52–63)	**71** (64–78)	**80** (73–89)	**93** (83–102)	**102** (91–112)	**111** (99–123)	**120** (106–133)	**131** (115–146)	**140** (122–157)
60-min	**30** (27–33)	**36** (33–40)	**45** (41–50)	**52** (47–58)	**62** (55–68)	**69** (62–76)	**76** (68–84)	**84** (74–93)	**94** (82–105)	**102** (89–114)
2-hr	**18** (16–20)	**22** (20–24)	**28** (25–30)	**32** (29–35)	**38** (34–42)	**43** (38–47)	**48** (43–53)	**53** (47–59)	**61** (53–67)	**66** (57–74)
3-hr	**13** (12–15)	**16** (14–18)	**20** (18–22)	**24** (21–26)	**28** (25–31)	**32** (28–35)	**36** (32–40)	**40** (35–45)	**46** (39–51)	**50** (43–57)
6-hr	**8** (8–9)	**10** (9–11)	**13** (11–14)	**15** (13–17)	**18** (16–20)	**21** (18–23)	**23** (20–26)	**26** (23–29)	**31** (26–35)	**34** (29–39)
12-hr	**5** (5–6)	**6** (5–7)	**8** (7–9)	**9** (8–10)	**11** (10–12)	**13** (11–14)	**15** (13–17)	**17** (15–19)	**20** (17–23)	**23** (19–26)
24-hr	**3** (3–3)	**4** (3–4)	**4** (4–5)	**5** (5–6)	**6** (6–7)	**7** (7–8)	**9** (8–9)	**10** (9–11)	**12** (10–13)	**13** (11–15)
2-day	**2** (2–2)	**2** (2–2)	**3** (2–3)	**3** (3–3)	**4** (3–4)	**4** (4–5)	**5** (4–5)	**6** (5–6)	**7** (6–7)	**7** (6–8)
3-day	**1** (1–1)	**1** (1–2)	**2** (2–2)	**2** (2–2)	**3** (2–3)	**3** (3–3)	**3** (3–4)	**4** (3–4)	**4** (4–5)	**5** (4–5)
4-day	**1** (1–1)	**1** (1–1)	**1** (1–2)	**2** (2–2)	**2** (2–2)	**2** (2–2)	**3** (2–3)	**3** (3–3)	**3** (3–4)	**4** (3–4)

7-day	**1**	**1**	**1**	**1**	**1**	**1**	**2**	**2**	**2**	**2**
	(1–1)	(1–1)	(1–1)	(1–1)	(1–1)	(1–2)	(2–2)	(2–2)	(2–2)	(2–3)
10-day	**1**	**1**	**1**	**1**	**1**	**1**	**1**	**1**	**2**	**2**
	(0–1)	(1–1)	(1–1)	(1–1)	(1–1)	(1–1)	(1–1)	(1–1)	(1–2)	(2–2)
20-day	**0**	**0**	**0**	**1**	**1**	**1**	**1**	**1**	**1**	**1**
	(0–0)	(0–0)	(0–1)	(1–1)	(1–1)	(1–1)	(1–1)	(1–1)	(1–1)	(1–1)
30-day	**0**	**0**	**0**	**0**	**0**	**1**	**1**	**1**	**1**	**1**
	(0–0)	(0–0)	(0–0)	(0–0)	(0–1)	(1–1)	(1–1)	(1–1)	(1–1)	(1–1)
45-day	**0**	**0**	**0**	**0**	**0**	**0**	**0**	**0**	**1**	**1**
	(0–0)	(0–0)	(0–0)	(0–0)	(0–0)	(0–0)	(0–0)	(0–1)	(0–1)	(1–1)
60-day	**0**	**0**	**0**	**0**	**0**	**0**	**0**	**0**	**0**	**0**
	(0–0)	(0–0)	(0–0)	(0–0)	(0–0)	(0–0)	(0–0)	(0–0)	(0–0)	(0–0)

[1] Precipitation frequency (PF) estimates in this table are based on frequency analysis of partial duration series (PDS).

Numbers in parenthesis are PF estimates at lower and upper bounds of the 90% confidence interval. The probability that precipitation frequency estimates (for a given duration and average recurrence interval) will be greater than the upper bound (or less than the lower bound) is 5%. Estimates at upper bounds are not checked against probable maximum precipitation (PMP) estimates and may be higher than currently valid PMP values.

Please refer to NOAA Atlas 14 document for more information.

Source: NOAA Atlas 14, volume 10, version 3.

TABLE 3.2
Precipitation frequency estimates for Downtown Los Angeles

	PDS-based point precipitation frequency estimates with 90% confidence intervals (in millimeters/hour)[1]									
	Average recurrence interval (years)									
Duration	1	2	5	10	25	50	100	200	500	1000
5-min	**47** (39–56)	**60** (50–73)	**78** (65–95)	**93** (77–114)	**114** (91–144)	**130** (101–169)	**147** (112–196)	**165** (122–226)	**190** (134–272)	**210** (143–312)
10-min	**33** (28–40)	**43** (36–52)	**56** (47–68)	**67** (55–82)	**82** (65–103)	**93** (73–121)	**106** (80–140)	**118** (87–162)	**136** (96–195)	**151** (102–224)
15-min	**27** (23–33)	**35** (29–42)	**45** (38–55)	**54** (44–66)	**66** (52–83)	**75** (59–98)	**85** (65–113)	**96** (70–131)	**110** (78–157)	**121** (83–180)
30-min	**18** (15–22)	**23** (19–28)	**30** (25–37)	**36** (30–44)	**44** (35–56)	**50** (39–65)	**57** (43–76)	**64** (47–87)	**73** (52–105)	**81** (55–120)
60-min	**13** (11–15)	**17** (14–20)	**22** (18–26)	**26** (21–31)	**31** (25–40)	**36** (28–46)	**41** (31–54)	**46** (34–62)	**52** (37–75)	**58** (39–86)
2-hr	**9** (8–11)	**12** (10–14)	**15** (13–19)	**18** (15–23)	**23** (18–29)	**26** (20–33)	**29** (22–39)	**33** (24–45)	**37** (26–54)	**41** (28–61)
3-hr	**8** (6–9)	**10** (8–12)	**13** (11–16)	**15** (13–19)	**19** (15–24)	**21** (17–28)	**24** (18–32)	**27** (20–37)	**31** (22–44)	**34** (23–50)
6-hr	**5** (4–6)	**7** (6–8)	**9** (7–11)	**11** (9–13)	**13** (10–16)	**15** (12–19)	**17** (13–22)	**19** (14–26)	**21** (15–30)	**23** (16–35)
12-hr	**3** (3–4)	**4** (4–5)	**6** (5–7)	**7** (6–8)	**8** (7–11)	**10** (7–12)	**11** (8–14)	**12** (9–16)	**13** (10–19)	**15** (10–22)
24-hr	**2** (2–2)	**3** (2–3)	**4** (3–4)	**4** (4–5)	**5** (5–6)	**6** (5–7)	**7** (5–9)	**8** (6–10)	**8** (6–11)	**9** (7–13)
2-day	**1** (1–1)	**2** (2–2)	**2** (2–3)	**3** (2–3)	**3** (3–4)	**4** (3–5)	**4** (4–5)	**5** (4–6)	**5** (4–7)	**6** (4–8)
3-day	**1** (1–1)	**1** (1–2)	**2** (2–2)	**2** (2–3)	**3** (2–3)	**3** (3–4)	**3** (3–4)	**4** (3–5)	**4** (3–6)	**5** (4–7)
4-day	**1** (1–1)	**1** (1–1)	**1** (1–2)	**2** (2–2)	**2** (2–3)	**3** (2–3)	**3** (2–4)	**3** (3–4)	**4** (3–5)	**4** (3–6)

7-day	**1**	**1**	**1**	**1**	**1**	**2**	**2**	**2**	**3**	**3**
	(0–1)	(1–1)	(1–1)	(1–1)	(1–2)	(1–2)	(2–2)	(2–3)	(2–3)	(2–4)
10-day	**0**	**1**	**1**	**1**	**1**	**1**	**2**	**2**	**2**	**2**
	(0–0)	(0–1)	(1–1)	(1–1)	(1–1)	(1–2)	(1–2)	(1–2)	(1–3)	(2–3)
20-day	**0**	**0**	**0**	**0**	**1**	**1**	**1**	**1**	**1**	**1**
	(0–0)	(0–0)	(0–1)	(0–1)	(1–1)	(1–1)	(1–1)	(1–1)	(1–2)	(1–2)
30-day	**0**	**0**	**0**	**0**	**1**	**1**	**1**	**1**	**1**	**1**
	(0–0)	(0–0)	(0–0)	(0–1)	(0–1)	(1–1)	(1–1)	(1–1)	(1–1)	(1–2)
45-day	**0**	**0**	**0**	**0**	**0**	**1**	**1**	**1**	**1**	**1**
	(0–0)	(0–0)	(0–0)	(0–0)	(0–1)	(0–1)	(0–1)	(1–1)	(1–1)	(1–1)
60-day	**0**	**0**	**0**	**0**	**0**	**0**	**1**	**1**	**1**	**1**
	(0–0)	(0–0)	(0–0)	(0–0)	(0–0)	(0–1)	(0–1)	(0–1)	(1–1)	(1–1)

[1] Precipitation frequency (PF) estimates in this table are based on frequency analys is of partial duration series (PDS).
Numbers in parenthesis are PF estimates at lower and upper bounds of the 90% confidence interval. The probability that precipitation frequency estimates (for a given duration and average recurrence interval) will be greater than the upper bound (or less than the lower bound) is 5%. Estimates at upper bounds are not checked against probable maximum precipitation (PMP) estimates and may be higher than currently valid PMP values.
Please refer to NOAA Atlas 14 document for more information.

Source: NOAA Atlas 14, vol. 6, version 2.

3.1.2 Rainfall Data

NOAA measures daily precipitation at their gauging stations throughout the United States. Also some universities collect rainfall data for research purposes. In New Jersey, for example, Rutgers University collects and analyzes rainfall and snowfall data. The NOAA, which was formerly called the US Weather Bureau and is run by the US Department of Commerce, operates 10,000 stations throughout the country. Table 3.3, downloaded from the National Weather Service, presents an example of NOAA climate that shows daily data for the month of September 2023 at JFK International Airport in New York City. The table shows that 8.05 in. (205 mm) of rain fell on

TABLE 3.3
Climate Data for JFK International Airport, New York, September 2023

Day	Max	Min	Avg	HDD	CDD	Pcpn	Snow	Snwg
1	76	59	67.5	0	3	0.00	0.0	0
2	75	60	67.5	0	3	0.00	0.0	0
3	90	65	77.5	0	13	T	0.0	0
4	89	70	79.5	0	15	T	0.0	0
5	93	74	83.5	0	19	0.00	0.0	0
6	93	74	83.5	0	19	0.00	0.0	0
7	92	73	82.5	0	18	0.00	0.0	0
8	86	73	79.5	0	15	0.01	0.0	0
9	84	73	78.5	0	14	T	0.0	M
10	80	69	74.5	0	10	1.11	0.0	0
11	82	70	76.0	0	11	0.40	0.0	0
12	81	69	75.0	0	10	0.00	0.0	0
13	79	70	74.5	0	10	0.02	0.0	0
14	80	63	71.5	0	7	0.00	0.0	0
15	71	58	64.5	0	0	0.00	0.0	0
16	78	61	69.5	0	5	0.00	0.0	0
17	73	56	64.5	0	0	T	0.0	0
18	71	62	66.5	0	2	0.70	0.0	0
19	76	59	67.5	0	3	0.00	0.0	0
20	76	56	66.0	0	1	0.00	0.0	0
21	74	57	65.5	0	1	0.00	0.0	0
22	73	57	65.0	0	0	0.00	0.0	0
23	63	58	60.5	4	0	0.96	0.0	0
24	65	62	63.5	1	0	0.56	0.0	0
25	62	57	59.5	5	0	0.35	0.0	0
26	61	53	57.0	8	0	0.01	0.0	0
27	67	51	59.0	6	0	0.00	0.0	0
28	67	54	60.5	4	0	0.62	0.0	0
29	61	57	59.0	6	0	8.05	0.0	0
30	68	56	62.0	3	0	0.22	0.0	0
Sum	2286	1876	-	37	179	13.01	-	0.0
Average	76.2	62.5	69.4	-	-	-	-	0.0
Normal	75.8	61.0	68.4	33	135	3.58	0.0	-

Source: National Weather Service NOAA, NOWData.

September 29 and that September was the wettest month on record with 13.01 inches (330 mm) of rainfall; that storm caused significant flooding and property damage in the city.

NOAA's Applied Climate Information System (ACIS) website includes data from over 6700 stations in the U.S. The rainfall data in the northeastern United States can also be downloaded for the current and previous months from the Cornell University Northeast Regional Climate Center website (http://www.nrcc.cornell.edu/page_nowdata.html).

Daily precipitation data are also collected by the Community Collaborative Rain, Hale & Snow (CoCoRaHS) Network. The data for any station operated by the CoCoRaHS throughout the United States can be downloaded from their website (http://www.cocorahs.org/).

Hourly rainfall depths are measured at selected recording precipitation stations in the United States. The hourly rainfall records in the northeastern states are kept at a climate center of Cornell University. These data may be purchased from the website Northeast Regional Climate Center of the Cornell University. (http://www.nrcs.cornell.edu/page_nowdata.html).

3.1.3 Rainfall Hyetograph

In a given storm, the instantaneous rainfall intensity is the slope of the cumulative (mass) rainfall depth at that time. The hyetograph, which represents variations in instantaneous rainfall intensity with time, is depicted as a continuous curve; see Figure 3.2 for an example. To simplify hydrologic analysis, this curve is divided into discrete segments, each representing the average rainfall intensity over a time increment. Hyetographs are more accurate than the average intensity during the storm period and are used in calculating runoff hydrographs, plots of variation of runoff rate with time. Storm hyetographs can be constructed from the data collected by precipitation stations that either measure hourly rainfall depth or have a recorder continuously measuring rainfall.

Since rainfall hyetographs vary from one storm to another, runoff hydrographs are usually constructed based on synthetic rainfall distribution. The SCS 24-hour rainfall distributions are among the most widely used synthetic hyetographs in the United States. These hyetographs were developed by the US Department of Agriculture and Soil Conservation Service (SCS), which was renamed the National Resources Conservation Service (NRCS) in 1994. There are four types of SCS unit hydrographs: type I, type IA, type II, and type III. Figure 3.3 presents these rainfall types, and Figure 3.4 shows the geographic location in the United States where each rainfall type applies. Figure 3.3 indicates that types II and III are nearly the same and that type IA and type II are the least and the most intense storms, respectively. An inspection of this table also indicates that nearly 50% of the 24 hr rainfall depth occurs within the middle 2 hours of the 24-hour storm period in type II and type III storms. The SCS 24-hour storm distributions apply to all storm frequencies, each of which has a different total rainfall depth. This assumes that the shape of the 24-hour rainfall hyetograph is independent of the return frequency.

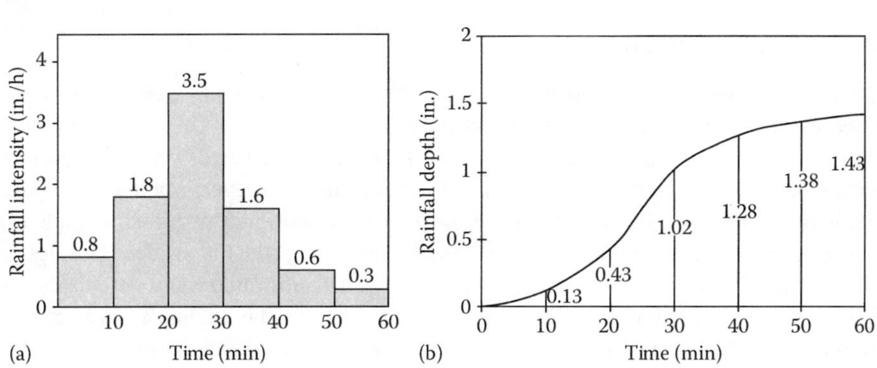

FIGURE 3.2 Rainfall hyetograph: (a) discrete hyetograph; (b) cumulative rainfall.

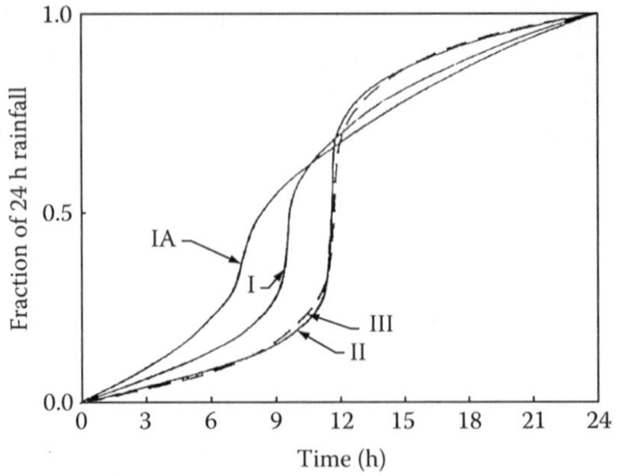

FIGURE 3.3 SCS 24-hour rainfall distributions.

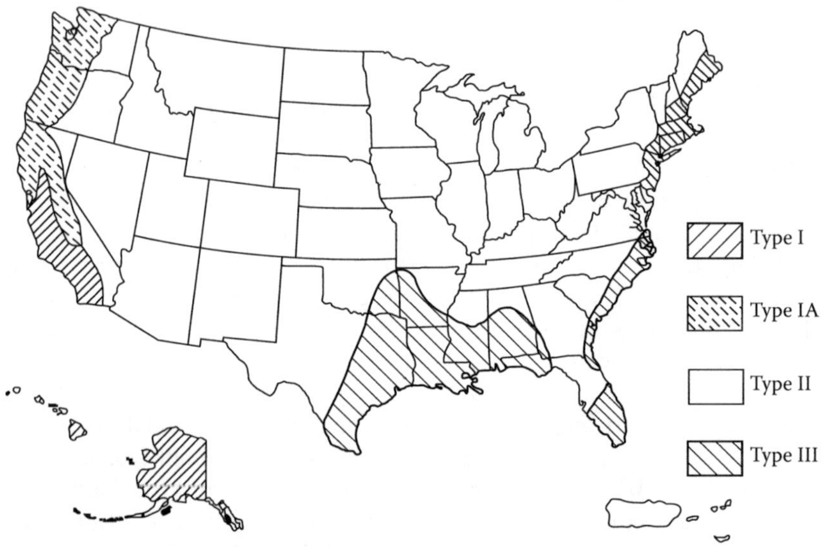

FIGURE 3.4 Approximate geographic locations for SCS rainfall distribution.

3.2 INITIAL ABSTRACTIONS

Rain falling on a pervious area is partly intercepted by vegetation such as tree canopies, partly fills surface depressions as puddles, and partly infiltrates into the ground. These retention elements at the beginning of a rainfall are known as initial abstraction. If the rain ends before the initial abstraction is satisfied, there will be no excess water to flow overland. However, when the rainfall exceeds the initial losses, water builds up on the catchment surface and flows overland, completely filling surface depressions on its path and gradually concentrating in gullies or swales. Through its route, the runoff continues to infiltrate into the ground. A part of the infiltrated water percolates downward through the soil to contribute to base flow and groundwater recharge. Figure 3.5 is a schematic representation of the disposition of stormwater during a uniformly distributed rainfall.

The initial abstraction of a vegetative cover is significant during light and moderate, short-duration storms. In fact, during such storms, the vegetation can fully retain the rainfall. This behavior is

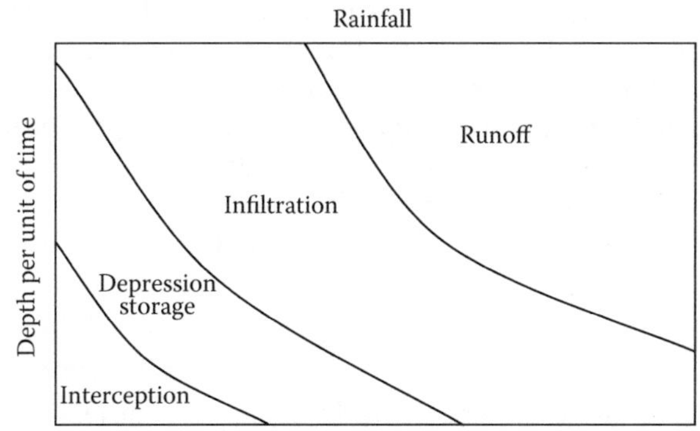

FIGURE 3.5 Schematic diagram of disposition of uniform rainfall.

evidenced by observations that no runoff occurs from lawns during storms of short duration. On an annual basis, large portions of rainfall infiltrates into the ground are retained by vegetation, and are lost to evaporation and evapotranspiration and therefore do not contribute to runoff.

Storm abstraction by vegetation continues until the vegetation becomes saturated. Beyond this point, vegetation has little retention effect; however, other losses such as surface retention in puddles or depressions and infiltration into the ground, continue to occur. In addition to these, evaporation also adds to losses; this latter effect is, however, negligible for short-duration storms.

3.2.1 Interception

Interception, as indicated, is a portion of abstraction by tree canopies, shrubs, and vegetation before the rain reaches the ground. This occurs as wetting of vegetation surface and retention of rain drops on the tree leaves. Interception is small in urban areas; however, it is quite significant in vegetated and forested lands. Development projects, including housing and roadway projects that involve clearing woods, significantly reduce if not totally eliminate interception, thereby increasing runoff.

A number of empirical formulae have been suggested for interception, and many of these are similar to that originally proposed by Horton in 1919, as follows:

$$Xi = a + b\,P^n \tag{3.1}$$

where a, b, and n are constants and P is the precipitation depth. In spite of its simplicity, a, b, and n are vegetation and soil dependent and cannot be readily determined. This severely limits practical applications of this equation. A more sophisticated interception equation is suggested by Brooks et al. (1991):

$$I = S\left(1 - e^{-P_e/S_d}\right) + KE_t \tag{3.2}$$

where

Xi = interception depth, mm (in.)

S = storage capacity of vegetation per unit of projected area, mm (in.)

P = rainfall depth, mm (in.)

E = evaporation rate, mm/h (in./h)

t = storm duration, h

K = leaf index area; the ratio of the upper surface area of intercepting leaves to the projected area of plant (or trees) on the ground

Storage capacity, S, is typically in the range of 1 mm (0.04 in.) for bare woods to 7 mm (0.3 in.) for evergreen trees and spruce. For crops, the interception during a 25 mm (1 in.) rain falling in 1 hour varies from 0.8 mm (0.03 in.) for corn to 4.1 mm (0.15 in.) for small grains and 8.4 mm (0.33 in.) for cotton. Therefore, trees and plants can intercept a large portion of rain during light or moderate short-duration storms. For P/S ratios of 3 or greater, the second term in the parentheses becomes increasingly small. Thus, interception depths can be approximated to be equal to S for rainfalls exceeding 4 mm to 25 mm (0.15 to 1 in.), depending on the vegetative cover.

Example 3.1

Estimated parameters in Equation 3.2 for a wooded area are $S = 5\text{ mm}$, $K = 3$, and $E = 0.25\text{ mm/h}$. Calculate the interception depth during a 30-minute storm of 50 mm/h (2 in./h) intensity.

Solution

$$P = 50 \times \frac{30\ min}{60\ min} = 25\ mm$$

Using Equation 3.2,

$$Xi = 5\left(1 - e^{-\frac{25}{5}}\right) + 3 \times 0.25 \times \frac{30\ min}{60\ min} = 5.34\ mm\ (0.21\ in.)$$

This example demonstrates that interception by trees and vegetation takes a large portion of the rain; the shorter the storm duration and the lighter the storm, the larger the interception-to-rainfall ratio will be.

3.2.2 Depression Storage

Depression storage is the water that is retained on surface depressions during a storm. Surface depressions act like miniature reservoirs that hold rainfall until they are filled. Depressions vary widely in area and depth. After the interception is satisfied, the rain falling on the ground begins to fill the depressions on the surface and infiltrate into the ground. The small depressions are filled first and overflow to larger depressions; once those are filled, the water begins to flow overland.

The water retained in depressions does not contribute to runoff; rather it infiltrates into the ground and evaporates after the storm. Depression storage depends on the type of soil cover, natural topography, and ground slope; its magnitude is commonly expressed in terms of the average depth of water over the drainage basin in millimeters or inches. Typical values of depression storage are reported to vary from 1 to 3.0 mm (0.04 to 0.12 in.) for paved areas and up to 7.5 mm (0.3 in.) in forested land (ASCE and WEF, 1992). However, the storage may be significantly higher on flat, uneven land and where the land has blind drainage, namely large depressions. For lawns, the depression storage varies from 3.0 to 5.0 mm (0.12 to 0.2 in.).

The volume of water in depression storage V_s (Linsley et al., 1982) can be expressed as:

$$V_s = S_d\left(1 - e^{-\frac{P_e}{S_d}}\right) \tag{3.3}$$

where

V = depression storage per unit area, mm (in.)
S_d = depression storage capacity, mm (in.)
P_e = depth of precipitation in excess of interception and infiltration

Equation 3.3 neglects any evaporation. In this equation, the terms in the parentheses imply that if depressions are filled, $P_e = 0$ and the depression storage will be equal to its capacity. This, of course, assumes that there will be no overland flow until all depressions are full. However, this condition only occurs when the largest depressions are located downstream. The assumption that depression storage subtracts the volume of rainfall from the initial storm period has been used with satisfactory results under normal conditions.

Experiments indicate that depression storage of an impervious area depends on its slope and varies from 3 mm (0.12 in.) for a 1% slope to 1.3 mm (0.05 in.) for a 3% slope. In natural basins, S_d varies from 10 to 50 mm (0.4 to 2 in.). For lawn and turf, depression storage, as indicated, is about 5 mm (0.20 in.).

The rate of depression storage during rainfall, V_s, can be expressed by

$$V_s = e^{-\frac{P_e}{S_d}}\left(I - f\right) \tag{3.4}$$

where

I = rainfall intensity, mm/h (in./h)
f = infiltration rate, mm/h (in./h)

Example 3.2

A 10-year, 10-minute storm in New Jersey has an average intensity of approximately 4.5 in./h (114 mm/h). Calculate the portion of rainfall that turns into runoff from a mildly sloped, semismooth paved area.

Solution

Calculations are performed both in CU and SI units. and the depression storage is estimated at 2.5 mm (0.1 in.).

The rainfall depth is

$$\frac{4.5 \times 10}{60} = 0.75\ in.$$

$$\frac{145 \times 10}{60} = 24.1\,mm$$

The net rainfall that becomes runoff is

$$0.75 - 0.1 = 0.65\ in.$$

$$19 - 2.5 = 16.5\ mm$$

The ratio of runoff to rainfall depth is

$$C = \frac{0.65}{0.75} = 0.87$$

$$C = \frac{16.5}{19} = 0.87$$

This ratio, as will be discussed later, is called the runoff coefficient. Since depression storage is independent of rainfall depth, the runoff coefficient is not constant; rather, it varies with the rainfall intensity and duration: The shorter the storm duration and the lighter the storm, the smaller the runoff coefficient will be. In practice, however, a constant runoff coefficient is employed to perform runoff calculations, which is unrealistic.

Example 3.3

A 0.5 acre (2025 m^2) subdivided lot contains 5000 ft^2 (465 m^2) of impervious surfaces, and the remainder is wooded/landscape. Estimate the percentage of rainfall that turns into runoff for a 10-minute storm of 5.7 in./h (145 mm/h) intensity. Assume the amounts of interception at 0.15 in. (3.8 mm) and depression storages for pervious and impervious areas at 0.4 in. (10 mm) and 0.1 in. (2.5 mm), respectively. Also, conservatively neglect evaporation and infiltration.

Solution

Calculation are presented in both CU and SI.

Rainfall depth is

$$\frac{5.7\times 10}{60}=0.95\,in.$$

$$\frac{145\times 10}{60}=24.1\,mm$$

Abstraction and depression storage amount to

$$\frac{\left[(43{,}560\times 0.5-5000)\times(0.15+0.40)+5000\times 0.1\right]}{12}=810.8\,ft^3$$

$$\frac{\left[(2025-465)\times(3.8+10)+465\times 2.5\right]}{1000}=22.69\,m^3$$

Abstraction depth is

$$\frac{810.8}{(43{,}560\times 0.5)}=0.037\,ft=0.45\,in.$$

$$\frac{22.69}{2025}=0.0112\,m=11.2\,mm$$

Net rainfall becoming runoff is

$$0.95-0.45=0.50\,in.$$

$$24.1-11.2=12.9\,mm$$

The ratio of runoff to rainfall is 53%

$$0.5/0.95=0.53\%$$

$$12.9/0.95=53\%$$

Note that infiltration during rainfall is ignored in this problem and that accounting for infiltration would produce a much smaller runoff-to-rainfall ratio than that calculated herein. See Example 3.4.

3.3 INFILTRATION

Infiltration is the passage of water through the surface soil. Many factors affect infiltration, and are classified as natural and surface factors. Natural factors are related to natural processes, such as precipitation, freezing, season, temperature, moisture, and, above all, soil texture. Surface factors are associated with soil cover. A bare soil forms a crust under the impact of raindrops, which, in turn, reduces infiltration. By preventing the soil from crust formation, grass cover increases infiltration.

Infiltration is different from percolation, which represents a downward flow of water through soil due to gravity. Although different, the two processes are closely related since infiltration cannot continue indefinitely unless percolation removes infiltrated water from the surface soil. Percolation occurs through the flow of water in noncapillary channels. The capillary water, the water absorbed into the soil particles, does not flow downward by gravity. The capillary suction distinguishes permeable soils such as sand from impermeable soils such as clay and is much smaller for the former than the latter. Typical capillary suction can be less than 1 cm (0.4 in.) for sand but over 5 m (15 ft) for clay.

The infiltration rate is equal to the percolation rate just below the ground surface, where the soil is saturated. The movement of water through soil is governed by Darcy's law:

$$q = K\frac{dh}{dz} \tag{3.5}$$

where

K is permeability or hydraulic conductivity and is a function of soil texture and moisture content
h is the piezometric head of pore water
z is the vertical coordinate taken positive downward in the preceding equation

The piezometric head is the total of pore water pressure and depth, z, as follows:

$$h = \frac{p}{d} + z \tag{3.6}$$

Negative pore water pressure indicates tension or suction. This occurs in unsaturated soil due to the capillary effect. Because of the capillary (suction) effect, both piezometric head and permeability are at a maximum when the soil is dry. The maximum rate at which the water can enter soil under a given set of conditions is called the infiltration capacity, f_p. The actual infiltration, f, equals f_p only when the effective rainfall intensity, that is, the rainfall intensity minus the rate of interception and the rate of depression storage, equals or exceeds f_p. As infiltration continues, the soil pores fill with water, the capillary suction diminishes, and the infiltration reaches its lower limit, which is governed by the gravity flow alone. Under this condition, the infiltration rate becomes equal to the percolation rate, which is also called the hydraulic conductivity, K. If the soil is stratified, the least pervious subsoil layer limits the infiltration.

A number of equations or models are available for estimating infiltration. The Green–Ampt model and Horton and Philip's equations are the most widely used in engineering practice. The validity of each of these models should be based on its consistency with the actual infiltration process.

3.3.1 Green-Ampt's Model

A physically based model was originally introduced by Green and Ampt in 1911 and was placed on a firm basis by Philip in 1954. The Green–Ampt model, also called the delta function model, is one of the most realistic models for infiltration. This model is widely employed in continuous simulation models such as the EPA's Stormwater Management Model (SWMM). However, as will be shown in this section, this model involves implicit equations and tedious iterative calculations, and it breaks down if improperly applied.

Consider that water is impounded to a depth of H_o over the ground surface. When the infiltration begins, the soil below the ground becomes saturated with water, but the lower soil remains unsaturated. This produces a sharp moisture gradient near the interface of moist and dry soils, resulting in a high infiltration rate. Figure 3.6 depicts the infiltration process for this case. As the infiltration continues, the interface, called the wetting front, moves downward. If the rain is sustained, the wetting front eventually reaches the water table.

Using straight-line approximation for the saturated soil between the soil surface and the wetting front and neglecting ponding depth, H_o, Darcy's equation (Equation 3.5) becomes

$$q = f_p = \frac{K\left[0-(L+\psi)\right]}{(0-L)}$$

or

$$f_p = K\left(1+\frac{\psi}{L}\right) \tag{3.7}$$

where

f_p = potential infiltration rate
E = soil suction at the wetting front L

Since suction is in a downward direction, the z will be positive if it is also downward in this equation. The cumulative infiltration, F, is equal to L multiplied by the initial moisture deficit:

$$F = L(\alpha - \theta i) = L\Delta\theta \tag{3.8}$$

where

θ_i = initial moisture content in the dry soil
$\Delta\theta$ = the soil moisture deficit
α = soil porosity

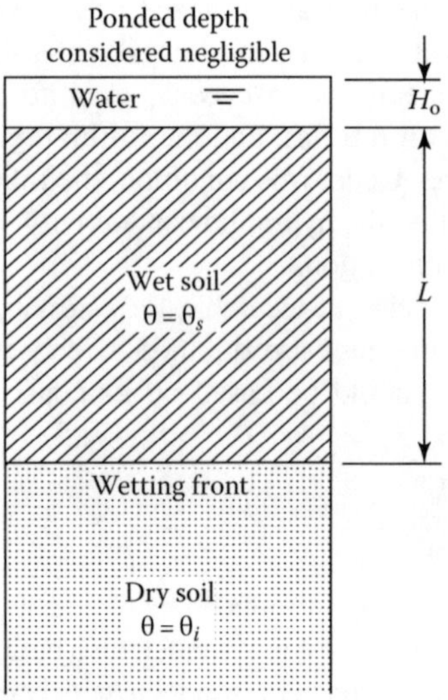

FIGURE 3.6 Schematic diagram of the Green–Ampt infiltration model.

Eliminating L between Equations 3.7 and 3.8 yields

$$f_{\mathrm{p}} = K\left[1+\left(\frac{\psi\Delta\theta}{F}\right)\right] \tag{3.9}$$

or, inversely,

$$F = \frac{\Psi\Delta\theta}{\left(f_{\mathrm{p}}/K-1\right)} \tag{3.10}$$

Table 3.4 gives typical Green–Ampt parameters for the US Department of Agriculture (USDA) soil texture classes. The preceding equations indicate that the initial infiltration capacity is larger than the hydraulic conductivity. However, as infiltration progresses, ψ diminishes, and the infiltration capacity decreases, eventually approaching the hydraulic conductivity. Note that the permeability values in Table 3.5 are significantly smaller than reliable information (see, e.g., Freeze and Cherry, 1979; Linsley et al., 1982; Todd, 1980). They are also far smaller than those in Tables 3.5 and 3.7.

The preceding equations for f and F are valid only when water is impounded on the ground and/or the rainfall intensity exceeds the infiltration capacity. Since Equations 3.9 and 3.10 have two variables, they cannot be solved directly. However, noting that

$$f_{\mathrm{p}} = \frac{dF}{dt} \tag{3.11}$$

TABLE 3.4
Green–Ampt Infiltration Parameters for USDA Soil Texture Classes[a]

USDA Soil Classification	Porosity θ	Effective Porosity	Permeability mm/h (in./h)	Wetting Front Suction Head Ψ mm (in.)
Sand	0.44	0.42	117.8 (4.64)	49.5 (1.95)
Loamy sand	0.44	0.40	29.9 (1.18)	61.3 (2.41)
Sandy loam	0.45	0.41	10.9 (0.43)	110.1 (4.33)
Loam	0.46	0.43	3.4 (0.13)	88.9 (3.50)
Silt loam	0.50	0.49	6.5 (0.26)	166.8 (6.57)
Sandy clay loam	0.40	0.33	1.5 (0.06)	218.5 (8.60)
Clay loam	0.46	0.31	1.0 (0.04)	208.8 (8.22)
Silty clay loam	0.47	0.43	1.0 (0.04)	273.0 (10.75)
Sandy clay	0.43	0.32	0.6 (0.02)	239.0 (9.41)
Silty clay	0.48	0.42	0.5 (0.02)	292.2 (11.50)
Clay	0.48	0.39	0.3 (0.01)	316.3 (12.45)

Note: Actual θ and $\Delta\theta$ vary by nearly 30% from the average values listed in this table; ψ values vary by up to 25-fold from the listed values.

Source: Rawls, W. J. et al., 1983, *Journal of Hydraulic Division, ACSE*, 109 (1): 62 ± 70; condensed and rounded to the second decimal place.

[a] See Figure 3.6.

TABLE 3.5
Typical Values for $f_c = K$ for USDA Soil Textures

	f_c	
Soil type	**mm/h**	**in./h**
Sand	>500	>20.0
Loamy sand	250	10.0
Sandy loam	100	4.0
Loam	40	1.6
Sandy clay loam	30	1.2
Silt loam	10	0.4
Clay loam	7.5	0.3
Sandy clay	5.0	0.2
Silt	2.5	0.1
Silty clay loam	2.0	0.08
Silty clay	1.0	0.04
Clay	<0.5	<0.02

Note: Values of f_c are derived from a soil permeability/textural triangle (Figure 3.11). For sandy loam through sandy clay soil covered with turf, the f_c values may be two- to three-fold larger than those shown in this table.

and combining Equations 3.8 and 3.10 and separating variables gives

$$\left[\frac{F}{(F+\Psi\Delta\theta)}\right]dF = Kdt \tag{3.12}$$

Integrating from $t = 0$ to $t = t$ results in the following equation:

$$Kt = F - \Delta\theta\psi ln\left(1+\frac{F}{\Psi\Delta\theta}\right) \tag{3.13}$$

This equation can be solved through an iterative process to calculate F at any time during the storm period when the rainfall intensity exceeds the infiltration capacity. Having F (the infiltration depth), f_p can be calculated using Equation 3.9.

If the rainfall intensity is initially less than the infiltration capacity, all the rainfall infiltrates until the initial moisture deficit is satisfied. In this case, the ponding does not occur at time $t = 0$, but at a time $t = t_p$, where

$$t_p = \frac{F_p}{I} \tag{3.14}$$

and

$$f = I \; for \; t \le t_p \tag{3.15}$$

Note that F_p, which is the infiltrated water depth before ponding occurs, cannot be calculated using Equation 3.10. Substituting $f_p = I$ in that equation yields a negative infiltration volume when $I < K$ and breaks down when $K = I$, both of which are unreal.

In the preceding case, the variation of infiltrated depth with time can be written as

$$K\left(t - t_p + t_{p'}\right) = F - \Psi\Delta, \ \ln\left[1 + \frac{F}{\Psi\Delta\theta}\right] \tag{3.16}$$

where, as indicated, t_p is the time at which ponding occurs and $t_{p'}$ is the time to infiltrate volume F_p, under initial surface ponding condition. The time $t_{p'}$ can be calculated by substituting F_p for F in Equation 3.13.

Under this condition, Equations 3.14 and 3.16 are to be employed together to calculate F in an iterative process. To expedite the calculation process, the Green–Ampt model is generally applied by incrementing F and solving for t in Equation 3.16 and then using Equation 3.9 to calculate f_p. The following example illustrates the tediousness of calculating a relatively simple rainfall distribution.

Example 3.4

A loamy, sandy soil has the following properties:

$K = 40$ mm/h
$\theta = 0.45$
$\theta_i = 0.15$
$\psi = 50$ mm water

Using the Green–Ampt model, calculate the time for soil surface to become saturated for a 60-minute rainfall with the following distribution:

$I = 30$ mm/h; 0–15 minutes
$I = 60$ mm/h; 15–60 minutes

Solution

$$\Delta\theta = 0.45 - 0.15 = 0.3$$

$$\Psi\Delta\theta = 50 \times 0.3 = 15$$

During the first 15 minutes, the rainfall intensity is less than the minimum K; therefore all the rainfall infiltrates:

$$F_{15} = \frac{30 \times 15}{60} = 7.5 \text{ mm}$$

In the t = 15- to 60-minute period, since the rainfall intensity is larger than hydraulic conductivity, ponding is possible. Use Equation 3.9 to relate f_p and F

$$f_p = 40 + \frac{600}{F}$$

The time to surface ponding is calculated by equating the infiltration capacity to the rainfall intensity in the preceding equation:

$$60 = 40 + \frac{600}{F_p}$$

$$F_p = 30 \ mm$$

Time to surface ponding is

$$t_p = \frac{F_p}{I} = \frac{30}{60} = 0.5h = 30\ min$$

Therefore, the ponding starts at

$$t_p = 30 + 15 = 45\ min = 0.75\ h$$

from the beginning of rainfall.

Cumulative infiltration at this time is

$$F = 7.5\ mm + 30\ mm = 37.5\ mm$$

Next, calculate $t_{p'}$ namely the time that it would take F = 37.5 mm to infiltrate at a potential rate from t = 0; using Equation 3.13:

$$40 \times t_p' = 37.5 - 15ln\left(1 + \frac{37.5}{15}\right)$$

$$t_p' = \frac{18.71}{40} = 0.468\ \text{h} = 28.1\ \text{min}$$

Express F as a function of time using Equation 3.16:

$$40\left(t - t_p + t_p'\right) = F - 15\ln\left(1 + \frac{F}{15}\right)$$

Input the previously calculated t_p and t_p values in this equation:

$$40(1 - 0.75 + 0.468) = F - 15ln\left(1 + \frac{F}{15}\right)$$

Simplify:

$$F - 15ln\left(1 + \frac{F}{15}\right) = 28.72$$

Solve iteratively:

$$F = 51\ mm$$

Since the amount of rainfall at t = 1 h is

$$7.5 + \frac{60 \times 45}{60} = 52.5\ mm$$

the ponding depth at the end of the storm (t = 1 h) is

$$52.5\ mm - 51\ mm = 1.5\ mm$$

3.3.2 Horton's Equation

In 1939/1940, Horton presented the following empirical equation for the infiltration capacity at a given time:

$$f_t = f_c + \left(f_o - f_c\right)e^{-\alpha t} \tag{3.17}$$

where

f_t is the infiltration capacity at time t
f_o and f_c are the initial and ultimate (or equilibrium) infiltration rates, respectively
e is the Napierian base
α is a decay constant, which depends on soil
t is time from beginning of rainfall

Integrating Equation 3.17 gives the cumulative infiltration depth:

$$F_t = f_c t + \left[\frac{(f_o - f_c)(1 - e^{-\alpha t})}{\alpha} \right] \tag{3.18}$$

Eliminating t between Equations 3.17 and 3.18 yields a direct relationship between f and

$$F_t = \frac{f_c \ln(f_o - f_c)}{\alpha} - \frac{f_c \ln(f - f_c)}{K} + \frac{(f_o - f)}{\alpha} \tag{3.19}$$

To solve Equation 3.19, f_o, f_c, and α need to be known. These parameters may be determined from a graph of measured infiltration rate over an extended period of time. The tail end of the graph can be extrapolated to calculate f_c. Two sets of f and t data can be taken from the plot and input into Equation 3.19 to solve for f_o and α. Although the process looks simple, it is tedious and somewhat inconclusive.

Similar to the Green–Ampt model, Equations 3.18 and 3.19 are valid only when the net rainfall intensity exceeds the infiltration capacity, specifically,

$$f = f_p, I \geq f_p \tag{3.20}$$

If the rainfall intensity is smaller than the infiltration capacity ($I < f_p$), the infiltration occurs at the rate of $f = I$.

Experience shows that the infiltration capacity for a given soil varies with the initial moisture content, organic matter of soil, vegetative cover, and season (Linsley et al., 1982). These effects result in tremendous variation in soil permeability, meaning widely varying f_o, f_c, and α values reported in the technical literature and some of the values do not agree with field observations. For instance, Butler and Davies reported 75 mm/h and 3 mm/h, respectively, as f_o and f_c for clay both of which, however, are exaggerated (see Chin, 2006).

The same researchers reported f_c = 12 mm/h for medium-textured soil and f_c = 25 mm/h for coarse-textured soils. While the former results grossly exaggerate the permeability of clay, the latter figures underestimate the permeability of medium- and coarse-textured soils. Additionally, Rawls et al. reported α = 0.64 min^{-1} for loamy sand referenced in Chin (2006) and others, indicating that soil loses 25% of its initial infiltration capacity within just 1 minute after rain and reaches its ultimate limit in less than 10 minutes, both of which appear unrealistic. Because of these inconsistencies, no reliable values for f_o and α are presented here. Only typical values of f_c, primarily soil-dependent, are provided for the USDA soil textures in Table 3.5.

The uncertainty in the Horton model stems from the reported α values, which exaggerate the drop in the infiltration with time. To adjust for this deficiency, Viessman et al. (1989) suggest relating f as a function of the cumulative infiltration rather than time. The EPA's SWMM uses this form of the Horton equation. Figure 3.7 depicts a typical variation of infiltration with time.

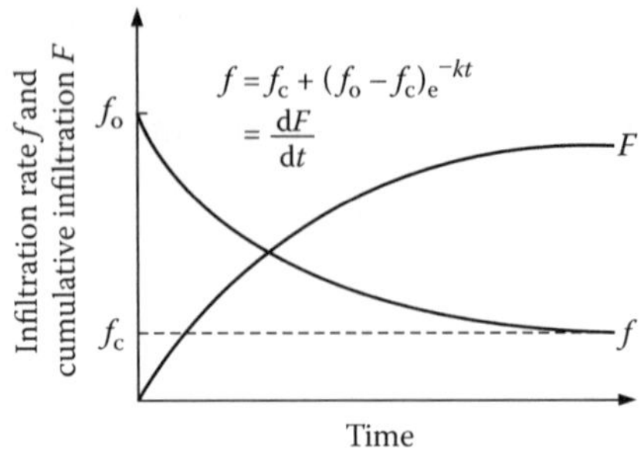

FIGURE 3.7 Infiltration curves.

Example 3.5

Assuming f_o = 55 mm/h, f_c = 25 mm/h, and α = 1.2 hour^{-1}, calculate infiltration rates at the end of one, two, and three hours during a three-hour storm having an intensity larger than infiltration. Also, calculate the cumulative infiltration at the end of the three-hour period.

Solution

$$f_p = f_c + (f_o - f_c)\, e^{-\alpha t}$$

$$f_c = 25 + 30\, e^{-1.2t}$$

f_p at the end of each hour interval is

$$f_1 = 34.04\ mm;\ f_2 = 27.72\ mm;\ f_3 = 25.82\ mm$$

$$F(t) = f_c t + \left(\frac{f_o - f_c}{\alpha}\right)\left(1 - e^{-\alpha t}\right)$$

$$= 25 \times 3 + \left(\frac{30}{1.2}\right)\left(1 - e^{-3.6}\right) = 99.3\ mm$$

Note: Under initially wet ground surface conditions, the infiltration rate would be nearly equal to hydraulic conductivity, f_c, and the cumulative infiltration = 3 × 25 = 75 mm.

Example 3.6

Rain falls at a rate of 1.5 in./h for one hour on a lawn in a silty loam soil. Calculate the amount of runoff within the rainfall period. Conservatively estimate $f = K$ and use Table 3.5.

Solution

From Table 3.5, K = 0.4 in./h. Since the soil is covered with lawn, estimate K = 1.0 in./h (2.5 times greater than bare soil). Estimate interception and depression storage for the lawn at 0.1 and 0.25 in., respectively.

Runoff depth = net rainfall = 1.5–1.0 – (0.1 + 0.25) = 0.15 in.

3.3.3 Philip's Equation

Philip (1958) presented the following equation

$$f = \frac{1}{2} s t^{-1/2} + K \tag{3.21}$$

where

f = instantaneous infiltration rate
s = an empirical parameter related to the progression of the waterfront
K = hydraulic conductivity at the surface
t = time

In this equation, $f = \infty$ for $t = 0$; however, it gradually decreases with time until $f = K$ at $t = \infty$. Integrating this equation results in

$$F = St^{1/2} + Kt \tag{3.22}$$

in which F is the cumulative depth of infiltration. The parameter s in this equation, similar to α in Horton's model, depends on soil properties and ground cover and varies widely.

Example 3.7

A 1000 m^2 residential lot comprises 350 square feet of impervious area, and the remainder is covered with grass. Assume

- Lawn interception = 2 mm
- Depression storage = 7 mm for grass and 1.5 mm for pavement
- Initial and equilibrium infiltration rates of 75 mm/h and 25 mm/h, respectively and a Horton decay constant of $\alpha = 1.0^{-1}$ hour

Calculate the portion of rainfall that turns into runoff for a 30 minutes of rainfall at 100 mm/h intensity using Horton's equation. Also, calculate the runoff-to-rainfall ratio.

Solution

The overall initial abstractions and depression storage for the site are calculated as follows:

$$Lawn\ area = 1000 - 350 = 650\ m^2$$

The combined depth of initial abstraction and depression storage is

$$\frac{[650 \times (2+7) + 350 \times 1.5]}{1000} = 6.4\ mm$$

Infiltration rate, Horton's equation, is

$$f = 25 + (75 - 25)\,e^{-1} = 25 + 50\,e^{-1}$$

he infiltration amount is

$$F(t) = f_c t + \left(\frac{f_o - f_c}{\alpha}\right)\left(1 - e^{-\alpha_t}\right)$$

For t = 30 minutes = 0.5 h,

$$F = 25 \times 0.5 + 50\left(1 - e^{-0.5}\right) = 12.5 + 50\,(1 - 0.607) = 32.2\ mm$$

Average infiltration over the entire lot is

$$\frac{(650 \times 32.2 + 350 \times 0.0)}{1000} = 20.9\ mm$$

Total losses = 6.4 + 20.9 = 37.3 mm
Rainfall depth = 100 × 30 min/60 min = 50 mm
Runoff depth = rainfall depth – combined losses

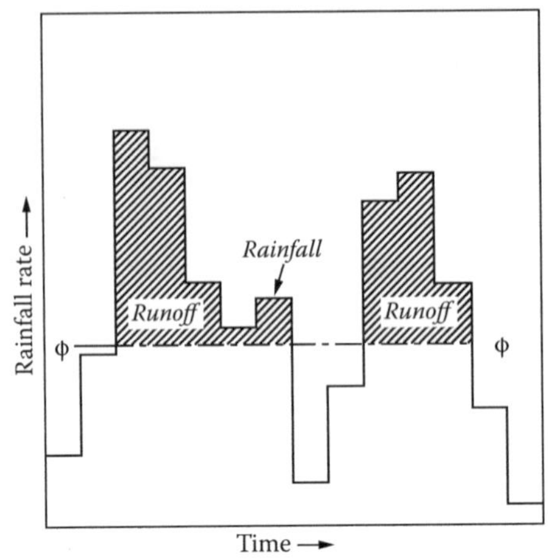

FIGURE 3.8 The φ infiltration index.

$$= 50 - 37.3 = 12.7\ mm$$

$$\frac{Runoff}{Rainfall} = \frac{12.7}{50} = 0.25$$

As seen, a large percentage of rainfall is dissipated through initial abstraction and infiltration. The losses are significantly larger for soils of higher permeability.

3.3.4 Infiltration Indexes

In practice, infiltration indexes are commonly employed to simplify calculations. Infiltration indexes generally assume that infiltration occurs at a constant rate during the storm. As such, the indexes underestimate the initial rate of infiltration and exaggerate its ultimate value. The best application of an index is to a large storm on wet soil where infiltration rate may be assumed to be relatively constant.

The most common index is the φ index. In this index, the total volume of rainfall abstraction is estimated and distributed uniformly during the storm period (Figure 3.8). The depth of precipitation above the index line then represents the overland runoff. A variation of this index is the W index, in which the initial abstractions are deducted from the early storm period.

To calculate the φ index for a given storm, the amount of observed runoff from the storm hydrograph is deducted from the total precipitation and the difference is divided by the storm duration. It should be noted, however, that φ obtained from a single storm may not be applicable to other storms.

3.4 MEASUREMENT OF INFILTRATION AND PERMEABILITY

Infiltration and permeability can be measured directly through in situ percolation tests. Soil samples can also be sent to a laboratory for permeability measurement. Permeability can also be estimated indirectly through soil gradation. These methods are discussed in this section.

3.4.1 Infiltrometers

Infiltration rates into soils can be measured with a ring infiltrometer, consisting of a metal tube 8 to 12 in. (20 to 30 cm) in diameter. This tube is driven 18 to 24 in. (45 to 60 cm) into the soil with 4 in. (10 cm) or so projecting above the ground. Water is poured into the tube, and the rate of drop is measured as an indicator of infiltration rate. As water enters the soil, the air escapes around the tube; as such, the measured water drop exaggerates the actual infiltration rate. To minimize this effect, a double-ring infiltrometer is used, and both rings are filled with water. In this device, the water that

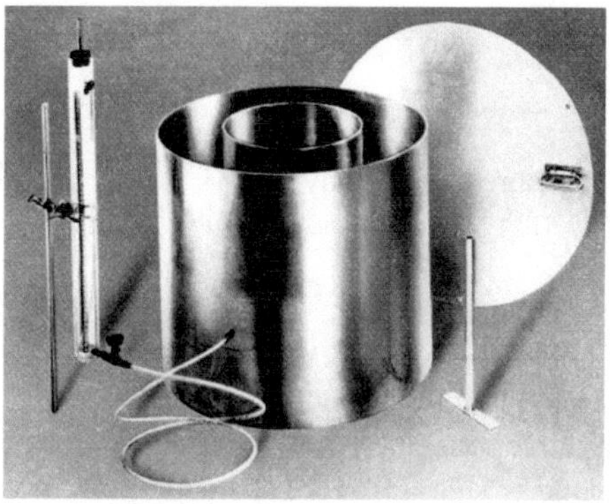

FIGURE 3.9 Rickly double-ring infiltrometer.

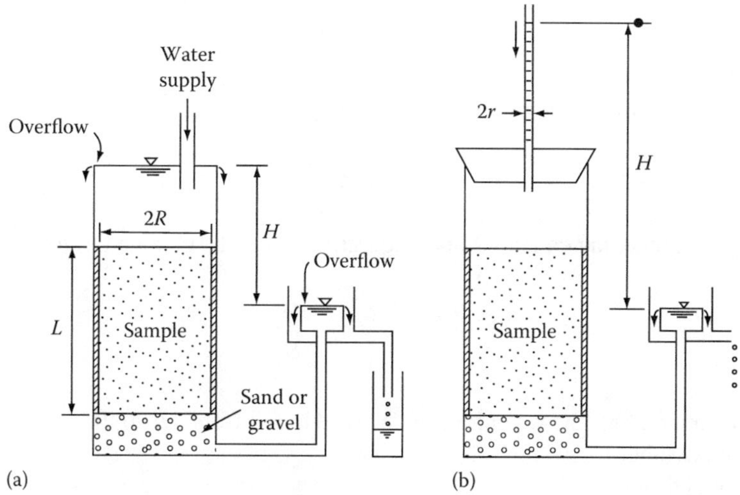

FIGURE 3.10 Permeameters for measuring *K*: (a) constant head, (b) falling head.

enters the inner ring tends to move down with a minimal lateral spread. Because of spatial variation of permeability, several measurements spread over the area of interest should be made to determine the average value of infiltration rate.

Figure 3.9 shows a double-ring infiltrometer (by Rickly Hydrological Company, Columbus, OH, USA). This instrument consists of two concentric rings and a driving plate with handles for the inner and outer rings. The outer and inner rings are 24 and 12 in. diameter, respectively. The rings are driven into the soil and partially filled with water. The volume of water poured in the inner tube to maintain the water level for a specific period is measured using a Mariotte tube. The information is converted into the infiltration rate using the data sheet provided by the manufacturer.

3.4.2 Permeameters

A permeameter is a laboratory device for measuring hydraulic conductivity. In this device, flow is maintained through a small sample of soil material, and flow rate and head are measured. Two types of permeameters are available: constant head and falling head.

The constant-head permeameter can measure the hydraulic conductivity of consolidated or unconsolidated soils under low head (Figure 3.10a). Water flows upward through the sample; by measuring its rate, the hydraulic conductivity, K, can be obtained from Darcy's law as follows:

$$K = \frac{(Q / \pi R^2)}{(H / L)} \tag{3.23}$$

where

Q = rate of flow (volume measured divided by time)
L = height of sample
H = constant head
R = radius of sample in permeameter

The soil should be fully saturated before taking any measurements. In practice, several measurements are made to obtain a reliable result.

In the falling head permeameter (Figure 3.10b), a slender, tall tube is filled with water, and the rate of fall of the water level in the tube is measured. The hydraulic conductivity can be obtained by noting that the measured rate of flow is

$$Q = \pi r^2 \frac{dH}{dt} \tag{3.24}$$

and that the flow rate through the sample follows Darcy's law:

$$Q = \pi R^2 K \frac{H}{L} \tag{3.25}$$

By equating the preceding equations and integrating, K can be obtained as follows:

$$K = \left(\frac{r}{R}\right)^2 \left(\frac{L}{t}\right) ln \left(\frac{H_1}{H_2}\right) \tag{3.26}$$

where

t= is the time interval for the water level in the tube to fall from H_1 to H_2
r = radius of tall tube
ln = natural logarithm
R and L = are as previously defined.

The laboratory measure of permeability is more accurate than in situ percolation testing; however, it has the drawback of disturbing soil structure and stratification.

3.4.3 Soil Gradation Analysis

The percolation rate depends primarily on the soil particle size. Soil porosity has little effect on this rate. In fact, clayey soils, are much more porous than both silt and sand but are far less permeable than either of them. Soils are commonly classified by their particle size distribution. Table 3.6 presents the USDA soil classification. Thus, percolation and infiltration are controlled by the soil texture, which is defined as the proportions by weight of clay, silt, loam, and sand after granular material larger than 2 mm is removed.

The USDA uses a soil texture triangle, which is shown in Figure 3.11. The figure shows that, for example, a soil composed of 30% clay, 50% sand, and 20% silt is classified as sandy clay loam. If more than 15% of the soil is larger than 2 mm, a prefix, such as gravely or stony, is added to the soil texture names in this figure.

Gradation analysis of a soil sample in a laboratory provides an indirect method of estimating permeability. In this method, the percentages of sand, silt, and clay are measured. Based on this analysis, the soil texture is determined using the USDA Soil Texture Triangle. Then, the soil permeability is estimated using the Soil Permeability/Textural Triangle in Figure 3.12. As shown in this figure, the soils are classified into six permeability ratings or classes, identified as K_0 through K_5; K_0 has the lowest permeability and K_5 the highest. The permeability ratings for these soil classes, which range from less than 5 mm/h (0.2 in./h) for K_0 soils to well over 500 mm/h (20 in./h) for K_5 soils, are provided in Table 3.7.

TABLE 3.6
USDA Soil Classification

Soil Name	Particle Size (mm)
Clay	<0.002
Silt	0.002–0.05
Sand	0.05–2.0
Gravel	>2.0

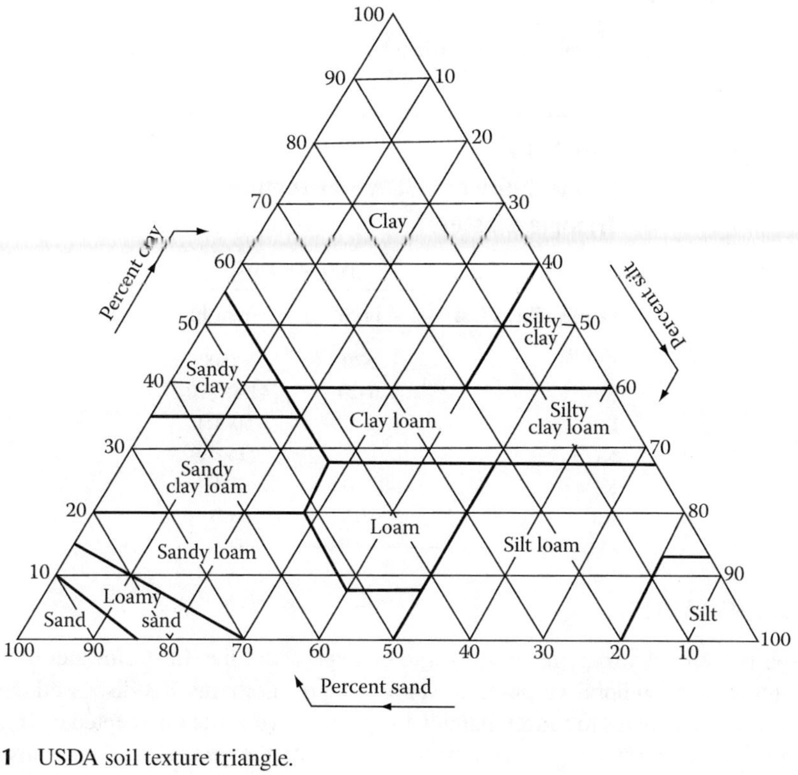

FIGURE 3.11 USDA soil texture triangle.

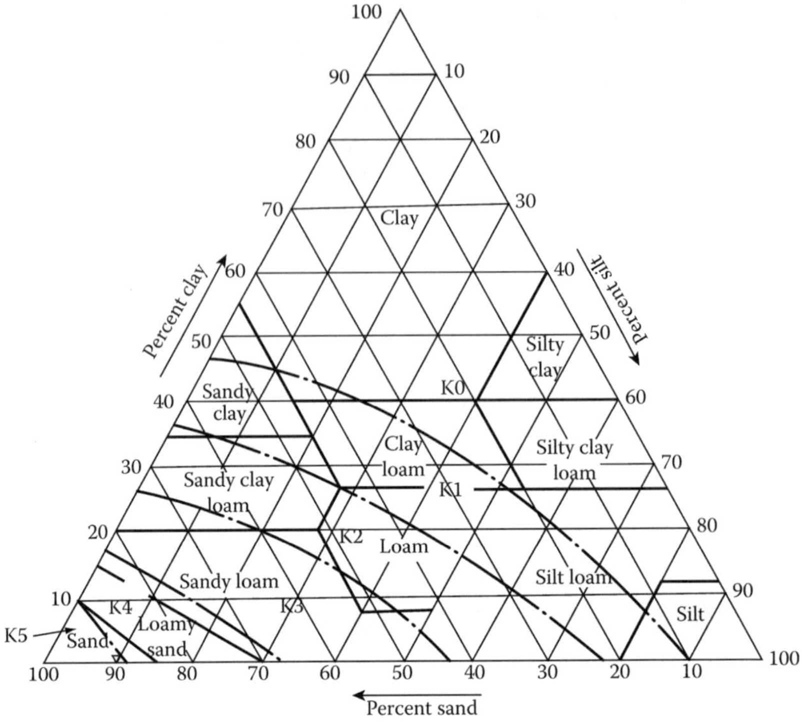

FIGURE 3.12 USDA soil permeability/textural triangle.

TABLE 3.7
Permeability of USDA Soil Texture Triangle Classes

	Permeability	
Permeability Class	**in./h**	**(mm/h)**
K5	>20	> (500)
K4	6–20	(150–500)
K3	2–6	(50–150)
K2	0.6–2	(15–50)
K1	0.2–0.6	(5–15)
K0	<0.2	(5)

3.5 HYDROGRAPHS

A hydrograph is graph of variations in discharge or stage with time. In stormwater management, a hydrograph represents temporal variation of runoff rate or discharge. As discussed earlier, only a portion of rainfall contributes to runoff, namely the portion that is not intercepted by tree leaves and vegetation, retained in surface depression, or infiltrated into the ground. The relationship between rainfall and runoff is an integral part of stormwater management calculations.

A hydrograph of a single storm is shown in Figure 3.13. A hydrograph is characterized by a rising limb, a peak, and a falling limb, also known as a recession. The rising limb reflects increased runoff as the rainfall is stored and flows overland, and the recession represents the withdrawal of stored water after a storm is ended. A hydrograph is also characterized by three parameters: peak flow,

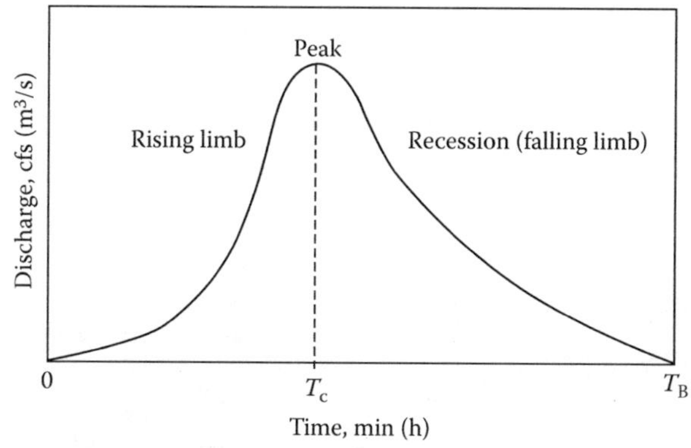

FIGURE 3.13 A typical storm hydrograph.

time to peak (also referred to as time of concentration, T_c), and time base, T_B. The time to peak, also referred to as time of concentration, represents the time it takes for the runoff from the entire catchment area to reach the point of discharge. These parameters are discussed in the following sections.

3.5.1 Time of Concentration Equations and Nomographs

Time of concentration is defined as the time for the runoff from a catchment area to reach equilibrium under steady rainfall. It is also defined as the longest travel time it takes the runoff to reach the discharge point of a catchment area. Travel time is a parameter most often used to characterize the response of a catchment area to rainfall. This parameter is a function of length, L, average catchment slope, S, and the catchment surface condition. The time of concentration is the sum of the overland flow time and the travel time in drainage channels along the flow route to the outlet. A number of empirical equations and nomographs have been proposed for time of concentration calculations is the most popular of which are the following.

3.5.1.1 Kirpich's Equation

Kirpich, in 1940, presented the following empirical formula in CU (Chow et al. 1988):

$$T_c = \frac{0.0078L^{0.77}}{S^{0.385}} \tag{3.27}$$

In metric units, this equation reads

$$T_c = \frac{0.21L^{0.77}}{S_o^{0.385}} \tag{3.28}$$

where

T_c = time of concentration, in minutes
L = flow length, ft (m)
S = average slope of flow path, ft/ft (m/m)

Kirpich's equation was originally developed from SCS data for seven rural basins in Tennessee with well-defined channels and slopes ranging from 3% to 10%. Figure 3.14 shows a graphical solution of Kirpich's equation that is widely used. For overland flow on the lawn, the T_c obtained from this figure should be multiplied by 2. The author finds that Kirpich's nomograph (Figure 3.14) gives unreasonably short times of concentrations, and therefore, he does not recommend its use. See Example 3.8.

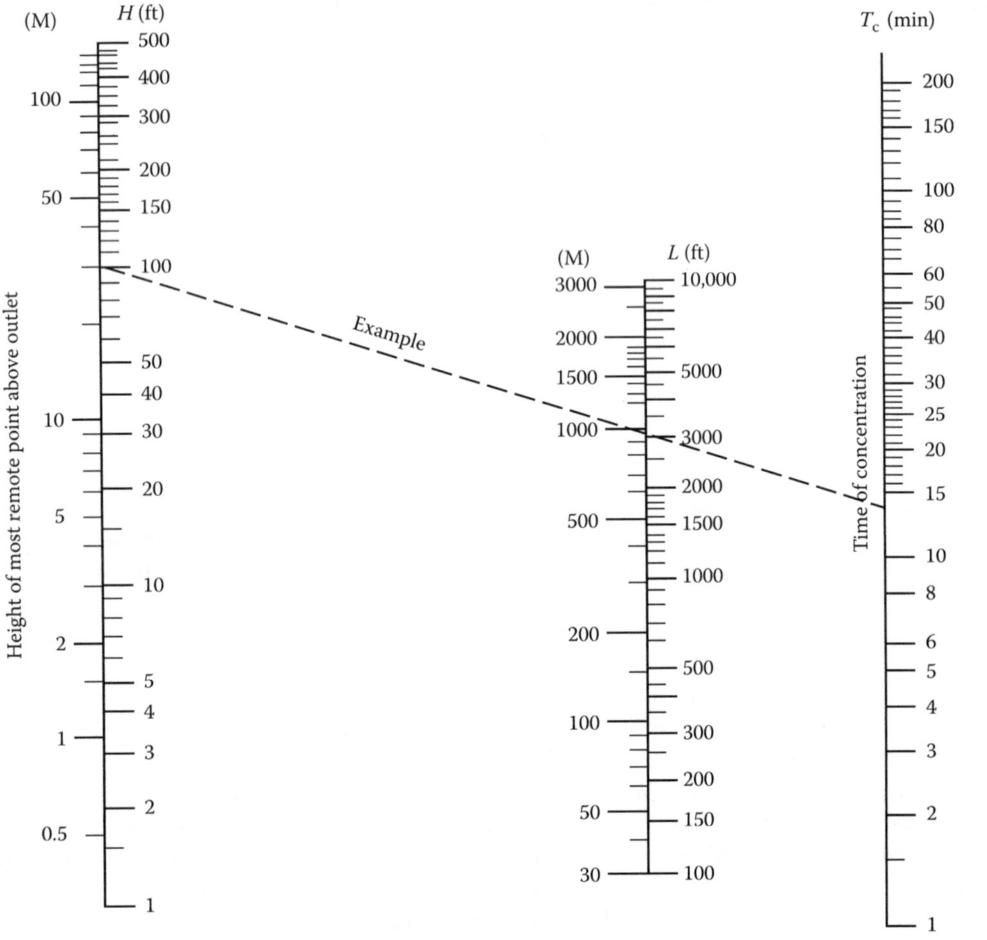

FIGURE 3.14 Kirpich's time of concentration nomograph.

Notes: Use nomograph for natural catchment areas with well-defined channels, for overland on bare soil, and for roadside grass swales. For overland flow on the lawn, multiply T_c by 2. For overland flow on pavement, multiply T_c by 0.4. For concrete channel, multiply T_c by 0.2.

3.5.1.2 Izzard's Equation

Based on laboratory experiments by the Bureau of Public Works on roads and turf surfaces, Izzard, in 1946, developed the following equation in CU:

$$T_c = \frac{41.025\left(0.0007I + C_r\right)L^{1/3}}{S^{1/3} \cdot I^{2/3}} \tag{3.29}$$

In SI, this equation reads as

$$T_c = \frac{527\left(2.8 \times 10^{-5} \times I + C_r\right)L^{1/3}}{S^{1/3} \times I^{2/3}} \tag{3.30}$$

TABLE 3.8
Values of C_r in Izzard's Equation

Surface	C_r
Very smooth asphalt	0.007
Tar and sand pavement	0.008
Concrete	0.012
Closely clipped sod	0.016
Tar and travel pavement	0.017
Dense blue grass	0.060

where

T_c = time of concentration, in minutes
I = rainfall intensity, in./h (mm/h)
L = length of flow path, ft (m)
S = slope of flow length, ft/ft (m/m)
C_r = retardance coefficient, ranging from 0.007 for very smooth pavements to 0.06 for dense turf (see Table 3.8).

3.5.1.3 Kirby's Equation

Kirby, in 1959, proposed the following equation in CU:

$$T_c = 0.83\left(\frac{Lr}{S^{1/2}}\right)^{0.467} \tag{3.31}$$

This equation in SI becomes

$$T_c = 1.45\left(\frac{Lr}{S^{1/2}}\right)^{0.467} \tag{3.32}$$

where

T_c = time of concentration, in minutes
L = flow length, ft (m)
r = retardance roughness coefficient, varying from 0.02 to 0.06 (see Table 3.9)
S = slope of catchment in ft/ft (m/m)

3.5.1.4 Garden State Parkway's Nomograph

In 1957, the New Jersey Highway Authority Garden State Parkway adopted the time of concentration nomograph shown in Figure 3.15. In this figure, T_c is shown as a function of the flow length, type of surface cover, and slope of the land. The surface cover ranges from woodland to pavement and anything in between. Based on the experience of the author, this figure yields a more realistic result than Kirpich's equation for natural land surfaces. This matter is illustrated by the following example. Note that neither in Figure 3.14 nor in Kirpich's, Izzard's, and Kirby's equations is the time of concentration proportional to the overland flow length. Therefore, contrary to common practice, the flow reach should not be divided into reaches because this would exaggerate the time of concentration.

TABLE 3.9
Retardance Roughness, *r*, in Kirby's Formula

Surface	*r*
Smooth pavement	0.02
Asphalt/concrete	0.05–0.15
Smooth, bare, packed soil, free of stones	0.10
Light turf	0.20
Poor grass on moderately rough ground	0.20
Average grass	0.40
Dense turf	0.17–.80
Dense grass	0.17–0.30
Bermuda grass	0.30–0.48
Deciduous timberland	0.60
Conifer timberland, dense grass	0.60

Source: Westphal, J. A., 2001, in L.W. Mays, ed. (2001) *Stormwater collection system design handbook*, McGraw-Hill, New York.

Example 3.8

A five-acre (2.0 ha) watershed is covered by 15% pavement and 85% dense grass. The watershed main flow path is 1100 ft (335 m) long and has an average slope of 3%. Assuming that the flow path is entirely vegetated, calculate the time of concentration using the Kirpich, Izzard, and Kirby equations and the New Jersey Garden State Parkway nomograph (Figure 3.15). Use a rainfall intensity of 4 in./h (100 mm/h).

Solution

a. Kirpich equation (Equation 3.27)

$$T_c = \frac{0.0078 \times 1100^{0.77}}{0.03^{0.385}} = 6.6\ min$$

Multiply the calculated t_c by 2 for grass cover:

$$T_c = 2 \times 6.6 = 13.2\ \text{min}\ .$$

b. Izzard equation
For dense grass, $C_r = 0.06$ (see Table 3.9)
Inputting $I = 4.0$ in./h and $C_r = 0.06$ in Equation 3.29.

$$T_c = \frac{41.025(0.007 \times 4 + 0.06) \times 1100^{1/3}}{\left(0.03^{1/3} \times 4^{2/3}\right)} = 34\ \text{min}$$

c. Kirby equation
From Table 3.10, $r = 0.3$ (maximum for dense grass).

$$T_c = 0.83\left(\frac{1100 \times 0.3}{0.03^{\frac{1}{2}}}\right)^{0.467} = 28.2\text{min, say, 28 min}$$

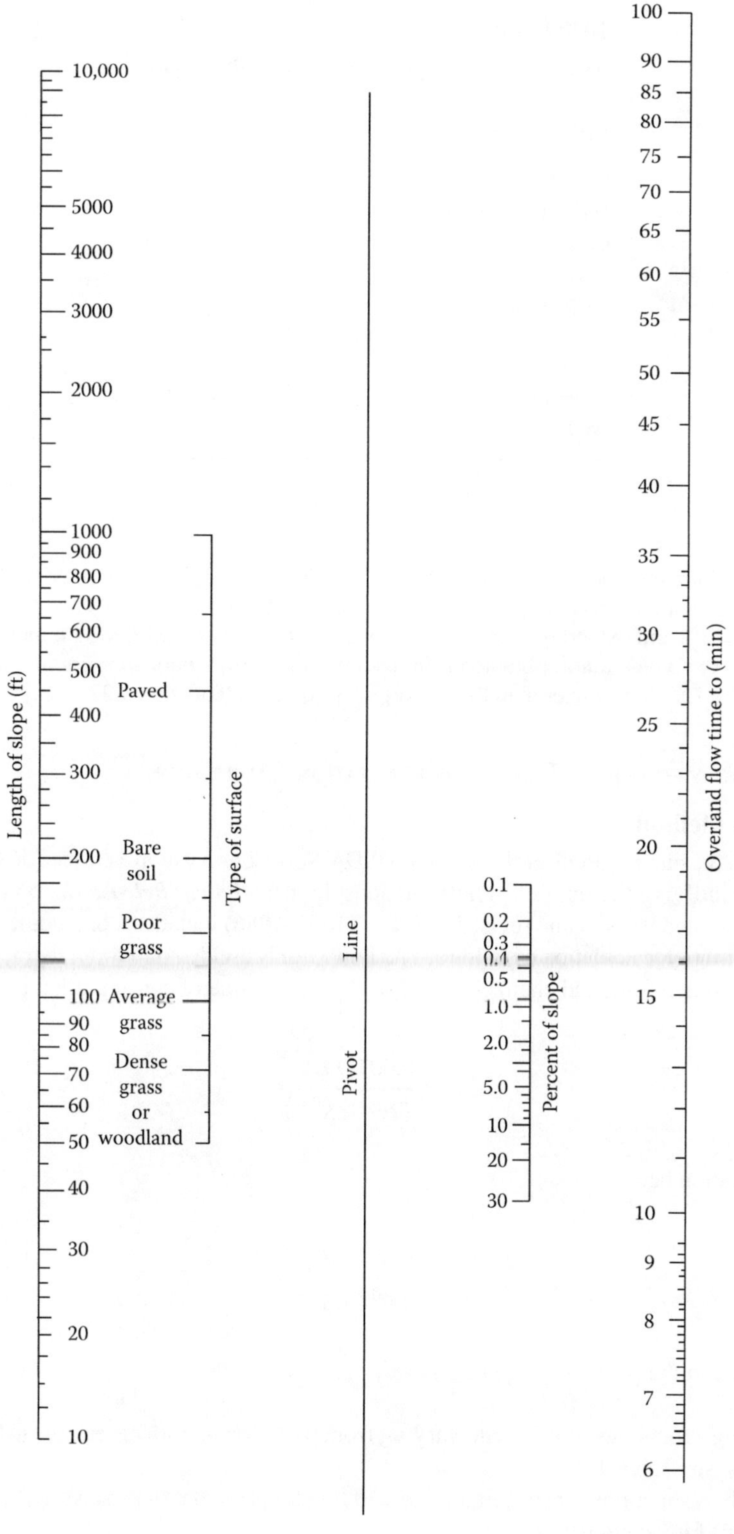

FIGURE 3.15 New Jersey Highway Authority-Garden State Parkway (1957). Time of concentration nomograph.

TABLE 3.10
Sample Manning's *n* Values for Sheet Flow

Surface Description	*n*
Smooth asphalt	0.011
Fallow (no residue)	0.050
Cultivated soil	0.06 ± 0.17
Grass	
Short grass	0.150
Dense grass	0.240
Bermuda grass	0.400
Range (natural)	0.130
Woods	0.400

d. From Figure 3.15:
Draw a straight line connecting 1100 ft on the length scale to dense grass on the "type of surface" line and extend to the pivot line. Then, from the point of intersection, draw another line to S = 3% and extend the time of the concentration line to read T_c = 34 minutes. Note that the times of concentration obtained from Izzard's and Kirby's equations are in fair agreement and nearly 2.5 times longer than that of Kirpich's equation (Equation 3.27 or Figure 3.14).

3.5.2 Other Methods of Time of Concentration Calculation

3.5.2.1 SCS Method

Based on tests on small agricultural plots, the USDA Soil Conservation Service (SCS) developed a method for quantifying the rainfall–runoff relationship in *Technical Release No. 55* (TR-55), which was originally issued in 1975 and revised in 1986. TR-55 (1986) includes a procedure for calculating time of concentration calculation as follows. The flow path is divided into three segments: sheet flow, shallow concentrated flow, and channel flow. The sheet flow time of concentration is calculated by

$$T_t = \frac{0.007(nL)^{0.8}}{\left(P^{0.5} \times S^{0.4}\right)} \tag{3.33}$$

In SI, this equation becomes

$$T_t = \frac{0.091(nL)^{0.8}}{\left(P^{0.5} \times S^{0.4}\right)} \tag{3.34}$$

The parameters in the preceding equations are

n = Manning's roughness coefficient, varying from 0.011 for smooth pavements to 0.4 for wooded areas (see Table 3.10)
L = Length of sheet flow, ft (m); limited to 150 ft (±50 m) in pervious surfaces and 100 ft (30 m) for pavements
P = 2- year, 24 hr precipitation, in. (mm)
S = land slope, ft/ft (m/m)
T = sheet flow travel time, hours

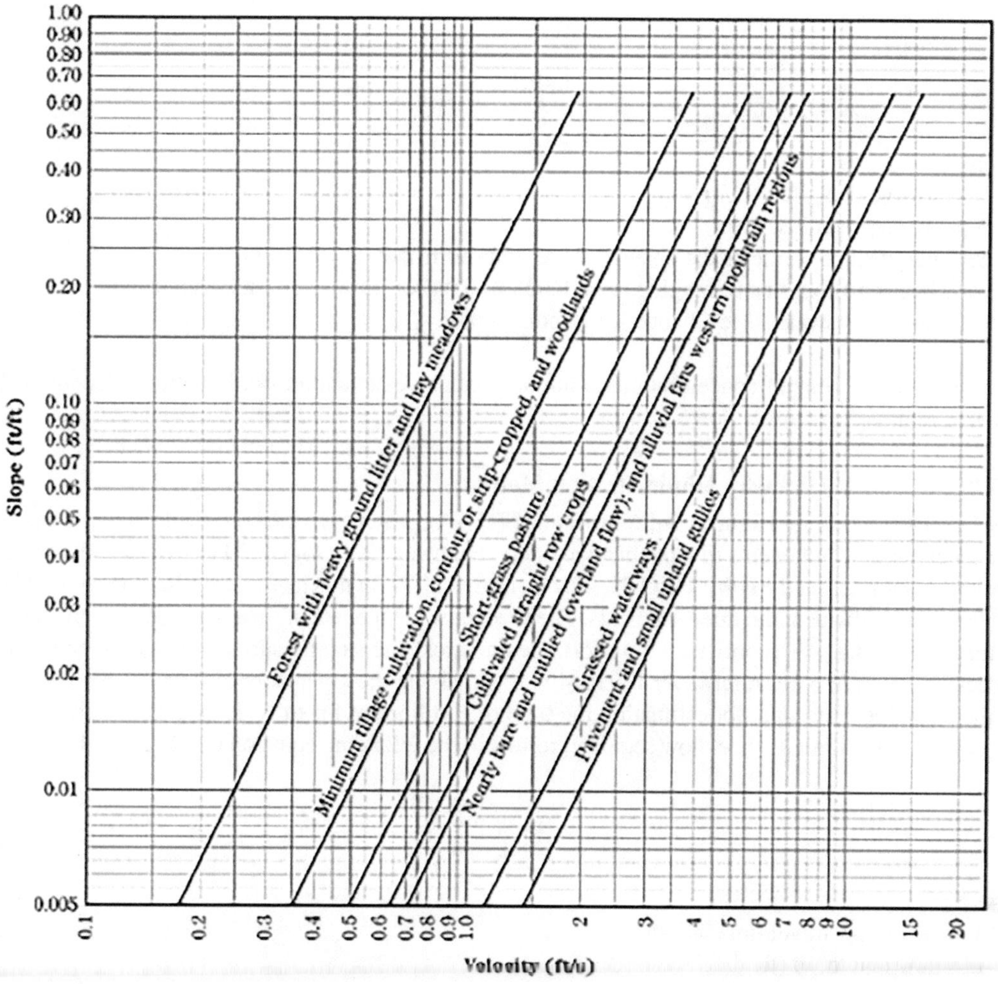

FIGURE 3.16 Shallow concentrated flow velocity.

Source: Adapted from the TR-55 method (NJDEP, July 2023).

The shallow concentrated flow lies between sheet flow and channel flow, if any. The average flow velocity for the sheet flow is derived from Figure 3.16, which is a refined form of the figure in the TR-55 manual, and the shallow concentrated flow time of concentration is calculated by

$$T_s = \frac{L}{60V} \tag{3.35}$$

where

T_s = time of concentration, in minutes
L = length of this flow segment, in ft, m
V = shallow concentrated flow velocity in ft/s, m/s

Channel flow begins where the flow concentrates into swales or pipes. The flow velocity for this segment is calculated using the Manning's formula, which as noted in Chapter 2, is expressed as

$$V = \frac{R^{2/3} S^{1/2}}{n} \quad SI \tag{3.36}$$

and

$$V = \frac{1.49R^{2/3}S^{1/2}}{n} \; CU \tag{3.37}$$

where

V = flow velocity, ft/s (m/s)
R = hydraulic radius, ft (m)
S = energy slope (same as the channel slope in uniform flow), ft/ft (m/m)
n = Manning's roughness, ranging from 0.025 to 0.04 for gravel and rock channels to over 0.10 for grass swales (see Chapter 4 for more detail)

The channel flow time of concentration, in minutes, can be calculated from Equation 3.35, where L is the length of the channel flow segment.

3.5.2.2 Federal Highway Administration Method

The U.S. DOT's Federal Highway Administration (FHWA), presented a method of calculating the time of concentration components in a publication titled "Highway Hydrology," FHWA HDS-2 Hydraulic Design Series no. 2, and FHWA HEC-22, 4th ed., March 2024. In this method, similar to that of SCS, the flow path is divided into three segments. The sheet flow component of time of concentration relates to the shallow mass of runoff with a uniform depth across the sloping surface. Its length generally exceed 25 m (80 ft) but is rarely over 130 m (400 ft). Applying the kinematic wave model, a derivative of the Manning's formula, for estimating the time of equilibrium for fully developed turbulent flow (representing the limit of sheet flow segment) results in the following equation (HDS-2, 2024):

$$T_t = \left(\frac{K_n}{I^{0.4}}\right)\left(\frac{nL}{S^{1/2}}\right)^{0.6} \tag{3.38}$$

where

T_t = sheet flow travel time, minutes
L = flow length, m (ft)
I = rainfall intensity of the design storm, mm/h (in./h)
S = surface slope, m/m (ft/ft)
K_n= empirical coefficient = 6.92 SI, 0.933 CU
n= Manning's roughness coefficient

Equation 3.38 may be applied to calculate the overall time of concentration in small catchment areas. Though incorrectly, it is also used to segmented sheet flow reaches to calculate T_t for the entire reach by adding the sheet flow travel time for the sub-reaches. The Manning's roughness coefficient in Equation 3.38 varies from 0.011 for smooth asphalt to woods with dense underbrush (see Table 3.1 in HDS-2).

After a distance of 100 m (300 ft), sheet flow tends to concentrate in rills and then gullies. The flow velocity in this reach, which is referred to as shallow concentrated flow, can be estimated from the following equation:

$$V_s = K_u k S^{0.5} \tag{3.39}$$

where

K_u= 1 SI; 3.28 CU
k = intercept coefficient (see Table 3.11)
S = slope, percent

TABLE 3.11
Intercept Coefficient *k* in Shallow-Concentrated Flow Velocity[a]

Land Cover/Flow Regime	*k*
Forest with heavy ground litter; hay meadow (overland flow)	0.076
Trash fallow or minimum tillage cultivation; contour or strip cropped; woodland (overland flow)	0.152
Short grass pasture (overland flow)	0.213
Cultivated straight row (overland flow)	0.274
Nearly bare and untilled (overland flow); alluvial fans in western mountain regions	0.305
Grassed waterway (shallow concentrated flow)	0.457
Unpaved area (shallow concentrated flow)	0.491
Paved area (shallow concentrated flow; small upland gullies	0.619

Source: FHWA Hydraulic Design Series no. 2, (HDS-2) 2nd ed., 2002, Chapter 6; HEC-22, 3rd ed., 2009 (revised August 2013), Chapter 3.

[a] Equation 3.39.

The flow velocity in the channel/pipe flow reach is calculated using the Manning formula (Equations 3.36 and 3.37), the same as in the SCS method. The overall time of concentration is then calculated by adding travel time in the previously indicated three segments as follows:

$$T_c = T_t + \left(\frac{L_s}{60V_s}\right) + \left(\frac{L_c}{60V_c}\right) \tag{3.40}$$

where L_s and L_c are the lengths of the shallow concentrated flow and channel/pipe flow segments, respectively, and other parameters are as defined previously.

It should be noted that the sheet flow time of the concentration equation in this method, though seemingly similar to the SCS method, is different from that method. In the SCS method the two-year rainfall depth is used to calculate the sheet flow travel time for all storm frequencies. However, in the FHWA method, like the Izzard equation, T_t is calculated using the storm intensity for a design storm, and as such, the sheet flow travel time varies with the storm frequency.

Since the rainfall intensity, *I*, depends on T_c, which is not initially known, the calculation of T_t requires an iterative process. Specifically, an initial estimate of T_c is assumed to obtain *I* from the rainfall IDF curve for the locality. The T_c is then calculated by adding the time of concentration components for sheet flow, shallow concentrated flow, and channel flow using Equations 3.38 through 3.40, respectively, and then is checked against the initial T_c. If they differ, the process is repeated until the two successive T_c estimates become equal. The following example illustrates the calculation process.

Like other methods presented in the previous section, the sheet flow time of concentration in the TR-55 and FHWA methods is not a linear function of the flow length. Therefore, dividing the sheet flow reach into sub-reaches, which is commonly practiced, will result in an overestimation of the sheet flow travel time. This matter will be demonstrated for the TR-55 method later in this chapter.

Example 3.9

The flow path characteristics for a 4 ha (10 acres) are as follows:

Flow Segment	Length, m	Slope	Segment Cover
Sheet flow	50	0.7%	Short grass
Shallow concentrated	85	1.2%	Short grass
Pipe flow	180	1.0%	450 mm (18 in.) concrete pipe

Calculate the time of concentration using:

a. TR-55 method
b. FHWA method (HEC-22)

The local 2-year, 24-hour storm is 84 mm (3.3 in.), and the rainfall intensity–duration relation for the design storm is listed below:

T_c, min	20	30	45	60	80
I, mm/h	110	82	63	50	42

Solution

a. TR-55 method

- Sheet flow, using Equation 3.34 and n= 0.24 (see Table 3.11):

$$T_t = \frac{0.091(0.24\times 50)^{0.8}}{\left(84^{0.5}\times 0.007^{0.4}\right)} = 0.527\ hr = 31.7\ min$$

- Shallow concentrated flow: For S = 1.2%, read the flow velocity for unpaved surface in Figure 3.16:

$$V = 0.8 \text{ ft/s} = 0.20 \text{ m/s}$$

$$T_s = \frac{85}{(60\times 0.20)} = 7.1\,min$$

- Pipe flow:
  Assuming full flow in pipe:

$$V = \frac{R^{2/3}S^{1/2}}{n}$$

$$R = \frac{D}{4} = \frac{450}{4} = 112.5\ mm = 0.113\ m$$

$$n = 0.012;\ see\ Table\,2.5\ in\ Chapter\,2$$

$$V = \frac{(0.113)^{2/3}\times(0.01)^{1/2}}{0.012} = 1.95 \text{ m/s}$$

$$T_{ch} = \frac{180}{60\times 1.95} = 1.5\ min$$

Time of concentration:

$$T_c = 31.1 + 7.1 + 1.5 = 39.7\ min,\ say,\ 40\ min$$

b. FHWA method

Assume T_c = 30 min as a trial; refine if necessary:

- Sheet flow travel time (Equation 3.38):
  Based on the IDF table, I = 82 mm/h

$$T_t = \left[\frac{6.92}{(82)^{0.4}}\right]\left[\frac{0.24\times 50}{(0.007)^{0.5}}\right]^{0.6} = 23.4\ min$$

- Shallow concentrated flow:

$$V = kS^{0.5}$$

$$k = 0.213, \textit{ short grass pasture overland flow (see Table 3.11)}$$

$$S = 1(1\%)$$

$$V = 0.213 \times (1)^{0.5} = 0.213\ m/s$$

$$T_s = \frac{85}{(60 \times 0.213)} = 6.7\ min$$

- Pipe flow travel time:
  Same as TR-55 method

$$T_c = 1.5\ min$$

- Time of concentration:

$$T_c = 23.4 + 6.7 + 1.5 = 31.6\ min, \textit{ round to 32 minutes}$$

The calculated T_c is not significantly different from the assumed value of 30 minutes, so no further trial is necessary.

Note that the results from the two methods are somewhat different and that the FHWA method gives more reasonable answers for shallow concentrated flow velocity than the SCS TR-55 method.

Example 3.10

Using the TR-55 method, calculate the time of concentration for a 10 acre watershed, having

100 ft sheet flow in woods, at 2% slope
375 ft shallow concentrated flow in grass, at 3% slope
500 ft channel flow: Manning's n = 0.06; flow area = 10 ft²; wetted perimeter = 12 ft, slope = 1%

The 2-year, 24-hour storm is 3.5 in.

Solution

Sheet flow, n = 0.4 woods

$$T_t = \frac{0.007(n \times L)^{0.8}}{\left(P_2^{0.5} \times S^{0.4}\right)} = \frac{0.007(0.4 \times 100)^{0.8}}{\left[3.5^{0.5} \times (0.02)^{0.4}\right]} = 0.34\ hour$$

Shallow concentrated flow (based on TR-55, 1986, V = 2.8 ft/s)

$$T_t = \frac{L}{3600V} = \frac{375}{3600 \times 2.8}$$
$$= 0.04\ hour \textit{ (rounded to second decimal place)}$$

For S = 0.03, Figure 3.16 (form NJDEP) reads V = 1.6 fps for shallow concentrated flow:

$$= 0.06\ hour \textit{ (rounded to second decimal place)}$$

Channel flow

$$V = \left(\frac{1.49}{n}\right) R^{\frac{2}{3}} S^{\frac{1}{2}}$$

$$V = \left(\frac{1.49}{0.06}\right) \times (0.83)^{\frac{2}{3}} \times (0.01)^{\frac{1}{2}} = 2.2 \; ft/s$$

$$T_t = \frac{500}{(2.2 \times 3600)} = 0.06 \; hour$$

$$T_c = 0.34 + 0.06 + 0.06 = 0.46 \; hour, \; round \; to \; 0.5 \; hour$$

3.5.3 Sheet Flow Length Analysis

Rain falling on a surface moves initially at a shallow depth, but the rate and velocity of flow keep increasing along the flow. At a point located at distance L from the onset, the flow rate and velocity are given by the following parametric equations:

$$q \approx I\,L \tag{3.41}$$

$$V \approx I\frac{L}{d} \tag{3.42}$$

where

- q = flow rate per unit width
- L = flow length from the high point
- I = rainfall intensity
- d = depth of flow

the sign "≈" represents the proportionality factor, which depends on the system of units (SI, CU).

The Reynolds number at the point is

$$R = \frac{Vd}{\upsilon} \tag{3.43}$$

Combining Equations 3.42 and 3.43 results in

$$R_e \approx I\frac{L}{\upsilon} \tag{3.44}$$

The original (1946) Izzard's equation for the time of concentration placed a limit on laminar sheet flow based on the product of rainfall intensity and length as follows:

$$I\,L < 500 \tag{3.45}$$

where I is the rainfall intensity in inches/hour and L is the sheet flow length in feet. This equation reflects a Reynolds number of approximately 1100. Chow (1964) reported that Izzard's equation is in reasonable agreement for sheet flow across very wide airport aprons. For a rainfall intensity of 1.5 in./h, which approximates a 2-year, 60-minute storm in New Jersey, the Izzard equation gives a maximum sheet flow length of 330 ft (100 m). In metric units, expressing I in mm/h and L in meters, Equation 3.45 becomes

$$IL \leq 3870 \tag{3.46}$$

McCuen and Spiess (1995) report that nL/√s is a better parameter than either L or IL as a limiting criterion for sheet flow. This parameter was later expanded to Equation 3.38 for sheet flow time of concentration (McCuen et al., 2002). Factors affecting sheet flow are identified in this book as follows.

To relate the sheet flow length with the surface slope and roughness, the flow velocity in Equation 3.42 is equated to velocity in Manning's formula in which the hydraulic radius is replaced by the flow depth in a broad sheet flow:

$$IL/d = (1/\mathrm{n})d^{2/3}S^{1/2} \tag{3.47}$$

Solving for L:

$$L \approx \left(d5/3\sqrt{S}\right)/(\mathrm{n}\, I) \tag{3.48}$$

This equation indicates that the sheet flow length is lumped in the parameter (nL/√S) x ($I/d^{5/3}$). This parameter includes not only nL/√S, but also the flow depth d and rainfall intensity I affect the sheet flow length. The proportionality factor depends on the system of units. In SI, Equation 3.48 is expressed as

$$L = 36\frac{\left(d^{5/3}\sqrt{S}\right)}{(n\, I)} \tag{3.49}$$

Rewriting this equation in terms of flow depth:

$$\mathrm{d} = 0.116\left(\frac{nI}{\sqrt{S}}\right)^{0.6} \times L^{0.6} \tag{3.50}$$

In these equations:

L = sheet flow length, m
d = flow depth, mm
I = rainfall intensity, mm/h
S = surface slope, m/m

The term in the parentheses on the right-hand side is identical to the second term in the parentheses of the FHWA equation for sheet flow time of concentration (Equation 3.38). This demonstrates that both the depth of sheet flow and time of concentration are related to rainfall intensity to power 0.6 and the slope of surface to power 0.3.

Combining Equation 3.46, which is based on Izzard's experiment, and Equation 3.50 yields the following equation for limiting sheet flow depth:

$$d_{\mathrm{sf}} = 16.5\left(\frac{n}{\sqrt{S}}\right)^{0.6} \tag{3.51}$$

This equation indicates that the sheet flow depth in $\left(\frac{n}{\sqrt{S}}\right)^{0.6}$ increases with increasing surface roughness and decreasing slope.

In CU, Equation 3.51 is expressed as

$$d_{sf} = 0.65\left(\frac{n}{\sqrt{S}}\right)^{0.6} \tag{3.52}$$

where d is in inches and S is in ft/ft.

For a smooth paved surface (n = 0.015) at a 2% slope, Equations 3.51 and 3.52 give d = 4.3 mm and 0.17 in., which appear reasonable. Substituting this depth in Equation 3.49, the sheet flow length for a rainfall intensity of 75 mm/h (3 in./h) is calculated at 54.5m (170 ft). For a grass-covered surface (n = 0,15) at 2% slope, Equation 3.51 shows a limiting depth of 17 mm (0.67 in.). Conservatively using a flow depth of 12.5 mm (0.5 in.), Equation 3.51 gives a sheet flow length of 77.2 m (253 ft) for a rainfall intensity of 50 mm/h (2in./h). These calculations indicate that the SCS method underestimates the length of sheet flow for both grass and paved covers, suggesting that the method underestimates the time of concentration, which in turn exaggerates the peak rates of runoff.

3.6 RUNOFF CALCULATION METHODS

In stormwater management practices, runoff is commonly calculated based on two methods: the rational method and the TR-55 method. A brief review and comparison of the rational and the SCS methods are presented by Pazwash (1989). An in-depth discussion of these two methods is presented in the following sections. Also presented in this book is a universal method developed by the author.

3.6.1 Rational Method

In 1898, Emil Kuichling introduced a simple formula for calculating peak rates of runoff for a catchment area, known as the rational formula, as follows:

$$Q = CIA\left(\textit{see footnote}\right), \text{CU} \tag{3.53}$$ *

$$Q = CIA/360, \text{ SI} \tag{3.54}$$

where

C = runoff coefficient, dimensionless
I = rainfall intensity, mm/h (in./h)
A = catchment area, hectares (acres)
Q = discharge, m³/s (cfs)

The rational method is based on the following assumptions of linearity and proportionality:

a) Runoff is proportional to rainfall intensity.
b) Abstractions and losses vary linearly with rainfall and are incorporated in the runoff coefficient.
c) The duration of storm must be equal to or longer than the time of concentration of catchment area.

The runoff coefficient varies from less than 0.1 for forested lands of mild slope to 0.95 for paved areas. Table 3.12 lists typical runoff coefficients for various land surface covers and developments. *The Standards for Soil Erosion and Sediment Control in New Jersey* (New Jersey Standards 2017)

* The Rational formula inherits an approximation as follows: 43,560 sf/acre × 1/(12 × 3600) = 1.088.

TABLE 3.12
Typical Runoff Coefficients, C[a]

Description of A	Runoff Coefficient
Business	
Downtown areas	0.70–0.95
Neighborhood areas	0.50–0.70
Residential	
Single-family areas	0.30–0.50
Multi-unit, detached	0.40–0.60
Multi-unit, attached	60–0.75
Residential, suburban	0.25–0.40
Apartment dwelling areas	0.50–0.70
Industrial	
Light areas	0.50–0.80
Heavy areas	0.60–0.90
Parks, cemeteries	0.10–0.25
Railroad yard areas	0.20–0.35
Unimproved areas	0.10–0.30
Pavement	
Asphalt or concrete	0.70–0.95
Brick	0.70–0.85
Roofs	0.75–0.95
Lawns, sandy soil	
Flat, 2%	0.05–0.10
Average, 2%±7%	0.10–0.15
Steep, 7% or more	0.15–0.20
Lawns, heavy soil	
Flat, 2%	0.13–0.17
Average, 2%-7%	0.18–0.22
Steep, 7% or more	0.25–0.35

Sources: ASCE and WEF, 1992, *Design and construction of urban stormwater management systems,* ASCE Manuals and Reports of Engineering Practice, no. 77, and *WEF Manual of Practice* FD-20; ASCE, 2006

a For storms of up to 10-year frequency. A somewhat higher *C* may be used for larger design storms.

recommends using the coefficients in this table for storms of less than 25 year frequency and applying an adjustment factor C_r to less frequent storms as follows:

$$T_r < 10\ years, \qquad C_r = 1.0$$

$$T_r = 25\ years, \qquad C_r = 1.1$$

$$T_r = 50\ years, \qquad C_r = 1.20$$

$$T_r = 10\ years, \qquad C_r = 1.25$$

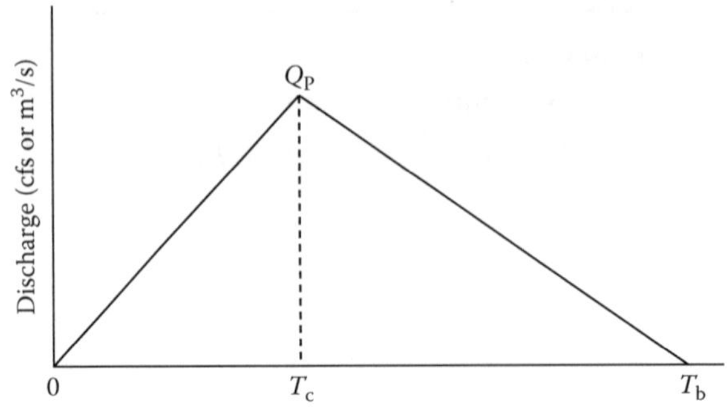

FIGURE 3.17 Rational method hydrograph.

TABLE 3.13
Base Times in the Rational Method

C	T_b/T_c
0.75–1.0	2.00
0.5–0.75	2.33
0.30–0.5	2.50
<0.3	2.67

The Rational formula inherits an approximation as follows: 43,560 sf/acre × 1/(12 × 3600) = 1.088.

Where there is more than one land use, the weighted runoff coefficient is calculated, and the resulting C is used in the rational formula. The calculation process is illustrated by Example 3.11.

The rainfall intensity in the rational formula is the average intensity of rainfall having a duration equal to the time of concentration of the catchment area. As was noted previously, the rainfall intensity varies with the storm frequency and storm duration Thus, for a given storm, the peak flow is calculated using the IDF curves for the locality.

TABLE 3.14
Suggested Runoff Coefficients for the Denver Area

	2 h Rainfall Depth, in. (mm)			
Land Use/Surface Cover	**1.2 (30)**	**1.7 (38)**	**2.0 (51)**	**3.1 (79)**
Lawns, sandy soil	0	0.05	0.10	0.20
Lawns, clayey soil	0.05	0.15	0.25	0.50
Paved areas	0.87	0.88	0.90	0.93
Gravel pavement	0.15	0.25	0.35	0.65
Roofs	0.80	0.85	0.90	0.93

Source: Urban Drainage, 2011, Denver.

Note: Values in this table apply for area averaging; the coefficients may not be valid for areas over 200 acres.

The rational formula might yield a larger peak runoff for a portion of a drainage area than the entire area if that subarea is mostly paved and has a much shorter time of concentration than the entire area. This condition can occur in pipe design where an upstream pipe reach receives runoff from a large paved area, while a downstream pipe has a larger but mostly pervious drainage area with a longer time of concentration than the upstream reach. In such cases, the downstream pipe reach should be sized for the larger discharge.

For a storm duration equal to the time of concentration, the runoff hydrograph reaches its peak at T_c and recedes to zero at a time base. The rational hydrograph is commonly plotted as a triangular hydrograph with a time base T_b 2.67 times longer than T_c. Figure 3.17 shows a typical rational method hydrograph. It is evident that for storm durations shorter than the time of concentration, the runoff from all of the catchment areas cannot reach the point of outlet.

It should be noted that for fully paved areas such as parking lots and building roofs, using Figure 3.17 results in a runoff volume greater than the rainfall amount, which is unrealistic. Pazwash (1992) suggests selecting a base time according to the impervious coverage or simply based on the runoff coefficient. Table 3.13 presents his suggested base times.

3.6.1.1 Drawbacks/Limitations of the Rational Method

The rational method assumes that the runoff is directly related to and is proportional to the rainfall intensity, C being the proportionality factor. As discussed earlier, such an assumption is unrealistic. For this reason, many jurisdictions limit the use of the rational method to small areas. ASCE and WEF (1992), for example, recommends that the rational method be used for areas less than 80 ha (200 acres). However, the New Jersey Department of Transportation (NJDOT) limits the application of the method to 20 acres (8 ha). The author agrees with the use of the rational method up 100 ha. The rational method is best suited and most accurate for mostly paved areas where interception is nonexistent, infiltration is negligible, and surface retention is small.

In an attempt to account t for the effect of rainfall frequency on the runoff coefficient, some jurisdictional agencies recommend adjusting the coefficient for less frequent storms. The indicated adjustment factors in the NJ Standards (2017) serve as an example. However, this does not address a shortcoming of the method, that it omits the impact of storm duration on the rainfall–runoff relation. The following example demonstrates the outcome of the omission.

Figure 3.1b shows a 1-year, 15-minute storm and a 25-year, 40-minute storm with the same intensity of approximately 2.9 in./h (~ 74 mm/h). However, these storms amount to 0.72 in. (18 mm) of rainfall for the former but 1.93 in. (49 mm) for the latter. Estimating the initial losses of 10 mm, which as will be indicated represents an average amount of losses for lawns, only 8 mm of the rainfall becomes available for infiltration and runoff for the former storm and 39 mm for the latter. Since the infiltration rate is the same for both storms, the former storm creates a far smaller peak discharge than the latter. In fact, for a moderate permeability of 50 mm/h (2 in./h), the former (10 min) storm creates no runoff. This exercise proves the dependency of the runoff coefficient on the rainfall duration.

To account for the effect of initial losses on the runoff coefficient, some agencies suggest adjusting the runoff coefficient based on rainfall depth rather than its frequency. This is reflected in Table 3.14, which lists suggested runoff coefficients in Denver, Colorado (see Maidment, 1993). While these coefficients may not be applicable to all localities, the C values for smaller rainfall depths are in fair to good agreement with visual observations, which indicate that the runoff coefficient varies with rainfall depth.

Example 3.11

Calculate the peak rates of runoff for 2- and 10-year frequency storms for the catchment area of Example 3.10 based on the rational formula. The catchment area is 90% grass-covered and 10% paved. Base calculations on the New Jersey IDF curves (Figure 3.1). Assume the time of concentration is the same as that calculated T_c using the FHWA method in Example 3.9.

Solution

For this site, select $C = 0.15$ for vegetated land and $C = 0.95$ for pavement (see Table 3.12). Calculate the composite C as follows:

$$C = 0.15 \times \left(\frac{90}{100}\right) + 0.95\left(\frac{10}{100}\right) = 0.23$$

Round T_c in Example 3.9 from 31.6 to 32 minutes. Interpolate rainfall intensities for a storm duration of 32 minutes in Figure 3.1b.

$$I_2 = 2.2\ in./h$$
$$I_{10} = 3.0\ in./h$$

Input these parameters in the rational formula:

$$Q = 0.23 \times 10 \times I = 2.3\ I$$

$$Q_2 = 5.1\ cfs\ (rounded\ to\ the\ first\ decimal\ place)$$

$$Q_{10} = 6.9\ cfs$$

3.6.2 Modified Rational Method

The rational method is widely used for estimating the peak flow for small drainage basins, urban developments, and roadway drainage systems. In this method, the peak runoff for a given storm frequency is calculated for storm duration equal to the time of concentration. Storms lasting longer than the time of concentration have smaller intensity and produce smaller peak flows than the triangular hydrograph; however, they generate greater volumes of runoff. Therefore, these storms create more discharge when routed through a detention basin. Thus, routing computations for a detention basin should be performed for a range of storm durations to determine the critical storm, the one that produces the largest discharge. This will be illustrated by Example 3.12.

A modified rational hydrograph (see Figure 3.18) is represented by a trapezoid peaking at T_c with a uniform discharge extending to the storm duration T_d and falling to zero at a time equal to $T_d + T_c$. The peak discharge in modified rational hydrographs is calculated using the rational formula with a rainfall intensity associated with the storm duration. The runoff volume in a modified rational hydrograph is calculated as follows:

$$V = \frac{1}{2}T_c Q + (T_d - T_c)Q + \frac{1}{2}T_c Q$$

This equation simplifies as

$$V = T_d Q \tag{3.55}$$

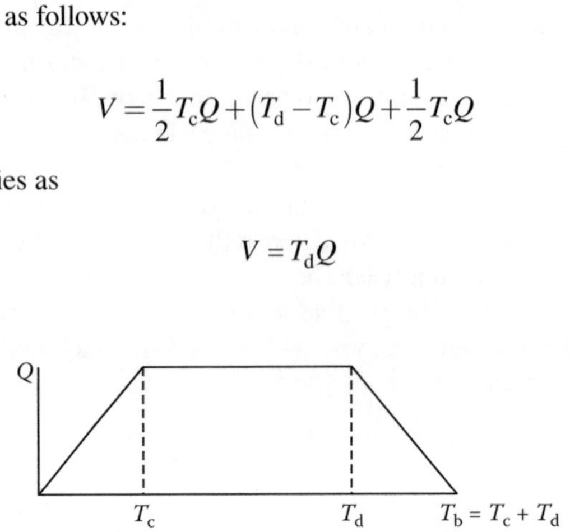

FIGURE 3.18 Modified rational hydrograph.

This equation shows that the runoff volume is independent of the time of concentration, rather, it depends solely on the storm duration. For T_d equal to T_c, the runoff hydrograph becomes a triangle, as shown in Figure 3.17.

Example 3.12

Calculate the runoff volume from a 45,200 sq. ft flat roof for 10-year storms of the following durations:

a. 10-minute storm duration, representing the time of concentration of roof runoff
b. 30-minute duration
c. 60-minute duration

Use the IDF curves in Figure 3.1b.

Solution

1. Rational formula

Use the IDF curves in Figure 3.1b.

$$A = 45,200 / 43,560\ sf = 1.038 \text{ acres}$$
$$Q = CIA = 0.95 \times 1.038 \times I = 0.986I$$

a. 10-minute storm

$$I = 5.8 in./h$$
$$Q = 0.986 \times 5.8 = 5.72\ cfs$$

Use an isosceles triangular hydrograph to calculate runoff volume:

$$V = 2 \times T_c \times Q / 2 = 10 \times 60 \times 5.72 = 3430 \text{ cf}$$

b) 30-minute storm

$$I = 3.2\ in./h$$

$$Q = 0.986 \times 3.2 = 3.15\ cfs$$

$$V = Q \times T_d = 3.15 \times 30 \times 60 = 5670\ cf$$

c) 60-minute storm

$$I = 2.0\ in./h$$

$$Q = 0.986 \times 2.0 = 1.97\ cfs$$

$$V = 1.97 \times 60 \times 60 = 7092\ cf$$

2. Direct volume calculations

Alternatively, the runoff volumes can be calculated from the product of area times rainfall depth, accounting for losses. For the 30-minute storm, the calculations are exemplified as follows:

$$\text{Rainfall depth} = 3.2 \text{ in /h} \times 30 / 60 = 1.6 \text{ in.}$$

$$\text{Effective rainfall depth} = 1.6 \times 0.95 = 1.52 \text{ in.}$$

where 0.08 in. (1.6–1.52) represents the initial losses, namely surface retention

$$\text{Runoff volume} = 45200 \times 1.52/12 = 5725 \text{ cf}$$

This is more accurate than the former result; the difference is due to an inherent approximation in the rational formula.

Exercise

Solve the preceding example for a 4200 m^2 roof of a commercial building for the following rainfall events:

a. 10-minute duration storm of 145 mm/h
b. 30-minute duration storm of 80 mm/h
c. 60-minute duration storm of 50 mm/h

3.6.3 SCS TR-55 Method

Based on tests on small agricultural plots, the Soil Conservation Service (presently the NRCS) developed a method for relating the rainfall-runoff relation. This method was presented in a report titled "Technical Release No. 55 (TR-55)", which was originally issued in 1975 and revised in 1986. The method, which is based on 24-hour storm events and time of concentrations, T_c, ranging from 0.1 hours to 10 hours, is applied to small and midsized catchment areas up to 5 square miles (13 km^2). The SCS method separates the rainfall into three components: initial abstraction, I_a; retention storage, F; and runoff depth, R (denoted as Q in the TR-55).

Figure 3.19 depicts these three components. The initial abstraction includes interception, initial infiltration, and depression storage. If the amount of rainfall is less than the initial abstraction, no runoff occurs. However, as will be indicated, the TR-55 method is based on 24-hour storm never displays this condition The retention, F, represents the continuing losses after the initial abstraction has occurred and is primarily due to the infiltration into the ground surface.

The SCS method is based on the hypothesis that the ratio of actual retention, F, to potential (maximum) retention, S, is equal to the ratio of runoff, R, to the potential runoff, namely $P - I_a$. This assumption is presented by the following equation:

$$\frac{F}{S} = \frac{R}{(P - I_a)} \tag{3.56}$$

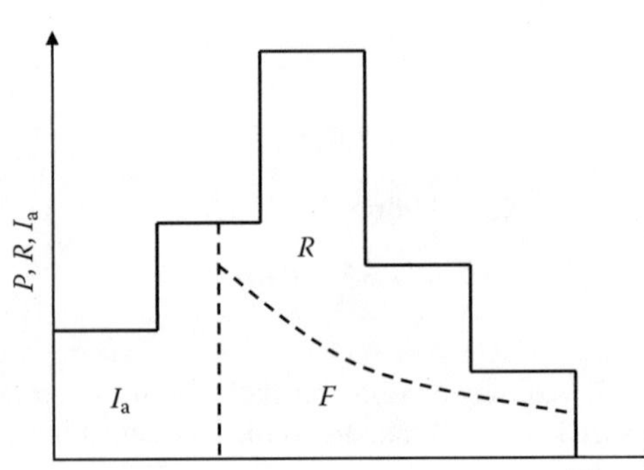

FIGURE 3.19 Components of the SCS model.

The potential retention, S, does not include I_a. According to Figure 3.21,

$$F = P - R - I_a \tag{3.57}$$

Combining Equations 3.56 and 3.57 yields

$$R = \frac{(P - I_a)^2}{(P - I_a) + S} \tag{3.58}$$

Using field data, the SCS method further assumes the initial abstraction, I_a, is equal to two tenth (0.2) of the maximum retention, S, namely:

$$I_a = 0.2\,S \tag{3.59}$$

Studies indicate that the preceding assumption exaggerates peak rates of runoff, especially for small to medium storms (Pazwash, 1989; Schneider and McCuen, 2005).

Eliminating I_a between Equations 3.58 and 3.59 results in

$$R = \frac{(P - 0.2S)^2}{(P + 0.8S)} \tag{3.60}$$

Since this equation is dimensionally homogeneous, it is equally applicable to SI and customary units.

This equation relates R and P, provided that S is known. To establish a relation for S, the SCS introduces the soil curve number, CN, by the following equation:

$$CN = \frac{1000}{(10 + S)} \tag{3.61}$$

where CN varies with soil group (to be defined later) and soil cover.

Inversely:

$$S = \frac{1000}{CN} - 10,\ in. \tag{3.62}$$

In SI units, these equations become

$$CN = \frac{1000}{(10 + 0.0394S)} \tag{3.63}$$

and

$$S = 25.4\left(\frac{1000}{CN} - 10\right),\ mm \tag{3.64}$$

The curve number (CN), as calculated by Equation 3.61 or 3.63, applies to normal antecedent moisture conditions, known as AMC II. In drier than normal soil conditions (AMC I) and in wetter

TABLE 3.15
Relative Amounts of Runoff for Different AMCs[a]

AMC Type	CN	Runoff in. (mm)	Runoff/Rainfall	Relative to AMC II
II	72	2.2 (55.9)	44%	100%
I	52	0.8 (20.3)	16%	36%
III	86	3.47 (88.1)	69%	158%

[a] *For 5 in. (127 mm)-24-hour storm.*

TABLE 3.16
Description of SCS Soil Groups

Group	Soil Type	Minimum Infiltration Rate in./h (mm/h)
A	Deep sand; deep loess; aggregated silts	0.30 (7.6)
B	Shallow loess; sandy loam	0.15–0.30 (3.8–7.6)
C	Clay loams, shallow sandy loam; organic soils; soils of high clay concentration	0.05–0.15 (1.3–3.8)
D	Soils that swell significantly when wet; heavy plastic clays, certain saline soils	0–0.05 (0–1.3)

than normal soil conditions (AMC III), the curve number would be adjusted as follows (Chow et al., 1988):

$$CN(I) = \frac{CN}{(2.38 - 0.014\ CN)} \tag{3.65}$$

$$CN(III) = \frac{CN}{(0.43 + 0.0057\ CN)} \tag{3.66}$$

Table 3.15 lists runoff depths for AMC I and AMC III relative to AMC II for a given storm event and CN value.

For the purpose of calculating runoff, the SCS classifies the soils into four hydrologic groups. These are group A through group D, where group A is the most permeable and group D the least. Table 3.16 presents a description of these soils and their permeability.

As indicated, the SCS relates the soil curve number to soil cover and hydrologic soil groups. Table 3.17 lists soil curve numbers for residential developments, and Table 3.18 presents soil curve numbers for virgin lands. CN tables for agricultural and forested lands may be found in the TR-55 manual (1986).

When impervious areas are not wholly connected, the runoff is partly dissipated through losses in pervious areas. In such a case, the composite CN would be somewhat smaller than that of fully connected impervious surfaces and may be calculated from Figure 3.20. This figure is applicable for cases where total impervious coverage in a catchment area is less than or equal to 30%.

3.6.4 SCS Peak Discharge Calculations

In the TR-55 method, the peak runoff, Q_p, is calculated using two different methods: the graphical method and the tabular hydrograph method.

TABLE 3.17
SCS Curve Numbers for Urban Areas

Cover Description		Curve Numbers for Hydrologic Soil Group			
Cover Type and Hydrologic Condition	**Average Percent Impervious Area**	**A**	**B**	**C**	**D**
Fully Developed Urban Areas (Vegetation Established)					
Open space (lawns, parks, golf courses, cemeteries, etc.)					
Poor condition (grass cover < 50%)		68	79	86	89
Fair condition (grass cover 50% to 75%)		49	69	79	84
Good condition (grass cover > 75%)		39	61	74	80
Impervious areas					
Paved parking lots, roofs, driveways, etc. (excluding right-of-way)		98	98	98	98
Streets and roads		98	98	98	98
Paved; curbs and storm sewers (excluding right-of-way)					
Paved; open ditches (including right-of-way)		83	89	92	93
Gravel (including right-of-way)		76	85	89	91
Dirt (including right-of-way)		72	82	87	89
Western desert urban areas					
Natural desert landscaping (pervious areas only)		63	77	85	88
Artificial desert landscaping (impervious week barrier, desert shrub with 1- to 2-in. sand or gravel mulch and basin borders)		96	96	96	96
Urban districts					
Commercial and business	85	89	92	94	95
Industrial	72	81	88	91	93
Residential districts by average lot size					
1/8 acre or less (town houses)	65	77	85	90	92
1/4 acre	38	61	75	83	87
1/3 acre	30	57	72	81	86
1/2 acre	25	54	70	80	84
1 acre	20	51	68	79	84
2 acres	12	46	65	77	82

Developing Urban Areas
Newly graded areas (pervious cover only). 77 86 91 94
Idle lands (CNs are determined using cover types similar to those in Table 3.18)

Note: Average runoff condition and Ia = 0.2S. The average percent impervious area shown was used to develop the composite CNs. Other assumptions are as follows: Impervious areas are directly connected to the drainage system, impervious areas have a CN of 98, and pervious areas are considered equivalent to open space in good hydrologic condition. CNs for other combinations of conditions may be computed using Figures 3.20a or b. CNs shown are equivalent to those of pasture. Composite CNs may be computed for other combinations of open space cover types. Composite CNs for natural desert landscaping should be computed using Figures 3.20a or based on the impervious area percentage (CN = 98) and the pervious area CN. The pervious area CNs are assumed to be equivalent to desert shrub in poor hydrologic conditions. Composite CNs to use for the design of temporary measures during grading and construction should be computed using Figure 3.20a or b based on the degree of development (impervious area percentage) and the CNs for the newly graded pervious areas.

TABLE 3.18
Runoff Curve Numbers for Virgin Lands[a]

Cover Type	Curve Numbers for Hydrologic Soil Group A	B	C	D
Brush- brush-week-grass mixture with brush the major element	30	48	65	73
Woods- grass combination	32	58	72	79
Woods	30	55	70	77

[a] Listed CN values are for soil cover in good condition. Conservatively, this condition should be assumed in performing runoff calculations for the predevelopment conditions.

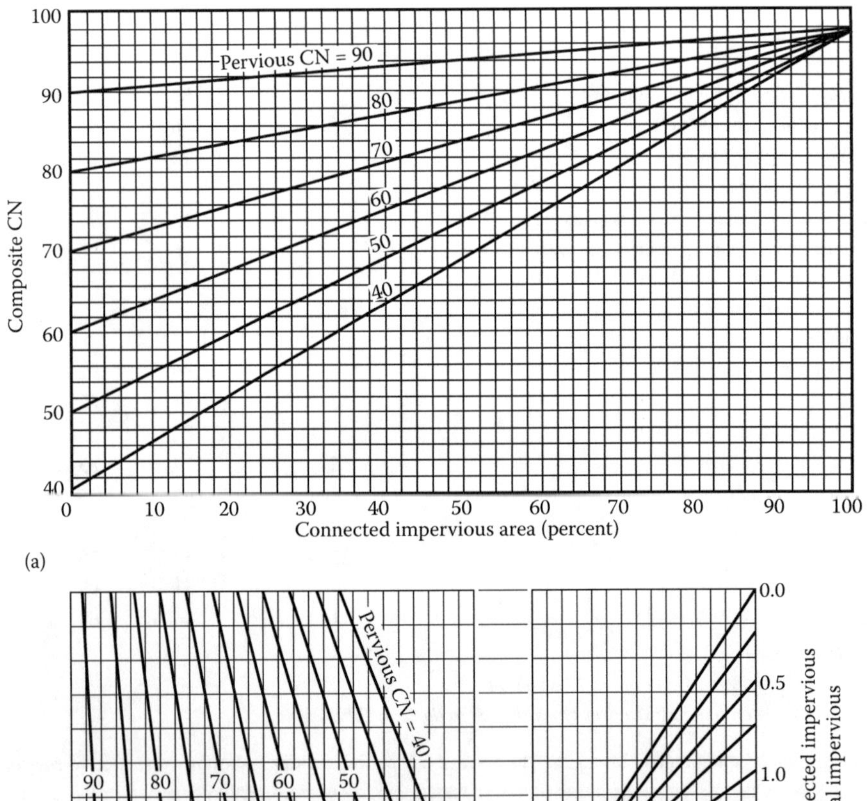

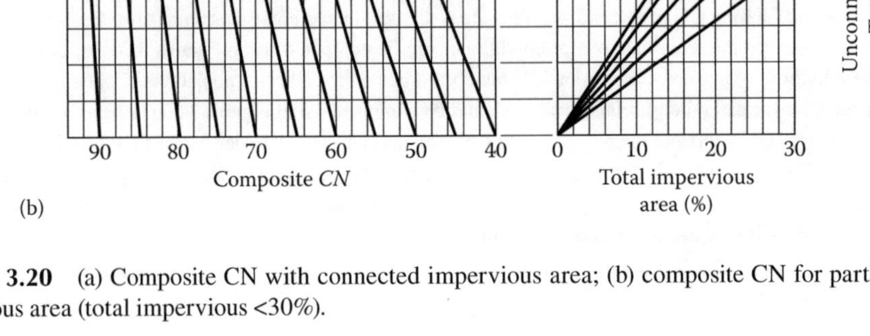

FIGURE 3.20 (a) Composite CN with connected impervious area; (b) composite CN for partly connected impervious area (total impervious <30%).

3.6.4.1 Graphical Method

The peak runoff in the graphical method is calculated using the following equation:

$$Q_p = q_u \times A \times R \times F_p \quad (3.67)$$

where

q_u = unit peak discharge in cfs/mi², abbreviated as csm
A = catchment area, in mi²
R = runoff depth from a 24-hour storm in inches
F_p = pond and swamp adjustment factor, dimensionless
Q_p = peak runoff, in cfs

The unit peak discharge, q_u, depends on T_c and travel time, T_t, and is calculated using the following equation:

$$log(q_u) = C_0 + C_1\ logT_t + C_2\ logT_c - C_3 \quad (3.68)$$

where C_0, C_1, and C_2 are related to I_a / P. The initial abstraction, I_a in turn is a function of the soil curve number and the relation is given in Table 3.19 (Note: I_a = 0.2S) The time of concentration calculation in the TR-55 method was discussed in Section 3.5.2.

To simplify calculations, TR-55 includes a graphical representation of Equation 3.68 for rainfall distribution types I, IA, II, and III. These storm hyetographs and a geographic map, where they apply, were presented in a previous section. A graph of q_u versus T_c and I_a / P for type III storms, which is typical of the East Coast of the United States including New England, New Jersey, Delaware, and eastern portions of the Carolinas, is shown in Figure 3.21. TR-55 contains graphs of q_u for other storm distributions. The SCS method of calculating the time of concentration was presented in Section 3.5, and this matter will be discussed further in the next section. Table 3.20 lists values of F_p against the percentage of a catchment area covered by ponds and swamps.

3.6.4.2 Tabular Hydrograph Method

Tabular hydrographs, as the name implies, give variations in runoff discharge with time. Included in the TR-55 are tables listing the unit discharges in cubic feet per section (cfs) per square mile per inch of runoff versus the time of concentration and travel time for values of I_a / P ranging from 0.1 to 1.0, T_c from 0.1 to 2.0 h, and T_t of 0 to 3 h. Multiplying the unit hydrograph ordinates in these tables by the watershed area in square miles, and the depth of runoff, R, in inches, will produce runoff hydrographs. A number of computer programs that facilitate calculations are available commercially. One such earlier program was called Quick TR-55 in MS-DOS by Haestad Methods in Waterbury, but Haestad Methods was acquired by Bentley Systems in 2009, and a new version of this program in Windows is available at Bentley. StormCAD is another software package available for performing the TR-55 calculations.

3.6.4.3 Unit Hydrograph Method

The SCS has developed a synthetic hydrograph that is widely used. This hydrograph, similar to the rational hydrograph, has a triangular distribution, as shown in Figure 3.22. In this figure,

D = excess rainfall period (not the same as storm duration)
L_a = lag time of catchment area; time from center of rainfall excess to the time to peak
T_p = time to peak = 0.5 D + 0.6 T_c (T_c = time of concentration)
Q_p = peak runoff rate, cfs

TABLE 3.19
I_a Values for Soil Curve Numbers

Curve Number	I_a (in.)	Curve Number	I_a (in.)
40	3.000	70	0.857
41	2.878	71	0.817
42	2.762	72	0.778
43	2.651	73	0.740
44	2.545	74	0.703
45	2.444	75	0.667
46	2.348	76	0.632
47	2.255	77	0.597
48	2.167	78	0.564
49	2.082	79	0.532
50	2.000	80	0.500
51	1.922	81	0.469
52	1.846	82	0.439
53	1.774	83	0.410
54	1.704	84	0.381
55	1.636	85	0.353
56	1.571	86	0.326
57	1.509	87	0.299
58	1.448	88	0.273
59	1.390	89	0.247
60	1.333	90	0.222
61	1.279	91	0.198
62	1.226	92	0.174
63	1.175	93	0.151
64	1.125	94	0.128
65	1.077	95	0.105
66	1.030	96	0.083
67	0.985	97	0.062
68	0.941	98	0.041
69	0.899		

The peak runoff rate, Q_p, is given by the following equations:

$$Q_p = 0.208 \frac{A \times R}{T_p} \ SI \tag{3.69}$$

$$Q_p = 484 \frac{A \times R}{T_p} \ CU \tag{3.70}$$

The constant 0.208 (484) is an average obtained from studies. This method requires calculating T_p. In practice, T_p is approximated by $T_p = (2/3)T_c$. (In HEC-22, issued in 1975 and revised in 1986 and again in (2013), p. 3–33, a coefficient of 2.08 is noted for 1 mm of unit hydrograph. This coefficient should read 0.208 author's note.)

In flat coastal areas of Maryland, Delaware, Virginia, and New Jersey, where sandy soils are prevalent, the preceding equations exaggerate the peak runoff. In these areas, the SCS unit hydrograph

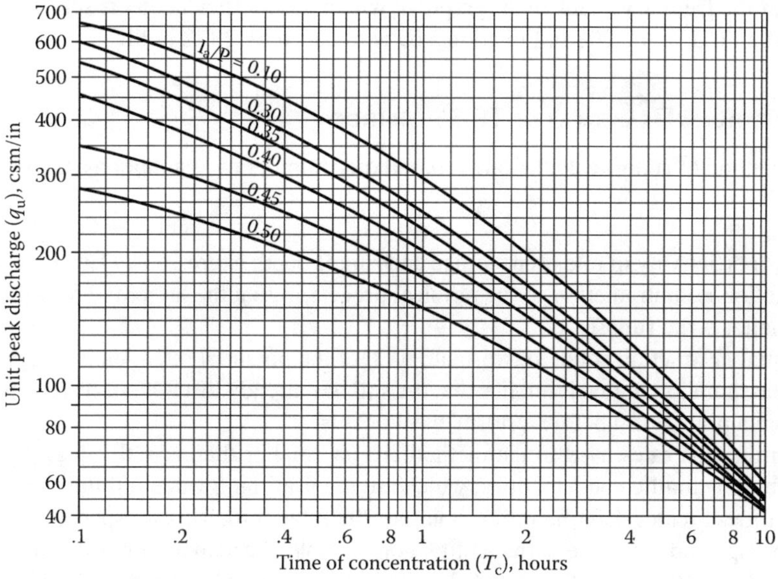

FIGURE 3.21 Unit peak discharge (q_u) for SCS type III rainfall.

TABLE 3.20
Pond and Swamp Adjustment Factor, F_p

Percentage of Pond and Swamp Area	F_p
0.0%	1.00
0.2%	0.97
1.0%	0.87
3.0%	0.75
5.0%	0.72

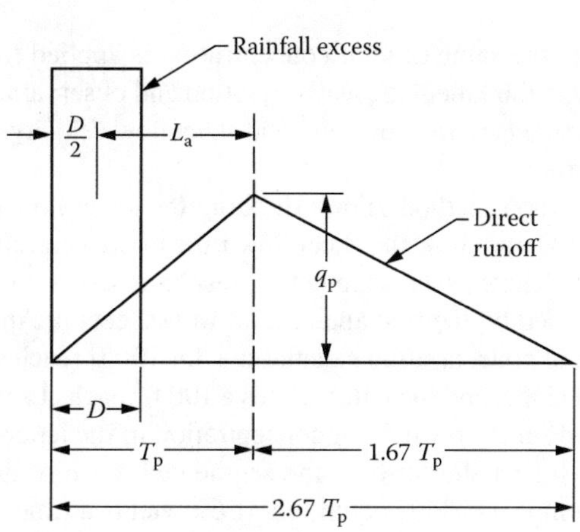

FIGURE 3.22 SCS synthetic triangular hydrograph.

is modified as the Delmarva Unit Hydrograph, in which the coefficient 284 is substituted for 484 in Equation 3.70 in CU, and 0.208 in Equation 3.69 is replaced by 0.122 in SI.

3.6.5 Limitations/Drawbacks of TR-55 Method

The TR-55 method of runoff calculations is based on a number of assumptions, the most notable of which are

a) The proportionality of retention storage and runoff depth expressed by Equation 3.56. This assumption is contradictory to the physical process in that the larger the retention storage, the smaller is the runoff.
b) An initial abstraction equal to two-tenth (0.2) of the potential retention is a highly conservative assumption in that it implies that the soil has already lost 80% of its retention capacity from a prior storm before a storm event occurs.
c) The infiltration rates used in the method are far smaller than actual values. Specifically, the actual hydraulic conductivities, which are the minimum value of infiltration rates, are many times greater than the infiltration rates in Table 3.16. This is especially the case for soils of high and moderate permeability. For example, the actual infiltration rate for sand is in the order of 30 in./h, rather than 0.3 in./h listed in that table. The effect of this unrealistic assumption will be elaborated in an example in Chapter 8.
d) The SCS classifies natural soils into four groups with a small variation in infiltration from one group to another. In reality, the soil permeability varies very widely for different soils (refer to Table 3.7 and Figure 3.12).
e) Many developed areas are classified as urban land. This soil classification has no defined hydrologic group. As such, a soil group has to be assumed in performing runoff calculations in these areas.
f) There is no justification for sheet flow length, which is limited to 100 ft in the current TR-55 time of concentration worksheet.
g) A comparison of the kinematic wave equations (Equation 3.38) with Equation 3.33 of the TR-55 method indicates the following discrepancies:
   - The exponent ($nL/\sqrt{s}$)in the TR-55 method deviates from the kinematic wave equation
   - The time of concentration is a function of the rainfall intensity, I, in the kinematic wave equation but is related to the 2-year rainfall depth, P_2, in the TR-55 method. The exponents for I and P in these equations also differ.
   - The 2-year storm is hypothetical. Its relation with sheet flow length cannot be determined.
h) In the TR-55 method, the same time of concentration is applied for any storm frequency. This is contradictory to the kinematic wave equation and observations, which indicate that, as rainfall intensity increases, the flow rate (and sheet flow velocity) increases, and the time of concentration shortens.
i) Worksheet 3 in the TR-55 method allows dividing the sheet flow reach to two segments. Since in Equations 3.33 and 3.34, the sheet flow time of concentration is a nonlinear function of the sheet flow length, such segmenting results in overestimation of the sheet flow time. This is exemplified by the first and second worksheets in Appendix 3A. The former table shows the time of concentration calculations for 100 ft reach separated to 60 ft at 2% slope and 40 ft at 1% slope, and the latter relates a 100 ft reach at a uniform slope of 2%. As shown, the calculated sheet flow time of concentration in the latter table, which represents a steeper surface, is 0.16 h shorter than the segmented reach of the smaller slope, which is unreal. Using a composite slope equal to 0.016% yields a time of concentration of 0.6, which is more realistic (see the third worksheet in Appendix 3A).

j) In the TR-55 manual, revised in 1986, the sheet flow length was limited to 300 ft. This length was changed to 150 ft for previous surfaces and 100 ft for pavements through memoranda. The current worksheet 3 that can be filled in online does not accept a sheet flow length longer than 100 ft long.

Because of these assumptions and limitations, the calculated peak runoff using the TR-55 does not relate to actual observations. In performing hydrologic analysis for a number of projects, the author has found that the TR-55 exaggerates peak flows. To exemplify, the author employed the TR-55 method to perform a dam breach analysis for a stream in Ho-Ho-Kus, New Jersey. The computed 100-year flood discharge was found to be nearly equal to the spillway capacity. However, no overtopping of the spillway was reported during Tropical Storm Floyd in September 1999, when the storm surpassed the 24-hour, 100-year storm by nearly 50%.

It is also to be noted that soil maps are commonly in 1 in. = 2000 ft or 1:20,000 (1 in. = 1767 ft) scale. As such, soil types cannot be accurately identified for small catchment areas. Considering this limitation, the author recommends that the TR-55 method not be employed for any area less than 10 acres (4 ha) and that soil permeability be measured by in situ or laboratory tests. Also for urbanized areas, where the soil has been already disturbed and has been classified as urban land, the author suggests basing the calculations on the soil type(s) occurring in the neighboring areas. Recently, the NRCS has developed a website for hydrologic soil groups and their names in some states, including New Jersey. This software allows the preparation of user-selected soil maps and soil reports for any US postal address. This partly resolves concerns with the soil map scale, where the NRCS soil maps are available online (see Case Study 3.1).

Example 3.13

A natural watershed is composed of 40% wood in HSG C and 60% grass–brush also in HSG C. Calculate the runoff depth for

a. 3.5 in. rainfall
b. 90 mm rainfall

Consistent with common practice, assume soil cover is in good condition.

Solution

Using Table 3.17,

CN = 70 for woods, HSG C
CN = 65 for grass-brush, HSG C

Calculate composite CN:

$$CN = 70 \times 0.4 + 65 \times 0.6 = 67$$

Potential abstractions:

$$S = \frac{1000}{CN} - 10 = 4.93\ in. = 125.1\ mm$$

$$\text{a. } R = \frac{(P - 0.2S)^2}{(P + 0.8S)} = \frac{(3.5 - 0.2 \times 4.93)^2}{(3.5 + 0.8 \times 4.93)} = 0.85\ in.$$

$$\text{b. R} = \frac{(90 - 0.2 \times 125.1)^2}{(90 + 0.8 \times 125.1)} = 22.2\ mm$$

Note that for this 24-hour storm, nearly 25% of rainfall turns to runoff. In reality, such covers retain the rainfall in its entirety and create zero runoff.

Example 3.14

The catchment area in the previous example is 20 acres. Calculate the peak runoff for the specified storms in that example. The storm distribution is type III, and the time of concentration is 0.6 h.

Solution

For, CN=67 in., I_a = 0.985 in. (see Table 3.19); A = 20 acres = 0.031 mi².

a. $P = 3.5\text{in .}; I_a / P = 0.28$

For T_c = 0.6 hours and I_a/P = 0.28, interpolate q_u using Figure 3.21:

$$q_u = 325 \text{ csm}$$

Using Equation 3.67, noting that $F_p = 1$

$$Q = 325 \times 0.031 \times 0.85 = 8.6\ cfs$$

b. P = 90 mm rainfall

$$I_a = 0.985 \times 25.4\ \text{mm/in.} = 25.0\ \text{mm}$$
$$I_a / P = 25/90 = 0.28$$

From Figure 3.21,

$$q_u = 325\,\text{cfs/mi}^2$$
$$R = 22.2\ mm = 0.87\ in.$$
$$Q = 325 \times 0.031 \times 0.87 = 8.8\ cfs$$
$$Q = 8.8/35.3 = 0.25\ \text{m}^3/\text{s}$$

3.6.6 WinTR-55 Method

There has been a considerable amount of literature presenting comments and discussions on TR-55, which was issued in January 1975 and revised in June 1986. To address these comments, the Natural Resources Conservation Service (NRCS) formed a committee in 1998 that ultimately produced WinTR-55; the user guide was issued in January 2009, and the NRCS also completely revised the TR-55 computer model for Windows.

The new WinTR-20 program is a driving tool to perform hydrologic analysis of small watersheds. The latest version of WinTR-55 was updated in August 2023, can be downloaded free of charge from the NRCS website or simply by searching for WinTR-55 version 1.10. WinTR-55 (2023) offers a number of improvements over the DOS-based TR–55. For instance, calculations can be performed in SI or CU for watersheds up to 25 square miles (65 km²), consisting of 1 to 10 subareas.

The software also calculates CN for subareas in which the pavements are partly or wholly (100%) drained to pervious area and allows customized CN to be used for unconnected impervious surfaces (0% to 100%).

WinTR-55 allows using default rainfall depths or user-defined rainfall depths 0–50 in. (0–1270 mm). The software accepts NRCS types I, IA, II, and III or user-defined rainfall distribution (hyetograph) and allows the use of custom dimensionless hydrographs such as Delmarva.* The software incorporates two average antecedent runoff conditions. Additionally, incorporating the built-in WinTR-20 computer model, it can perform channel or storage routing based on the Muskingum–Cunge method and perform structure routing based on the storage-indication (level pool) method. Pipe and weir are the structure types that WinTR-55 models.

WinTR-55 limits the sheet flow to 100 ft (30 m); however, it allows using $n = 0.8$ to calculate the sheet flow time of concentration in wood-dense underbrush soil cover. Similar to TR-55, this software accepts a time of concentration of $0.1 \le t_c \le 10$ hours. WinTR-55 is applied in the following case study.

CASE STUDY 3.1

Figure 3.23 shows a drainage area map of a 26 acre watershed in Monmouth County in New Jersey. This area, designated as area 2, is hydraulically separated from the downstream watershed by a roadway that rises over 2 ft above a depression in this area. Figure 3.24 depicts a depression behind the roadway. According to the Monmouth County soil survey map (Figure 3.25), the soil in this watershed is Lakehurst sand, which is HSG A. A copy of the soil survey map is provided as Figure 3.25.

Runoff calculations for the watershed are performed using the WinTR-55 computer model (WinTR-55.exe). A 0.6-acre area of the watershed, which drains to and is fully retained by a depression on the northwesterly side of the watershed, is excluded from the calculations. Table 3.21 presents calculations for the runoff curve number. The paved area within this watershed, which is wholly drained to the wooded area, is inputted as 100% unconnected in a subtable that does not appear in the WinTR-55 printouts.

Table 3.22 includes times of concentration calculations for the watershed comprising sheet flow and shallow concentrated flow but no channel flow. Presented in Table 3.23 are the 24-hour rainfall and runoff depths for 2-, 10-, and 100-year storms (note that the 2-year storm rainfall depth is required for T_c calculations). Table 3.24 lists the computed peak flows for the said storms. As shown, the 2-year storm does not produce any runoff, and the peak runoff during the 10-year storm is very small. To account for the detention/retention effect of the depression and the infiltration loss thereof, routing calculations are performed. Note that WinTR-55 cannot calculate infiltration or routing through a multistage outlet structure.

Table 3.25 lists the stage–storage–discharge relation for the depression. The storage calculations are prepared based on topographic data shown in Figure 3.24, and the listed discharges are calculated using an estimated infiltration rate of 8 in./h through the depression. Note that the estimated infiltration rate is conservative in that for sandy soils, such as that at the site, the actual infiltration may be greater than 20 in./h. In fact, a percolation test in the vicinity of the depression area shows an infiltration rate of 19 in./h.

$$Q = \text{area} \times 8\ \text{in.} / (12 \times 3600) = 1.852 \times 10^{-4} \times \text{area}$$

The inundation depth and retention volume in the depression are computed by routing the runoff hydrograph (calculated using WinTR-55) through the depression. See Table 3.25. For brevity, only the results for the 100-year storm are presented herein:

* The Delmarva unit hydrograph was developed by the New Jersey Department of Agriculture for runoff estimation on agricultural lands located within the coastal zone of flat topography (<5% slope). The method is characterized by a peak rate factor of 284 instead of 484 in Equation 3.70. The hydrograph was adopted by the State Soil Conservation Committee on July 12, 2004 through Technical Bulletin 2004-2.0 (http://www.state.nj.us/agriculture//pdf/delmarvabulletin.pdf) Hydrograph 2004-2.0 doc. The state of Maryland has included this hydrograph in Appendix D.14 of its stormwater design manual, vol. II.

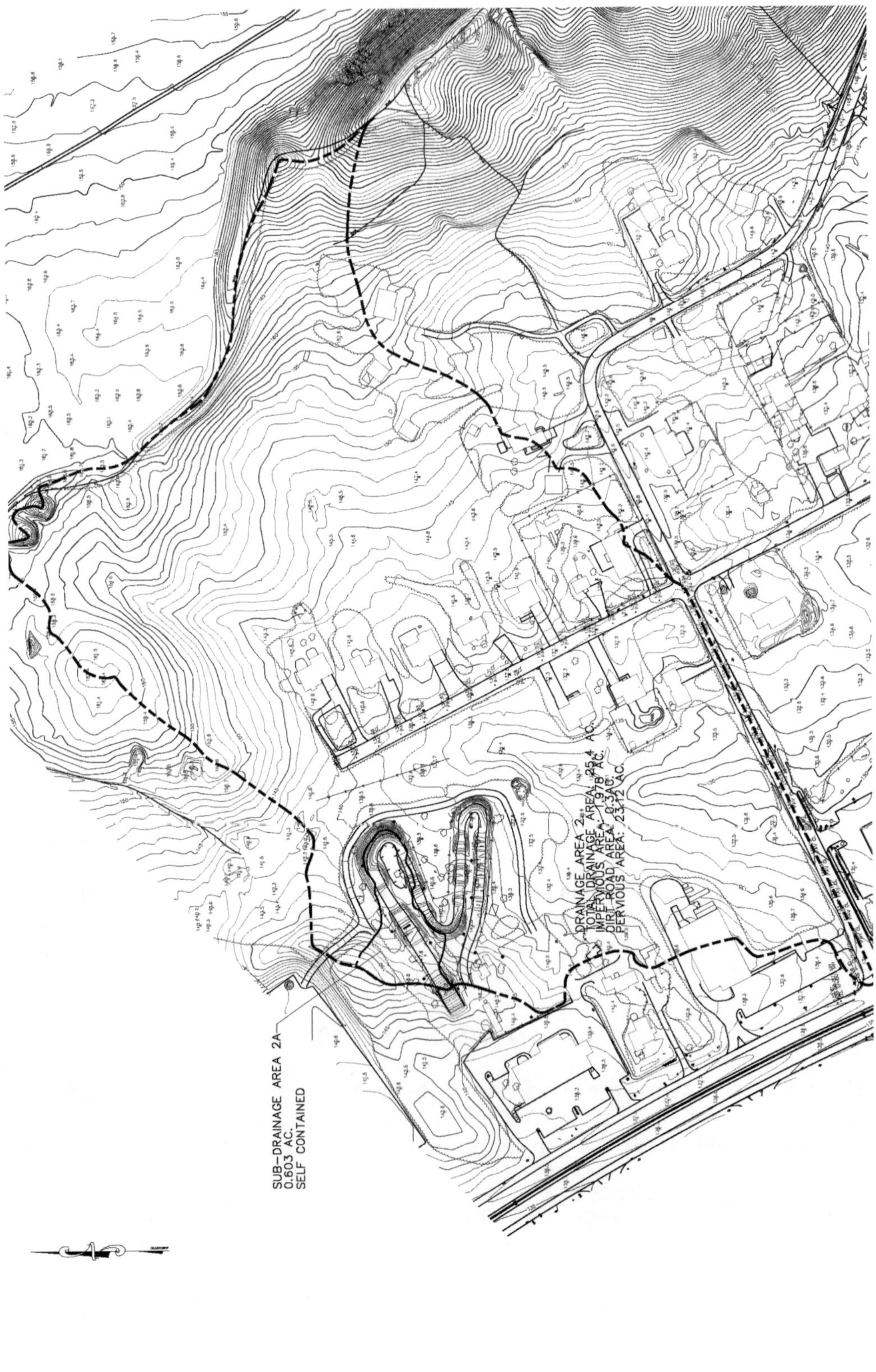

FIGURE 3.23 A drainage area map of a 26.0-acre watershed in Monmouth County, New Jersey.

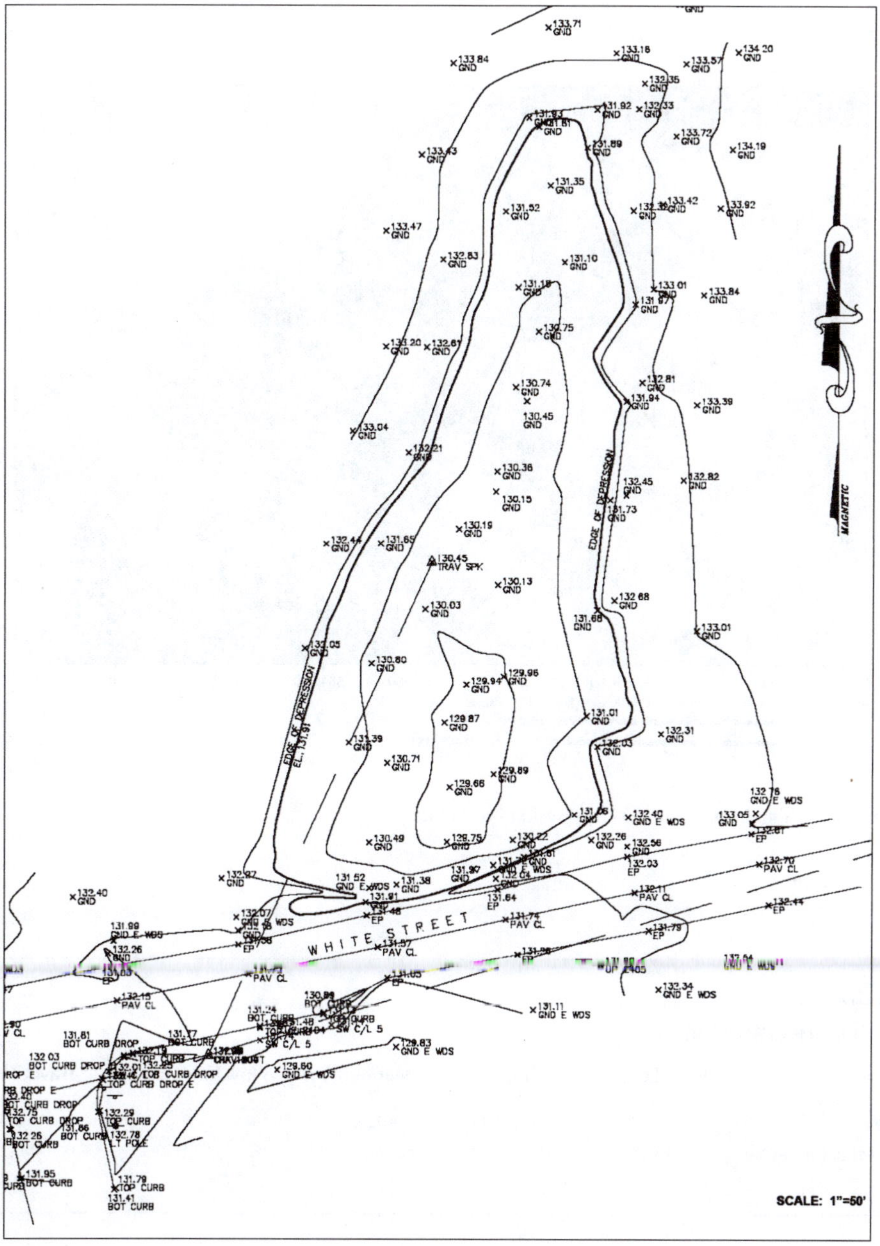

FIGURE 3.24 Topographic map of the depression.

TABLE 3.21
CNs Calculated for A2

Land Use	Hydrologic Soil Group	Subarea Acres	CN
Woods (good)	A	23.12	30
Paved (100% unconnected)	A	1.98	33[a]
Dirt road (100% unconnected)	A	0.30	30 (72)
		25.40	30

[a] *8% imperviousness; 100% unconnected.*

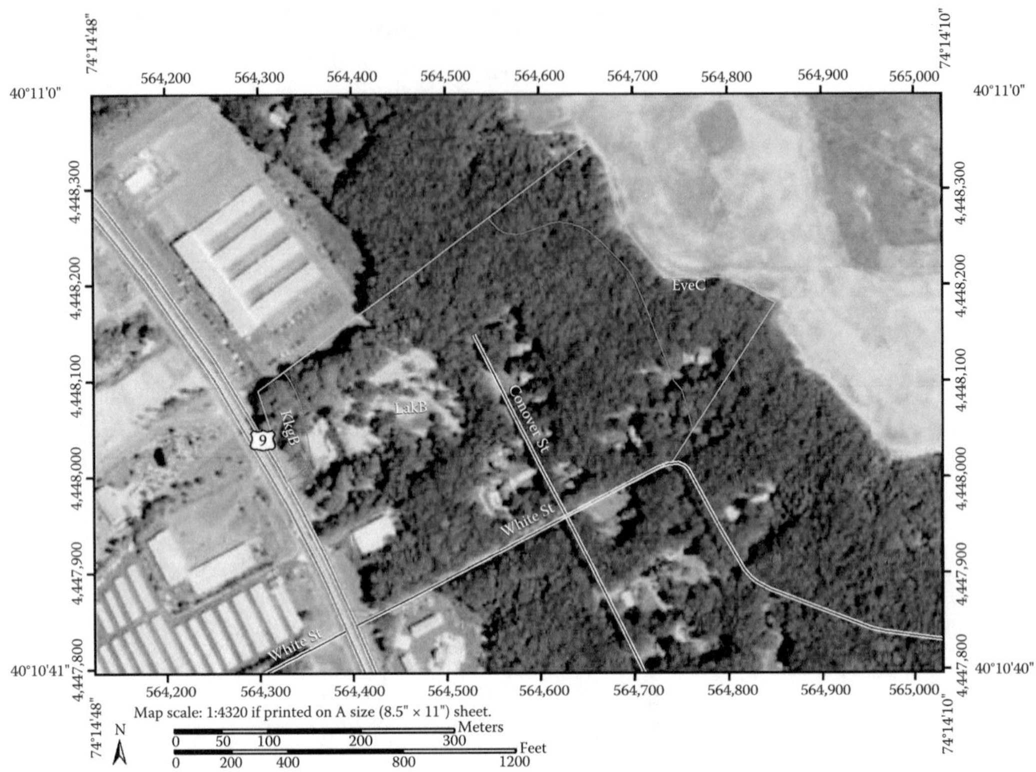

FIGURE 3.25 The Monmouth County soil survey map.

TABLE 3.22
Times of Concentration

Flow Type	Flow Length, ft	Slope ft/ft	Manning's *n*	Flow Velocity ft/s	Travel Time Hour
Sheet flow	100	0.0260	0.40		0.313
Shallow conc. (unpaved)	1450	0.0186	(N/A)	2.8	0.183
					0.496

TABLE 3.23
24-Hour Rainfall and Runoff Depths

Storm Freq. (yrs.)	Rainfall Depth (in.)	Runoff Depth
2	3.4	0
10	5.2	0.005
100	8.9	0.65

TABLE 3.24
Computed Peak Flows

Storm Freq. (yrs.)	Peak Runoff (cfs)
2	0
10	0.05
100	4.79

TABLE 3.25
Stage–Storage–Discharge – Infiltration Relation

Elev.	Area ft²	Avg. Area ft²	Δ Vol. ft³	Vol. ft³	*Q* cfs
129.75	0			0	0
		765	191		
130.00	1530			191	0.28
		5370	5370		
131.00	9210			5561	1.71
		13,445	12,235		
131.57	17,680			17,796	3.27

Max. water surface elev. = 131.32 ft
Max. retention storage ≐ 9839 cf < 17,796
Max. discharge (infiltration) = 2.26 cfs < 3.27

These results indicate that the 100-year storm does not overtop the roadway. Note that the listed discharge in Table 3.25 represents the maximum infiltration through the inundated area, which is less than the infiltration capacity of the depression when full.

3.7 UNIVERSAL RUNOFF METHOD

The shortcomings of the rational and SCS methods were discussed in previous sections. As noted, the rational method is based on assumptions of linearity and proportionality that do not properly account for initial losses and contradict actual observations (see Table 3.14). The SCS method is also based on a number of assumptions and approximations that do not agree with the physical rainfall–runoff process.

The universal model that this author presented in a 2009 paper and expanded in the first (2011) and second (2016) editions of the book and in Pazwash (2013) provides a realistic representation of the rainfall–runoff relation. This physically based model directly accounts for initial losses and infiltration. It requires estimating initial abstractions based on the soil cover and performing either a soil gradation analysis or a permeability test to determine infiltration.

3.7.1 Lag Time between Rainfall and Runoff

Figure 3.26 depicts a hyetograph of a uniformly distributed rainfall and initial losses, which include interception, surface depression, and infiltration for a pervious area. Figure 3.26 may be modified by

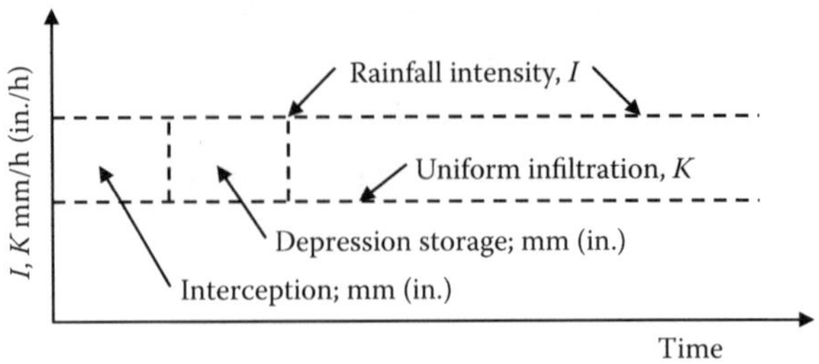

FIGURE 3.26 Representation of the universal rainfall–runoff relation.

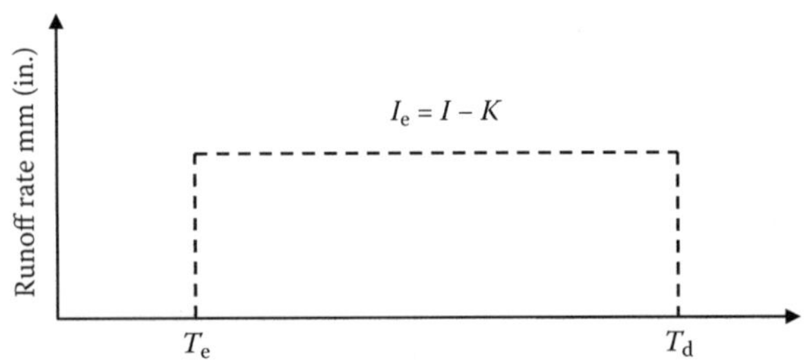

FIGURE 3.27 Excess rainfall from a pervious area.

removing the interception and infiltration lines and adjusting the surface depression, also referred to as depression storage. While the rainfall intensity and infiltration rate are expressed in mm/h (in./h) in this figure, the initial abstractions, including interception and depression storage, have a length scale and are expressed in mm (in.). The infiltration curve in the figure is conservatively shown as having a constant rate equal to hydraulic conductivity, which is the minimum rate of infiltration and which occurs under the condition of surface saturation. For simplicity, interception and depression storage are shown as rectangular blocks in the figure.

Accounting for the initial losses, the net rainfall excess will be uniformly distributed beyond a lag time, which is shown in Figure 3.27. This figure indicates that no runoff occurs until the initial losses are satisfied. The lag time from the start of rainfall to the onset of runoff is expressed by

$$T_e = (X_t + S_d) + KT_e \tag{3.71}$$

or

$$T_e = \frac{L}{(I - K)} \tag{3.72}$$

where

- I = rainfall intensity, mm/h (in./h)
- X_t = interception by trees and vegetation, mm (in.)
- S_d = depression storage, mm (in.)
- $L = X_t + S_d$ = initial abstractions (excluding infiltration), mm (in.)
- K = permeability, mm/h (in./h)
- T_e = time for runoff to begin (lag time between the beginning of rainfall and start of runoff), hours
- T_d = storm duration, hours

TABLE 3.26
Typical Values of Interception and Depression Storage (Initial Losses)

Surface Cover	X_t Interception mm (in.)	S_d = Depression Storage mm (in.) Steep Slope	S_d = Depression Storage mm (in.) Mild Slope	Initial Losses $L = X_t + S_d$
Woodlands	2.5–7.5 (0.1–0.3)	7.5 (0.30)	20 (0.5)	10–25 (0.4–1.0)
Grass-lawn	2.0–5.0 (0.08–0.2)	5 (0.20)	10 (0.4)	7.0–15 (0.3–0.6)
Landscape	2.5 (0.1)	20 (0.8)	50 (2)	25–50 (1–2)
Porous pavement, pavers[a]	0	2.0 (0.08)	3.0 (0.12)	2.0–3.0 (0.08–0.12)
Pavement	0	1.0 (0.04)	2.0 (0.08)	1.0–2.0 (0.04–0.08)

[a] Pavers with more than 20% openings may be considered as a pervious surface, having the same infiltration as grass in soils of high permeability. However, in soils of low permeability, pavers have a higher infiltration rate than lawn.

Typical values of interception and depression storage are listed in Table 3.26. In this table, the term "Initial losses" is denoted as L, which is the sum of interception and depression storage. The values in this table are based on observations and reports covered in Sections 3.2.1 and 3.2.2.

The author suggests that permeability be determined based on either the results from a permeability test or the USDA Soil Texture Permeability Ratings K_0 to K_5 in Table 3.8. The initial losses for pervious areas will vary based on antecedent moisture conditions, applying a factor of less than 1 to their potential values listed in Table 3.26.

$$T_e = M_c \frac{L}{(I - K)} \tag{3.73}$$

where M_c = antecedent moisture factor is the portion of initial abstractions (interception and depression storage capacity) that is still available after a previous storm. The author suggests $M_c = 0.5$ as a conservative estimate of the initial abstractions for pervious surfaces. When $I \le K$, the permeability rate becomes limited to the rainfall intensity and the preceding equation indicates that no runoff occurs.

3.7.2 Runoff Volume and Discharge from Pervious Surfaces

Figure 3.28 shows the depth of excess rainfall, that is, runoff from a pervious area. According to this figure, the runoff, R, is calculated by

$$R = (T_d - T_e) \cdot (I - K) \tag{3.74}$$

or

$$R = I(T_d - T_e) - K(T_d - T_e) \tag{3.75}$$

where R = depth of runoff, mm (in.).

In Equation 3.75, the first term on the right-hand side represents the net rainfall depth after the initial losses are satisfied, and the second term accounts for infiltration losses from the pervious surface after the runoff begins. For $I \le K$, the infiltration becomes limited to rainfall intensity, and Equation 3.74 gives $R = 0$.

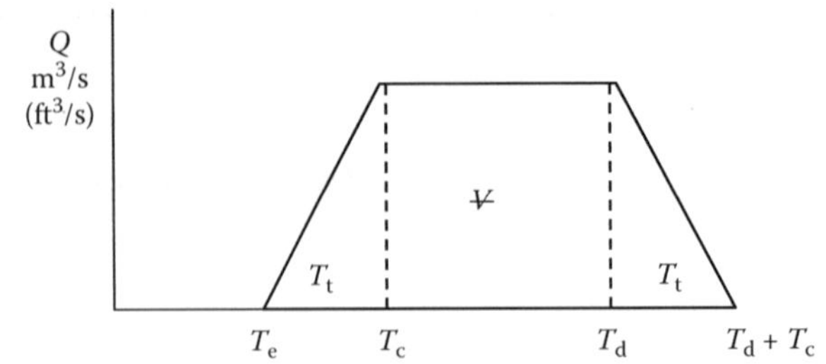

FIGURE 3.28 Peak runoff distribution.

The runoff volume is calculated from the product of runoff depth times and the area of pervious surface, as follows:

$$V_p = 0.001\,AR = 0.001\,A \cdot (I - K) \cdot (T_d - T_e)\ SI \tag{3.76}$$

$$V_p = \frac{AR}{12} = 0.083\,A \cdot (I - K) \cdot (T_d - T_e)\ CU \tag{3.77}$$

where V_p = runoff volume, m³ (ft³) and A = area of pervious surface, m² (ft²)

Given uniform variation of excess rainfall, the peak discharge can be approximated by a trapezoidal geometry shown in Figure 3.28. In this figure, T_c is the time of concentration of the catchment area, and the area under the trapezoid equals the runoff volume given by Equations 3.76 and 3.77. In this figure, $T_c = T_e - T_t$, where T_t is the travel time through the catchment area and the peak rate of runoff is given by

$$\mathrm{Q} = \frac{\forall}{3600(T_d - T_e)} \tag{3.78}$$

where the factor 3600 represents seconds per hour. Combining Equations 3.76 through 3.78 results in

$$Q = 0.278 \times 10^{-6} \times A(I - K)\,SI \tag{3.79}$$

$$Q = 2.31 \times 10^{-5} \times A(I - K)\ CU \tag{3.80}$$

where Q = peak discharge m³/s (cfs) and other parameters are as defined previously.

For a catchment area composed of different types of pervious surfaces, such as lawn, woods, and landscape, Equations 3.76 through 3.78 can be amended as follows:

$$T_e = \frac{M_c \sum AxL}{\sum A(I - K)} \tag{3.81}$$

$$V = 0.001\left[\sum A(I - K)\right](T_d - T_e)SI \tag{3.82}$$

$$V = 0.083\left[\sum A(T_d - T_e)\right]CU \tag{3.83}$$

$$Q_p = V / \left[3600(T_d - T_e)\right] \tag{3.84}$$

where

ΣA = all pervious surfaces combined
K = soil permeability relating to each A

and other parameters were described previously.

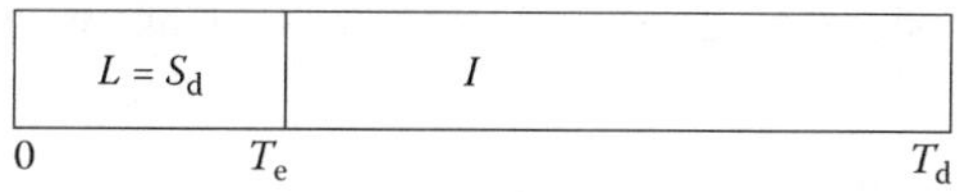

FIGURE 3.29 Rainfall–runoff relation for impervious surfaces.

3.7.3 Lag Time and Runoff Equations for Impervious Surfaces

For an impervious surface, the rainfall–runoff relation is depicted in Figure 3.29. In this case, the lag time is given by the following equation:

$$T_e = \frac{L}{I} \tag{3.85}$$

where
$L = S_d$
I = rainfall intensity, mm/h (in./h)

Referring to Figure 3.29, the runoff depth is given by the equation:

$$R = I(T_d - T_e)mm(in.) \tag{3.86}$$

where
T_d = rainfall duration, h
T_e = lag time calculated from Equation 3.85, h

In this case, equations for runoff volume and peak discharge are as follows:

$$V_i = c_x IA(T_d - T_e)SI / CU \tag{3.87}$$

$$Q_i = V / [3600(T_d - T_e)]\ SI / CU \tag{3.88}$$

where

$$c_x = 0.001\text{ SI}; c_x = 0.083\text{ CU}$$

3.7.4 Unconnected Pervious and Impervious Surfaces

For a catchment area consisting of pervious and paved surfaces, the lag time and runoff volume depend on weather or not pervious and impervious surfaces are connected. When the surfaces are unconnected, the lag time, runoff volume, and peak discharge are calculated separately for each surface using Equations 3.73 through 3.89. Then, the composite runoff volume and peak discharges for the entire area are obtained respectively by adding the volumes and discharges from pervious and impervious areas, as follows:

$$V = V_p + V_i \tag{3.89}$$

$$Q = Q_p + Q_i \tag{3.90}$$

The calculation process is illustrated by an example later in this section.

3.7.5 Equations for Connected Composite Surfaces

This section presents calculations for composite surfaces both for the cases that pervious areas drain to impervious surfaces and impervious areas drain to pervious surfaces. The pervious surfaces may consist of lawn, woods, brush, or bare soil and impervious areas may include asphalt or concrete pavement or roofs.

3.7.5.1 Pervious Areas Drain to Impervious Areas

Where one type or various pervious areas drain to a single or a number of impervious surfaces, the lag time, volume, and peak runoff from pervious and impervious areas are first calculated separately. Then, the composite lag time, runoff volume, and peak runoff from the site are calculated as follows:

Pervious areas

$$T_e = \frac{\sum M_c Li \cdot A}{\sum A(I - K)} \tag{3.91}$$

$$V_p = C_x I[\sum A(I - K)](T_d - T_p)\text{SI/CU} \tag{3.92}$$

$$Q_p = V_p / 3600(T_d - T_p)\text{SI/CU} \tag{3.93}$$

or

$$Q_p = c_x[\sum A\ (I - K)]/3600$$

b. Impervious areas

$$T_e = \frac{\sum L \cdot A}{I \sum A} \tag{3.94}$$

$$V_i = c_x I(\sum A)\ (Td - Te) \tag{3.95}$$

$$Q_i = V_p / 3600\ (T_d - T_p) \tag{3.96}$$

$$V = V_p + V_i$$

3.7.5.2 Impervious Areas Drain to Pervious Areas

Where the runoff from an impervious surface is directed to a pervious surface, that runoff may be partly or wholly retained by the pervious surface and infiltrated over time. In this case, the equation for the lag time can be derived from a graph similar to Figure 3.26, with the following result:

$$T_e = \frac{\sum M_c L \cdot A}{\sum A(I - K)} \tag{3.97}$$

In this equation, $K = 0$ for impervious surfaces. Similarly to the previous case, an antecedent moisture factor M_c may be applied to pervious surfaces.

The runoff volume and peak discharge are given by the following equations:

$$V = c_x 1\left[\sum A(I - K)\right]'(T_d - T_e)\text{SI}/CU \tag{3.98}$$

$$Q = \frac{V}{\left[3600(T_d - T_e)\right]}\text{SI/CU} \tag{3.99}$$

or

$$Q = c_x\left[\sum A(I - K)\right]/3600$$

and all parameters are as described previously, *Mc* for pavements may be taken as 1.0.

Definitions:

A = area in m^2 (ft^2)
I = rainfall intensity, mm/h (in./h)
K = permeability, mm/h (in./h)
S_d = surface retention, mm (in.)
X_i = plants interception, mm (in.)
$L_i = X_i + S_d$ initial losses for plants, mm (in.)
$L_i = S_d$ surface retention for pavements, mm (in.)
R = runoff depth, mm (in.)
V = runoff volume, m^3 (ft^3)
Q = discharge, m^3/s (ft^3/s)
M_c = antecedent moisture condition factor; M_c = 1.0 impervious; M_c = 0.5 pervious
T_e = runoff lag time, h
T_d = rainfall duration, h
c_x = 0.001 SI, 1/12 = 0.083 CU

Example 3.15

A 1500 m^2 residential lot is currently wooded. Based on a soil gradation analysis, the site soil is characterized as sandy clay loam. Calculate the runoff volume and the peak runoff for the following rainfalls:

a. 10 year – 10 minutes storm of 150 mm/h intensity
b. 10 year – 60 minutes storm of 50 mm/h intensity

Solution

Using Table 3.26, estimate initial losses L_p at 20 mm. According to Figure 3.12, sandy clay loam has K_2 permeability rating; K = 15–50 mm/h (see Table 3.7).

Estimate K = 30 mm/h.

a. 10- minutes, 150 mm/h storm

Using Equations 3.73 through 3.76 to calculate the lag time, runoff depth, and runoff volume, respectively

$$T_e = \frac{0.5\times 20}{(150-30)} = 0.0833\ \text{h}$$

$$R = (0.1667-0.0833)(150-30)$$

$$R = 10.04\ \text{mm}$$

$$V = 0.001 R\cdot A = 15.07\ \text{m}^3$$

Peak runoff, using Equation 3.79 is

$$Q = 0.278\times 10^{-6}(150-30)\times 1500 = 0.05\ \text{m}^3/\text{s}$$

Alternatively, use Equation 3.78:

$$Q = \frac{V}{3600(T_d - T_e)} = \frac{15.07}{[3600(0.1667-0.0833)]}$$

$$Q = 0.05\ \text{m}^3/\text{s}$$

b. 60- minute, 50 mm/h rainfall

$$T_e = 0.5 \times 20/(50-30) = 0.5\text{h} = 30\text{ min}$$

Runoff begins 30 minutes past the start of rainfall. Runoff depth is

$$R = (1-0.5)(50-30) = 10.0\text{ mm}$$

Runoff volume is

$$V = 0.001R \cdot A = 0.001 \times 10 \times 1500 = 15.0\text{ m}^3$$

Peak rate of runoff is

$$Q = 0.278 \times 10^{-6} \times (50-30) \times 1500 = 0.0125\text{ m}^3/\text{s}$$

Note: This example shows that, contrary to the rational method and the TR-55 method, the runoff volume may not increase with the storm duration. The example further reveals that the peak rate of runoff does not increase proportionally to the rainfall intensity.

Example 3.16

Calculate the peak and volume of runoff for the single-family home shown in the following figure during a 30-minute storm having 3.25 in./h (83 mm/h) intensity. Perform the calculations for the following cases:

a. Direct discharge from pavements, and
b. Roof and driveway drain to lawn and landscape Roof.

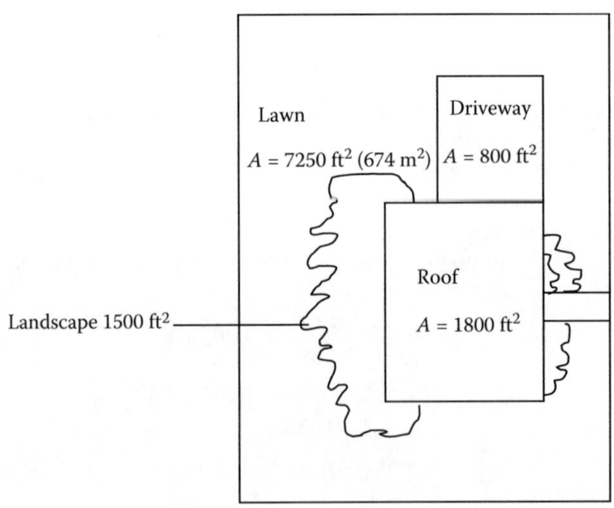

Given:

L_i = 0.5 in. (12.5 mm) lawn
L_i = 1.5 in. (38 mm) landscape
S_d = 0.12 in. (3 mm) driveway
S_d = 0.04 in. (1 mm) roof

Landscape:

A = 1500 ft² (139 m²)
K = 6 in./h (150 mm/h)

Use the equations in Section 3.7.5.1 to prepare calculations and spreadsheets separately for "a" and "b".

CASE A: UNCONNECTED PERVIOUS AND IMPERVIOUS AREAS

Pervious areas

A	*L*	*K*	*I*	*Me*	M_c *LA*	*A(I-K)*
7250	0.5	3	3.25	0.5	1812.5	1812.5
1500	1.5	6	3.25	0.5	1125	-4125
				Totals	2937.5	–2312.5

ΣA (*I*-K)< 0

No runoff from lawn and landscape

Impervious areas

A	*L*	*I*	*Mc*	*Mc LA*	*AI*
1800	0.04	3.25	1.00	72	5850
800	0.08	3.25	1.00	64	2600
2600		Totals		136	8450

$Ti = \frac{136/}{8450} = 0.02$ h	*V* = 3.25 × 2600 × (0.5– 0.02)/12
V = 340.8 cf*	*Q* = V/3600 × (0.5 – 0.021)]
Q = 0.196 cfs	

CASE B: CONNECTED PERVIOUS AND IMPERVIOUS AREAS

A	*L*	*K*	*I*	M_C	M_C *LA*	*A(I-K)*
7250	0.5	3.0	3.25	0.5	1812.5	1812.5
1500	1.5	6	3.25	0.5	1125	-4125
1800	0.04	0	3.25	1.0	72	5850
800	0.08	0	3.25	1.0	64	2600
				Totals	3073.5	6137.5

$T_{e=3073.5/6127.5=0.50}$ h=30 min

V= 0 cf	no runoff from home
Q= 0 cfs	

*Calculated using $c_X = 1/12\ (0.083)$

3.7.6 Applying the Universal Method to Nonuniform Rainfall

For a nonuniformly distributed storm, the rainfall can be divided into discrete segments of uniform rainfall intensity; then the universal runoff equations are applied to each segment. If initial losses are satisfied for any part of the drainage area during the first segment, then no losses other than infiltration will be included in the runoff calculations for that part, in other distributed storm, the rainfall can be divided into discrete segments of uniform rainfall intensity; then the universal runoff equations are applied to each segment. If initial losses are satisfied for any part of the drainage area during the first segment, then no losses other than infiltration will be included in the runoff calculations for that part in other segments. The following example illustrates the application of the method for a rainfall divided into two segments.

Example 3.17

Calculate the volume and peak runoff in Example 3.16 for a 30-minute storm falling at 4.0 in./h for the first 15 minutes and 2.5 in./h for the next 15 minutes. Perform calculations for case B (pervious areas and pavements hydraulically connected).

Solution

a. The rainfall depth during the first 15 minutes is 4 in. × 15/60 = 1.0 in. Therefore, the initial losses for lawn and pavements are fully satisfied during the first part of rainfall. Also, accounting for antecedent moisture factor $M_c = 0.5$, all of the available initial losses for landscape are met (Note: 1.5 × 0.5 = 0.75 in.).

A ft²	*L* in.	*K* in./h	*I* in./h	M_c	M_cLA ft² in.	$A(I-K)$ ft² in./h
7250	0.50	3.0	4	0.5	1812.5	7250
1500	1.50	6.0	4	0.5	1125.0	–3000
1800	0.04	0	4	1.0	72.0	7200
800	0.08	0	4	1.0	64.0	3200
Total					3073.5	14,650

$$T_e = \frac{\Sigma M_c LA}{\Sigma A(I-K)} = \frac{3073.5}{14,650} = 0.21\,h$$

$$V = 0.083 \times \Sigma A(I-K)\cdot(T_d - T_e)$$

$$= 0.083 \times 14,650(0.25-0.21) = 48.9\ ft^3$$

$$Q = \frac{V}{3600}\cdot(T_d - T_e) = \frac{48.9}{3600(0.25-0.21)} = 0.34\ \text{ft}^3/\text{s}$$

b. For the second 15 minutes, $I = 2.5$ in./h.

During this period, the losses are solely due to infiltration. Runoff and infiltrations calculations for this period are summarized in the following table:

A ft²	*L* in.	*K* in./h	*I* in./h	M_c	M_cLA ft² in.	$A(I-K)$ ft² in./h
7250	0	3.0	2.5	0.5	0	-3625
1500	0	6.0	2.5	0.5	0	-5250
1800	0	0	2.5	1.0	0	4500
800	0	0	2.5	1.0	0	2000
Total					0	-2375

Since $\Sigma A(I-K) < 0$, *there is no runoff during the second period.*

Note: The depth of rainfall in this storm is equal to 1.625 in., which is the same as that in the previous example. However, this nonuniform storm produces a larger peak but smaller runoff volume than the uniform rainfall distribution of the previous problem. It is evident that neither the rational method nor the TR-55 method can perform such calculations.

3.8 STORMWATER MANAGEMENT MODELS

Stormwater management models can be classified into two types: flow quantity and flow quantity–runoff quality. The former type, which provides no information on stormwater quality, includes the rational method, SCS method, universal method (presented in this book), HEC-1 and HEC-HMS TR–20, Pond pack, and HydroCAD. This type also includes a number of empirical synthetic hydrographs and regression equation empirical hydrographs, such as the 1938 Snyder hydrograph (which was included in the previous editions of this book) and the USGS regression equations and StreamStats. The latter type generally includes comprehensive computer models, which calculate not only the flow rates but also estimate amounts of pollutants carried by runoff.

The most widely used comprehensive model in the United States is the EPA's SWMM, developed in the 1980s. The EPA also developed WinSLAMM (Source Loading and Management Model) for Windows in 2007. Another model is SWAT which was developed by the US Department of Agriculture Research Services and Texas A&M AgriLife Research in 2000. The computer simulation models, together with the USGS methods, are presented in the following sections.

3.8.1 HEC-1/HEC-HMS

HEC-1 and its Windows version, HEC-HMS, are US Army Corps of Engineers programs that designed to simulate the runoff in stream and rivers based on rainfall and the watershed geographic and hydrologic parameters. HEC-HMS provides most of the HEC-1 capabilities but has dropped some of the less-used functions while adding others. HEC-1 was written in Fortran and until 1984 could only be run on a mainframe computer. The HEC-HMS can also perform routing computations through detention basins. However, more user-friendly software such as PondPack and HydroCAD are commonly employed in stormwater management practices.

3.8.2 TR-20

TR-20 is a project formulation hydrology that was first released by the SCS in 1964. The Windows version of TR-20, which was released in 2004 and has since been updated a number of times, performs hydrologic flow routing through detention basins and channel reaches. The model requires input information for rainfall distribution and dimensionless hydrograph, time of concentration, and stage-storage-discharge table. As such, these calculations must be performed by the user before they are inputted in TR–20. Thus, the model is not user-friendly and takes a considerable amount of time to understand and apply. As such, many engineers use other models such as HydroCAD, PondPack or EPA SWMM. The new WinTR-20 provides a complete graphical interface to the TR-20 hydrological analysis, and its version 10, released October 2024, is available at: https://wintr20.software.inf.

3.8.3 PondPack

PondPack is a Window-based hydrological modeling software that can analyze pre- and post-developed watershed conditions. This modeling software was initially developed by Haestad Methods, a company founded by John Haestad in Waterbury, CT, in 1979. Haestad Methods was the first company in any industry to move software from large mainframe computers to PCs. Haestad Methods' PondPack software was extensively used by engineers and government agencies.

This software computes stage–storage–discharge relation and performs routing computations using TR-55 and the rational method. On August 2, 2004, Bentley Systems (https://www.

bentley.com) acquired Haestad Methods. This company then expanded the modeling capabilities of PondPack to include interconnected ponds with or without backwater Bentley's "OpenFlows Storm software includes PondPack, Storm CAD, and Culvert Master. Bentley has over ten offices in the United States. and one in Burlington, ON, Canada. One of his offices is at the original Haestad Methods in Waterbury.

3.8.4 HydroCAD

HydroCAD is a computer-aided design stormwater modeling software for performing a wide range of commonly used hydrological and hydraulic computations. In 1985, the Applied Microcomputer Systems (AMS) began development of HydroCAD; the model was updated before it was released in 2001, and the company changed its name to HydroCAD Software Solutions LLC (https://www.hydrocad.net).

HydroCAD performs hydrological and routing calculations using the SCS, NRCS, and rational methods and the SBUH (Santa Barbara Unit Hydrograph). It also allows using local rainfall data or the built-in rainfall library. It is also capable of culvert hydraulics, dam breach simulation, land-use analysis, and pollutant loading calculation. The program runs on Windows PC and provides on-screen routing diagrams and preface reports.

3.8.5 SWMM

SWMM is a dynamic rainfall–runoff simulation model. This model can be used for a single event or long-term (continuous) simulation of runoff quantity and quality from urban areas. SWMM was first developed in 1971 and has since undergone several major upgrades. The latest edition, version 5.2.4, which was released on August 7, 2023, runs under Windows. EPA SWMM was developed by the Water Supply and Water Resources Division of the US EPA's National Risk Management Laboratory with assistance from CDM, Inc. The updated SWMM is widely used for planning, analysis, and design of urban stormwater runoff; combined sewers; and drainage systems both nationally and internationally. The model applications include designing and sizing drainage systems; controlling site runoff through green infrastructure practices such as rain gardens, grass swales, roof downspout disconnection, rain barrels, and cisterns; and mapping flood plains of streams. SWMM is a FEMA-approved model for the National Flood Insurance Program.

The SWMM allows the user to perform water quality simulations and view the results in formats including color-coded drainage area maps, time series tables and graphs, profile plots, and statistical frequency analysis. The runoff component of the SWMM generates rainfall–runoff relation for a collection of sub-catchment areas and transports the runoff through pipes and channel conveyance systems, storage/treatment devices, pumps, and regulators. It computes the flow rate, flow depth, and quantity of runoff in each pipe and channel and tracks the quantity and quality of runoff generated within each sub-watershed during a simulation period or multiple time intervals.

3.8.6 WinSLAMM

The original SLAMM was developed in the 1970s to investigate the relationships between sources of urban runoff pollutants and runoff quality. WinSLAMM version 5.0—the Windows version—released in December 2022, is the only urban stormwater quality model that evaluates runoff volume and pollution loading for individual source areas within each land use for each rainfall event. The new version includes a wide variety of source conditions and outfall control practices such as infiltration basins, wet ponds, porous pavement, street cleaning, and grass swale. Unlike many other models, WinSLAMM (https://www.winslamm.net) is applicable to frequent and relatively small flows. As such, this model is most suited for water quality analysis. It can predict the sources of runoff pollutants and flows for the rainfalls that are of interest to water quality analysis and is capable of evaluating the effectiveness of stormwater control measures.

3.8.7 SWAT (Soil and Water Assessment Tool)

Soil and Water Assessment Tool (SWAT) simulates the impact of land development practices on the quantity and quality of surface and groundwater. The model is applicable not only to small watersheds but also to large complex river basins. It can quantify the environmental impacts of land use, land management practices, and climate change on flooding. This model is supported by the USDA Agricultural Research Services at the Grassland, Soil, and Water Research Laboratory in Temple, Texas. Unlike traditional methods that use design storms, this model allows a user to predict continuous impacts of flow. The model requires climatic (rainfall and temperature) data, topographic information, and historical and future land use patterns. The city of Austin in Texas has used this model to study the Walnut Creek Watershed in Austin at the creek's confluence with the Colorado River (Lopez et al., 2012). The last update of this model, SWAT Season 7, was released on February 16, 2024.

3.8.8 USGS Regression Equation for Peak Flow Calculations

The USGS has developed and compiled regression equations for estimating the peak flows for 2- through 500-year storms throughout the United States. The equations are available for both rural and urban areas for every state.

The rural equations are based on the watershed and climatic characteristics within specific regions in each state. The regression equations are in the following general form:

$$Q_T = kA^a B^b C^c \tag{3.100}$$

where

Q_T = T-year rural peak flow
k = regression constant
a, b, and c are regression coefficients
A, B, and C: basin characteristics

For urban areas, the regression equations include seven factors, one of which is the peak flow for rural areas. The most significant factor in these equations is the basin development factor (BDF) which is a measure of the efficiency of damage basins. Though verified, these equations may still deviate 35% to 50% from field measurements.

3.8.9 StreamStats

The USGS has also prepared a computer program known as StreamStats (https://water.usgs.gov/osw/streamstat) that provides information not only for peak discharges but also characteristics such as drainage area, length, and slope for any stream basin. The program has been fully implemented in a number of states, and efforts are underway to achieve the long-term goal of national coverage. New Jersey is among states where the program is fully operational.

The information for a watershed can be readily accessed by inputting the latitude and longitude of a point of interest in a watershed or surfing for the location on the interface map. Appendix 3B presents the application of this program to a stream in New Jersey. As this appendix shows, the StreamStats software output includes a drainage area map of the stream and basin information such as the stream length and slope, percentage of the area covered by urban use, and peak discharges for 2- through 500-year storms.

PROBLEMS

3.1 An undisturbed rock sample has an oven-dry weight of 425.30 g. After saturation with kerosene, its weight is 476.19 g. It is then immersed in kerosene and displaces 196.07 g of kerosene. What is the porosity of the sample?

3.2 A mildly sloped, undisturbed catchment area is covered with approximately 50% wood and 50% grass. For a 60-minute storm of 50 mm/h intensity, calculate:

a. Interception using the Brooks et al. method. Estimate $S = 4$ mm and neglect evaporation during the storm

b. Depression storage using Equation 3.3 with $S_d = 20$ mm.

3.3 Solve Problem 3.2, using: $I = 2$ in./h, $S = 0.15$ in., and $S_d = 0.8$ in.

3.4 Calculate infiltration volume for sandy clay loam soil, having $f_o = 2.5$ in./h, $f_c =$ K=1 in./h during a 2-hour storm of $I = 1.6$ in./h intensity. Use:

a. Horton equation with $\alpha = 1$ hour $^{-1}$

b. Horton equation and Table 3.5

3.5 Work Problem 3.4 for the following parameters: $f_o = 62.5$ mm/h, $f_c = 25$ mm/h, and $I = 40$ mm/h.

3.6 The infiltration rate in a small area was found to be 100 mm/h at the beginning of a 6-hour storm, and it decreased to an equilibrium of 20 mm/h. A total of 250 mm of water was infiltrated during the storm period. Calculate the value of α in Horton's equation.

3.7 Average rainfall intensities during each hour of a 4-hour storm over a 00 ha watershed were 30, 50, 25, and 15 mm/h, respectively. If the infiltration φ index for the storm was 20 mm/h,-calculate the direct runoff from the watershed.

3.8 The rainfall intensities of a 60-minute storm falling on a 10-acre basin were as follows:

Time, min	15	30	45	60
Intensity, in./h	3.0	2.5	4.0	1.5

a. Calculate the total rainfall depth in inches.

b. Calculate the φ index if the net rainfall (overland flow) from the basin was measured to be 1.5 in.

c. Calculate the volume of runoff from the basin.

3.9 Rain falls uniformly on sandy loam soil at a rate of 30 mm/h for 2 hours. Using the Green-Ampt method, calculate:

a. Time for the soil surface to become saturated

b. The amount of infiltration at the end of the storm period Assume initial moisture content at 0.25 and fp = 50 mm/h.

3.10 Solve Problem 3.9 for a 2-hour storm of 1 in./h intensity; $f_p = 2.0$ in./h.

3.11 In Problem 3.9, rain falls at a rate of 15 mm/h during the first hour and 40 mm/h during the second hour. Calculate the time to surface saturation and the infiltration amount.

3.12 A falling head permeameter, similar to that shown in Figure 3.10b, is used to measure the permeability of a soil sample. The head, H, is measured to drop from 400 to 150 mm in 90 seconds. Calculate the soil permeability. The permeameter dimensions are sample length, $L = 120$ mm; $r = 10$ mm, and $R = 100$ mm.

3.13 Calculate the soil permeability of Problem 3.12 for a head drop of 14 in. to 6 in. in 90 seconds. The perimeter dimensions are $L = 5$ in.; $r = 0.4$ in.; and $R = 4$ in.

3.14 A 25-acre wooded area receives 2 inches of rain uniformly in 1 hour. If the time of concentration is 30 minutes, what is the peak flow rate at the watershed outlet? Base your calculations on the rational method.

3.15 Rework Problem 3.14 for 10 ha and 50 mm rainfall.

3.16 The rain falls uniformly for 1 hour at the rate of 2 in./h on a 100-acre watershed composed of: 25 acres, $C = 0.3$; 20 acres, $C = 0.4$; 40 acres, $C = 0.6$; and 15 acres, $C = 0.90$. Calculate the peak rate of runoff, assuming a time of concentration of 1 hour. What will be the peak discharge, if the time of concentration is 45 minutes?

3.17 Rain falls uniformly at a rate of 50 mm/h for 1 hour on a 40 ha watershed which comprises 10 ha having $C = 0.3$; 10 ha with $C = 0.4$; and 20 ha with $C = 0.6$. Calculate the peak rate of runoff assuming a time of concentration of 1 hour.
If the time of concentration is 45 minutes, what will be the peak discharge? Plot a hydrograph for this storm.

3.18 A watershed contains two subareas with the following hydrologic characteristics:
   a. Subarea 1: $A = 80$ acres, $C = 0.50$, $T_c = 30$ min
   b. Subarea 2: $A = 100$ acres, $C = 0.35$, $T_c = 40$ min
   c. The 25-year rainfall intensities of the 30- and 40-minute duration storms are 3.6 and 3.0 in./h, respectively. Calculate the peak rate of runoff from each subarea and the composite discharge from the watershed.

3.19 In Problem 3.18, subareas 1 and 2 are 32 and 40 ha, respectively. Calculate the 25-year peak runoff from
   a. Each subarea
   b. The watershed
   c. The 25 -year rainfall intensities of the 30- and 40-minute duration storms are 90 and 75 mm/h, respectively.

3.20 A 150-acre watershed includes two subareas: Subarea A, which includes 30% of the watershed, has a time of concentration of 20 min; subarea B has mild slope comprising 70% of the watershed with a time of concentration of 60 min. The abstraction can be taken as 1 in./h. Calculate the 25-year peak flow. Use the following IDF relation:

$$I = \frac{\left(30\text{x}T^{0.22}\right)}{\left(t_d + 18\right)^{0.75}}$$

where I = rainfall intensity, in./h, T = return period in years; and t_d = rainfall duration in minutes. Assume a linear discharge at the watershed outlet. State any other assumptions used.

3.21 Name the factors on which the soil curve number, CN, in the TR-55 method depends.

3.22 A drainage area has a composite soil curve number of 55. Based on the SCS method, how much rain must fall before any runoff occurs?

3.23 A natural watershed has the following characteristics:
   a. Woods in good condition, soil group B, covering 40% of the area
   b. Grass-brush in good condition, soil group C, covering 60%
   c. Calculate the runoff depth, in centimeters, for an 11.5 cm rainfall in 24 hours. Use the SCS type III storm distribution.

3.24 Calculate runoff depth in inches in Problem 3.23 for a 24 hour-5 inch rainfall.

3.25 Using the TR-55 method, calculate the time of concentration for a drainage area having the following characteristics:
   a. Sheet flow: dense grass, length $L = 100$ ft; slope $S = 2.0\%$, 2 year-24 hour rainfall $P_2 = 3.5$ in.
   b. Shallow concentrated flow: unpaved, length $L = 250$ ft, slope $S = 1\%$
   c. Stream flow: Manning's $n = 0.06$, flow area $A = 9$ ft^2, wetted perimeter $P = 11$ ft, slope $S = 0.7\%$, length $L = 700$ ft.

The watershed area is 10 acres.

3.26 Calculate the time of concentration of the drainage area in Problem 3.25 using the FHWA method. Base your calculation on rainfall intensity of 10-year storm frequency in Figure 3.1b.

3.27 The drainage area described in Problem 3.25 is composed of 40% wood, 30% grass, and 30% pavement. For this area, calculate the depths of runoff and the peak rates of flow for the 2-, 10-, and 100-year storms of 24-hour duration. The 24-hour storms in the area are $P_2 = 3.5$ in., $P_{10} = 5.3$ in., and $P_{100} = 8.7$ in. The storm is type III distribution and the soil is group C.

3.28 Use the rational method to calculate the peak rates of flow for the drainage area described in Problem 3.27 for the storms of 2-, 10-, and 100-year frequency. Use the New Jersey rainfall intensity-duration-frequency (IDF) curves of Figure 3.1b for this area. Base your calculations on the following runoff coefficients:

a. $C = 0.95$ pavement
b. $C = 0.30$ lawn
c. $C = 0.25$ wooded area

3.29 A 10 ha watershed is composed of 60% woods, 20% lawn, and 20% pavement. The soil is uniformly hydrologic group B. Calculate the depth, volume, and peak runoff for a 24-hour, 150 mm type III storm. Assume a time of concentration of 45 minutes.

3.30 The drainage area of Problem 3.27 is composed of loamy sand soil and the pavements drain onto lawn and wooded areas. Calculate the volume and the peak runoff for the storms of 2-, 10-, and 100-year frequency and 45-minute duration based on the Universal Method. Use the IDF curves in Figure 3.1b.

3.31 Calculate the peak and volume of runoff in the previous problem if the pavements are not drained onto lawns and woodland.

3.32 A mildly sloped 10 ha watershed includes 4 ha of woodland, 3.5 ha of lawn, and 2.5 ha of pavement. The soil is sandy loam. Using the Universal Method, calculate the volume and peak runoff rate for a 45-minute storm of 75 mm/h intensity for the following cases:

a. Pavements are hydraulically separate from pervious areas.
b. Pavements are drained to lawn and woodland.

3.33 A single-family home includes a 130 m^2 dwelling, 80 m^2 driveway and paved patio, 275 m^2 lawn, and 150 m^2 landscape. Initial abstraction and hydraulic conductivity of lawn and landscape are as follows:

a. $L = 10$ mm, $K = 50$ mm/h- lawn
b. $L = 30$ mm, $K = 75$ mm/h- landscape

Calculate the lag time, the runoff volume, and peak runoff for a storm of 60 mm/h, lasting 1 hour using the Universal Method for the following cases:

a. Roof and driveway drain directly to street
b. Roof downspouts end at the landscape
c. Roof and driveway drain to lawn and landscape

3.34 Solve Problem 3.33 for a rainfall intensity of 40 mm/h.

3.35 Rain falls at a rate of 50 mm/h for 30 minutes and 40 mm/h for the next 30 minutes. Calculate the lag time, runoff volume, and peak runoff from the dwelling in Problem 3.33 for cases a, b, and c.

3.36 Calculate the peak and volume of runoff for the single-family home shown below during a 30-minute storm having 3.2 in./h intensity. Perform the calculations using the Universal Method for the following cases:

a. Direct discharge from pavements
b. Roof and driveway drain to lawn and landscape

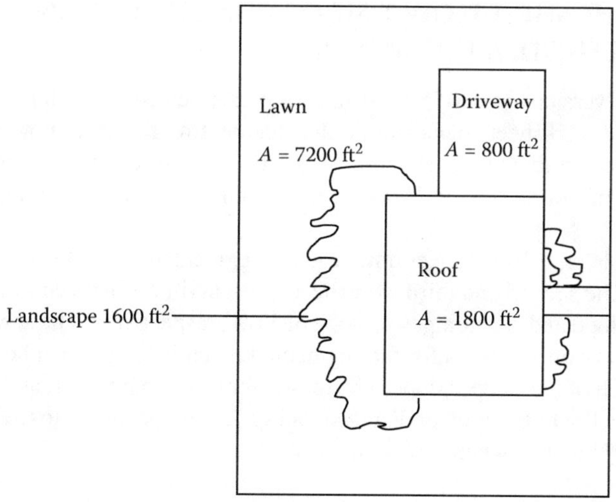

Given:

$L_i = 0.4$ in. lawn
$L_i = 1.2$ in. landscape
$S_d = 0.08$ in. driveway
$S_d = 0.04$ in. roof

Landscape:

$A = 1600\ \text{ft}^2$
$K = 5$ in./h

APPENDIX 3A: THE SHEET FLOW TIME OF CONCENTRATION IN THE TR-55 METHOD: A COMMENTARY

The following three versions of TR-55 Worksheet 3 present calculations for the sheet flow time of concentration. These worksheets demonstrate that segmenting the sheet flow reach, as commonly practiced and allowed in the TR-55 method, is improper because it always results in an overestimation of the time of concentration. Worksheet 3 should be revised, eliminating this erroneous procedure.

In the first worksheet, a 100 ft sheet flow reach is segmented to a 60 ft reach at 2% and a 40 ft reach at 1% grade. The second and third worksheets present time of concentration calculations for a 100 ft reach at 2% slope and at a composite slope of 1.6%, respectively. These worksheets show that the calculated time of concentration for the segmented reach is longer than both a composite slope but also a 100 ft reach of the steeper slope. The reason for this anomaly is that the time of concentration equation is not a linear function of distance; rather, it is proportional to the 0.8 power of length. (Note that, e.g., $100^{0.8} = 39.8$, whereas $2 \times 50^{0.8} = 45.7$.)

U.S. Department of Agriculture
Natural Resources Conservation Service

FL-ENG-21B
06/04

TR 55 Worksheet 3: Time of Concentration (T_c) or Travel Time (T_t)

Project: ______ Designed By: ______ Date: ______

Location: ______ Checked By: ______ Date: ______

Check one: Present Developed

Check one: T_c T_t through subarea ______

NOTES: Space for as many as two segments per flow type can be used for each worksheet. Include a map, schematic, or description of flow segments.

Sheet Flow (Applicable to T_c only)	Unit	Segment ID			
1. Surface description (Table 3-1)					
2. Manning's roughness coeff., n (Table 3-1)		0.80	0.80		
3. Flow length, L (total L ≤ 100 ft)	ft	60	40		
4. Two-year 24-hour rainfall, P_2	in	3.5	3.5		
5. Land slope, s	ft/ft	0.020	0.010		
6. $T_t = \frac{0.007\,(nL)^{0.8}}{P_2^{0.5}\,s^{0.4}}$ Compute T_t	hr	0.40	+ 0.38	=	0.77

Shallow Concetrated Flow	Unit	Segment ID			
7. Surface description (paved or unpaved)					
8. Flow length, L	ft				
9. Watercourse slope, s	ft/ft				
10. Average velocity, V (Figure 3-1)	ft/s				
11. $T_t = \frac{L}{3600\,V}$ Compute T_t	hr		+	=	

Channel Flow	Unit	Segment ID			
12. Cross sectional flow area, a	ft^2				
13. Wetted perimeter, P_w	ft				
14. Hydraulic radius, $r = \frac{a}{P_w}$ Compute r	ft				
15. Channel Slope, s	ft/ft				
16. Manning's Roughness Coeff., n					
17. $V = \frac{1.49\,r^{2/3}\,s^{1/2}}{n}$ Compute V	ft/s				
18. Flow length, L	ft				
19. $T_t = \frac{L}{3600\,V}$ Compute T_t	hr		+	=	

20. Watershed or subarea T_c or T_t (add T_t in steps 6, 11, and 19 ... hr **0.77**

U.S. Department of Agriculture
Natural Resources Conservation Service

FL-ENG-21B
06/04

TR 55 Worksheet 3: Time of Concentration (T_c) or Travel Time (T_t)

Project: ______________________ Designed By: ______________ Date: ________

Location: ______________________ Checked By: ______________ Date: ________

Check one: Present Developed

Check one: T_c T_t through subarea ______________________

NOTES: Space for as many as two segments per flow type can be used for each worksheet. Include a map, schematic, or description of flow segments.

Sheet Flow (Applicable to T_c only)

		Segment ID			
1. Surface description (Table 3-1)					
2. Manning's roughness coeff., n (Table 3-1)		0.80			
3. Flow length, L (total L $\leq$ 100 ft)	ft	100			
4. Two-year 24-hour rainfall, P_2	in	3.5			
5. Land slope, s	ft/ft	0.020			
6. $T_t = \frac{0.007\,(nL)^{0.8}}{P_2^{0.5}\,s^{0.4}}$ Compute T_t	hr	0.60	+	=	0.60

Shallow Concetrated Flow

		Segment ID			
7. Surface description (paved or unpaved)					
8. Flow length, L	ft				
9. Watercourse slope, s	ft/ft				
10. Average velocity, V (Figure 3-1)	ft/s				
11. $T_t = \frac{L}{3600\,V}$ Compute T_t	hr		+	=	

Channel Flow

		Segment ID			
12. Cross sectional flow area, a	ft^2				
13. Wetted perimeter, P_w	ft				
14. Hydraulic radius, $r = \frac{a}{P_w}$ Compute r	ft				
15. Channel Slope, s	ft/ft				
16. Manning's Roughness Coeff., n					
17. $V = \frac{1.49\,r^{2/3}\,s^{1/2}}{n}$ Compute V	ft/s				
18. Flow length, L	ft				
19. $T_t = \frac{L}{3600\,V}$ Compute T_t	hr		+	=	

20. Watershed or subarea T_c or T_t (add T_t in steps 6, 11, and 19 .... hr 0.60

U.S. Department of Agriculture
Natural Resources Conservation Service

FL-ENG-21B
06/04

TR 55 Worksheet 3: Time of Concentration (T_c) or Travel Time (T_t)

Project: ______________________ Designed By: ______________ Date: ________

Location: ______________________ Checked By: ______________ Date: ________

Check one: Present Developed

Check one: T_c T_t through subarea ______________________

NOTES: Space for as many as two segments per flow type can be used for each worksheet. Include a map, schematic, or description of flow segments.

Sheet Flow (Applicable to T_c only)

	Unit	Segment ID	Segment ID		Total
1. Surface description (Table 3-1)					
2. Manning's roughness coeff., n (Table 3-1)		0.80			
3. Flow length, L (total L ≤ 100 ft)	ft	100			
4. Two-year 24-hour rainfall, P_2	in	3.5			
5. Land slope, s	ft/ft	0.016			
6. $T_t = \frac{0.007\,(nL)^{0.8}}{P_2^{0.5}\,s^{0.4}}$ Compute T_t	hr	0.65	+	=	0.65

Shallow Concetrated Flow

	Unit	Segment ID	Segment ID		Total
7. Surface description (paved or unpaved)					
8. Flow length, L	ft				
9. Watercourse slope, s	ft/ft				
10. Average velocity, V (Figure 3-1)	ft/s				
11. $T_t = \frac{L}{3600\,V}$ Compute T_t	hr		+	=	

Channel Flow

	Unit	Segment ID	Segment ID		Total
12. Cross sectional flow area, a	ft^2				
13. Wetted perimeter, P_w	ft				
14. Hydraulic radius, $r = \frac{a}{P_w}$ Compute r	ft				
15. Channel Slope, s	ft/ft				
16. Manning's Roughness Coeff., n					
17. $V = \frac{1.49\, r^{2/3}\, s^{1/2}}{n}$ Compute V	ft/s				
18. Flow length, L	ft				
19. $T_t = \frac{L}{3600\,V}$ Compute T_t	hr		+	=	

20. Watershed or subarea T_c or T_t (add T_t in steps 6, 11, and 19 .... hr **0.65**

APPENDIX 3B: USGS STREAMSTATS: A CASE STUDY

This study concerns an unnamed stream along Elm Avenue in Bogota, New Jersey. Figure 3B.2 presents the watershed parameters of the stream computed by the StreamStats software.

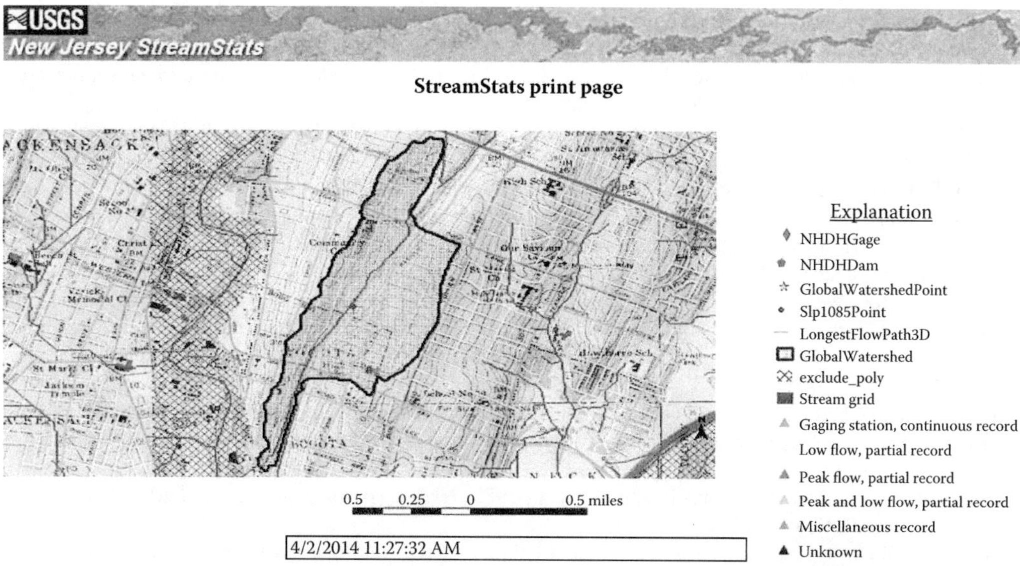

FIGURE 3B.1 StreamStats drainage area map of the stream.

USGS
New Jersey StreamStats

StreamStats ungaged site report

Date: Wed Apr 2 2014 12:04:46 Mountain daylight time
Site location: New_Jersey
NAD27 latitude: 40.8703 (40 52 13)
NAD27 longitude: −74.0331 (−74 01 59)
NAD83 latitude: 40.8704 (40 52 14)
NAD83 longitude: −74.0327 (−74 01 58)
Drainage area: 0.54 mi2

Peak flows region basin characteristics

100% peak glaciated piedmont region 2009 5167 (0.54 mi2)

Parameter	Value	Regression equation valid range	
		Min	Max
Drainage area (square miles)	0.54 (below min value 1.27)	1.27	56.4
Percent storage (percent)	0 (below min value 0.62)	0.62	11.6
Stream slope 10 and 85 method (feet per mi)	36	9.37	176
Basin population density (persons per square mile)	8640	645	13492

Warning: Some parameters are outside the suggested range. Estimates will be extrapolations with unknown errors.

Peak flows region streamflow statistics

Statistic	Flow (ft^3/s)	Prediction error (percent)	Equivalent years of record	90-Percent prediction interval	
				Minimum	Maximum
PK2	101		1		
PK5	162		2		
PK10	209		3		
PK25	275		4		
PK50	329		5		
PK100	385		5		
PK500	524		6		

FIGURE 3B.2 StreamStats ungauged stream report.

REFERENCES

ASCE/EWRI, 2006, *Standard guidelines for the design of urban stormwater systems*, ASCE/EWRI 45-05, *Standard guidelines for installation of urban stormwater systems*, ASCE/EWRI 46-05, *Standard guidelines for the operations and maintenance of urban stormwater systems*, ASCE/EWRI 47-05.

ASCE and WEF, 1992, *Design and construction of urban stormwater management systems*, American Society of Civil Engineers Manuals and Reports of Engineering Practice No. 77 and Water Environment Federation Manual of Practice FD-20.

Brooks, K.N., Elliott, P.R., Gregersen, H.M. and Thames, J.L., 1991, *Hydrology and the management of watershed*, Iowa State University Press, Ames, IA.

Chin, D.A., 2006, *Water resources engineering*, 2nd ed., Chapter 5, Pearson Prentice Hall, Upper Saddle River, NJ.

Chow, V.T., 1964, *Handbook of applied hydrology*, McGraw-Hill, New York.

Chow, V.T., Maidment, D.R. and Mays, L.W., 1988, *Applied hydrology*, Chapter 15, McGraw-Hill, New York.

FHWA (Federal Highway Administration, US Department of Transportation), March 2024a, *Highway Hydrology*, Chapter 3, 3rd ed., FHWA, Hydraulic Design Series No. 2, (HDS-2), Publication No. FHWA-HIF-24-00.

FHWA (Federal Highway Administration, US Department of Transportation), HEC-22, 3rd ed., 2009, revised August 2013; *Urban drainage design manual*, 4th ed., February 2024, FHWA HEC-22, Publication No. FHWA-HIF-24-006.

Freeze, R.A. and Cherry, J.A., 1979, *Groundwater*, Chapter 2, Prentice Hall, Upper Saddle River, NJ.

Izzard, C.F., 1946, *Hydraulics of runoff from developed surfaces, report 26*, Hwy Research Board, Washington, DC, pp. 129±146.

Linsley, R.K., Jr., Kohler, M.A. and Paulhus, J.L.H., 1982, *Hydrology for engineers*, 3rd ed., Chapters 3–4, McGraw-Hill, New York.

Lopez, J.M., Glick, R., and Gosselink, L., January–February 2012, Taking a SWAT at changing urban cycles, a combined approach to evaluate changes in flooding, erosion and aquatic life, *Stormwater*, pp. 39–46.

Maidment, D.R., editor in chief, 1993, *Handbook of hydrology*, Chapter 28, McGraw-Hill, New York.

Mays. L.W., editor in chief, 1996, *Water Resources Handbook, chapter 26 Urban Stormwater Management*, McGraw Hill, New York.

McCuen, R.H., Johnson, P.A. and Regan, R.M., 2002, *US department of transportation, FHWA, highway hydrology*, 2nd ed., FHWA, Hydraulic Design Series No. 2, Publication No. FHWA-NHI02-001.

McCuen, R.H. and Spiess, J.M., 1995, Assessment of kinematic wave time of concentration, *ASCE Journal of Hydraulic Engineering*, 121 (3): 256–266.

New Jersey Department of Environmental Protection, Trenton, NJ, July 2023, *New Jersey stormwater best management practices manual, February 2004*, last updated October 9, 2024, https://dep.nj.gov/stormwater/bmp-manual/.

Pazwash, H., May 22–26, 1989, *Comparison of rational and SCS-TR55 methods for urban stormwater management*, Channel Flow and Catchment Runoff, Proceedings of the International Conference for centennial of Manning's Formula and Kuichling's Rational Formula, University of Virginia, Charlottesville, pp. 156–165.

Pazwash, H., August 2–6, 1992, *Simplified design of multi-stage outfalls for urban detention basins*, Proceedings of Water Resources Sessions at Water Forum 92, Baltimore, MD, pp. 861–866.

Pazwash, H., August 16–20, 2009, *Universal runoff model, paper R13*, StormCon, Anaheim, CA.

Pazwash, H., 2011 *Urban storm water management*, 1st ed., CRC Press, Taylor & Francis Company, Boca Raton, FL, London and New York.

Pazwash, H., May 20–22, 2013, *Universal runoff model, comparison with rational and SCS methods*, World Environmental and Water Resources, Congress 2013, Cincinnati, OH.

Pazwash, H., 2016, *Urban storm water management*, 2nd ed., CRC Press, Taylor & Francis Company, Boca Raton, FL, London and New York.

Rawls, W.J., Brakensiek, D.L., and Miller, N., 1983, Green-Ampt infiltration parameters from soil data, *Journal of Hydraulic Division, ACSE*, 109 (1): 62±70.

Rickly Hydrological Company, 1700 Joyce Avenue, Columbus, OH, 43219, 1-800-561-9677, http://www.rickly.com/MI/Infiltrometer.htm.

Schneider, L.E. and McCuen, R.H., 2005, Statistical guidelines for curve number generation, *Journal of Irrigation and Drainage Engineering, ASCE*, 1311 (3): 282–290.

Standards for Soil Erosion and Sediment Control in New Jersey, July 2017, Adopted by the New Jersey State Soil Conservation Committee.

Todd, D.K., 1980, *Groundwater hydrology*, 2nd ed., Chapter 3, John Wiley & Sons, New York.

Urban Drainage and Flood Control District, revised 1991, *Urban storm drainage criteria annual, Denver, CO*, https://denvergov.org>do.

USDA (US Department of Agriculture, NRCS), January 2009, *Small watershed hydrology WinTR-55 user guide*, WinTR-55, version 1.10 Windows 7, last updated November 2023.

USDA (US Department of Agriculture, Soil Conservation Service, June 1986, *Urban hydrology for small watersheds*, NCRS, Technical Release 55 (TR 55).

Viessman, W., Jr., Lewis, G.L. and Knapp, J.W., 1989, *Introduction to hydrology*, 3rd ed., Harper & Row, New York.

Westphal, J.A., 2001, Hydrology for drainage system design and analysis, in L.W. Mays, ed. *Stormwater collection system design handbook*, McGraw-Hill, New York.

4 Design of Storm Drainage Systems

This chapter presents procedures for the design of storm drainage elements, including inlets, storm drains, culverts, and erosion control measures at conduit outlets and lined swales. Inlets, culverts, and swales are primarily intended to provide a discharge capacity adequate to carry the calculated design flow.

4.1 INTRODUCTION TO ROADWAY DRAINAGE ANALYSIS

Management of stormwater runoff is an important aspect of every urban development roadway project. Traffic safety depends on surface drainage. Proper drainage can minimize, if not eliminate, the occurrence of the hazardous phenomenon of hydroplaning.

Roadway runoff can be managed in different ways, depending on whether the streets and roads are curbed or not. Conveyance of runoff from curbed streets and roads is discussed in this chapter. Also presented in this chapter is the design of roadside swales to carry runoff. For streets with curbs and gutters, runoff is collected and conveyed by a drainage system comprising inlets and pipes. Sufficient numbers of inlets is the key to effective removal of stormwater from pavements. Expensive drainage systems often flow below their capacity because of insufficient inlets.

The design of any drainage system begins with runoff calculations. These calculations were discussed in a previous chapter. For curbed streets and roads, the design is followed by gutter flow analysis, inlet calculations, and sizing of drainage pipes. The following sections discuss the flow spread in gutters, procedures for calculating the efficiency of inlets, and design of storm drains. A section also presents flow equations developed by the author for roadways with zero longitudinal profile.

4.1.1 Gutter Flow

The hydraulic capacity of a gutter depends on its geometry and the longitudinal grade. Applying the Manning formula to a triangular shaped gutter, which is a typical curbed gutter section (Appendix 4A) yields

$$Q = S_x^{5/3} S^{1/2} T^{8/3*} \tag{4.1}$$

where

$K = 0.375$ SI
$K = 0.56$ CU
S_x = cross slope for roadway, m/m (ft/ft)
S = longitudinal slope of gutter m/m (ft/ft)
T = flow spread, m (ft)
Q = gutter flow m^3/s (cfs)

See Appendix 4A for derivation of this equation

DOI: 10.1201/9781003460121-4

Expressing Q in terms of the flow depth, d, at the curb gives

$$Q = \left(\frac{K}{n}\right)\left(\frac{S^{1/2}}{S_x}\right)d^{8/3} \tag{4.2}$$

where

$$d = TS_x,\ m/ft \tag{4.3}$$

Various forms of nomographs are available for estimating the flow in triangular gutters in terms of the depth of flow. One such version is shown in Figure 4.1a in SI and Figure 4.1b in CU. However, to improve accuracy, gutter flow calculations should be performed directly using either Equation 4.1 or Equation 4.2.

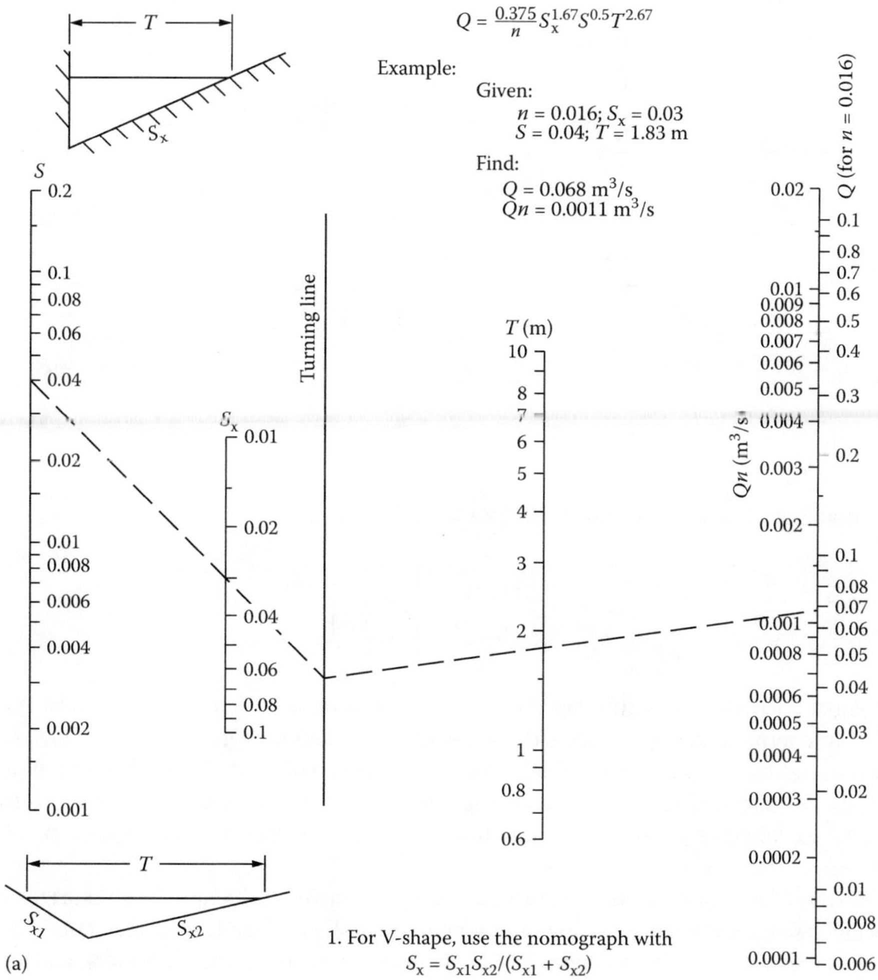

FIGURE 4.1A Flow in triangular gutter sections (SI).

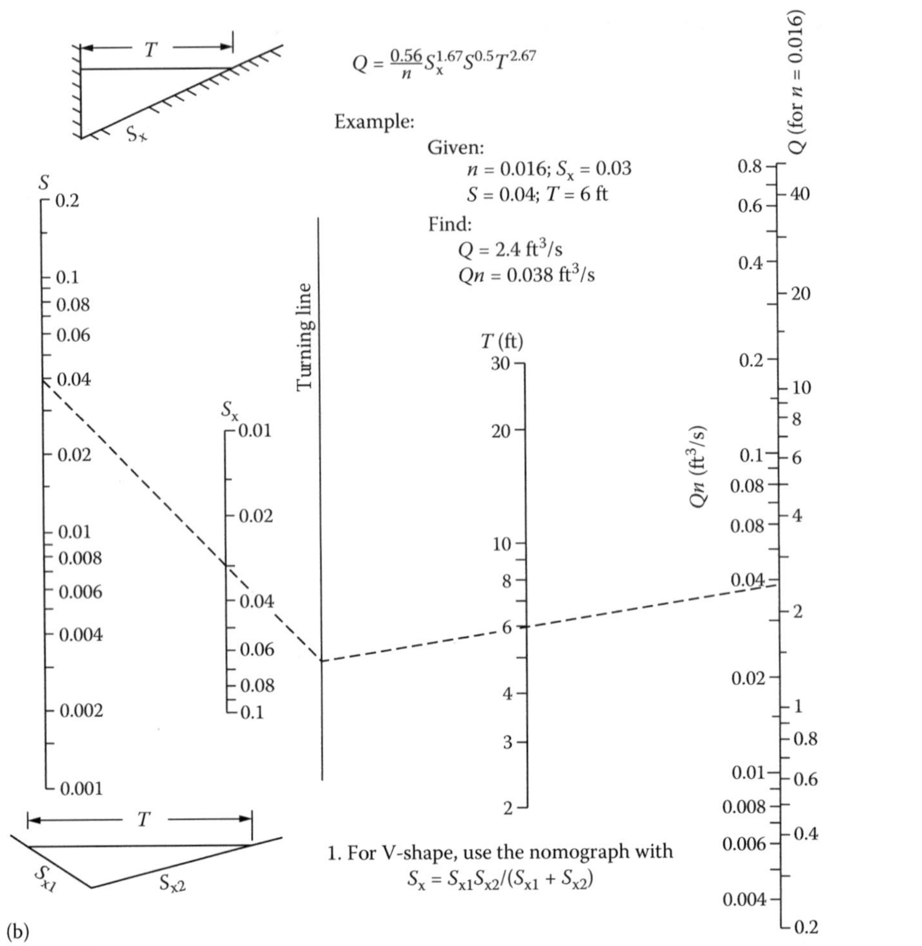

FIGURE 4.1B Flow in triangular gutter sections (CU).

Rewriting Equation 4.1 in terms of the flow spread, T, gives

$$T = \left[\frac{(n/K)^{0.375}}{S_x^{0.625}}\right] \times \left(\frac{Q}{S^{0.5}}\right)^{0.375} \tag{4.4}$$

This equation is dimensionally homogeneous and can be used both in metric and customary units. The first term on the right-hand side is generally constant for a given project. Thus, substituting that constant for the first term in the preceding equation will simplify tabulation of flow spread along roadway gutters; note that the preceding equation breaks down for roadways at zero grade. Spread flow calculations for such a condition are derived in Section 4.5 and Appendix 4B in this chapter.

The n values for various surface textures are listed in Table 4.1. Table 4.2 summarizes the flow spread calculations for a range in flow and longitudinal slope of gutters in meters and feet. The calculations are based on a cross slope of 4%, which is a common grade for gutters, and $n = 0.013$, which represents a smooth gutter surface pavement. Highlighted in this table are the conditions in which the spread exceeds 2 m (6 ft).

TABLE 4.1
Typical Manning's *n* Values for Street and Pavement Gutters

Type of Gutter or Pavement	Manning's *n*
Concrete gutter, troweled finish	0.012
Asphalt Pavement	
Smooth texture	0.013
Rough texture	0.016
Concrete Pavement	
Float finish	0.014
Broom finish	0.016

TABLE 4.2
Flow Spread in Gutters

	Longitudinal Slope, *S*%					
Q L/s (m^3/s)	1	2	3	4	5	7
in Meters						
30 (0.03)	1.35	1.18	1.10	1.04	1.0	0.93
60 (0.06)	1.75	1.53	1.42	1.35	1.29	1.21
90 (0.09)	**2.03**	1.79	1.65	1.57	1.50	1.48
120 (0.12)	**2.26**	1.99	1.84	1.75	1.67	1.65
150 (0.15)	**2.46**	**2.16**	**2.00**	1.90	1.82	1.79

	Longitudinal Slope, *S*%					
Q cfs	1	2	3	4	5	7
			in Feet			
1	4.32	3.80	3.52	3.33	3.20	3.00
2	5.61	4.92	4.56	4.32	4.15	3.89
3	**6.53**	5.73	5.31	5.03	4.83	4.53
4	**7.27**	**6.38**	5.92	5.61	5.38	5.05
5	**7.91**	**6.94**	**6.43**	**6.10**	5.85	5.49

Note: $n = 0.013$, $S_x = 4\%$.

These tables show that the smaller is the longitudinal slope, the wider is the flow spread. It is interesting to note that for longitudinal slopes of 2% or less, the spread exceeds 2 m for discharges larger than 120 L/s (6 ft for discharges greater than 3.5 cfs). Thus, in a roadway with a 2 m (6 ft) wide shoulder, the spread may extend to traffic lanes on roads at small slope. To reduce the length of spread on roadways, composite gutters are sometimes used along roadways. Figure 4.2 shows composite gutter sections having a 0.6 m (2 ft) gutter section with 50 mm (2 in.) depression. A further discussion of the flow spread will be given later in this chapter.

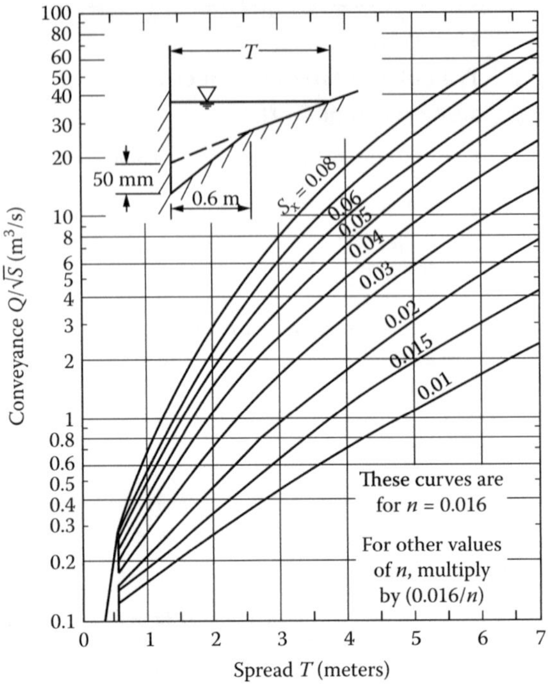

FIGURE 4.2 Discharge versus spread for a composite gutter section in SI.

Source: Figure 3 in HEC-12 transformed from CU.

Example 4.1

A smooth asphalt roadway includes four 12 ft wide lanes, two 4 ft median shoulders. and two 6 ft wide side shoulders (one at each side). The roadway is elevated and receives no offroad runoff. The road is at 1.5% longitudinal slope, and the traffic lanes and the side shoulders are at 2% and 4% cross slopes, respectively. A Jersey barrier is placed at the center of the roadway. The first set of inlets are placed at side shoulders 350 ft from the road high point.

a. Calculate the 25-year peak rate of runoff to the first inlet on each shoulder. Base your calculations on $T_c = 10\,\text{min}$, $I = 6.7$ in./h.
b. Calculate the flow spread on shoulder at the inlet for $n = 0.013$.

Solution

a. $\Sigma w = 2 \times 12 + 4 + 6 = 34$ ft
   $A = 34 \times 350 = 11{,}900$ sf $= 0.273$ acres
   $Q = CIA = 0.950 \text{ x } 273 \times 6.7 = 1.74$ cfs

b. $T = \left[\dfrac{Qn}{\left(0.56 \times S_x^{5/3} S^{1/2}\right)}\right]^{3/8}$

$$T = \left(\frac{1.74 \times 0.013}{0.56 \times 0.04^{5/3} \times 0.015^{1/2}}\right)^{0.375} = 70.57^{0.375} = 4.93 \text{ ft}$$

Example 4.2

Rework Example 4.1 for 3.75 m wide lanes, 1.3 m center shoulder, 1.8 m side shoulder, and $I = 170$ mm/h. The first sets of inlets are placed 100 m from the high point.

Solution

a. $\Sigma w = 2 \times 3.75 + 1.3 + 1.8 = 10.6$ m

$A = 10.6 \times 100 = 1060\ \text{m}^2 = 0.106$ ha

$Q = 2.78 \times 10^{-3} ACI = 2.78 \times 10^{-3} \times 0.95 \times 0.106 \times 170 = 0.0476\ \text{m}^3/\text{s} = 47.6$ L/s

b. $$T = \left[\frac{Qn}{\left(0.375 \times 0.04^{5/3} \times 0.015^{1/2}\right)}\right]^{0.375}$$

$$T = \left(\frac{0.0476 \times 0.013}{0.375 \times 0.00468 \times 0.1225}\right)^{0.375} = 1.49\ \text{m}$$

4.2 TYPES OF INLETS

Figure 4.3 depicts four commonly used types of inlets, namely grate inlet, curb opening inlet, combination inlet, and slotted drain. It should be noted that curb openings and combination inlets are not common in some countries. Figures 4.4 and 4.5 show an inlet along a curb in Budapest, Hungary, and an inlet at a parking lot median in a roadside service center in Austria, respectively. The latter photo also shows a basket under an inlet grate to intercept floatable objects and debris.

To keep bottles and cans from entering curb openings, a new type of curb piece designated as an Eco-curb piece is being used on new curb opening inlets in New Jersey. The New Jersey Department of Environmental Protection (NJDEP) has mandated that all existing curb pieces with openings less than 2 in. wide be replaced with the new curb piece. Campbell Foundry in Harrison, New Jersey, is one of the companies that makes Eco-curb pieces for inlets. Figure 4.6a and 4.6b show a new curb piece and the traditional curb piece, respectively.

A trench drain is a modified type of slotted drain. This type of drain commonly consists of a rectangular flume and a continuous grate. A variety of trench drains are commercially available.

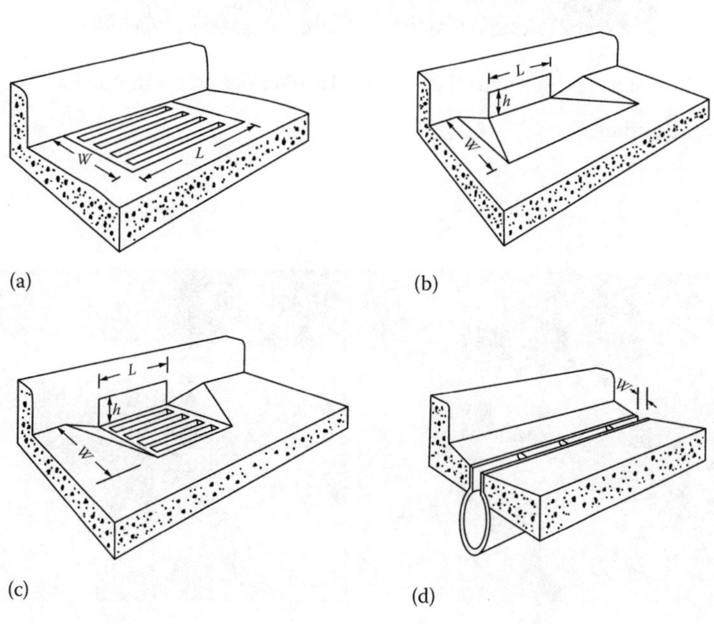

FIGURE 4.3 Types of inlets: (a) grate inlet; (b) curb-opening inlet; (c) combination inlet; and (d) slotted drain inlet.

FIGURE 4.4 An inlet along a curb in Budapest, Hungary.

Source: Photo by the author.

FIGURE 4.5 An inlet at a parking lot median, in a roadside service center in Austria.

Source: Photo by the author.

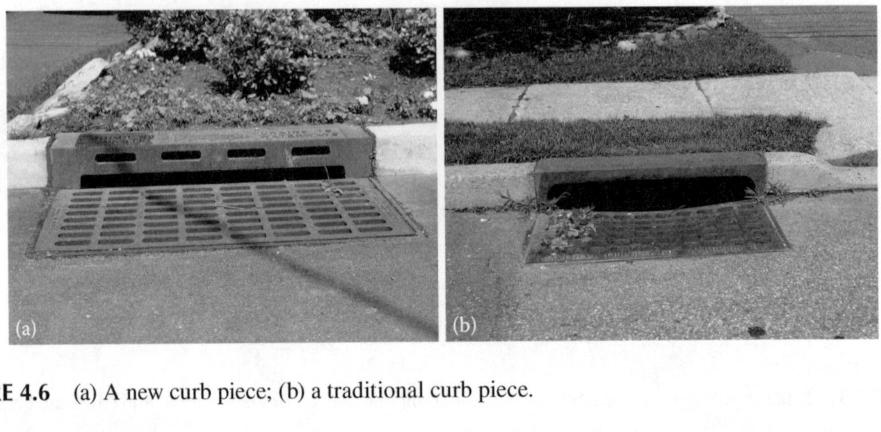

FIGURE 4.6 (a) A new curb piece; (b) a traditional curb piece.

For instance, ACO makes trench drains in 0.5 m and 1 m long sections (see Figure 4.7a) also highway drains. These drains, which are made of polymer concrete, come in 4 ft long, 8 in. inside width sections with 1/16 in. per ft (0.5%) internal longitudinal slope (refer to Figure 4.7b). ACO highway drains provide significantly larger capacity than original ACO drains. They are also nearly four times less expensive than the older models, and as such, are most suitable and cost-effective for roads of small longitudinal slope. Following the author's recommendation, these drains were installed along State Route 30 in Magnolia, Camden County, under a New Jersey DOT roadway improvement project.

To provide access for maintenance, ACO highway drains are also available with ductile iron grates bolted at four corners. ACO also manufactures flume outlets for connection into a collector line or catch basin. These flumes are available in 8 and 12 in. (20–30 cm) schedule 40 vertical or horizontal outlets with iron access grates. Figure 4.8 depicts grates manufactured by Campbell Foundry in Harrison and Kearny, New Jersey, another company that makes various sizes of grates for trench drains (http://www.campbellfoundrry.com). Neenah Foundry in Neenah, Wisconsin is another manufacturer of inlet grates.

4.3 INLET DESIGN

The hydraulic capacity, that is, the portion of gutter flow intercepted by an inlet, depends on the gutter flow, the inlet type, and its location. Inlets at grades have different hydraulic characteristics from inlets at low points.

The US Department of Transportation, Federal Highway Administration (FHWA), has prepared two comprehensive publications for drainage design: *Drainage of Highway Pavements*, Hydraulic Engineering Circular No. 12 (HEC-12), and *Urban Drainage Design Manual* (HEC-22). The former publication, dated 1984, has been revised on May 1, 2003, and the latter, last updated 2013, has been archived and superseded by its fourth edition, in February 2024 that, however, omits some of the information of the third edition.

4.3.1 Grate Inlets at Grade

Grates provide effective means of draining roadway and parking lot pavements. Inlet grates come in a variety of sizes and geometrics. Grates are selected considering three basic elements: hydraulic efficiency, traffic safety (vehicular, bicycle, and pedestrian), and debris clogging. Structural strength, cost, durability, and vandalism are among other elements to be considered. Grates are designated as P-1-7/8 and so on; the letter P stands for parallel bar grates and the number indicates the bar spacing in inches. There are also curved vane (CV) grates and reticuline or honeycomb patterns. Figures 4.9 and 4.10 show P-1-7/8 and reticuline grates, respectively.

Grates are effective pavement drainage elements where debris is not a problem. Grate inlets intercept gutter flow in part through their front edge and partly from their side. All the water passing over the upstream edge of the grate (called frontal flow) will be intercepted if the gutter is long enough and the flow velocity is low enough to prevent a splash-over. At steep areas where the gutter flow velocity exceeds a threshold, splash-over occurs. The splash-over velocity depends on the type of grate. The ratio of the frontal flow to total gutter flow, E_o, for a uniform gutter slope, is given by

$$E_o = \frac{Q_W}{Q} = 1 - \left(1 - \frac{w}{T}\right)^{2.67} \quad (4.5)$$

where

Q = total gutter flow, m³/s (cfs), calculated using Equation 4.1
w = width of grate (or depressed gutter), m (ft)
Q_W = frontal flow; flow in width w, m³/s (cfs)
T = total flow spread in gutter, m (ft)

(a)

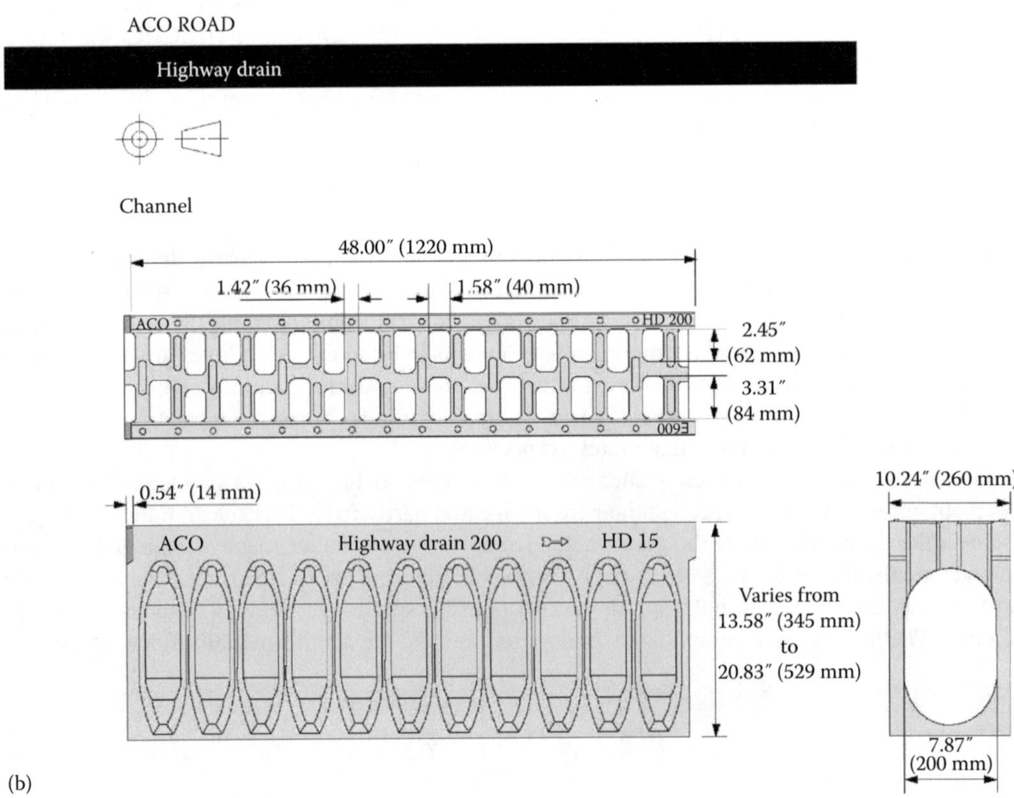

(b)

FIGURE 4.7 (a) ACO trench drain; (b) highway drain by ACO.

Standard heavy duty trench covers/grates are designed for A.A.S.H.T.O. HS20-44 highway loading and are suitable for use in most cargo handling areas. For any application other than standard truck traffic please contact our engineering department for design and material recommendations.

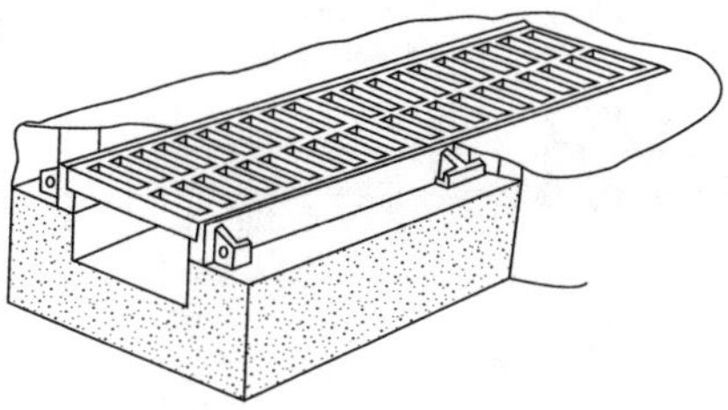

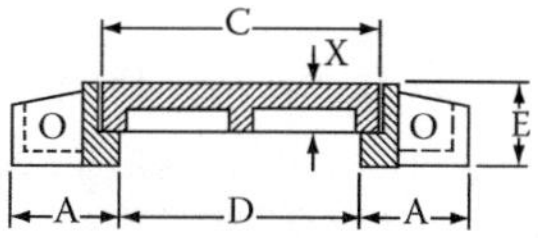

Trench frame and cover

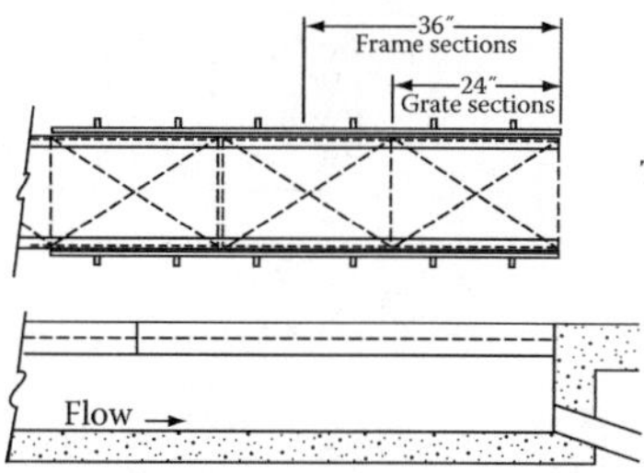

Standard heavy duty

Pattern number		Dimensions in in.					Grate type
Cover	Grate	A	C	D	E	X	
4500	4525*	3 ¾	7 ¾	6	2 ¼	2	1
4500A	4525A*	3 ¾	9 ¾	8	2 ¼	2	2
4501	4526*	3 ¾	11 ¾	10	2 ¼	2	2
4501A	4526A*	3 ¾	14	12	2 ¼	2	2
4501B	4526B	3 ¾	15 ¾	14	2 ¼	2	2
4501C	4526C	3 ¾	17 ¾	16	2 ¼	2	2
4503	4528	5	19 ¾	18	4	2 ¾	3
4503A	4528A	5	21 ¾	20	4	2 ¾	2
4504	4529	5	25 ¾	24	4	2 ¾	3
4505	4530	5	31 ¾	30	4	2 ¾	3
4506	4531	5	38	36	4	2 ¾	3
4507	4532	5	44	42	4	2 ¾	3
4508	4533	5	50	48	4	2 ¾	3

* Available with A. D. A. compliant openings (openings ½″ or less in width)

FIGURE 4.8 Campbell Foundry trench drains.

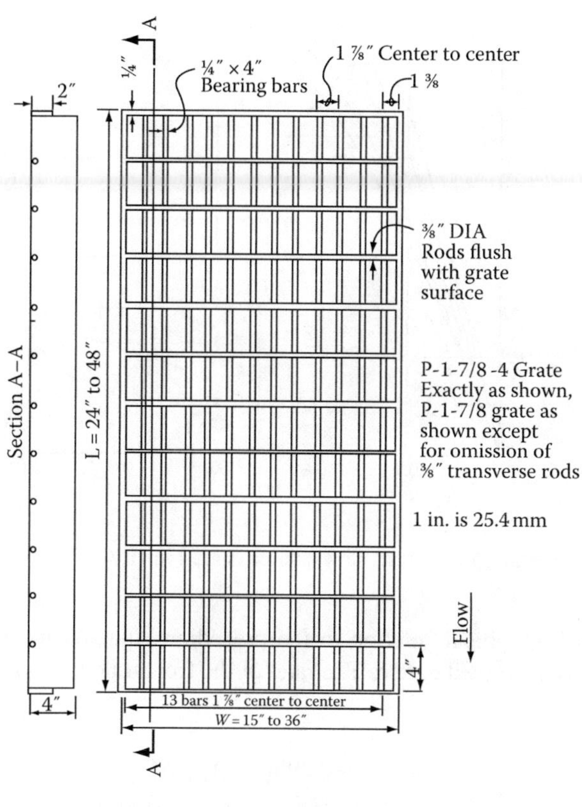

FIGURE 4.9 Campbell Foundry P-1–7/8 and P-1–7/8–4 grates.

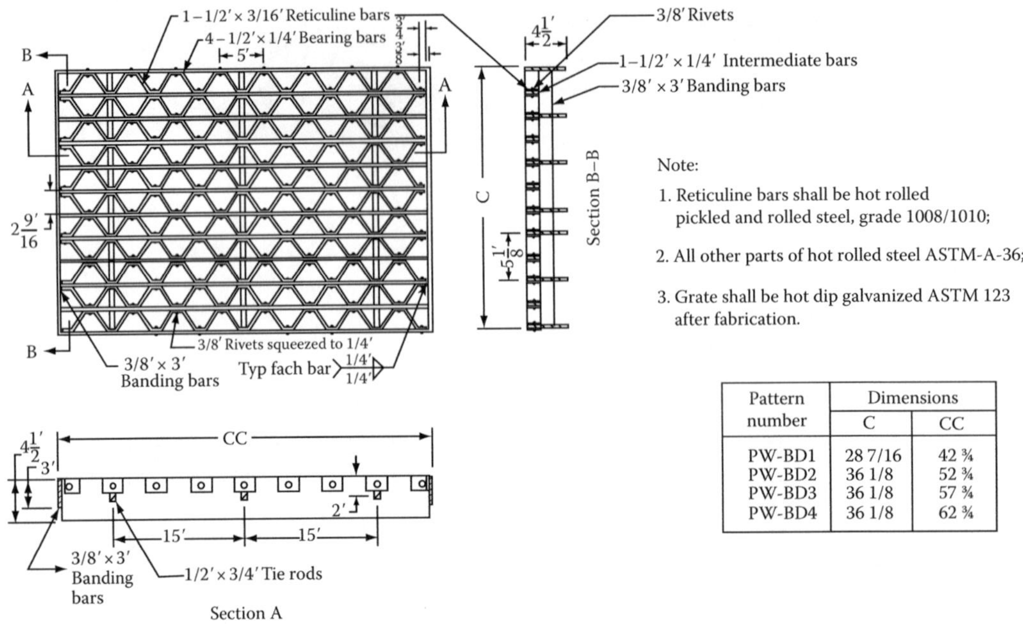

Pattern number	Dimensions	
	C	CC
PW-BD1	28 7/16	42 ¾
PW-BD2	36 1/8	52 ¾
PW-BD3	36 1/8	57 ¾
PW-BD4	36 1/8	62 ¾

FIGURE 4.10 A Campbell Foundry reticuline grate.

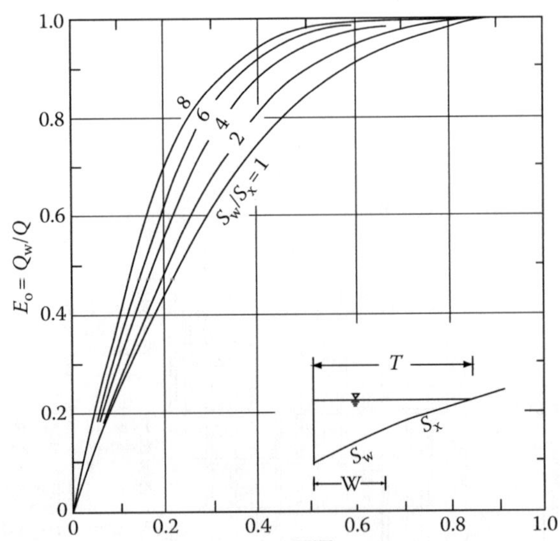

FIGURE 4.11 Ratio of frontal flow to gutter flow (source chart 4 in HEC-12).

Figure 4.11 shows a graphical solution to the preceding equation in dimensionless form for straight cross slope and depressed gutters The ratio of the frontal flow to gutter flow intercepted by an inlet is given by

$$R_f = 1 - 0.295(V - V_o) \quad SI \tag{4.6}$$

$$R_f = 1 - 0.09(V - V_o) \quad CU \tag{4.7}$$

TABLE 4.3
Splash-Over Velocity of Common Inlet Grates, V_o

Grate Type	Length, ft (m)	Splash-Over Velocity, ft/s (m/s)
Reticuline	2 (0.6)	4.2 (1.25)
	4 (1.2)	7.0 (2.10)
Curved vane	2 (0.6)	5.9 (1.75)
	4 (1.2)	9.0 (2.70)
P-1-1/8	2 (0.6)	6.3 (1.90)
(P-30)	4 (1.2)	9.1 (2.75)
P-1-7/8	2 (0.6)	8.1 (2.40)
(P-50)	4 (1.2)	11.5 (3.45)

Note: P-30 and P-50 represent parallel bars spaced 30 and 50 mm on center, respectively.

where

V = velocity of flow in gutter in m/s (ft/s)

V_o = gutter velocity where splash-over first occurs in m/s (ft/s)

The splash-over velocity depends on the type and length of inlet grate. Table 4.3 lists typical values of the splash-over flow velocity of various types of grates in metric and customary units. This table, for example, shows that the splash-over velocity for a 4 ft long grate varies from 7 ft/s for a reticuline inlet to over 11 ft/s for a P-1–7/8 inlet. Thus, if the gutter flow velocity is less than these figures, the second term in Equations 4.6 and 4.7 disappears and R_f becomes 1.0. The gutter flow velocity can be calculated by the following equation:

$$V - \frac{Q}{\left(S_x T^2 / 2\right)} \tag{4.8}$$

Alternatively flow velocities can be calculated using the following equations:

$$V = \frac{0.75 S^{0.5}}{n S_x^{0.67} T^{0.67}} \text{ SI} \tag{4.9}$$

$$V = \frac{1.12 S^{0.5}}{n S_x^{0.67} T^{0.67}} CU \tag{4.10}$$

Figure 4.12 depicts the splash-over velocity for a large number of grates in SI units. In addition to frontal flow, water also enters a grate as side flow. This flow depends on the cross slope of the gutter, the length of the grate, and flow velocity. The ratio of side flow intercepted to the total side flow, referred to as side flow interception efficiency, can generally be calculated by

$$R_s = \left[1 + \left(K V^{1.8} / S_x L^{2.3}\right)\right]^{-1} \tag{4.11}$$

where:

L = length of grate, in m (ft)

V = the gutter flow velocity, in m/s (ft/s)

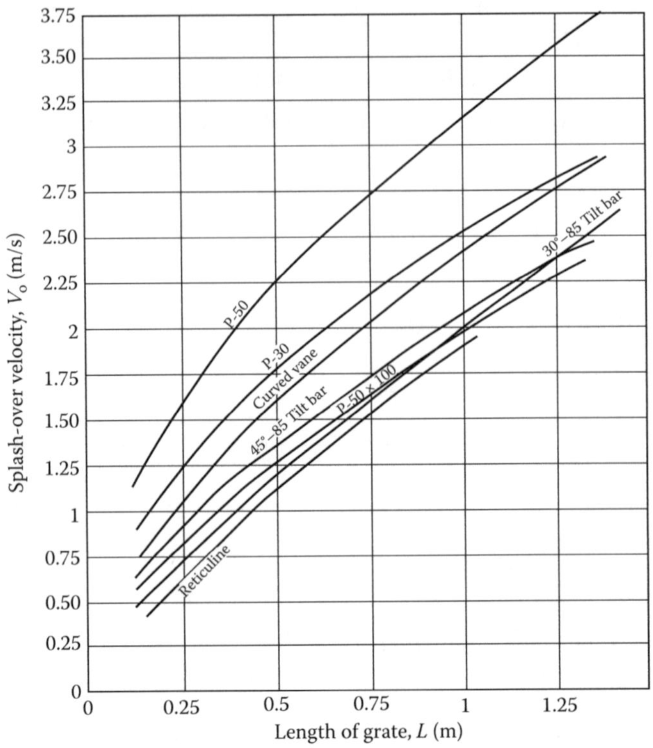

FIGURE 4.12 Splash-over velocity for various grates (SI units). See FHWA HEC-22 (2024).

$K = 0.0828$ SI
$K = 0.15$ CU
S_x = shoulder cross slope, m/m (ft/ft)

Thus, the overall efficiency and the intercepted flow of a grate are expressed by the following two equations, respectively:

$$E = R_f E_o + R_s (1 - E_o) \tag{4.12}$$

and

$$Q_i = EQ \tag{4.13}$$

Note that for a roadway at 4% or greater slope and grates of limited length, the side flow is negligible. Likewise, the intercepted flow of curb openings in combination inlets can be ignored for steep roads. In such cases, the preceding equation simplifies as

$$E = R_f E_o \tag{4.14}$$

Figures 4C.1 and 4C.2 in Appendix 4C give the overall efficiency of P1–1/8 (0.6 m × 0.6 m, 2 ft × 2 ft) and P1–7/8 (0.6 m × 1.2 m, 2 ft × 4 ft) grates for a 4% cross slope for a large range in gutter flow and longitudinal slope. These figures show that a 0.6 m × 0.6 m (2 ft × 2 ft) grate inlet cannot intercept 75% of a 0.09 m³/s (3 cfs) gutter flow under any roadway longitudinal slope. The figures also demonstrate that, all in all, inlets are not always as effective as many engineers perceive; they must be spaced close enough to effectively remove runoff from streets and roadways.

Example 4.3

Given $T = 6$ ft, $S_x = 0.04$, $S = 0.03$, $n = 0.016$, and no bicycle traffic, calculate interception capacity for the following grates:

a. 2 ft × 2 ft P-1–1/8 grate
b. 2 ft × 4 ft (long) P-1–1/8-4 grate
c. 2 ft × 2 ft reticuline grate

Base your calculations for the condition of no spread beyond the shoulder.

Solution

First calculate gutter flow capacity:

$$Q = \left(\frac{0.56}{n}\right) S_x^{1.667} S^{1/2} T^{8/3}$$

$$Q = \left(\frac{0.56}{0.016}\right) \times 0.04^{1.667} \times (0.03)^{1/2} \times 6^{2.667} = 3.37 \text{ cfs}$$

Calculate gutter flow velocity:

$$V = Q/A = 3.37/\left(1/2 \times 6^2 \times 0.04\right) = 4.67 \text{ ft/s}$$

Next, calculate the frontal flow to gutter flow with ratio

$$\frac{W}{T} = \frac{2}{6} = 0.33$$

$$E_o = \frac{Q_w}{Q} = 1 - \left(1 - \frac{W}{T}\right)^{2.67} = 1 - (1 - 0.33)^{2.67} = 0.66$$

Alternatively, enter Figure 4.11 with $W/T = 0.33$ and $S_x/S = 1.33$ to read $E_o = 0.66$.
The frontal flow interception efficiency of grates can be calculated using Equation 4.7 in combination with Table 4.3.

a. For P-1-1/8, $V_o = 6.3$ ft/s $> V$ $R_f = 1.0$
b. For P-1-1/8-4 (four long) grate, $V_o = 9.1$ ft/s $> V$ $R_f = 1.0$
c. For reticuline, 2 ft long grate, $V_o = 4.2 < V$

$$R_f = 1 - 0.09(4.67 - 4.2) = 0.96$$

Then calculate side flow interception efficiency. For grates (a) and (c) where $L = 2$ ft, the side flow interception can be calculated using Equation 4.11, as follows:

$$R_s = \left(\frac{1 + 0.15 \times 4.67^{1.8}}{0.04 \times 2^{2.3}}\right)^{-1} = 0.076$$

Likewise, for grate (b), where $L = 4.0$ ft,

$$R_s = 0.288$$

From Equation 4.12:

a. $E = 1 \times 0.66 + 0.076(1 - 0.66) = 0.66 + 0.026 = 0.686$
$Q_i = 3.37 \times 0.686 = 2.31$ cfs

b. $E = 1 \times 0.66 + 0.288(1 - 0.66) = 0.758$
   $Q_i = 3.37 \times 0.758 = 2.55$ cfs
c. $E = 0.96 \times 0.66 + 0.076(1 - 0.66) = 0.66$
   $Q_i = 2.22$ cfs

These calculations indicate that less than 4% of the overall efficiency of the 2 ft long parallel grate and reticuline inlets is due to side interception and that this efficiency is approximately 13% for a 4 ft long parallel grate. It is also seen that the 2 ft parallel grate provides nearly 5% more capacity than the reticuline grate and that increasing the length of grate from 2 to 4 ft increases the capture capacity by a mere 10%. Note that the preceding results for the 2 ft × 2 ft and 2 ft × 4 ft P-1–1/8 grates can be directly derived from Figures 4C.1b and 4C.2b.

4.3.2 Curb Opening Inlets

Curb opening inlets do not interfere with traffic and in the absence of a curb piece are also less vulnerable to clogging than grate inlets. In the past, these inlets were commonly used without any curb piece. However, as previously indicated, to address the current stormwater management regulations, a curb piece is installed to prevent floatable objects from entering the drainage pipes. See Figure 4.6. Also as indicated previously, curb inlets offer little capacity in intercepting the runoff on steep slopes unless they are depressed.

The flow through a curb opening inlet occurs as a weir flow before the opening is fully inundated and as an orifice when the opening is submerged. For a curb inlet at level grade (no depression) the weir flow discharge is given by

$$Q = C_w L d^{1.5} \tag{4.15}$$

$$C_w = 1.66\ SI;\ C_w = 3.0\ CU$$

where
L = length of curb opening in m (ft)
d = depth of water at the curb opening in m (ft)

Under submerged conditions, the flow equation becomes

$$Q = C_o L h \left(2 g h_o\right)^{1/2} \tag{4.16}$$

where
C_o = orifice coefficient = 0.67 SI and CU
h = height of opening, m (ft)
$h_o = (d - h/2)$ = water depth to the center of the opening, m (ft)
$g = 9.81$ m/s^2 (32.2 ft/s^2)

For an inlet at depression, the equation for orifice flow remains unchanged, while the weir flow equation modifies as

$$Q = C_w \left(L + 1.8 W_o\right) d^{1.5} \tag{4.17}$$

where
W_o is the width of gutter depression in m (ft)
C_w = 1.25 SI
C_w = 2.3 CU

According to HEC-22 (FHWA, 2013) the length of curb opening inlet that needed to fully intercept the gutter flow on a straight cross slope is given as

$$L_{100} = KQ^{0.42}S^{0.3}\left(nS_x\right)^{-0.6} \tag{4.18}$$

where

L_{100} = required curb opening length to intercept 100% of gutter flow, in m (ft)
$K = 0.82$ SI (this factor reads as 0.076 in HEC-22, 2013).
$K = 0.60$ CU
Q = gutter flow m³/s (ft³/s)

The same publication proposes the following equation for the interception efficiency of a curb-opening inlet of length L:

$$E = 1 - \left(1 - \frac{L}{L_{100}}\right)^{1.8} \tag{4.19}$$

Using this equation, the length of curb opening inlet required to achieve a 75% efficiency can be calculated as $L = 0.54L_{100}$.

Note that Equation 4.16 can be used to calculate the capacity of a grate inlet fully submerged by substituting the opening area of inlet, A, for Lh in that equation.

4.3.3 Slotted Inlets

Slotted inlets offer little interference to traffic and can be used in curbed or uncurbed sections. They are particularly effective in airport runways, parking lots, garage entries at downhill driveways, docks, ports, and roads with small longitudinal slope where the runoff would impound on the gutter. For slotted inlets in sag locations, the inlet capacity can be calculated using the weir flow equation for a depth of up to 6 cm (0.2 ft) and the orifice flow equation for depths larger than 0.12 m (0.4 ft). Within these depths, the flow is in a transient condition. The inlet capacity for the weir flow condition can be calculated using the same equation as for the curb opening (Equation 4.15). In that equation, the depth at curb measured from the normal cross slope is substituted for depth of water, d, and the weir coefficient, C_w, varies with flow depth and slot length with a typical value of

$$C_w = 1.4\ SI\ C_w = 2.5\ CU$$

The orifice flow condition follows the equation:

$$Q = 0.8LW\left(2gd\right)^{0.5} \tag{4.20}$$

where

L= length of slot, m (ft)
W= width of slot, m (ft)
g = acceleration of gravity, 9.81 m/s² (32.2 ft/s²)
d= depth of water at slot inlet, m (ft)

This equation is valid when the depth of water is greater than the opening which results in inundation of slot. For smaller flow depths, the weir flow equation should be used.

As was previously indicated, a number of trench drains are commercially available. One such drain is the ACO drain, which comes in sections 0.5 and 1 m long and 95 mm (3.75 in.) and upward widths and various depths. Another one is the ACO highway drain, which was previously described. Capacity and technical information on this and other trench drains can be obtained from the manufacturers' specifications.

4.3.4 Combination Inlets

Combination inlets combine the benefits of grate and curb openings. However, this capacity is not considerably greater than the capacity of grates alone on roadways at 4% or greater slope. This is particularly the case for new curb pieces with reduced openings. Therefore, the interception capacity of combination inlets is commonly calculated neglecting the curb pieces.

4.3.5 New Jersey Inlets

Many states have specific inlet grates. In New Jersey, for example, four types of inlets are commonly used for streets, roads, and parking areas. These are inlet types A and E with 2 ft × 4 ft and 4 ft × 4 ft grates, respectively, and combination inlet types B and D with 2 ft × 4 ft grates plus curb pieces. The only difference between type B and type D inlets is that the former inlet has larger inside dimensions than the latter. These inlets are depicted in Figure 4.13. Also available are 2 ft × 2 ft and smaller inlets for lawns and gardens.

NJDOT (2015) developed the following equation for the interception capacity of grate inlets

$$Q_I = \frac{Kd^{1.54} S^{0.233}}{S_X^{0.276}} \tag{4.21}$$

$$K = 2.98\ SI,\ \ K = 16.88\ CU$$

where

d = depth of water in gutter, m (ft)
S = longitudinal slope, m/m (ft/ft)
S_X = cross slope, m/m (ft/ft)
Q = flow intercepted by the grate, m³/s (cfs)

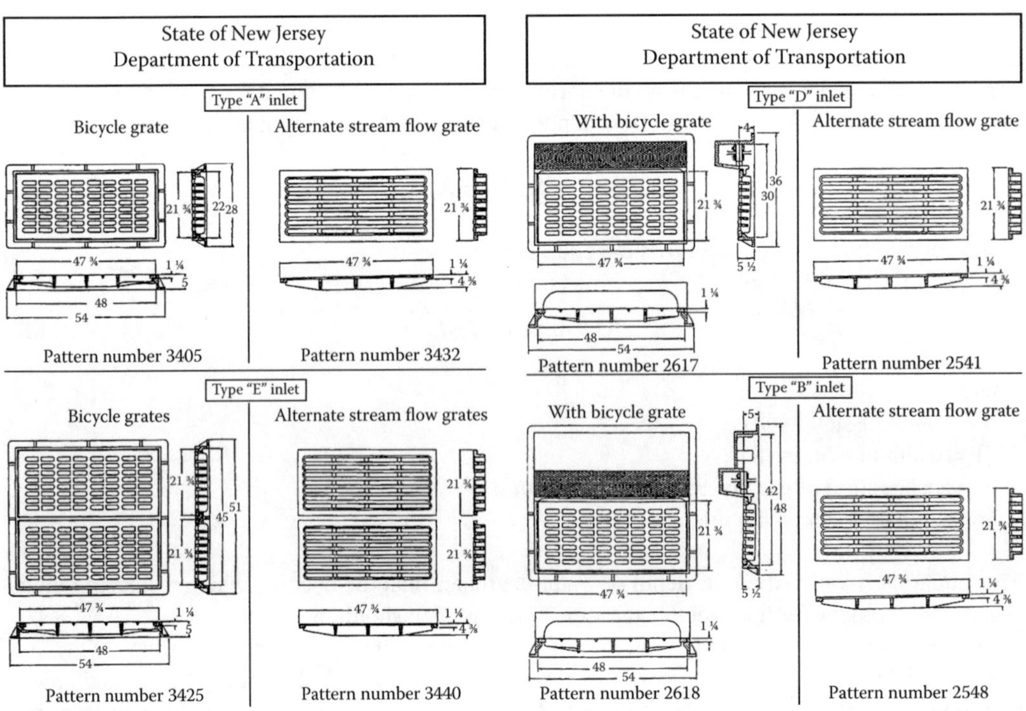

FIGURE 4.13 New Jersey type A, B, D, and E inlets.

The type A inlet is similar to the FHA type P-1–7/8, 2 ft × 4 ft inlet grate. Using Equation 4.21 yields results reasonably close to, though somewhat smaller, than that shown in Figure 4C.2b for a 2 ft × 4 ft grate. The NJDOT suggests using Equation 4.21 for inlet grate types A, B, and D without any modification. However, combination inlets such as D and B provide a somewhat larger interception capacity than type A inlets in small longitudinal slopes. In using these inlets, the author suggests applying an adjustment factor of 1.05 to the K in Equation 4.21 only when $S \leq 2\%$.

Eliminating the depth of flow, d, between Equations 4.21 and 4.2, results in

$$Q_I = 5.23\frac{S_X^{0.302}}{S^{0.056}}(nQ)^{0.5775} \ SI \tag{4.22}$$

$$Q_I = 23.59\frac{S_X^{0.302}}{S^{0.056}}(nQ)^{0.5775} \ CU \tag{4.23}$$

where Q = gutter flow, m³/s (cfs).

These equations indicate that the flow interception capacity of an inlet increases with the gutter flow and cross slope of the gutter but decreases with the longitudinal slope. The ratio of the inlet flow to the gutter discharge is termed the inlet efficiency. This ratio represents the portion of the flow in gutter that is captured by an inlet; the remainder is the flow that bypasses an inlet and contributes to the gutter flow to the next downstream inlet. The efficiency of an inlet is written as

$$E = \frac{Q_I}{Q} = 5.23\frac{S_X^{0.302}}{S^{0.056}}\frac{n^{0.5775}}{Q^{0.4225}} \ \text{SI} \tag{4.24}$$

$$E = \frac{Q_I}{Q} = 23.6\frac{S_X^{0.302}}{S^{0.056}}\frac{n^{0.5775}}{Q^{0.4225}} \ \text{CU} \tag{4.25}$$

These equations demonstrate that the efficiency of an inlet decreases with increasing gutter flow. This relation is illustrated by Table 4.4, which is prepared based on Equation 4.24 and shows the

TABLE 4.4
Efficiency of "A" Inlet at Grade[a]

Q m³/s	S = 1%	S = 2%	S = 4%	S = 6%	S = 8%
		S_o = 4%, n = 0.016			
0.04	92	88	85	83	82
0.06	77	74	71	70	69
0.09	65	63	60	59	58
0.120	58	55	53	52	51
0.150	52	50	48	47	47

Q cfs	S = 1%	S = 2%	S = 4%	S = 6%	S = 8%	S = 10%
			S_x = 4%, n = 0.016			
2	79	76	73	72	70	70
3	67	64	62	62	59	59
4	59	57	55	53	53	52
5	54	52	50	49	48	47
6	50	48	46	45	44	44

[a] For types [a] B° and [a] D° inlets, the listed efficiencies in this table may be adjusted using a factor of 1.05 for $S \leq 2\%$.

inlet efficiency versus gutter flow and roadway slope. As this table indicates, an A inlet at slopes of 2% or greater provides less than 75% efficiency for a 0.06 m³/s gutter flow. Table 4.4, in, CU shows that A inlets placed at over 2% slope capture less than 75% of gutter flows of 2 cfs or more.

4.3.6 Grates on Sags

The flow at an inlet at low point occurs as weir flow until the water builds up to a certain depth. Beyond that depth, the inlet functions as an orifice. For bicycle-safe grates, NJDOT reports that inlets behave as a weir when flow depth is less than 9 in. (230 mm) and as an orifice for greater depths. For a conservative design, this author suggests using 15 cm (6 in.) depth as the criteria for differentiating these types of flow. The flow for each type can be calculated as follows:

a. Weir flow:

$$Q_i = C_w P d^{1.5} \tag{4.26}$$

$$C_w = 0.37(2g)^{0.5}, \text{ (see HEC-22)}$$

$$\text{Or } C_w = 1.64\ SI, 2.97\ CU;\ d \le 15\ cm\ (0.5\ ft)$$

where

C_w = adjusted weir coefficient
P = perimeter around the open area of grate, m (ft)
d = the depth of the approach flow, m (ft)

The perimeter for inlets at curbs and off curbs can be calculated from the following equations, respectively:

$$P = 2w + l$$

$$P = 2(w + l)$$

where w is the width and ℓ is the length of inlet grate in m (ft). NJDOT specifies a perimeter length of 5.18 ft (1.58 m) for A and B inlets and 6.96 ft (2.12 m) for C and D inlets.

b. Orifice flow:

$$Q_i = C_o A_o (2gd)^{0.5}\quad d \ge 15\ cm\ (0.5\ ft) \tag{4.27}$$

where

C_o = orifice coefficient
A_o = clear opening area of the inlet grate in m² (ft²)
d = depth of approach flow in m (ft)

This equation is applicable to both metric and customary units, with C_o = 0.6 and g = 9.81 m/s² (32.2 ft/s²).

NJDOT specifies clear opening areas (A_o) as 1.45 ft² (0.135 m²) for A, B modified, B1 modified, and B2 modified inlets and 2.90 ft² (0.269 m²) for B, D and E inlets (including B, B1, B2. C., D1, D2, E, and ES). B and D inlets have curb pieces that allow water to enter the inlets when the grate is partly clogged by debris. For inlets without curb pieces, A_o in the preceding equation should be reduced by 50% to account for partial clogging. The capture efficiency of a curb inlet is increased significantly by depressing the gutter at the inlet. Also, it is recommended to use curb opening

E inlets but preferably, combination inlets on sags. A procedure for proper design of inlets and drains has been presented by Pazwash and Boswell (2003).

4.4 INLET SPACING

The spacing of inlets on curbed gutters is determined based on the rate of flow in the gutter, the capacity of inlet, and the allowable spread. The first two parameters have been discussed already. The allowable spread is selected based on the degree of safety for the traffic. NJDOT (2015) has adopted the following criteria:

- Interstate highways and freeways: full shoulder
- Land service roads: shoulder + one-third width of lane
- Ramps: one-third width of ramp, for all roadways
- Acceleration/deceleration lanes: one-half width of lane, for all roadways

A proper procedure for spacing of inlets is

a. Calculate flow in the gutter. Include overland flow tributary to the roadway.
b. Calculate the spread in the gutter. Place the first inlet at the location where the spread approaches the allowable limit.
c. Calculate the capacity of the inlet and its efficiency. Adjust the inlet location, if necessary, to meet the required efficiency and calculate the bypassing flow.
d. Add the bypassing flow to the tributary runoff to the next inlet.
e. Repeat steps b through d to the end of the system.

Figure 4.14 shows a sample worksheet, prepared by the author, that can be used to tabulate the results in three sections.

Section 1 presents gutter flow calculations in six columns. The first five columns list the inlet number, the area tributary to an inlet, the runoff coefficient, the time of concentration, and rainfall intensity determined using local IDF curves. Then the rational formula (Q = CIA) is applied to the listed figures in columns 2, 3, and 5 to calculate the flow in the gutter.

The next section of the table includes the flow calculations in the gutter. The inputs for columns 7 and 8 are the flow bypassing the previous inlet (nonexistent for the first inlet) and the total flow in the gutter, respectively. The next two columns list the longitudinal slope and cross slope of the gutter. Column 11 presents the calculated spread width based on the listed slopes in the previous two columns using Equation 4.4. The calculated spread width is then compared with the allowable spread, T_a. If T is larger than T_a, then the inlet is relocated closer to the previous inlet, and Q and T are recalculated in columns 2 through 11.

The last section of the table examines the inlet efficiency versus its allowable limit. In column 13, the inlet capacity, the intercepted flow, is determined using appropriate figures in Appendix 4C. Alternatively, the intercepted flow can be calculated using acceptable local or state DOT equations, such as NJDOT's Equation 4.21. The ratio of intercepted inlet flow to the gutter flow is input in column 14, and this ratio is compared with the allowable efficiency. If the efficiency is acceptable, then the flow bypassing the inlet is calculated by subtracting the intercepted inlet flow from the gutter flow and entering the result in column 16 and also column 7 for the next inlet. However, if the inlet efficiency is not satisfactory, the inlet spacing is reduced and calculations are repeated from column 2 through column 15. The iteration process can be significantly reduced, or even eliminated, by examining Table 4.2 for flow spread, Appendix 4C for applicable figures for inlet capacity, or Table 4.4 for inlet efficiency.

The NJDOT (2015) drainage design in Roadway Design manual specifies 400 ft (125 m) as the maximum spacing between inlets. The same manual specifies that inlets have a minimum efficiency

Allowable spread: T_a = ft/m Manning's n = ________

Gutter flow						Gutter spread						Inlet efficiency			
Inlet no.	A	C	T_c	I	Q	Q_B	Q_T	S	S_x	T	$T < T_a$	Q_I	$E = Q_I/Q$	$E < E_A$	$Q_B = Q - Q_I$

FIGURE 4.14 Worksheet for tabulating calculated inlets spacing (blank form).

of 75%. The Residential Site Improvements Standards (NJDCA 2023), enacted in New Jersey in 1997, also specify the maximum spacing between inlets in a development as 400 ft. The same standards specify 6 cfs as the maximum capacity of a B inlet. As shown in the previous section, this figure far exceeds the interception capacity of an inlet at grade.

Example 4.4

In Example 4.1, apply the NJDOT equation for inlet capacity to calculate:

a. The flow captured by the first inlet on each shoulder
b. The efficiency of the inlet
c. The maximum spacing to the second set of inlets, accounting for the bypass flow from the first set.

Solution

In Example 4.1, Q = 1.74 cfs, S = 1.5%, S_x = 4%, T = 4.93 ft.

a. $d = TxS_x = 4.93 \times 0.04 = 0.197$ ft

$$Q_I = \frac{16.88d^{1.54}S^{0.233}}{S_x^{0.276}} = 16.88 \times 0.197^{1.54} \times \frac{0.015^{0.233}}{0.04^{0.276}} = 1.26\ cfs$$

Add 5% accounting for interception by the curb piece, Q = 1.32 cfs.

b. $E = \frac{Q_I}{Q} = \frac{1.32}{1.74} = 76\%$

Bypass flow = 1.74–1.32 = 0.42 cfs

c. Calculate gutter flow for 75% inlet efficiency by an iteration process.
First try: T = 5.0 ft

$$Q=\frac{0.56}{0.013}\times 0.04^{1.667}\times 0.015^{0.5}\times 5^{2.667}=0.02466\times 5^{2.667}=1.80\ cfs$$

$$d=S_xT=0.04\times 5=0.2\ ft$$

Calculate flow intercepted by the second set of inlets:

$$Q_i=\frac{16.88\times 0.20^{1.54}\times 0.015^{0.233}}{0.04^{0.276}}=15.425\times 0.2^{1.54}=1.29\ cfs$$

Add 5% for side interception

$$Q_i=1.05\times 1.29=1.35\ cfs$$

$$E=\frac{Q_i}{Q}=\frac{1.35}{1.80}=0.75\ OK$$

Allowable roadway runoff:

$$Q=1.80-0.42=1.38\ cfs$$

Maximum tributary area:

$$A=\frac{Q}{CI}=\frac{1.38}{0.95\times 6.7}=0.217\ acres=9444\ sf$$

$$L=\frac{9444}{34}=278\ ft$$

Place second set of inlets 275 ft from the first inlets.
Show that for a roadway with n = 0.016, the spacing of a second set of inlets in the preceding example can be increased to 360 ft.

Example 4.5

Solve the previous example using 2 ft × 4 ft P-1–1/8 FHA grates and Figure 4C.2b (n = 0.016).

Solution

Enter Figure 4C.2b with Q = 1.74 cfs, and S = 1.5% to read

$$Q_i=1.5\ cfs$$

$$E=\frac{Q_i}{Q}=\frac{1.5}{1.74}=0.86:86\%$$

$$Bypass\ runoff=1.74-1.5=0.24\ cfs$$

Note: Due to a higher roughness coefficient, the flow spread is wider; therefore, the inlet has greater efficiency than the previous example. Also, the FHA P1–1–1/8 grate provides a larger opening area than the NJ type B bicycle-safe grates. As such, the second set of inlets can be spaced further than that of the previous example.

First try: T = 6 ft (entire side shoulder)

$$Q=(0.56/0.013)\times 0.04^{1.667}\times 0.015^{0.5}\times 6^{2.667}=2.38\ cfs$$

Enter the same figure with Q = 2.38, S_o = 0.015 to read:

$$Q_i=1.97\ cfs$$

$$E = \frac{1.97}{2.38} = 82.8\%$$

Contributing roadway runoff $Q = 2.38 - 0.24 = 2.14$ cfs

$$Area = \frac{2.14}{(0.95 \times 6.7)} = 0.336\ acre = 14{,}645\ sf$$

$$L = \frac{14{,}645}{34} = 430\ ft$$

Set the second set of inlets at the maximum allowable spacing of 400 ft from the first set.

Example 4.6

Using the calculated discharge in Example 4.2, calculate:

a. The flow captured by the first inlet on each shoulder
b. The efficiency of the inlet
c. The maximum spacing to the second inlet for 75% efficiency

Base your calculations on 0.6 m × 1.2 m FHA P-30 grate ($n = 0.016$).

Solution

Given

$$Q = 0.0476\ \text{m}^3/\text{s};\ \text{say},\ 0.048\ \text{m}^3/\text{s}$$

$$S_x = 4\%, S = 1.5\%, \Sigma W = 10.6\ \text{m}$$

Check efficiency: Enter Figure 4C.2a $Q = 0.048$ m³/s and $S = 1.5\%$ and interpolate between $Q = 0.03$ and $Q = 0.05$ to read

$$Q = 0.041 \quad \text{and} \quad E = 86\%$$

Check efficiency:

$$E = 0.041/0.048 = 85.4\% \quad \text{OK}$$

Bypass flow:

$$Q_b = 0.048 - 0.041 = 0.007\ \text{m}^3/\text{s}$$

Try $T = 2$ m for the second set of inlets.

Calculate Q using Equation 4.1:

$$Q = (0.375/0.016) \times 0.04^{5/3} \times 0.015^{1/2} \times 2^{8/3}$$

$$Q = 0.085\ \text{m}^3/\text{s}$$

Enter the same figure with $Q = 0.085$ and $S = 0.015$, to read:

$$E = 81\% > 75\% \quad \text{OK}$$

Contributing roadway runoff:

$$Q = 0.085 - 0.007 = 0.078\ \text{m}^3/\text{s}$$

$$\text{Area} = Q / \left(2.78 \times 10^{-3} \times CI\right)$$

$$\text{Area} = 0.078 / \left(2.78 \times 10^{-3} \times 0.95 \times 170\right) = 0.1737\ \text{ha} = 1737\ \text{m}^2$$

$$L = 1737 / 10.6 = 163.9\ \text{m}$$

Set the second inlets at a maximum allowable distance of 120 m (<400 ft).
Note that Figures 4C.1 and 4C.2 are not accurate enough for efficiency calculations.

4.5 INLETS ON ROADWAYS AT 0% GRADE

Roadways traversing through flat lands can be at zero or nearly zero longitudinal grade. A section of the New Jersey Turnpike traversing through Meadowlands is an example where the roadway is at zero or nearly zero longitudinal slope. The gutter flow in such roadways is not governed by Equation 4.1 because the longitudinal slope is zero. Rather, the water impounds on the shoulder as depicted in Figure 4.15 and flows toward the inlet by gravity. This figure presents the flow parameters for a road consisting of lanes of total width *W*, shoulder width *W*′, and inlet spacing *L*. It is assumed that no offroad runoff enters the shoulders. The flow enters the shoulder laterally from the traffic lanes; thus, the flow increases linearly toward inlets. However, the flow depth increases away from inlets and reaches a maximum, y_{max}, midway between two consecutive inlets. Pazwash (2011) developed the following procedure for calculating the inlet spacing for roadways at zero % grad.

Approximating the flow on the shoulder as a one-dimensional flow, this author has developed equations for flow spread (Pazwash, 2011). Appendix 4B presents the deviation for the water surface equation with the results:

$$y^{19/3} = y_o^{19/3} + Kn^2 S_x^2 q^2 \left[\left(\frac{L}{2}\right)^3 - x^3\right] \tag{4.28}$$

where
- q is the gutter flow per unit length of roadway
- y_o is the depth of flow at the inlet
- x and y are the distance from the water divide and corresponding depth of water, respectively

$$K = 21.3\ SI; K = 10\ CU$$

The depth of flow at the inlet can be calculated using the weir flow equation:

$$y_o = \left(\frac{qL}{C_w P}\right)^{2/3} \tag{4.29}$$

where
- $P = l + 2W$ (l = length, W = width of inlet)
- C_w = weir coefficient

At $x = 0$, where the depth of water attains its maximum, the equation reads as

$$y_{max}^{19/3} = y_o^{19/3} + KS_x^2 n^2 q^2 \left(\frac{L}{2}\right)^3 \tag{4.30}$$

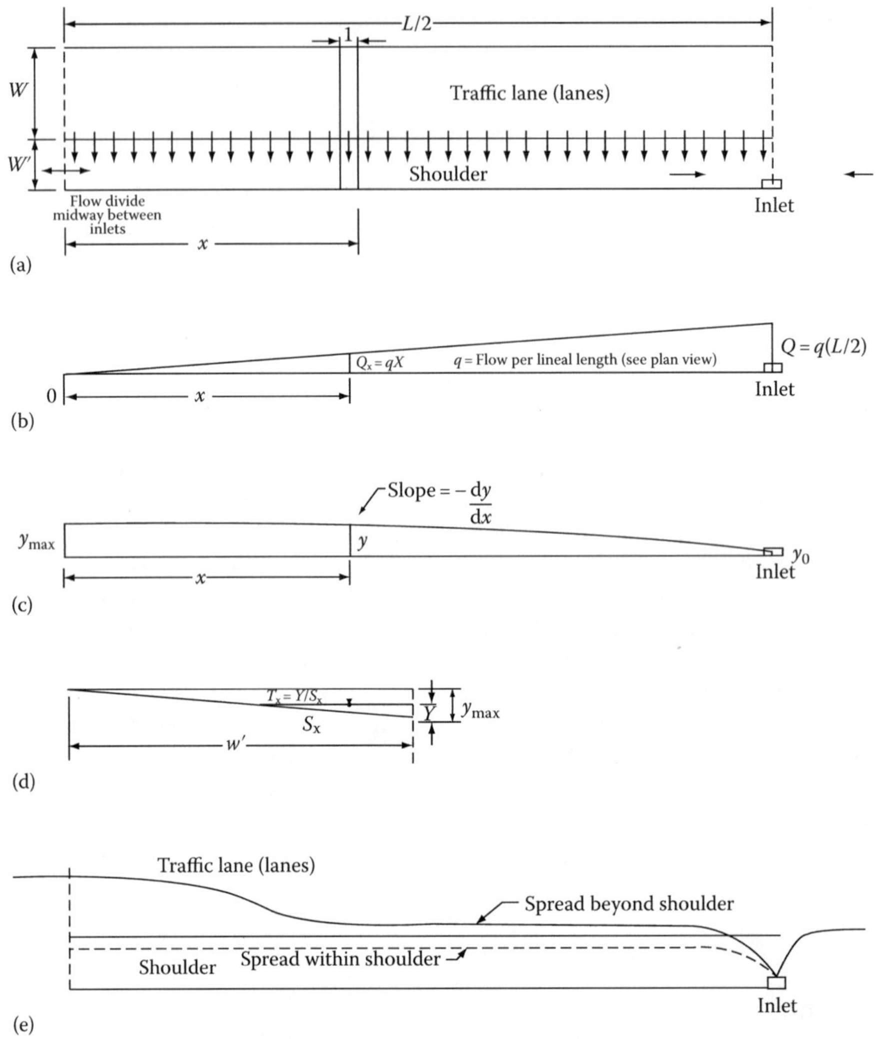

FIGURE 4.15 Gutter flow for a roadway of horizontal profile: (a) plan view; (b) cutter flow vs. distance; (c) variation of depth of flow; (d) flow cross-section; and (e) flow spread.

Rewriting the preceding equation in terms of L:

$$L = 2\left\{ \frac{\left[\left(y_{\max} \right)^{19/3} - \left(y_{o} \right)^{19/3} \right]}{K S_x^2 n^2 q^2} \right\}^{1/3} \tag{4.31}$$

The maximum inlets spacing L can be calculated by setting $y_{max} = S_x W'$, where W' is the shoulder width (assumed to be the allowable spread). Since y_o varies with L, this equation has to be solved by trial and error to determine the maximum spacing between inlets that would keep the water spread within the shoulder. To facilitate calculations, the term containing y_o may be neglected in the preceding equation, resulting in

$$L_{max} = 2\left[\frac{\left(S_x W'\right)^{19/3}}{K S_x^2 n^2 q^2}\right]^{1/3} \tag{4.32}$$

The preceding equations neglect the effects of debris and clogging on the inlets. In the case of uniformly spaced inlets, the full clogging of one inlet will increase the length of flow spread from one-half of the inlets' spacing to full spacing from the clogged inlet. Thus, the maximum spacing between inlets calculated by Equation 4.32 should be reduced by one-half. Further, to account for the depth of flow at inlets, the author recommends that the maximum spacing between inlets be additionally reduced by 20%. Thus,

$$L_{max} = 0.8\left[\frac{\left(S_x W'\right)^{19/3}}{K S_x^2 n^2 q^2}\right]^{1/3} \tag{4.33}$$

Example 4.7

Calculate the maximum spacing between inlets for a flat roadway having the following parameters:

$$S_x = 4\%$$

$$W = 22\ ft(6.8\ m);\ W' = 8\ ft(2.5\ m)$$

$$n = 0.016$$

$$I = 6.7\,\text{in./h}(17\ \text{mm/h}); 25-\text{year}, 10-\text{minute storm in New Jersey}$$

Solution

The maximum spacing for the inlets in this case is calculated in both CU and SI as follows:

$$\Sigma W = 22 + 8 = 30\ \text{ft}; 6.8 + 2.5 = 9.3\ \text{m}$$

$$q = 30 \times 6.7 / 43{,}560 = 4.61 \times 10^{-3}\text{cfs/ft}$$

$$q = 2.78 \times 10^{-3} \times 9.3 \times 170 / 10^4 = 4.40 \times 10^{-4}\text{m}^3/\text{m}$$

$$y_{max} = S_w W' = 0.03 \times 8 = 0.24\ \text{ft}; 0.03 \times 2.5 = 0.075\ \text{m}$$

$$L_{max} = 107\ \text{ft};\ 34\ \text{m}$$

This result shows that inlets must be closely spaced even in the absence of any offsite runoff. When the offroad runoff is present and significant, it is more feasible to use trench drains such as ACO highway drains or slot drains than grate inlets in flat roadways and pavements.
Note, q is calculated using a runoff coefficient = 1.

4.6 DESIGN OF STORM DRAINS

Storm drainpipes are designed to provide sufficient capacity to convey the calculated peak rate of runoff. Commonly, the pipes are designed based on an inlet location plan, which logically should be prepared following inlet spacing calculations. In many development projects, however, an inlet location plan is prepared before any inlet efficiency calculations are performed. The peak rate of runoff to each inlet is generally calculated using the rational formula:

$$Q = 2.78 \times 10^{-3}\ CIA\ \ SI \tag{4.34}$$

$$Q = CIA\ \ CU \tag{4.35}$$

The parameters in this equation as defined in Chapter 3 are
A = area, hectares (acres)
C = runoff coefficient
I = rainfall intensity, mm/h (in./h)
Q = peak runoff, m^3/s (cfs)

For the first reach of pipe, it is assumed that the pipe receives all of the runoff tributary to the most upstream inlet. This is a conservative assumption considering that inlets might not fully capture the gutter flow. The design flow for the pipe downstream of a second inlet is calculated by adding the runoff from the tributary area to that inlet and the discharge in the preceding pipe. The calculation continues to the most downstream pipe that terminates at a stream, detention basin, receiving drainage system, or body of water.

The results can be tabulated. The USDOT and many state highway agencies, including the NJDOT (2015), have tables for pipe sizing calculations. Figure 4.16 is a spreadsheet prepared by the author for recording the results calculated for inlet and pipe reach description, flow calculations, pipe sizing calculations, and the depth of cover over pipes.

The inlet calculations section, which follows the inlet and pipe description section, organizes runoff discharges calculated using the rational method. This section, which shows the gutter flow at each inlet, is similar to the calculations presented in Figure 4.14. The third section presents calculations for design flow of each pipe and compares that flow with the capacity of the pipe, which is calculated using the Manning formula. The Manning's n values for various types of pipes can be found in Table 2.3 (see Chapter 2). Also included in this section are the calculations for the flow velocity assuming full flow condition in each pipe.

The tables in Appendix 2A list the flow capacity factors for round pipes and elliptical concrete pipes; the actual velocity depends on the ratio of the design discharge to the pipe capacity. Table 2.6 and Figure 2.5 facilitate the calculations for partial flow in round pipes. Figure 2A.1 shows the part-full flow parameters for elliptical pipes.

The last section in the table lists the invert elevations at the upstream and downstream ends of a pipe reach and the depth of cover over the pipe, accounting for the wall thickness of the pipe. The calculated cover is then compared with the minimum required cover specified by the pipe manufacturer for a given loading on the pipe to ensure that the pipe has adequate cover. The following case study illustrates the calculation process.

Note that calculations in Figure 4.16 provide a preliminary design of drainage pipes. In final design, the hydraulic grade line (HGL) calculations are also performed to determine that no surcharge from any inlet occurs. These calculations include not only frictional losses but also entrance and exit losses and local losses at structures and bends. Accounting for these losses, the flow in many pipes that according to the preliminary calculations would be part full could be under pressure due to the backwater effect. The calculations process is illustrated in Case Study 4.2.

CASE STUDY 4.1

This study relates to a drainage improvement project in the borough of New Milford in New Jersey. Figure 4.17 shows the existing drainage system in the project area. Because of inadequacy of the drainage pipes and an obstruction due to a sanitary sewer crossing through MH#2, the existing manhole just to the east side of the Lake Street-New Milford Avenue intersection was experiencing periodic flooding. To mitigate flooding at this manhole, a new drainage system was

Type of pipes: ______ Manning's n: ______

Pipe section				Inlet calculations					Storm drain calculations										Elevations and cover			
No	From	To	Length (ft)	Acres A (acres)	Runoff coef C	Conc. time (min)	Rain intensity (in/h)	Inlet Q (cfs)	Sum A × C	Total time (min)	Rain intensity (in/h)	Pipe discharge Q (cfs)	Slope (%)	Diameter (in)	Full flow capacity (cfs)	Normal depth (in)	Normal velocity (ft/sec)	Travel time (min)	Ground elev at upper end (ft)	Invert elev upper end (ft)	Invert elev lower end (ft)	Cover at upper elevation (ft)

FIGURE 4.16 Worksheet for recording storm drain design calculation results

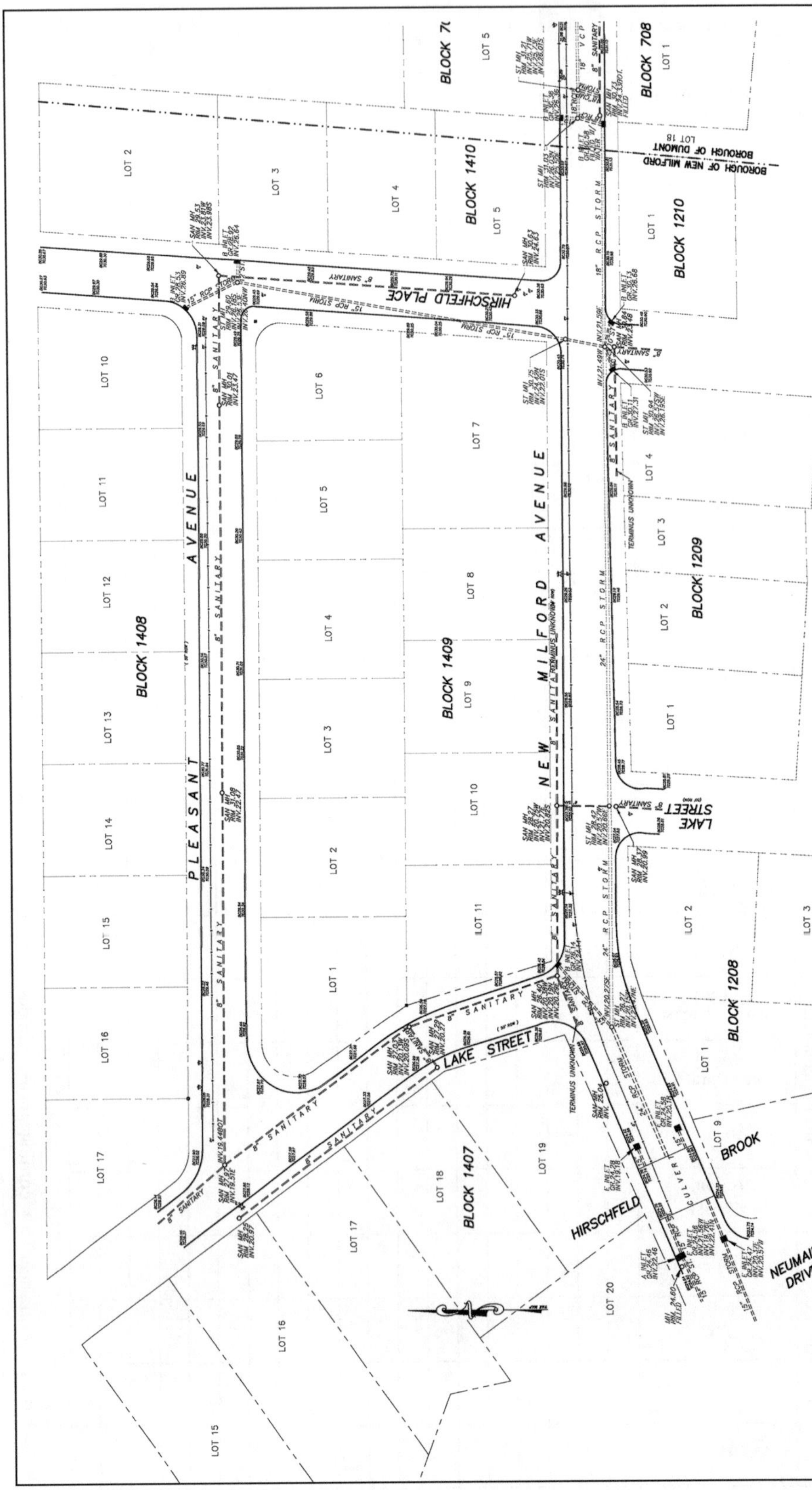

FIGURE 4.17 Existing condition drainage system map, New Milford Avenue.

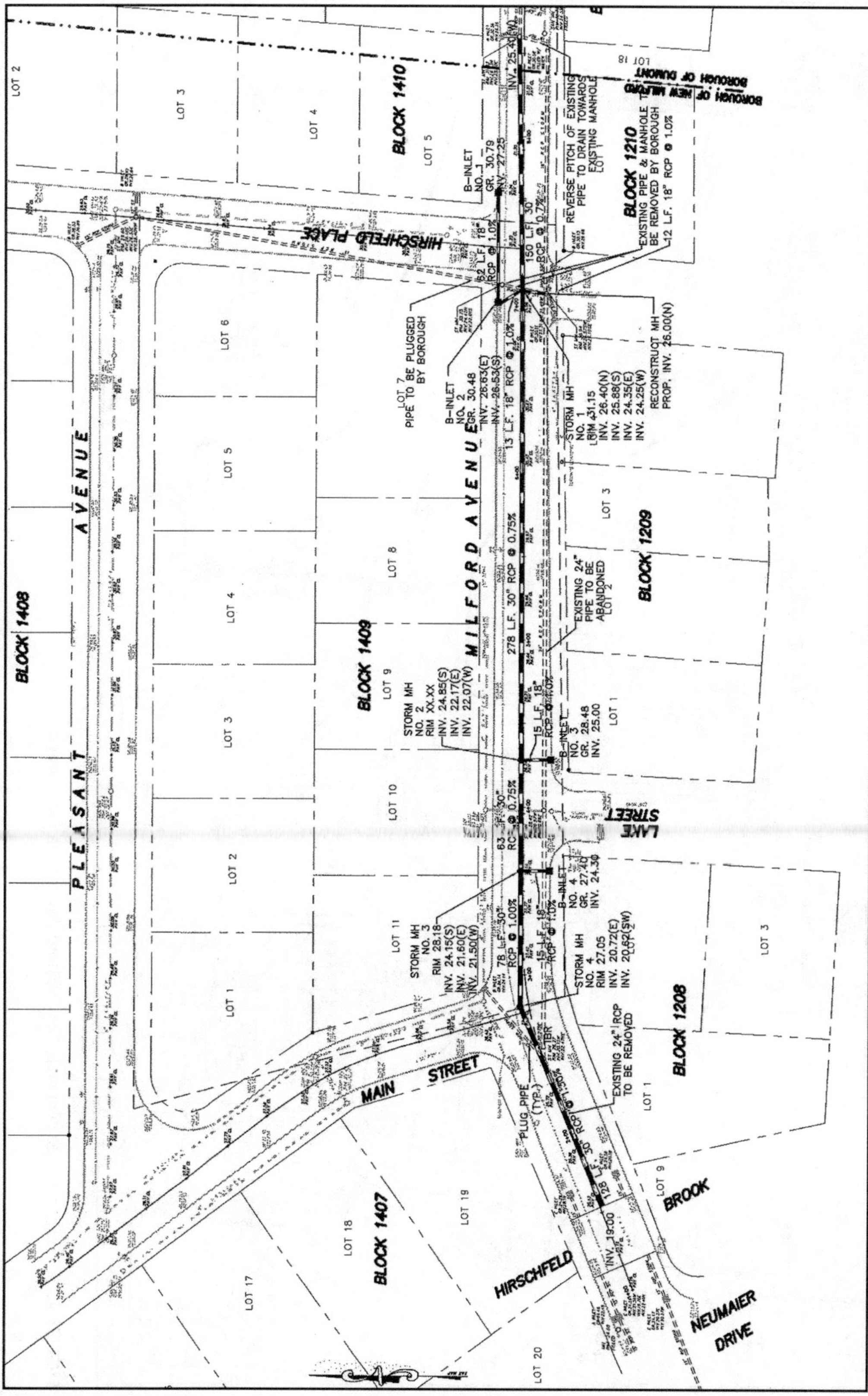

FIGURE 4.18 Proposed drainage system, New Milford Avenue.

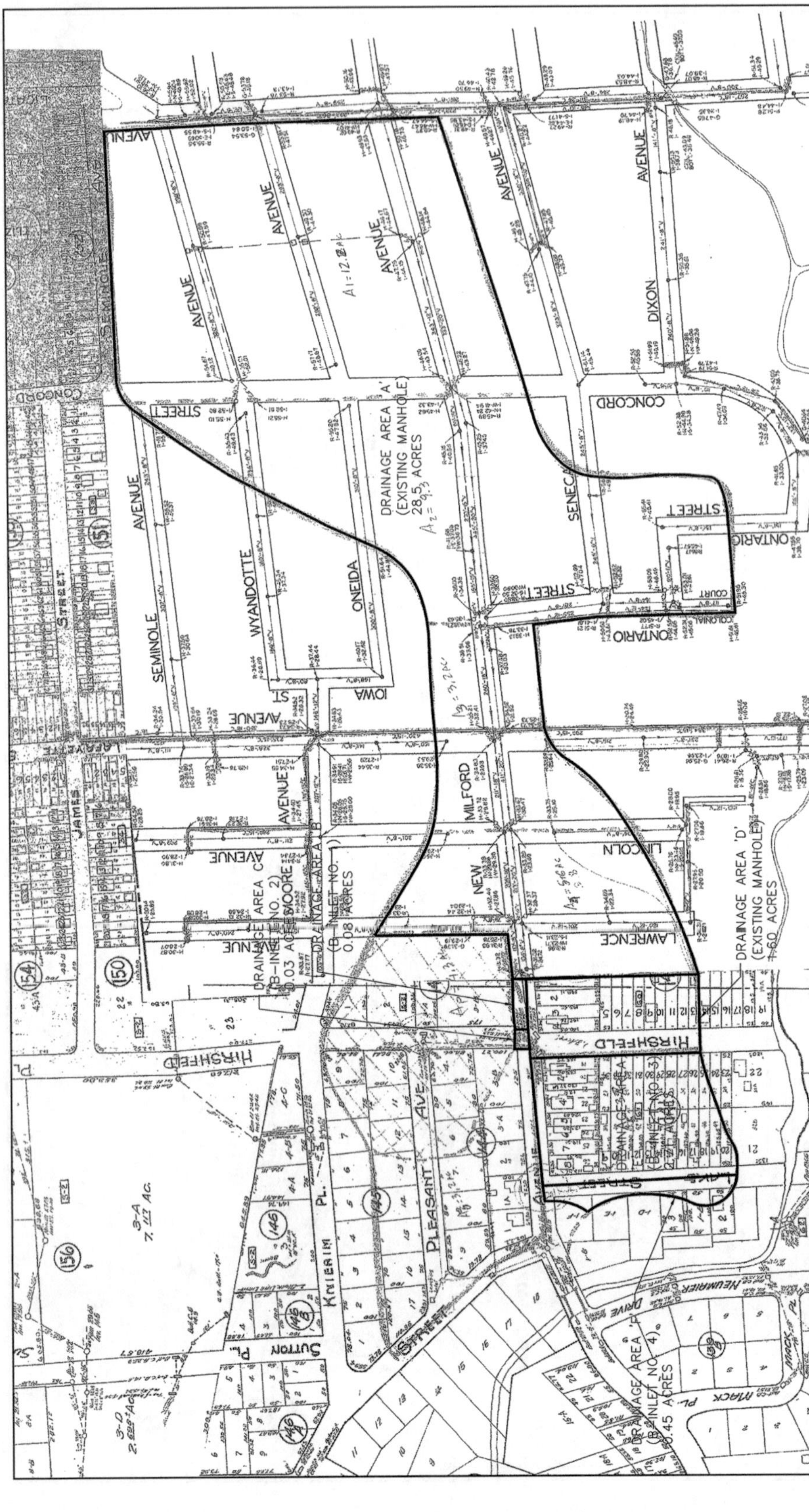

FIGURE 4.19 Inlet drainage area map, New Milford Avenue drainage system.

Pipe section				Inlet calculations					Storm drain calculations										Elevations and cover			
No	From	To	Length (ft)	Acres A (acres)	Runoff coef C	Conc. time (min)	Rain intensity (in/h)	Inlet Q (cfs)	Sum $A \times C$	Total time (min)	Rain intensity (in/h)	Pipe discharge Q (cfs)	Slope (%)	Diameter (in)	Full flow capacity (cfs)	Normal depth (in)	Normal velocity (ft/sec)	Travel time (min)	Ground elev at upper end (ft)	Invert elev upper end (ft)	Invert elev lower end (ft)	Cover at upper elevation (ft)
1	EX MH	MH 1	150	28.50	0.46	60	2.4	31.46	13.110	60.00	2.4	31.46	0.70%	30	37.17	20.83	6.51	0.38	31.03	25.40	24.35	3.1
2	CB 1	CB 2	40	0.08	0.9	10	6.7	0.48	0.072	10.00	6.7	0.48	1.00%	18	11.38	2.61	3.19	0.21	30.57	27.25	26.85	1.8
3	CB 2	MH 1	16	0.03	0.9	10	6.7	0.18	0.099	10.21	6.7	0.66	1.00%	18	11.38	3.11	3.48	0.08	30.48	26.75	26.59	2.2
4	EX MH	MH 1	12	1.60	0.46	20	4.6	3.39	0.736	20.00	4.6	3.39	1.00%	18	11.38	6.66	5.54	0.04	30.94	26.00	25.88	3.4
5	MH 1	MH 2	334					0.00	13.945	60.38	2.4	33.47	0.75%	30	38.48	21.60	6.74	0.83	31.15	24.25	21.74	4.4
6	CB3	CB4	38	2.71	0.46	30	3.6	4.49	1.247	30.00	3.6	4.49	2.00%	18	16.09	6.39	7.70	0.08	28.45	25.00	24.24	2.0
7	CB4	MH2	16	0.45	0.46	10	6.7	1.39	1.454	30.08	3.6	5.23	1.00%	18	11.38	8.64	5.54	0.05	27.54	24.14	23.98	1.9
8	MH2	MH3	88					0.00	15.399	61.21	2.4	36.96	1.00%	30	44.43	20.66	7.78	0.19	28.18	21.64	20.76	4.0
9	CB5	MH3	22	10.43 cfs from Lake St from Borough improvement calculations								10.43	2.00%	18	16.09	10.53	7.83	0.05	26.14	21.80	21.36	2.8
10	MH3	CULV	128						15.399	61.40	2.4	47.39	1.30%	30	50.66	25.80	8.88	0.24	27.05	20.66	19.00	3.9

FIGURE 4.20 Storm drain design (design storm = 25 years, Manning's **n** = 0.012).

Pipe reach				Head losses											
From	To	L (ft)	D (ft)	Q (cfs)	V (ft/s)	$V^2/2g$ (ft)	R	h_f	K_s	h_s (ft)	K_b	h_b (ft)	h_e (ft)	h_{xt} (ft)	H_t (ft)
MH3	CUL	128.0	2.5	47.4	9.7	1.46	0.625	1.45	0.0	0.0	0.0	0.0	0.7	1.46	3.62
MH2	MH3	88.0	2.5	37.0	7.5	0.9	0.6	0.6	0.3	0.3	0.0	0.0	0.3	0.0	1.3
MH1	MH2	334.0	2.5	33.5	6.8	0.7	0.6	1.9	0.3	0.2	0.0	0.0	0.4	0.0	2.5
EX MH	MH1	150.0	2.5	31.5	6.4	0.6	0.6	0.7	0.3	0.2	0.0	0.0	0.3	0.0	1.3

Note:

1. $h_f = (29n^2L/R^{1.33})V^2/2g$ Friction loss
2. $h_s = K_sV^2/2g$ Structural loss ($K_s = 0.3$ pressure flow, $K_s = 0$ open channel flow)
3. $h_b = K_bV^2/2g$ K_b – Bend factor = 0, no bend
4. $h_e = KV^2/2g$ Entrance loss ($K = 0.5$ square edge)
5. $h_{xt} = V^2/2g$ Exit loss
6. $H_t = h_f + h_s + h_b + h_e + h_{xt}$ Total loss

FIGURE 4.21 Calculated head losses at the New Milford Avenue project site (**n** = 0.012).

designed. The proposed system is shown in Figure 4.18. As shown, the existing 24 in. pipe would be replaced by a 30 in. pipe and this pipe would be dropped by over 2 ft at manhole (MH2) in order to clear the sanitary sewer that traverses through the manhole in front of the 18 in. pipe.

The drainage area to the proposed system is delineated based on a review of the USGS Hackensack quadrangle and site visits. Figure 4.19 shows the tributary drainage areas to the new inlets. Based on a review of tax maps, Google aerial maps, and site visits, the runoff coefficient was estimated at $C = 0.46$. Figure 4.20 presents calculations for the storm drain pipes, which are designed based on a 25-year frequency storm.

CASE STUDY 4.2

Calculate the hydraulic grade line (HGL) in the proposed 30-inch pipe of Case Study 4.1.

SOLUTION

First, frictional and structural losses in the 30 pipe reaches and structures (manholes and inlets) are calculated. The losses in structure are due to changes in velocity in the structure and also because of changes in flow direction at bends. Figure 4.21 summarizes head loss calculations for the proposed 30 in. pipes and manholes thereon.

Frictional losses in the pipes are calculated based on Manning's formula, which as shown in Chapter 2 may be written as:

$$h_f = \left(29n^2L / R^{1.33}\right)V^2 / 2g \quad (4.36)$$

where n is Manning's roughness and L and R are the length and hydraulic radius of each pipe reach, respectively. The losses in structures and bends are calculated applying an adjustment factor to the velocity head. The losses also include entrance and exit losses. The former loss occurs at each structure and depends on the type of connection (a coefficient $K_e = 0.5$, which is typical for square edge is conservatively used herein). However, the exit loss only occurs where the pipe terminates at a stream, lake, or any stagnant body of water, and $K_{ex} = 1.0$.

Figure 4.22 presents calculations for hydraulic grade line for the proposed 30 in. pipe along New Milford Avenue. The water surface elevation at MH3 on the 30 in. pipe is calculated based on inlet control at the manhole and outlet control due to losses in the first reach of pipe (refer to

Pipe reach		Q	H_t	T_w	$T_w + T_t$	H_w	Inv. el.	I.C. elev.	HGL	T.O.S.	CL
From	To	(cfs)	Total head loss (ft)	Tailwater elev. (ft)	(ft)	Inlet control headwater (ft)	Upstream inv. elev. (ft)	Inlet control elev. (ft)	Headwater elev. (ft)	Top of structure elev. (ft)	Clearance
MH3	CUL	47.4	3.6	21.4	25.0	5.1	20.7	26.5	26.5	27.1	0.60
MH2	MH3	37.0	1.3	26.5	27.3	3.7	21.6	25.3	27.3	28.2	0.9
MH1	MH2	33.5	2.5	27.3	29.8	3.3	24.3	27.6	29.8	31.2	1.4
EX MH	MH1	31.5	1.3	29.8	31.1	3.0	25.4	28.4	31.1	31.0	−0.1

Note:
1. H_t = Total losses (See Figure 4.23)
2. T_w = Tailwater or $(D + d_c/2)$ (whichever greater for first pipe)
3. I.C. elev. = H_w + Inv. el.
4. HGL = $T_w + H_t$ or I.C. elev. (whichever is greater)
5. H_w is calculated using orifice flow equation

FIGURE 4.22 Hydraulic grade line calculations.

Figure 4.21). Calculations for inlet–outlet control for the pipe are elaborated in an example in the next section (see Example 4.9).

Figure 4.22 indicates that flow in each pipe reach is under the backwater effect of downstream pipe and that all pipes are completely full. The table also shows that the HGL rises nearly 0.1 ft (less than 1 in.: the calculations are rounded to the first decimal place) at the existing manhole. However, no surcharge would actually occur at this manhole as the 30 in. pipe has significantly more capacity than the existing 24 in. pipe that precedes this manhole and therefore will receive less flow than the calculated 47.4 cfs in the previous example.

4.7 HYDRAULIC DESIGN OF CULVERTS

A culvert is a short reach of conduit employed to convey the flow of a stream, drainage channel, or swale under a roadway or railway. The flow to a culvert can occur with its inlet and outlet faces either being fully submerged or partly open. Also, one of the faces can be partly open while the other face is submerged.

The capacity of a culvert depends on the flow that the inlet face of the culvert can accept as well as the flow that the culvert can pass under the force of gravity. Depending upon which one of these capacities limits the discharge, the culvert flow is labeled accordingly. Specifically, the culvert flow is termed as inlet control if the inlet capacity is smaller than the gravity-driven capacity. On the other hand, the outlet control condition prevails if the inlet capacity is not the limiting factor. Figure 4.23 depicts flow conditions where the inlet capacity is less than the gravity-flow capacity, and thus the flow occurs as inlet control. Figure 4.24 shows outlet control cases where water can enter the culvert at a greater rate than the culvert can freely convey.

Regardless of the submergence at the inlet or outlet facade of a culvert, both the inlet and outlet control capacities should be calculated and the smaller of the two used in design. The inlet control capacity is basically the capacity of the opening face, which acts as a weir flow when partly full and as an orifice when fully submerged. The flow equations for these cases are

$$Q = C_w L H^{1.5} \tag{4.37}$$

$$Q = A(2gh)^{0.5} \tag{4.38}$$

where

L= span of culvert, m (ft)
H = head above culvert invert, m (ft)
h = head above center of opening, m (ft)

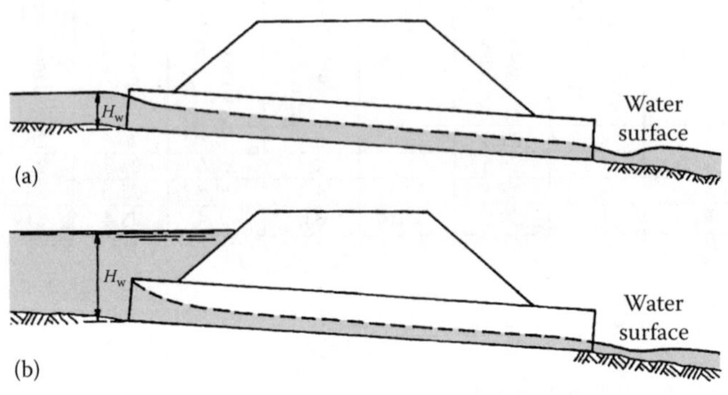

FIGURE 4.23 Inlet control schemes: (a) inlet façade open, (b) inlet face submerged.

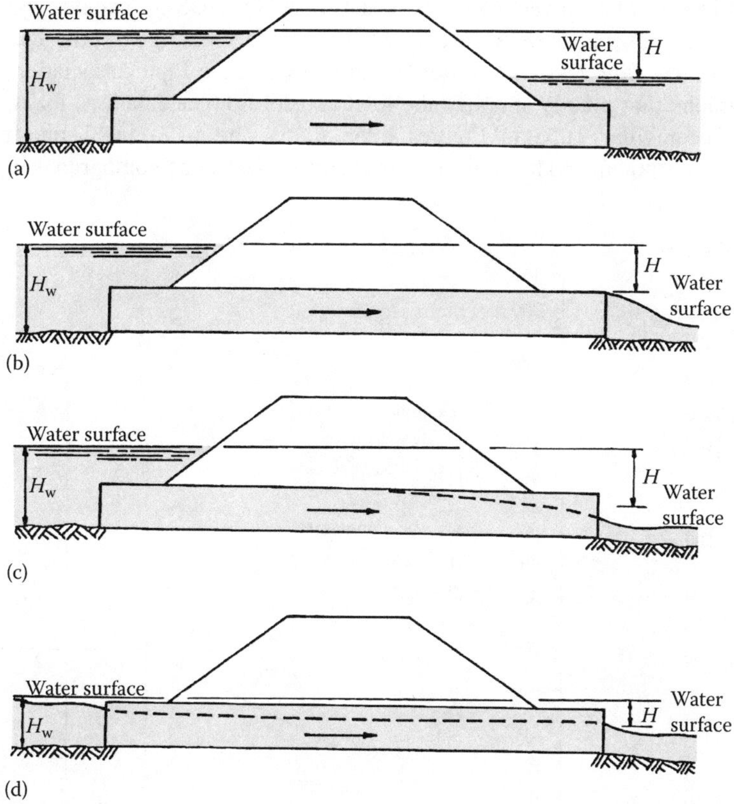

FIGURE 4.24 Outlet control schemes: (a) both inlet and outlet facades submerged; (b) inlet submerged, outlet full; (c) inlet submerged, outlet partly open; (d) inlet and outlet both partly open.

A – area of barrel, m^2 (ft^2)
C_w= weir coefficient
C_o= orifice coefficient

The weir coefficient C_w and orifice coefficient C_o vary with the type of culvert, its geometry, and entrance face configurations.

The outlet control capacity can be calculated using the Manning's formula for frictional losses and accounting for the entrance and exit losses and calculate the energy as follows:

$$H = \left(\sum K_\ell + \frac{16.62n^2 L}{R^{4/3}}\right)\left(\frac{V^2}{2g}\right) \text{ SI} \tag{4.39}$$

$$H = \left(\sum K_\ell + \right)\left(\frac{V^2}{2g}\right) \text{ CU} \tag{4.40}$$

In these equations,
V = full flow velocity in culvert, m/s (ft/s)
n = Manning's roughness coefficient
L= length of culvert, m (ft)
R= hydraulic radius, m (ft)
ΣK_ℓ = sum of entrance and exit loss coefficients

The Hydraulic Design of Highway Culverts (FHWA, 2012), HDS-5, includes rigorous parametric equations for various makes of culverts and their inlet face configurations. The same publication and the *Concrete Pipe Design Manual* (American Concrete Pipe Association, ACPA, 2007) include nomographs that greatly simplify the inlet control flow calculations for round, elliptical, arch, and box (rectangular) culverts. Figures 4.25a, 4.26a, and 4.27a, in SI, present inlet control capacity of round, elliptical, and box culverts, respectively. The same nomographs in CU are shown in Figures 4.25b–4.27b.

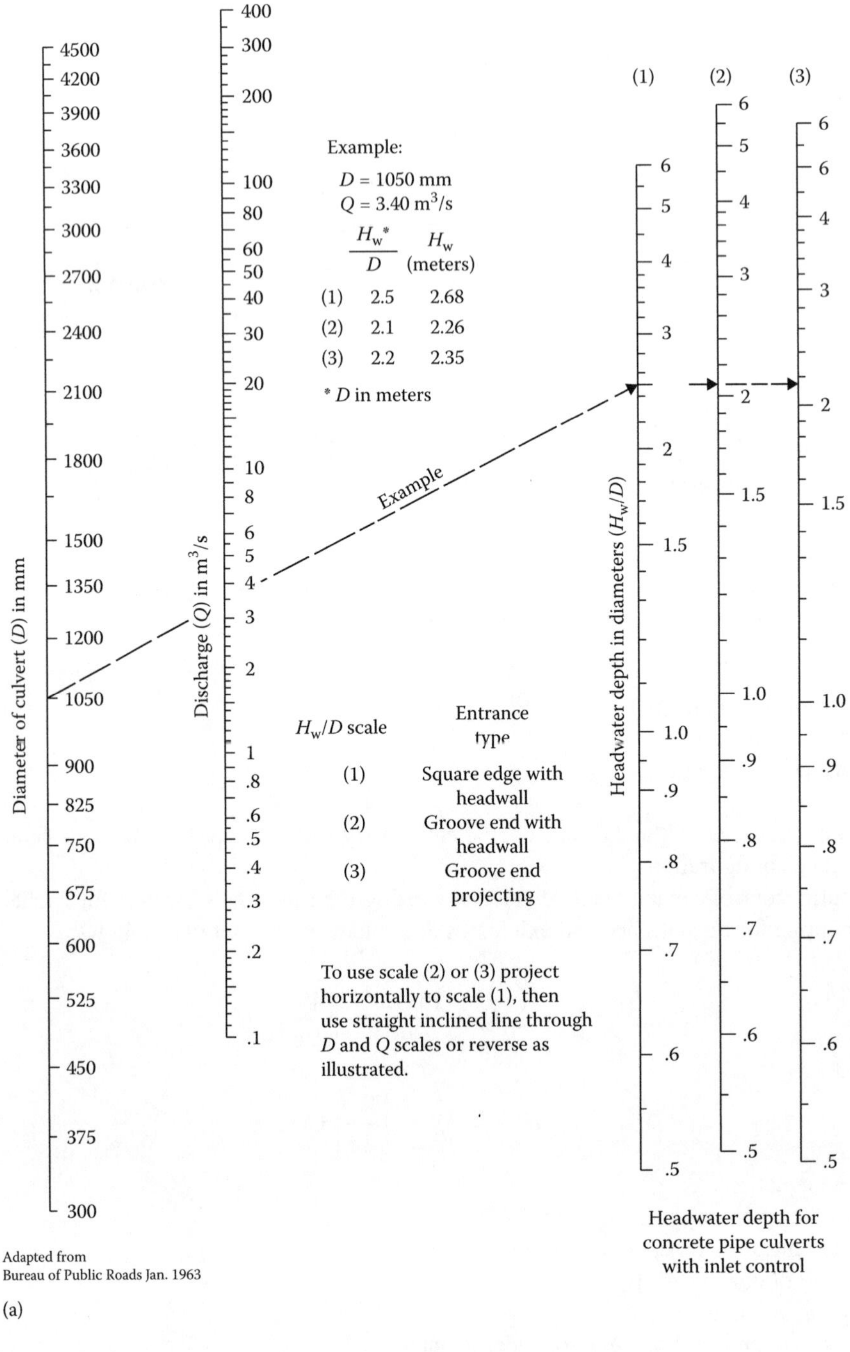

FIGURE 4.25A Inlet control nomograph, round culverts (SI).

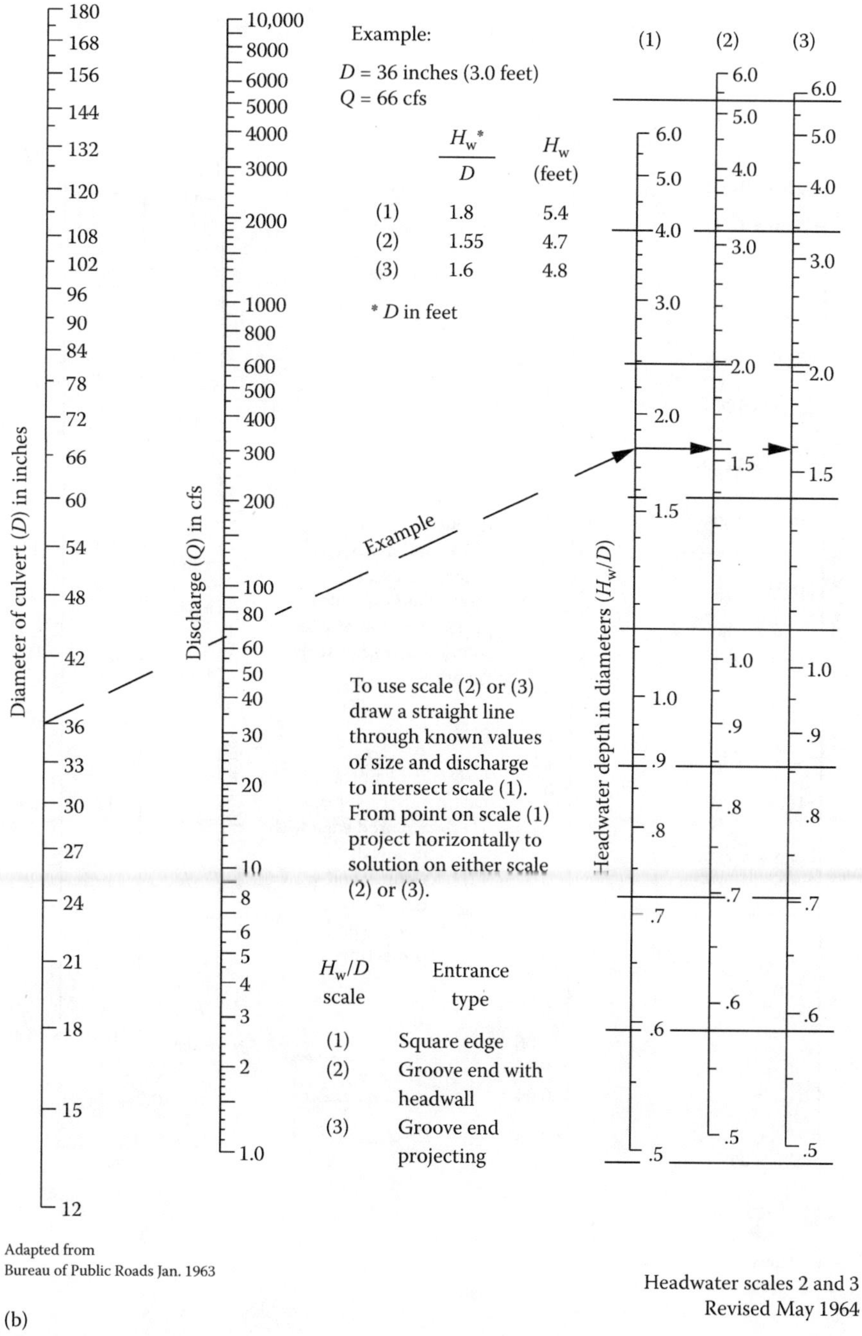

(b)

FIGURE 4.25B Inlet control nomograph, round culverts (CU).

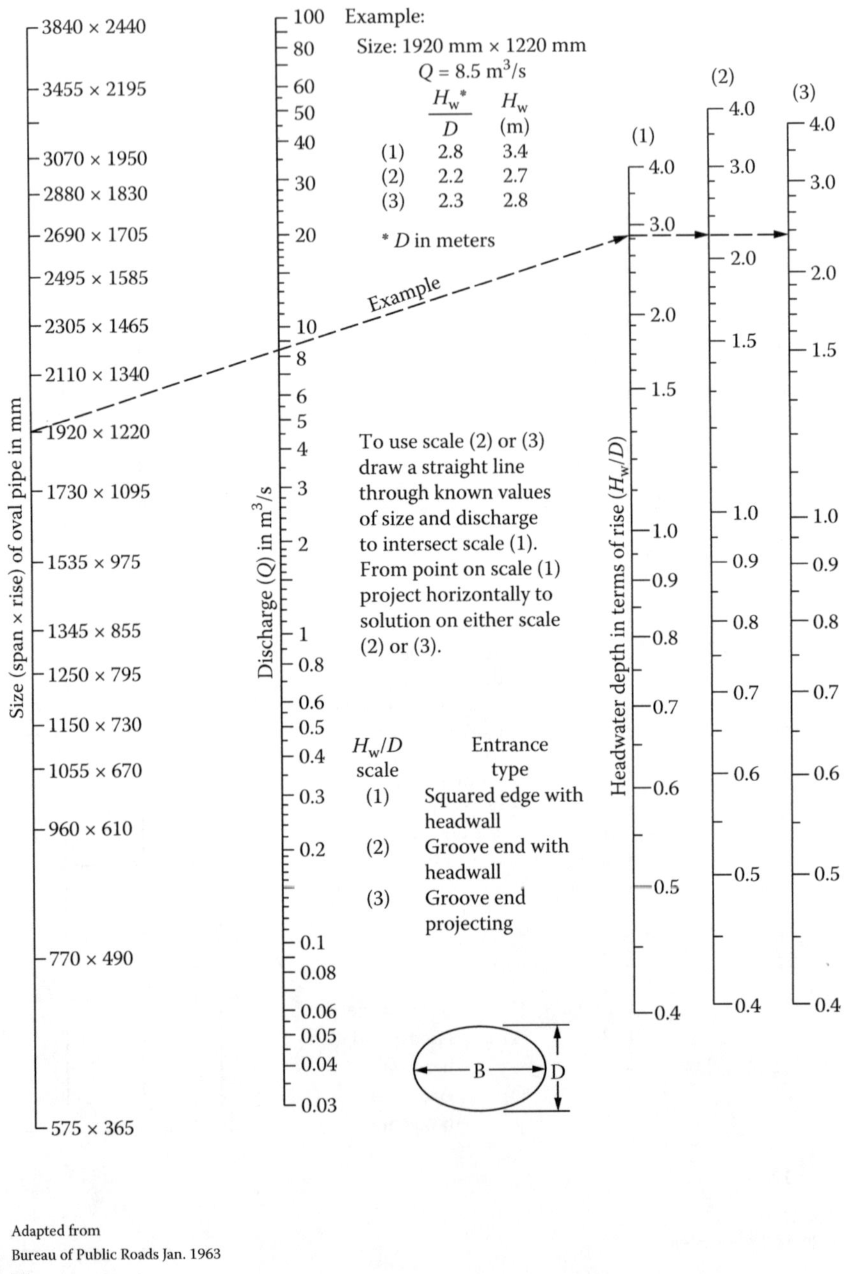

FIGURE 4.26A Inlet control nomograph, elliptical pipes (SI).

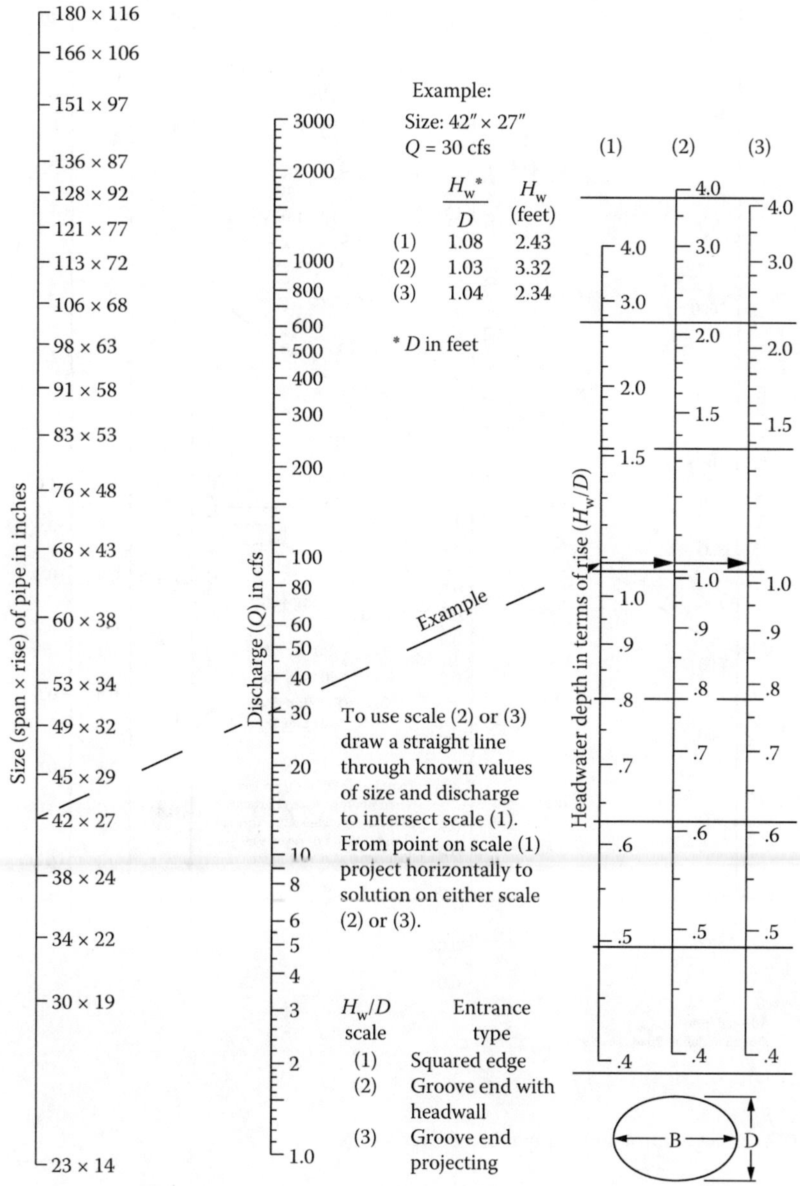

Adapted from
Bureau of Public Roads Jan. 1963
(b)

FIGURE 4.26B Inlet control nomograph, elliptical pipes (CU).

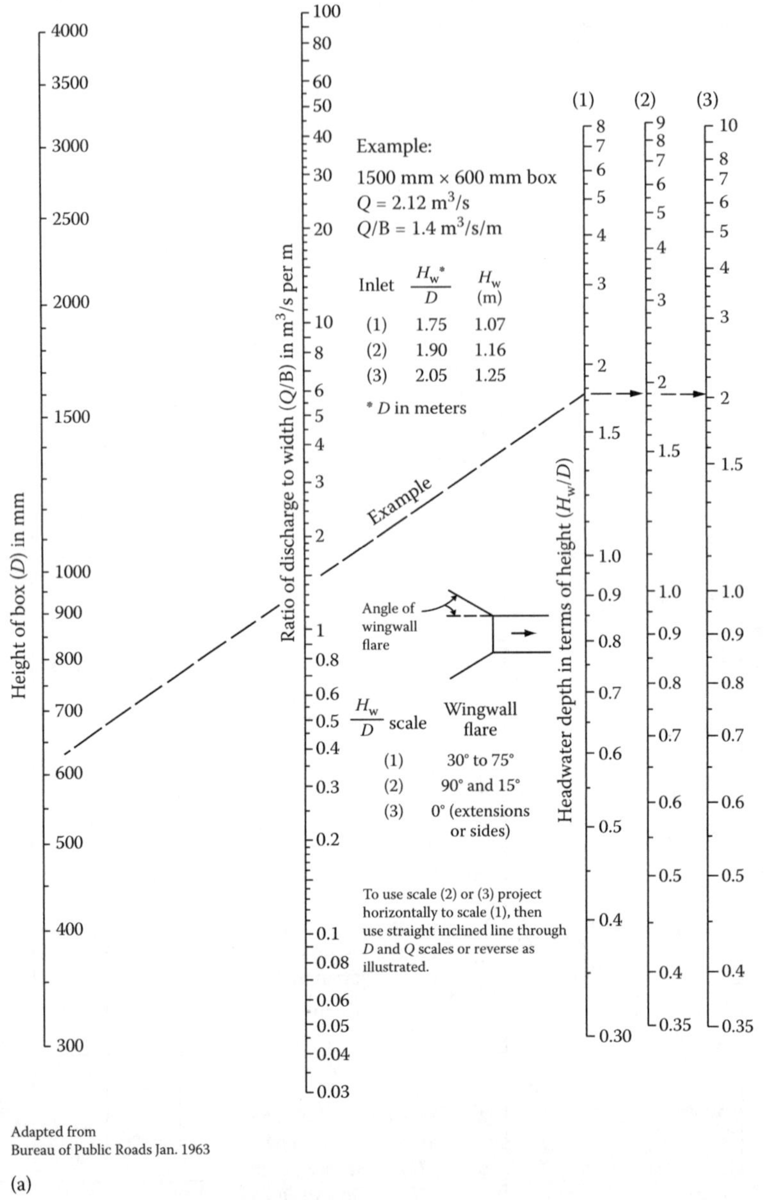

FIGURE 4.27A Inlet control nomograph, box culverts (SI).

Table 4.5 lists typical values of Manning's n for concrete, HDPE, and corrugated metal conduits, which are commonly used. The entrance loss coefficient can vary from 0.2 to 0.9 depending on the entrance conditions; the former coefficient represents a tapered inlet face, the latter a CMP (corrugated metal pipe) culvert projecting from fill. Table 4.6 presents entrance loss coefficients for concrete and CMP culverts for various inlet face configurations. The exit loss coefficient is almost always equal to 1.0. Thus, under normal conditions, ΣK_ℓ for a single culvert can be estimated at 1.5.

HDS-5 (FHWA, 2012) and the *Concrete Pipe Design Manual (2007)* by American Concrete Pipe Association and ASCE and WEF, 2011) contain charts to facilitate calculating conduit capacity under outlet control. Figure 4.28 is a sample of charts for round pipes. To improve accuracy, however, Equations 4.39 and 4.40 should be used for head loss calculations.

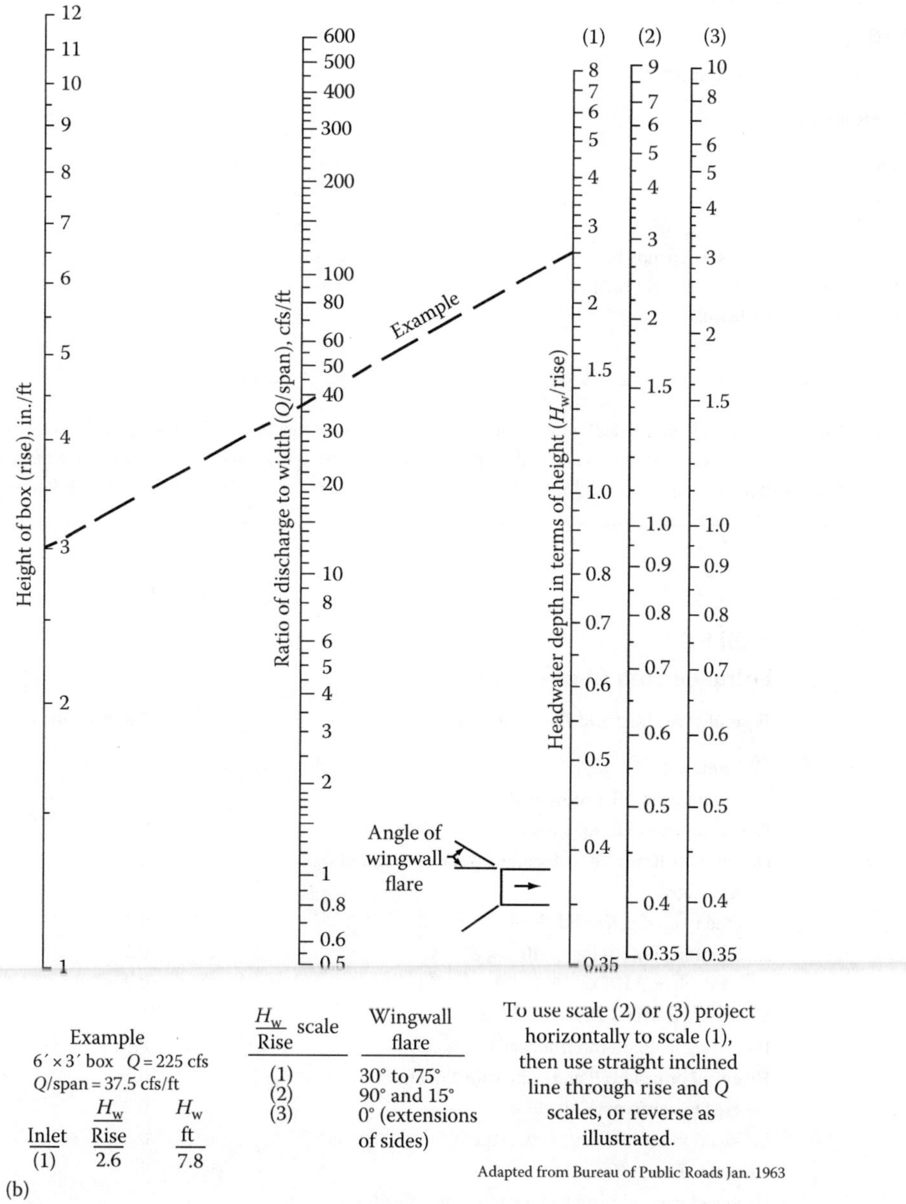

FIGURE 4.27B Inlet control nomograph, box culverts (CU).

Culvert design parameters for carrying a given flow under a roadway or embankment are commonly performed in an iterative process. This process consists of selecting a culvert size and type and then performing calculations to determine the required head for both the inlet and outlet control flow conditions; the larger of the two is used in design. Commonly, the required headwater for inlet control is estimated based on inlet control charts, such as Figures 4.25 through 4.27; the head losses for the outlet control are calculated using Equations 4.39 and 4.40 (see Figure 4.29).

TABLE 4.5
Manning's *n* for Various Culverts

Conduit Type		Manning's *n*
Concrete pipe	Smooth walls	0.012–0.013
Concrete box	Smooth walls	0.012–0.015
HDPE	Smooth interior	0.012
Corrugated metal pipes *n* depends on corrugation size and barrel size; smaller *n* values for larger barrels	2 -2/3 × 1/2 in.	0.022–0.027
	6 × 1 in.	0.023–0.028
	6 × 2 in.	0.033–0.035
	9 × 2-1/2 in.	0.033–0.037

Note: The Manning's *n* values listed in this table are recommended design values for new culverts. For concrete pipes and culverts with deteriorated walls and poor joints, *n* values may be as high as 0.018. Corrugated metal pipes, in addition to corrosion, also experience deformation and shape change that can significantly reduce their hydraulic capacity.

TABLE 4.6
Entrance Loss Coefficients

Type of Structure and Design of Entrance	Coefficient, K_e
Concrete Pipe	
Projecting from fill, groove end	0.2
Projecting from fill, sq. cut end	0.5
Headwall or headwall and wingwalls Groove end of pipe	0.2
Square-edge	0.5
Mitered to conform to fill slope	0.7
End section conforming to fill slope[a]	0.5
Beveled edges 33.7° or 45° bevels	0.2
Corrugated Metal Pipe and Arch	
Projecting from fill (no headwall)	0.9
Headwall or headwall and wingwalls Square-edge	0.5
Mitered to conform to fill slope	0.7
End section conforming to fill slope[a]	0.5
Concrete Box	
Headwall parallel to embankment (no wingwalls)	
Square-edged on three edges	0.5
Rounded on three edges to radius of 1/12 barrel dimension	0.2
Wingwalls at 30° to 75° to barrel	
Square-edged at crown	0.4
Crown edge rounded to radius of 1/12 barrel dimension	0.2
Wingwalls at 10° to 30° to barrel	
Square-edged at crown	0.5
Wingwalls parallel (extension of sides)	
Square-edged at crown	0.7

Source: **FHWA,** *Hydraulic Design of Highway Culverts* **(HDS-5), 3rd ed., 2012.**

[a] End section conforming to fill slope made of either metal or concrete are the sections commonly available from manufacturers.

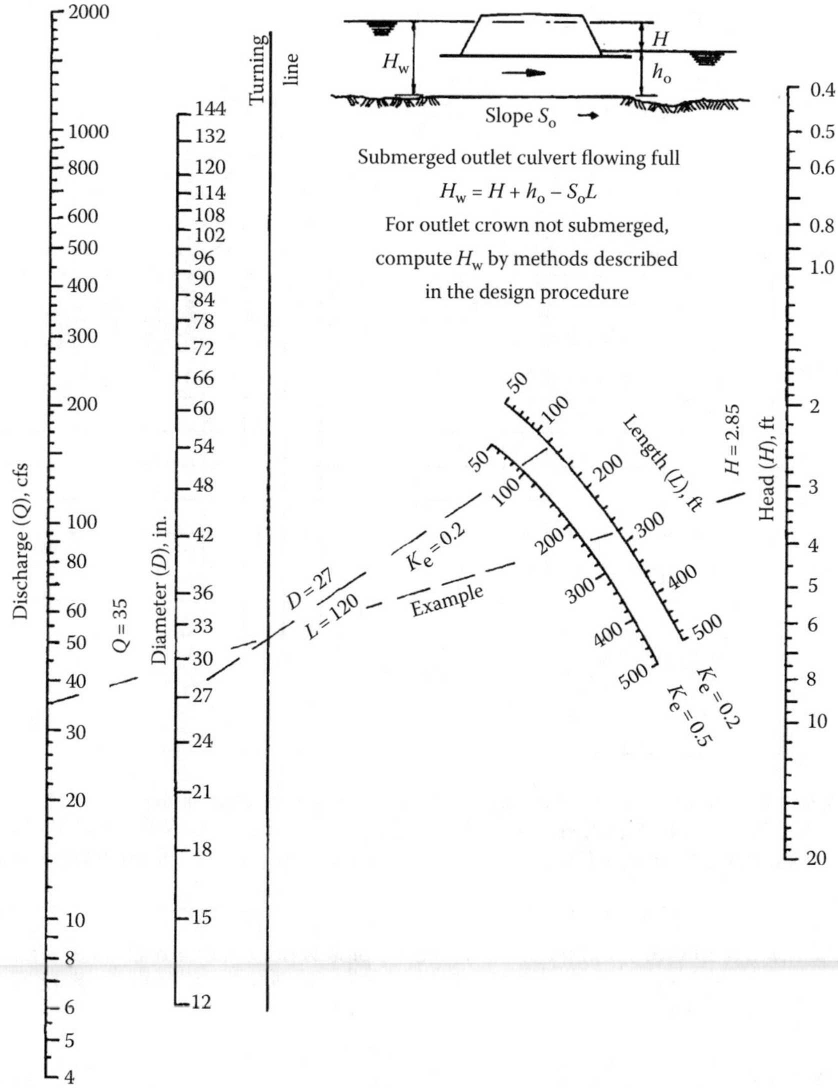

Adapted from Bureau of Public Roads Jan. 1963

FIGURE 4.28 Outlet control discharge headwater relation for circular concrete pipe culverts flowing full, n = 0.012 (CU).

Example 4.8

The proposed 30 in. pipe along New Milford Avenue (Case Study 4.1) terminates at the New Milford Bridge over Hirschfeld Brook (Figure 4.18). A review of the state and FEMA flood maps shows the 10-year flood elevation of Hirschfeld Brook to be at approximately elev. 17.7 ft at the bridge. Calculate the inlet control capacity of the 30 in. pipe.

Solution

The 30 in. pipe is designed based on a 25-year frequency storm. Considering that the floods in the brook lag hours behind rainfalls, the 10-year flood level of Hirschfeld Brook can conservatively be assumed as the backwater elevation at the pipe outlet. However, since the invert of the pipe

Hydrologic and channel information	Sketch
Q_1 = ______ T_{w1} = ______ Q_2 = ______ T_{w1} = ______ (Q_1 = Design discharge, say Q_{25} Q_2 = Check discharge, say Q_{50} or Q_{100})	Station: ______ 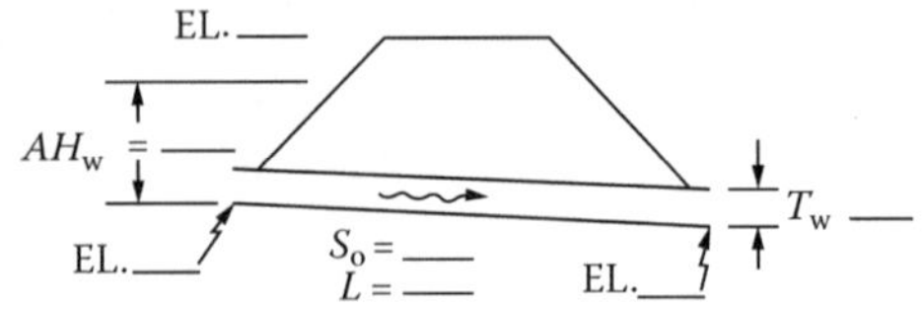 Mean stream velocity = ______ Max. stream velocity = ______

Culvert description (entrance type)	Q	Size	Headwater computation: Inlet cont.		Outlet control $H_w = H + h_o - LS_o$								Controlling H_w	Outlet velocity	Cost	Comments
			$\frac{H_w}{D}$	H_w	K_e	H	d_c	$\frac{d_c+D}{2}$	T_w	h_o	LS_o	H_w				

Summary and recommendations:

FIGURE 4.29 Hydraulic tabulation for culverts. Parameters are derived as follows: **H_w/D**, inlet control nomograph; **K_e**, Table 4.6; **H**, Equations 4.39 and 4.40; **d_c**, critical depth, Figures 4D.1 and 4D.2 or Table 2.2 for circular culverts; **h_o**, **T_w** or (**d_c** + **D**)/2, whichever is greater. **S_o** and **L** are as defined in the sketch.

is well above the 10-year flood level, the proposed pipe is not under a backwater effect of the brook. Therefore, headwater can be calculated based on the rim and invert elevations of MH#3. According to Figure 4.18, these elevations are, respectively,

Rim elev. = 27.05 ft
Inv. elev. = 20.66 ft

Therefore, the headwater depth is

$$H_w = 27.05 - 20.66 = 6.39 \text{ ft}$$

$$D = 30 \text{ in.} = 2.5 \text{ ft}$$

$$\frac{H_w}{D} = 2.6$$

Drawing a straight line connecting these values for pipe diameter and H_w / D in Figure 4.25b gives an inlet control capacity of Q= 55 cfs.

Example 4.9

Calculate the inlet capacity of the existing 30 in. pipe in the preceding example using the orifice flow equation.

Solution

Applying the orifice flow equation to MH-3 results in

$$Q = 0.6 \times \left(\pi \times \frac{2.5^2}{4} \right) \times (2gH)^{0.5} \tag{4.41}$$

where H is the available head to the center of the 30 in. pipe and is given as

$$H = 27.05 - (20.66 + 1.25) = 5.14 \ ft$$

where 27.05 and 20.66 are the rim elevation of MH-3 and the invert elevation of the 30 in. pipe, respectively (see Figure 4.20).

$$Q = 0.6 \times 4.91 \times (2 \times 32.2 \times 5.14)^{1/2} = 53.6 \text{ cfs}$$

This result is very close to 55 cfs that was obtained using Figure 4.25b in Example 4.8.

Example 4.10

The calculated 25-year peak runoff from a 38.85 acre tributary to the 30 in. pipe in Example 4. is calculated at 47.4 cfs. Calculate the water surface elevation at manhole MH#3. The pipe is 128 ft long and, as indicated, is not under a backwater effect of Hirschfeld Brook.

Solution

Calculate headwater under inlet and outlet control conditions using Figure 4.25b and Equation 4.40, respectively.

a. Inlet control condition
   Drawing a straight line through given data for pipe diameter and discharge and extending to H_w / D line for square edge line gives a conservative value of $H_w / D = 2.15$. Therefore $H_w = 2.15 \times 2.5 = 5.38$ ft*

$$W.S.\ elev. = 20.66(inv.elev.) + 5.38 = 26.04 \approx 26 \ ft$$

(using orifice flow equation with $C_o = 0.6$ gives $H_w = 5.32$ ft (4.07+1.25).

b. Outlet control conditions

The entrance loss coefficient for this pipe square edge with headwall is $K_e = 0.5$ (see Table 4.6).

$$A = \left(\frac{\pi}{4} \right) D^2 = 4.91 \ ft^2$$

$$V = \frac{47.4}{4.91} = 9.65 \text{ ft/s}$$

$$\frac{V^2}{2g} = \frac{9.65^2}{2 \times 32.2} = 1.45$$

$$R = \frac{D}{4} = 0.625 \ ft$$

Use $n = 0.012$ (Table 4.5).

$$H = \left(1.5 + \frac{39 \times 0.012^2 \times 128}{0.625^{4/3}}\right) \times 1.45$$

$$H = (1.5 + 1.0)1.45 = 3.62\ ft$$

The preceding calculations are arranged in the following table. In this table, the backwater effect is input as

$$T_w = \frac{(d_c + D)}{2}$$

Where d_c may be derived from Table 2.2 (critical depth for circular pipes) as follows:

$$K_c = \frac{Q}{\sqrt{g d^{2.5}}}$$

$$K_c = \frac{47.4}{\left[(32.2)^{1/2} \times 2.5^{2.5}\right]} = 0.845$$

Hydrologic and channel information

Q_1 = 47.4 cfs T_{w1} = None

Q_2 = ______ T_{w2} = ______

(Q_1 = Design discharge, say Q_{25}; Q_2 = Check discharge, say Q_{50} or Q_{100})

Sketch

Station: ______

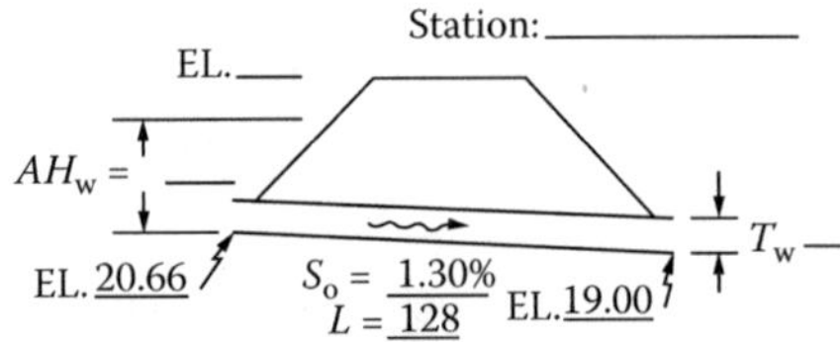

Mean stream velocity = ______
Max. stream velocity = ______

Culvert description (Entrance type)	Q	Size	Headwater computation										Controlling H_w	Outlet velocity	Cost	Comments
			Inlet cont.		Outlet control $H_w = H + h_o - LS_o$											
			$\frac{H_w}{D}$	H_w	K_e	H	d_c	$\frac{d_c+D}{2}$	T_w	h_o	LS_o	H_w				
RCP square-edge	47.4	30″	2.15	5.38	0.50	3.62	2.26	2.38	–	2.38	1.66	4.34	5.38	9.65		

Summary and recommendations:

$H_o = T_w$ or $(d_c + D)/2$, whichever is greater

Inlet control
Available head at inlet face = 6.39 ft > 5.50 ft

FIGURE 4.30 Worksheet to record Example 4.10 headwater calculations.

$$\frac{d_c}{D} = 0.905\,(see\ Table\,2.2\ in\ Chapter\,2)$$

$$d_c = 2.26\ ft$$

Alternatively, using Figure 4D.1b gives

$$d_c = 2.25\,ft$$

$$h_o = \frac{(2.26 + 2.5)}{2} = 2.38$$

$$LS_o = 128 \times 1.3\% = 1.66\ ft$$

$$H_w = H + h_o - LS_o = 3.62 + 2.38 - 1.66 = 4.34\ ft$$

Since H_w for inlet control is larger than H_w for outlet control, the 30 in. pipe is under inlet control. According to Figure 4.20, the invert of the pipe is laid 6.39 ft below the rim of MH3; therefore, the available head is sufficient and water would not surcharge from the manhole. The calculations may be organized in a table (see Table 4.29).

4.8 EROSION CONTROL AT OUTFALLS

Erosive flow can occur at the outlet of a pipe in a stream or channel; the potential for erosion is determined based on either the permissible velocity or permissible shear stress for the ambient soil. Allowable velocity is commonly used for outlet conduit protection. Table 4.7 includes allowable velocities for various soils. To determine soil stability, these velocities are compared with the design velocity in the conduit. The design velocity is that which occurs during the erosion control design storm, which is generally the 25-year storm. When design velocity exceeds the allowable velocity, the conduit outlet must be protected to prevent erosion and scour.

The protection measure commonly consists of a stone riprap section extending from the culvert outlet to where the channel is stable or to reach to reduce the flow velocity below permissible values in the channel. Following a recommendation by the National Academies of Science in 1970, riprap aprons are designed based on, at a minimum, the 25-year flood. Erosion can also be controlled using a scour hole. Design calculations for riprap apron and scour hole are presented in the following sections.

TABLE 4.7
Permissible Velocities for Various Soils

	Allowable Velocity	
Soil Texture	**m/s**	**ft/s**
Sand	0.5[a]	1.5
Sandy loam	0.8	2.50
Silt loam, loam	0.9	3.00
Sandy clay loam, graded loam to gravel	1.1	3.50
Clay loam	1.2	4.00
Clay, fine gravel	1.5	5.00
Cobbles	1.7	5.50
Shale (nonweathered)	1.8	6.00

Source: Soil Erosion and Sediment Control Standards in New Jersey, July 2017

[a] *Rounded to the nearest tenth.*

4.8.1 Riprap Aprons

Where the outlet terminates either at a well-defined channel or a flat area such as the bed of a detention basin, a riprap apron is used to protect the soil from erosion. The length of an apron, L_a, is calculated based on empirical equations, developed by the EPA (1976). In SI L_a, and the size of stone, d_{so}, are as follows:

$$L_a = \frac{3.2q}{D_o^{1/2}} + 7D_o \; for \; T_w < \frac{D_o}{2} \tag{4.42}$$

$$L_a = \frac{5.43q}{D_o^{1/2}} \; for \; T_w \geq \frac{D_o}{2} \tag{4.43}$$

$$d_{so} = \frac{3.5}{T_w} q^{1.33} \tag{4.44}$$

In CU, these equations become

$$L_a = \frac{1.8q}{D_o^{1/2}} + 7D_o \; for \, T_w < \frac{D_o}{2} \tag{4.45}$$

$$L_a = \frac{3q}{D_o^{1/2}} \; for \, T_w \geq \frac{D_o}{2} \tag{4.46}$$

$$d_{so} = \frac{0.2}{T_w} q^{1.33} \tag{4.47}$$

where

d_{so} = median stone diameter in cm (in.)
D_o= culvert height in m (ft)
$q = Q/W_o$
Q = culvert discharge, m³/s (cfs)
W_o= maximum culvert width
T_w= tailwater depth above the invert of the culvert, m (ft)

Where T_w cannot be estimated, $T_w = 0.2D_o$ is used.

In a well-defined channel, the bottom width of the apron should be at least equal to the bottom width of the channel; the lining should extend at least 1 ft above the tailwater elevation but no lower than two-thirds of the vertical conduit dimension above the culvert invert. In addition, the riprap channel section should satisfy the following conditions: The side slopes should be 2:1 (2 H, 1 V) or flatter, the bottom grade should be level (0% slope), and there should be no drop at the end of the apron or at the terminus of the culvert. When Reno mattress or erosion control blankets are used in lieu of riprap stone, the side slopes may be increased to 1:1.

Where there is no well-defined channel downstream of the apron, W at the end of the apron should be as follows:

$$W = 3D_o + L_a \; for \, T_w < \frac{D_o}{2} \tag{4.48}$$

$$W = 3D_o + 0.4L_a \; for \; T_w > \frac{D_o}{2} \tag{4.49}$$

The preceding equations are valid in both SI and CU. The width of apron at the culvert outlet should be at least three times the culvert width (Figure 4.31). The size of the stone in these cases is calculated using Equations 4.44 and 4.47 in SI and CU, respectively.

For two or more culverts closely spaced, the length of the riprap can be sized for one culvert and the width should accommodate all culverts. When the culvert spacing equals or exceeds one-fourth of the width dimensions, the riprap length and stone size, calculated for one culvert, are increased by 25%. The bottom grade of the apron should be constructed on a flat grade 0. The SCS has prepared graphs of riprap apron design parameters. However, based on this author's experience, these charts show shorter riprap lengths than those of the equations, and therefore, are not included in this section.

4.8.2 Preformed Scour Holes

Where a flat apron is either impractical or would be too large, existing holes are modified to control erosion. In new outfalls, scour holes are installed to achieve the same objective. Figure 4.32 shows a layout and section view of a scour hole. The depth of scour holes, y, commonly varies from $D_o/2$ to D_o, where D_o is the vertical diameter of the pipe/culvert.

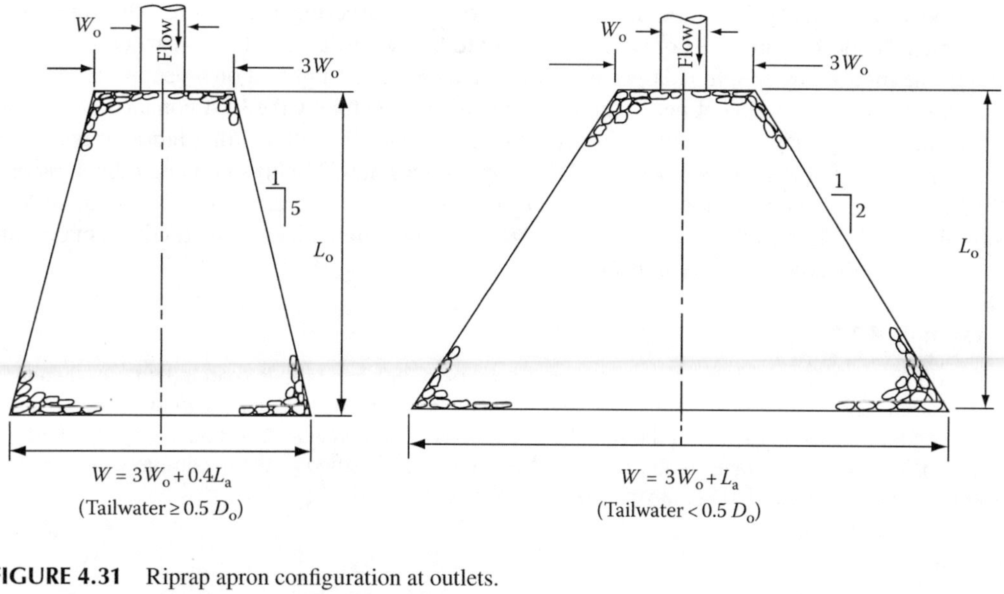

FIGURE 4.31 Riprap apron configuration at outlets.

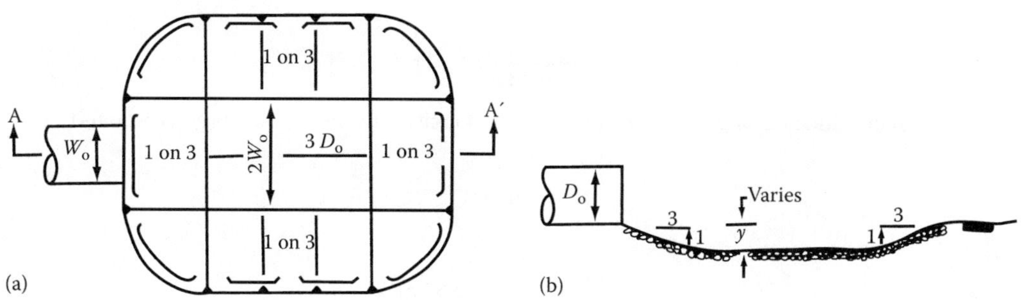

FIGURE 4.32 Preformed scour hole: (a) plan, (b) section.

The median stone diameter, d_{so}, can be calculated using the following equations:

$$d_{so} = \frac{2.8}{T_w} q^{4/3} \text{ when } y = \frac{D_o}{2} \tag{4.50}$$

$$d_{so} = \frac{1.8}{T_w} q^{4/3} \text{ when } y = D_o \tag{4.51}$$

In CU, these equations become respectively

$$d_{so} = \frac{0.15}{T_w} q^{4/3} \text{ when } y = \frac{D_o}{2} \tag{4.52}$$

$$d_{so} = \frac{0.1}{T_w} q^{4/3} \text{ when } y = D_o \tag{4.53}$$

The parameters and units in these equations are the same as those defined for riprap apron.

The riprap stone should be fieldstone or rough unhewn quarry stone. Stone should be angular, hard, and of such quality that it does not disintegrate due to weathering or exposure to flowing water. The riprap should be composed of well-graded mixture down to 25 mm (1 in.) particles such that 50% of the mixture by weight is larger than the calculated d_{so} and placed on a geotextile fabric or stone filter or a combination of both. The NJ Standards (2017) follow the EPA equations for riprap dimensions and specify a riprap thickness equal to three times d_{so} with no filter fabric or two times d_{so} when a filter fabric is placed beneath the riprap. Some states, Michigan among others, recommend a minimum riprap thickness of 1.5 d_{so} or 6 in., whichever is greater. The standards in New York State (2016, Section 5.8) and many other states call for riprap thickness of 1.5 times maximum stone size plus thickness of filter or bedding.

Example 4.11

Runoff from a residential development is conveyed by a 750 mm pipe into an infiltration–detention basin. The 25-year discharge in the pipe is calculated at 0.5 m³/s. Calculate dimensions of the required riprap apron to protect the sand bed in the basin from erosion based on the NJ Standards (2017). The design tailwater depth, which in New Jersey is specified as the maximum water level at the detention basin during a 2-year storm is calculated as 0.6 m.

Solution

$$A = \pi \times 0.75^2 / 4 = 0.442 \text{ m}^2$$

$$V = \frac{Q}{A} = \frac{0.5}{0.442} = 1.13 \text{ m/s}$$

Since design velocity is larger than the permissible velocity for sand, riprap apron is required.

$$T_w = 0.6 \text{ m} > \frac{D_o}{2}$$

$$q = \frac{Q}{D_o} = \frac{0.5}{0.75} = 0.667 \text{ m}^2/\text{s}$$

$$L_a = \frac{5.43q}{D_o^{1/2}} = \frac{5.43 \times 0.667}{(0.75)^{0.5}} = 4.18\text{ m, use } 4.2\text{ m}$$

$$W = 3D_o + 0.4L_a = 3 \times 0.75 + 0.4 \times 4.18 = 3.92\text{ m, use } 4.0\text{ m}$$

Using Equation 4.44:

$$d_{50} = \left(\frac{3.5}{0.6}\right) \times 0.667^{1.33} = 3.39\text{ cm}$$

$$L_a = \frac{5.43q}{D_o^{1/2}} = \frac{5.43 \times 0.667}{(0.75)^{0.5}} = 4.18\text{ m, use } 4.2\text{ m}$$

$$W = 3D_o + 0.4L_a = 3 \times 0.75 + 0.4 \times 4.18 = 3.92\text{ m, use } 4.0\text{ m}$$

Using Equation 4.44:

$$d_{50} = \left(\frac{3.5}{0.6}\right) \times 0.667^{1.33} = 3.39\text{ cm}$$

Use 7.5 cm (3 in.) min. size. Thickness of riprap 3 × 7.5 = 22.5 cm, no filter fabric.

Example 4.12

A 36 in. pipe terminates at an unlined drainage channel. The 25-year discharge from the pipe is calculated at 49 cfs. Design stone riprap to protect the channel from erosion. Base your calculations on the EPA method, assuming that the tailwater lies above the center of the pipe.

Solution

Given Q = 49 cfs and D_o = 36 in.=3 ft
Using Equations 4.46 and 4.47:

$$q = \frac{Q}{D_o} = \frac{49}{3} = 16.33$$

$$L_a = 3 \times \frac{16.33}{3^{0.5}} = 28.3\text{ ft (use 28 ft)}$$

Conservatively assume $T_w = D_o/2 = 1.5$ ft.

$$d_{50} = \frac{0.15 \times 16.33^{1.33}}{1.5} = 4.1\text{ in. (use 4.5 in.)}$$

Construct channel banks at 2:1 slope and place riprap on channel banks for a distance of 28 ft. from the pipe outfall and to 2 ft above the bed, which is equal to two-thirds of the pipe vertical diameter.

Example 4.13

The discharge pipe from an infiltration–detention basin terminates at a stream. The pipe is 450 mm in diameter, and the 100-year discharge from the basin is computed at 0.55 m³/s. To reduce the area of disturbance at the stream, a scour hole is provisioned at the outfall point. The depth of scour hole is 22.5 cm. The tailwater at the stream, which may be taken as the 2-year flood elevation, is 0.9 m deep. Size the scour hole.

Solution

Top length of scour hole,

$$L = 3D + \left(3\frac{D}{2}\right) \times 2 = 6D = 6 \times 0.450 = 2.7\ m$$

$$d_{50} = \frac{0.15 \times 16.33^{1.33}}{1.5} = 4.1\ \text{in.}(\text{use } 4.5\ \text{in.})$$

Top width of scour hole, $W = 5D = 2.25$ m
Bottom length of scour hole, $L' = 3D = 1.35$ m
Bottom width of scour hole, $W' = 2D = 0.9$ m
Stone size is

$$q = \frac{Q}{D} = \frac{0.55}{0.45} = 1.222\ \text{m}^2/\text{s}$$

$$d_{so} = \left(\frac{2.8}{T_w}\right) q^{1.33}$$

$$d_{so} = \left(\frac{2.8}{0.9}\right) 1.222^{1.33} = 4.06\ cm$$

Use three layers of 7.5 cm (3 in.) stone, which is the minimum size specified in the NJ Standards and by some other states.

4.9 DRAINAGE CHANNELS

Drainage channels, like pipes, are used to convey the runoff from developments and roadways. Channels, also referred to as swales, are used along the toe of slopes. To prevent erosion, channels may need to be protected with lining. The lining can be rigid or flexible. Rigid linings include cast-in-place concrete, stone masonry, and grouted riprap. Gabion walls and mattresses are classified as semirigid linings. Rigid linings were widely used in the past; however, the new trend is to use flexible linings. As such, rigid linings are not discussed in this text.

Flexible linings include riprap stone, gravel mulch, grass, synthetic mats, and fiberglass roving and the like. In humid areas such as the northeast part of the United States, grass lining is the most effective means of protecting swales in mildly sloped terrains. In addition to being cost-effective, grass swales are easy to maintain and aesthetically pleasing. By removing silt and suspended sediment from runoff, grass swales also improve water quality. Grass lining can be best achieved by sodding, with sods laid parallel to the flow direction and secured with pins or staples. If the runoff is diverted during the grass growth or the grass is protected until the vegetative cover is established, seeding is satisfactory. In arid climates, in the absence of irrigation, and in steep terrains where high-flow velocities are likely to occur, other lining such as gravel, riprap stone, or gabion mattress should be considered.

Swales covered with flexible lining such as riprap stone or grass are generally designed based on the resistance of the cover to erosion, disregarding the properties of the native soil. See for example FHWA (1967). An exception to this design procedure is presented in HEC-15 (FHWA, 2005), to be discussed later. In designing grass-lined swales, the ability of equipment to mow the grass is an important factor to consider. Over time, trapezoidal or V-shaped sections become parabolic, and as such parabolic grass swales are more practical than other shapes.

Drainage channels, whether lined or unlined, are designed based on two different concepts: maximum permissible velocity and permissible shear stress, also known as the tractive force method. The former method is empirical in nature and considers the channel to be stable against erosion as long as the actual flow velocity is less than a critical one. However, in the tractive force method, the shear stress is the basis of the stability criterion.

4.9.1 Permissible Velocity

The permissible velocity of soils was introduced earlier as a criterion for determining the need for, rather than the design of, riprap apron at a conduit outfall. The use of a definite permissible velocity for a given soil is too simplistic, as the said velocity for noncohesive material varies not only with the size of soil particle but also with the compactness of the soil. For cohesive soils, the permissible velocity depends on a plasticity index and soil compactness. Figure 4.33 shows the USSR data for permissible velocity of noncohesive soils as a function of grain size, in millimeters and US standard sieve sizes; Figure 4.34 shows the same data for cohesive soils (Chow, 1959). The velocities in Figures 4.33 and 4.34 relate to approximately 1 m deep flow. For other flow depths, a correction factor should be applied to the velocities shown in the figures. Table 4.8 lists the flow depth adjustment factor using the USSR data.

It should be noted that clear water has a larger potential for erosion than a sediment-laden flow. Thus, the permissible velocities for sediment-laden are somewhat higher than those in sediment-free flow Table 4.9 presents permissible velocities in channels with clear water and those carrying colloidal and noncolloidal silts. Since stable channels are expected to be sediment free, Figure 4.33, and the no detritus column of this table provide a more conservative basis for estimating the permissible velocity of channels than the right two columns in the table.

Channel side slopes and channel bends also affect the channel stability. These parameters, however, will be treated in the tractive force method, which is more widely used than the permissible velocity for design of stable channels.

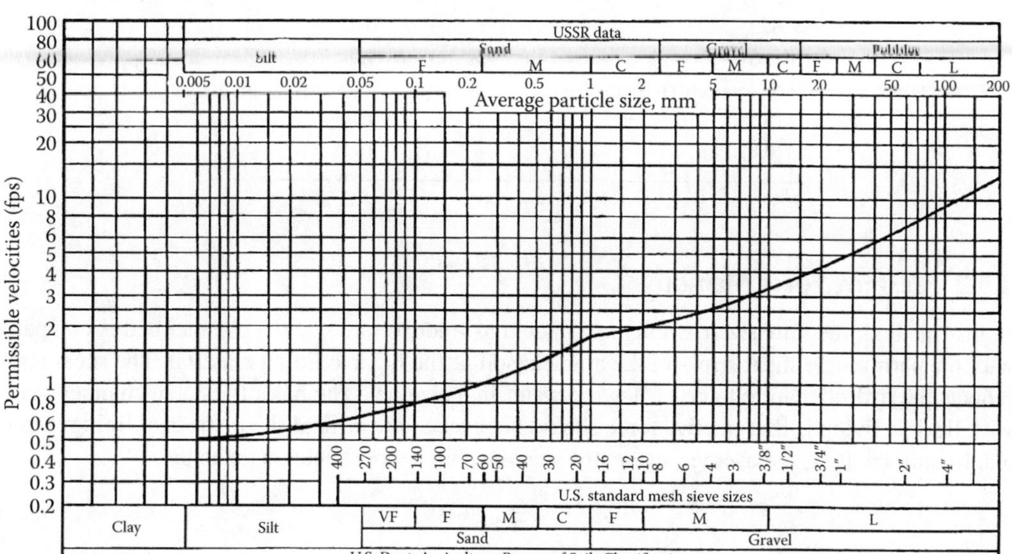

FIGURE 4.33 USSR data on permissible velocities of noncohesive soils: CD coarse; FD fine; LD large; MD medium; VFD very fine.

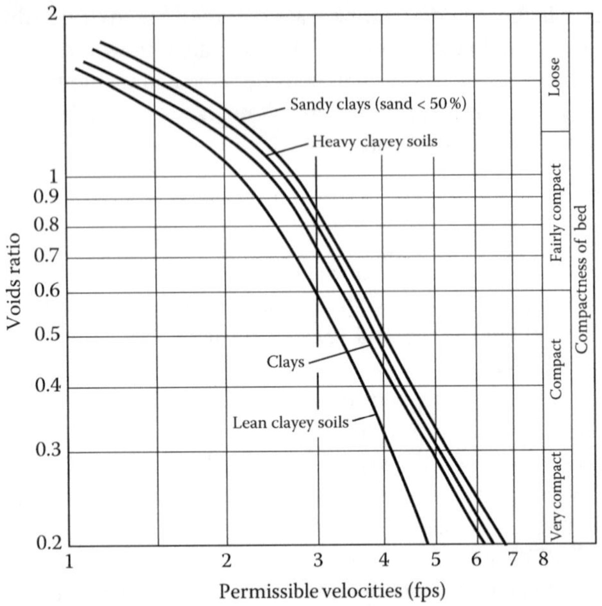

FIGURE 4.34 Curves showing USSR data for permissible velocity of cohesive soils.

TABLE 4.8
Corrections of Permissible Velocity with Depth for Both Cohesive and Noncohesive Material

Flow Depth, m (ft)	Adjustment Factor
0.3 (1)	0.8
0.6 (2)	0.92
1 (3.3)	1.0
1.5 (4.9)	1.09
2 (6.6)	1.16
3 (9.8)	1.25

4.9.2 Tractive Force Method

In this method, the flow shear stress, also known as tractive force, at the channel bed is compared with the permissible shear stress of the ambient soil or the flexible lining material. The shear stress reflects the hydrodynamic force of flowing water in a channel. The basis for stable channel design with flexible living is that tractive force should not exceed the critical shear stress of lining materials. In uniform flow, the average shear stress on the channel perimeter is given by

$$\tau = \gamma RS \tag{4.54}$$

where

γ = unit weight of water, N/m^3 (lb/ft^3)
R = hydraulic radius, m (ft)
S = average bed slope, m/m (ft/ft)
τ = shear stress, in SI units = N/m^2 (Pascal); lb/ft^2 (CU)

TABLE 4.9
Permissible Velocities for Various Soil Textures in Unlined Channels with Clear and Silty Water

Original Material Excavated for Canals	Mean Velocity for Straight Canals of Small Slope, after Aging, with Flow Depths Less than 3 ft (0.9 m)					
	Clear Water, No Detritus		Water Transporting Colloidal Silts		Water Transporting Noncolloidal Silts, Sands, Gravels, or Rock Fragments	
	ft/s	m/s	ft/s	m/s	ft/s	m/s
Fine sand (noncolloidal)	1.5	0.46	2.5	0.76	1.5	0.46
Sandy loam (noncolloidal)	1.75	0.53	2.5	0.76	2.0	0.61
Silt loam (noncolloidal)	2.0	0.61	3.0	0.91	2.0	0.61
Alluvial silt (noncolloidal)	2.0	0.61	3.5	1.07	2.0	0.61
Ordinary firm loam	2.5	0.76	3.5	1.07	2.25	0.69
Volcanic ash	2.5	0.76	3.5	1.07	2.0	0.61
Stiff clay (very colloidal)	3.75	1.14	5.0	1.52	3.0	0.91
Alluvial silt (colloidal)	3.75	1.14	5.0	1.52	3.0	0.91
Shales and hardpans	6.0	1.83	6.0	1.83	5.0	1.52
Fine gravel	2.5	0.76	5.0	1.52	3.75	1.14
Graded, loam to cobbles (when noncolloidal)	3.75	1.14	5.0	1.52	5.0	1.52
Graded silt to cobbles (when colloidal)	4.0	1.22	5.5	1.68	5.0	1.52
Coarse gravel (noncolloidal)	4.0	1.22	6.0	1.83	6.5	1.98
Cobbles and shingles	5.0	1.52	5.5	1.68	6.5	1.98

Source: USDA (2007), NEH Part 654, Chapter 8 Threshold channel design, Table 8.3

Note that the permissible shear stress can be related to the permissible velocity. Eliminating the slope, S, between Equation 4.54 and the Manning formula results in

$$V_p = \frac{0.01 R^{1/6} \tau_p^{1/2}}{n}, SI \tag{4.55}$$

$$V_p = \frac{0.189 R^{1/6} \tau_p^{1/2}}{n}, CU \tag{4.56}$$

where τ_p and V_p are the permissible shear stress and the permissible velocity, respectively.

These equations indicate that V_p is not a constant for a given channel lining; rather, it varies with the hydraulic radius. Thus, permissible shear stress, which depends on hydraulic conditions, is more appropriate than permissible velocity for the selection of a lining material. For this reason, the concept of permissible velocity is losing popularity in engineering practice. In fact, many jurisdictional agencies recommend using the tractive force method instead in stable channels made of noncohesive material and cohesive material of low plasticity (NJ Standards, 2017).

Shear stress is distributed nonuniformly along the wetted perimeter of a channel. Distribution of shear stress in a trapezoidal channel tends toward zero at the corners with a maximum at the center

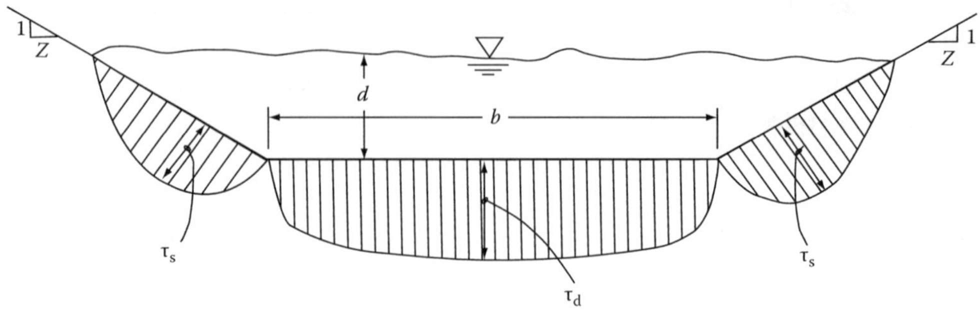

FIGURE 4.35 Variation of shear stress in trapezoidal channel.

bed of the channel. The maximum shear stress at the sides occurs nearly at the lower third of the slope. Figure 4.35 depicts the shear stress variation in a trapezoidal channel section.

The maximum shear stress for a straight channel is given as

$$\tau_d = \gamma y S \tag{4.57}$$

where y is the maximum depth of flow. Note that the depth of flow in rectangular and trapezoidal channels was denoted as y in Chapter 2. In swales, the flow depth is commonly nonuniform and the average depth is designated as D_a here to differentiate from y.

In stable channel design, a safety factor is applied to the shear stress at the bottom of the channel using the following equation:

$$\tau_p \geq SF\tau_d \tag{4.58}$$

where

SF = safety factor, typically 1.0 to 1.5
τ_p = permissible shear stress of channel cover

4.9.3 Bare Soil and Stone Lining

Permissible shear stress for noncohesive soils and stone linings is listed in Table 4.10. The table shows the permissible shear stress of fine-grained soils of d_{75} smaller than 1.3 mm (0.05 in.) is relatively constant and may be conservatively estimated at 1.0 N/m^2 (0.02 lb/ft^2). For noncohesive soils of $1.3 < d_{75} < 15$ mm (0.05–0.6 in.), the permissible shear stress can be approximated using the following equation:

$$\tau_p = Kd_{75} \tag{4.59}$$

where

τ_p = permissible shear stress N/m^2 (lb/ft^2)
K = 0.75 SI, 0.4 CU
d_{75} = 75% finer soil size, mm (in.)

The same equation can be applied to coarse and very coarse gravel and stone riprap substituting d_{50} for d_{75}:

$$\tau_p = Kd_{50} \quad d_{50} > 15\,mm \tag{4.60}$$

TABLE 4.10
Permissible Shear Stresses for Noncohesive Lining Materials

Lining Category	Lining Type	Permissible Shear Stress N/m²	Permissible Shear Stress lb/ft²
Bare soil cohesive (PI = 10)	Clayey sands	1.8–4.5	0.037–0.095
	Inorganic silts	1.1–4.0	0.027–0.110
	Silty sands	1.1–3.4	0.024–0.072
Bare soil cohesive (PI ≥ 20)	Clayey sands	4.5	0.094
	Inorganic silts	4.0	0.083
	Silty sands	3.5	0.072
	Inorganic clays	6.6	0.140
Bare soil noncohesive (PI < 10)	Finer than coarse sand D_{75} < 1.3 mm (0.05 in.)	1.0	0.02
	Fine gravel D_{75} = 7.5 mm (0.3 in.)	5.6	0.12
	Gravel D_{75} = 15 mm (0.6 in.)	11.0	0.24
Gravel mulch	Coarse gravel D_{50} = 25 mm (1 in.)	19.0	0.40
	Very coarse gravel D_{50} = 50 mm (2 in.)	38.0	0.80
Rock riprap	D_{50} = 0.15 m (0.5 ft)	113.0	2.40
	D_{50} = 0.30 m (1.0 ft)	227.0	4.80

Source: FHWA, *Design of Roadside Channels with Flexible Linings* (HEC-15), 3rd ed., September 2005.

Figure 4.36 shows the permissible shear stress of cohesive materials. This figure indicates that the permissible shear stress of cohesive material such as clay depends on the soil plasticity index and its compactness (porosity, void ratio). The third edition of HEC-15 (FHWA, 2005) presents the following formula for permissible shear stress of cohesive soils:

$$\tau_p = \left(C_1 PI^2 + C_2 PI + C_3\right)\left(C_4 + C_5 e\right)^2 C_6 \tag{4.61}$$

where

PI = plasticity index

e = void ratio (volume of void to volume of solids)

C_1 through C_6 are coefficients that depend on the soil type. (These coefficients are published in HEC-15 Figure 4.18 [FHWA, 2005] as well as Appendix 4D.)

The use of Equation 4.61 requires measurement of void ratio in addition to plasticity and soil type.

4.9.4 Side Slope Stability

In riprap-lined channels with side slopes steeper than 3:1, side slope stability must also be analyzed. The permissible shear stress on the side of a channel is given by

$$\tau_s = K_1 \tau_d \tag{4.62}$$

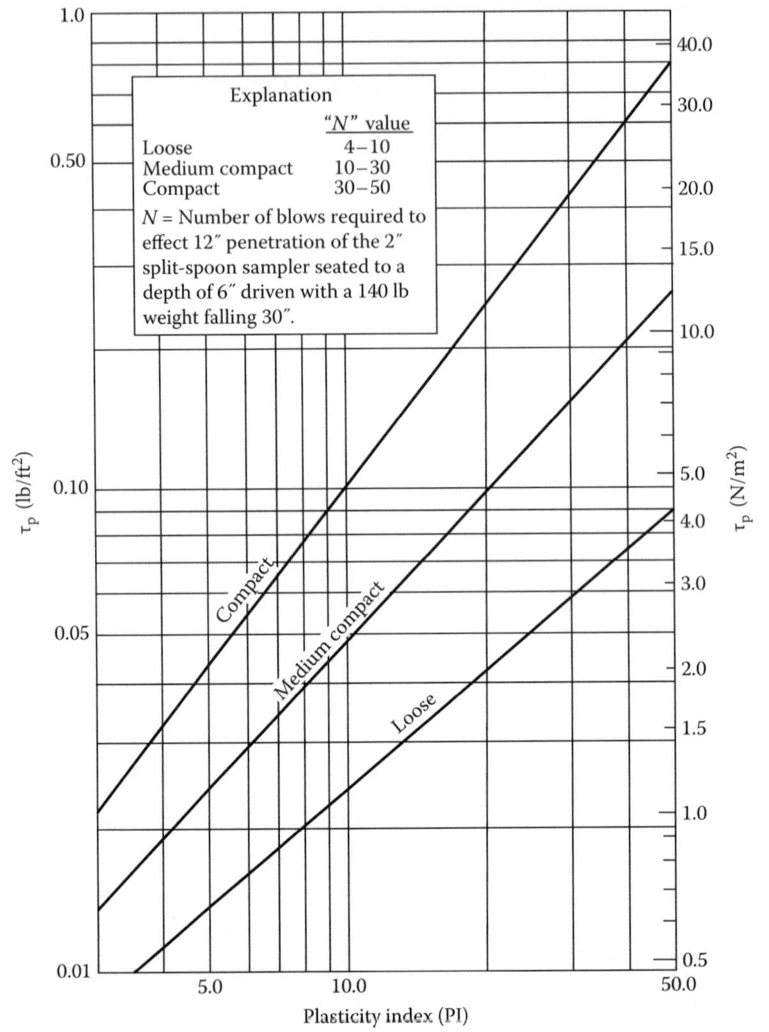

FIGURE 4.36 Permissible shear stress of cohesive soils.

Source: **FHWA, Urban Drainage Design Manual, HEC-22, 3rd Ed., August 2013.**

where

τ_s = side shear stress, N/m^2 (lb/ft^2)

τ_d = bottom shear stress, N/m^2 (lb/ft^2)

K_1 = ratio of channel side shear stress to bottom shear stress

For parabolic and triangular channels with rounded bottom, there is no sharp discontinuity along the perimeter. Thus, it can be assumed that the shear stress on the sides is equal to the bottom shear stress given by Equation 4.57, that is, $K_1 = 1$. For trapezoidal channels, the factorK_1 depends on the side slope, *m*, and the ratio of bottom width to depth (*b*/*d*) of the channel. The relationship between τ_s/τ_d and side slope, m, can be approximated by the following equation:

$$K_1 = 0.77m < 1.5$$
$$K_1 = 0.67 + 0.066m1.5 < m < 5 \qquad (4.63)$$
$$K_1 = 1m > 5$$

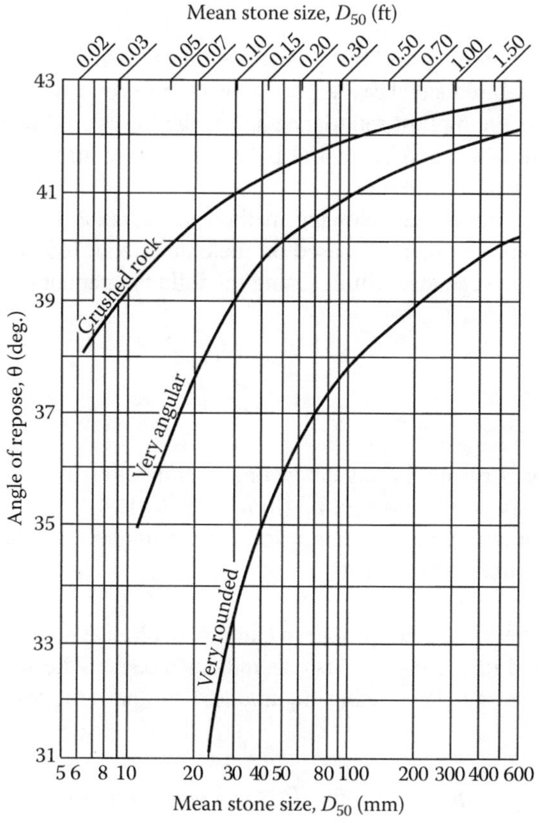

FIGURE 4.37 Angle of repose of riprap stones.

Source: HEC-15, 3rd ed. September 2005

Equation 4.63 can also be applied to triangular channels with sharp angles. For a channel lined with stone, the side stability is also affected by the angle of repose of the stone. Specifically, the ratio of tractive forces (permissible shear stresses) on the side and the bottom depends on the side slope and the angle of repose of the riprap stone and is given by

$$K_2 = \left(1 - \frac{\sin^2 \psi}{\sin^2 \theta}\right)^{1/2} \tag{4.64}$$

where θ and ψ are the angle of repose and the angle of side slope, respectively. The angle of repose depends on the size and shape of the stone and is plotted in Figure 4.37. The preceding equation indicates the side angle should be smaller than the angle of repose. The median stone size for the side slopes is calculated using the following equation:

$$(d_{50})sides = \frac{K_1}{K_2}(d_{50})bottom \tag{4.65}$$

where

(d_{50}) bottom for stability, see Table 4.10
K_1= ratio of side shear stress to bed shear stress (Equation 4.63)
K_2= ratio of tractive forces on the sides and bottom (Equation 4.64)

4.9.5 Grass Lining

The permissible shear stress of grass lining depends on the density and height of grass, known as retardance degree or retardance class. Vegetal covers are grouped into five retardance classes, A through E; class A has the highest retardance and E the lowest. Table 4.11 serves as a guide in selecting the vegetal retardance, and Table 4.12 provides specific information about the retardance class of various grass covers.

The HEC-15 (2005) presents the following method for calculating the stability of grass lined channels. This method, unlike others, is based on the combined effect of soil shear stress and the effective shear stress of the vegetative lining, using the following equation

$$\tau_p = \left[\frac{\tau_p, \text{soil}}{(1-C_f)}\right]\left(\frac{n}{n_s}\right)^2 \tag{4.66}$$

where

τ_p = permissible shear stress on the vegetative lining, N/m² (lb/ft²)
τ_p, soil = permissible soil shear stress, N/m² (lb/ft²), see table 4.10
C_f = grass cover factor (see Table 4.13), n_s = soil grain roughness coefficient
n = overall lining roughness coefficient

The permissible shear stress, τ_p, soil for cohesive and noncohesive soils applicable to this method is the same as those listed in Table 4.10. Also, as indicated earlier, the shear stress of noncohesive and cohesive soils can be calculated using Equations 4.59 and 4.61, respectively. Soil roughness coefficient can be held as

$$n_s = 0.016 \; for \; d_{75} \le 1.3 \; mm \; (0.05 \; in.)$$

For soils of larger grain size, the following equation is used:

$$n_s = \alpha_1 (d_{75})^{1/6} \tag{4.67}$$

where

n_s = soil grain roughness
α_1 = 0.015 SI; 0.026 CU
d_{75} = 75% finer stone size, mm (in.)

TABLE 4.11
Vegetal Retardance Selection Guide

Average Height of Grass, cm (in.)	Retardance Class: Grass Density: Good	Retardance Class: Grass Density: Fair
75 (30)	A	B
27.5–60 (11–24)	B	C
15–25 (6–10)	C	D
5–15 (2–6)	D	D
5 (<2)	E	E

TABLE 4.12
Retardances classes of Vegetal Covers

Retardance Class	Cover[a]	Condition
A	Weeping love grass	Excellent stand, tall, average 760 mm (30 in.)
	Yellow bluestem (*Bothriochloa ischaemum*)	Excellent stand, tall, average 910 mm (36 in.)
B	Kudzu	Very dense growth, uncut
	Bermuda grass	Good stand, tall, average 300 mm (12 in.)
	Native grass mixture (little bluestem, bluestem, blue gamma, and other long and short Midwest grasses)	Good stand, unmowed
	Weeping lovegrass	Good stand, tall, average 610 mm (24 in.)
	Lespedeza sericea	Good stand, not woody, tall, average 480 mm (19 in.)
	Alfalfa	Good stand, uncut, average 280 mm (11 in.)
	Weeping lovegrass	Good stand, unmowed average 330 mm (13 in.)
	Kudzu	Dense growth, uncut
	Blue gamma	Good stand, uncut, average 280 mm (11 in.)
C	Crabgrass	Fair stand, uncut 250 to 1200 mm (10 to 48 in.)
	Bermuda grass	Good stand, mowed, average 50 mm (6 in.)
	Common lespedeza	Good stand, uncut, average 280 mm (11 in.)
	Grass-legume mixture – fall, spring (orchard grass, redtop, Italian ryegrass, and common lespedeza)	Good stand, uncut, 150 to 200 mm (6 to 8 in.)
	Centipede grass	Very dense cover, average 150 mm (6 in.)
	Kentucky bluegrass	Good stand, headed, 150 to 300 mm (6 to 12 in.)
D	Bermuda grass	Good stand, cut to 60 mm (2.5 in.) height
	Common lespedeza	Excellent stand, uncut, average 110 mm (4.5 in.)
	Buffalo grass	Good stand, uncut, 80 to 150 mm (3 to 6 in.)
	Grass-legume mixture – fall, spring (orchard grass, redtop, Italian ryegrass, and common lespedeza)	Good stand, uncut, 100 to 130 mm (4 to 5 in.)
	Lespedeza sericea	After cutting to 50 mm (2 in.) height, very good stand before cutting
E	Bermuda grass	Good stand, cut to height, 40 mm (1.5 in.)

Source: US Soil Conservation Service, *Handbook of Channel Design for Soil and Water Conservation*, SCS-61, Stillwater Outdoor Hydraulic Laboratory, Oklahoma, June 1954, revised August 1966.

[a] *Covers classified have been tested in experimental channels. Covers were green and generally uniform.*

TABLE 4.13
Grass Cover Factor C_f

Growth Form	Excellent	Very Good	Good	Fair	Poor
Sod	0.98	0.95	0.90	0.84	0.75
Bunch	0.55	0.53	0.50	0.47	0.41
Mixed	0.82	0.79	0.75	0.70	0.62

Source: HEC-15, 3rd ed., 2005

The overall roughness coefficient, n, for grass linings varies with both vegetative retardance class and the flow shear stress. The latter parameter affects the roughness coefficient through bending the grass stem, which in turn reduces the stem height relative to the flow depth. The coefficient n in HEC-15 (2005) is given as

$$n = \alpha C_n \tau_o^{-0.4} \tag{4.68}$$

where
τ_o = mean boundary shear stress ($\tau_o = \gamma RS$)
C_n = grass roughness coefficient; dimensionless and the same in SI and CU
α = unit conversion constant 1.0 SI; 0.213 CU

The grass roughness coefficient depends on the density stiffness of grass and the grass cover condition and is given by

$$C_n = \beta C_s^{0.1} h^{0.528} \tag{4.69}$$

where
C_s = density-stiffness coefficient
h = stem height, m (ft)
β = unit conversion factor, 0.35 SI, 0.237 CU

Table 4.14 presents the density stiffness coefficient for various grass cover conditions and Table 4.15 lists the calculated roughness coefficient, C_n for a range of grass factors and stem height based on Equation 4.69. As Table 4.15 shows for grass heights, which range from 7.5 to 22.5 cm (3–9 in.), and grass cover conditions, the C_n for roadside grass channels varies from 0.1 to 0.3, with 0.2 being an average.

4.9.6 Manning's Roughness Coefficient for Linings

Manning's roughness, n, depends on the channel roughness and its relative magnitude to the flow depth. For a given lining, the channel roughness will increase with a decrease in flow depth. In urban developments and roadside channels, the flow depth typically ranges from 0.15 to 1 m (0.5 to 3.3 ft).

TABLE 4.14
Density–Stiffness Coefficient, C_s

Grass Condition	Excellent	Very Good	Good	Fair	Poor
C_s (SI)	580	290	106	24	8.60
C_s (CU)	49	25	90	2.0	0.73

TABLE 4.15
Roughness Coefficient C_n

Stem Height m (ft)	Excellent	Very Good	Good	Fair	Poor
0.075 (0.25)	0.168	0.157	0.142	0.122	0.111
0.150 (0.50)	0.243	0.227	0.205	0.177	0.159
0.225 (0.75)	0.301	0.281	0.254	0.219	0/197

For riprap, cobble, and gravel lining, Blodgett (1986) proposed the following equation for n which is included in HEC-15 (FHWA, 2005):

$$n = \frac{\alpha D_a^{1/6}}{7.05 + 16.4 \log\left(\frac{D_a}{d_{50}}\right)} \quad (4.70)$$

where

D_a = average flow depth in channel, m (ft)
d_{50}= mean riprap, gravel size, m (ft)
α = unit conversion factor = 1 SI; 0.82 CU

The average flow depth is defined as the ratio of waterway area to the top width (A/T). The preceding equation gives very conservative Manning's n. For an average depth of 0.5 m (1.5 ft) and 100 mm (4 in.) stone, for example, Equation 4.70 gives $n = 0.048$, which is far greater than values reported in the literature (e.g., see Table 5.1 in HEC-22, 2013).

The NJ Standards (2017) include graphs of the relationship between n and stone size based on the following equation:

$$n = 0.0395 (d_{50})^{1/6} \quad (4.71)$$

where d_{50} = stone size in feet.

Expressing d_{50} in millimeters and inches, Equation 4.71 becomes

$$n = 0.015 (d_{50})^{1/6} \text{ SI} \quad (4.72)$$

$$n = 0.026 (d_{50})^{1/6} \text{ CU} \quad (4.73)$$

These equations, which are the same as Equation 4.67 with d_{50} substituted for d_{75}, disregard the effect of relative roughness, the ratio of flow depth to stone size, and they underestimate n values for shallow channels. For example, for a 6 in. (150 mm) riprap lining, Equation 4.73 gives $n = 0.035$, which is unrealistic for such shallow flow depth. It is the author's recommendation that Equations 4.71, 4.72, and 4.73 be applied only when the depth of flow is at least five times greater than the stone size, d_{50}.

USDA-NRCS (1992) offers the following equation for the Manning's n for gravel and stone-lined channels:

$$n = \frac{\alpha y^{1/6}}{\left[14 + 21.6 \log\left(y / d_{50}\right)\right]} \quad (4.74)$$

where

y= flow depth, m (ft)
d_{50}= 50% finer stone size, m (ft)
α = 1.22 SI; 1 CU

Apart from differences in coefficients between Equations 4.70 and 4.74, the former equation is based on the average flow depth and the latter on the flow depth. Since the average flow depth (area divided by the top width) is always smaller than the flow depth in trapezoidal or parabolic channels, a direct comparison between these equations cannot be made. See Example 4.20. Both Equations 4.70 and 4.74 yield n values that are more realistic than Equation 4.71. Table 4.16 list n values calculated using Equation 4.74. However, it should be noted that Equation 4.74 still shows n values that are somewhat smaller than those in HEC-15 (2005).

For vegetal covers, as indicated, Manning's n varies with density and height of grass, that is the retardance class, and can be calculated using Equation 4.68. The HEC-15 (FHWA, 2005) method of calculating n was discussed in Section 4.9.3. Two other methods are discussed herein.

Figure 4.38 shows, in CU, the relationship between Manning's roughness n and the product of flow velocity, V, and hydraulic radius, R, for various vegetal retardance classes. Since the flow velocity and hydraulic radius are related to the roughness coefficient by Manning's formula, the design of grass channel using this figure requires iterative calculations. Specifically, for a trial n, the velocity or hydraulic radius is calculated using Manning's formula, and is refined iteratively until it fits the appropriate retardance curve on this figure (see Example 4.15).

The NJ Standards (2017) contains design charts for vegetal retardance classes A to E that eliminate any iterative calculations. These charts present flow velocity and Manning's n as a function of hydraulic radius, R, and channel slope, S, for each retardance class. Figure 4.39 here shows the chart for retardance D, that is, the design criterion for calculating the discharge capacity of grass-lined channels in New Jersey.

In practice, design calculations for well-maintained manmade grass swales are conservatively performed assuming retardance D or E, which correspond to dormant season or freshly mowed grass, respectively. To ensure stability against erosion, the flow velocity obtained from Figures 4.38

TABLE 4.16
Manning's n Values for Stone Linings

		Flow Depth, m (ft)		
Lining Category	**Stone Size, mm (in.)**	**0.15/(0.5)**	**0.5/(1.5)**	**1.0/(3.3)**
Gravel	25 (1)	0.029/0.029	0.026/0.026	0.025/0.025
	50 (2)	0.037/0.037	0.031/0.031	0.029/0.029
Cobbles	100 (4)	-	0.037/0.038	0.034/0.035
Riprap	150 (6)	-	0.043/0.044	0.038/0.039
	200 (8)	-	-	0.042/0.043
	300 (12)	-	-	0.048/0.049

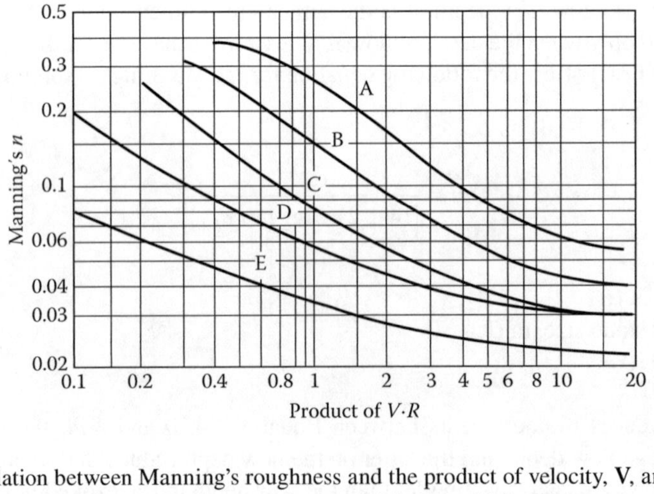

FIGURE 4.38 Relation between Manning's roughness and the product of velocity, V, and hydraulic radius, R. Curves A to E represent vegetal retardance: A for very high, B for high, C for moderate, D for low, and E for very low.

Source: Chow, V. T., Handbook of Applied Hydrology, McGraw-Hill, New York (1964).

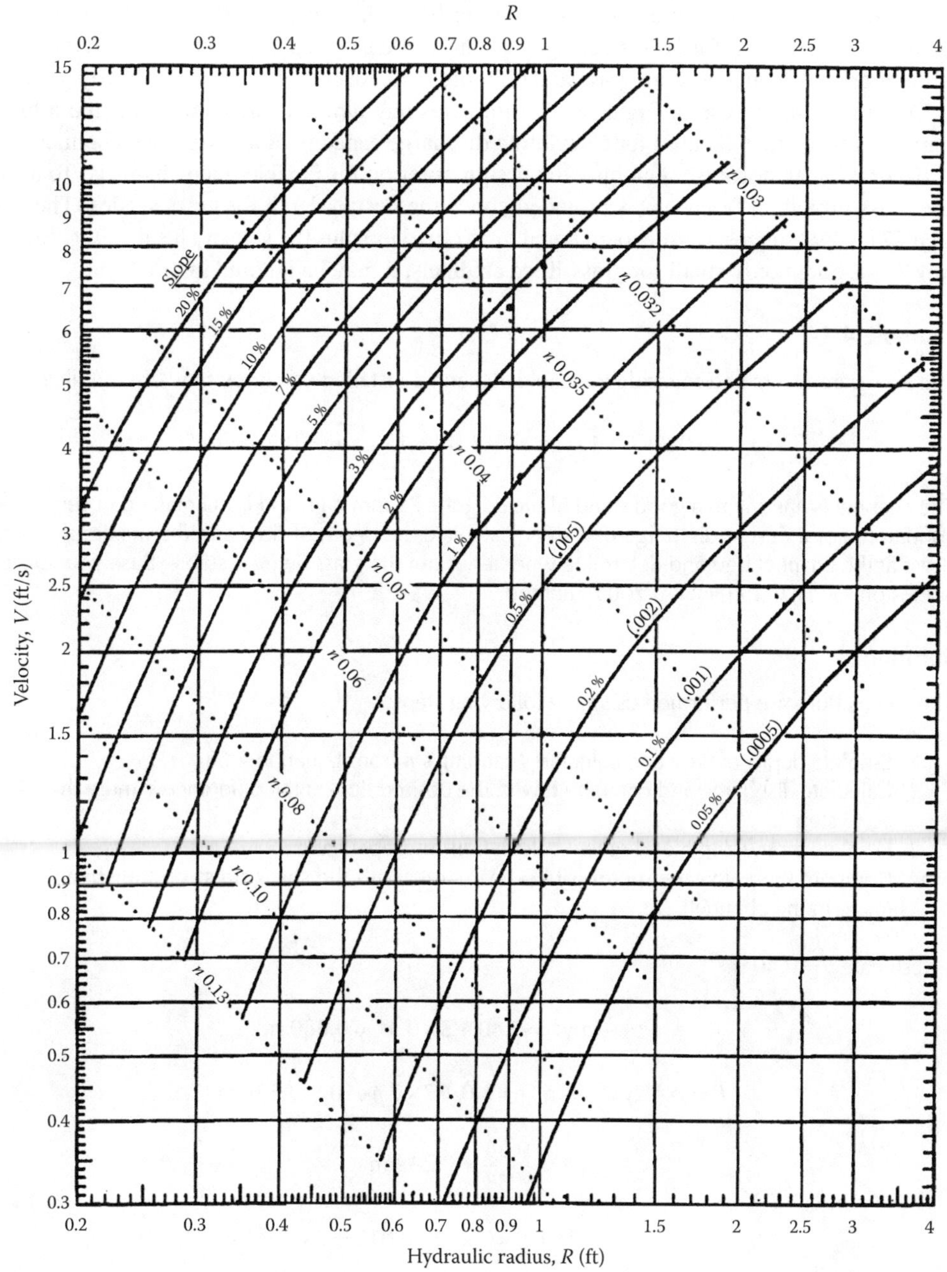

FIGURE 4.39 Manning's **n** Relation with **R**, **S**, and **V** for low vegetal retardance.

Source: NJ Standards, Figure A6-3.

and 4.39 are compared with the flow velocity in Table 4.9, those given by U.S. Soil Conservation Service (SCS, 1966), or those specified in soil erosion standards of many states. The reference list in this chapter contains multiple sources of SCS permissible velocity tables such as Table 7.6 in Chow (1959) and Table 9.3 in the ASCE manual 77 in ASCE and WEF (1992).

In general, the velocities in grass swales should be limited to 1 m/s (3 ft/s) for common grass cover and 1.5 m/s (5 ft/s) for very sturdy grass. The swale slopes should also be limited to 5% for common grass cover such as crab grass and no more than 10% for exceptionally sturdy grass mixtures. As indicated, the use of Figure 4.39 eliminates any iterative calculations for the allowable velocity, which can then be used for calculating discharge capacity of a grass-lined channel.

A 10- or 25-year storm is commonly the design criterion for flexible lined channels. To account for the grass growth, a freeboard is provided above the design depth for grass swales. The ASCE manual 77 in 1992 suggests a freeboard equal to 15 cm (6 in.) plus the velocity head, $V^2/2g$. However, since $V^2/2g$ is commonly small for grass-lined channels, it may be ignored in practice.

Example 4.14

The flow parameters of a roadside trapezoidal channel installed on clayey sand are as follows:

$$S_o = 0.012;\ b = 1.0\ m;\ m = 3;\ PI = 15; ande = 0.5\ for\ soil$$

The channel is lined with a good stand of mixed grass 75 mm high and is maintained in very good condition. For a design discharge of 1 m³/s, calculate the depth of flow and the maximum shear stress at the channel bed and determine whether or not the grass lining is stable. Base your calculations on the HEC-15 (FHWA, 2005) method.

Solution

The calculations are performed using the following steps:

1 Estimate depth of flow and calculate Manning's n from Equation 4.68.
2 Calculate discharge and compare it with the desired flow. If the difference is more than 5%, go to step 3.
3 Estimate a new depth and repeat calculations in steps 1 and 2.
4 Compare the calculated permissible shear stress with the shear stress on the channel bed (max. in the channel).

Estimate y = 0.40 m

$$A = by + my^2 = 0.40 + 3\times 0.4^2 = 0.880\ \text{m}^2$$

$$P = b + 2y\sqrt{}\left(1+\text{m}^2\right) = 1.0 + 2\times 0.4\sqrt{10} = 3.530\ \text{m}$$

$$R = \frac{0.88}{3.53} = 0.249\ \text{m}$$

$$\tau_o = \gamma RS = 9.81\times 10^3 \times 0.249 \times 0.012 = 29.4\ \text{N/m}^2$$

Calculate Manning's n from Equations 4.69 and 4.68 in that order:

$$C_n = 0.142\ \text{see Table 4.15}$$

$$n = \pm C_n \ddot{A}_o^{-0.4} = 1\times 0.142 \times 29.4^{-0.4} = 0.037$$

$$Q = \frac{1}{n} AR^{2/3} S^{1/2} = \frac{1}{0.037} \times 0.88 \times 0.249^{2/3} \times (0.012)^{1/2}$$

$$Q = 1.03 \text{ m}^3/\text{s} \approx 1.0 \text{ m}^3/\text{s}$$

The trial depth is good. Next calculate the maximum shear stress on the channel bed:

$$\tau_d = \gamma y S = 9.81 \times 10^3 \times 0.40 \times 0.012 = 47.1 \text{N}/\text{m}^2$$

Calculate the permissible shear stress of the soil from Equation 4.61:

$$\tau_p = \left(C_1 PI^2 + C_2 PI + C_3\right)\left(C_4 + C_5 e\right)^2 \times C_6$$

Coefficients C_1 through C_6 may be obtained from Table 4D.1 for SC soil:

$$C_1 = 1.07,\ C_2 = 14.3,\ C_3 = 47.7,\ C_4 = 1.42,\ C_5 = -0.61,\ C_6 = 4.8 \times 10^{-3}$$

$$\tau_p, \text{soil} = \left(1.07 \times 15^2 + 1.43 \times 15 + 47.7\right)\left(1.42 - 0.61 \times 0.45\right)^2 \times 4.8 \times 10^{-3}$$

$$\tau_p, \text{soil} = 1.92 \text{N}/\text{m}^2 \text{soil}$$

Next calculate the permissible shear stress on the grass using Equation 4.66:

$$\tau_p = \left[\frac{\tau_p, \text{soil}}{(1 - C_f)}\right]\left(\frac{n}{n_s}\right)^2$$

$$n = 0.037$$

$$n_s = 0.016$$

$$d_{50} < 1.3 \text{ mm}$$

$$C_f = 0.79 \text{ (see Table 4.13)}$$

$$\tau_p = \left[\frac{1.92}{1 - 0.79}\right] \times \left(\frac{0.037}{0.016}\right)^2 = 48.9 \text{ N}/\text{m}^2$$

$$\tau_p > \tau_d = 47.1$$

The grass is stable and the safety factor in this case is 48.9/47.1=1.03.

Example 4.15

For a grass swale at 2% slope, calculate Manning's n and R for a stable channel based on Figure 4.38.

Solution

Consider retardance D for the dormant season and conservatively assume a permissible velocity of 4 ft/s. Determine the design R using the following procedure:

1. Assume an *n* and determine the corresponding *VR* from Figure 4.38, curve D.
2. Calculate *R* based on the selected maximum permissible velocity ($R = VR/V_p$)
3. Calculate *VR* using the Manning formula:

$$\text{VR} = \frac{1.49R^{5/3}S^{1/2}}{n}\ \text{CU}$$

and check this value against the value of *VR* obtained in step 1.

4. Refine the assumed *n* and repeat steps 1 through 3 until the calculated *VR* is equal to the *VR* obtained from the *n* versus *VR* curve in Figure 4.38.

The trial process in CU are summarized in the following table:

Trial No.	n	VR	$R = VR/4$	$1.49\,R^{5/3}\,S^{1/2}/n$
1	0.050	1,5	0. 375	0.82
2	0.045	2.0	0.50	1.47
3	0.040	2.8	0.7	2.91

The design value for the channel section is interpolated as $R = 0.68$ ft and $n = 0.041$ for V= 4 ft/sec. Note this method is not sufficiently accurate and not recommended.

Example 4.16

Using the NJ Standards (Figure 4.39), find the hydraulic parameters for a grass-lined channel to carry a discharge of 35 cfs at 2% slope, based on

a. Permissible velocity of 4 ft/s
b. Hydraulic radius of 0.68 ft, obtained in the previous example

Solution

Entering Figure 4.39 with V = 4.0 ft/s and S = 0.02, the values of n and R can be obtained directly, without a need for any trial, as follows:

$$n = 0.041$$

$$R = 0.69\ ft$$

The channel area $A = Q/V = 8.75$ ft^2.

b. Entering Figure 4.39 with R = 0.68 ft, nearly equal to 0.69, A = 8.7 ft^2.

Note: The NJ Standards method yielded results which closely agree with those using Figure 4.38. However, this method eliminates the need for trial and error.

Example 4.17

Design a channel lined with 10 cm stone riprap to carry a discharge of 1.5 m^3/s at 1% slope. Base your design on NRCS's Equation 4.74 for calculating Manning's *n*. Perform your design for the following channel geometries:

a. Trapezoidal section of a 3:1 side slope
b. Parabolic section

Solution

Given Q = 1.5 m³/s, S = 1%.
First calculate hydraulic parameters of channels based on the permissible shear stress for d_{50} = 0.1 m, using Equation 4.60:

$$\tau_p = 0.75 \times 100\,mm = 75\,\mathrm{N/m^2}$$

The maximum shear stress occurs at the channel bed, which according to Equation 4.57 is

$$\tau = \gamma y S$$

Conservatively apply a safety factor of K=1.5:

$$\tau_p = K\tau\tau_d = 75/1.5 = 50\,\mathrm{N/m^2}$$

The maximum channel depth is

$$y = \frac{50}{(9.81 \times 10^3 \times 0.01)} = 0.5\ m$$

Both sections should be designed based on this depth.

From Equation 4.74:

$$n = 1.22y^{1/6} / \left[14 + 21.6\log(y/d_{50})\right]$$

$$n = 1.22 \times 0.5^{1/6} / (14 + 21.6\log 5) = 0.037$$

Simplify Manning formula for Q = 1.5 m³/s, S = 0.01, and n = 0.037:

$$AR^{2/3} = 1.50 \times 0.037 / (0.01)^{1/2} = 0.555$$

Trapezoidal sections, 3:1 side slope
Express area, A, and hydraulic radius, R, in terms of flow depth, y, and

$$A = by + my^2 = 0.5b + 0.75$$
$$R = A/P = (0.5b + 0.75)/(b + 2 \times 0.5\sqrt{10})$$
$$= (0.5b + 0.75)/(b + 3.16)$$

Solve the preceding equations for $AR^{2/3}$ for a trial value for b and refine it by an iterative process. The following table summarizes the calculations.

Trial No.	b, m	A, m²	R, m	$AR^{2/3}/n$
1	1.00	1.25	0.300	0.560
2	0.90	1.20	0.296	0.532
3	0.98	1.24	0.300	0.555

Therefore,

$$A = 1.24\ m^2$$

$$V = \frac{Q}{A} = \frac{1.50}{1.24} = 1.21\text{m/s}$$

Add depth by 0.15 m for free board; channel depth = 0.65 m

$$Top\ width\ B = 0.98 + 2 \times 3 \times 0.65 = 4.88\ m$$

For a parabolic section (refer to Figure 2.4 in Chapter 2), For this section:

$$A = (2/3)Ty = 0.333T,\ \text{where } T = \text{top width}$$

$$P = T + 8y^2/3T = T + 0.667/T$$

$$R = A/P = 0.333T/(T + 0.667/T)$$

Solve for T by a trial process (see table):

Trial No.	T, m	A, m²	R, m	$AR^{2/3}$
1	4.0	1.332	0.320	0.623
2	3.0	1.000	0.310	0.458
3	3.6	1.199	0.317	0.557

The wet section is 3.6 m wide and 0.5 m deep

$$V = 1.5/1.199 = 1.25\text{ m/s}$$

Adding 0.15 m as freeboard, the center depth will be

$$y = 0.5 + 0.15 = 0.65\text{ m}$$

Considering that the top width in parabolic sections varies in proportion to square root of depth, the top width of the section will be

$$T = 3.6(0.65/0.5)^{1/2} = 4.1\text{ m, top width}$$

While both sections are practical, the parabolic section is more efficient. Its top width is approximately 0.8 m narrower than the trapezoidal section.

Example 4.18

Using Equation 4.70, calculate the Manning's n value for the channels in the previous problem.

Solution

Using Equation 4.70, calculate Manning's n for the channels in the previous problem.

a. Trapezoidal section
   For the flow depth of 0.5, the top width and mean depths are

$$T = b + 2my = 0.98 + 2 \times 3 \times 0.5 = 3.98 \text{ m}$$

$$D = A / T = 1.24 / 3.98 = 0.312 \text{ m}$$

$$d_{50} = 0.1 \text{ m}$$

Then, rewriting Equation 4.70 as

$$n = \frac{1 \times 0.312^{1/6}}{7.05 + 16.4 \log\left(\frac{0.312}{0.1}\right)} = 0.054$$

b. Parabolic section
Based on the calculated top width and area for this channel, the mean depth and n are

$$D = A / T = 1.199 / 3.6 = 0.333 \text{ m}$$

$$n = \frac{1 \times 0.333^{1/6}}{7.05 + 16{,}4 \log\left(\frac{0.333}{0.1}\right)} = 0.053$$

The n calculated using Equation 4.70 deviates by over 50% from the n = 0.037 calculated based on Equation 4.74. As such, the NRCS method yields a larger channel section.

Example 4.19

Design the trapezoidal grass-lined channel of Example 4.14 based on the NJ Standards (Figure 4.39).

Solution

Given S = 1.2%, Q = 1 m^3/s = 35.3 cfs, and b = 1 m = 3.28 ft.,
Since both velocity and hydraulic radius are unknown, the problem requires an iterative solution.
First try:
Assume R = 1.0 ft
Step 1. Entering the figure with R = 1.0 and S = 0.012 gives

$$V = 4.25 \text{ ft/s}$$

$$n = 0.037$$

Step 2. Calculate area:

$$A = \frac{Q}{V} = \frac{35.3}{4.25} = 8.30\ ft^2$$

Calculate flow depth for the above area:

$$3.28y + 3y^2 = 8.3$$

$$y = 1.20\ ft\ (rounded\ to\ second\ decimal\ place)$$

$$P = 3.28 + 2\sqrt{10} \times 1.20 = 10.87\ ft$$

$$R = \frac{A}{P} = \frac{8.30}{10.87} = 0.76\ ft$$

Second try:
$R = 0.85$ ft
Figure 4.39 gives $V = 3.5$ ft/s; $n = 0.040$.
Calculate $A = Q/V = 35.3/3.5 = 10.09$ ft²
Calculate flow depth:

$$A = 3.28y + 3y^2 = 10.09$$

$$y = 1.37\ ft$$

$$P = 3.28 + 2\sqrt{10} \times 1.37 = 11.94\ ft$$

$$R = \frac{A}{P} = 0.845\ ft \approx 0.85\ ft$$

Therefore:

$$R = 0.85\ ft = 0.26\ m$$

$$V = 3.5\ \text{ft/s} = 1.07\ \text{m/s}$$

$$n = 0.040$$

$$A = 10.09 ft^2 = 0.94\ m^2$$

$$y = 1.37\ ft = 0.418\ m$$

Note: The calculated channel depth in this method differs by nearly 2% from that calculated 0.40 m using the more laborious HEC-15 (FHWA, 2005).

4.9.7 Channel Bends

Because of a change in flow direction, the flow induces centrifugal forces in channel bends. The result is a rise in the water surface at the outer bend and a drop at the inner bend. This rise at the outer bend, termed superelevation, can be estimated from the following equation:

$$\Delta y = \frac{V^2 T}{g R_c} \tag{4.75}$$

where
V = mean velocity, m/s (ft/s)
T = water surface width of the channel, m (ft)
g = gravitational acceleration, 9.81 m/s² (32.2 ft/s²)
R_c = mean radius of the bend, m (ft)

Note: The NJDOT Roadway Design Manual, Sec.10, uses the bottom width rather than top width in Equation 4.75, which underestimates the rise in water surface elevation at the outer bend.

To account for superelevation, an extra free board must be provided in a channel bend. Also, because of centrifugal forces, a larger shear stress is exerted on the outer bend. To account for this increase, a factor greater than 1 is applied to the shear stress in the straight channel section, as follows:

$$\tau_b = K_b \tau_d \tag{4.76}$$

TABLE 4.17
Bend Factor in Channels

R_c/T	10	9	8	7	6	5	4	3	2
K_b	1.05	1.13	1.20	1.30	1.41	1.53	1.67	1.83	2.0

where K_b is the bend factor and τ_b and τ_d are the shear stresses at the bend and straight section, respectively. The bend factor depends on the ratio of the bend radius, R_c, to the top width of the water surface, T. For R_c/T ratios between 2 and 10, K_b is given by

$$K_b = 2.38 - 0.206\left(\frac{R_c}{T}\right) + 0073\left(\frac{R_c}{T}\right)^2 \tag{4.77}$$

$$K_b = 2.0 \; for \frac{R_c}{T} \le 2.0$$

$$K_b = 1.05 \; for \frac{R_c}{T} \ge 10$$

Table 4.17 lists K_b values for R_c/T ratios between 2 and 10. The increased shear stress persists to a distance downstream of the bend. This distance, L_b, can be calculated using the following equation:

$$L_b = \frac{K_u R^{7/6}}{n_b} \tag{4.78}$$

where

n_b = Manning's roughness in the channel bend
R = hydraulic radius of channel, m (ft)
K_u = a constant, 0.74 SI, 0.60 CU

Contrary to intuition, the length L_b in the preceding equation does not depend on the bend curvature.

Example 4.20

A trapezoidal channel of 1 m bottom width and 3:1 side slope is lined with 15 cm (6 in.) thick riprap. The channel is at 1% slope and includes a straight section and a bend with a centerline radius of 18 m (59 ft). For a discharge of 1.5 m³/s,

a. Calculate the maximum shear stress in the straight reach and in the bend.

b. Determine if the lining is stable.

c. Calculate the minimum distance past the bend where the channel protection must extend.

d. Super elevation at the outer edge of channel.

Solution

The solution will be performed in SI; the solution in CU is left as an exercise to interested readers.
Step 1. Estimate n

From Table 4.16, estimate $n = 0.043$ (assuming $y \approx 0.5$ m)
Using Manning's formula:

$$AR^{2/3} = \frac{Qn}{S^{1/2}} = \frac{1.5 \times 0.043}{(0.01)^{1/2}} = 0.645$$

$$b = 1 \text{ m}$$

$$A = y + 3y^2;$$

$$R = \frac{(y + 3y^2)}{1 + 2\sqrt{10}y}$$

Substitute A and R in the Manning formula:

$$\frac{(y + 3y^2)^{5/3}}{(1 + 2\sqrt{10}y)^{2/3}} = 0.645$$

Simplify:

$$f(y) = \frac{(y + 3y^2)^{5/2}}{(1 + 2\sqrt{10}y)} = 0.645^{1.5} = 0.518$$

Solve for y by trial and error. The following table summarizes calculations:

Trial No.	y	$f(y)$
1	0.50	0.4197
2	055	0.572
3	0.54	0.538
4	0.53	0.507

Interpolate between $y = 0.54$ and 0.53 to obtain $y = 0.534$ m.
Check n using Equation 4.74.

$$n = \left[1.22 \times (0.534)^{1/6}\right] / \left[14 + 21.6 \log(0.534 / 0.15)\right] = 0.0424;\ say\ 0.042$$

The selected n = 0.043 was good. Therefore,

$$y = 0.534 \text{ m}$$

$$A = 0.534 + 3 \times 0.534^2 = 1.389\ m^2$$

$$R = \frac{(0.534 + 3 \times 0.534^2)}{(1 + 2 \times 0.534 \times 10^{1/2})} = 0.317 \text{ m}$$

a. Shear stress on the channel bed in straight section:

$$\tau_d = \gamma y S = 9.81 \text{ kN/m}^3 \times 0.534 \times 0.01 = 0.052 \text{ kPa} = 52 \text{ Pa}$$

Shear stress at bend:

$$\text{Top width}, T = 1 + 2 \times 3 \times .0.534 = 4.209; say\ 4.21 m$$

$$R_c / T = 18 / 4.21 = 4.3$$

From Table 4.17, by interpolation (or Equation 4.77)

$$K_b = 1.63$$

$$\tau_b = K_b \tau_d = 1.63 \times 52 = 85 \text{ Pa}.$$

b. Estimate the permissible shear stress from Equation 4.59:

$$d_{50} = 150 \text{ mm}$$

$$\tau_p = 0.75 \times 150 = 112.5 \text{Pa}$$

The d_{50} = 15 cm (6 in.) stone is stable.
The influence distance of bend can be calculated using Equation 4.78:

$$L_b = \frac{K_u R^{7/6}}{n}$$

$$L_b = \frac{0.74 \times 0.317^{7/6}}{0.042} = 4.61 \text{m} (15.1 \text{ft})$$

b. Super Elevation

$$V = 1.5 / 1.389 = 1.08 \; m/s$$

$$\Delta y = 1.08^2 \times \frac{4.21}{9.81 \times 18} = 0.028 \; m = 3 \; cm$$

4.9.8 Composite Lining

In practice, occasionally two different linings are employed in a single channel. An example is a channel that experiences low flows where concrete, riprap, or gravel is used in the bottom and a more cost-effective lining, such as grass, is used in the upper section. Figure 4.40 shows a grass-lined swale with gravel bed in Saint-Sauveur, Canada.

Flow calculations in a composite channel involve the use of an equivalent Manning's n for the entire perimeter of the channel. The equivalent, also referred to as effective roughness coefficient, n_e, is calculated using the following equation:

$$n_e = \left[\frac{P_L}{P} + \left(1 + \frac{P_L}{P}\right) \left(\frac{n_s}{n_L}\right)^{3/2} \right]^{2/3} n_L \tag{4.79}$$

where

P_L = perimeter of low flow lining
P = total flow perimeter
n_s = Manning's n for the side slope lining
n_L = Manning's n for the low flow lining

When vegetation is used as side slope lining, a transition lining should be used adjacent to the low flow channel to prevent erosion until the vegetative lining is established.

The stability calculations for composite channel are similar to those described for grass lining with additional steps as follows:

FIGURE 4.40 Grass-lined swale with gravel bed in Saint-Sauveur, Canada.

Source: Photo by the author (2014).

a. Calculating the effective n (n_e) using Equation 4.79.
b. Calculating the shear stress at the maximum depth τ_d and the sear stress on the side slope (Equations 4.57 and 4.62).
c. Comparing the shear stresses τ_d and τ_s to the permissible shear stresses τ_p for each lining. If τ_d or τ_s is greater than the τ_p for the respective lining, a different type of lining should be investigated.

4.10 OTHER LININGS

This section discusses gabion boxes and Reno mattresses (which are among the earliest methods of channel protection) and some of the newest methods such as turf reinforcement mats and erosion control blankets.

4.10.1 Gabion Baskets and Mattresses

Gabion is a riprap stone enclosed tightly in wire baskets made of zinc-coated (galvanized) or PVC-coated steel wire woven in the form of a rectangular container and reinforced on corners with heavier gage wire (see Figure 4.41). The baskets are installed along the channel sides, connected together and filled with well graded, durable stone. The baskets vary from 15 cm (6 in.) thick mattresses to 1 m (3 ft) thick box-like gabions. The wire mesh binds the stones together, preventing their movement. Gabion baskets provide significantly more resistance to erosion from flowing water than riprap. Figure 4.42 shows gabion walls at a drainage outfall. The outfall (designed by the author) is equipped with a Tideflex check valve to prevent water in the tidal channel from backing up into the drainage system during high tides.

To simplify calculations, gabion baskets and mattresses are commonly designed based on permissible velocity rather than permissible shear stress. Table 4.18 gives the maximum permissible velocity of gabion baskets and mattresses.

Over the past five decades, prefabricated concrete blocks and interlocked concrete mats were introduced to the market. In addition to the ease of installation, these blocks and mats have a more pleasing appearance than gabions. Also unlike the gabion baskets, which are vulnerable to collapse because of wire mesh damage, the concrete blocks are stable. Keystone Concrete Block and Supply Company in Scranton, PA, USA, which was established during the first half of the nineteenth century, was the first concrete block manufacturer in the United States. Now many companies make

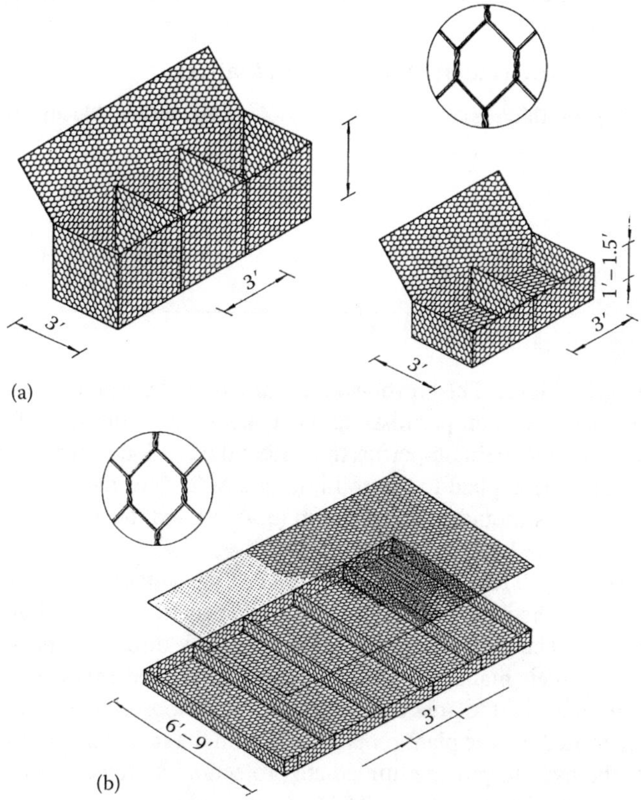

FIGURE 4.41 Gabion baskets (a) and mattress (b).

FIGURE 4.42 Gabion walls at drainage outfall at a tidal channel.

Source: Photo by the author.

various types of blocks and mats. ArmorLoc, ArmorFlex, Allan Block, Versa-Lok, and Verdura are some of the concrete blocks and erosion control mats that are available on the market.

4.10.2 Turf Reinforcement Mats

Turf reinforcement refers to a means of providing a structure to the soil/vegetation matrix that helps the establishment of vegetation and protects the vegetation once established. Turf can also be reinforced using gravel mulch. The latter method involves adding coarse to very coarse gravel into soil

TABLE 4.18
Maximum Flow Velocity for a Gabion Mattress

Basket/Mattress Thickness, cm (in.)	Permissible Velocity, m/s (ft/s)
15 (6)	2.0 (6)
23 (9)	3.5 (11)
30 (12)	4.3 (14)

Source: NJ Standards (2017).

and seeding the soil–gravel layer. The gravel–soil mixture provides a non-degradable lining. Gravel mulches are designed based on their permissible shear stress (see Table 4.10). The density, size, and gradation of the gravel are the main properties that affect the erosion control performance of gravel mulch. The gravel should be applied to the soil at a rate of 25% of the mixture. The thickness of mixture for fine graded soils should be 95–100 mm (3–4 in.) and the gravel should be applied at 6 kg/m^2 per centimeter depth (3 lb/ft^2 per inch).

A turf reinforcement mat (TRM) consists of a long-lasting synthetic fiber, filament, or netting of sufficient thickness. The mats provide strength and void space to retain soil and to establish grass roots within this matrix. The mats are placed along the flow direction and secured with staples at intervals specified by their manufacturer. Mats can be installed first and covered with topsoil and then seeded. Alternatively, the area can be covered with topsoil and seeded before the mat is installed. In the former method, the plant roots grow within the mat, whereas in the latter, the grass stems grow through the mat. To provide immediate protection, a growth medium may be added to the TRM to form an intimate bond with the TRM matrix, seed, and soil.

A number of companies manufacture TRMs, such as Green Armor by Profile EVS (https:///profleevs.com). This TRM consists of a three-dimensional matrix composed of thermally fused nylon filaments that provide superb anchors for root reinforcement. Enka Solutions also produces a range of these mats under its EnkaMat line (https://enkasolutiond.com): EnkaMat 7010, EnkaMat ASV 7020, and EnkaMat R45. A roll of EnkaMat R45 is 8 ft wide by 90 ft long (2.44 m × 27.45 m) and 0.75 in. (19 mm) thick and weighs 80 lbs (36.3 kg). EnkaMat products are especially suited for dry slopes.

LandLok, manufactured by Propex Global, is another brand-name TRM made of geosynthetics. This TRM is available in Landlok 300, which is a second-generation TRM, and Landloks 435, 450, and 1051, which are first-generation TRMs. Landlok 300 comes in 8.5 ft wide by 106 ft long (2.6 m × 32.3 m) rolls. Landlock 450 is 0.5 in (12.7 mm) thick, has tensile strength of 425 × 350 lb/ft (6.2 × 5.1 kN/m), and is rated for 18 ft/s (5.5 m/s) flow velocity. Landloks 450 and 300 are, respectively, stitch-bonded and woven-bonded and which last up to 10 years and 25 years, also respectively. Propex also makes Pyramat 50 and 75, which are high-performance mats (HP-TRMs) suitable for high-velocity flows that last up to 50 years.

Armormax is another Propex HP-TRM that is expected to last 50 years or longer. The US Army Corps of Engineers used this product to control erosion at their Rapid Repair Levee Break Laboratory in Vicksburg, Mississippi. More than 4500 m^2 (5000 yd^2) of Armormax were installed by January 2012, and less than 2 months after its application, the entire eroded area was covered with thick grass vegetation.

4.10.3 Erosion Control Blankets

An erosion control blanket (ECB) is a degradable product composed of natural or polymer fibers that are physically or chemically bonded together to form a uniform continuous mat. ECBs are stiffer, thicker, and denser than open-weave textile linings such as jute net, woven paper net, and straw blanket, which were common in the past.

Similar to TRMs, ECBs are placed on the channel banks parallel to the flow direction and secured with staples at specified spacings. A 30 to 60 cm (1 to 2 ft) overlap is provided to ensure full coverage. Depending on type, ECBs degrade in one to three years; but long-term protection is provided by the established vegetation. Curled wood mat, reinforced polypropylene fiber, and coconut fiber are among erosion control blankets.

In addition to TRMs, Propex makes ECBs including Landloks C2, CS2, S1, and S2. Landlok S1 and S2 consist of 100% wheat straw mechanically bound and covered on one side by netting. Both S1 and S2 are 7 mm (0.28 in) thick. The former resists 1.5 m/s (5 ft/s) flow velocity and the latter 1.8 m/s (6 ft/s) velocity before vegetation is grown. Futerra EnviroNet is another ECB that is manufactured by Profile EVS.

4.10.4 Properties of ECBs and TRMs

TRMs and ECBs are collectively known as rolled erosion control products (RECPs). The density, stiffness, and thickness are the main properties of RECPs in terms of their erosion control performance. These physical properties are measured by a series of standard tests that are referred to as index tests (see Table 4.19).

The Manning's n values for RECPs are a function of the shear stress. The roughness factor of each product is commonly measured by a full-scale laboratory test and is included in the manufacturer's specification. Table 4.20 presents the erosion control properties of the Landlok S2 ECB; the permissible shear stress and velocity are based on conditions of no vegetation growth.

ECBs and TRMs reduce tractive force of flowing water before it reaches the underlying soil. Therefore, to control erosion, the applied shear stress at the soil surface should be less than the permissible shear stress of the soil. As the shear stress on the erosion control mats or blankets increases beyond their limits, the lining is detached from the soil, and the flow directly contacts the soil surface. This limit, defined as the shear stress that results in 12.5 mm (0.5 in.) soil loss. To prevent this condition, a safety factor is applied to permissible shear stresses of TRMs and ECBs as specified by the manufacturer. Depending on the application, the safety factor varies from 1.5 to 2.

4.10.5 Design of RECP Lined Channels

The Manning's n and the permissible shear stress of RECPs (TRMs and ECBs) vary, and therefore, a single equation cannot be established for these products. To design a channel lined with any RECP, a table of n values versus applied shear stresses—upper, middle, lower— should be obtained from the manufacturer including n corresponding to each. The upper and lower shear stress must be equal to twice and one-half of the middle shear stresses, respectively.

The manufacturer's information is used to calculate n from the following equation:

$$n = a\tau_o^b \tag{4.80}$$

TABLE 4.19
Index Tests for RECPs (ECBs and TRMs)

Property	Index Test	Description
Density	ASTM D 6475	Standard test method for mass per unit area for erosion control blankets
	ASTM D 6566	Standard test method for measuring mass per unit area of turf reinforcement mats
	ASTM D 6567	Standard test method for measuring the light penetration of turf reinforcement mats
Stiffness	ASTM D 4595	Test method for tensile properties of geotextile by the wide-width strip method
Thickness	ASTM D 6525	Standard test method for measuring nominal thickness of erosion control products

TABLE 4.20
Properties of Landlok ECB S2 by Propex Global

Properties	Test Method	CU	SI
Origin of material		100%	100%
% US Manufactured			
Physical Properties			
Mass/Unit Area	ASTM D64A75	8.0 oz/sy	271 g/m[1]
Thickness	ASTM D6525	0.28	7 mm
Light Penetration (% Passing)	ASTM D6567	22%	
Water Absorption	ASTM D1117	450%	
Mechanical Properties			
Tensile Strength	ASTM D6818	57.6 × 57.6 Ib/ft	0.8 × 0.8 kN/m
Elongation	ASTM D6818	15% × 20%	
Performance			
Velocity (Unvegetated)	ASTM D6460	6 ft/s	1.8 m/s
Shear Stress (Unvegetated)	ASTM D6460	1.8 Ib/ft[1]	86 Pa
Manning's n (Unvegetated)	Calculated	0.032	
Rusle Cover Factor (C Factor)	ASTM D6459	0.04	
Roll Sizes		8.0 ft × 112.5 ft	2.4 m × 34.3m
		16.0 ft × 112.5 ft	4.9 m × 34.3 m

Notes: The values in the table are effective 05/01/2023 but are subject to change without notice.

[1] *Values represent testing at time of manufacture and are shown as typical values.*

where

τ_o = mean boundary shear stress N/m^2 (lb/ft^2)

Coefficient a and exponent b depend on n at the mid-range of the applied shear and n corresponding to the range of shear stresses, respectively.

These coefficients are calculated from the following equations:

$$a = n_m / \tau_m^b \tag{4.81}$$

$$b = -\sqrt{\left[\ln\left(n_m / n_\ell\right)\ln\left(n_u / n_m\right)\right]} / 0.693 \tag{4.82}$$

where

τ_m = middle shear stress

n_m = n value corresponding to τ_m

n_ℓ = n value corresponding to lower τ (τ_ℓ)

n_u = n value corresponding to upper τ (τ_u)

The permissible shear stress on an RECP lining is calculated both for the underlying soil and the RECP. In the case of TRMs, the presence of vegetation affects erosion resistance properties.

RECPs dissipate shear stress before it reaches the soil surface. To control erosion, the shear stress at the soil surface should be less than the permissible shear of the soil. As the shear stress on the surface of RECPs increases, the lining detaches from the soil and the current erodes the soil. The effective shear stress on the soil surface is related to the RECP shear stress and the design shear stress in the channel by the following equation:

$$\tau_e = \left(\tau_d - \tau_\lambda / 4.3\right) / \left(\pm / \tau_\lambda\right) \tag{4.83}$$

where

τ_e = effective shear stress on the soil N/m² (lb/ft²)
τ_λ = shear stress on the RECP that results in 12.5 mm (0.5 in.) of erosion
τ_d = design shear stress N/m² (lb/ft²)
α = conversion constant: 6.5 (SI), 0.14 (CU)

For stability, the permissible shear stress for the RECP should be at least equal to the design shear stress, the shear stress on the bed. Likewise, the effective shear stress should be no more than the effective shear stress on the soil, τ_λ. Substituting τ_p for τ_d and τ_p,soil for τ_e in Equation 4.83 results in the following equation for the permissible shear stress of the RECP:

$$\tau_p = (\tau_\lambda / \alpha)(\tau_p,\text{soil} + \alpha / 4.3) \tag{4.84}$$

The following example illustrates the design procedure for an RECP lined channel.

Example 4.21

To control erosion in a trapezoidal earthen channel an RECP of the following performance data is selected:

Roughness Rating

Applied Shear, N/m²	n Value
45	0.039
90	0.035
180	0.030

$$\tau_\lambda = 80\ \text{N/m}^2\ (\text{sher on lining at 12.5 mm soil loss})$$

The channel is at 2.0% slope, the bottom width and side slopes are 1.0 m and 3:1, respectively, and the soil is clayey sand of PI = 16, e = 0.45. Determine if the RECP lining is satisfactory as a temporary lining for a discharge of 0.70 m³/s.

Solution

Estimate depth at 0.35 m.
Calculate hydraulic radius, shear stress, and Manning's n as follows:

$$A = by + my^2 = 1 \times 0.35 + 3 \times 0.35^2 = 0.7175\ \text{m}^2$$

$$P = b + 2y\sqrt{(m^2+1)} = 1 + 2 \times 0.35\sqrt{10} = 3.21$$

$$R = 0.223\ \text{m}$$

$$\tau_o = \gamma RS = 9.81 \times 10^3 \times 0.223 \times 0.020 = 43.8\text{N/m}^2$$

$$b = -\sqrt{[\ln(0.035/0.039)\ln(0.030/0.035)]} / 0.693$$

$$b = -0.186$$

$$a = n_m / \tau_m^b = 0.035 / (90)^{-0.186} = 0.081$$

$$n = a\tau_o^b = 0.081 \times 43.8^{-0.186} = 0.040$$

Then calculate the discharge:

$$Q = (1/0.040) \times 0.7175 \times (0.223)^{2/3} \times (0.020)^{1/2} = 0.946 \text{m}^3/\text{s}$$

The calculated flow differs by 35% greater than the design discharge, try a smaller flow depth.

$$y = 0.31 \text{ m}$$

$$A = 0.31 + 3 \times \times 0.31^2 = 0.598 \text{ m}^2$$

$$P = 1 + 2 \times 0.31\sqrt{10} = 2.96 \text{ m}$$

$$R = 0.202 \text{ m}$$

$$\tau_o = \gamma RS = 9.81 \times 10^3 \times 0.202 \times 0.02 = 39.6 \text{ N/m}^2$$

$$n = 0.081 \times 39.6^{-.0.158} = 0.041$$

$$Q = (1/0.041) \times 0.598 \times (2.02)^{2/3} \times (0.627 \times 0.207)^{2/3} = 0.71 \text{m}^3/\text{s}$$

The calculated flow differs by 1.6% (less than 5%) of the design flow. Proceed with calculating the permissible shear stresses of the soil and the RECP using Equations 4.61 and 4.85, respectively.

$$\tau_p, \text{soil} = \left(C_1 PI^2 + C_2 PI + C_3\right)\left(C_4 - C_5 e\right)^2 C_6$$

See Appendix 4E for C values.

$$\tau_p, \text{soil} = \left(1.07 \times 16^2 + 14.3 \times 16 + 47.7\right)\left(1.42 - 0.61 \times 0.45\right)^2 \times 0.0048 = 3.47 \text{ N/m}^2$$

$$\tau_p = \left(\tau_{\text{»}} / \pm\right)\left(\tau_p, \text{soil} + \pm/4.3\right) = (80/6.5)(3.47 + 6.5/4.3) = 61.3 \text{ N/m}^2$$

Calculate the maximum shear stress on the channel bottom.

$$\tau_d = \gamma y S = 9.81 \times 10^3 \times 0.31 \times 0.02 = 60.8 \text{ N/m}^2$$

The selected RECP is acceptable as a temporary lining until the vegetation is established. Note that the permanent channel stability for the established grass has to be evaluated separately.

PROBLEMS

4.1 Calculate the flow spread in a 2 m wide asphalt gutter for a flow of 0.08 m³/s. The gutter is at 2% longitudinal and 4% cross slopes, respectively.

4.2 Calculate the spread in a 6 ft wide asphalt gutter for a flow of 2.5 cfs. The gutter is at 1.5% longitudinal slope and 4% cross slope.

4.3 A 2 m wide smooth asphalt shoulder is at 4% cross slope and 0.5% longitudinal slope. Does the spread extend beyond the shoulder for a flow of 0.09 m³/s?

4.4 A 6 ft wide smooth asphalt shoulder is at 4% cross slope. For a flow of 2.5 cfs, does the spread extend beyond the shoulder if the longitudinal slope is 0.5%?

4.5 Calculate the flow spread on a 2 m wide shoulder at 4% cross slope for gutter flows of 0.03 m³/s, 0.06 m³/s, and 0.09 m³/s. The roadway and shoulder have an asphalt cover (n = 0.016). Perform calculations for a longitudinal slope of 3%.

4.6 Solve Problem 4.5 for a 6 ft wide shoulder and discharges of 1, 2, and 4 cfs.

4.7 In Problem 4.6, calculate the flow capture and efficiency of a " B" inlet using the NJDOT method.

4.8 Recalculate flow capture and efficiency in Problem 4.7 using the HEC-22 Charts in Appendix 4C.

4.9 A grate inlet is to be placed at a roadway sag flush with a 15 cm curb to capture the gutter flow, calculated at 0.15 m^3/s. Calculate the minimum required length of a 60 cm wide grate. For grate in sags assume that 50% of the grate opening and 25% of the grate perimeter are clogged by debris.

4.10 Calculate the required grate opening area in Problem 4.9 for a discharge of 6 cfs and 2 ft wide grate. The curb is 6 in. high. Consider weir flow through grate, in this case.

4.11 In Example 4.2 of this chapter, calculate the maximum spacing of the first set of inlets from the high point for a 75% efficiency.

4.12 A pond discharges through a 1.2 m × 1.2 m square box culvert at a rate of 2.5 m^3/s. The culvert is very long, Manning's n is 0.014, and $S = 0.005$. Calculate the elevation of water level in the pond at the inlet face of the culvert. The exit loss in the pond and the entrance loss can be neglected. Assume that the outlet is unsubmerged.

4.13 Rework Problem 4.12 if a 1200 mm circular culvert is used in lieu of the box culvert. Use $n = .013$.

4.14 Solve Problem 4.12 for a discharge of 85 cfs and 48 in. culvert.

4.15 Solve Problem 4.12 if the culvert is 300 ft long and the culvert outlet is submerged up to its crown. Account for inlet and exit losses in this case.

4.16 A 60 in. RCP culvert carries a stream under a roadway. The culvert is 100 ft long; at 2% slope, and its upstream invert lies 10 ft below the edge of road. Calculate the capacity of the culvert for the following conditions:
The culvert outlet is unsubmerged.
The outlet is submerged 2 ft above its crown.

4.17 A box culvert is to carry a discharge of 10 m^3/s. The culvert is 20 m long and has 0.4% slope. The depth of water at the upstream face of the culvert is not to rise more than 1 m above its crown. Select a suitable size culvert. The downstream face of e culvert is unsubmerged.

4.18 For a channel in alluvial silt, the Manning's n value and the maximum permissible velocity are 0.02 and 2 ft/s, respectively. Calculate the corresponding permissible tractive force if the channel is at 0.9%.

4.19 Calculate the discharge and the cross-sectional area of a channel excavated in a noncohesive soil having permissible shear stress of 5 Pa, angle of repose of 32°, and $n = 0.025$. The channel slope is 0.4%.

4.20 Design a parabolic swale lined with Bermuda grass to carry a discharge of 35 cfs, at 2% slope. Base your design on retardance class D and the NJ Soil Erosion Standards.

4.21 Design the grass swale of Problem 4.20 for a discharge of 1m^3/s, using Figure 4.38 (n vs. VR relation).

4.22 Design a trapezoidal swale, lined with riprap stone of $d_{50} = 15$ cm, to carry a discharge of 1.5 m^3/s. The channel has an average slope of 1.5%. Base your design on the permissible shear stress method and Equation 4.70 for n, and a channel of 3:1 side slope.

4.23 Solve Problem 4.22 for $d_{50} = 6$ in. and 50 cfs discharge and
a. Equation 4.70
b. NRCS Equation 4.74

4.24 A trapezoidal grass-lined swale is to carry a discharge of 1.5 m^3/s at 1.5% slope. Th e soil is clayey sand of PI = 20 and $e = 0.5$. The swale is 1.25 m wide at the bottom and has 3:1 side slope. For a mixed-grass lining in very good condition, 75 mm thick, design the swale using the HEC-15 (FHWA, 2005) method.

4.25 A trapezoidal channel of bottom width of 2 m and side slopes of 2:1 has a bend with a radius of 15 m. For a discharge of 8 m^3/s, water depth at the inner wall of the bend is 1.0 m. Calculate water depth at the outer bank around this bend. The channel is at mild slope.

4.26 Calculate the water depth at the outer bank of a trapezoidal channel of 2:1 side slope and 6 ft bottom width at a 50 ft radius bend. The channel discharge is 250 cfs and the depth of water at the inner bend is 3 ft.

4.27 Calculate the length of the channel past the bend in Problem 4.26, which requires protection. Base your calculations on a Manning's n of 0.040.

4.28 The trapezoidal channel of Problem 4.25 is lined with d_{50} = 15 cm stone and is at 1.3% slope. Determine whether or not the stone is stable. Base your analysis on permissible shear stresses listed in Table 4.10.

4.29 Calculate the effective Manning's n for the trapezoidal channel in Problem 4.24 if it is lined with concrete at the bottom and 75 mm high grass at the sides. Assume grass cover to be in good condition. Base your calculations for n = 0.016 for concrete bed.

4.30 In Problem 4.24, if the slope of the channel is 3%, will the grass be stable?

4.31 An RECP having the following listed roughness rating is used to control erosion in a trapezoidal earthen channel of 3:1 side slope:

Roughness Rating applied Shear (N/m^2)	n Value
40	0.042
80	0.039
160	0.035

$$\tau_\ell = 75\text{N}/\text{m}^2$$

The bottom width and the channel slope are 0.9 m and 1.5%, respectively. The soil is clayey sand (SC) having PI = 16 and e = 0.5. Determine if the RECP lining is satisfactory for a discharge of 0.5 m^3/s.

APPENDIX 4A: DERIVATION OF THE GUTTER FLOW EQUATION

In a gutter cross-section, the top width considerably exceeds the flow depth at the curb (25 times for a 4% cross slope). As such, a direct application of the Manning's formula to gutter cross section does not accurately reflect the flow in the gutter. To be accurate, the flow should be calculated by the sum of elemental flows through the section.

Figure 4A.1 illustrates flow in a triangular curb gutter. The partial flow through an element of width, dx, can be calculated from the following equation:

$$dQ = Vydx \tag{4A.1}$$

where V is the average flow velocity and y is the depth of flow within the segment. Using the Manning formula in SI units, the average flow velocity within the segment can be calculated by

$$V = \frac{y^{2/3} S^{1/2}}{n} \tag{4A.2}$$

where n is the Manning's roughness of the gutter section and S is the longitudinal slope of the gutter. According to Figure 4A.1,

$$\mathrm{d}x = \frac{\mathrm{d}y}{S_x} \tag{4A.3}$$

Substituting V and dx from Equations 4A.2 and 4A.3 into Equation 4A.1 yields

$$\mathrm{d}Q = \left(\frac{1}{n}\right)\left(\frac{S^{1/2}}{S_x}\right) y^{5/3} \mathrm{d}y \tag{4A.4}$$

Integrating the preceding equation within the limits of the flow depth from 0 to d across gutter, the total flow can be calculated as

$$Q = \int_0^d \frac{1}{n} y^{5/3} \frac{S^{1/2}}{S_x} \mathrm{d}y$$

$$Q = 0.375 \left(\frac{S^{1/2}}{S_x}\right) \frac{d^{8/3}}{n} \tag{4A.5}$$

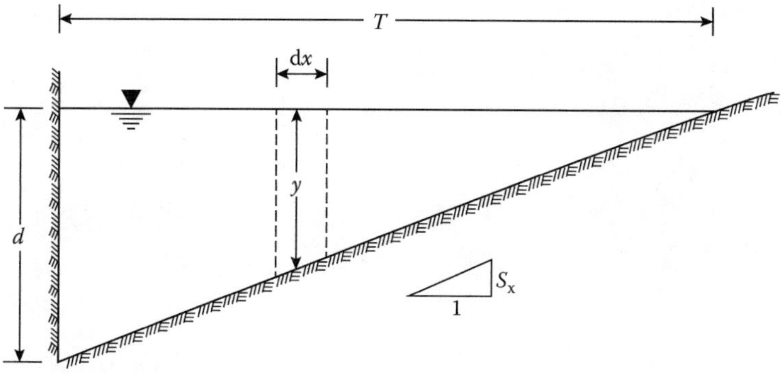

FIGURE 4A.1 Triangular gutter flow.

Substituting for $d = TS_x$ in the preceding equation gives

$$Q = 0.375 S^{1/2} S_x^{5/3} \frac{T^{8/3}}{n} \tag{4A.6}$$

In customary units, the gutter flow equation can be derived by simply applying 1.49 as a factor instead of 1 in Equation 4A.2, resulting in

$$Q = 0.56 \left(\frac{S^{1/2}}{S_x} \right) \frac{d^{8/3}}{n} \tag{4A.7}$$

and

$$Q = 0.56 S^{1/2} S_x^{5/3} \frac{T^{8/3}}{n} \tag{4A.8}$$

APPENDIX 4B: FLOW EQUATIONS FOR INLETS ON ROADWAYS AT 0% GRADE

The flow at a distance of x from the midway point between any two inlets can be calculated from the following equation:

$$Q_x = qx \tag{4B.1}$$

where q represents the flow per lineal length of roadway (refer to Figure 4B.1). Using the rational method, the unit discharge is related to the rainfall intensity, I, mm/h (in./h) by the following equations:

$$q = \frac{2.78 \times 10^{-3}\left(W + W'\right)}{10{,}000} \times ISI \tag{4B.2}$$

$$q = \frac{W + W'}{43{,}200} \times ICU \tag{4B.3}$$

where W and W' are the width of roadway and gutter, respectively.

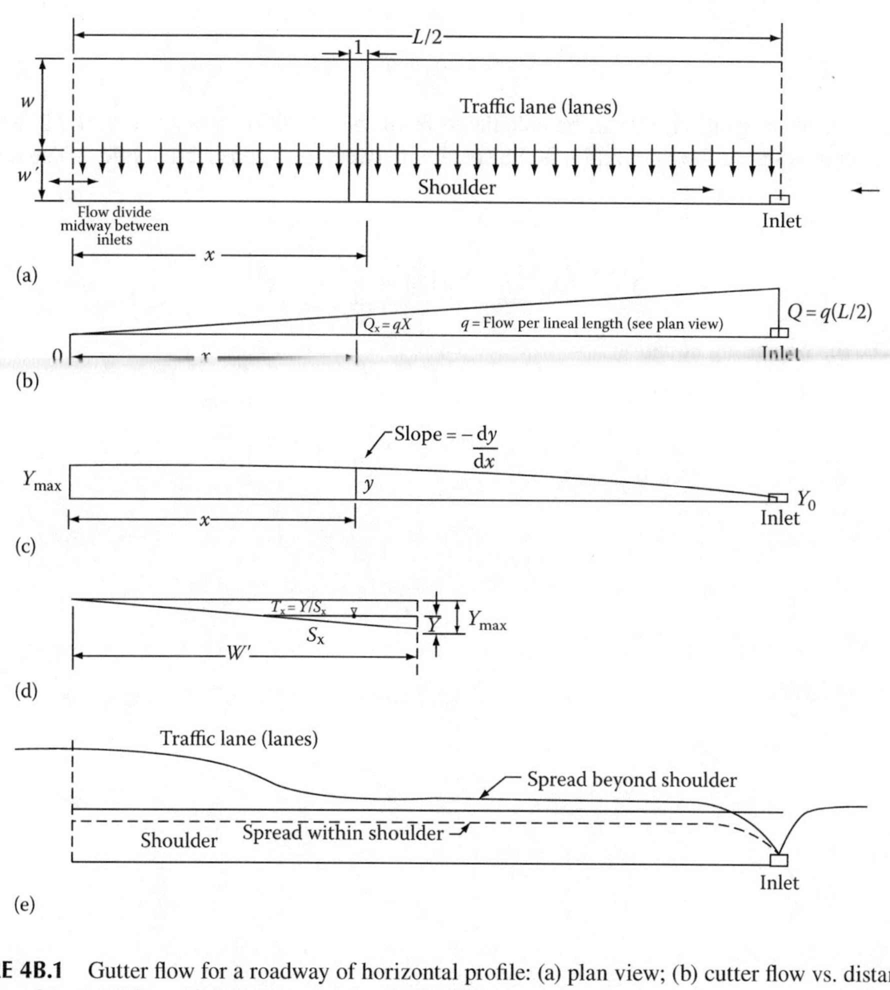

FIGURE 4B.1 Gutter flow for a roadway of horizontal profile: (a) plan view; (b) cutter flow vs. distance; (c) variation of depth of flow; (d) flow cross-section; and (e) flow spread.

A runoff coefficient of 1.0 is used for paved roadway in these equations. Using Manning's formula, Q_x can be related to the gutter flow parameters shown in Figure 4B.1 as follows:

$$A = \frac{y^2}{2S_x};\ R = \frac{y}{2}\ (\textit{refer to flow cross section, d})$$

$$Q = AR^{2/3}\frac{S^{1/2}}{n} = 0.315y^{8/3}\frac{S^{1/2}}{S_x n}\ \text{SI} \tag{4B.4}$$

where S is the energy slope.

Eqating Equations 4B.1 and 4B.4 and raising to the second power yields

$$q^2x^2 = 0.1y^{16/3}\frac{S}{S_x^2 n^2}$$

Approximating the energy slope S by dy/dx in the preceding equation,

$$y^{16/3}dy = -10.08S_x^2 n^2 q^2 x^2 dx$$

Integrating:

$$y^{19/3} = -21.3S_x^2 n^2 q^2 x^3 + C \tag{4B.5}$$

The constant of integration, C, can be calculated from the condition $y = y_o$ at $x = L/2$ where y_o is the depth of water at the inlet and L is the spacing between two consecutive inlets. The resulting equation is

$$y^{19/3} = y_o^{19/3} + 21.3S_x^2 n^2 q^2\left[\left(\frac{L}{2}\right)^3 - x^3\right] \tag{4B.6}$$

In CU, the preceding equation reads as

$$y^{19/3} = y_o^{19/3} + 10.0S_x^2 n^2 q^2\left[\left(\frac{L}{2}\right)^3 - x^3\right] \tag{4B.7}$$

APPENDIX 4C: HYDRAULIC DESIGN CHARTS FOR INLETS

Figures 4C.1 and 4C.2 give the overall efficiency of a 0.6 m × 0.6 m and 0.6 m × 1.2 m, 2 ft × 2 ft, and 2 ft × 4 ft grates for a 4% cross slope for a large range in gutter flow and longitudinal slope.

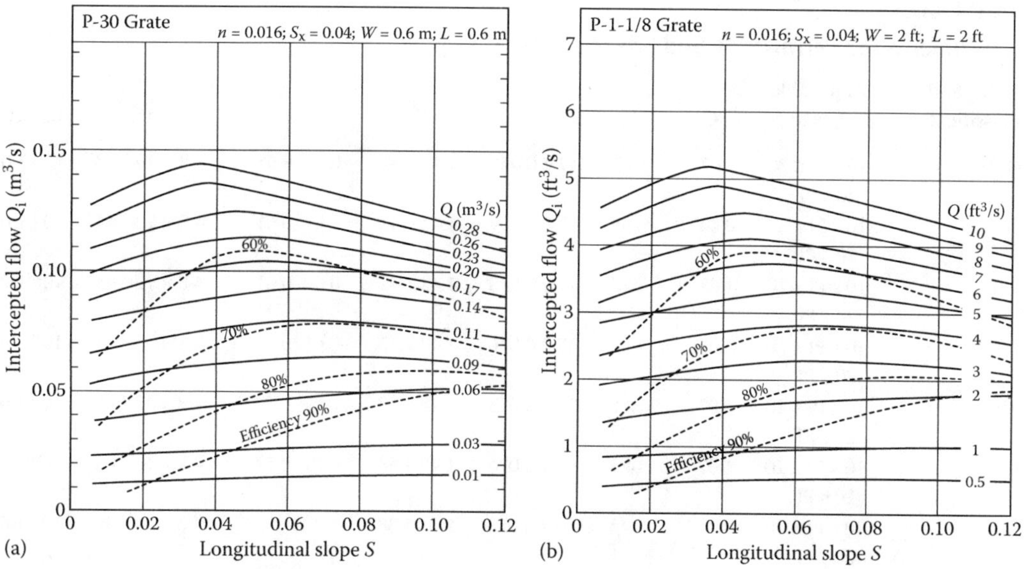

FIGURE 4C.1 (a) Interception capacity of a 0.6 m × 0.6 m, P-30 grate (SI units). (b) Interception capacity of a 2 ft × 2 ft, P1–1/8 grate (CU).

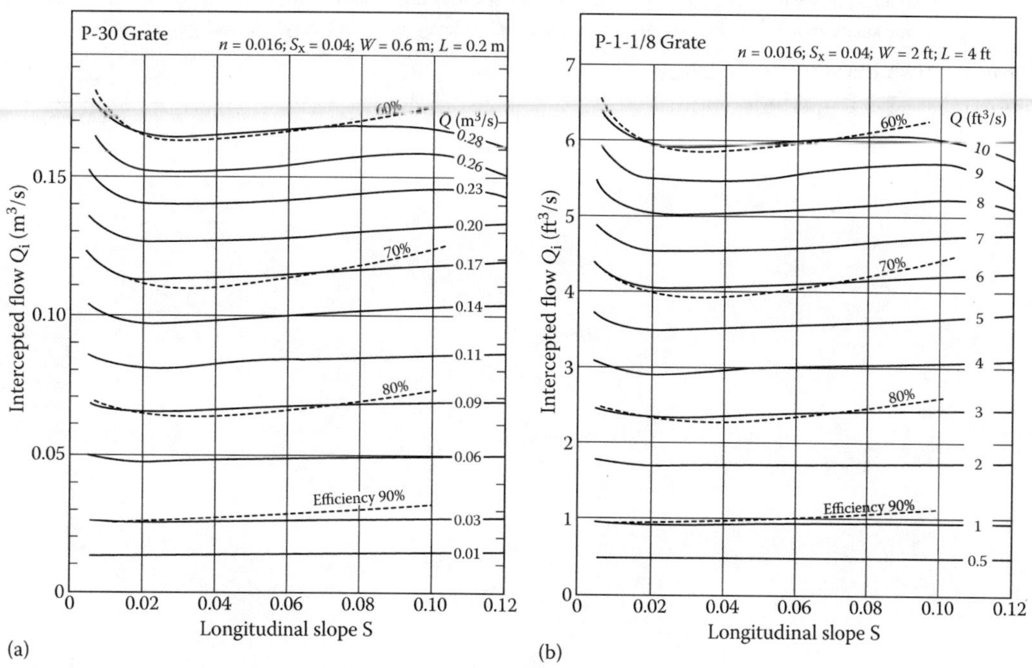

FIGURE 4C.2 (a) Interception capacity of a 0.6 m × 1.2 m P-30 grate (SI units). (b) Interception capacity of a 2 ft × 4 ft, P1–1/8 grate (CU).

APPENDIX 4D: PERMISSIBLE SHEAR STRESS OF COHESIVE MATERIAL IN HEC-15

TABLE 4D.1
Coefficients for Permissible Soil Shear Stress

ASTM Soil Classification	Applicable Range	C_1	C_2	C_3	C_4	C_5	C_6 (*SI*)	C_6 (*CU*)
GM	$10 \le PI \le 20$ $20 \le PI$	1.07	14.3	47.7 0.076	1.42 1.42	-0.61 -0.61	$4.8 \times 10^{\pm3}$ 48	10^{-4} 1.0
GC	$10 \le PI \le 20$ $20 \le PI$	0.0477	2.86	42.9 0.119	1.42 1.42	-0.61 -0.61	$4.8 \times 10^{\pm3}$ 48	10^{-3} 1.0
SM	$10 \le PI \le 20$ $20 \le PI$	1.07	7.15	11.9 0.058	1.42 1.42	-0.61 -0.61	$4.8 \times 10^{\pm3}$ 48	10^{-4} 1.0
SC	$10 \le PI \le 20$ $20 \le PI$	1.07	14.3	47.7 0.076	1.42 1.42	-0.61 -0.61	$4.8 \times 10^{\pm3}$ 48	10^{-4} 1.0
ML	$10 \le PI \le 20$ $20 \le PI$	1.07	7.15	11.9 0.058	1.48 1.48	-0.57 -0.57	$4.8 \times 10^{\pm3}$ 48	10^{-4} 1.0
CL	$10 \le PI \le 20$ $20 \le PI$	1.07	14.3	47.7 0.076	1.48 1.48	-0.57 -0.57	$4.8 \times 10^{\pm3}$ 48	10^{-4} 1.0
MH	$10 \le PI \le 20$ $20 \le PI$	0.0477	1.43	10.7 0.058	1.38 1.38	-0.373 -0.373	$4.8 \times 10^{\pm3}$ 48	10^{-3} 1.0
CH	$20 \le PI$			0.097	1.38	-0.373	48	1.0

Note: CH = inorganic clays of high plasticity, fat clays; CL = inorganic clays of low to medium plasticity, gravelly clays, sandy clays, silty clays, lean clays; GC = clayey gravels, gravel–sand–clay mixtures; GM = silty gravels, gravel–sand silt mixtures; MH = inorganic silts, micaceous or diatomaceous fine sands or silts, elastic silts; ML = inorganic silts, very fine sands, rock flour, silty or clayey fine sands; SC = clayey sands and–clay; SM =silty-sand sand–silt.

Source: FHWA HEC-15.

REFERENCES

American Concrete Pipe Association (ACPA), 2007, *Concrete pipe design manual*, 19th printing, American Concrete Pipe Association.

ASCE and WEF, October 1992, *Design and construction of urban stormwater management systems*, American Society of Civil Engineers Manuals and Reports of Engineering Practice No. 77 and Water Environment Federation Manual of Practice FD-20.

Blodgett, J.C., 1986, *Rock riprap design for protection of stream channels near highway structures, Vol. 1D. Hydraulic characteristics of open channels*, USGS, Water Resources Investigation Report 86-4127.

Campbell Foundry Company, Harrison, Kearny, New Jersey, http://www.campbell.com.

Chow, V.T., 1959, *Open channel hydraulics*, McGraw-Hill, New York.

Chow, V.T., 1964, *Handbook of applied hydrology*, McGraw-Hill, New York.

EPA (US Environmental Protection Agency), 1976, *Erosion and sediment control, surface mining in the Eastern U.S.*, EPA, EPA-62515-76-006, Washington, DC.

FHWA (US Department of Transportation), 1967, *Use of riprap for bank protection*, Washington, DC.

FHWA (US Department of Transportation), March 1984, *Drainage of highway pavements*, Hydraulic Engineering Circular No. 12 (HEC-12), Archived. FHWA, *HEC-12*, 2020, PDH online course C343, Instructure Vincent D. Reynolds.

FHWA (US Department of Transportation), September 2005, *Design of roadside channels with flexible linings*, Hydraulic Engineering Circular No. 15 (HEC-15), 3rd ed.

FHWA (US Department of Transportation), 2012, Hydraulic design of highway culverts, Hydraulic Design Series (HDS-5), Washington, DC, 3rd ed., April.

FHWA (US Department of Transportation), August 2013, *Urban drainage design manual*, 3rd ed., FHWA, Hydraulic Engineering Circular No. 22 (HEC-22)., Archived., *HEC-22*, 4th ed., 2024, by R.K. Kilgore, p. 313, online.

NJDCA, 2023, *Residential site improvements standards*, New Jersey Administrative Code, Title 5, Chapter 21, NJ Department of Community Affairs, adopted January 1996, last updated 2024, https://www.nj.gov/dca/codes/offices/rsis.shtml.

NJDOT (New Jersey Department of Transportation), April 2015, *Drainage design, section 10*, Roadway Design Manual.

NJ State Soil Conservation Committee, July 2017, *Standards for soil erosion and sediment control in New Jersey*, NJ State Soil Conservation Committee.

NY State Department of Environmental Conservation, November 2016, *New York state standards and specifications for erosion and sediment control, August 2005*, NY State Department of Environmental Conservation, Albany, NY.

Pazwash, H. and Boswell, S.T., 2003, Proper design of inlets and drains for roadways and urban developments, *ASCE EWRI Conference, World Water and Environmental Resources Congress*, June 22±26, Philadelphia, PA.

Pazwash, H. 2011, *Urban Storm water management*, p. 434. CRC Press, Boca Raton, FL.

US Soil Conservation Service, 1966, *Handbook of channel design for soil and water conservation, March 1947, revised June 1954, converted to metric system August 1966*, Stillwater Outdoor Hydraulic Laboratory.

USDA, NRCS, 1992, *Engineering field handbook*, Washington, DC.

USDA, August 2007, *Threshold channel design, NEH (National engineering handbook)*, Part 654, Chapter 8, U.S. Department of Agriculture, NRCS (Natural Resources Conservation System.

US Soil Conservation Service, 1966, Handbook of channel design for soil and water conservation, SCS-61, March, 1947, revised June 1964, converted to metric system August 1966 Stillwater Outdoor Hydraulic Laboratory, Oklahoma.

5 Stormwater Management Regulations

This chapter covers federal regulations to improve the quality of water in national streams and lakes. The regulations include the Clean Water Act (CWA) of 1972 and the National Pollutants Discharge Elimination System (NPDES) Phases I and II and EISA (Energy Independence and Security Act of 2007) Section 438. Following the enactment of CWA and NPDES, states adopted their own stormwater management regulations to address the issue of runoff quality. A Brief descriptions are given of the regulations for the states of New Jersey, Maryland, and New York.

5.1 CLEAN WATER ACT OF 1972

Prior to the enactment of the Clean Water Act of 1972, sanitary sewage, combined sewage, and industrial wastewater were directly discharged into state open waters without receiving any treatment. To restore and maintain the chemical and biological integrity of the nation's waters, the U.S. EPA enacted the CWA, which prohibited discharge of dredged or fill material as well as untreated wastewater from municipal and industrial sources into US water bodies without a NPDES permit. These permits (discussed in the following three subsections) fall under Section 404 of CWA. A June 5, 2024, Stormwater report by the Water Environment Federation (WEF), indicates that according to the EPA estimated that more than \$630 billion was needed to achieve the Clean Water Act goals such as repairing or upgrading stormwater and sewer infrastructure over the next 20 years.

5.1.1 NPDES Phase I

The CWA somewhat improved the quality of our nation's waters, but it did not address the nonpoint source pollutants carried with stormwater runoff. Following an extensive study conducted between 1979 and 1983 under Nationwide Urban Runoff Program (NURP), the EPA amended the 1972 CWA to the Water Quality Act of 1987 (Pub. Law 100-4), which is also known as the National Pollution Prevention Act, phase I of the US EPA Stormwater Program.

To accelerate compliance with the 1987 Act, the EPA started to take actions in 1990 to promulgate numeric water quality criteria for states that had not adopted sufficient standards for toxic pollutants, specifically the NPDES Phase I. As was indicated in Chapter 1, Phase I covered any construction activity disturbing five acres or more of land. The program affected medium and large municipal separate storm sewer systems. MS4s that served populations of 100,000 or more. In addition, the permit coverage under NPDES Phase I included 10 categories of industrial activities.

For wetlands, regulatory guidance requiring compensatory mitigation has been set by various agencies, including the Corps of Engineers, the EPA, the US Fish and Wildlife Services, and the National Marine Fisheries Service since 1990. Because of a general deficiency in the effectiveness of compensatory wetlands at mimicking the natural ones, the EPA and the Corps of Engineers, on March 2, 2006, proposed revisions to compensatory mitigation regulations. On April 10, 2008, the EPA and the Corps published a final rule that improved and consolidated existing regulations and established equivalent standards for all types of mitigation under the CWA section 404. This rule, which became effective on June 9, 2008, was intended to provide greater consistency and ecological effectiveness of mitigation projects. The rule made changes to where and how the compensatory mitigation is required.

DOI: 10.1201/9781003460121-5

Despite the NPDES Phase I, degraded water bodies still existed. In the 1998 National Water Quality Inventory, a biennial summary of state surveys of water quality, approximately 40% of the surveyed water bodies in the United States were still impaired by pollution and did not meet water quality standards (EPA, 2000a). A leading source of the impairment was polluted runoff. A 2000 inventory report by states, territories, and interstate commissions was even more grim. Of the 33% of the US water assessed for this national inventory, nearly 40% of streams, 45% of lakes, and 50% of estuaries were not clean enough for fishing and swimming (EPA, 2000b).

5.1.2 NPDES Phase II

To lessen the adverse impacts of urbanization on water quality, the EPA published Phase II of the stormwater program in the Federal Register on December 8, 1999 (vol. 64, no. 235, NPDES). This program, which was implemented under Section 402 of the CWA, became effective on March 10, 2003, and extended the Phase I coverage to include the following two additional classes of stormwater discharges nationwide:

Operators of small MS4s located in "urbanized areas" in any MS4 not already covered under.
Phase I. MS4s in Phase II include those with populations of 50,000 or more and population density of at least 1000 people per square mile.
Operators of construction sites that will disturb at least one acre of land.

These classes were required to obtain a Phase II MS4 stormwater permit and a Construction General Permit (CGP). The CGP regulates the discharge of stormwater runoff from construction sites. Under the new program, the number of permittees in the NPDES system increased from 100,000 to more than 500,000. Phase II also covers 11 industrial categories that discharge to an MS4 or to the waters of the United States. Any category, except construction can be certified for a condition of no exposure if its industrial material and operations are not exposed to stormwater; these industries are exempt from stormwater general permits.

Phase II MS4s are covered by general permit, and all regulated MS4s are required to perform certain steps to receive the permit. All must develop and implement a Stormwater Management Program (SWMP) or Stormwater Pollution Prevention Plan (SWPPP) to reduce stormwater runoff contamination and prohibit illicit discharge and provide adequate long-term operation and maintenance measures. MS4s must also train staff to protect stormwater when performing daily municipal activities.

As of 2007, approximately half of the rivers, lakes, and bays under EPA oversight were still unsafe for fishing and swimming. These were very significant numbers given that urban/suburban areas cover a small percentage of US land. Nearly one-half of the lake and river impairment was due to stormwater runoff from construction activities. A National Rivers and Streams Assessment Report (EPA, 2013) indicates that of the 1.19 million miles (1.92×10^6 km) of the nation's surveyed rivers and streams, 46% are in poor biological condition, 23% in fair condition, and only 21% in good condition. Up to 60% of the nation's existing water pollution problems are estimated to be attributed to stormwater nonpoint sources.

To assist MS4s with the preparation of SWPPP, the EPA in 2007 developed a user-friendly pamphlet titled "Developing a Stormwater Pollution Prevention Plan (SWPPP)." This publication was last updated on September 2, 2025, The updated version is user-friendly and includes examples for small and medium-sized construction sites (<5 acres). Most MS4s are covered under individual Statewide General Permits.

Since NPDES Phase II, approximately 7250 MS4s operate under NPDES, and following its passage, many states developed their own stormwater discharge regulations, such as New Jersey's NJDPES and New York's SPDES. Most states administer their own PDES, although the EPA is the NPDES permitting authority for Massachusetts, New Hampshire, New Mexico, the District of Columbia and US territories (except the Virgin Islands), some federal facilities, and tribal lands in various states. The NPDES permits are generally valid for a period of five years.

A four-page EPA (2023) summary report updated on August 11, 2023, contains information about NPEDS program areas, technical resources, New and Notables. Each subject in this user-friendly pamphlet can be readily downloaded. Appendix 5A includes a copy of this report. One of the nine NPDES program areas in this publication is stormwater. The NPDES Stormwater Program, last updated July 10, 2025, covers stormwater discharge from:

- Stormwater Pollution of Emerging Concern
- Construction activities
- Industrial activities
- Municipal sources
- Transportation Sources
- Oil and Gas Stormwater Permitting
- EPA Residual Designation Authority
- Stormwater Maintenance
- Long-Term Stormwater Planning
- Stormwater Planning

EPA also issued a CGP that went into effect in 2003 and expired on July 1, 2008, with updates since then in 2009, 2012, 2017, and 2022. The current CGP (EPA, 2022), which became effective on February 17, 2023 and was last updated on June 13, 2025, provides permit coverage in Massachusetts, New Hampshire, New Mexico, and the District of Columbia as well as Denali National Park, oil and gas activities in Oklahoma, and Federal facilities in Colorado, Delaware, Vermont, and Washington, DC. Key requirements for CGP (2022) in part include developing a SWPPP and submitting a Notice of Intent (NOI) to the EPA via the NPDES e-Reporting tool, implementing soil erosion and sediment control practices, and conducting inspections during the entire construction period.

In 2009, EPA promulgated national regulations and standards for Construction and Development Effluent. These regulations, known as the C&D, were renewed in 2014 and 2015 and last updated on April 28, 2025. They cover stormwater runoff from construction sites with one acre (21,047 m^2) or more of land disturbance under the NPDES program. The regulations also require a 50 ft (15 m) buffer from state open waters.

5.1.3 EISA Section 438

To address stormwater runoff from federal facilities, Congress enacted EISA Section 438 of 2007. Under Section 438 of EISA, which became law in December 2007, federal agencies are required to reduce stormwater runoff from federal development and redevelopment projects to protect water resources. Federal agencies can employ a variety of stormwater management practices including green infrastructure and low-impact development to achieve runoff reduction. Section 438 requirements for federal development projects reads as follows:

> The sponsor of any development or redevelopment project involving a federal facility with a footprint that exceeds 5000 square feet (465 m^2) shall use site planning, design, construction, and maintenance strategies for the property to maintain or restore, to the maximum extent technically feasible, the predevelopment hydrology of the property with regard to the temperature, rate, volume, and duration of flow. Section 438 of EISA establishes a far more stringent stormwater requirement on federal facilities than any regulation governing private projects in any state in the nation.

The intention of Section 438 is to preserve or restore the hydrology of a federally owned facility during construction or reconstruction. More specifically, Section 438 is intended to ensure that aquatic biota, stream channel stability, and natural aquifer recharge of receiving water are not adversely impacted by changes in runoff temperature, volumes, durations, and rates due to federal projects (EWRI December, 2009). EPA also prepared technical guidance on implementing the stormwater runoff requirements for federal projects under Section 438 on December 1, 2009, which was updated on November 5, 2024.

Subsequent to EISA, President Obama signed Executive Order (EO) 13514 on October 2009 "on Federal leadership. in environmental, energy and economic performance." Pursuant to this executive order, the EPA prepared "Technical guidance on implementing the stormwater runoff requirements for federal projects" on December 1, 2009. EO 13514 was replaced by EO 13693 and signed by President Obama on March 19, 2015. This EO was called Planning for Federal Sustainability in the Next Decade but was revoked with the publication of the EO 13834 on May 17, 2018, by President Trump. That in turn was revoked and replaced by EO 14057, which was signed by President Biden on December 8, 2021. This EO reestablished the federal government as a leader in sustainability and clean energy. These changes in executive orders, however, did not affect EISA Section 438.

5.2 CURRENT STORMWATER MANAGEMENT REGULATIONS

Following the enactment of the NPDES programs, states prepared their own NPDES program and their own stormwater management rules in states other than those where the EPA is their permitting authority. The regulations generally include measures for controlling runoff quantity and water quality. The standards for water quality storms generally require treating a certain storm, called a water quality storm, for pollutant removal.

Earlier practices specified the water quality storm as 0.5 in. (12.5 mm) of rainfall. This was based on the premise that capturing and treating the early part of rainfall, which carries a higher concentration and/or mass of contaminants, is more effective than treating the later portion of the rainfall.

Both here and around the world, the rainfall is seasonal. In Southern California, the bulk of the rainfall occurs from November to March, with the heaviest precipitation in January and February. The first heavy storms of the season transport more pollutants than those later in the season. This phenomenon, known as first flush, is measured by the concentration and mass of the pollutants carried in the initial runoff. The mass first flush is the event where 80% of the pollutant load is discharged in the first 30% of the runoff. However, the concept of mass first flush is less relevant in the northeastern United States than elsewhere because the rainfall is fairly uniformly distributed throughout the year (Kayhanian and Stenstrom, 2008).

The concept of first flush has been fading away and giving way to water quality storms. In New Jersey, for example, the water quality storm is defined as 1.25 in. of rainfall in a 2-hour period; this storm represents the 90th percentile of rainfalls in one year. In New York State, the water quality storm is also the 90% annual rainfall, which is either 0, 9, or 1 inch depending on location. NPDES Phase II and some states define the water quality storm as an 85% annual rain event.

The pollutant removal criteria selected by different states are typically tied to the quality of receiving waters. Many states use total suspended solids (TSS) as a surrogate pollutant parameter. It is presumed that effective TSS removal (typically 70% to 80%) will remove other pollutants. Many states including New Jersey aim for 80% TSS removal for new pavements; the goal in states such as California, Maryland, and New York is to treat a certain percent of the water quality runoff volume. To achieve the intended goal, many states specify capturing a certain percentage (typically 85% to 90%) of the stormwater depth over a one- or two-hour period, which amounts to 1 in. (2.5 cm) or more of rainfall.

The percentage of removal may not be a good measure of BMP performance because pollutants in stormwater are complex mixtures of chemical and physical substances. Chemicals adhere to solid particles, which in turn attach to larger objects. Therefore, achieving a regulatory agency's specified

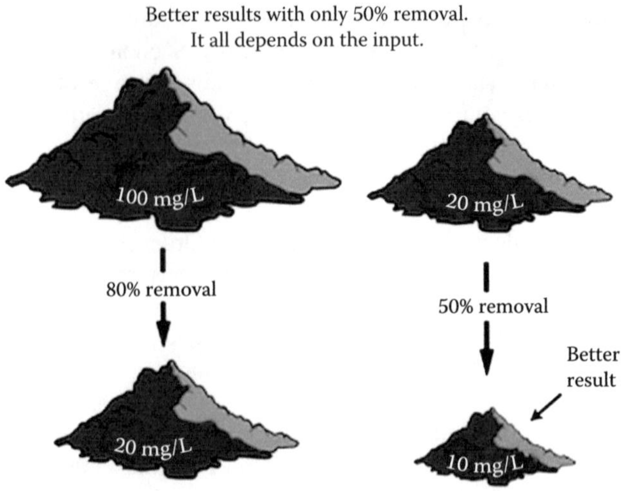

FIGURE 5.1 Input–output pollutant concentration relation

Source: EPA.

percentage of TSS removal might not necessarily reflect a BMP's actual performance. To determine the actual percentage of TSS removal, the quantity of pollutants entering and leaving a system has to be directly measured. Though it appears simple, it is not. Removing TSS from a wet pond, for example, not only varies from site to site but also is dependent on rainfall duration and intensity.

The impact on the environment does not solely depend on the percentage of removal (output loading) of given pollutants; it is also a function of the input loading. A simplified example shown in Figure 5.1 indicates that removing 50% of a lower pollutant load (20 mg/L) is twice as effective as removing 80% of a higher pollutant concentration (100 mg/L). This example, though an oversimplification, indicates that reducing the pollutant load is as important as, if not more important than, the effectiveness of a BMP. While the performance of a BMP varies with location, rainfall pattern, and land uses, the average outflow concentration is useful when comparing different types of BMPs.

Equally important is the volume of runoff. Reducing runoff volume plays a significant role in overall pollutant load reduction. Practices such as pervious pavements, disconnection of runoff from roofs and pavements, infiltration, bioretention and the like, which capture a portion of and infiltrate it into soil, reduce the total outflow pollutant load. Thus, inflow concentration, runoff loading, and outflow concentration are all key elements in pollutant removal.

5.3 NJDEP STORMWATER MANAGEMENT REGULATIONS

The stormwater management regulations in the state of New Jersey were first adopted on February 2, 2004 in the *NJ Stormwater Best Management Practices Manual.* The regulations expired in February 2009 and were extended for one year through February 2010. On April 19, 2010, NJDEP issued amended stormwater management rules). The manual was updated in 2014, 2016, 2023, and 2024. The current manual which was issued in July 2023 and updated on October 9, 2024 (NJDEP, 2024) includes nine chapters and four appendices which in part has been updated through 17, 2023. The current regulations are available online, and each chapter, subchapter, and appendix can be readily downloaded for review and print. For this update and future updates, one can visit www.njstormwater.org.

Effective March 2, 2021, New Jersey required all public and private major development projects to incorporate green infrastructure into stormwater management technique to further reduce pollution in

TABLE 5.1
NJPDES Permitting Regulations

Criterion	Requirement
Water quality	1.25 in. of rainfall in 2 hours
	80% TSS removal for new pavement
	50% TSS removal for reconstruction
	0% TSS removal for roofs and pervious areas
Peak flow rate	Reductions below predevelopment values for 2-, 10-, and 100-year storms
	50% for 2-year storm
	25% for 10-year storm
	20% for 100-year storm
Groundwater recharge	Maintaining predevelopment annual recharge or infiltrating the increased runoff volume for the 2-year storm

New Jersey waterways. The New Jersey Pollutant Discharge Elimination System (NJPDES) governs any development that either disturbs at least one acre or creates 0.25 acres or more of new impervious area. Land disturbance is defined as the placement of impervious surfaces, exposure or movement of soil or bedrock, and clearing or removal of vegetation. A new impervious area means any net increase in impervious surfaces on site, any replacement of an existing drainage system with a system of larger capacity, or any proposed collection and discharge of runoff into a regulated area from an existing impervious surface, the runoff from which currently occurs as sheet flow into vegetation

While the one-acre land disturbance criterion follows the federal NPDES Phase II rule, the 0.25 acre increase in impervious area is outside of the federal regulations. Any project with runoff that enters a water body requires a state stormwater permit from the state, and this applies to include any water course with a drainage area of 50 acres (20 ha) or larger. The stormwater management regulations in New Jersey require provision of measures for runoff quantity, stormwater quality, and groundwater recharge. These requirements are summarized in Table 5.1 and are described in more detail in the next few sections.

5.3.1 Runoff Quantity Requirements

Because of extensive annual flood damage in New Jersey, the runoff quantity regulations in this state are more stringent than those of many others in the nation. The regulations require performing hydrologic and hydraulic analyses to demonstrate no adverse increase in both runoff rates and volumes or create certain reductions in the peak rates of runoff as described next:

1. Demonstrate that the post-construction runoff hydrographs for the 2-, 10-, and 100-year frequency storms do not exceed at any point and time respective hydrographs for preconstruction runoff, or
2. Demonstrate that the post-construction peak runoff rates are no greater than the preconstruction peak runoff rates for the storms of 2-, 10-, and 100-year frequency at any time. In addition, any increased volume or change in the peak timing of the runoff will not increase flood damage at downstream offsite, or
3. Design stormwater management measures so that the post-development runoff rates for the 2-, 10-, and 100-year storm events are 50%, 75%, and 80% of the respective preconstruction peak runoff rates. The reductions apply only to the portion of the site that is disturbed by the project. Specifically, any offsite runoff entering the site and portions of the site that remain intact are not subject to the previously indicated reductions.

In practice, most major developments find it more practical to undertake the latter measures to address condition 3. Tidal areas are exempt from the runoff quantity unless the increased volume of stormwater would aggravate flood-related damage downstream of the point of discharge. Agricultural projects that meet the definition of major development in N.J. A.C. 7:8–1.2 need to submit an application to the Soil Conservation District for review and approval.

5.3.2 Stormwater Quality Standards

The water quality standards only apply when the project creates 0.25 acre or more of new impervious area, as defined previously. The standards mandate measures to reduce the post-construction TSS load in stormwater runoff generated from the water quality storm as follows: 80% for any new impervious area (90% where the runoff is discharged to special water resources such as C1 streams) or 50% for any impervious surface to be reconstructed.

In the NJDEP regulations, building roofs, uncompacted gravel, and porous pavements are not considered impervious surfaces for water quality purposes, brick and concrete pavers with 25% or more openings also count as pervious areas. Lawns and landscapes are not subject to any TSS removal requires, and any re-pavement that does not require land disturbance, such as milling and repaving, is considered maintenance and does not require any TSS removal. The water quality design storm in New Jersey is 1.25 in rainfall, nonuniformly distributed in 2 hours. This rainfall represents the 90th percentile of average annual storms. Figure 5.2 shows the rainfall intensity curve, and Table 5.2 presents the cumulative rainfall depths and intensities of the water quality storm.

Table 5.3 lists the permissible TSS removal rates for certain BMPs when designed in accordance with the New Jersey Stormwater Best Management Practices Manual, 2024. If more than one BMP is used in series, the composite TSS removal is calculated by

$$R = A + B - (A \times B)/100 \tag{5.1}$$

where

R = combined TSS removal for two systems

A and B = TSS removal rate for each BMP

See Chapter 4 in the 2024 manual (first issued February 2004).

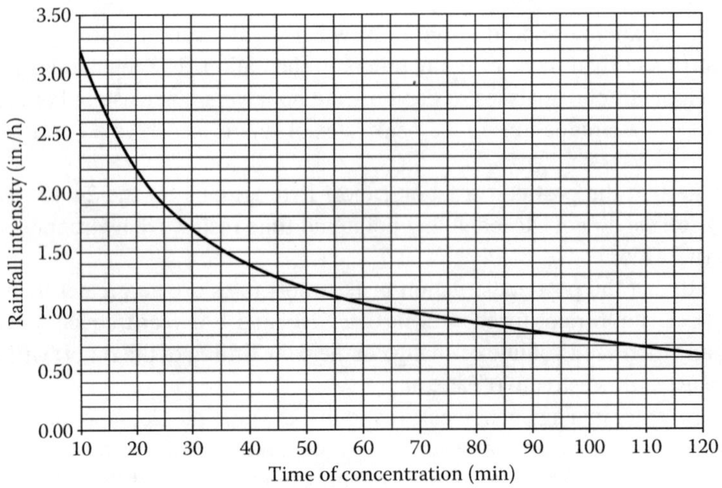

FIGURE 5.2 NJDEP 1.25 in./2 h water quality storm rainfall intensity curve.

TABLE 5.2
NJDEP Water Quality Storm, Temporal Distribution

Time (min)	Cumulative Rainfall (in.)	Incremental Rainfall (in.)	Intensity (in./h)
0	0.0000	0.0000	0.00
5	0.0083	0.0083	0.10
10	0.0166	0.0083	0.10
15	0.0250	0.0084	0.10
20	0.0500	0.0250	0.30
25	0.0750	0.0250	0.30
30	0.1000	0.0250	0.30
35	0.1330	0.0330	0.40
40	0.1660	0.0330	0.40
45	0.2000	0.0340	0.41
50	0.2583	0.0583	0.70
55	0.3583	0.1000	1.20
60	0.6250	0.2667	3.20
65	0.8917	0.2667	3.20
70	0.9917	0.1000	1.20
75	1.0500	0.0583	0.70
80	1.0840	0.0340	0.41
85	1.1170	0.0330	0.40
90	1.1500	0.0330	0.40
95	1.1750	0.0250	0.30
100	1.2000	0.0250	0.30
105	1.2250	0.0250	0.30
110	1.2334	0.0084	0.10
115	1.2417	0.0083	0.10
120	1.2500	0.0083	0.10

Note: Intensity = (incremental rainfall/5 min) × 60 min/h.

TABLE 5.3
TSS Removal Rates for BMPs

BMP	TSS Removal Rate%
Bioretention systems (small scale or large)	80–90
Constructed wetlands	90
Infiltration basins (small scale or large)	80
Wet ponds	50–90
Extended detention basins	40–60
Vegetative filter strips	60–80
Sand filler	80
Grass Swale	≤50
MTDs	50 or 80
Pervious Pavement	80
Sub-surface gravel wetlands	90% (also 90% Nitrogen)

If a site contains two or more drainage areas, the required TSS removal for each area should be met. However, if some of the drainage areas converge to a common discharge point, the TSS calculations are based on the weighted TSS removal; the NJDEP does not accept two similarly manufactured treatment devices used in series. Also, when using an infiltration basin, NJDEP requires that the MTD be placed before the basin. The regulations also call for the minimal use of nutrients through the application of green infrastructure measures and the reduction in phosphorous pollutants to the maximum extent practicable (MEP).

5.3.3 Groundwater Recharge Standards

All major developments must address groundwater recharge requirements as follows. The infiltration from the site and stormwater management features maintains the average annual predevelopment groundwater recharge or the stormwater management measures infiltrate the difference between the pre- and post-construction runoff volumes for 2-year storms.

The New Jersey Geological Survey has developed a method for performing pre- and post-annual groundwater recharge volume calculations. This method is described in "Geological Survey Report, No. 32 (1993), *A Method of Evaluating Groundwater Recharge Areas in New Jersey* and is known as the GSR-32 method. The GSR-32 is a spreadsheet that calculates the pre- and post-construction recharge depths and volumes for any municipality in New Jersey based on soil cover and stormwater infiltration systems. The spreadsheet can be downloaded free of charge from the NJGS website (http://www.state.nj.us/dep/njgs/). Figure 5.3 exemplifies its application for a project involving improvements to the Garden State Parkway in New Jersey (see Case Study 7.4). Sheet 1, Figure 5.3a, shows a reduction of 36,900 cubic feet (1044 m^3) owing to the proposed impervious surfaces. Figure 5.3b presents information about the retention system (which consisted of Cultec chambers in stone trench) and indicates a recharge volume of 91,254 cubic feet.

Sites with contaminated soil are exempt from groundwater recharge requirements, and the state discourages placing groundwater recharge systems in sites with limestone substrata. Groundwater recharge requirements also do not apply to urban redevelopment areas and the previously developed portions of a site, any area that had already been graded, filled, paved, or occupied by structures. The most feasible means of addressing groundwater recharge is infiltrating the roof runoff, which is considered clean. However, for rooftop garages and industrial buildings where contaminants can be deposited on the roofs, a pretreatment is required.

5.3.4 Runoff Calculation Methods

The current NJ stormwater management regulations manual (July 2023, last updated October 9, 2024) only specifies the use of NRCS data to calculate the peak and volume of stormwater runoff. The NRCS uses updated NWS 24-hour rainfall data in lieu of the outdated SCS TR-55 (1986). Also in 2012 New Jersey replaced Type III storm with two new rainfall distribution regions, C and D. Region C covers the counties of Sussex, Warren, Hunterdon, Somerset, Mercer, Burlington, Camden, Gloucester, Atlantic, Salem, Cumberland and Cape May, and region D covers Bergen, c, Atlantic, Salem, Cumberland, and Cape May, and region D covers Bergen, Hudson, Essex, Passaic Morris, Union, Middlesex, Monmouth, and Ocean counties (NJ Bulletin 2012). The rainfall distribution for these regions is incorporated in the WinTR-55. Table 5.4 includes NOAA 24 h precipitation depths for all counties in New Jersey.

New Jersey N.J.A.C. 7:8–5,6 requires that the NOAA precipitation depths for the 2-, 10- and 100-year storms be adjusted for the current (present) and projected (future, proposed) conditions in meeting the runoff quantity standards. The adjustment factors vary from county to county and are listed in two tables in Chapter 5 of the manual. These tables show adjustment factors ranging from 0.99 to 1.15 for the current conditions and from 1.16 to 1.50 for the projected conditions.

The current regulations specify the following procedures for calculating the runoff volume for the cases in which pervious and impervious areas are separated or connected. In the former case,

New Jersey Groundwater Recharge Spreadsheet Version 2.0 November 2003	Annual Groundwater Recharge Analysis (based on GSR-32)		
	Select Township ↓	Average Annual P (in)	Climatic Factor
	MIDDLESEX CO., PERTH AMBOY CITY	47.8	1.53

Project Name:	GSP INTERCHANGE 88/89
Description:	This is a test application
Analysis Date:	09/01/03

Pre-Developed Conditions

Land Segment	Area (acres)	TR-55 Land Cover	Soil	Annual Recharge (in)	Annual Recharge (cu.ft)
1	0.64	Open space	Downer	15.9	36,902
2	0.13	Woods	Woodstown	12.9	6,080
3					
4					
5					
6					
7	0				
8	0				
9	0				
10	0				
11	0				
12	0				
13	0				
14	0				
15	0				
Total =	0.8			Total Annual Recharge (in)	Total Annual Recharge (cu-ft)
				15.4	42,982

Post-Developed Conditions

Land Segment	Area (acres)	TR-55 Land Cover	Soil	Annual Recharge (in)	Annual Recharge (cu.ft)
1	0.64	Impervious areas	Keyport	0.0	-
2	0.13	Open space	Woodstown	12.9	6,082
3					
4					
5	0				
6	0				
7	0				
8	0				
9	0				
10	0				
11	0				
12	0				
13	0				
14	0				
15	0				
Total =	0.8			Total Annual Recharge (in)	Total Annual Recharge (cu.ft)
				2.2	6,082
				Total Impervious Area (sq.ft)	27,878

Annual Recharge Requirements Calculation ↓

% of Pre-Developed Annual Recharge to Preserve =	100%	
Post-Development Annual Recharge Deficit=	*36,900*	(cubic feet)

Recharge Efficiency Parameters Calculations (area averages)

RWC=	2.75	(in)	DRWC=	0.00	(in)
ERWC =	0.65	(in)	EDRWC=	0.00	(in)

Procedure to fill the Pre-Development and Post-Development Conditions Tables

For each land segment, first enter the area, then select TR-55 Land Cover, then select Soil. Start from the top of the table and proceed downward. Don't leave blank rows (with A=0) in between your segment entries. Rows with A=0 will not be displayed or used in calculations. For impervious areas outside of standard lots select "Impervious Areas" as the Land Cover. Soil type for impervious area are only required if an infiltration facility will be built within these areas.

FIGURE 5.3A Groundwater recharge calculations: sheet 1 of 2 (Case Study 7.4).

Project Name	Description	Analysis Date	BMP or LID Type
GSP INTERCHANGE 88/89	This is a test application	09/01/03	

Recharge BMP Input Parameters

Parameter	Symbol	Value	Unit
BMP Area	ABMP	2112.0	sq.ft
BMP Effective Depth, this is the design variable	dBMP	8.4	in
Upper level of the BMP surface (negative if above ground)	dBMPu	48.0	in
Depth of lower surface of BMP, must be>=dBMPu	dEXC	56.4	in
Post-development Land Segment Location of BMP, Input Zero if Location is distributed or undetermined	SegBMP	2	unitless

Root Zone Water capacity Calculated Parameters

Parameter	Symbol	Value	Unit
Empty Portion of RWC under Post-D Natural Recharge	ERWC	1.08	in
ERWC Modified to consider dEXC	EDRWC	0.00	in
Empty Portion of RWC under Infilt. BMP	RERWC	0.00	in

Recharge Design Parameters

Parameter	Symbol	Value	Unit
Inches of Runoff to capture	Qdesign	0.26	in
Inches of Rainfall to capture	Pdesign	0.34	in
Recharge Provided Avg. over Imp. Area		16.0	in
Runoff Captured Avg. over imp. Area		16.0	in

BMP Calculated Size Parameters

ABMP/Aimp	Aratio	0.03	unitless
BMP Volume	VBMP	1,478	cu.ft

Parameters from Annual Recharge Worksheet

Post-D Deficit Recharge (or desired recharge volume)	Vdef	36,900	cu.ft
Post-D Impervious Area (or target Impervious Area)	Aimp	68,607	sq.ft
Root Zone Water Capacity	RWC	4.61	in
RWC Modified to consider dEXC	DRWC	0.00	in
Climatic Factor	C-factor	1.53	no units
Average Annual P	Pavg	47.8	in
Recharge Requirement over Imp. Area	dr	15.9	in

System Performance Calculated Parameters

Annual BMP Recharge Volume		91,254	cu.ft
Avg BMP Recharge Efficiency		100.0%	Represents % Infiltration Recharged
%Rainfall became Runoff		78.3%	%
%Runoff Infiltrated		42.7%	%
%Runoff Recharged		105.0%	%
%Rainfall Recharged		82.2%	%

CALCULATION CHECK MESSAGES

Volume Balance-> **Solve Problem to satisfy Annual Recharge**

dBMP Check---> **OK**

dEXC Check---> **OK**

BMP Location---> **OK**

OTHER NOTES

Pdesign is accurate only after BMP dimensions are updated to make rech volume= deficit volume. The portion of BMP infiltration prior to filling and the area occupied by BMP are ignored in these calculations. Results are sensetive to dBMP, make sure dBMP selected is small enough for BMP to empty in less than 3 days. For land Segment Location of BMP if you select "impervious areas" RWC will be minimal but not zero as determined by the soil type and a shallow root zone for this Land Cover allowing consideration of lateral flow and other losses.

How to solve for different recharge volumes: By default the spreadsheet assigns the values of total deficit recharge volume "Vdef" and total proposed impervious area "Aimp" from the "Annual Recharge" sheet to "Vdef" and "Aimp" on this page. This allows solution for a single BMP to handle the entire recharge requirement assuming the runoff from entire impervious area is available to the BMP.
To solve for a smaller BMP or a LID-IMP to recharge only part of the recharge requirement, set Vdef to your target value and Aimp to impervious area directly connected to your infiltration facility and then solve for ABMP or dBMP. To go back to the default configuration clik the "Default Vdef & Aimp" button

FIGURE 5.3B Groundwater recharge calculations: sheet 2 of 2.

TABLE 5.4
24-Hour Rainfall Depth–Frequency Data for New Jersey Counties (in inches)

County	1-Year	2-Year	5-Year	10-Year	25-Year	50-Year	100-Year
Atlantic	2.72	3.31	4.30	5.16	6.46	7.61	8.90
Bergen	2.75	3.34	4.27	5.07	6.28	7.32	8.47
Burlington	2.77	3.36	4.34	5.18	6.45	7.56	8.81
Camden	2.73	3.31	4.25	5.06	6.28	7.34	8.52
Cape May	2.67	3.25	4.22	5.07	6.34	7.47	8.73
Cumberland	2.69	3.27	4.25	5.09	6.37	7.49	8.76
Essex	2.85	3.44	4.40	5.22	6.44	7.49	8.66
Gloucester	2.71	3.29	4.24	5.05	6.29	7.36	8.55
Hudson	2.73	3.31	4.23	5.02	6.19	7.20	8.31
Hunterdon	2.28	3.38	4.26	5.00	6.09	7.02	8.03
Mercer	2.74	3.31	4.23	5.01	6.19	7.20	8.33
Middlesex	2.76	3.35	4.30	5.12	6.36	7.43	8.63
Monmouth	2.79	3.38	4.38	5.23	6.53	7.66	8.94
Morris	2.94	3.54	4.47	5.24	6.37	7.32	8.35
Ocean	2.81	3.42	4.45	5.33	6.68	7.87	9.20
Passaic	2.87	3.47	4.42	5.23	6.43	7.47	8.62
Salem	2.69	3.26	4.20	5.00	6.22	7.28	8.45
Somerset	2.76	3.34	4.25	5.01	6.15	7.13	8.21
Sussex	2.68	3.22	4.02	4.70	5.72	6.60	7.58
Union	2.80	3.39	4.35	5.17	6.42	7.49	8.69
Warren	2.78	3.34	4.18	4.89	5.93	6.83	7.82

Notes: The average point rainfall amounts listed in the table were developed from data contained in NOAA Atlas 14 Volume 2. Point rainfall estimates for specific locations may be obtained from the Precipitation Frequency Data Server located at http://www.nws.noaa.gov.ohd/hdsc/.

For most hydrologic design procedures, the rainfall amounts listed above may be rounded to the nearest tenth of an inch.

the runoff volume from pervious and impervious areas is computed separately and added to calculate the composite runoff volume from the site. Also, the hydrographic data for the pervious and impervious areas, which may have different concentration times, shall be added to calculate the composite runoff from the site. In the case, where the runoff from impervious areas travels as a sheet flow over previous areas, part of the runoff infiltrates into the soil. In this case, the overall hydrograph can be computed based on the composite CN using Figure 3.20b (in Chapter 3), provided that the impervious surface is less than 30% of the total area. Otherwise, runoff calculations should be based on a reduced soil curve number. However, since such a procedure has not been verified through experiments, it will not be discussed herein. See also Chapter 8.

5.3.5 Standards for Stormwater Management Structures

Structural stormwater management systems must take into account the existing site conditions, such as wetlands, flood plains, slopes, depth to seasonal high-water table, drainage pattern, and presence of limestone. They must require minimal maintenance and be readily accessible to facilitate maintenance and repairs. For instance, trash racks should be installed at the inlet and outlet structures, and racks should be parallel bars with on-center spacing of no greater than one third the diameter of the orifice or the width of the weir but no less than 1 in. and no more than 6 in. Effective systems should also have an orifice no smaller than 2.5 in. in diameter at the intake of the outlet structure and have a minimum of 2 ft (60 cm) separation between the bottom of an infiltration basin and the spring high water table.

5.3.6 Green Infrastructure Stormwater Strategies

The NJ Standards emphasize the use of green stormwater strategies as prime solutions for meeting the state's soil erosion, groundwater recharge, runoff quantity, and runoff quality standards. Appendix A in the manual includes a form titled "Low Impact Development Checklist (LID)" that must be filled in and included with the test results and calculations report in any application for a stormwater management project.

5.3.7 Municipal Stormwater Management Review

Appendix D, titled "Model Municipal Stormwater Control Ordinance," dated July 2023, includes sample ordinances to assist municipalities in revising their stormwater content ordinance to N.J.A.C.7:8 adopted March 2, 2020, and updated July 17, 2023. It is important that municipalities carefully review the amended rules before adopting any portion of this draft ordinance. The sample ordinances represent the minimum standards that municipalities must follow. A municipality may choose stronger or other measures to address local water quality or flooding issues. For example, municipalities may choose to define a "major development" with a smaller area of disturbance and/or a smaller area of regulated impervious covers.

5.3.8 Suggestions for Improving the NJDEP Regulations

The current stormwater manual, revised through October 2024, excludes the rational formula for calculating the peak and volume of runoff. This method is simple and provides more accurate results when used by an experienced engineer. The manual requires that all runoff calculations be performed using the TR-55 method, which in spite of the vast variations in soil permeability only includes four soil groups from most to least permeable, which cannot be accurately defined in small areas. Also, as will be indicated in Chapter 8, this method is unsuitable for green roofs, which act as microretention basisn.

The NJ manual goes one step further and employs the TR-55 method to compute peak flow reduction of green roofs. As will be indicated in Chapter 8, these calculations yield erroneous results in that they exaggerate the retention volume and underestimate the peak rate of runoff during heavy storm events. It is evident that once the soil media in the green roof becomes saturated, the green roof discharges with no attenuation. The New York State Stormwater Management Design Manual uses a version of the rational formula with a runoff coefficient of 0.95 to calculate the volume of water quality storm for the design of green roofs.

The NJ Standards regarding TSS removal are generic in that they disregard the rates of pollutant generation; one site can produce twice as much pollutant as another, and removing 50% TSS from the latter results in more improved water quality removing 80% from the former. A truck-stop parking area generates many times more pollutants, particularly oil and grease, than driveways and streets in a residential development. The treatment target should also include the pollution generation rate. The regulations assume that lawn and landscape areas are virgin land and pollutant-free, but as indicated in Chapter 1, fertilizers, pesticides, and other lawn chemicals can degrade the water quality of runoff more than that of a paved area such as a driveway.

The practicality of maintenance is also ignored. The Standards accept the use of a 2.5 in. (63 mm) orifice in both aboveground and underground detention basins, but a 2.5 or even 3 in. (7.5 cm) orifice is highly vulnerable to clogging by leaves, small debris, and even a tennis ball. While in an open detention basin, the clogging can be noticed, in an underground system where the outlet structure is hidden, the clogging will be left undetected, leaving the detention basin partly filled with water before the next storm event occurs. Considering practicality and maintenance, the author recommends using a 6 in. (15 cm) orifice as the smallest opening in any underground detention basin. Following his recommendations, hundreds of underground detention and infiltration basins were so constructed in New Jersey.

Regarding water quality, it is more prudent to use a device that is inexpensive and can be readily maintained than another system that is costly to maintain and hidden from view. The author has observed NJDEP-approved underground sand filters become rendered totally ineffective due to lack of maintenance. There are cases where the sand filter was ignored years after installation. Once the sand is clogged, the runoff that enters the system is discharged from the overflow weir without receiving any treatment. Figure 5.4 shows a sand filter that remained unmaintained for five years after its installation and was totally clogged. Installing inlet filters that are easy to maintain will be far more prudent than a sophisticated water treatment device that is left idle.

For extremely flat sites, the 100 ft sheet flow limit for calculating the time of concentration, $T_{c,}$ is too short. The sheet flow lengths of 200 feet (60 m) for pervious areas and 100 feet (30 m) for impervious surfaces (post-development) are more reasonable for performing T_c calculations (see Maryland stormwater management regulations in the next section).

The NJ Standards also entirely exclude single-family homes from any water quality, and more importantly, any runoff quantity control. A single-family residence in an affluent community in New Jersey can disturb one acre (3859 m^2) of virgin land and create a quarter acres (1020 m^2) of pavement yet be free of any stormwater management regulations. It should be realized that most residential units in any suburban community consist of single-family homes. While each home may have little impact on stormwater runoff, collectively they create an enormous increase in the peak and volume of runoff and significantly degrade stormwater quality.

In fact, single-family homes are prime culprits in flooding in New Jersey and many other states as well. It is time that the regulations address not only a development's size but also its per capita environmental impacts. To be successful, it is imperative to focus on not just large projects but rather on the urban life as a whole. Maryland is one of few states where the jurisdictional criterion for stormwater management is 5000 ft^2 (465 m^2) of disturbance. As such, the regulations cover most single-family homes.

In his profession, the author has had the opportunity to review and comment on hundreds of development projects in many municipalities, mostly in northern New Jersey. To control runoff, he has been requiring a net zero increase in runoff rate/volume not only from all single-family homes, which were exempt from the NJDEP regulations, but also any improvement resulting in more than 200 ft^2 (20 m^2) of paved area. Depending on site and soil conditions, his recommendations to prevent

FIGURE 5.4 Grates on an unmaintained sand filter.

Source: Photo by the author.

any increase in runoff have included dry wells, underground retention–infiltration chambers, rain gardens, and rain tanks for collecting roof runoff. The author also recommends using concrete paver blocks (pavers) and disconnecting roof leaders from driveways for reducing the runoff volume and improving water quality.

5.4 STATE OF MARYLAND STORMWATER MANAGEMENT REGULATIONS

Maryland was one of the first states that enacted statewide stormwater management regulations. The Maryland Department of the Environment (MDE) developed a manual titled *The 2000 Maryland Stormwater Design Manual*, volumes I & II, in April 2000. In addition, the MDE issued Model Stormwater Management Ordinance in July 2005 that was supplemented in 2007. The state amended the stormwater design manual on May 4, 2009, and the new regulations were enacted in May 2010. The manual contains two volumes: volume I includes five chapters, and volume II includes five appendices covering supplementary material. examples, specifications, and NPDES requirements.

In Maryland, the stormwater management rules apply to any construction activity disturbing 5000 ft^2 (465 m^2) or more of land. Chapter 1 in volume I of the manual is introductory and presents performance standards (14 in all) for stormwater management practices, including minimizing the generation of stormwater, pretreatment, meeting groundwater recharge, removing 80% of TSS and 40% of the total annual post-development phosphorus load, controlling peak flows, preparation of a stormwater pollution plan for certain industrial areas, and provisions for stormwater discharges from hotspots and redevelopment areas. One of the new requirements in the manual is to implement environmental site design (ESD) to the MEP (maximum extent practicable) to mimic predevelopment conditions. Chapter 2 in Volume I contains "Unified Stormwater Sizing Criteria." These rules, which include runoff quantity control, water quality control, and groundwater recharge, are summarized in Table 5.5 and briefly discussed in this section.

TABLE 5.5
Maryland's Unified Stormwater Sizing Criteria

Sizing Criteria	Description of Stormwater Sizing Criteria
Water quality volume (WQ_v) (acre-feet) or (cu-ft)	$WQ_v = [(P)(R_v)(A)]/12$ P = rainfall depth in inches and is equal to 1.0 in. in the eastern rainfall zone and 0.9 in. in the western zone (Figure 2.1 in the manual) R_v = volumetric runoff coefficient A = area in acres (or ft^2)
Recharge volume (Re_v) (acre-feet)	Fraction of WQ_v, depending on predevelopment soil hydrologic group $Re_v = [(S)(R_v)(A)]/12$ S = soil-specific recharge factor in inches
Channel protection storage volume (Cp_v)	Cp_v = 24 hour (12 hour in USE III and IV watersheds) extended detention of post-developed 1-year, 24-hour storm event Not required for direct discharges to tidal waters and the Eastern Shore (see Figure 5.5)
Overbank flood protection volume (Q_p)	Controlling the peak discharge rate from the 10-year storm event to the predevelopment rate (Q_{p10}) is optional; consult the appropriate review authority. For Eastern Shore: Provide peak discharge control for 2-year storm events (Q_{p2}). Control of the 10-year storm event (Q_{p10}) is not required.
Extreme flood volume (Q_f)	Consult with the appropriate reviewing authority. Normally, no control is needed if development is excluded from 100-year floodplain and downstream conveyance is adequate

Source: The 2009 Maryland Stormwater Design Manual, Table 2.1.

5.4.1 Water Quality Volume

The water quality volume, WQ_v, is the storage needed to capture and treat the runoff from 90% of the average annual rainfall. This 90% rainfall event is 1 in. in the eastern zone and 0.9 in. in the western zone. The eastern zone covers over two-thirds of the counties in the eastern half of the state. The water quality volume is calculated by

$$WQ_v = (P)(R_v)(A)/12 \tag{5.2}$$

where
$R_v = 0.05 + 0.009\ (I)$
A= area in acres
P= 90% of average annual rainfall (0.9 or 1 in.)
I= percent impervious cover on site
WQ_v= water quality volume, acre-feet

In SI, the preceding equation reads as

$$WQ_v = 10\cdot(P)(R_v)(A) \tag{5. 3}$$

where
P= 22.9 or 25.4 mm
A= area in hectares
WQ_v = water quality volume, m^3

A minimum of WQ_v = 0.2 in./acre (0.5 mm/ha) must be met at sites or drainage areas with less than 15% impervious cover. The water quality requirement can be met by providing a 24 h drawdown time of WQ_v in an extended detention basin or a 24-hour drawdown of a portion of WQ_v in conjunction with a stormwater pond or wetland system (refer to Chapter 3 of the Manual).

5.4.2 Recharge Volume

The criterion for groundwater recharge is to maintain the predevelopment average annual recharge of the hydrologic soil present at the site. The recharge volume, Re_v, is calculated from the following equation:

$$Re_v = (S)(R_v)(A)/12 \tag{5.4}$$

where
S = soil-specific recharge factor (see Table 5.6)
Re_v and A are as defined for water quality criteria (Table 5.5)

In SI, Equation 5.4 reads as

$$Re_v = 10\cdot(S)(R_v)(A) \tag{5.5}$$

where
A= Area in hectares
Re_v= recharge volume, m^3

Re_v, which is smaller than WQ_v, is inclusive in it. Specifically, satisfying WQ_v by a structural practice such as infiltration basin or bioretention, or nonstructural practice (e.g., buffers, disconnection

TABLE 5.6
Soil-Specific Recharge S

Hydrologic Soil Group	in.	(mm)
A	0.38	(9.65)
B	0.26	(6.60)
C	0.13	(3.30)
D	0.07	(1.78)

of roof runoff, or combination thereof) also satisfies the recharge requirement. However, when utilizing separate measures for water quality and groundwater recharge, Re_v may be subtracted from WQ_v when sizing the water quality BMP.

If more than one HSG is present at the site, the composite S is calculated based on the proportion of the total site area within each group. The composite S figure will be applied to calculate percent volume runoff for structural measures and percent impervious area for nonstructural measures. Acceptable nonstructural measures include filter strips to treat parking lot and rooftop runoff, sheet flow discharge to stream buffers, and grass channels to treat roadway runoff.

Groundwater recharge requirements do not apply to redevelopment projects or any portion of a site designated a hotspot. In addition, the local review agencies may relax the recharge volume requirements if a site is situated on unsuitable soil such as marine clay, karst, or an urban redevelopment area. In such situations, nonstructural practices should be used for a portion of the site (percent area method) to the MEP. The use of the percent volume and percent area methods is clarified in the next example.

Example 5.1

A 30-acre residential development site comprises 60% hydrologic group B and 40% group C soil and has 35% impervious cover. Calculate the required Re_v using the percent volume and percent area methods.

Solution

First, calculate composite S based on Table 5.6:

$$S = 0.26 \times 0.60 + 0.13 \times 0.4 = 0.208$$

a. Percent volume method:

$$R_v = 0.05 + 0.09(I) = 0.05 + 0.009 \times 35 = 0.365$$

$$Re_v = (S)(R_v)\frac{(A)}{12}$$

$$Re_v = 0.208 \times 0.365 \times \frac{30}{12} = 0.19\ ac - ft$$

b. Percent area method:

$$Re_v = (A_i) \times imp.\ percentage$$

$$A_i = 30 \times 0.35 = 10.5\ ac$$

$$Re_v = 0.208 \times 10.5 = 2.18\ acres$$

The recharge requirement may be met by either retaining 0.19 ac-ft of runoff using an structural method or disconnecting 2.18 acres of impervious area and directing them to a nonstructural system such as vegetated buffer.

5.4.3 Channel Protection Storage Volume

To protect downstream channels from erosion, 24-hour extended detention of the 1-year, 24-hour storm event shall be provided. An exception is the Eastern Shore, where Cp_v is not required. Also, in Use III and IV watersheds, only 12 hours of extended detention time are required. Table 5.7 lists 24-hr rainfall depths of 1-, 2-, 10-, and 100-year storms in all counties in Maryland. See Figure 5.5 for the county map.

To meet the Cp_v requirement, detention ponds and underground vaults are commonly used. The rationale for this requirement is that runoff will be stored and released in such a gradual manner that erosive velocities will not be created in downstream channels. Because of large storage requirements, infiltration is not recommended for Cp_v.

The Cp_v requirement does not apply to direct discharges to the tidal waters and the Eastern Shore. It is also not required when the 1-year post-development peak discharge is less than 2 cfs. Included in the manual is a simplified solution for addressing the required detention time. Offsite

TABLE 5.7
Rainfall Depths of the 1, 2, 10, and 100-year, 24-hour Storm Events in Maryland

County	Rainfall Depth			
	1 Year, 24 h (in.)	2 Year, 24 h (in.)	10 Year, 24 h (in.)	100 Year, 24 h (in.)
Allegany	2.4	2.9	4.5	6.2
Anne Arundel	2.7	3.3	5.2	7.4
Baltimore	2.6	3.2	5.1	7.1
Calvert	2.8	3.4	5.3	7.6
Caroline	2.8	3.4	5.3	7.6
Carroll	2.5	3.1	5.0	7.1
Cecil	2.7	3.3	5.1	7.3
Charies	2.7	3.3	5.3	7.5
Dorchester	2.8	3.4	5.4	7.8
Frederick	2.5	3.1	5.0	7.0
Garrett	2.4	2.8	4.3	5.9
Harford	2.6	3.2	5.1	7.2
Howard	2.6	3.2	5.1	7.2
Kent	2.7	3.3	5.2	7.4
Montgomery	2.6	3.2	5.1	7.2
Princ George's	2.7	3.3	5.3	7.4
Queen Anne's	2.7	3.3	5.3	7.5
St. Mary's	2.8	3.4	5.4	7.7
Somerset	2.9	3.5	5.6	8.1
Talbot	2.8	3.4	5.3	7.6
Washington	2.5	3.0	4.8	6.7
Wicomico	2.9	3.5	5.6	7.9
Worcester	3.0	3.6	5.6	8.1

Source: Maryland Stormwater Design Manual, 2009.

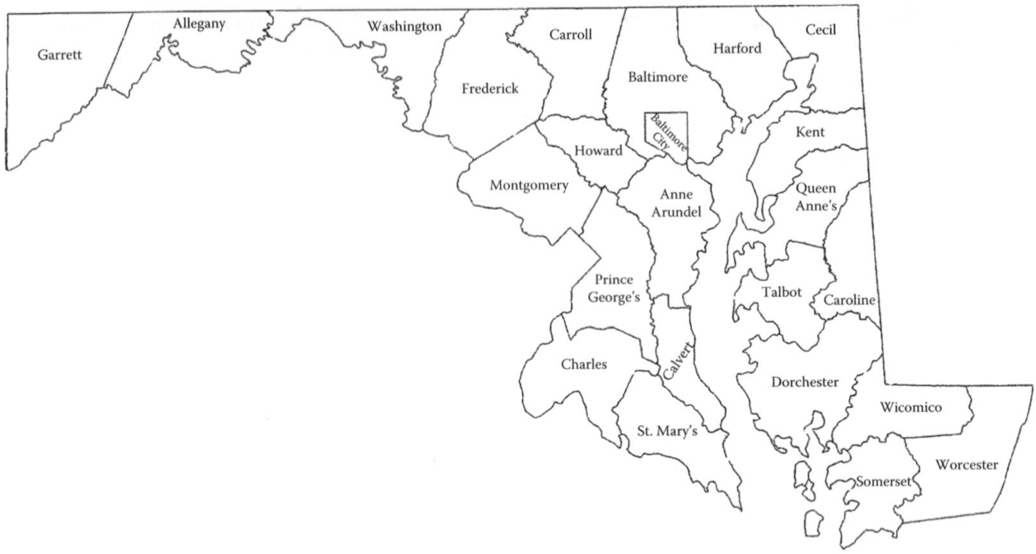

FIGURE 5.5 Map of Maryland counties.

areas should be modeled as present land use in good condition for Cp_v calculations. If a site consists of multiple drainage areas, Cp_v may be distributed proportionally to each drainage area, and the Cp_v for the entire drainage area is addressed.

5.4.4 Overbank Protection Volume

The criterion for overbank protection is a 10-year, 24-hour storm event Q_{p10}, except for the Eastern Shore, where the criterion is a 2-year storm event, Q_{p2}. Calculations for the Q_{p2} and Q_{p1} are based on TR-55 or TR-20, using rainfall depths shown in Table 5.7. In computing the predevelopment runoff, the land should be assumed meadow in good cover condition. Offsite areas should also be assumed as present land use conditions with good vegetative cover. The length of sheet flow, in T_c calculations, is limited to 150 ft for predevelopment conditions and 100 ft for post-development conditions.

Detention basins/ponds and underground chambers/vaults are commonly employed to address Q_{p2} and Q_{p10}. Similarly to Cp_v, overbank protection does not apply to direct discharges to tidal waters.

5.4.5 Extreme Flood Volume

The objectives of the extreme flood criterion (Q_f) are

a. Prevent flood damage from large storm events
b. Avoid expanding the boundaries of pre-development 100-year Flood Emergency Management Agency (FEMA)
c. Protect the BMP control structures.

The criterion for extreme flood Q_f is either a 100- or 10-year storm event, depending on whether or not the area downstream of the site is situated within or out of the 100-year floodplain, respectively. In the latter case, hydraulic/hydrologic analyses may be required to demonstrate no adverse impact on downstream structures. The analyses typically extend to the first downstream tributary of equal or greater drainage area than that of the development or to any downstream dam, bridge, highway culvert, or point of restricted stream flow. In performing the 100-year flood analysis, offsite areas should be modeled as ultimate (full development) conditions.

5.4.6 BMP Design

Chapter 3 of Maryland's manual includes performance criteria for urban BMP design, such as general feasibility, conveyance, pretreatment, treatment, environment, and maintenance. An introductory note explains that in small watersheds, increases in temperature due to development have a primary impact on the quality of receiving waters. To minimize this effect, BMP designs in such watersheds are required to minimize permanent pools, limit extended detention times for C_p to 12 h, maintain existing forested buffer, and bypass available base flow.
The manual identifies the following BMPs:

a. Five types of ponds, P-1 through P-5: P-1 is a micro pool extended detention pond, P-2 a wet pond, P-3 a wet extended detention pond, P-4 a multiple pond system, and P-5 a pocket pond. Each type is defined with descriptive illustration and its features specified, such as riser, valves, drains, buffers, and setbacks, landscaping, and maintenance.
b. Four types of stormwater wetlands, W-1 through W-4: these represent shallow wetland, extended detention shallow wetland, pond/wetland system, and pocket wetland, respectively. Each type is illustrated by a figure and its application is defined. The manual also requires that 25% of the total WQ_v be retained in deep water zones with a minimum depth of 4 ft (1.2 m), that at least 35% of the surface area of wetlands be less than 6 in. (15 cm) deep, and that 65% of the total surface area be shallower than 18 in. (45 cm).
c. Infiltration trenches/basins: Infiltration systems are required to exfiltrate the entire WQ_v, less the pretreatment volume, through the bed of the systems. Porosity, $n = V_v/V_t$, of 0.40 is allowed for the design of stone trenches (reservoirs) for infiltration practices.
d. Filtration systems: The manual classifies filtering systems into six groups: F-1 through F–6. These are surface sand filter, underground sand filter, perimeter sand filter, organic filter, pocket sand filter, and bioretention, respectively. These systems are not to be designed to meet Wp_v and Q_p. Rather, filter practices are generally combined with a separate facility to provide those controls.

The filtration surface area is designed based on the following listed permeabilities, K, and using the following equation:

$$A = WQ_v \,/\, KT \tag{5.6}$$

where K for

sand = 3.5 ft/day (1.1 m/day)
peat = 2.0 ft/day (0.6 m/day)
leaf compost = 8.7 ft/day (2.65 m/day)
bioretention soil = 0.5 ft/day (0.15 m/day)
WQ_v = water quality volume, ft^3 (m^3)
T = drain time, days
A = filtration area ft^2 (m^2)

A maximum 40-hour infiltration time is recommended for sand bed and two days for bioretention soils. Dry or wet pretreatment equivalent to at least 25% of the calculated WQ_v is required prior to the filtration system. A typical pretreatment system consists of a sedimentation basin that has a length-to-width ratio of 2:1 and a minimum surface area calculated by

$$A_s = E'(Q_o / V_f) \quad (5.7)$$

where

A_s = sedimentation basin area, ft² (m²)
Q_o = discharge rate from basin = WQ_v/24 hour, ft³/day (m³/day)
V_f = fall velocity of particles, ft/s (m/s)
E' = sediment trapping efficiency constant
and, for $I \leq 75\%$, use:
V_f = 0.0004 ft/s (0.00013 m/s) for particle size = 20 μm
and, for $I > 75\%$, use:
V_f = 0.033 ft/s (0.01 m/s) for particle size = 40 μm
I = percent impervious

The reason for using a larger settling velocity for $I > 75\%$ is that sites with greater than 75% imperviousness have a higher percentage of coarse-grained sediments.

The sediment trapping efficiency constant, E', is related to the sediment trap efficiency, E, by the following equation:

$$E' = -\ln\left[1-(E/100)\right] \quad (5.8)$$

For E = 90%, E' = 2.30.
Substituting for E' and V_f in Equation 5.7 give .

$$I \leq 75\%$$

$$A_s = 0.066\ WQ_v, \quad (5.9)$$

$$A_s = 0.217\ WQ_v,\ m^2\ (5.10)$$

$$I > 75\%$$

$$A_s = 0.0081\ WQ_v,\ ft^2 \quad (5.11)$$

$$A_s = 0.027\ WQ_v,\ m^2 \quad (5.12)$$

The criterion for a filtering system is that the entire treatment system (including pretreatment) holds at least 75% of WQ_v prior to infiltration. Minimum filter bed thickness is typically 18 in. (45 cm) for infiltration basins and 12 in. (30 cm) for sand filters.

e. Swales are designed to treat the full WQ_v and may be dry swale or wet swale, designated as O-1 and O-2, respectively. Dry swale is basically a vegetated open channel, and wet swale has an expanded basin with wetland vegetation and constricted outlet. Figure 5.6 shows a schematic plan view of a wet swale.

Open-channel swales have to be designed for 10-year storms, and the storms' peak flow velocities have to be nonerosive. The channels must have moderate side slopes, flatter than 3:1 and no steeper than 2:1. Additionally, a minimum ponding time of 30 minutes is recommended for WQ_v treatment, but the maximum allowable time is less than 48 hours. An underdrain system is necessary in dry swales to meet the maximum ponding time requirement.

Pretreatment storage of 0.1 in. (2.5 mm) of runoff per impervious acre is required. This storage is usually provided by installing check dams at pipe inlet and/or driveway crossings. For lateral inflow, a pea gravel diaphragm of gentle side slopes should be provided along the top of the channel. Direct discharge of concentrated flow should be prevented.

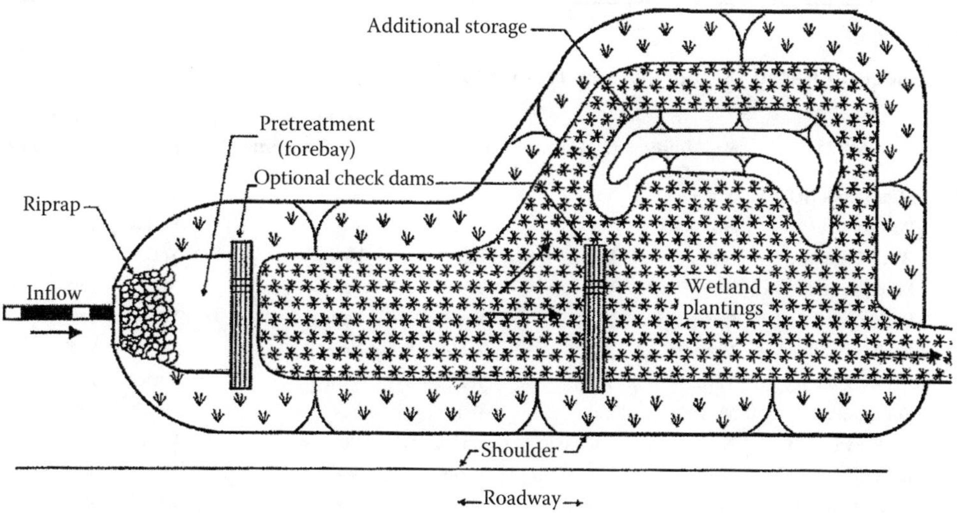

FIGURE 5.6 Schematic plan view of a wet swale.

Chapter 4 of the manual discusses selecting BMPs for a site based on watershed; terrain; the physical feasibility of stormwater treatment suitability; and community, environmental, location, and permitting factors.

One of the watershed factors is development in the intensely developed area of the Maryland critical area (a zone extending 1000 ft from mean high tide and the landward edge of tidal wetlands). BMPs in these areas must comply with the "10% rule." This rule requires that post-development phosphorus load be reduced to 10% below predevelopment loads. Other watershed factors are cold water streams, sensitive streams, wetlands and reservoir protection, and shellfish/beaches.

Terrain factors include low-relief areas containing karst and carbonaceous rock. Stormwater treatment suitability includes meeting Re_v, C_{pv}, Q_{p2}, and Q_{p10}, as well as being able to accept hotspot runoff. Stormwater treatment suitability is summarized in Table 5.8, adapted from Table 4.3 in the manual.

Physical feasibility factors include soils, water table, drainage area, slope restriction, and ultra-urban sites. Community and environmental factors entail ease of maintenance, community acceptance, construction cost, habitat quality, and others. The manual includes a table each of relevant BMP selection for physical feasibility factors, community and environmental factors, and location and permitting factors checklists.

5.4.7 Environmental Site Design (ESD)

Chapter 5 in the current manual (adopted May 4, 2009) has been expanded considerably following the state's Stormwater Management Act of 2007. The Act requires establishing a comprehensive plan for stormwater management approval, implementing ESD to the MEP, and ensuring that structural practices (discussed in Chapter 3 of the manual) are used only where absolutely necessary. Many stormwater management strategies that seek to replicate natural hydrology; these are known as better site design, low-impact development, green infrastructure, or sustainable site design. For consistency, ESD was adopted as a more generic classification for use in Maryland. The Act defines ESD as using small scale stormwater management practices, nonstructural techniques or better site planning to mimic natural hydrologic, runoff characteristics and minimize the impact of land development on natural resources. Under this definition, ESD includes optimizing conservation of

TABLE 5.8
BMP Selection Based on Stormwater Treatment Suitability

Code	BMP List	Re_v Ability	C_{pv} Control	Q_p Control	Additional Safety Concerns	Space	Accept Hotspot Runoff
P-1	Micro-pool ED	No[a]	Yes	Yes	No	Yes	Yes[c]
P-2	Wet pond	No[a]	Yes	Yes	Yes	Varies	Yes[c]
P-3	Wet ED pond	No[a]	Yes	Yes	Yes	Yes	Yes[c]
P-4	Multiple pond	No[a]	Yes	Yes	Yes	No	Yes[c]
P-5	Pocket pond	No[a]	Yes	Yes	Varies	Yes	Yes[c]
W-1	Shallow wetland	Varies[b]	Yes	Yes	No	No	Yes[c]
W-2	ED wetland	Varies[b]	Yes	Yes	Varies	Varies	Yes[c]
W-3	Pond/wetland	Varies[b]	Yes	Yes	Yes	No	Yes[c]
W-4	Pocket wetland	No	Varies	Varies	No	Varies	Yes[c]
I-1	Infiltration trench	Yes	Varies	Varies	No	Yes	No[c]
I-2	Infiltration basin	Yes	Varies	Varies	No	Varies	No[c]
F-1	Surface sand filter	Varies[b]	Varies	Varies	No	Yes	Yes[d]
F-2	Underground SF	No	No	No	Varies	Yes	Yes
F-3	Perimeter SF	No	No	No	No	Yes	Yes
F-4	Organic filter	Varies[b]	Varies	Varies	No	Yes	Yes[d]
F-5	Pocket sand filter	Varies[b]	Varies	Varies	No	Yes	Yes[d]
F-6	Bio-retention	Yes	Varies	Varies	No	Varies	Yes[d]
O-1	Dry swale	Yes	No	No	No	Varies	Yes[d]
O-2	Wet swale	No	No	No	No	Varies	No

Source: Table 4.3 of the *Maryland Stormwater Design Manual*, updated through May 4, 2009, volumes I and II, prepared by Center for Watershed Protection, Ellicott City, and the Maryland Department of the Environment, Baltimore, MD.

a Structures that require impermeable liners or that intercept groundwater cannot be used for groundwater recharge.

b Re_v can be provided by exfiltration (see Section 3.4 in the manual).

c Not allowed unless pretreatment to remove hydrocarbons, trace metals, and toxicants is provided.

d Yes, but only if the bottom of the facility is lined with impermeable filter fabric that prevents leachate infiltration.

natural features (e.g., drainage patterns, soil, vegetation); minimizing impervious surfaces (e.g., pavement, concrete channels, roofs); slowing down runoff to maintain discharge timing and to increase infiltration and evapotranspiration; and using other nonstructural practices or innovative technologies approved by MDE.

Studies indicate that stream quality and watershed health diminish when impervious cover exceeds 10% and become severely degraded beyond 25%. Thus, fundamental principles of ESD during the planning process help minimize adverse impacts of imperviousness. Table 5.9 provides a summary of site development strategies.

The manual also covers procedures for the site development phase, review of site development plans, and final plan design and review. To achieve the goal of maintaining predevelopment runoff characteristics, the Act established that "The standard for characterizing pre-development runoff characteristics for new development projects is woods in good hydrologic condition." ESD must be implemented to the MEP to mimic predevelopment conditions, and at minimum, ESD must be used to address both Re_v and WQ_v requirements. The manual stipulates that channel protection obligations are met when ESD practices are designed according to the reduced runoff curve number method described in Section 5.4.8.

TABLE 5.9
Better Site Design Development Strategies

Technique	Strategy
Using narrower, shorter streets, rights-of-way, and sidewalks	Streets can be as narrow as 22 ft in neighborhoods serving low traffic volumes; open space designs and clustering will reduce street lengths; rights-of-way can be reduced by minimizing sidewalk width, providing sidewalks on one side of the road, and reducing the border width between the street and sidewalks
Cul-de-sacs	Allow smaller radii for turnarounds as low as 33 ft; use a landscaped island in the center of the cul-de-sac, and design these areas to treat stormwater runoff
Open vegetated channels	Allow grass channels or biofilters for residential street drainage and stormwater treatment
Parking ratios, parking codes, parking lots, and structured parking	Parking ratios should be interpreted as maximum number of spaces; use shared parking arrangements; minimum parking stall width should be less than 9 ft and stall length less than 18 ft; parking garages are encouraged rather than surface lots
Parking lot runoff	Parking lots are required to be landscaped and setbacks are relaxed to allow for bioretention islands or other stormwater practices in landscaped areas
Open space	Flexible design criteria should be provided to developers who wish to use clustered development and open space designs
Setbacks and frontages	Relax setbacks and allow narrower frontages to reduce total road length; eliminate long driveways
Driveways	Allow for shared driveways and alternative impervious surfaces
Rooftop runoff	Direct to pervious surfaces
Buffer systems	Designate a minimum buffer width and provide mechanisms for long-term protection
Clearing and grading	Clearing, grading, and earth disturbance should be limited to that required to develop the lot
Tree conservation	Provide long-term protection of large tracts of contiguous forested areas; promote the use of native plantings
Conservation incentives	Provide incentives for conserving natural areas through density compensation, property tax reduction, and flexibility in the design process

Source: Center for Watershed Protection (1998).

The required ESD runoff depth and storage volume are calculated using the following equations:

$$R_E = (P_E)(R_v) \tag{5.13}$$

$$\text{ESD}_v = (P_E)(R_v)(A)/12 \tag{5.14}$$

where

R_E = required runoff depth
P_E = rainfall target for reduced runoff curve numbers (tabulated for HSG A through D) in inches. Appendix 5B in the manual includes tables for reduced curve numbers:
$R_v = 0.05 + 0.009(I)$, dimensionless runoff coefficient
ESD_v = required storage volume, ft^3
I = impervious cover percent
A = the drainage area in square feet

In SI, the preceding equations become

$$R_E = (P_E)(R_v) \tag{5.15}$$

$$ESD_v = 0.001(P_E)(R_v)(A) \tag{5.16}$$

where

R_E= required runoff depth, mm
P_E= rainfall target, mm
A= the drainage area, m^2
R_v= dimensionless runoff coefficient
ESD$_v$= required storage volume, m^3

5.4.7.1 ESD Stormwater Management Requirements

Stormwater treatment requirements in the 2009 manual are as follows:

ESD practices must be used to treat the runoff from 1 in. of rainfall (P_E = 1 in., 25 mm) on all new developments where stormwater management is required. For Cp_v, ESD practices must be used to the MEP for all sites where the peak discharge, q_i, is more than 2 cfs. Cp_v must be based on 1-year, 24-hour design storm runoff calculated using reduced CN (RCN; see the tables in Appendix 5B in the manual). If the RCN for a drainage area reflects woods in good condition, then Cp_v is met for that area. If the targeted rainfall is not met, any remaining Cp_v requirement must be addressed using structural practices in Chapter 3 of the state manual (Section 5.4.6). The runoff stored in ESD practices can be subtracted from the overbank flood protection and extreme flood volumes (Q_{p2}, Q_{p10}, Q_f), where these are required.

5.4.8 Addressing ESD

The required runoff depth, R_E, and storage volume, ESD, can be addressed using alternative surfaces and nonstructural or microscale practices.

5.4.8.1 Alternative Surfaces

Alternative surfaces include

- Green roofs
- Permeable pavements
- Reinforced turfs

5.4.8.1.1 Green Roofs

In Maryland's 2009 Design Manual, the green roof contribution to ESD was based on the reduced curve numbers (RCNs), which depended on the green roof thickness. Because of discrepancies between observations and the calculations, MDE took a step-based approach in green roof design. MDE understands that a different procedure for analyzing the performance of green rooftops needs to be developed. In the interim, MDE recommends that designers and plan reviewers use the reduced CNs based on retention depth of the green roof tops (Table 5.10).

5.4.8.1.2 Permeable Pavements

Permeable pavements, similar to green roofs, reduce the soil curve number based on the thickness of the sub-base. Effective RCNs for pervious pavements are listed in Table 5.11. Design requirements and soil conditions for pervious pavements are as follows:

TABLE 5.10
Reduced CNs for Green Roofs

Retention depth of green roof in in. (cm)	Effective RCN*
0.6 (1.5 cm)	94
1.0 (2.5 cm)	92
1.2 (3.0 cm)	90
1.4 (3.5 cm)	88
1.6 (4.0 cm)	86
1.8 (4.5 cm)	84
2.0 (5 cm)	82
2.2 (5.5 cm)	81
2.4 (6 cm)	77

TABLE 5.11
Effective RCNs for Permeable Pavements

	Hydrologic Soil Group			
Subbase	**A**	**B**	**C**	**D**
6 in. (15 cm)	76[a]	84[a]	93[b]	–
9 in. (22.5 cm)	62[c]	65[c]	77[c]	–
12 in. (30 cm)	40	55	70	–

[a] Design shall include 1–2 in. minimum over-drain (inv. 2 in. below pavement base) per 750 s.f. of pavement area.

[b] Design shall include 1–2 in. minimum over-drain (inv. 2 in. below pavement base) per 600 s.f. of pavement area.

[c] Design shall include 1–3 in. minimum over-drain (inv. 2 in. below pavement base) and a 0.50 in. underdrain at subbase invert.

Treatment: All permeable pavement systems shall meet the following conditions:

First, applications that exceed 10,000 ft^2 (930 m^2) must be designed as infiltration practices using the design methods for infiltration trenches using porosity (η) of 30% and an effective area of the trench (A_t) equal to 30% of the pavement surface area. Then, a sub-base layer of clean, open graded, washed aggregate with η = 30% (1.5 to 3 in. stone is preferred) is to be used below the pavement surface; the sub-base can be 6, 9, or 12 in. (15, 22.5, or 30 cm) thick. *Filter cloth shall not be used between the sub-base and soil subgrade.* If needed, a 12 in. (30 cm) layer of washed concrete sand or pea gravel (1/8 to 3/8 in. stone) can be used as a bridging layer between the sub-base reservoir and subsurface soils.

In terms of soils, permeable pavements must not be installed in HSG D or on areas of compacted fill, and the underlying soil types and conditions have to be field verified prior to final design. For applications that exceed 10,000 ft^2 (930 m^2), underlying soils have an infiltration rate (f) of 0.50 in. per hour or greater. This rate can be initially determined from NRCS soil textural classification and subsequently confirmed by geotechnical tests in the field. The invert of the sub-base reservoir must be at least 4 ft (1.2 m) (2 ft or 0.6 m on the lower Eastern Shore) above the seasonal high-water table.

5.4.8.1.3 Reinforced Turfs

Reinforced turf consists of interlocking modular units with interstitial areas for planting gravel or grass. These systems are suitable for light traffic areas and can be installed in all soil groups but work best in sandy soils. Reinforced turf should not be used in hotspots that generate higher concentrations of hydrocarbons, trace metals, or toxicants.

5.4.8.2 Nonstructural Practices

Disconnecting impervious cover and treating runoff closer to its source is the next step in implementing ESD. These nonstructural practices include N-1 Disconnection of rooftop runoff, N-2 Disconnection of non-rooftop runoff, and N-3 Sheet flow to conservation areas.

5.4.8.2.1 Disconnection of Rooftop Runoff, N-1

The ESD sizing factor for rooftop runoff disconnection depends on the length of pervious area downstream of downspouts and is specified in Table 5.12.

5.4.8.2.2 Non-Rooftop Runoff Disconnection

The minimum length of flow path through vegetated areas should be 10 ft, and the maximum contributing impervious flow path 75 ft. ESD sizing for non-rooftop disconnection depends on the ratios of impervious and pervious flow paths to the vegetated buffer areas and varies from 0.2 to 1.0 in. (5 to 25 mm).

5.4.8.2.3 Sheet Flow to Conservation Areas

Conservation areas (the author notes that these are called vegetated buffers in many jurisdictions, including New Jersey) are an effective means of treating stormwater runoff. The ESD, P_E, depends on the width of the conservation area and ranges from 0.6 in. for 50 ft buffer to 1.0 in. for 100 ft buffer. The conservation area must be 20,000 square feet or larger, and its minimum effective width is 50 ft.

5.5.8.3 Microscale Practices

Microscale practices are small water-quality treatment or flow-control measures to capture runoff from discrete impervious areas, and Maryland classifies them as M-1: rainwater harvesting, M-2: submerged gravel wetlands, M-3: landscape infiltration, M-4: infiltration berms, M-5: dry wells, M-6: micro-bioretention, M-7: rain gardens, M-8: swales, M-9: enhanced filters.

TABLE 5.12
ESD Sizing Factors for Rooftop Disconnection

Disconnection Flow Path Length, ft (m)		P_E
Eastern Shore	**Western Shore**	**in. (mm[a])**
15 (4.9)	12 (3.0)	0.2 (5)
30 (9.8)	24 (7.9)	0.4 (10)
45 (14.8)	36 (11.8)	0.6 (15)
60 (19.7)	48 (15.7)	0.8 (20)
75 (24.6)	60 (19.7)	1.0 (25)

[a] Rounded to whole numbers.

Microscale practices used for new development shall promote runoff reduction and water quality treatment through infiltration, filtration, evapotranspiration, rainwater harvesting, or a combination of these techniques. Microscale filters used for new development must be designed to promote recharge (e.g., enhanced filter) and be planted as part of the landscaping plans.

Since a number of these practices will be discussed in Chapters 7 and 8, only a brief description together with design equations for sizing of some of the microscale practices will be reviewed herein.

5.4.8.3.1 Rainwater Harvesting (Cisterns and Rain Barrels)

The Maryland manual recommends designing rain barrels and cisterns for at least 0.2 in (5 mm) of rainfall. This suggests that two 65 gallons (250 liter) rain tanks should be used to retain 5 mm of rainfall from every 1000 ft^2 (100 m^2) roof area.

5.4.8.3.2 Landscape Infiltration

Landscape infiltration utilizes on-site vegetative planting areas to capture, store, and treat stormwater runoff. Rainwater is initially stored filters through the planting soil and gravel media below, and then it infiltrates into native soils. These practices can be integrated within the overall site design by utilizing a variety of landscape features for storage and treatment of stormwater runoff. Storage may be provided in constructed planters made of stone, brick, or concrete, or in natural areas excavated and backfilled with stone and topsoil.

Landscape infiltration can be best implemented in residential and commercial land uses. Residential areas with compact housing such as clustered homes and townhouses can utilize small green spaces for landscape infiltration. Because space in these instances prevents structural pretreatment, the drainage area to these practices should be limited to less than 10,000 ft^2 (930 m^2). Larger drainage areas may be allowed where soil testing is performed and pretreatment forebays can be implemented. Successful application is dependent upon soil type and groundwater elevation. Landscape infiltration is not recommended in HSG C and D soils. Figure 5.7 depicts plan view of a landscape infiltration.

When landscape infiltration systems are designed in accordance with the guidance in the manual, their ESD sizing criteria are given by

$$P_E = (20\,\text{in.})(A_f)/DA(\text{in.}) \tag{5.17}$$

$$P_E = (50\,\text{cm})(A_f)/DA(\text{mm}) \tag{5.18}$$

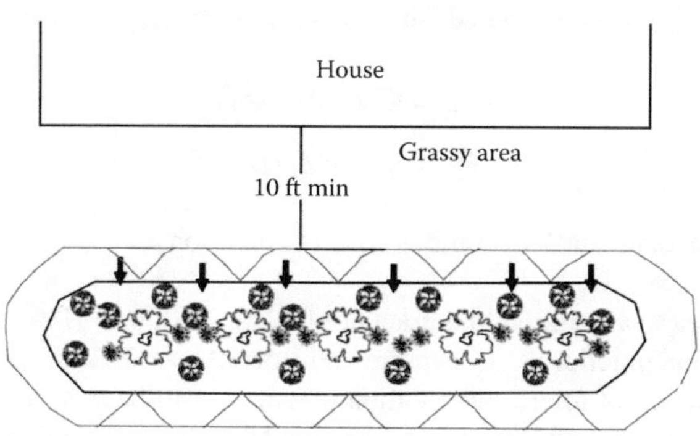

FIGURE 5.7 Plan view of landscape infiltration (depression).

where A_f and DA are the surface area of landscape infiltration and the drainage area tributary thereof and P_E *is in inches* in CU and cm in SI. The infiltration area should be at least 2% of the contributing drainage area. Landscape infiltration design criteria are as follows:

- The drainage area to any individual practice shall be 10,000 ft² or less.
- Landscape infiltration facilities located in HSG B (i.e., loams, silt loams) shall not exceed 5 ft (1.5 m) in depth. Facilities located in HSG A (i.e., sand, loamy sand, sandy loam) shall not exceed 12 ft (3.6 m) in depth.
- Landscape infiltration facilities shall be designed to fully dewater the entire ESD_v within 48 hours. Temporary storage of the ESD_v may be provided above the facility.
- A 12 to 18-in. (30-45 cm) layer of planting soil shall be provided as a filtering media at the top of the facility.
- A minimum 12-in. (30 cm) layer of gravel is required below the planting soil.
- A 12-in. (30 cm) layer of clean sand shall be provided at the bottom to allow for a bridging medium between the existing soils and stone within the bed.
- The storage volume for the ESD_v shall be determined for the entire system including the temporary ponding area, the soil, and the sand and gravel layers in the bottom of the facility. Storage calculations shall account for the porosity (n = 0.40) of the gravel and soil media.
- Pretreatment measures shall be implemented along the main stormwater runoff collection system where feasible. These include installing gutter screens, a removable filter screen on rooftop downspout pipes, a sand layer or pea gravel diaphragm at the inflow, or a two to three-inch surface mulch layer.

5.4.8.3.3 Micro-bioretention

Micro-bioretention practices capture and treat runoff from discrete impervious areas by passing it through a filter bed mixture of sand, soil, and organic matter. Filtered stormwater is either returned to the conveyance system or partially infiltrated into the soil. Micro-bioretention practices are versatile and may be adapted for use anywhere there is landscaping. Micro-bioretention is a multifunctional practice that can be easily adapted for new and redevelopment applications in commercial and industrial projects. Stormwater runoff is stored temporarily and filtered in landscaped facilities shaped to take runoff from various sized impervious areas. Micro-bioretention provides water quality treatment and aesthetic value, and it can be applied as concave parking lot islands, linear roadway or median filters, terraced slope facilities, residential cul-de-sac islands, and ultra-urban planter boxes.

The P_E values determined by Equation 5.19 may be applied to the ESD sizing criteria when micro-bioretention systems are designed according to the guidance that follows. Re_v requirements are also met when the P_E meets or exceeds the soil-specific recharge factor (see Table 5.6).

$$P_E = 15\,\text{in.}\left(A_f / DA\right) \tag{5.19}$$

$$P_E = 38\,\text{cm}\left(A_f / DA\right) \tag{5.20}$$

Micro-bioretention practices shall meet the following conditions:

- The drainage area to any individual practice shall be 20,000 ft² (1860 m²) or less.
- Micro-bioretention practices shall capture and store at least 75% of the ESD_v.
- The surface area (A_f) of micro-bioretention practices shall be at least 2% of the contributing drainage area. A P_E value based on Equation 5.19 (or 5.20) shall be applied to the contributing drainage area. Temporary storage of the ESD_v may be provided above the facility with a surface ponding depth of 12 in. (30 cm) or less.

- Filter beds shall be between 24 and 48 in. (50 cm and 1 m) deep.
- Filter beds shall not intercept groundwater. If designed as infiltration practices, filter bed inverts shall be separated at least 4 ft vertically (2 ft on the lower Eastern Shore) from the seasonal high- water table.
- A surface mulch layer (maximum 2 to 3 in. thick) should be provided to enhance plant survival and inhibit weed growth.

Setbacks:

- Micro-bioretention practices should be located down gradient and set back at least 10 ft (3 m) from structures. Micro-bioretention variants (e.g., planter boxes) that must be located adjacent to structures should include an impermeable liner.
- Micro-bioretention practices shall be located at least 30 ft (10 m) from water supply wells and 25 ft (7.5 m) from septic systems. If designed to infiltrate, then the practice shall be located at least 50 ft (15 m) from confined water supply wells and 100 ft (30 m) from unconfined water supply wells.

Figure 5.8 shows a variation of micro-bioretention employed at a parking lot median.

5.4.8.3.4 Rain Gardens

A rain garden is a shallow, excavated landscape feature or a saucer-shaped depression that temporarily holds runoff for a short. Rain gardens typically consist of an absorbent planted soil bed, a mulch layer, and planting materials such as shrubs, grasses, and flowers. An overflow conveyance

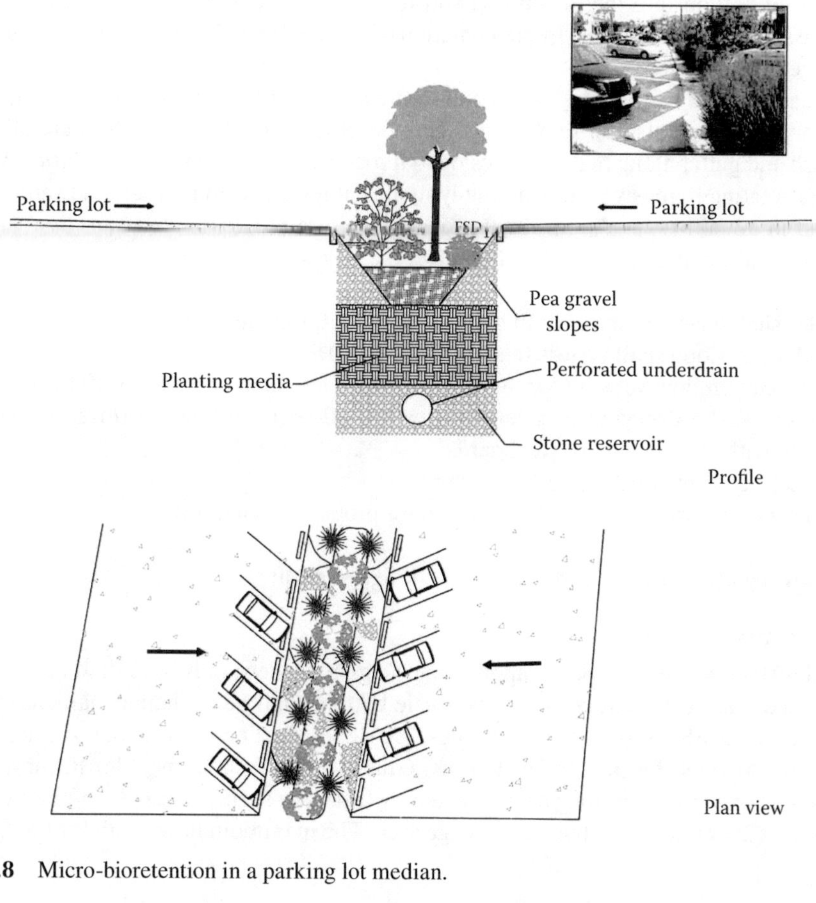

FIGURE 5.8 Micro-bioretention in a parking lot median.

system is included to pass larger storms. Captured runoff from downspouts, roof drains, pipes, swales, or curb openings temporarily ponds and slowly filters into the soil over 24 to 48 hours.

The ECS sizing criterion for rain gardens is the P_E calculated by the following equation:

$$P_E = (10\,\text{in.})(A_f / DA),\ \text{in.} \tag{5.21}$$

$$P_E = (25\,cm)(A_f / DA),\ \text{cm} \tag{5.22}$$

Rain gardens are designed based on the following conditions:

The drainage area to a rain garden serving a single lot in a residential subdivision should be 2000 ft^2 (190 m^2) or less, and the maximum drainage area to a rain garden for all other applications is 10,000 ft^2 (900 m^2). Micro-bioretention or bioretention should be considered when these requirements are exceeded. The surface area (A_f) of rain gardens must be at least 2% of the contributing drainage area. P_E based on Equation 5.21/5.22 must be applied to the contributing drainage area. Temporary storage of the ESD_v can be provided above the facility with a surface ponding depth of 6 in. or less. Excavated rain gardens work best where HSG A and B are prevalent; in areas of HSG C and D, at-grade applications or soil amendments should be considered. A minimum 6–12 in. (15–30 cm) layer of planting soil must be provided, and a mulch layer 2–3 in. (5–7.5 cm) deep is applied to the planting soil to maintain soil moisture and to prevent premature clogging.

5.4.8.3.5 Swales

Swales are channels that provide conveyance, water quality treatment, and flow attenuation of stormwater runoff. Swales remove pollutants through vegetative filtering, sedimentation, biological uptake, and infiltration into the underlying soil media. Three design variants are covered in this section: grass, wet, and bioswales. Implementation of each is dependent upon site soils, topography, and drainage characteristics.

Swales can be used for primary or secondary treatment on residential, commercial, industrial, or institutional sites and also for retrofitting and redevelopment. The linear structure allows use in place of curb and gutter along highways, residential roadways, and property boundaries. Wet swales are ideal for treating highway runoff in low-lying or flat terrain with high groundwater. Bioswales can be used in all soil types due to the use of an underdrain. Grass swales are best suited along highway and roadway projects. Swales shall meet the following criteria:

- Swales shall have a bottom width between 2 and 8 ft (0.6 and 2.4 m).
- The channel slope shall be less than or equal to 4.0%.
- The maximum flow velocity for the ESD_v, shall be less than or equal to 1.0 fps (0.3 m/s).
- Swales shall be designed to safely convey the 10-year, 24-hour storm at a nonerosive velocity with at least 6 in. of freeboard.
- Channel side slopes shall be 3:1 or flatter.
- A thick vegetative cover shall be provided for proper functioning.

The following criteria apply to each specific design variant:

5.4.8.3.6 Grass swales

Grass swales shall be used for linear applications (e.g., roadways) only, and shall be as long as the treated surface. The surface area (A_f) of the swale bottom shall be at channel gardens (P_E = 10 in. × A_f/DA) is used for ESD sizing of grass swales side slopes shall be 3:1 or flatter. A thick vegetative cover shall be provided for proper functioning. Grass swales shall be used for linear applications (e.g., roadways) only and be as long as the treated surface. The surface area (A_f) of the swale bottom shall be at least 2% of the contributing drainage area. The maximum flow depth for ESD_v treatment should be 4 in. (10 cm), and the channel should have a Manning's n of 0.15. This can be accomplished by either maintaining vegetation height equal to the flow depth or using energy dissipaters like check dams, infiltration berms, or riffle/pool combinations. Equation 5.21 for rain

5.4.8.3.7 Bioswales

The surface area (A_f) of the bioswale bottom must be at least 2% of the contributing impervious area, and P_E based on Equation 5.19 (P_E = 15 in. × A_f/DA) must be applied to the contributing drainage area. Bioswales must be designed to temporarily store at least 75% of the ESD_v. A 2 to 4 ft (0.6–1.2 m) deep layer of filter media shall be provided in the swale bottom. Underdrains are recommended for all applications and is essential in HGS C or D soils. Figure 5.9 is a schematic cross-section of a bioswale.

5.4.8.3.8 Wet swales

Wet swales must store at least 75% of the ESD_v. A P_E equivalent to the volume captured and treated must be applied to the contributing drainage area. Wet swales should be installed in areas with a high groundwater table, and check dams or weirs may be used to enhance storage.

5.4.9 Redevelopment

5.4.9.1 Introduction

Redevelopment is defined as any construction, alteration, or improvement performed on sites where the existing land use is commercial, industrial, institutional, or multifamily residential and existing site impervious area exceeds 40%. The term site is defined as a single tract, lot, or parcel of land, or a combination of tracts, lots, or parcels of land that are in one ownership or are contiguous and in diverse ownership where development is to be performed as part of a unit, subdivision, or project. When calculating site imperviousness, the local approving agency may allow lands protected by forest preservation, conservation easements, or other mechanisms to be subtracted from the total site area. This will create incentive to preserve and protect natural resources in redevelopment projects.

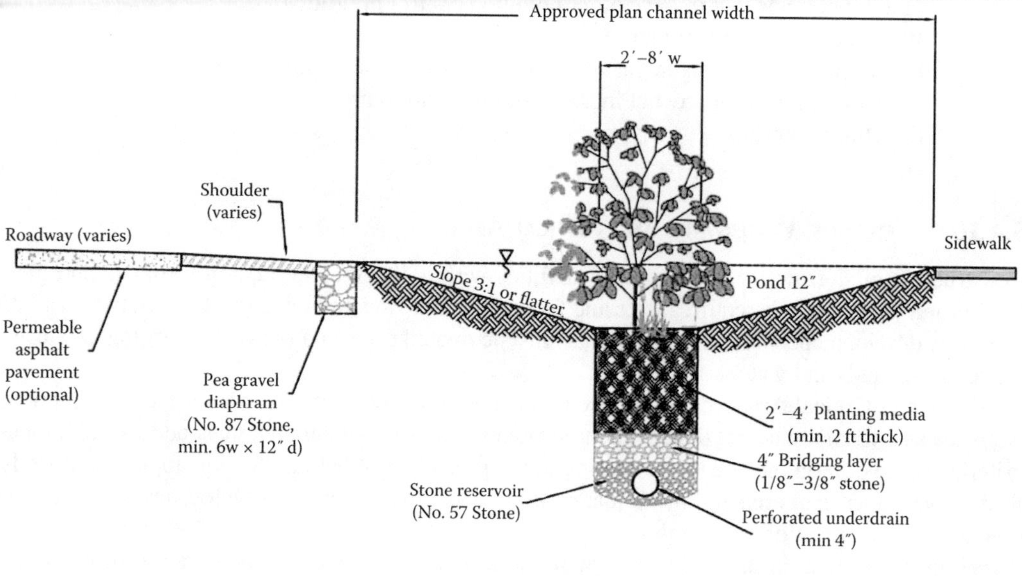

FIGURE 5.9 Section view of a bioswale.

5.4.9.2 Redevelopment Policy

The 40% site impervious area threshold will determine whether a project in Maryland will be regulated as new development or redevelopment. When redevelopment requirements apply, all existing impervious areas located within a project's limit of disturbance (LOD) must be part of the management plan. Because redevelopment projects present a wide range of constraints and limitations, the following policy allows for flexibility and an evaluation of options that can work in conjunction with broader watershed goals and local initiatives:

1. Stormwater management shall be addressed for redevelopment according to the following criteria:
   a. Reduce existing impervious area within the LOD by at least 50%
   b. Implement ESD practices to the MEP to provide water quality treatment for at least 50% of existing impervious area within the LOD
   c. Use a combination of impervious area reduction and ESD implementation for at least 50% of existing impervious areas
2. Alternative stormwater management measures may be used provided that the developer satisfactorily demonstrates to the approving authority that impervious area reduction and ESD have been implemented to the MEP. Alternative stormwater management measures include but are not limited to
   a. An onsite structural BMP
   b. An offsite structural BMP to provide water quality treatment for an area equal to or greater than 50% of existing impervious areas
   c. A combination of impervious area reduction, ESD implementation, and on-site or off-site structural BMP for an area equal to or greater than 50% of existing impervious area within the LOD
3. An approving agency may develop separate programmatic policies for providing water quality treatment for redevelopment projects when the preceding requirements cannot be met. These policies shall be reviewed and approved by MDE and may include but are not limited to
   a. Retrofitting existing structural BMPs
   b. Stream restoration
   c. Trading policies that involve other pollution control programs
   d. Watershed management plans
      i. Stormwater management shall be addressed according to new development requirements for any net increase in impervious area
      ii. Other criteria

5.4.10 Wetlands, Waterways, and Critical Areas

Construction of stormwater management facilities within 100-year floodplain wetlands and their buffers and tidal wetlands requires state and federal permits. In addition to these restrictions, runoff from new developments and redevelopments must be treated prior to directly discharging into jurisdictional wetlands and waters of the state.

Maryland's Critical Area Act designated all land within 1000 ft of tidal waters and adjacent tidal wetlands as critical areas. All development located within the critical area must address additional criteria. These criteria include those relating to the protection of habitat that are applied uniformly throughout the critical area and those that relate to water quality and imperviousness and are specific to land classifications as follows.

Within the critical area, land is designated as either intensely developed area, limited development area, or resource conservation area (IDA, LDA, and RCA, respectively) based on uses that existed at the time the local programs were adopted. IDAs are areas of concentrated development where there is little natural habitat. Any new development and redevelopment projects within an IDA must include stormwater management practices to reduce post-development phosphorus loads to at least 10% below predeveloped levels, commonly known as the 10% rule.

LDAs are regions where development density is low to moderate and wildlife habitat is not dominated by agriculture, wetlands, forests, or other natural areas. Similarly, RCAs are characterized by the dominance of agriculture or protected resources like forests or wetlands. Within these areas, any new development or redevelopment must address standard water quality requirements, conserve natural areas, and incorporate corridors to connect wildlife and plant habitat. To accomplish these goals, impervious surfaces and lot coverage are generally limited to 15% of the property or project area. Alternative surfaces may be used to meet this limitation. There are also strict limits on clearing existing woodland or forests. All clearing of these areas requires at least a 1:1 replacement.

To protect habitat, a forested buffer is required on all new development in all three land designations. Extending a minimum of 100 ft (30 m) from the mean high-water line of tidal waters or the landward edge of tidal wetlands and tributary streams, this buffer acts as a water quality filter and protects important riparian habitat within the critical area. This distance may be expanded to include adjacent sensitive areas like hydric or highly erodible soils or steep slopes. If a sensitive area exists within a subdivision in the RCA, the minimum width of the buffer is 200 ft. Disturbance associated with new development is generally prohibited within the buffer, and, accordingly, stormwater practices (e.g., microscale practices, structural facilities) cannot be located within it.

5.5 STATE OF NEW YORK STORMWATER REGULATIONS

5.5.1 Introduction

The State of New York Stormwater Management Design Manual was first published in 2010, by the State Department of Environmental Conservation. This manual was updated in 2015, and the current manual in draft was issued in May 2022. The current manual contains twelve chapters, nine appendices and covers 496 pages. Table 5.13 presents the Unified Stormwater Sizing Criteria in the current manual.

5.5.2 Water Quality Volume

The water quality volume, abbreviated as WQ_v, represents the runoff from 90% of the average annual rainfall and is the design criteria for sizing stormwater management systems to improve water quality. The WQ_v is directly related to the amount of impervious cover and is calculated by the following equation:

$$WQ_v = (P)(R_v)(A)/12 \text{ CU} \tag{5.23}$$

In metric units, this equation reads as:

$$WQ = 10PR_vA \text{ SI} \tag{5.24}$$

where

WQ_v = water quality volume, in ac-ft (m^3)
P = 90% rainfall event number, in. (mm)
$R_v = 0.05 + 0.009(I)$, where I is percent impervious cover
A = site area (contributing area), acres (ha)

TABLE 5.13
New York State Stormwater Sizing Criteria

Water Quality (WQ_v)	Standards
	90% rule: $WQ_v = [(P)(R_v)(A)]/12$ $R_v = 0.05 + 0.009(I)$ I = impervious cover (percent) Minimum $R_v = 0.2$ P = 90% rainfall event number A = site area in acres
Runoff reduction volume (RR_v)	RR_v (ac-ft) = reduction of the total WQ_v by application of green infrastructure techniques and SWPs to replicate predevelopment hydrology. The minimum RR_v is defined as the specified reduction factor (S) provided that objective technical justification is determined.
Channel protection (Cp_v)	Cp_v = 24-hour extended detention of post developed 1-year, 24-hour storm event, remaining after runoff reduction. Where conditions allow, runoff reduction of total Cp_v is encouraged. Option for sites larger than 50 acres: Distributed runoff control- geomorphic assessment to determine the bank full channel characteristics and thresholds for channel stability and bedload movement.
Overbank flood (Q_p)	Control the peak discharge from the 10-year storm to 10-year predevelopment rates.
Extreme storm (Q_f)	Control the peak discharge from the 100-year storm to 100-year predevelopment rates. Safely pass the 100-year storm event
Alternative Methods	Design, construct, and maintain systems sized to capture, reduce, reuse, treat, and manage rainfall on-site, and prevent the offsite discharge of the precipitation from all rainfall events less than or equal to the 95th percentile rainfall event, computed by an acceptable continuous simulation model.

Note: Channel protection, overbank flood, and extreme flood requirements may be waived in some instances if the conditions specified in this chapter are met. For SMPs involving dams, follow Appendix A, Guidelines for Design of Dams for safe passage of the design flood.

WQ_v treatment must be able to capture and treat the full WQ_v to meet 80% TSS removal and 40% TP removal. The practices must have acceptable longevity in the field and have a pretreatment mechanism.

Acceptable water quality treatment practices are as follows:

a. Stormwater ponds: Practices have either a permanent pool of water or a combination of permanent pool and extended detention capable of treating the WQ_v. The manual describes five different types of ponds denoted as P-1 through P-5 (see Table 5.4 in the manual).
b. Stormwater wetlands: Practices include stagnant shallow marsh areas and can incorporate small permanent pools and extended detention storage to achieve the full WQ_v.
c. Infiltration practices: Practices capture and temporarily store the WQ_v before allowing it to infiltrate into the soil.
d. Filtering practices: Practices that capture and temporarily store the WQ_v and pass it through a filter bed of sand, organic matter soil, or other acceptable treatment media.
e. Open channel practices: Practices are explicitly designed to capture and treat the full WQ_v within dry or wet cells formed by check dams or other means.

For any offsite area entering an onsite stormwater management facility, treatment should be provided based on its current condition. If water quality treatment is provided offsite, the facility needs only to treat onsite runoff.

5.5.3 WQ_v Treatment Practices

The WQ_v will be treated by an acceptable practice that meets the following criteria:

1. It can capture and treat the full water quality volume (WQ_v).
2. It is capable of 80% TSS removal and 40% TP removal.
3. It has acceptable longevity in the field.

It has a pretreatment mechanism.
Acceptable water quality treatment practices are as follows:

1. Stormwater ponds: Practices have either a permanent pool of water or a combination of permanent pool and extended detention capable of treating the WQ_v. The manual presents description for five different types of ponds denoted as P-1 through P-5 (see Table 5.4 in the manual).
2. Stormwater wetlands: Practices include stagnant shallow marsh areas and may also incorporate small permanent pools and extended detention storage to achieve the full WQ_v.
3. Infiltration practices: Practices capture and temporarily store the WQ_v before allowing it to infiltrate into the soil.
4. Filtering practices: Practices capture and temporarily store the WQ_v and pass it through a filter bed of sand, organic matter soil, or other acceptable treatment media.
5. Open channel practices: Practices are explicitly designed to capture and treat the full WQ_v within dry or wet cells formed by check dams or other means.

For any offsite area entering an on-site stormwater management facility, treatment should be provided based on its current condition. If water quality treatment is provided offsite, the facility needs only to treat on-site runoff.

5.5.4 Runoff Reduction Volume

Runoff can be reduced through infiltration, groundwater recharge, and reuse and recycle of 100% of post-development water quality volume. The minimum required runoff reduction volume (PR_V) for newly constructed impervious areas is calculated by the following equation:

$$PR_V = PxR_V xA_i xS/12 \tag{5.25}$$

where

P= 90% rainfall event (Figure 5.10)
R_v= 0.05 +0.09I, (I = % impervious)
A_i = total area of impervious cover
S= Hydrological Soil Group (HSG) special reduction factor as follows
S = 0.55 HSG A
S = 0.40 HSG B
S = 0.30 HSG C
S = 0.20 HSG D

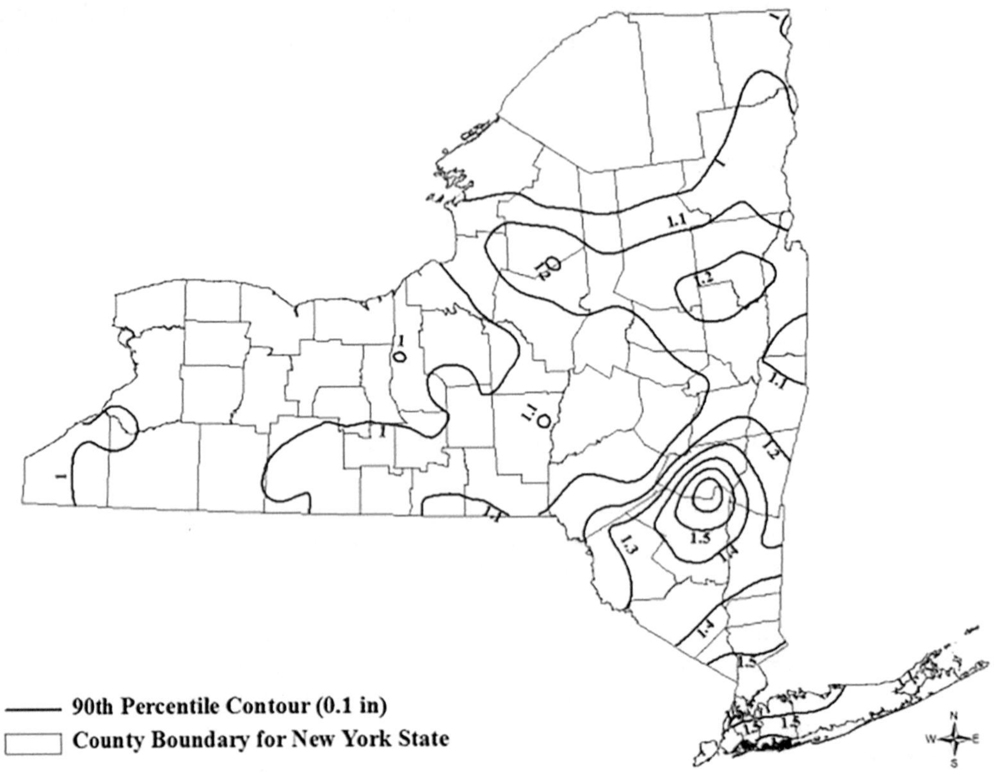

FIGURE 5.10 Ninety percent rainfall event in New York State.

Infiltration techniques such as rain gardens, rain tank/cistern, dry well, green roof and pervious pavements are among the measures that reduce runoff volume. For watersheds requiring phosphorous treatment, the minimum PR_V should be calculated using the one-year 24- hour event, rather than the 90% rainfall.

5.5.5 Stream Channel Protection Volume

The stream channel protection volume (Cp_v) requirement is intended to protect stream channels from erosion. This goal is accomplished by providing 24-hour extended detention of the 1-year, 24-hour storm event. The depth of this storm as depicted in a figure in the NY State Stormwater Management Design Manual (2015) ranges from 1.8 to 2.9 inches (46–74 mm) across the state. For sites that discharge to trout water, only 12 hours of extended detention are required to meet this criterion. Regardless of meeting the required retention time, the minimum size of an orifice in basins provided for channel protection is 3 in (75 mm). Cp_V is required for the entire site. If a site has multiple design points, Cp_V should be calculated and met at each of those points. For developments greater than 50 acres (20 ha) with impervious cover greater than 25%, it is recommended that a detailed geomorphic assessment be performed to determine the appropriate level of control. Appendix I in the manual provides guidance on how to conduct this assessment.

The Cp_v requirement does not apply in certain conditions, including the following:

- The entire Cp_v volume is recharged at a site.
- The site discharges directly into tidal waters or fourth-order or larger streams. Within New York State, streams are classified using: New York State Codes Rules and Regulations (NYCRR).
- The sites where the one-year, 24-hour peak discharge is less than 2 cfs (56.6 L/s).

However, this classification system does not provide a numeric stream order. The methodology identified in this manual is consistent with Strahler-Horton methodology. Specifically, a small branch is designated as a first-order stream. When two branches merge, a second order is created downstream of the confluence. Third-order stream refers to those downstream of the confluence of the second-order streams and so on.

Detention ponds or underground vaults are methods to meet the Cp_v requirement (and subsequent Q_{p10} and Q_f criteria). Note that although these practices meet water quantity goals, they are unacceptable because of poor pollutant removal, and they need to be coupled with ponds, wetlands, and infiltration and filtration practices. The Cp_v requirement may also be provided above the water quality (WQ_v) storage in a wet pond or stormwater wetland

In determining channel protection storage volume, TR-55 and TR-20 (or approved equivalent) should be used to determine peak discharge rates for the 1-year, 24-hour storm event, and offsite areas should be modeled as "present condition." The length of overland flow used in T_c calculations is limited to no more than 100 ft for post-development conditions and 150 ft for the predevelopment conditions. However, in flat areas (less than 1% average slope), the maximum length of overland flow is extended to 250 feet).

5.5.6 Overbank Flow Control

The primary purpose of the overbank flooding design criterion (Q_P) is to prevent an increase in the frequency and magnitude of out-of-bank flooding generated by urban development. Overbank control requires storage to attenuate the post-development 10-year, 24-hour peak discharge rate (Q_p) to predevelopment rates. The 10-year 24-hour storm depth ranges from 3.25 in (82,5 mm) in the northwesterly parts of the state to 5.25 in, (133 mm) in southeasterly parts of the state and Long Island. The 2015 Design Manual includes a figure showing the rainfall contour lines in 0.25 in, (64 mm) increments, The overbank flood control requirement does not apply in certain conditions, including where the site discharges directly into tidal waters or fourth-order (fourth downstream) or larger streams.

Design calculations for overbank flood control are performed on the following basis:

- TR-55 and TR-20 (or approved equivalent) will be used to determine peak discharge rates.
- When the pre-development land use is agriculture, the curve number for the predeveloped condition shall be "taken as meadow."
- Off-site areas should be modeled as "present condition" for the 10-year storm event.
- The length of overland flow used in T_c calculations is limited to no more than 150 ft for pre-development conditions and 100 ft for post-development conditions. On areas of extremely flat terrain (<1% average slope), this maximum distance is extended to 250 ft for pre-development conditions and 150 ft for post-development conditions.

5.5.7 Extreme Flood Control

The extreme flood control criterion is a 100-year, 24-hour storm throughout the state of New York. One-hundred-year flood control requires storage to attenuate the post-development 100-year,

24-hour peak discharge rate (Q_f) to predevelopment rates. According to a figure in the 2015 stormwater manual, this storm varies from 5 in. to 8 in. (127–203 mm) in the state. Extreme flood control must be addressed for the entire site. If a site contains multiple design points, the 100-year flood storage requirement must be provided for each.

The 100-year storm control requirement can be waived if

- The site discharges directly to tidal waters or fifth-order or larger streams, lakes where the increase in flows will not impact the stream bank or channel stability. However, the point of discharge must be adequately protected against erosion and scour.
- Development is prohibited within the ultimate 100-year floodplain.
- A downstream analysis reveals that 100-year control is not needed.

Detention structures involving dams must provide safe overflow of the design flood. The flow rates and floodplain extents referred to herein should not be confused with those developed by FEMA for use in the NFIP.

The design criteria for this storm event are the following:

- The Q_f is calculated using the same methods discussed for overbank flood control, Q_p.
- When determining the storage required to reduce 100-year flood peaks, offsite areas are modeled under current conditions.
- When calculating the storage required to safely pass the 100-year flood, offsite areas are modeled assuming ultimate (fully developed) conditions.

5.5.8 Downstream Analysis

A downstream analysis for overbank and extreme flood control is recommended for sites greater than 50 acres to size facilities in the content of a larger watershed. The analysis will help ensure that storage provided at a site is appropriate when combined with upstream and downstream flows. For example, detention at a site may in some instances exacerbate flooding problems within a watershed. This section provides brief guidance for conducting this analysis, including the specific points along the downstream channel to be evaluated and the minimum elements to be included in the analysis.

Downstream analysis can be conducted using the 10% rule. That is, the analysis should extend from the point of discharge downstream to the point on the stream where the site represents 10% of the total drainage area. For example, the analysis limits for a 50-acre site would include points on the stream from the point of discharge to the nearest downstream point with a drainage area of 500 acres. The required elements of the downstream analysis are as follows:

- Compute pre-development and post-development peak flow and velocities for design storms (e.g., 10 year and 100 years) at all downstream confluences with first-order or higher streams up to and including the point where the 10% rule is met. These analyses should include scenarios both with and without stormwater quantity control practices in place, where applicable.
- Evaluate hydrologic and hydraulic effects of all culverts and/or obstructions within the downstream channel.
- Assess water surface elevations to determine if an increase in water surface elevations will impact existing buildings and other structures.

Overbank and extreme flood requirements may be waived if both of the following conditions are met:

a. Peak flow rates increase by less than 5% of the predeveloped condition for the design storm (e.g., 10 year or 100 year).
b. No downstream structures or buildings are impacted.

5.5.9 Rainfall Data and Hydrological Analysis

Rainfall data for runoff reduction volume (PR_V), channel protection (Cp_V), overbank flood control (Q_P) and extreme flood control (Q_F) should be extracted from NRCC and NRCS joint collaborative website (https://precip.eas.cornell.edu) or NOAA Atlas 14 website (https://hdsc.nws.noaa.gov/hdsc/pfds/). Hydrologic and Hydraulic modeling software including TR-20 (by IJSDA), HydroCAD, PondPack, StormCAD, HydraFlow and Hydrologic Engineering Center HEC-HMS may be used for performing hydrologic calculations and flood routing.

5.5.10 Stormwater Hotspots

A stormwater hotspot is defined as a land use or activity that generates higher concentrations of hydrocarbons, trace metals, or toxicants that are found in typical stormwater runoff. These activities may include municipal, industrial, transportation, or institutional that, in addition to generating higher concentrations of pollutants, have a risk of spills, leaks, or illicit discharge. If a site is designated a hotspot, measures should be taken to protect surface waters and groundwater from contamination. An effective measure is to isolate the hotspot from the rest of the watershed by

- Diverting the runoff upgradient from the Hotspot and conveying it to a separate stormwater practice if needed.
- Capturing the runoff from Hotspot at the source

The runoff from the hotspot should then be properly treated. The measure and level of treatment should be selected based on the type and level of pollutants. Depending on the pollutant concentrations, the hotspots are classified as Level 1 for moderate pollutant concentrations or Level 2 for severe concentrations. To prevent groundwater contamination, infiltration is restricted in Level 1 Hotspot and prohibited in Level 2.

5.5.11 Redevelopment Projects

Redevelopments, like new projects, are subject to four criteria: water quality treatment, channel protection, overbank flooding, and control of extreme storms. If a project proposes no increase in impervious area or if it does not change hydrology that would increase the discharge rate, the overbank flooding and control of extreme flows do not apply. The water quality criteria are met using the following options:

1. The project reduces the existing impervious coverage by a minimum of 25%.
2. If the 25% reduction in impervious area is not met, the project captures and treats a minimum of 25% of the water quality volume (WQ_v) from the redevelopment area through standard practices. The plan should target areas with the greatest pollution generation potential, such as parking lots and service stations.
3. The project proposes the use of alternative practices to treat 75% of WQ_v from the disturbed area as well as additional runoff, if any, beyond the disturbed area.
4. The plan proposes a combination of impervious area reduction and standard or alternative practices that provide a weighted average equal to or more than options 2 or 3. The level of water quality treatment must be acceptable for redevelopment applications, Proprietary

practices include wet vaults, underground infiltration systems, and manufactured water quality devices (hydrodynamic gravity or water separators and media filters).

PROBLEMS

5.1 Which agency passed the Clean Water Act (CWA) of 1972? What did the Act cover?

5.2 What are MS4s, and what type of permit do they require?

5.3 What is a construction general permit (CGP)? When did the current CGP go into effect?

5.4 What does NPDES stand for? When did the NSPES phase II become effective and what does it require?

5.5 What does TSS mean? What is the required TSS removal in New Jersey? What is the requirement in your state/your area?

5.6 What does EISA Section 438 cover?

5.7 What are the new trends (green infrastructure) for stormwater quantity control?

5.8 What is defined as a major development in New Jersey?

5.9 Are major developments required to provide peak flow reductions in New Jersey? What are those reductions?

5.10 What is the water quality requirement in New Jersey? Are all development projects required to provide stormwater treatment?

5.11 Calculate the volume of water quality storm for a 10-acre development with an impervious coverage of 30% in New Jersey. What is the volume for a 4.0 ha site? State any assumptions made.

5.12 Calculate the water quality runoff volume for a 10-acre development having the following impervious/pervious coverages:
15% building
25% parking
60% lawn/landscape

How much of the runoff is subject to water quality treatment? Hint: Lawns and landscapes generate little runoff during a water-quality storm.

5.13 The runoff from the building and parking area of Problem 5.12 is routed through an extended detention basin that provides 40% TSS removal. Is this adequate to address the state of New Jersey's stormwater management regulations? If not, what is the required TSS removal deficiency? What can be done to address the required TSS removal for this site?

5.14 Calculate the required groundwater recharge in Problem 5.12 using option 2 in the NJDEP regulation. The site soil is hydrologic group B, and the 2-year, 24-hour storm is 3.5 in.

5.15 What are the water quantity sizing criteria in the state of Maryland? Calculate the required water quality volume for an 8-acre residential site in the eastern zone. The impervious coverage on the site is 2.4 acres.

5.16 Rework Problem 5.15 for a 2 hectares site with 30% impervious coverage.

5.17 Calculate the water quality volume WQ_v of the residential project in Problem 5.15 if the project was in New York City.

5.18 Calculate the recharge volume for the residential site of Problem 5.15. The site soil is hydrologic group B.

5.19 Calculate the channel protection volume, Cp_v, for the site of Problem 5.18. The site is located in Frederick County. Base your calculations on a time of concentration of 30 minutes for the predevelopment conditions and 15 minutes for post-development.

5.20 Calculate overbank flood protection volume, Q_p, for Problem 5.15. The site is situated in Frederick County.

5.21 Calculate the required recharge for Example 5.1 in metric units. For simplicity, round the site area to 12 hectares.

APPENDIX 5A: NATIONAL POLLUTANT DISCHARGE ELIMINATION SYSTEM (NPDES)

About
NPDES

- **Permit basics/FAQs** <https://epa_gov/nodes/npdes-permit:basics>
- **State program information** <https://epa.gov/npdes/npedes-state-program-information>
- **Permit limits** <https://epa.gov/npdes/npdes-permit-limits>
- **Program management and oversight** https://epa.gov/npdes/npdes-program-management-and-oversight
- **NPDES regulations** <https://epa.gov/npdes/npdes-regulations>
- **Other applicable federal laws** https://epa.gov/npdes/other-federal-laws-apply-npdes-permit-program>
- **Clean Water Act** <https://epa.gov/laws-regulations/summary-dean-water-act>
- **Flexibilities under CWA related to COVID-19** <https://epa.gov/npdes/flexibilities-under-clean-water-act>
- **Keys to permitting success** <https://epa.gov/npdes/keys-permitting-success>
- **Public Participation in the NPDES Permit Issuance Process (pdf):** <https://epa.gov/system/files/documents/202205/public%20participation%20brochure%20v2.pdp (786.67 KB)

Program
Areas

- **Animal feeding operations** <https://epa.gov/npdes/animal-feeding-operations-afos>
- **Aquaculture** <https://epa.gov/npdes/aquaculture>
- **Industrial wastewater** <https://epa.gov/npdes/industrial-wastewater>
- **Municipal wastewater** <https://epa.gov/npdes/municipal-wastewater>
- **National pretreatment program** <https://epa.gov/npdes/national-pretreatment-program>
- **Pesticide permitting** <https://epa.gov/npdes/pesticide-permitting>

- **Stormwater** <https://epa.gov/npdes/npdes-stormwater-program>
- **Vessel discharges** <https://epa.gov/vessels-marinas-and-ports/commercial-vessel-discharge-standards>
- **All NPDES program areas** <https://epa.gov/vessels-marinas-and-ports/commercial-vessel-discharge-standards>

Technical Resources

- **Resources for Permit Writers** <https://epa.gov/node/118313>
- **Training** <https://epa.gov/npdes/npdes-training>
- **Tools** <https://epa.gov/npdes/npdes-tools>
- **Electronic Notice of Intent (eNOI** <https://epa.gov/npdes/electronic-reporting-epas-npdes-general-permits>)
- **Applications and forms** <https://epa.gov/npdes/npdes-applications-and-forms-epa-applications>
- **Find NPDES permits in your state or territory** <https://epa.gov/npdes-permits>

Explore the NPDES Permit Writers' Clearinghouse, a searchable database of permit writing resources
<https://epa.gov/npdes/permit-writers-cleringhouse>

Rainfall Erosivity Factor Calculator for Small Construction Sites <https://lew.epa.gov/>

Join NPDES News ☑: <https://lp.constantcontact.com/su/bvf6dae/npdes> to receive updates on webinars, training, and publications.

New and Notable

- **Case Studies of Successful Approaches to Off-Site Stormwater Management** <https://epa.gov/npdes/altemative-site-stormwater-management-case-studies>
- **Understanding Nutrient Management Plans Learning Modules** <https://epa.gov/npdes/understanding-nutrient-management-plans>
- **Off-Site Stormwater Management MS4 Compendium: Compendium Part 7: Off-Site Stormwater Management (pdf)** <https://eja.gov/system/files/documents/2023-01/off-site-stormwater-management-ms4-compendium.pdf> (1.08 MB)
- **Off-Site Stormwater Management** <https://epa.gov/npdes/altemative-site-stormwater-management>
- **Addressing PFAS Discharges in NPDES Permits and Through the Pretreatment Program and Monitoring Programs (pdf)** <https://epa.gov/system/files/documents/2022-12/npdes_pfas_state%20memo_december_2022.pdf (315 KB)
- **Technical Assistance Opportunity for CSO Communities** <https://epa.gov/npdes/combined-sewer-overflow-technical-assistance>
- **Low-Flow Statistics Tools Handbook** <https://epa.gov/npdes/low-flow-statistical tooks-handbook>
- **FAQs for Control Authorities on Dental Rule** <https://epa.gov/npdes/frequently-asked-questions-control-authorities-dental-rule-40-cfr-part-441>
- **Final Rule Published in Federal Register: Public Notification Requirements for CSOs to the Great Lakes Basin** <https://eja.gov/npdes/public-notification-requirements-combined-sewer-overflows-great-lakes-basin>
- **Online Training: Permit Writing Tips and Best Practices** <https://epa.gov/npdes/npdes-training#past>
- **Compendium of State Approaches for Manure Management** <https://epa.gov/node/250091>
- **MS4 General Permit Remand Rule** <https://epa.gov/npdes/stormwater-rules-and-notices>
- **Public Participation in the NPDES Permit Issuance Process (pdf)** <https://epa.gov/system/files/documents/2022-06/public%20participation%20brochure%20v2.pdf (786.67 KB)

Contact Us <https://epa.gov/npdes/forms/contact-us-about-npdes> to ask a question, provide feedback, or report a problem.

LAST UPDATED ON AUGUST 11, 2023

REFERENCES

EPA (Environmental Protection Agency), June 2000a, *The quality of our nation's waters, a summary of the national water quality inventory: 1998 report to Congress*, Office of Water, Washington, DC, EPA-841-S-00-001.

EPA (Environmental Protection Agency), 2000b, *National water quality inventory report*, EPA-841-R-02-001, http://www.epa.gov/305b.

EPA (Environmental Protection Agency), December 2009, *Technical guidance on implementing the stormwater runoff requirements for federal projects under section 438 of the energy independence and security act*, last updated November 5, November 5, 2024.

EPA (Environmental Protection Agency), February 2022, *Construction general permit*, (CGP), last updated April 15, 2025.

EPA (Environmental Protection Agency), 2013, *National rivers and streams assessment 2008–2009, a collaborative survey*, EPA, draft, EPA/841/D-13/001, February 28.

EPA (Environmental Protection Agency), *Developing a stormwater pollution prevention plan SWPPP*, EPA, last updated September 2, 2025.

EPA (Environmental Protection Agency), 2025, *National pollutant discharge elimination system (NPDES), stormwater discharge from municipal sources*, EPA, last April 22.

EPA (Environmental Protection Agency), 2025, *National pollutant discharge elimination system (NPDES), NPDES around the nation*, EPA, p. 4, May 14.

EWRI Currents, 2009, Federal facilities face a new stormwater hurdle, *Newsletter of the Environmental and Water Resources Institute of the ASCE*, 11 (1): 1–2. August.

Kayhanian, M. and Stenstrom, M.K., March–April 2008, First flush characterization for stormwater treatment, *Stormwater*, March–April, pp. 32–45.

Maryland Department of the Environment, *Maryland Stormwater Design Manual*, Vol. I and II, 2009, adopted May 4, 2010, prepared by Center for Watershed Protection, Ellicott City, and the Maryland Department of the Environment, MD, http://www.mde.state.md.us.

New York State Stormwater Management Design Manual, 2015, *Department of environmental conservation, draft – May 2022*, New York State Stormwater Management Design Manual.

NJ Stormwater Best Management Practices Manual, NJDEP, Trenton, NJ, February 2004, July 17, 2023, last updated October 9, 2024.

NJDEP, Stormwater Management, N.J.A.C.7:8, July 17, 2023.

NJDEP, NJ Bulletin 2012, No. NJ 210-12-1 Subject, September 10, 2012.

NJGS (New Jersey Geological Survey), 1993, *GSR-32*, Geological Survey Report No. 32, http://www.state.nj.us/dep/njgs/N.J.A.C.7:15.

SCS TR-55, 1986, *Urban hydrology for small watersheds, technical release 55 method (TR-55)*, Soil Conservation Service (SCS) a part of U.S.DA. TR-55; first issued 1975.

6 Manufactured Water Treatment Devices

This chapter discusses manufactured water quality devices that are employed to improve the quality of stormwater runoff. These devices are neither intended for nor are they are they capable of runoff quality control. The water quality devices vary from simple inlet inserts to sophisticated filter media structures. Also, included in this chapter are bioretention and bio-infiltration cells that in addition to treating water provide some runoff reduction.

6.1 AN OVERVIEW

In practice, a variety of structural stormwater management systems are employed to improve water quality including vegetated swales, extended detention basins, wet ponds, infiltration basins, bioretention basins, constructed wetlands, and sand filters. In addition, there exists a variety of source control, also referred to as green infrastructure stormwater management measures, such as porous pavements, rain gardens, and vegetative buffers.

As was indicated in the previous chapter, traditional stormwater management systems, including detention basins and wet ponds do not always fully meet the water quality requirements. As such, during the past four decades, various water quality devices have been manufactured to address either fully or partly the federal and state water quality regulations. This chapter discusses manufactured water quality devices (also referred to as water treatment devices) and presents a brief comparison of these devices with structural water quality systems. Structural and nonstructural management systems will be discussed in the next two chapters of this book.

Among structural stormwater management systems, detention basins and ponds, if properly designed, can serve as effective measures for runoff quantity control; they can also improve water quality, though not as effectively as they control flow rate. Infiltration basins and bioretention basins are most suited for water quality. Though some structural BMPs are reported to remove up to 95% of the total suspended sediment, their actual effectiveness is site-dependent and can also vary from storm to storm.

The effectiveness of structural BMPs in removing total phosphorus and nitrate is far smaller than that of their TSS removal. Some chemicals, such as chlorine derived from road deicing, cannot be removed by any structural BMP. Additionally, the absorption of some pollutants, such as metals, by soil may cause soil toxicity, turning infiltration basins into contaminated sites over a period of a few decades. Apart from inadequate efficacy, structural BMPs are expensive. Their cost includes not only the construction but also the land in the case of open detention-retention and infiltration basins, as well as their maintenance.

A manufactured device is either a prefabricated or cast-in-place stormwater treatment or other technologies to remove pollutants from runoff. Manufactured treatment devices are intended to capture sediments, floatable objects, hydrocarbons, and other pollutants in stormwater runoff before discharging to a drainage conveyance system or another stormwater management facility. These devices are suited for small drainage areas and particularly for small pavement areas such as parking lots or gas stations. For larger sites, the devices can be used to pretreat runoff before discharging it to larger-capacity structures such as detention basins, ponds, and underground retention/infiltration basins.

In general, water treatment devices are designed based on the peak discharge. However, if a device is placed past a detention system, it can be sized based on the attenuated flow from the system. Such an application, which reduces the size and cost of the device, is best suited for filter media devices. Media

DOI: 10.1201/9781003460121-6

filters (and sand filters as well) have been accepted as BMPs for water quality, and both have been approved by some jurisdictional agencies including the NJDEP for 80% TSS removal.

Sand filters comprise a horizontal bed of sand underlain by aggregate and a perforated drainpipe. The silt deposit eventually occludes the sand and inhibits percolation. To maintain the filter, the water over the sand has to be vacuumed and the top layer, if not the entire sand bed, removed and replaced by clean sand. On the other hand, media filter maintenance is straightforward and simply involves replacing media filters when they are found to be losing their draining capacity. Also, the cost of maintaining sand filters is nearly twice as high as that of media filters. According to data reported by the Virginia Planning Board Commission, the average annual cost of maintaining sand filters was estimated at $2000 per acre of impervious area compared with $1000 for media filters (Doerfer, 2008). Today, the maintenance cost would be nearly twice as much.

The Water Environment Research Foundation (WERF), a nonprofit organization that conducts research on stormwater and wastewater, established that a single BMP might not remove all pollutants of concern and recommends combining BMPs that address the stormwater management requirements (WERF, 2005). First, the extent to which a BMP reduces the volume of runoff should be considered. The consideration for volume reduction should include the storage of runoff for reuse. Next, the selection should carefully consider the rate of runoff that can be treated by a BMP. Many manufactured devices have a bypass system to divert flows higher than their capacity.

An important consideration in selecting a manufactured device (and any BMP) is its overall cost compared with other devices. In analyzing the overall cost, not only the initial installation cost but also the long-term costs of maintenance should be evaluated for the project, keeping in mind that a given BMP does not cost the same at different sites. It should also be noted that stormwater treatment technology is rapidly changing. Today's leading technology may not be the best approach in the long run.

There are a large number of national and state publications on the performance of stormwater management BMPs. An excellent EPA publication on the subject is "National Management Measures to Control Nonpoint Source Pollution from Urban Areas," which was released in November 2005. This publication (EPA No. 841-B-05-004) includes such subjects as watershed assessment and protection, site development, construction site erosion and sediment control, pollution prevention, new development runoff treatment, operation and maintenance, and evaluation of program effectiveness. Other excellent publications are:

- EPA (2025): "Source Water Protections," August 11, 2025
- "Critical Assessment of Stormwater Treatment Control and Selection Issues," published by WERF, 2005
- "Stormwater Best Management Practices Handbook Portal," published by California Stormwater Quality Association, November 2009
- "Stormwater Best Management Practices Manual," published by the New Jersey Department of Environmental Protection, last updated October 9, 2024
- "Construction Site Monitoring Program Guidance Manual," published by Caltrans Division of Environmental Analysis, CTSW-RT-11–255–20.1, August 2013

Another major source is the International Stormwater BMP Database, a nonprofit organization that has a main office in Alexandria, VA, and another office in Denver, CO. This organization provides useful information on stormwater management practices. The database, which was established by the EPA and the American Society of Civil Engineers (ASCE) in 1996, has since grown considerably and includes other partners, such as the WERF, the Federal Highway Administration (FHWA), the American Public Works Association, the Environmental and Water Resources Institute (EWRI) of the ASCE, and services that provide performance data and BMP evaluation. The database can be accessed online (http://www.bmpdatabase.org). The International Stormwater BMP Database Summary Statistics in 2020 was prepared by the WERF. This summary, in 8 pages, can be downloaded from https://www.waterrf.org.

6.2 CERTIFICATION OF WATER QUALITY DEVICES

Many manufactured water treatment devices were introduced to the market following the EPA Phase II stormwater management rule, which became final on December 8, 1999. Stormceptor, CDS (Continuous Deflective Separation), and Vortechnics were among the first devices that came to the market in the mid-1990s.

The CDS, an Australian device, for example, was introduced to the US market in 1996. The number of manufactured devices has since grown to three dozen or so and is still growing. Manufactured devices were initially faced with objections from institutions. Entities like ASCE and EPA began to consider them in their third-party independent evaluation during the early 2000s. Many jurisdictional agencies only accepted manufactured devices as a supplement to structural or nonstructural BMPs and as a last resort. This attitude has gradually changed as the experiments with the manufactured devices proved their effectiveness and efficiency.

In selecting a water quality device (and any water quality BMP as well) for a project, the practitioner needs to have reliable information about its performance, maintenance requirements, and life cycle. In addition, because of the site-specific nature of stormwater management practice, BMPs must be tailored to suit a given site. Ideally, the selection should be based on test data collected within the same state or a region of the same or similar hydrologic nature. However, dependable data specific to the locality are not always available for manufactured water treatment devices.

Also, reported data by manufacturers are based on test protocols that may vary from one manufacturer to another. The claimed total suspended sediment removal for a device may not specify the type and size of sediment tested. Particle size is important in pollutant removal. As indicated in Chapter 1, large particles settle much faster than fine particles; as such, tests that utilize sand show a far larger TSS removal rate than silt. However, fine particles, which transport pollutants such as metals, constitute the sediments of concern, take a long time to settle. Also, due to variations, the natural environment does not lend itself to a standardized test for soil removal. Different geologic conditions generate differently sized particles carried in runoff. In addition, sediment size can vary from storm to storm; an intense rainfall can produce and transport larger particles and significantly more sediment than a tight rain.

Another issue in evaluating test data is whether the study was conducted in a laboratory or a field. Laboratory tests show how a device performs under a set of strictly controlled testing protocols. As such, the effectiveness of different devices can be closely compared on that basis. However, since the size, type, and concentration of particles entering a manufactured treatment device (MTD) vary from one site to another, the results of laboratory testing may or may not reflect the actual field performance of a device.

Field tests have the benefit of representing the device's performance under actual circumstances. To be conclusive, however, field tests should be conducted within a large range of storm conditions and varying sites. Not all states have undertaken the task of preparing some kind of document that local stormwater practitioners can use. New Jersey is one of the states that has developed a rigorous testing and certification procedure for manufactured water quality devices. This procedure is briefly described in the following section.

6.2.1 NJDEP Approval Process for MTDs

The approval procedure for certifying manufactured treatment devices in New Jersey was established in 1997 and updated in January 2021. Information for the certification process of an MTD is posted online. Excerpts of the posted information are as follows:

> This document outlines the process for a stormwater manufactured treatment device (MTD) to be approved by the New Jersey Department of Environmental Protection (NJDEP) in compliance with the Stormwater Management rules, N.J.A.C.7:8. Prior to an MTDs entrance into the NJDEP process, the MTD must obtain verification through the New Jersey Corporation for Advanced Technology (NJCAT). The process for NJCAT Verification is available at www.njcat.org titled "Procedure for Obtaining Veri-

fication of a Stormwater Manufactured Treatment Device from New Jersey Corporation for Advanced Technology: For use in accordance with the Stormwater Management Rules, N.J.A.C. 7:8." In addition to these process documents, there are protocols for sedimentation and filtration MTDs that must be used for approval; the protocols are available at www.njstormwater.org.

Upon successful completion of the technical and regulatory standards and the completion of the reporting of those standards in the NJCAT Verification Report, NJCAT will provide NJDEP's Stormwater Management Unit a link to its website to access the Verification Report. In addition to the Verification Report link, NJCAT will supply the MTD name, the MTD manufacturer name, and the respective TSS percent removal rate and the following information: Description of Technology, Laboratory Test Setup, Performance Claims, supporting Documentation, Design Limitations, Maintenance Plans, Statements of Compliance, and a Verification Appendix. The Verification Appendix highlights and translates the design specifications found in the rest of the Verification Report to the design engineer.

Formal representation of a NJDEP approval will then be established on the NJDEP stormwater website at www.njstormwater.org. Upon approval, the MTD can be used for compliance with the Stormwater Management rule as long as the conditions of the NJCAT Verification are met.

Any MTD which is not approved by the NJDEP may be used but it does not get any credit for stormwater treatment. Such devices include catch basin inserts, oil, and treatment. Such devices include catch basin inserts, and oil and grease, and grit separators. Currently, MTDs are classified as Green Infrastructure (GI) MTDs and non-GI MTDs. All of the NJDEP-approved hydrodynamic separators and some of the filtration devices are accepted for 50% TSS removal. To be considered as GI MTD, the device must treat the stormwater close to its source through (a) infiltration into the soil and (b) filtration by vegetation and soil. All GI MTDs are approved for 80% TSS removal. The latest updated Stormwater Management only specifies that the design flow calculations are based on the NRCS (formerly SCS method) and using the projected 2 year storm.

An updated list of certified GI and non-GI manufactured treatment devices may be downloaded from the NJDEP Stormwater MTD website. The list includes two tables, for GI and non-GI MTDs. Information for models, capacities, and maintenance requirements for each MTD can be downloaded by clicking on the certification for that MTD. It should be noted that the currently approved MTDs exclude those whose certification has expired and has not been renewed. Those devices, in part, include CDS (Continuous Deflective Separator) by Contech, Vortex Systems, Terre Kleen by Terre Hill Stormwater Systems, and Jellyfish® by Imbrium.

6.3 TYPES OF MANUFACTURED DEVICES

Manufactured treatment devices come in various types, providing different levels of water quality improvements. MTDs may be classified into three main groups

1. Catch basin filters.
2. Hydrodynamic separation water treatment structures.
3. Filter media water quality devices. These devices provide the greatest total suspended solids removal and can also partially remove other pollutants such as phosphorus and hydrocarbons. Filter media devices may be more effective than some structural water treatment systems, such as grass swales.

In addition to the these MTDs, bioretention and bio-infiltration cells have been introduced in recent decades as green infrastructure. These cells in addition to improving water quality serve as a measure for reducing the quantity of stormwater runoff.

6.3.1 Catch Basin Inserts

Catch basin inserts come in a number of varieties. One type consists of a series of perforated trays, with the top tray serving as the initial sediment trap and the underlying trays are composed of finer

TABLE 6.1
List of Catch Basin Inserts

Product Name	Manufacturer & Email
StormKleen Catch Basin Filter	Cleanway USA (www.cleanwayusa.com)
SOP Curb Inlet Filter	SOP Technologies Environmental Solutions (www.soptechint.com)
FlexStorm Pure Inlet Filter	Advanced Drainage Systems (ADS) Pipe (www.adspipe.com)
Bio Clean Catch Basin Filter	Contech Engineered Solutions (www.conteches.com, info@conteches.com)
Ultra-HydroKleen Catch Basin Filter	Ultratech Intentional, Inc. (https://spillcontainment.com)
Ultra-Urban Filter with Smart Sponge Filtration	AbTech Industries (www.abtechindustries.com)
FloGard +Plus Catch Basin Insert Filter	Oldcastle Infrastructure
Grate Inlet Skimmer Box	(www.oldcastleinfrastructure.com)
Aqua Guardian Catch Basin Insert	Aqua Shield Inc. (www.aquashieldinc.com)
Storm Clean Catch Basin Filtration	Clean Way USA (www.cleanwayusa.com)
REM Trinton Drop Inlet	Revel Environmental Manufacturing Inc. (www.remfilters.com)
Blocksom Filter[a]	Blocksom & Co. (www.blocksom.com)
Ultra Urban Filter	AbTech Industries (www.abtechindustries.com)

a The Blocksom filter is a mat placed over an inlet grate. This mat is used as a sediment control measure during construction (see Chapter 9). Note that since catch basin insert manufacturers change hands, the listed names in this table are subject to change.

filters. Another type employs filter fabric to remove pollutants. Yet another type includes a plastic or metal basket that fits directly into the catch basin. The most popular among such devices are the basket type that includes inserts and hoods. Table 6.1 lists some of the inlet inserts manufactured in the United States.

Among them, the Ultra-Urban Filter, manufactured by AbTech Industries (Scottsdale, AZ., USA) and the FloGard® filter, by KriStar Enterprises. were two of the early catch basin inserts. AbTech was founded in 1996, KriStar Enterprises was founded in 1993 and acquired by Oldcastle Infrastructure (Atlanta, GA, USA) in January 2014. This filter was then renamed FloGard +Plus and is now one of the most widely used catch basin inserts in the United States. Oldcastle also makes filters for trench drains.

Both Ultra-Urban and FloGard +Plus filters include a screen, a filter liner, and a non-leaching oil-absorbent material or sponge contained in a pouch or similar removable strainer. Each of these filters has an overflow weir to pass flows beyond the capacity of the filter to a downstream storm drain. These filters offer simple means of removing trash, floatable objects, leaves and grass needles and a large portion of sediment and oil and grease from stormwater runoff. Inlet filters have not received certification from the NJDEP; however, many states have approved them for capturing sediments and debris from the runoff entering a drainage system. They are well suited for small parking lots and also for developed areas where there is no room for a water treatment BMP, or it is unfeasible to make major modifications to existing drainage structures. These filters can be readily inserted into existing inlets on parking lots, streets, and roads. The author highly recommends the use of these filters on inlets that drain to storm drains or underground chambers and other types of underground retention-infiltration basins. By removing silt, leaves, trash, and oil and grease from runoff, inlet filters prolong the useful life of underground retention-infiltration basins. He had recommended the use of the inlet filters for 50% TSS removal in non-major projects that were and still are free from approval by the NJDEP in New Jersey.

Sponges in the previously indicated inlet filters are very effective in removing hydrocarbons. For example, the Smart-Sponge used in the Ultra-Urban Filter can remove three times its own weight in hydrocarbons, inhibit mildew and mold, and transform hydrocarbons into a stable solid per the

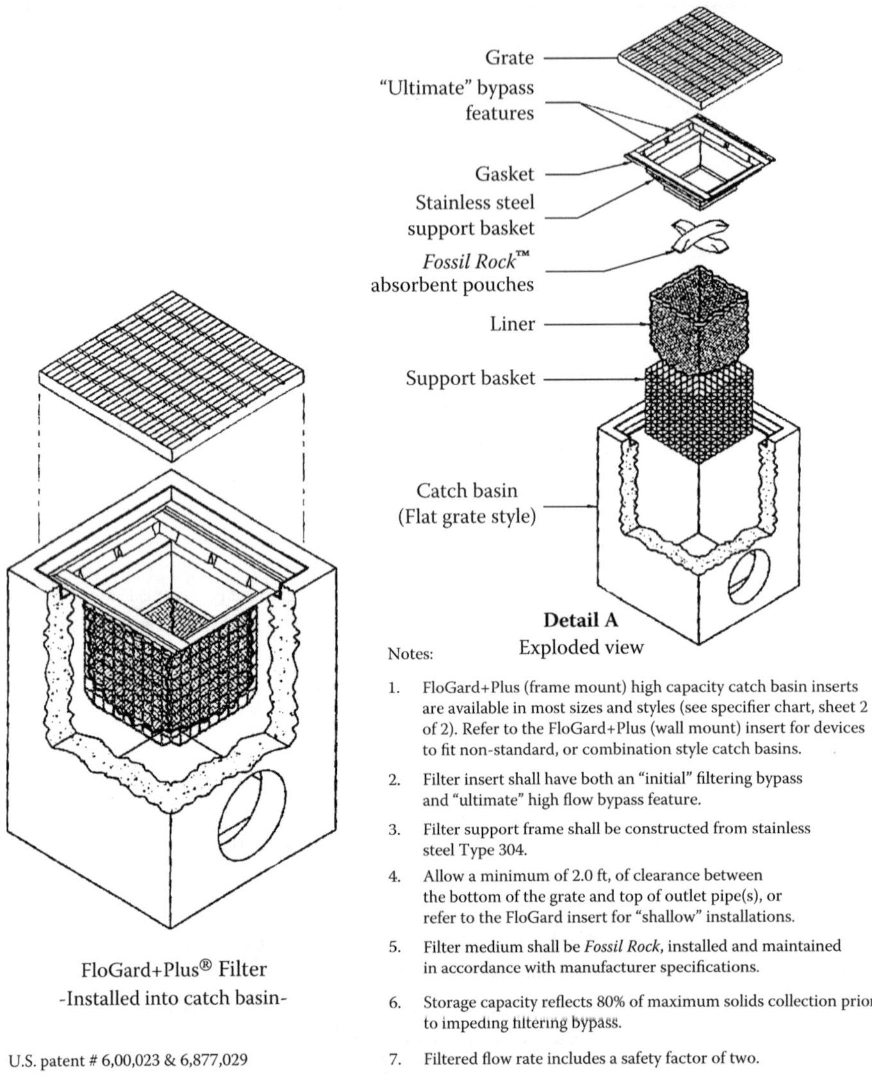

FIGURE 6.1 FloGard +Plus inlet filter components.

EPA's toxicity characteristic leach procedure (TCLP). A modified type of Smart Sponge called Smart is further capable of destroying bacteria on contact.

Catch basin filters not only serve as low-cost BMPs (installation cost ranges from less than \$1000 to \$3500 depending on the inlet size) but they are also easy to maintain. Figure 6.1 shows the components of a FloGard +Plus filter, and Figure 6.2 shows an Aqua-Guardian catch basin insert. Figure 6.3 shows a simple screen filter installed in a lawn inlet atop an underground chamber to capture leaves, grass cuts, debris, and coarse sediment. For more information about these and other water quality devices manufactured by these companies, the reader is referred to their manufactures' websites.

Inlet filters manufactured by Blocksom & Co. are made of natural fibers and can be used on inlet grates and curb openings. These filters can be readily attached to inlets without the need to remove grates. Blocksom filters come in 27 in. × 30 in. mats (69 cm × 76 cm), 27 in. × 21 in. (69 × 64 cm) mats, and 27 in. × 75 ft (22.9 m) rolls, and all are invariably 1.5 in. (3.8 cm) thick. Figure 6.4 shows a 27 in. × 30 in. mat. Blocksom inlet filters offer an economical means of removing silt, floatable

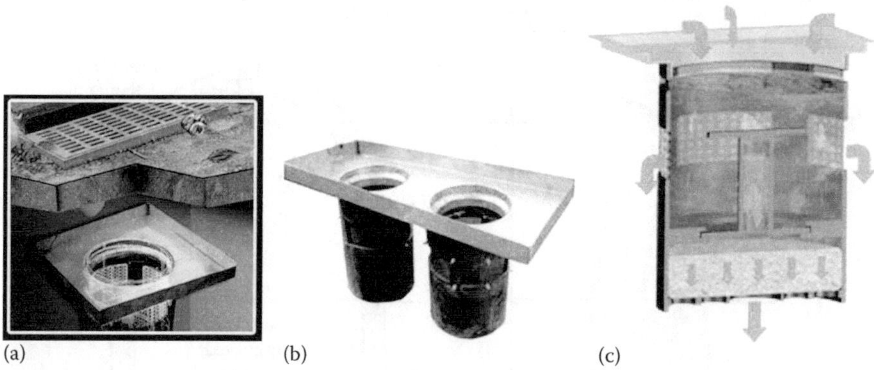

FIGURE 6.2 Aqua Guardian catch basin insert: (a) single insert, (b) double insert, (c) cross-section.

FIGURE 6.3 Lawn inlet screen filter.

FIGURE 6.4 Blocksom filter 27 × 30 in. (68 × 76 cm) mat.

objects, and debris and can be cleaned by sweeping the surface and sides. They are also durable, are not damaged by vehicular traffic, and are especially suited during construction. For more information, visit http://www.blocksom.com.

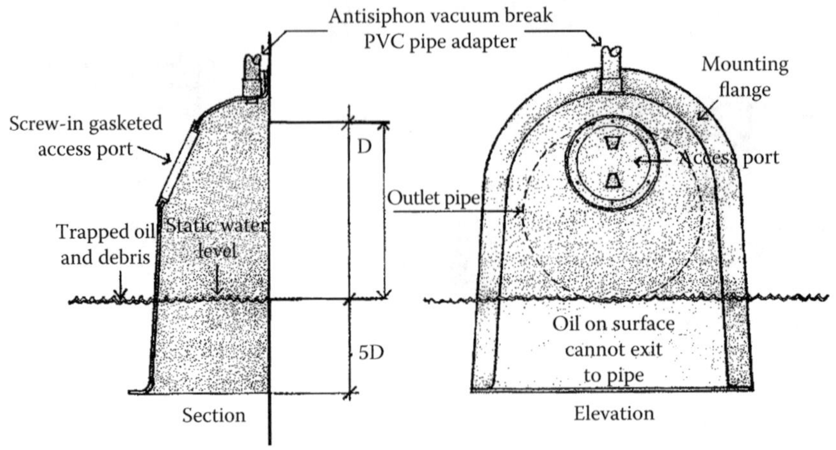

FIGURE 6.5 Snout water quality inlet hood.

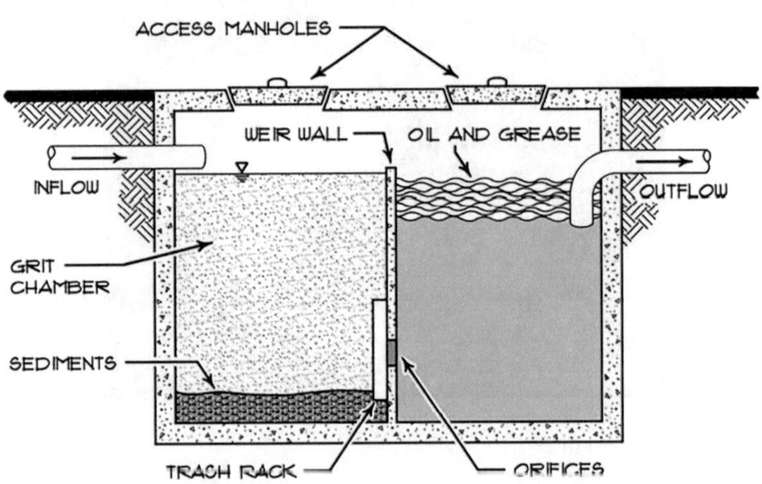

FIGURE 6.6 Grit and oil separator.

Source: ***Maine Stormwater Management Design Manual*****, May 2016, Figure 9.2.**

Snout is one of the oil-water separator hoods that is widely employed in the United States. These hoods, which are manufactured by Best Management Products Inc, can separate oil and other debris from water. Snout is attached to a catch basin wall over any type of discharge pipe to trap floatable objects, trash, sediment, oil and grease. Snout is equipped with an antisiphon flow vent to improve hydraulics and prevent contaminants from being drawn downstream. A clean-out port is also provided for easy access to the pipe.

Snout is made of marine-grade fiberglass which is a strong, yet lightweight, plastic composite. These hoods are easy to install and less expensive to maintain than inlet filters. However, since they are hidden underground, Snouts are more vulnerable to maintenance neglect than catch basin inserts and are less effective in removing suspended solids and oil, and grease. Figure 6.5 shows a SNOUT and its functioning. More information on Snout can be found at its website (https://www.bmpinc.com), which allows the user to size a Snout that suits any inlet.

A simple device for removing grit and oil and grease in urban areas is shown in Figure 6.6. This sediment and grease separator is practical for treating stormwater runoff from small parking lots, fueling stations or commercial loading areas.

6.3.2 Hydrodynamic Separation Water Treatment Devices

As was indicated, many water treatment devices have been introduced to the market over the past three decades. Table 6.2 is a partial list of hydrodynamic separation water treatment devices, also referred to as hydrodynamic sedimentation devices, some of which are or were approved by NJDEP. These devices are most effective in removing heavy particles, floatable objects, and some oils from runoff. However, unlike media filters, they are not fully capable of removing very fine particles, phosphorus, and heavy metals.

Some of the devices listed in Table 6.2, such as Aqua-Swirl and Stormceptor, separate suspended solids from inflowing flow by swirl action and sedimentation. In each of these, the water enters the device through a tangential inlet pipe that produces a swirl flow pattern that causes suspended solids to settle. The settlement occurs during each storm event and between successive storms. A combination of gravitation and hydrodynamic drag forces motivates the solids to drop out of the flow and migrate to the center of the chamber where the velocities are the lowest. During high flows, runoff entering these devices is directed into a treatment chamber through a secondary inlet, allowing the capture of floatable objects and debris. The bypass avoids high velocity or turbulence in the treatment chamber, which helps prevent the washout of previously captured pollutants.

All of the devices listed in Table 6.2 except Storm Vault, to be discussed later, were approved by the NJDEP for 50% TSS removal. Stormceptor, Stormceptor OSR, and CDS were among the earliest hydrodynamic separators that were introduced to the market. Stormceptor STS and OSR, both by Imbrium Systems, received the NJDEP certifications on September 1, 2011, which expired on September 1, 2013. CDS was introduced to the US market in 1997 and received its last certification from NJDEP on March 17, 2017. However, the certification of this device, like many others, was not renewed after it expired.

CDS units include a stainless screen with 2400 μm (2.4 mm) openings mesh that effectively removes floatable objects and any sediment particles coarser than 2.4 mm. Though this device was approved for 50% TSS removal by NJDEP, the Washington State Department of Ecology has approved CDS in line units for 80% TSS removal of particles coarser than 125 microns. Table 6.3 presents the model number and capacities of the CDS units as included in the Contech website. The treatment capacity of CDS in-line devices for removing particles 125 micron or larger varies from 0.5 cfs (14 L/s) for the smallest unit to 63 cfs (1780 L/s) for the largest, 12 ft (3.66 m) unit. The table also shows the minimum

TABLE 6.2
Partial List of Hydrodynamic Sedimentation Devicesa

Product Name	Company and Website
Barracuda, BayFilter, BaySeparator	Advanced Drainage System (www.adspipe.com)
Cascade Separator; CDS; Downstream Defender; STS, Stormceptor, Debris Separating Baffle Box, Ocean Guardian	Contech Engineered Solutions (www.conteches.com)
First Defense Optimum Vortex Separator; Storm Scape	Hydro International (www.hydro-int.com)
Storm Vault; Jensen Deflective Separator	Jensen Precast Inc. (www.Jensenprecast.com)
SciCloneX	Contech/Engineered Solutions (www.conteches.com)
StormPro	Environmental 21, LLC (www.env21.com)
Dual–Vortex Separator (DVS); StormCapture, Nutrient Separating Baffle Box (NSBB)	Oldcastle Infrastructure www.oldcastleinfrastructure.com
Terre Kleen	Terre Hill Concrete Products www.terrehill.com

TABLE 6.3
CDS Models and Capacities

		Treatment Capacity (cfs)			Estimated Maximum Peak Conveyance Flow** (cfs[3])	Minimum Sump Storage Capacity* (yd[4])	Minimum Oil Storage Capacity* (gal)
	CDS MODEL[1]	75 microns	125 microns[2]	Trash & Debris			
PRECAST	CDS1515–3	0.34	0.5	0.7	6	06	34
	CDS2015–4	0.5	0.7	1.0	10	0.9	61
	CDS2015–5	0.5	0.7.	1.0	10	1.5	83
	CDS2020–5	0.7	1.1	16	14	16	99
	CDS2025–5	1.1	1.6	22	14	16	116
	CDS3020–6	1.4	2.0	2.8	20	2.1	184
	CDS3025–6	1.7	2.5	36	20	2.1	210
	CDS3030–6	2.0	3.0	4.2	20	2.1	236
	CDS 3035–6	2.6	3.8	5.3	20	2.1	263
	CDS4030–8	3.1	4.5	6.3	30	5.6	426
	CDS4040–8	4.1	6.0	8.4	30	5.6	520
	CDS4045–8	5.1	7.5	10.5	30	5.6	568
	CDS5640–10	6.1	9.0	12.6	50	8.7	758
	CDS5653–10	9.5	14.0	19.6	50	8.7	965
	CDS5668–10	12.9	19.0	26.6	50	8.7	1172
	CDS5678–10	17.0	25.0	35.0	50	8.7	1309
	CDS9280–12	27.2	40.0	56.0	Offline	16.8	N/A
	CDS9290–12	35.4	52.0	72.0	Offline	16.8	N/A
	CDS92100–12	42.8	63.0	88.0	Offline	16.8	N/A

Large cast-in-place models available upon request.

Notes:

1. Alternative PSD/D_{50} sizing is available upon request.
2. 125 micron flows are based on the CDS Washington State Department of Ecology approval for 80% removal of a particle size distribution (PSD) having a mean particle size (D_{50}) of 125 microns.
3. Estimated maximum peak conveyance flow is calculated using conservative values and may be exceeded on sites with lower inflow velocities and sufficient head over the weir.
4. Sump and oil capacities can be customized to meet site needs.

sump storage capacity, the oil storage capacity, and the peak flow conveyance. CDS units offer greater drop inlet options than many others, accept multiple inlets and outlets at a variety of angles, can be placed in series and interconnected, and also give the option of internal and external bypasses. Figure 6.7 shows one of the smaller CDS devices with multiple inlets and outlets. The offline CDS units provide a treatment capacity of up to 638 cfs (18 m^3/s) for sediments 125 micron or coarser.

Because of a larger range in treatment capacity, CDS units are used extensively in urban stormwater projects. Another reason for their high application is that CDS units are among the least expensive treatment devices for a given discharge. The city of Laguna Beach in California, for example, selected a CDS hydrodynamic separator as the most cost-effective BMP for removing gross pollutants from runoff (Wieske and Penna, 2002). The city installed 3 cfs capacity CDS units offline to pretreat runoff before diverting it into the city combined sewer or returning it into the mainline storm drain. The author has specified the use of CDS water treatment devices for several projects, some of which are discussed in Chapter 7.

Dual Vortex Separator (DVS) by Oldcastle Infrastructures another hydrodynamic separator of high efficiency. Table 6.4 lists the model number, sizes, and treatment capacities of this device as approve by

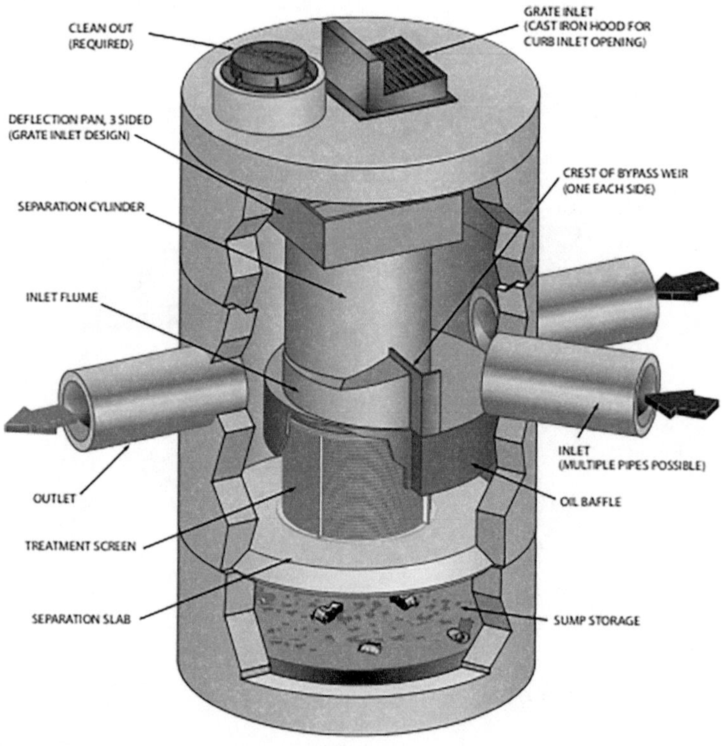

FIGURE 6.7 A small CDS unit with multiple inlets and outlets.

TABLE 6.4
NJDEP Certified Flow Rates of DVS

DVS Model	Manhole Diameter (ft)	Maximum Treatment Flow Rate (cfs)	Treatment Chamber Depth, ft
DVS-36	3	0.56	3.00
DVS-48	4	1.00	3.50
DVS-60	5	1.56	4.50
DVS-72	6	2.25	5.50
DV5S4	7	3.06	6.50
DVS-96	8	4.00	7.50
DVS-120	10	6.25	9.00
DVS-144	12	9.00	10.50

NJDEP on July 20, 2020, and Figure 6.8 depicts an elevation view of one of its smallest models. However, this device, too, has been removed from the NJDEP's current list of approved non-Gi treatment devices.

First Defense Optimum Vortex Separator is another hydrodynamic separation water treatment device that was last certified by the NJDEP on July 19, 2021. This device, which is manufactured by Hydro International of Portland, Maine comes in sizes ranging from 3 feet (92 cm) to 10 feet (3.05 m) in diameter. The treatment capacity of this device, as approved by the NJDEP, varies from 1.02 cfs (28.9 L/s) for a 3 ft diameter model to 11.33 cfs (321 L/s) for a 10 ft manhole.

Most hydrodynamic separation water treatment devices settle sediment due to a swirling action. The Terre Kleen stormwater device by Stormwater Systems, a subsidiary of Terre Hill Concrete

FIGURE 6.8 An elevation view of DVS.

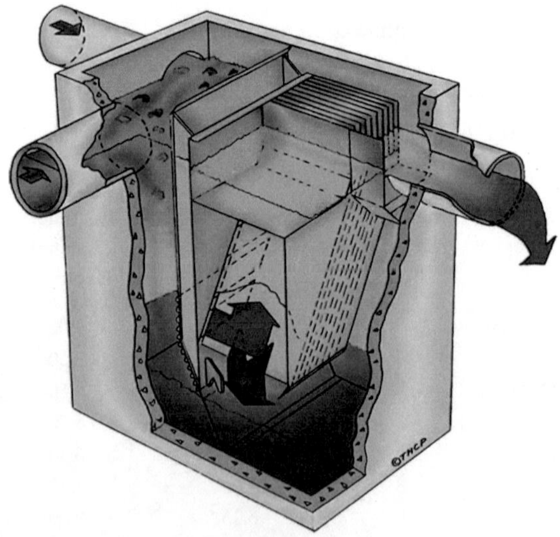

FIGURE 6.9 Terre Kleen™ stormwater treatment device.

Products in Terre Hill, Pennsylvania, functions on the basis of settling of particles in a laminar flow over an inclined plate. This process is similar to the one that has been practiced for many decades by the water treatment industry. To increase the area of contact, a number of parallel plates are placed in a chamber. The water enters the chamber from one or two pipes, and as it fills the chamber and flows slowly upward through the plates, suspended matter slides downward and settles to the bottom. The clean water that rises to the top of the chamber exits from an outlet pipe. Figure 6.9 shows the filtration process. To remove hydrocarbons, oil absorbent booms are added to the primary chamber. The booms float among the filter and settle when they become saturated with oil. This device received certification from NJDEP; on February 17, 2017 however, is lapsed.

The SciCloneX Hydrodynamic Separator by Bio Clean Environmental Inc. (Oceanside, CA, USA) now by Contech is one of the highly efficient non-Gi hydrodynamic devices certified by NJDEP for 50% TSS removal. Appendix 6A includes the certification letter for this hydraulic separator device and its models, sizes, and supporting calculations. Since the certification of this device predated the current N.J.A.C. 7:8 (last updated July 17, 2023), its sizing was based on the rational method using a runoff coefficient of 0.99 (which exaggerates the peak flow) and a rainfall intensity of 3.2 in./h, of the two-hour water quality storm (see Figure 5.2).

To prevent pollutant washout from becoming a source of contamination during high flows, the NJDEP mandates that all stormwater manufactured treatment devices be used as offline systems unless it can be demonstrated that the device will not release silt-sized pollutants during intense rain events. Based on the test results, only some of devices listed in Table 6.2 were approved for online applications. CDS was approved for online or offline application for sizes ranging from 4 to 12 ft (1.2–3.66 m) in diameter for 50% TSS removal. Downstream Defender was also approved for online application.

The use of internal bypass in these devices eliminates the need for a diversion structure, offering a compact design. It is expected that some other devices follow suit in being accepted for online use. Generally, the hydrodynamic sedimentation units are placed before detention basins, wet ponds, and, particularly, aboveground or underground infiltration and retention-infiltration basins. However, for underground detention systems (such as solid pipes) the treatment device may be placed past the detention system so that they are sized for attenuated flow.

Storm Vault by Jensen Infrastructure was the only hydrodynamic device with no filter media that was approved for 80% TSS removal by the NJDEP. This device was initially introduced by Contech. However, because of its large size and high cost, Contech stopped making the device before its permit expired on December 31, 2016. Since then, this device has been manufactured by Jensen Precast, Inc. Now, Jensen makes Storm Vault Bio-filtration with Sierra Blend, which approved as an online GI treatment device for 80% TSS removal by the NJDEP. The certification for this device was issued on June 21, 2021, and is valid for five years.

6.3.3 Media Filtration Water Quality Devices

To be approved for 80% TSS removal and to address the phosphorous removal requirement, a number of stormwater treatment devices containing media filtration cartridges have been introduced to the market since the turn of the century. Media filtration devices are generally far more expensive than catch basin inserts and hydrodynamic separation systems. They also require more maintenance and have significantly smaller flow capacity than hydrodynamic devices. To be cost-effective, media filter systems are generally placed after a detention basin or pond. Thus, they are utilized as post-treatment devices for attenuated flow rates. This type of application will be illustrated in a case study later in this chapter.

The filter surface area of a manufactured media filter (and sand filter, as well) is calculated using Darcy's equation as follows:

$$A_f = \frac{3600\ QxD}{K\left(D + \dfrac{d}{2}\right)} \tag{6.1}$$

where

A_f == filter surface area m^2 (ft^2)

Q = average flow rate through the fil$Q$ter associated with the water quality storm, m^3/s (cfs)

K = hydraulic conductivity of the filter, m/h (ft/h)

D = thickness of filter medium, m (ft)

d = maximum water depth, m (ft)

The term $(D+d/2)/D$ in Equation (6.1) represents the average hydraulic gradient. The average flow rate, Q, is related to the design water quality volume, WQ_v, as follows:

$$Q = \frac{WQ_v}{T}$$

where T denotes the drawdown time (or dewatering time) in the filter.

Table 6.5 lists some of filter media devices including those approved by NJDEP for 80% TSS removal. Mesia filter devices are mostly approved for treating runoff from small pavement areas: 0.13 acre (525 m^2 for Media Filtration System, 0.11 to 0.255 acres (445–1030 m^2) for Stormwater Management StormFilter (depending on cartridge height), 0.3 acres (1215 m^2) for Up-Flow Filter, and 0.7 acres (2840 m^2) for BayFilter. An exception was Jellyfish, which was approved for up to 4.63 cfs (132 liters) flow rate. This indicates that Jellyfish can treat up to 2 acres (0.81 ha).

The StormFilter, manufactured by Stormwater Management (in Portland, ME, USA), was the first filter media water treatment device introduced to the market. This company changed its name to Stormwater 360 and was later acquired by Contech Stormwater Solutions, Inc. in 2008. StormFilter is manufactured as a single unit or a linear grate storm filter. In a single unit, a number of cartridges (filters) are installed in a concrete chamber. The water enters the chamber, passes through the media filters, and starts filling the cartridges. Air below the hood is purged through a one-way check valve as the water rises. When water reaches the top of the float, buoyant forces pull the float free and allow filtered water to drain. After the storm, the water level starts to drop until it reaches the scrubbing regulators. Then, air rushes through the regulator, releases water and creates air bubbles that agitate the surface of the filter media, causing accumulated sediment to drop to the vault floor. This surface cleaning mechanism helps restore the filter's permeability between storm events. The linear grate StormFilter consists of a multichamber unit that can hold up to 29 filter cartridges This system receives sheet flow runoff through surface grates.

The StormFilter chambers range in size from 8'x11' to 8'x22' (2.44 × 3.36 to 2.4 × 6.71 m). Each chamber accommodates detached cartridges 27, 18, and 12 in. (68.6, 45.7, 30.5 cm) tall requiring nearly 3, 2, and 1.5 ft (91.4, 61, 45.7 cm) drops between the inflow and outflow, respectively. The largest chamber can hold sixty 27 in. high cartridges and treat up to 3 cfs (85 L/s) This device received certification from the NJDEP on December 19, 2016. Under this certification, which has expired, the 12, 18, and 27 in. cartridges were approved for 10, 15, and 2.5 GPM treatment, respectively.

Unlike this filter, the flow in many other filters, such as Media Filtration® by Oldcastle, Up-Flo Filter by Hydro International, and Jellyfish from Imbrium Systems, occurs in an upward direction. For example, the Media Filtration (formerly CDS Filter Media) system comes with 22 and 12 in.

TABLE 6.5
Filtration Systems Approved for 80% TSS Removal

Product Name	Manufacturer's Name and Website Address
Kraken High-Capacity Filter; Jellyfish Filter	Kraken by Bio Clean Environmental Services and Jellyfish by Imbrium are now by Contech Engineered Solutions (www.conteches.com)
BayFilter	BaySaver Technologies LLC, Mt Airy Maryland Now by ADS Pipe, https://www.adspipe.com
StormKleener Cartridge System	Lane Enterprises, www.lane-enterprises.com
Hydroworks HydroDome (HD)	Hydroworks LLC, www.hydroworks.com
PerkFilter Media Filtration Vault	Oldcastle Infrastructure
HydroChain Vortex Filter (HCVF)	Xerexes Corp. www.zerexes.com.hydrocahin-f

(55.9 30.5 cm) filters, requiring 2.3 and 1.5 ft. (70.1 and 45.7 cm) drop. Jellyfish, however, requires less than an 18 in. (45 cm) head (drop between inflow and outflow pipes) regardless of its size (model).

Since 2005, Contech has acquired companies such as CDS and Vortechs, Contech acquired the Jellyfish® filter from Imbrium Systems Inc. (in Rockville, Maryland) which also made Stormceptor in 2013. Contech also manufactures Infiltration StormFilter; this structure includes a perforated manhole (like a dry well) at the bottom and an upper compartment that includes StormFilter cartridges. The water entering the structure is filtered by StormFilter units, drains down to a perforated manhole, and infiltrates into the ground. Contech also makes an oil-water separator device known as VortClarex.

Jellyfish filter received its first certification by NJDEP on May 1, 2012, that expired on December 1, 2016. In Jellyfish, an insert deck divides the structure into an upper and a lower chamber. Water is treated in the lower chamber that contains a permanent pool of water. Flow enters the lower chamber tangentially and is directed around and under the cartridges in this chamber. Each 12 in. (30.5 cm) diameter cartridge consists of ninety-one 54 in. long filtration tentacles. Water is treated in the filtration tentacles, flows upward, and enters into a backwash pool created by a 6 in. (15 cm) weir. During rainfall, filtered water overflows the weir and enters the outlet pipe. After the runoff subsides, filtered water in the backwash pool drains down through the Jellyfish cartridge tentacles and brings the sediment that was accumulated on the tentacles to the sump. Water draining from the backwash pool displaces water from the lower chamber and the displaced water exits through the Jellyfish drawdown cartridge that is outside the backwash pool weir. Figure 6.10 depicts these three functions of a Jellyfish filter. Due to their self-cleansing capability, Jellyfish filters are fairly easy to clean and less expensive to maintain than some other stormwater filtering devices currently on the market.

Jellyfish cartridges come in 15, 27, 40, and 54 in. (28, 68, 101.6 and 137.2 cm) size. These cartridges are rated for high flow treatments of 0.05, 0.08, 0.13, and 0.18 cfs (1.4, 2.23, 3.4, and 5 L/s),

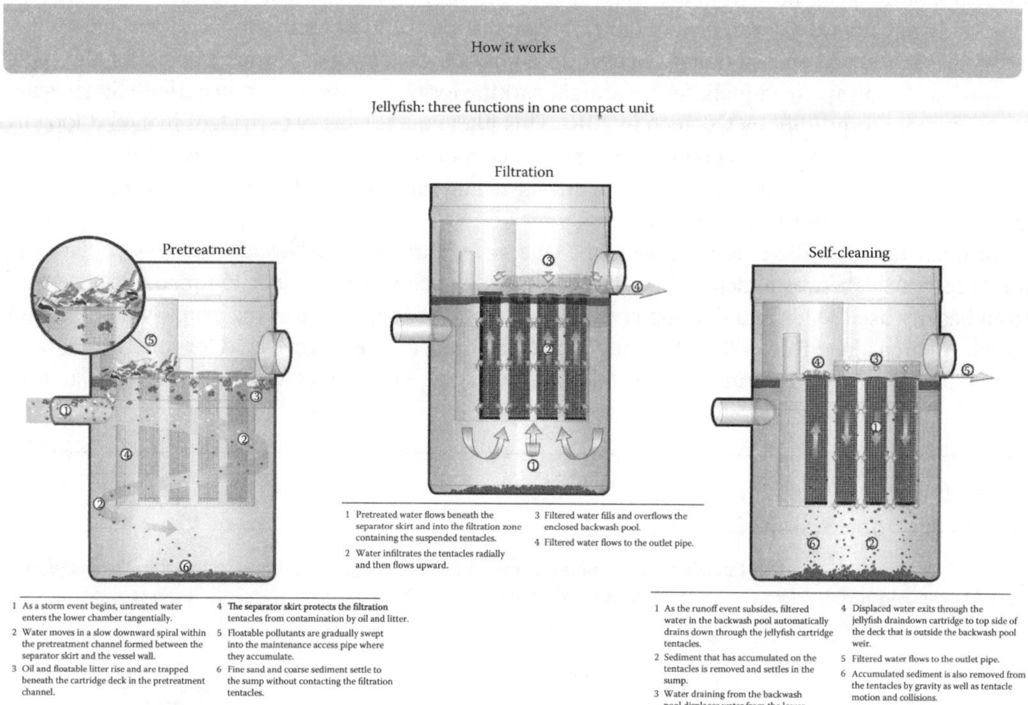

FIGURE 6.10 Jellyfish filter treatment (three functions).

respectively. A 54 in. (137.2 cm) high cartridge was approved for 80 GPM (3.8 L/min) by the NJDEP, which is the highest rated treatment element. Jellyfish are made in eight sizes, ranging from 4 ft to 12 ft (1.22–3.66 m) in diameter and provide treatment capacities of 0.27 to 5.26 cfs (2 to 149 L/s) using 1 to 27 pieces of 54 in. cartridges. Jellyfish also comes in a 16 ft (4.88 m) vault that can hold 40 cartridges and provide 7.85 CFS (222 L/s) treatment capacity. According to Imbrium, Jellyfish removes up to 99% of total trash, 89% TSS, 77% total phosphorus (TP), 51% total nitrogen (TN), over 80% total copper (TCu), over 50% total zinc (TZn) and over 15% total nephelometric turbidity (NTU).

Jellyfish filters incorporate Imbrium's oxide-coated Sorptive Media, which has a high phosphorus removal capability, including dissolved phosphorus which is a major cause of algae blooms and reduction of dissolved oxygen in ponds and bays. Sorptive Filter has received approval from the Maryland Department of the Environment (MDE) for use in sand filters, bioretention cells, and other microfilters throughout the state (Imbrium News Release, September 21, 2009).

BayFilter, manufactured by BaySaver Technologies, Inc. (in Mount Airy, Maryland), is another media filter treatment device. It was NJDEP rated for 30 GPM (115 L/min) per cartridge and appears to be a cost-effective device both in terms of initial cost and maintenance. BayFilter has been acquired by Advance Drainage Systems. The Bio Clean high-capacity Kraken is a newer stormwater treatment device that used to be manufactured by Bio Clean Environmental Services in Oceanside, CA, which was acquired by Forterra in 2019, which in turn was acquired by Quikrete Holding Inc. in Atlanta, Georgia, in 2022. Kraken (now by Contech which incorporated Bio Clean products in March 2022) can accommodate 10, 20, and 30 in. (25.4, 50.8, 76.2 cm) high cartridges. According to the NJDEP certification letter issued on November 18, 2022, these cartridges are approved for 0.011, 0.024, and 0.038 cfs (5, 11 and 17 GPM; 0.3, 0.7 and 1.1 L/s), respectively. Table 6.6 shows the models and treatment capacities of Kraken as approved by the NJDEP.

Because of their high costs and small capacities, media filter treatment devices are installed after a detention system. As such, they need to be sized for attenuated outflow from the system, which is a fraction of the peak flow from the water quality storm. Such an application is demonstrated in a case study in Chapter 7. Examples of the early applications of filter media applications are:

At the San Diego International Airport, to address the stormwater requirements for the rapid expansion of runways, terminals, and overnight parking for planes, the airport installed a Stormwater Management StormFilter by Contech in 2010. This filter contained 179 cartridges in a 48 ft long by 24 ft wide (14.63 m × 7.32 m) concrete chamber. The StormFilter controlled the discharge of heavy metals, petroleum products that leak from airplanes and service vehicles, and any trash, litter, and debris into San Diego Bay (Aird, 2012).

In Lake Tahoe, at the border between the states of California and Nevada, the loss of clarity of this blue water lake was a major issue. The sediment, which was finer than 16 μm and is generated from heavily used roads and developments in the Lake Tahoe Basin, was the primary cause of turbidity. An area along Route 28 in Incline Village was one of the areas selected for runoff treatment. A Jellyfish filter from Imbrium was installed in this area. This device was selected because of its

TABLE 6.6
Kraken Filtration System Capacities

Cartridge Height (in)	Cartridge Maximum Treatment Flow Rate (MTFR) (CFS)	Maximum Allowable Drainage Area (acres)
30	0.038	0.032
20	0.024	0.20
10	0.011	0.009

smaller footprint and lighter cartridges than other water treatment devices. The use of this device has reportedly stabilized the loss of clarity of water, which was continually a problem over the past 30 years (Goldberg, 2012).

6.4 BIORETENTION AND BIO-INFILTRATION

Bioretention and bio-infiltration cells, which were introduced to the market in the early part of the 21st century, function as not only water treatment devices but serve as small retention-infiltration structures. Because of such capability, this type of manufactured device is discussed in a separate section from the other types of water treatment devices, namely inlet filters, hydrodynamic and filter media devices. A bioretention structure (cell) typically includes a 2.3–3.0 ft (0.8–1.0 m) thick engineered soil mix, a filter element, and plants. The engineered soil mix generally consists of 50% construction sand, 30% topsoil, and 20% organic matter by volume. Table 6.7 lists some of bioretention and biofiltration systems that have been classified as Green Infrastructure MTDs and approved for 80% TSS removal by the NJDEP (and many other agencies, as well).

Bioretention cells can be coupled with catch basins to collect runoff from street gutters and parking lots. In this application, bioretention cells may be planted with salt-tolerant grasses and bushes to survive road salt. Experiments at the University of Maryland indicate that bioretention cells are effective in reducing the flow, trapping trash and floatable objects from runoff, and lowering suspended solids, phosphorus, and metal concentrations.

Filterra Bioretention was one of the earlier such systems manufactured in the United States. This system was developed by its parent company, Americast, in 2000. In 2008, the city of Virginia Beach in Virginia installed the Bacterra™ System from Filterra Bioretention Systems at Mount Trashmore Park to reduce the quantity of bacteria and pollutants discharging into the Lynnhaven River watershed. Filterra was the first bioretention cell that was approved for 80% TSS removal by the NJDEP in July 2014. Contech acquired the Filterra bioretention system from Americast and received a new certification from NJDEP on May 19, 2015. Now Contech manufactures Filterra HC, which provides far more treatment capacity than the original Filterra. Appendix 6B includes a part of the NJDEP certification issued on February 12, 2021, for Filterra HC.

Bio Clean Modular Wetlands System was another Green MTD infrastructure that received certification from NJDEP on June 11, 2022. This bio- infiltration MTD offers one of the highest treatment capacities. Table 6.8 lists the model number and the pavement area that this system can treat, and Figure 6.11 shows the layout, components and operation of one of its models.

The following case studies present the application of MTDs to address the stormwater quality requirements for two projects in New Jersey and New York.

The runoff volume and the storage volume of the retention system are calculated as follows:

TABLE 6.7
NJDEP-Certified Bioretention/Bio-Infiltration MTDs

MTD Name	Manufacturer
Aqua-Ponic Stormwater Biofiltration System	AquaShield Inc (www.aquashield.com)
Bio Clean Modular Wetlands	Bio Clean Environmental Services, Inc, a Forterra Company (now by Contech)
Filterra HC Bioretenrion System	Contech Engineered Solutions (www.conteches.com)
Focal Point High PerformanceModular Biofiltration	Convergent Water Technology (www.convergentwater.com)
StormVault Biofiltration System with Sierra Blend	Jensen Precast Corp. (https://jensenprecast.com)

TABLE 6.8
Bio Clean Modular Wetlands Linear System

Model Number	Maximum Treatment Flow Rate (cfs)	Maximum Allowable Drainage Area (acres)	Number of Pre-Filter Cartridges
MW-L-4-4	0.13	0.47	0.5
MW-L-4-6.5	0.17	0.65	1
MW-L-4-8	0.28	1.04	1
MW-L-6-8	0.35	1.32	1.5
MW-L-8-8	0.56	2.07	2
MW-L-8-12	0.83	3.11	3
MW-8-16	1.11	4.15	4
MW-L-8-20	1.39	5.19	5

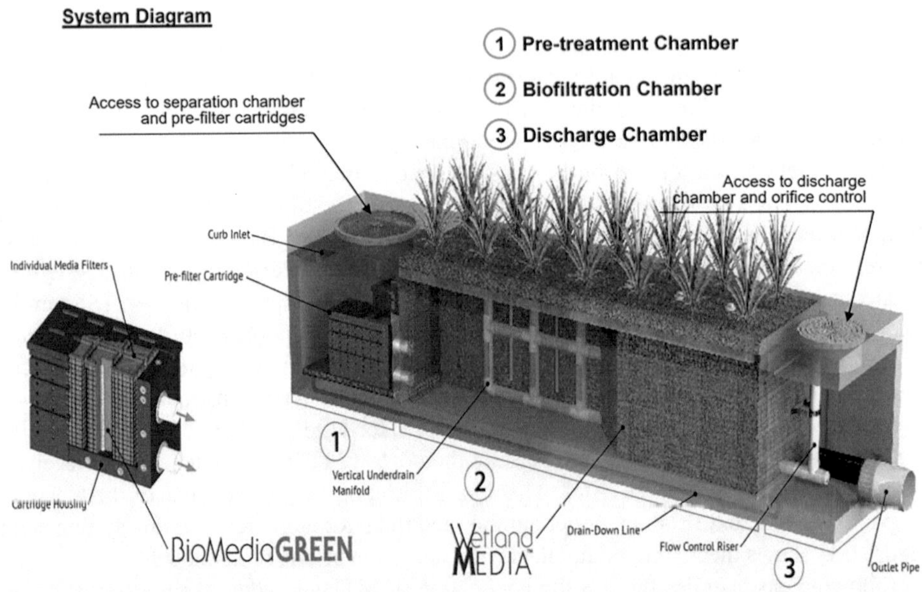

FIGURE 6.11 Layout and operation of a Modular Wetland system.

CASE STUDY 6.1

This case presents stormwater management measures for a parking lot expansion in the borough of Paramus, New Jersey.

PROJECT DESCRIPTION

Figure 6.12 depicts the pre- and post-improvement layouts of the parking lot in Paramus, New Jersey. Shown in this figure are the limits of the parking lot expansion, a preexisting paved basketball court that was to be removed, and the drainage facilities within the immediate vicinity of the site. The parking lot drains to an 18 in. RCP to the west and an 18 in. CMP to the east of the lot. According to the map, the net increase in impervious area measures to be 3020 ft^2 (280 m^2).

RETENTION SYSTEM CALCULATIONS

To offset the increased runoff, a retention system was provided to fully retain the runoff from a large portion of the expanded parking area. This system receives runoff from 4150 ft^2 (386 m^2) of pavement, which is larger than the net increase in impervious area. The installation of the system did not alter the preexisting drainage pattern in that the runoff from the site would continue to be discharged by the 18 in. concrete pipe to the wooded area north of the parking lot.

The retention system was designed for a 10-year, 60-minute storm event. In New Jersey, this storm has 2 in./h intensity and amounts to 2 in (50 mm) of rainfall. The system includes 60 linear feet (18.3 m) of 24 in. perforated HDPE pipe in a 6 ft by 4.58 ft (1.83 × 1.4 m) stone trench. Figure 6.13 shows the cross-section view of the system. A 6 in. (15 cm) HDPE overflow pipe is provisioned for the discharge of storms in excess of the retention capacity of the system.

Runoff volume = area × C × rainfall depth = 4150 × 0.95 × 2/12 in./ft = 657 ft^3 (18.60 m^3), where 0.95 represents the runoff coefficient (ratio of runoff to rainfall volume).

Retention volume:

$$\textit{Void vol.} = \left(6 \times 4 - \pi \times \frac{2.25^2}{4}\right) \times 60 \times 0.4 = 481\, ft^3 \left(13.62\ m^3\right)$$

$$\textit{Total storage} = 188 + 481 = 669 ft^3 \left(18.94 m^3\right) > 657\ ft^3 \left(18.60 m^3\right)$$

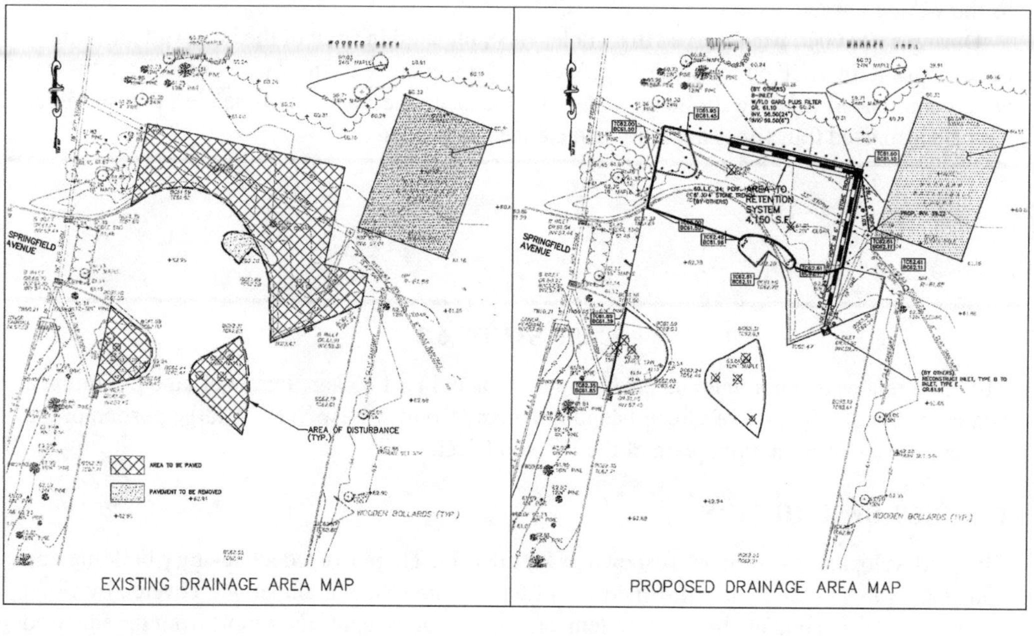

FIGURE 6.12 Pre- and post-construction site layouts and D.A. maps.

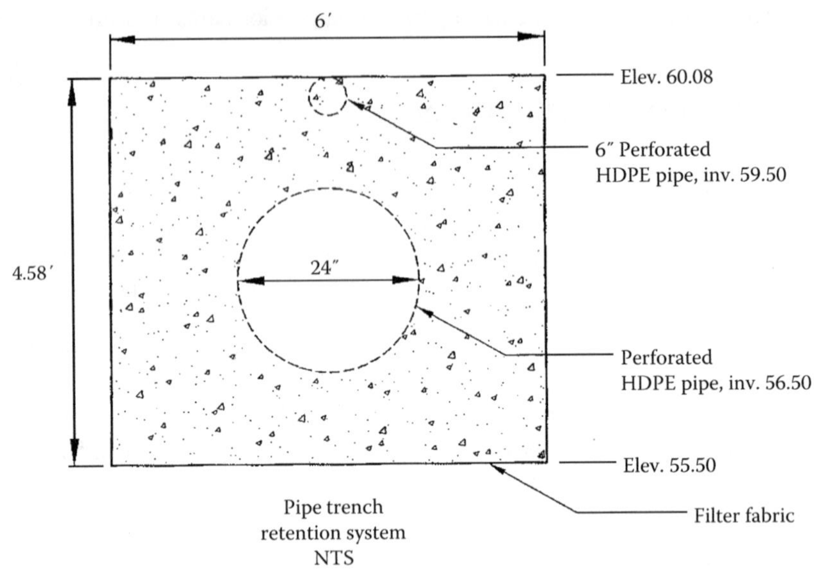

FIGURE 6.13 Pipe trench retention system.

In these calculations, 2.25 ft and 4 ft represent the outer diameter of the pipe and the effective depth of retention system below the 6 in. overflow pipe, respectively. Water quality and maintenance project was not subject to any water quality rule. However, to avoid silt, leaves, oils, and debris causing premature clogging of the retention trench, the new catch basin was equipped with a FloGard +Plus filter. In addition to prolonging the useful life of the retention system, the filter also improved the quality of stormwater runoff discharged to the wooded area.

During a site visit, we observed that silt had accumulated in front of the headwall and within the existing 18 in. RCP on the westerly side of the project area. To improve drainage conditions, we recommended that the drainage system be cleaned. We also recommended that the inlet filter be inspected four times annually and cleaned as needed.

CASE STUDY 6.2

This case relates to a commercial development project in a 1.45 acre parcel of land in Starbuck Island, part of the Village of Green Island in Albany County, New York. A large portion of the property lies within the floodplain of the Hudson River.

PROJECT DESCRIPTION

The predevelopment site layout is shown in Figure 6.14. There existed a one-story building and concrete pavement on the northwest corner of the site, and the remainder was covered by wood and grass. The site had no drainage system, and due to topography, the runoff from the site used to flow in all directions, entering private properties to the north and south, the Hudson River to

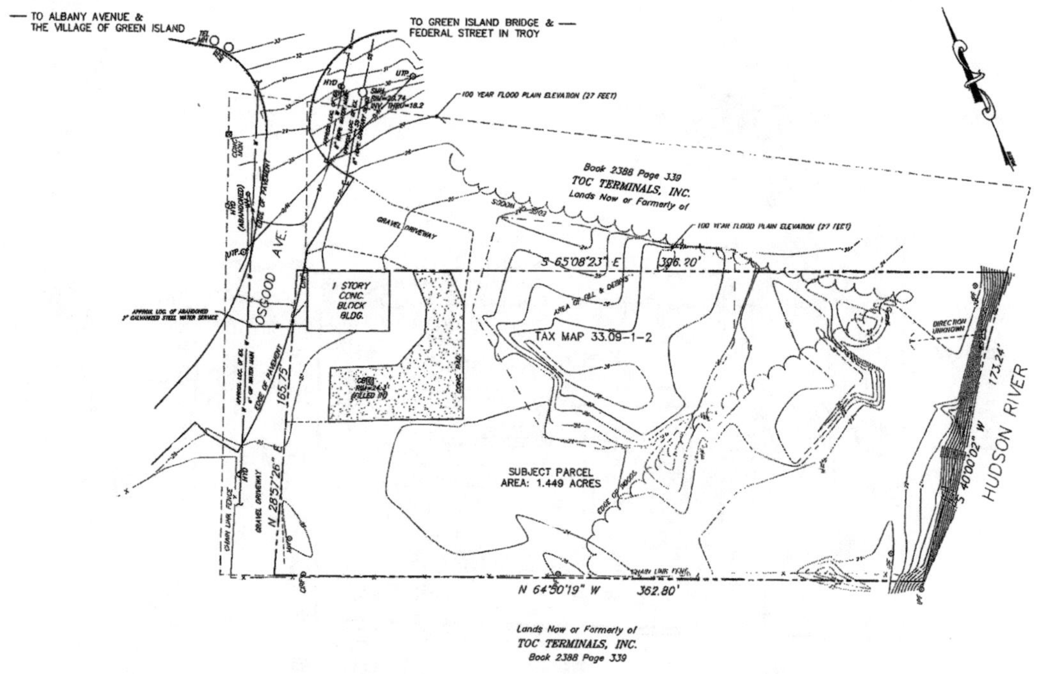

FIGURE 6.14 Predevelopment site layout.

the east, and Osgood Avenue to the west. A portion of the site was also draining to a depression at the center of the site.

The proposed project included an office building, a self-service car wash, and associated parking stalls and driveways. Figure 6.15 shows the proposed site layout and the limits of the area to be disturbed by the project. According to the FEMA flood insurance rate map of Green Island, the 100-year flood lies at an elevation of 27.0 ft at the site. The site of the office building, the car wash, and all parking stalls was raised above the 100-year flood elevation by constructing a retaining wall along portions of northerly and easterly property lines and placing fill behind walls.

DRAINAGE DESIGN

The runoff from the driveway and the parking areas is collected by a drainage system consisting of four inlets and two manholes and is routed through an underground detention basin comprising 290 ft (88.4 m) of 30 in. (750 mm) HDPE pipes. The roof runoff from the office building is discharged through roof leaders onto the undisturbed woodland behind the building.

Notably, the New York Department of Environmental Conservation (NYDEC) does not require a detention system to regulate the peak rates of runoff to the Hudson River. However, the objective of the proposed detention system is to attenuate the peak flows to reduce the size of StormFilter (the water quality device), which is designed to meet with the NYDEC water quality standards.

Consistent with the New York State drainage standards (NYDEC, 2015), the drainage system is designed based on the 10-year frequency storm. Flow calculations are based on the rational formula using runoff coefficients of 0.3 for pervious areas and 0.95 for impervious

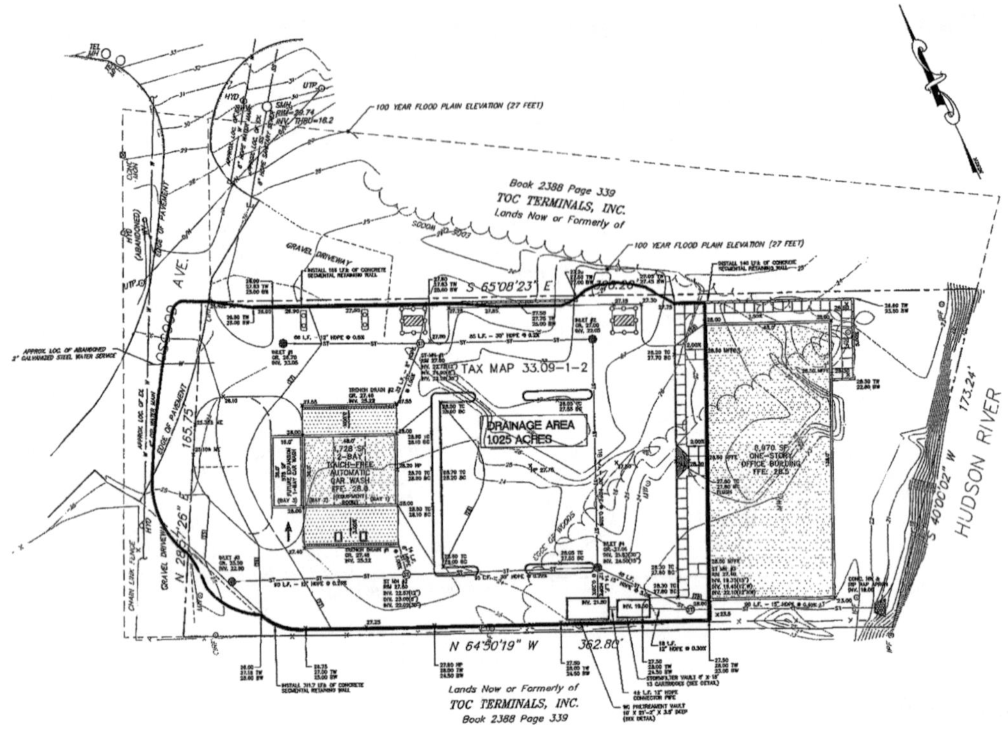

FIGURE 6.15 Post-development site layout and overall drainage area map.

surfaces. The drainage system collects runoff from the entire 0.85 acres (3440 m^2) paved area, routes it through a detention system, treats it with StormFilter, and discharges it on the overbanks of the Hudson River. The outlet control protection is designed based on the "New York Guidelines for Urban Erosion and Sediment Control" (effective at the time), conservatively neglecting any attenuation effect of the detention basin. Though calculations indicate no need for a riprap, a 5 ft × 5 ft (1.5 m × 1.5 m) stone spreader is provisioned at the outfall to disburse the runoff.

WATER TREATMENT DESIGN

Water quality system is designed based on the NY state water quality standards, which as described in a previous chapter call for treatment of the runoff volume generated during the water quality storm, which is 90% of the average daily annual rainfall (New York, 2015). Based on the water quality map of New York State (Figure 5.10 of this book), the 90% rainfall event at the site is approximately 1 in. (25 mm). For this rainfall event, the water quality volume, denoted as WQ_v, is calculated based on 1.025 acres (4148 m^2), the entire area of disturbance due to the project. This area includes 0.85 acres (3440 m^2) of impervious cover, inclusive of pavements and building roofs, and 0.175 acres (708 m^2) of lawn. The water quality volume calculations are as follows:

$$WQ_v = \frac{(PR_v A)}{12}$$

$$A = 1.025 \; acres \; P = 1 \; in.$$

$$I = \frac{(0.95 \times 0.85 + 0.3 \times 0.175)}{1.025} = 0.82 \ imperviousness\ ratio$$

$$R_v = 0.05 + 0.009 \times 0.82 = 0.788$$

$$WQ_v = 0.0673\ ac - ft = 2930\ ft^3$$

To treat this volume of runoff, a system comprising water quality pretreatment chamber and a StormFilter is used. The pretreatment structure is shown in Figure 6.16. This structure consists of a 21 ft, 2 in. (9.94 m) long reach (effective length) of a 10 ft wide by 3.5 ft high (3.28 m × 1.15 m) box culvert providing 740 cf (21 m^3) of storage capacity and as such it meets with the NYDEC Stormwater Design Manual (2015) which required a minimum pretreatment storage of 25% of WQ_v. The storage volumes of the pretreatment chamber and the detention system, which consisted of 290 lineal feet (98.4 m) of 30 in. (750 mm) pipe, are calculated as follows:

$$\text{Pretreatment chamber } V = 22.167 \times 10 \times 3.5 = 740\ ft^3\left(21\ m^3\right)$$

$$\text{Detention pipe } V = 290 \times \left(\pi \times \frac{2.5^2}{4}\right) = 1424\ ft^3\left(40.3\ m^3\right)$$

Four inlets/manholes = 4 × 4 × 2.5 ft (deep) each; V = 160 ft^3 (4.53 m^3)

Total volume = 1424 + 160 = 1584 ft^3 (44.85 m^3)

25% of WQ_v = 733 ft^3 (20.76 m^3)

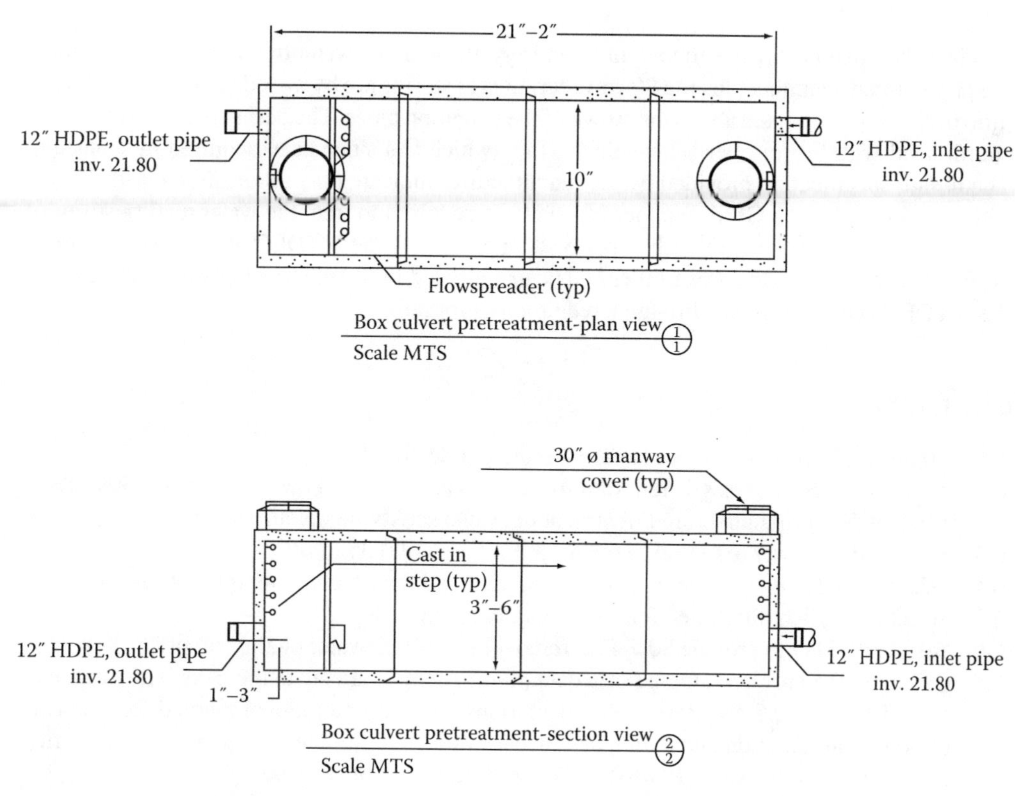

FIGURE 6.16 Pretreatment chamber

StormFilter sizing based on the NYDEC design methodology

Project Name:	Center Island South, LLC	Inpute	XXX
Date:	2/10/2005	Result	XXX

Site characteristic input

Design storm, P (inches)	1.00
Total area, A_T (acres)	1.03
Impervious area, A_I (acres)	0.84
Percent of WQv to be temporarily stored in system	75%

WQ_V calculations

Percent impervious cover, I	82%
Volumetric runoff coefficient, RV	0.79
Water quality volume, WQ_V (ac-ft)	0.067
Water quality volume, WQ_V (cu-ft)	2930
Required live storage volume (75% of WQ_V) (cu ft)	2198
Required pretreatment volume (25% of WQ_V) (cu ft)	733

StormFilter design constants (per MDE manual)

Filter bed depth, D_f (ft)	0.58
Coeff. of perm. of filter media, k (ft/day)	8.7
Avg. height of water above filter bed, H_f (ft)	0.75
Design filter bed drain time, T_f (days)	1.67
Surface area of stormfilter cartridge (sq ft)	7.1

Sizing calculations

Surface area of equivalent filter bed (sq ft)	88.0
Number of filter cartridges required	13
Maximum filtration rate (cfs)	0.22

System design

Water surface elevation (ft)	4.50

Volume StormFilter

Vault size	8 × 16
Live storage volume provided (cu ft)	375

Pretreatment structure

Vault size	10×12^{-2}
Total storage provided (cu ft)	740
Live storage volume (cu ft)	425

Storage summary

Total live storage provided in system (cu ft)	798
Additional live storage required (cu ft)	1574
Diameter of storage pipe (in)	30
Length of storage pipe required (ft)	290

**All volumes are based upon 4.5′ water surface elevation

The water quality storm is routed, in its entirety, through the water treatment system, namely the pretreatment chamber plus the StormFilter. Larger storms are bypassed through a 15 in. (375 mm) overflow pipe located above the crown of the detention pipes. The design of StormFilter was coordinated with Stormwater Management, Inc. (which has since been acquired by Contech). The design specification provided by the manufacturer indicate that a StormFilter with 13 cartridges, each with a 7.5 GPM treatment capacity, is required to treat the water quality storm. It should be noted that StormFilter had not been approved by the NYDEC at the time of project design. However, since the StormFilter had been approved for 80% TSS removal in New Jersey, the NYDEC accepted its use through a reciprocity program.

PROBLEMS

6.1 When would you use a manufactured treatment device?

6.2 A project in New Jersey is required to treat a water quality flow of 1.7 cfs for 50% TSS removal. What manufactured treatment device(s) satisfy this requirement?

6.3 What is the answer to Question 6.2 for a discharge of 0.05 m^3/s?

6.4 Select an MTD that can provide 80% TSS removal for 1.5 cfs water quality runoff.

6.5 Select a Jellyfish filter rated for 0.03 m^3/s treatment.

6.6 Select an MTD to provide 80% TSS removal for 0.5 cfs water quality runoff.

6.7 Calculate the required storage of a detention system to attenuate the peak water quality from 1.5 cfs to 0.5 cfs. Base your calculations on modified rational method for a storm of 30 minute duration and a time of concentration of 15 minutes. If 48 in. solid pipes are used to provide the storage, what is the required length of the pipe?

6.8 A commercial project results in 0.5 acre of new pavement and 0.4 acre of milled and resurfaced existing pavement. Calculate the required TSS removal and peak water quality runoff for this project based on the NJDEP Stormwater Management Rules. Base your calculations for a 10-minute time of concentration and select an MTD to address the water quality requirements for the project.

6.9 Solve Problem 6.8 for 2000 m^2 of new pavement and 1500 m^2 of repaved surfaces.

6.10 If the existing pavement in Problem 6.8 has to be reconstructed, what will be the required TSS removal? What MTD(s) would you choose to address the TSS removal requirement?

6.11 Solve Problem 6.9 if the existing pavement has to be reconstructed. Base your selection using the manufacturer's data.

6.12 To reduce the size of the MTD in Problem 6.10, the runoff is routed through an underground system that comprises solid pipes that reduce the peak water quality runoff by 70%. What is the peak water quality flow to be treated for TSS removal and witch device(s) can you use in this case.

APPENDIX 6A: CERTIFICATION OF SCICLONE HYDRODYNAMIC SEPARATOR

State of New Jersey

DEPARTMENT OF ENVIRONMENTAL PROTECTION

DIVISION OF WATERSHED PROTECTION AND RESTORATION
BUREAU OF NJPDES STORMWATER PERMITTING & WATER QUALITY MANAGEMENT
P.O. Box 420 Mail Code 401-02B
Trenton, New Jersey 08625-0420
609-633-7021 / Fax: 609-777-0432
www.njstormwater.org

PHILIP D. MURPHY
Governor

SHEILA Y. OLIVER
Lt. Governor

SHAWN M. LATOURETTE
Commissioner

June 10, 2021

Zachariha J. Kent
Vice President of Product Management
Bio Clean Environmental Inc.
398 Via El Centro
Oceanside, CA 92058

Re: MTD Lab Certification
SciCloneX Hydrodynamic Separator by Bio Clean Environmental, Inc.
On-line Installation

TSS Removal Rate 50%

Dear Mr. Kent:

The Stormwater Management rules under N.J.A.C. 7:8-5.2(f) and 5.2(j) allow the use of manufactured treatment devices (MTDs) for compliance with the design and performance standards at N.J.A.C. 7:8-5 if the pollutant removal rates have been verified by the New Jersey Corporation for Advanced Technology (NJCAT) and have been certified by the New Jersey Department of Environmental Protection (NJDEP). Bio Clean Environmental, Inc. has requested an MTD Laboratory Certification for the SciCloneX Hydrodynamic Separator (SciCloneX HDS).

The project falls under the "Procedure for Obtaining Verification of a Stormwater Manufactured Treatment Device from New Jersey Corporation for Advance Technology" dated January 25, 2013. The applicable protocol is the "New Jersey Laboratory Testing Protocol to Assess Total Suspended Solids Removal by a Hydrodynamic Sedimentation Manufactured Treatment Device" dated January 25, 2013.

NJCAT verification documents submitted to the NJDEP indicate that the requirements of the protocol have been met or exceeded. The NJCAT letter also included a recommended certification TSS removal rate and the required maintenance plan. The NJCAT Verification Report dated May 2021 with the Verification Appendix for this device is published online at http://www.njcat.org/verification-process/technology-verification-database.html.

The NJDEP certifies the use of the SciCloneX HDS by Bio Clean Environmental, Inc. at a TSS removal rate of 50% when designed, operated and maintained in accordance with the information provided in the Verification Appendix and the following conditions:

New Jersey is an Equal Opportunity Employer
Printed on Recycled Paper and Recyclable

1. The maximum treatment flow rate (MTFR) for the manufactured treatment device (MTD) is calculated using the New Jersey Water Quality Design Storm (1.25 inches in 2 hrs) in N.J.A.C. 7:8-5.5.

2. The SciCloneX HDS shall be installed using the same configuration reviewed by NJCAT and shall be sized in accordance with the criteria specified in in item 6 below.

3. This SciCloneX HDS cannot be used in series with another MTD or a media filter (such as a sand filter), to achieve an enhanced removal rate for total suspended solids (TSS) removal under N.J.A.C. 7:8-5.5.

4. Additional design criteria for MTDs can be found in Chapter 11.3 of the New Jersey Stormwater Best Management Practices (NJ Stormwater BMP) Manual which can be found on-line at www.njstormwater.org.

5. The maintenance plan for a site using this device shall incorporate, at a minimum, the maintenance requirements for the SciCloneX HDS, which is attached to this document. However, it is recommended to review the maintenance manual at https://biocleanenvironmental.com/wp-content/uploads/2021/03/SciCloneX-Operation-Maintenance-Manual_3-23-2021-v1.pdf for any changes to the maintenance requirements.

6. Sizing Requirements:

The example below demonstrates the sizing procedure for the SciCloneX HDS:

Example: A 0.25-acre impervious site is to be treated to 50% TSS removal using a SciCloneX HDS. The impervious site runoff (Q) based on the New Jersey Water Quality Design Storm was determined to be 0.79 cfs.

Maximum Treatment Flow Rate (MTFR) Evaluation:

The site runoff (Q) was based on the following:

time of concentration = 10 minutes
i=3.2 in/hr (page 21, Fig. 5-10 of Chapter 5 of the NJ Stormwater BMP Manual)
c=0.99 (curve number for impervious)
Q=ciA=0.99x3.2x0.25=0.79 cfs

Given the site runoff is 0.79 cfs and based on Table 1 below, the SciCloneX HDS Model SCX-3 with a MTFR of 1.02 cfs would be the smallest model approved that could be used for this site that could remove 50% of the TSS from the impervious area without exceeding the MTFR.

The sizing table corresponding to the available system models is noted below. Additional specifications regarding each model can be found in the Verification Appendix under Table A-1 and Table A-2.

Table 1 SciCloneX HDS Models

SciCloneX HDS Model	Manhole Diameter (ft)	Maximum Treatment Flowrate, MTFR (cfs)
SCX-3	3	1.02
SCX-4	4	1.82
SCX-5	5	2.84
SCX-6	6	4.09
SCX-7	7	5.56
SCX-8	8	7.27
SCX-10	10	11.36
SCX-12	12	16.35
SCX-14	14	22.25

Be advised a detailed maintenance plan is mandatory for any project with a Stormwater BMP subject to the Stormwater Management Rules, N.J.A.C. 7:8. The plan must include all the items identified in the Stormwater Management Rules, N.J.A.C. 7:8-5.8. Such items include, but are not limited to, the list of inspection and maintenance equipment and tools, specific corrective and preventative maintenance tasks, indication of problems in the system, and training of maintenance personnel. Additional information can be found in Chapter 8: Maintenance and Retrofit of Stormwater Management Measures.

If you have any questions regarding the above information, please contact Lisa Schaefer of my office at lisa.schaefer@dep.nj.gov.

Sincerely,

Gabriel Mahon

Gabriel Mahon, Chief
Bureau of NJPDES Stormwater Permitting & Water Quality Management
Division of Watershed Protection and Restoration
New Jersey Department of Environmental Protection

Attachment: Maintenance Plan

cc: Richard Magee, NJCAT

OPERATION & MAINTENANCE

The SciCloneX™ Separator is designed to remove high levels of trash, debris, sediments and hydrocarbons. Its efficient design and construction maximize longevity and minimize maintenance requirements. The simple design of the system allows for unimpeded access for quick and easy maintenance. The SciCloneX™ Separator is able to effectively capture and store sediment with no maintenance or loss of treatment capacity for several years based on annual average loading in most regions.

Yet, as with all stormwater BMPs inspection and maintenance on the SciCloneX™ Separator is necessary. Stormwater regulations require that all BMPs be inspected and maintained to ensure they are operating as designed to allow for effective pollutant removal and provide protection to receiving water bodies. It is recommended that inspections be performed multiple times during the first year to assess site-specific loading conditions.

This is recommended because pollutant loading can vary greatly from site to site. Variables such as nearby soil erosion or construction sites, winter sanding of roads, amount of daily traffic and land use can increase pollutant loading on the system. The first year of inspections can be used to set inspection and maintenance intervals for subsequent years. Without appropriate maintenance a BMP can exceed its storage capacity which can negatively affect its continued performance in removing and retaining captured pollutants.

System Diagrams:

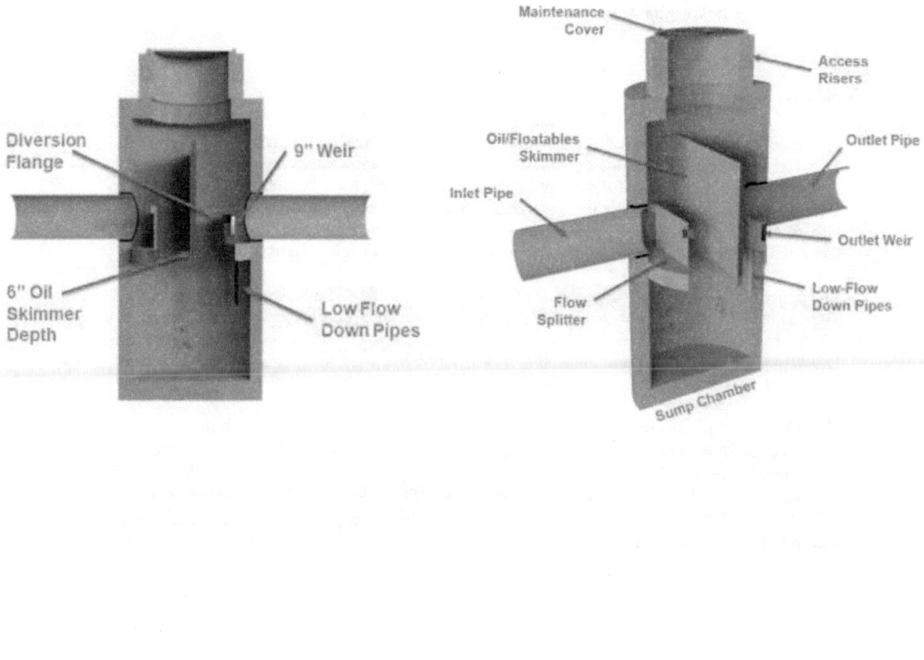

APPENDIX 6B: CERTIFICATION OF FILTERRA HC BIORETENTION SYSTEM

State of New Jersey
DEPARTMENT OF ENVIRONMENTAL PROTECTION
DIVISION OF WATER QUALITY
Bureau of Stormwater Permitting
401 East State Street
P.O. Box 420 Mail Code 401-02B
Trenton, NJ 08625-0420
Tel. (609) 633-7021 • Fax (609) 777-0432
www.nj.gov/dep/dwq/bnpc_home.htm

PHILIP D. MURPHY
Governor

SHEILA Y. OLIVER
Lt. Governor

SHAWN M. LATOURETTE
Acting Commissioner

February 12, 2021

Derek M. Berg
Director – Stormwater Regulatory Management - East
Contech Engineered Solutions LLC
71 US Route 1, Suite F
Scarborough, ME 04074

Re: MTD Lab Certification
Filterra® HC Bioretention System
Off-line Installation Approved

TSS Removal Rate 80%

Dear Mr. Berg:

The Stormwater Management rules under N.J.A.C. 7:8-5.5(b) and 5.7(c) allow the use of manufactured treatment devices (MTDs) for compliance with the design and performance standards at N.J.A.C. 7:8-5 if the pollutant removal rates have been verified by the New Jersey Corporation for Advanced Technology (NJCAT) and have been certified by the New Jersey Department of Environmental Protection (NJDEP). Contech Engineered Solutions LLC has requested a Laboratory Certification for the Filterra® HC Bioretention System (Filterra® HC.)

The project falls under the "Procedure for Obtaining Verification of a Stormwater Manufactured Treatment Device from New Jersey Corporation for Advance Technology" dated January 25, 2013. The applicable protocol is the "New Jersey Department of Environmental Protection Laboratory Protocol to Assess Total Suspended Solids Removal by a Filtration Manufactured Treatment Device" dated January 25, 2013.

NJCAT verification documents submitted to the NJDEP indicate that the requirements of the aforementioned protocol have been met or exceeded. The NJCAT letter also included a recommended certification TSS removal rate and the required maintenance plan. The NJCAT Verification Report with the Verification Appendix (dated January 2021) for this device is published online at http://www.njcat.org/uploads/newDocs/NJCATFilterraTechnologyVerificationReportFinal._.pdf.

The NJDEP certifies the use of the Filterra® HC stormwater treatment unit by Contech Engineered Solutions LLC at a TSS removal rate of 80% when designed, operated, and maintained in accordance with the information provided in the Verification Appendix and the following conditions:

1. The maximum treatment flow rate (MTFR) for the manufactured treatment device (MTD) is calculated using the New Jersey Water Quality Design Storm (1.25 inches in 2 hrs) in N.J.A.C. 7:8-5.5. The MTFR is calculated based on a verified loading rate of 3.12 gpm/ft^2 of effective filtration treatment area.

2. The Filterra® HC stormwater treatment unit shall be installed using the same configuration reviewed by NJCAT, and sized in accordance with the criteria specified in item 7 below.

3. This device cannot be used in series with another MTD or a media filter (such as a sand filter) to achieve an enhanced removal rate for total suspended solids (TSS) removal under N.J.A.C. 7:8-5.5.

4. Additional design criteria for MTDs can be found in the New Jersey Stormwater Best Management Practices (NJ Stormwater BMP) Manual, which can be found online at www.njstormwater.org.

5. The maintenance plan for a site using this device shall incorporate, at a minimum, the maintenance requirements for the Filterra® HC. A copy of the maintenance plan is attached to this certification. However, it is recommended to review the maintenance website at https://www.conteches.com/Portals/0/Documents/Maintenance%20Guides/Filterra%20HC%20OM%20Packet.pdf for any changes to the maintenance requirements.

6. For an MTD to be considered "green infrastructure" (GI) in accordance with the March 2, 2020 amendments to the Stormwater Management rules at N.J.A.C. 7:8, the MTD must meet the GI definition noted at amended N.J.A.C. 7:8-1.2. Specifically, the MTD shall (1) treat stormwater runoff through infiltration into subsoil; and/or (2) treat stormwater runoff through filtration by vegetation or soil; or (3) store stormwater runoff for reuse.

   The Filterra® HC filters stormwater runoff through an engineered biofiltration soil media and, thus, meets the definition of GI. Filterra® HC can be configured with or without a precast vault. Installations that will not include a precast vault will additionally need to comply the NJDEP Stormwater BMP Manual conditions regarding separation from the seasonal high water table and, if infiltration is proposed as an outlet, minimum vertical saturated hydraulic conductivity of the subsoil. Installations without a precast vault that do not rely on infiltration are required to maintain at least a one-foot separation from the seasonal high water table measured from the lowest point of the system. Installations without a precast vault that utilize infiltration are required to have the most hydraulically restrictive soil layer below the MTD meet the minimum tested vertical saturated hydraulic conductivity of one inch per hour and have at least two feet of separation from the seasonal high water table measured from the lowest point of the system.

7. Sizing Requirement:

The example below demonstrates the sizing procedure for the Filterra® HC:

Example: A 0.25-acre impervious site is to be treated to 80% TSS removal using the Filterra® HC. The impervious site runoff (Q) based on the New Jersey Water Quality Design Storm was determined to be 0.79 cfs.

The selection of the appropriate model of Filterra® HC is based upon both the maximum inflow drainage area and the MTFR. It is necessary to calculate the required model using both methods and to use the largest model determined by the two methods.

<u>Inflow Drainage Area Evaluation:</u>

The drainage area to the Filterra® HC in this example is 0.25 acres. Included in Table 1 below, all of the Filterra® HC models are designed with a maximum allowable drainage area greater than 0.25 acres. Specifically, the Filterra® HC with a 4'x4' media bay and a maximum allowable drainage area of 0.40 acres would be the smallest model able to treat runoff without exceeding the maximum allowable drainage area.

<u>Maximum Treatment Flow Rate (MTFR) Evaluation:</u>

The site runoff (Q) was based on the following:

time of concentration = 10 minutes
i = 3.2 in/hr (page 5-8, Fig. 5-3 of the NJ Stormwater BMP Manual)
c = 0.99 (runoff coefficient for impervious)
Q = ciA = 0.99 x 3.2 x 0.25 = 0.79 cfs

Given the site runoff is 0.79 cfs and based on the MTFR's listed in Table 1 below, the Filterra® HC with a 16'x8' media bay and an MTFR of 0.889 cfs would be the smallest model that could be used to treat the impervious area without exceeding the MTFR. If using more than one unit for treating runoff, the units should be configured such that the flowrate to each unit does not exceed the design MTFR for each unit and ensuring the entire 0.25 acre area is treated.

The MTFR evaluation results will be used since that method results in the highest minimum configuration determined by the two methods.

Be advised a detailed maintenance plan is mandatory for any project with a Stormwater BMP subject to the Stormwater Management rules, N.J.A.C. 7:8. The plan must include all of the items identified in the Stormwater Management rules, N.J.A.C. 7:8-5.8. Such items include, but are not limited to, the list of inspection and maintenance equipment and tools, specific corrective and preventative maintenance tasks, indication of problems in the system, and training of maintenance personnel. Additional information can be found in Chapter 8: Maintenance and Retrofit of Stormwater Management Measures.

If you have any questions regarding the above information, please contact me at (609) 633-7021.

Sincerely,

Gabriel Mahon, Chief
Bureau of Stormwater Permitting

Attachment: Maintenance Plan
cc: Chron File
Richard Magee, NJCAT
Vince Mazzei, NJDEP – Water & Land Management
Nancy Kempel, NJDEP– BSTP
Keith Stampfel, NJDEP – DLRP
Dennis Contois, NJDEP – DLRP

The sizing table corresponding to the available system models is noted below:

Table 1. Filterra® HC MTFRs and Maximum Allowable Drainage Areas

	Available Filterra® Media Bay Sizes (feet)	Effective Filtration Treatment Area (ft^2)	Treatment Flow Rate (cfs)	Maximum Allowable Drainage Area (ac)
Standard Configuration Filterra and Filterra Biosape Vaults	4x4	16	0.111	0.40
	4x6 or 6x4	24	0.167	0.60
	4.5x7.83 or 7.83x4.5 (Nominal 4x8/8x4)	35.24	0.245	0.89
	6x6	36	0.250	0.91
	6x8 or 8x6	48	0.333	1.21
	6x10 or 10x6	60	0.417	1.51
	6x12 or 12x6	72	0.500	1.81
	7x13 or 13x7	91	0.632	2.29
	14x8	112	0.778	2.82
	16x8	128	0.889	3.22
	18x8	144	1.000	3.62
	20x8	160	1.111	4.03
	22x8	176	1.222	4.43
Peak Diversion Filterra Vaults	4x4	16	0.111	0.40
	4.5x5.83 (Nominal 4x6)	26.24	0.182	0.66
	6x4	24	0.167	0.60
	6x6	36	0.250	0.91
	6x8	48	0.333	1.21
	6x10 or 10x6	60	0.417	1.51
	7x10	70	0.486	1.76
	8x10.5	84	0.583	2.11
	8x12.5	100	0.694	2.52
	Custom and/or Filterra Bioscape	Media Area in ft^2	0.00694 * (Media Area in ft^2)	0.0252 * (Media Area in ft^2)

REFERENCES

AbTech Industries, 4110 North Scottsdale Road, Suite #235, Scottsdale, AZ 85251, Ph. (480) 874-4000, https://www.abtechindustries.com.

Advanced Drainage Systems, Inc., 4671 Raindrop Way, Hillard, OH 43026, Ph. (800) 733-7473, https://www.ads-pipe.com.

Aird, J., September 2012, Separation devices for stormwater runoff, *Stormwater* pp. 48-53.

AquaShield Inc., 2733 Kanasita Drive, Chattanooga, TN 37343, Ph. 888-344-9044, https://www.aquashieldinc.com.

BaySaver Technologies, Inc., 1010 Deer Hollow Dr. Mount Airy, MD 21771, https://stormwater.com > baysaver . . .

Best Management Products, Inc., 9 Matthews Drive, Unit A1-A2, East Haddam, CT 06423, Ph. 1 (860) 434-0277, https://www.bmpinc.com.

Bio Clean Environmental Services, 2972 San Lois Rey, Oceanside, CA 92054, Ph. 760-433-7640, https://www.biocleanenvironmental.net.

Contech Engineered Solutions, 900 Centre Pointe Drive #400, West Chester, OH, 45069, Ph. 1 (800) 338-1122, https://www.Conteches.com.

Doerfer, J., July/August 2008, Filter maintenance sand filters vs. media filters, *Stormwater Solutions*, p. 8.

EPA, November 2005, *National management measures to control non-point source pollution from urban run-off*, Publication No. EPA-841-B-05-004, p. 518. https://www.epa.gov/owow/nps/urbanmm/index.html.

EPA, August 11, 2025, *Source water protection/EPA*, https://www.epa.gov.

Filterra Bioretention Systems, *A division of Americast, now by Contech engineered solutions*, https://conteches.com.

Goldberg, S., June 2012, Separation devices for stormwater runoff, *Stormwater* pp. 52–58.

Hydro International, 94 Hutchins Drive, Portland, ME 04102, Ph. 1 (207) 756-6212, https://www.hydro-int.com.

Imbrium Systems Inc, 605 Global Way, Suite 113, Linthicum, MD, 21090. 407 Fairview Drive, Whitby, ON L1N 3A9, Canada, Ph. 1(416) 960-9900

Jensen Precast, 9895 Double R Blvd, Reno, NV 89521, Ph. (775) 352-2700, https://www.jensenprecast.com.

KriStar OEM., Anaheim, CA, https://www.kristar.com.

Maine Department of Environmental Protection Maine Stormwater Management Design Manual. Technical Design Manual, Vol. III, May 2016.

NJDEP (New Jersey Department of Environmental Protection), February 2004, 2010, July 17, 2023, *Stormwater best management practices manual*, NJDEP, last updated October 9 2024.

NJDEP (New Jersey Department of Environmental Protection), N.J.A.C. 7:8, July 17, 2023, *Stormwater management*, NJDEP.

NYDEC (New York Department of Environmental Conservation), January 2015, *New York State stormwater management design manual*, NYDEC. last updated Draft May 2022.

Terre Kleen, Terre Hill Concrete Products, Box 10, 485 Weaverland Valley Road, Terre Hil, PA 17581, https://www.terrehill.com.

WERF, 2005, *Critical assessment of stormwater treatment and control selection issues*, https://werf.org.

Wieske, D. and Penna, L.M., 2002, Stormwater strategy, *ASCE Civil Engineering*, 72 (2): 62–67.

7 Structural Stormwater Management Systems

Detention, retention, and infiltration basins are conventional stormwater structures that are employed to control the peak rates and/or the volumes of runoff, and to improve the water quality. Constructed wetlands, bioretention basins, and vegetated swales are also used in stormwater management practices – more for improving water quality than peak flow attenuation. Also employed are sand filters and vegetated buffers, though mostly for improving water quality. This chapter presents simplified design procedures, examples, and case studies for various types of structural stormwater management systems. Also included in this chapter are recommended design criteria for stormwater management systems.

7.1 DETENTION BASINS/WET PONDS

Detention basins are storage facilities that impound stormwater runoff temporarily and release it slowly through their outlet structures. Figure 7.1 depicts a grass-lined detention basin with a center fountain designed by the author. Figure 7.2 shows a typical section of a detention basin-outlet structure system. Detention basins are intended to fully drain following a rainfall event and can be constructed aboveground or underground. Open detention basins not only regulate the peak rate of runoff but also serve as a means of enhancing the runoff quality.

Wet ponds are differentiated from detention basins in that the former are impounded at all times, while the latter remain dry between storm events. Both of these facilities are widely employed to control peak rates of runoff and can be quite effective if properly designed and well maintained. By prolonging the discharge, detention basins and wet ponds also improve water quality through partial settling of suspended sediment from runoff. This effect depends on retention time, the time it takes the stored water to be released. A detention basin that prolongs the release time is known as an extended detention basin. The longer the retention time, the greater the treatment efficacy of a detention basin. The efficiency can be increased by incorporating a presenting chamber in an underground basin or a forebay in an aboveground basin to trap coarse sediments.

The effectiveness of a detention basin for removing pollutants depends on site conditions and the types of developments. In general, detention basins may be more effective in commercial developments than residential ones since the runoff can have a higher concentration of sediments in the former than in the latter. In wet ponds, the storage above the normal pool level creates a detention effect, and the stagnant water helps improve water quality. In fact, a wet pond can serve as a highly effective best management practice. In addition to improving the water quality and controlling the peak runoff, wet ponds provide aesthetic amenity, aquatic habitat for fish, and, occasionally, recreation (Figure 7.3). In coastal areas where the water table is high, ponds are the most common practice to address stormwater management requirements. In South Carolina's coastal counties, wet ponds are not only a practical option but also the most popular system for meeting stormwater runoff requirements. Thousands of ponds have been constructed since 1987 under NPDES Phase I and Phase II.

Removal of pollutants in wet ponds occurs through various factors. The main factor is gravity, which results in the settling of suspended matter and pollutants attached thereto. As discussed in Chapter 1, the settling velocity of particles in a stagnant water body decreases exponentially with the particle size. As such in a wet pond and extended detention basin, the water quality effect for a given sediment type depends on the residence time, the time that takes water equal in volume to that of runoff to be discharged from the pond. Because of the mixing with the stagnant water, a wet pond is more effective than a detention basin of equal detention time.

DOI: 10.1201/9781003460121-7

FIGURE 7.1 Grass-lined detention basin with center fountain.

Source: Photo by the author.

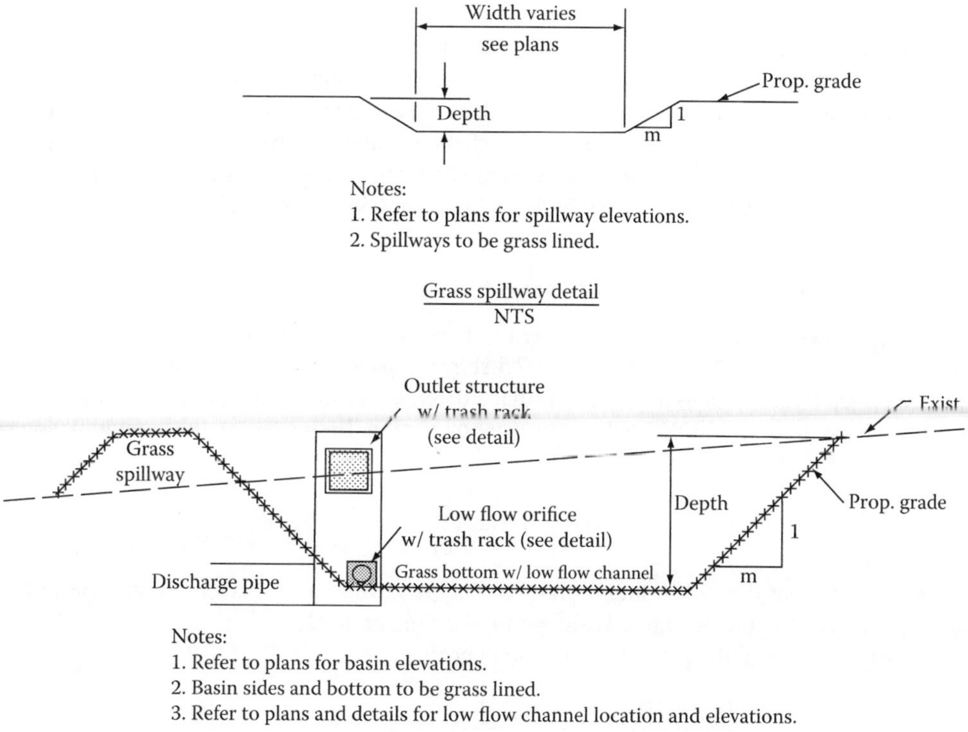

FIGURE 7.2 Typical detention basin cross-section.

Aquatic plants, if present, and microorganisms also improve water quality by uptake of nutrients and breakdown of organic contaminants. Aquatic vegetation also removes metals. In areas of high-water tables where aboveground detention basins and underground detention-infiltration basins are not practical, wet ponds may serve as one of the best management practices.

Detention basins and wet ponds are designed through an iterative process. The process begins with using a trial storage volume and preparing a grading plan of the basin. This is followed by preliminary sizing of the outlet openings and developing a stage storage- discharge relation for the

FIGURE 7.3 Wet pond.

Source: Photo by the author.

selected detention–outlet structure system. Then, routing computations are performed to evaluate the functioning of the system. Finally, outlet openings are refined until the desired discharges from the basin/pond are attained. This iterative process can be accelerated by first estimating the required detention storage. Example and case studies in this chapter illustrate the estimation procedure and the design process. Detailed information about design and construction of urban stormwater management systems can be found in an ASCE/WEF publication (1992) and ASCE (2006).

7.1.1 Flow Routing through Detention Basins

The runoff that enters a detention basin is partly discharged through an outlet structure and partly impounds in the basin (see Figure 7.4). Figure 7.5 shows typical inflow and outflow hydrographs through a detention basin. This figure indicates that a detention basin not only attenuates the runoff hydrograph but also prolongs discharge from its catchment area. In this figure, the continuity equation during a time interval d*t* is given by

$$I(t) - O(t) = \frac{\Delta S}{dt} \tag{7.1}$$

where *I* and *O* represent the inflow and outflow discharge at time, t, and ΔS is the change in the volume of water stored in the detention basin within the time period, dt.

In finite difference form, the preceding equation becomes

$$\left[\frac{(I_1 + I_2)}{2}\right] \cdot \Delta t - \left[\frac{O_1 + O_2}{2}\right] \cdot \Delta t = \Delta S = S_2 - S_1 \tag{7.2}$$

where subscripts 1 and 2 represent the quantity of variables I, O, and S at the beginning and at the end of Δt, respectively. Separating the known and unknown values of variables, the preceding equation can be written as

$$\left(\frac{2S_2}{\Delta t} + O_2\right) = (I_1 + I_2) + \left(\frac{2S_1}{\Delta t} - O_1\right) \tag{7.3}$$

Since the inflow hydrograph is known (obtained from runoff calculations) and the outflow discharge and storage volume are also known at the beginning of a period, the preceding equation gives the value of the term on the left side. To calculate the unknown O_2 and S_2, a storage-outflow

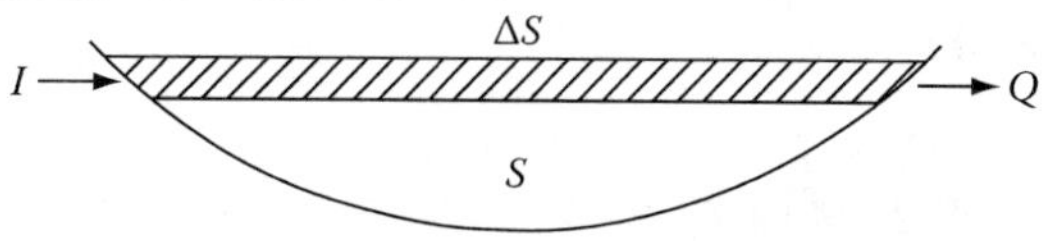

FIGURE 7.4 Flow routing through a detention basin.

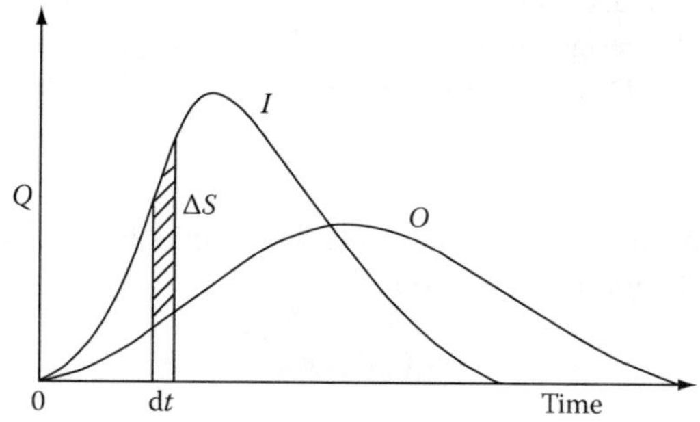

FIGURE 7.5 Typical inflow and outflow hydrographs through a detention basin.

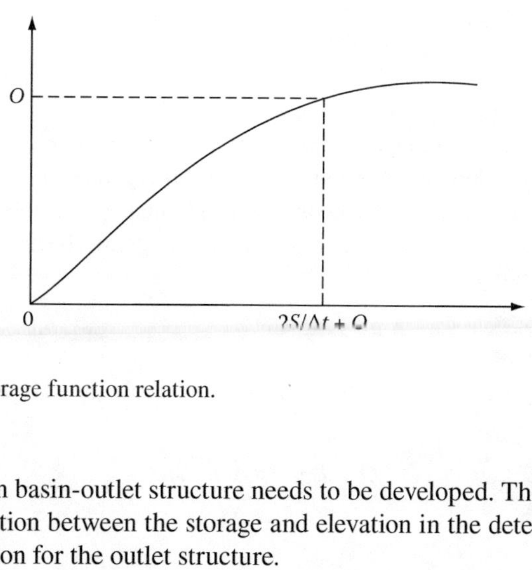

FIGURE 7.6 Outflow–storage function relation.

function for the detention basin-outlet structure needs to be developed. This function can be developed by preparing a relation between the storage and elevation in the detention basin and between the discharge and elevation for the outlet structure.

Specifically, for a given water surface elevation, the amounts of storage and discharge are calculated using the grading plan of the detention basin and the geometry of the outlet structure, respectively. Based on these calculations, a plot of O, versus $2S/\Delta t + O$ is prepared. This plot, sketched in Figure 7.6, together with Equation 7.3 can then be used to determine the outflow discharge at the end of the period.

In practice, Equation 7.3 is commonly solved using so-called reservoir routing, also known as level pool or hydrologic routing method. In this method, discharge and storage are related to water surface elevation or stage, and the variable $2S/\Delta t + O$ is also related to stage. Thus, by solving Equation 7.3, the stage and in turn discharge and storage can be determined. Equation 7.3 is based on an implicit assumption that variables I, O, and S are linearly related with time. Therefore, to arrive at a fairly accurate result, Δt must be small and properly selected. As has been discussed before a number of software programs are available to perform hydrologic routing. These include PondPack, StormCAD, and HEC-1.

7.1.2 Outlet Structure Design

An outlet structure forms an integral part of detention basin design. A detention basin is effectively utilized when the openings and stages of its outlet structures are properly selected. An outlet structure may be single-stage or multistage. A single-stage outlet comprises one opening, and multistage outlets include two or more openings commonly of varying size and geometry and at different levels. To attenuate runoff from storms of various frequencies, a multistage outlet structure should be employed. A simplified procedure for the design of detention basins with multistage outlets is presented by Pazwash (1992). Example 7.1 will illustrate the design process. Orifices, weirs, and grates are the most commonly used openings in an outlet structure. Discharges from these openings are discussed as follows.

7.1.2.1 Orifice

An orifice is an opening that is submerged at its inlet face, outlet face, or both. The flow through an orifice is calculated from the following equation:

$$Q = C_o A(2gh)^{1/2} \tag{7.4}$$

where

C_o = orifice coefficient
A = cross-sectional area of orifice, m² (ft²)
h = head above orifice center for free flow outlet or the difference in water surface elevation for submerged outlet, m (ft)
g = gravitation acceleration, 9.81 m/s² (32.2 ft/s²)
Q = discharge, m³/s (cfs)

This equation is dimensionally homogeneous and, therefore can be used in both SI and CU. The orifice coefficient is nearly independent of the depth of submergence and is approximated as $C = 0.6$ in practice. Orifices are commonly employed as low-level openings for the control of more frequent storms such as 1- to 2-year storms. Low-level orifice prolongs discharge from detention basins in order to improve water quality.

7.1.2.2 Rectangular Weir

For a rectangular weir (Figure 7.7), the ideal discharge can be calculated from the following equation:

$$Q = L\sqrt{2g}\int_0^H h^{1/2}\,dh$$

which integrates to

$$Q = \frac{2}{3}\sqrt{2g}LH^{3/2} \tag{7.5}$$

However, because of flow contractions at the sides and drawdown of the water surface on the weir, the effective area is smaller than LH. To account for this effect, a coefficient of discharge, C_d, is applied to the ideal discharge in the preceding equation. Thus, the actual discharge is given by

$$Q = C_d\left(\frac{2}{3}\right)\sqrt{2g}\;LH^{1.5} \tag{7.6}$$

For convenience, the term $C_d\,(2/3)\sqrt{2g}$ in the preceding equation is replaced by a coefficient, C_w, resulting in

$$Q = C_w\;LH^{1.5} \tag{7.7}$$

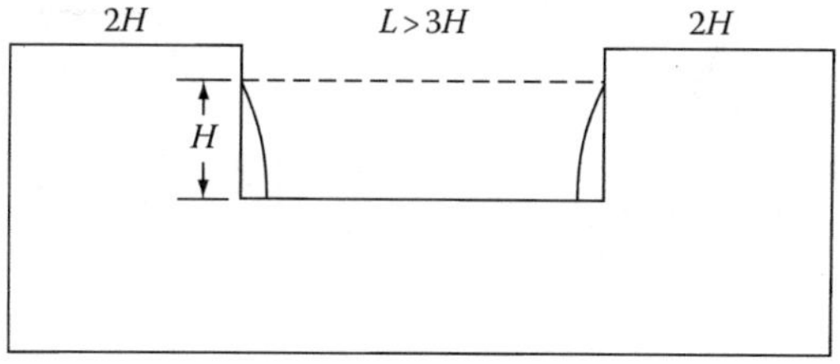

FIGURE 7.7 Side contractions in a rectangular weir.

where

C_w = weir coefficient
L= length of weir, m (ft)
H= depth of water above the weir crest, m (ft)

Experiments by Rehbock in Karlsruhe Hydraulic Laboratory in Germany showed that C_d, and C_w were somewhat dependent on H/P, where P is the depth of water below the weir crest (Pazwash, 2007). In practice, however, this dependence is neglected, and Equation 7.7 is expressed as

$$Q = 1.8\, LH^{1.5} \quad (SI) \tag{7.8}$$

$$Q = 3.3\, LH^{1.5} \quad (CU) \tag{7.9}$$

The preceding equations are accurate for $H/L < 0.4$, which is not far from the usual operating range of outlet structures.

For rectangular weirs with side contractions where the weir is narrower than the structure, there will be a lateral contraction of water nappe. The effect of side contractions, shown in Figure 7.7, has been found to be a reduction of the width of nappe by 0.1 H on each side. Accounting for contractions on both sides, the effective weir length of L- 0.2 H is used in the above equations.

It should be noted that a rectangular opening in an outlet structure functions as a weir as long as it is not submerged; once submerged, it behaves as an orifice. Under that condition, discharge should be calculated using Equation 7.4.

7.1.2.3 Triangular Weir

Triangular weirs, also known as V-notch weirs, serve as effective means of flow control. These weirs have the advantage that they can function for very small heads and also provide a very wide range in flow. The vertex angle of V-notch weirs commonly ranges from 15° to 90°. The ideal discharge through an elemental area da, shown in Figure 7.8, is given

$$dQ = \sqrt{2gh} \times dA \tag{7.10}$$

where

$$dA = 2x \times dh$$

$$x = h\ tan\left(\frac{\theta}{2}\right)$$

Substituting for h and dA in Equation 7.10 and integrating from h = 0 to h = H gives ideal discharge through a V-notch weir:

$$Q = \left(\frac{8}{15}\right)\sqrt{2g}\tan\left(\frac{\theta}{2}\right)H^{5/2} \tag{7.11}$$

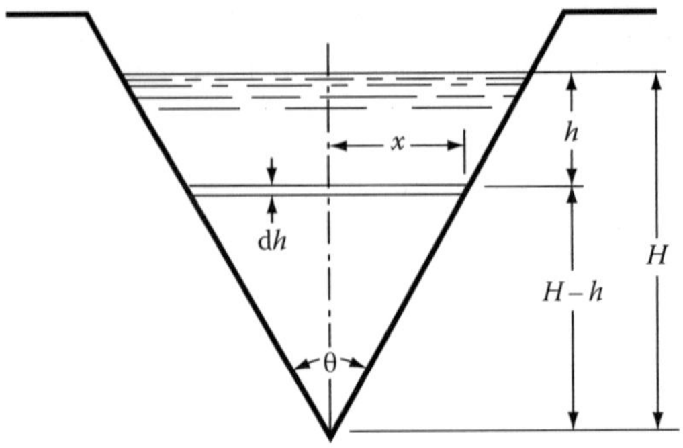

FIGURE 7.8 Triangular weir.

Introducing a coefficient of discharge C_d to account for side contractions, the preceding equation becomes

$$Q = C_d \left(\frac{8}{15}\right)\sqrt{2g}\tan\left(\frac{\theta}{2}\right)H^{5/2} \tag{7.12}$$

This equation is dimensionally homogeneous and as such it applies to both SI and CU. The coefficient C_d ranges from 0.58 to over 0.62 depending on the angle of vertex and the head over the weir; the smaller the head, the larger the C_d. Conservatively, using C_d = 0.585, the preceding equation simplifies as

$$Q = 1.38\, tan\left(\frac{\theta}{2}\right)H^{2.5}, (SI) \tag{7.13}$$

$$Q = 2.50\, tan\left(\frac{\theta}{2}\right)H^{2.5}, (CU) \tag{7.14}$$

For 90° and 30° weirs, the above equation simplifies as:

$$Q = KH^{2.5} \tag{7.15}$$

$$K = 1.38, (SI);\ \ 2.50, (CU)\ for\, 90°$$

$$K = 0.37, (SI);\ \ 0.67, (CU)\ for\, 30°$$

For heads smaller than 15 cm (0.5 ft), the discharge coefficient increases due to incomplete contraction. However, this variation may be ignored, still using Equations 7.13 and 7.14 in detention basin design.

7.1.2.4 Cipolletti Weir

A Cipolletti weir (Figure 7.9) is a trapezoidal weir with 1:4 (1H, 4V) side slopes that was developed to compensate for the side contractions in a rectangular weir. Thus, the weir in Equations 7.8 and 7.9 can be used disregarding side contractions.

7.1.2.5 Broad-Crested Weir

A broad-crested weir (Figure 7.10) is a weir with its width along the flow at least three times the depth of water over the weir. If the weir is high enough to create a backwater effect, critical depth

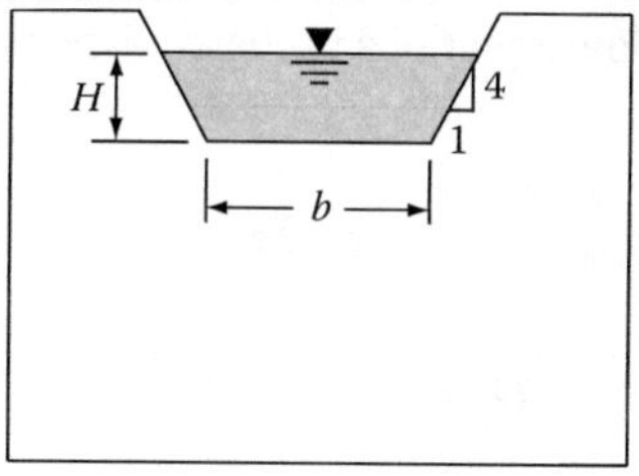

FIGURE 7.9 Cipolletti weir.

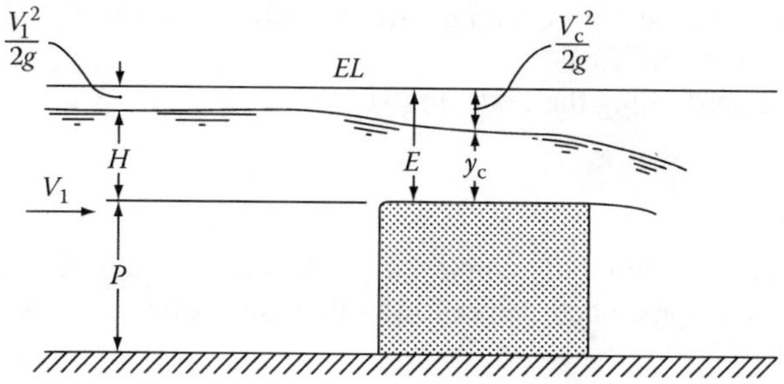

FIGURE 7.10 Broad-crested weir.

occurs on the crest of the weir. As shown in Chapter 2, the flow equations for a critical flow in a rectangular channel are

$$y_c = \frac{2}{3}E$$

$$V_c = \sqrt{gy_c}$$

$$Q = Ly_cV_c = \left(\frac{2}{3}\right)^{1.5}\sqrt{g}LE^{1/2} \tag{7.16}$$

For high weirs (that is, large *P*/*H*; refer to Figure 7.10), the velocity of approach becomes small, and *E* can be approximated by *H*, the depth of approach flow. In this case, the preceding equation becomes

$$Q = 1.7\ LH^{1.5}\ (SI) \tag{7.17}$$

$$Q = 3.09\ LH^{1.5}\ (CU) \tag{7.18}$$

7.1.2.6 Overflow Grates

Overflow grates are installed atop outlet structures to serve as an emergency measure to release water beyond the design storm. Flow-through grates were discussed in Chapter 4. As indicated, discharge from a grate occurs as weir flow at small depths of water and as orifice flow when the grate becomes fully submerged. Flow equations in these cases are given by the following equations respectively:

$$Q = C_w L H^{1.5} \tag{7.19}$$

$$Q = C_o f_c A (2gH)^{1/2} \tag{7.20}$$

where

C_w = weir coefficient = 1.8 SI, 3.3 CU
L = overall perimeter length of grate, m (ft)
C_o = orifice flow coefficient = 0.6 (square edge)
f_c = clogging factor to account for opening covered with leaves, usually taken as $f_c = 0.66$
A = opening area of grate, m² (ft²)
H = depth of water approaching the grate, m (ft)

7.1.2.7 Standpipes

Standpipes (Figure 7.11) are commonly employed for discharge from sediment basins during construction. Flow into a standpipe occurs as spillway flow for shallow depths. In this case, the discharge is given by

$$Q = C_L \pi D h^{1.5} \tag{7.21}$$

where C_L is the spillway coefficient (C_L = 1.8 SI; 3.3 CU), and h and D are the depth of water over the standpipe and pipe diameter, respectively. When the depth of water over the pipe exceeds 0.5 D, the inlet to standby pipe becomes fully submerged and the discharge follows the orifice equation:

$$Q = C_o \left(\frac{\pi D^2}{4} \right) \times (2gh)^{1/2} \tag{7.22}$$

where C_o is the orifice coefficient C_o = 0.6.

7.1.2.8 Hydro-Brake, Fluidic-Cone

Hydro-Brake is a device that functions similarly to an orifice at the low head. However, when submerged, the device creates a rotational flow in the conic section, which forms an air-cored vortex in

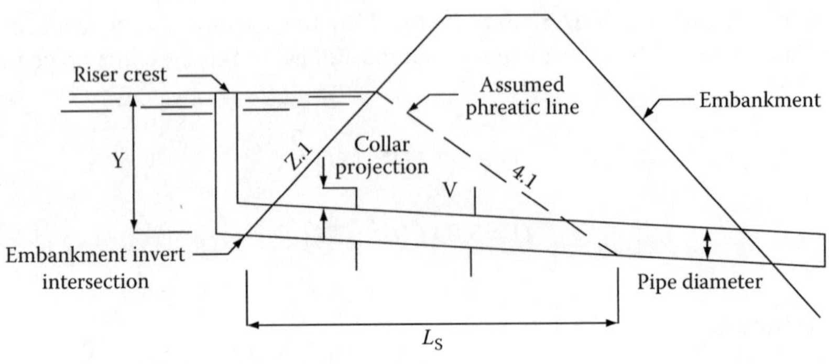

FIGURE 7.11 Standpipe in a wet pond (temporary during construction, permanent after).

the outlet section. The cone fills the center of the outlet and obstructs the flow. As a result, a Hydro-Brake with a relatively large opening performs like a small orifice. Hydro-Brake Fluidic-Cones come in 9 to 24 in. (23–60 cm) conic diameters and 3 to 10 in. (7.5–25 cm) outlet diameters. This device controls the design flow rate from 0 to 120 L/s (0–4.23 cfs) for design heads of 0.0 to 3 m (0–10 ft). For more information about this device, visit www.contechstormwater.com.

7.1.2.9 Thirsty Duck

Thirsty Duck (a trademark of Thirsty Duck LP) is a buoyant flow control device that controls outflow from a detention basin. This device uses the principle of buoyancy to suspend a discharge control device, such as orifice or weir, at a constant depth below the water surface. By maintaining a constant headwater over the control mechanism, this device provides a constant discharge. Using Thirsty Duck, the discharge from a detention basin can be kept at the maximum allowable limit and thus the storage volume of detention basins can be minimized.

The device does not have any active mechanism or component. As with any other stormwater control device, clogging due to debris can cause the structure to malfunction. The device can be accessed for clean-out during routine maintenance. The structure housing the Thirsty Duck device can be equipped with a locking hatch cover to prevent vandalism. Thirsty Duck LP. can be contacted at https://www.thirstyducksussex.com.

Example 7.1

Runoff calculations for a 3.29 acre residential development are performed based on the rational method. The existing peak rates of runoff for 2-, 10-, and 100-year storms are

$$Q_2 = 3.82 \text{ cfs}$$
$$Q_{10} = 5.40 \text{ cfs}$$
$$Q_{100} = 7.63 \text{ cfs}$$

The local stormwater management regulations require that the post-development peak rates of runoff from the site be 50%, 25%, and 20% smaller than those of predevelopment for the 2-, 10-, and 100-year storms, respectively. To meet these regulations and to provide a prolonged retention time to improve water quality, an outlet structure is provisioned comprising:

3 in. orifice at elevation 341.50 ft
12 × 12 in. opening at elevation 343.50 ft
Inlet grate (Campbell no. 3220) at elevation 344.75 ft

Calculate the elevation–discharge rating for this structure.

Solution

Discharge through openings is calculated as follows:

a. 3 in. orifice

$$Q_o = C_o A(2gh)^{1/2} = 0.6\times 0.049(64.4)^{1/2}(\text{El}-\text{El}_o) = 0.236(\text{El}-341.63)^{1/2}$$

b. 12 in. × 12 in. primary opening. Flow through this opening occurs as weir flow until the opening is submerged and as orifice flow when the water level rises above the opening.

$$Q_p = C_o L H^{1/5} = 3.3\times 1\times H^{1.5} = 3.3(\text{EL}-343.50)^{1.5} \quad \text{weir flow}$$

$$Q_p = 0.6\times A(2gH)^{1/2} = 0.6\times 1\times 1\times(64.4)^{1/2}(\text{EL}-344.0)^{0.5} \quad \text{orifice flow}$$

$$Q_p = 4.81(\text{EL} - 344)^{0.5} \quad \text{orifice flow}$$

c. Overflow grate. This grate has a perimeter length of 9.50 ft and 1 in. wide by 5-1/8 in. long slots with a total opening area of 345 in.^{2} (2.396 ft^2). Because of the small width of openings, the flow over the grate may be calculated using an orifice flow equation for depths of flow over 3 in.:

$$Q_g = CA(2gh)^{1/2} f_c$$

where $C = 0.6$ and $f_c = 0.66$ is the suggested clogging factor:

$$Q_g = 7.6\, h^{1/2}$$

Table 7.1 shows the discharge–elevation rating table for the structure. This table indicates that the 100-year water level in the detention basin will not rise above elevation 344.85 ft. At this elevation, the discharge is nearly equal to the allowable discharge for the 100-year storm, which is 7.63 × 0.8 = 6.10 cfs.

TABLE 7.1
Stage-Discharge Relation for the Outlet Structure

EL (ft)	Q_o[a]	Q_p	Q_g	Q_t
341.50	0			0
342.00	0.14			0.14
343.00	0.28			0.28
343.50	0.32	0		0.32
344.00	0.36	1.17		1.53
344.50	0.40	3.40		3.80
344.75	0.42	4.17	0	4.59
345.00	0.43	4.81	3.81[a]	9.05

[a] A clogging factor equal to 0.66 is applied.

7.2 PRELIMINARY SIZING OF DETENTION BASINS

This section presents simple procedures for estimating the required storage volume of detention basins. The calculations cover the rational and modified rational methods and the SCS TR-55 method, as well as the universal runoff method developed by the author.

7.2.1 Rational and Modified Rational Method Estimation

Assuming that the discharge from a detention basin varies linearly with time, the inflow–outflow hydrographs may be shown by Figure 7.12. In this figure, Q_p and O_o are peak inflow and peak discharge from the detention basin, respectively. The area between inflow–outflow hydrographs to the left of Q_o represents the water stored in the basin. Based on this figure, the required detention storage to attenuate the peak flow from Q_p to Q_o, the allowable discharge from the basin, is given

$$S = \left(\frac{1}{2}\right) \times \left(Q_p - Q_o\right) T_b \times 60 \tag{7.23}$$

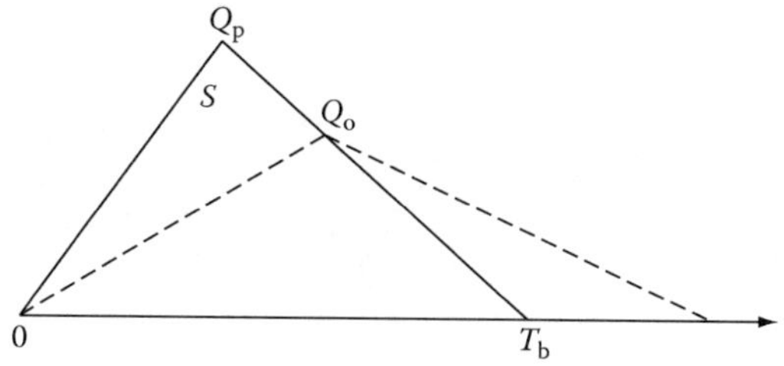

FIGURE 7.12 Detention storage estimation, rational method.

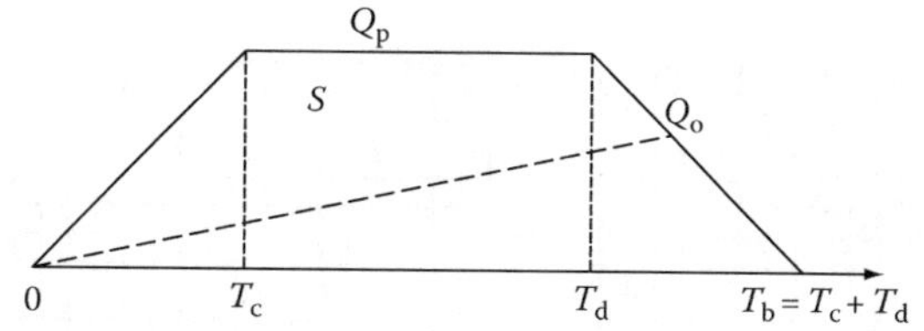

FIGURE 7.13 Detention storage estimation, modified rational method; $\mathbf{Q_o}$ = allowable discharge.

where T_b is the base time of inflow hydrographs, in minutes, and the factor 60 represents the number of **seconds** per minute.

Inversely, the outflow discharge for a given detention storage may be calculated as

$$Q_o = Q_p - \frac{2S}{(60\,T_b)} \tag{7.24}$$

Since Equations 7.23 and 7.24 are dimensionally homogeneous, they are applicable to both CU and SI. As indicated by these equations, the storage is related to the time base of the hydrograph, which is dependent on the time of concentration. For storms lasting longer than the time of concentration, assuming that discharge from a detention basin varies linearly with time, the inflow–outflow hydrographs can be presented by Figure 7.13. According to this figure, which represents the case of a modified rational hydrograph, the required detention storage to attenuate the peak inflow to the allowable discharge, Q_o, is calculated as follows:

$$S = \left(Q_p T_d - \frac{1}{2} Q_o T_b\right) 60 \tag{7.25}$$

where $T_b = T_d + T_c$.

Using the modified rational method, the required storage should be calculated for storm durations ranging from the time of concentration to the duration at which the peak runoff equals the allowable discharge from the detention basin. Figure 7.14 shows the calculation process in graphical form.

7.2.2 SCS TR-55 Method Estimation

The SCS's TR-55 method (1986) includes a figure for estimating the required storage volume of detention basins. Figure 7.15 shows the nondimensional relation between runoff volume and discharge ratios for type I through type III storms: one curve for types I and IA and another for types II and III. In this figure, the ratio of storage volume to runoff volume is plotted against the ratio of the peak outflow to peak inflow discharge. This figure indicates that for a given outflow-to-inflow ratio, types II and III storms require a significantly larger storage volume than types I and IA. Using this

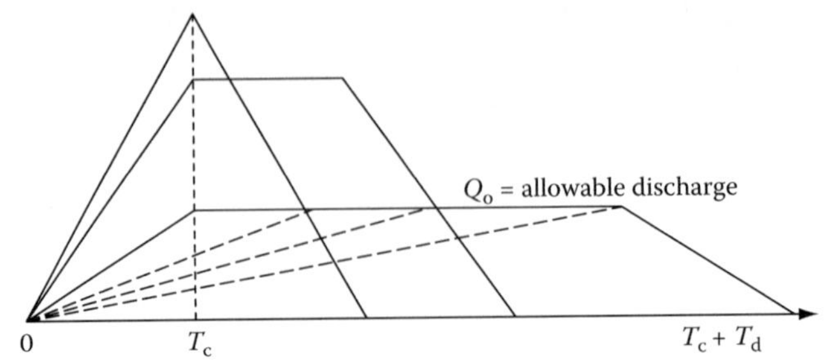

FIGURE 7.14 Required detention storage estimation, modified rational method.

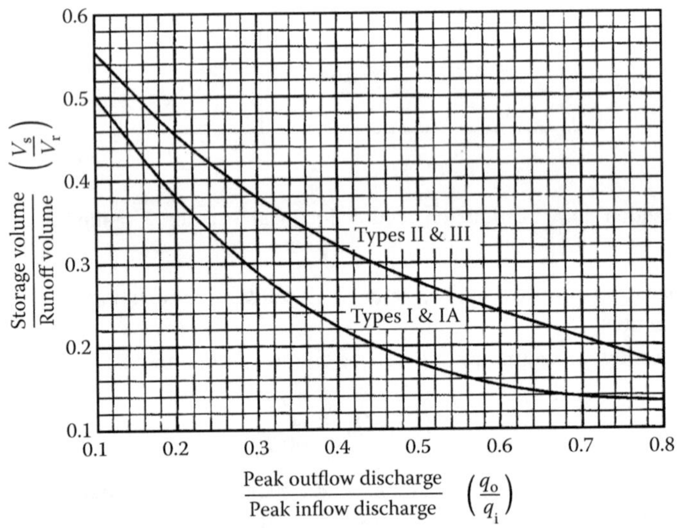

FIGURE 7.15 Approximate detention storage volume for types I, IA, II, and III storms.

figure allows for estimating storage for a given discharge or, inversely, estimating outflow discharge from a detention basin of a given storage volume (see Example 7.2). The required storage volume can be calculated as follows:

$$V_s = V_r\left(\frac{q_o}{q_i}\right) \tag{7.26}$$

where

V_s = detention storage volume, m³ (ft³)
V_r = runoff volume, m³ (ft³)
q_o = outflow (allowable) discharge, m³/s (cfs)
q_i = inflow discharge, m³/s (cfs)

7.2.3 Universal Method of Storage Volume Estimation

In the universal method (Pazwash, 2009, 2011, 2016), approximating the discharge variation with time by a straight line, the inflow–outflow hydrographs may be represented by Figure 7.16. It can be readily shown that the required detention storage in this method is given as

$$S = \left[(T_d - T_e)Q_i - \left(\frac{1}{2}\right)(T_d + T_c - 2T_e)Q_o\right]60 \tag{7.27}$$

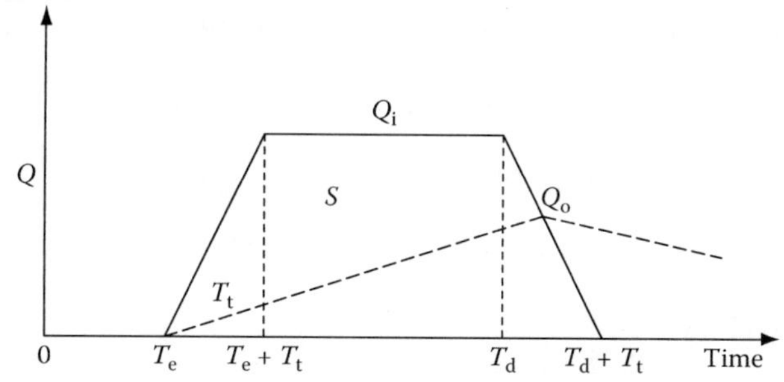

FIGURE 7.16 Routed inflow–outflow hydrograph in the universal method.

For a storm duration $T_d = T_c$, the preceding equation simplifies as follows:

$$S = (Q_i - Q_o)(T_c - T_e)60 \tag{7.28}$$

The parameters T_e, T_t, and T_d in this method are the lag time between the rainfall and the onset of runoff, travel time, and storm duration, respectively. Similar to the modified rational method, the calculations are performed for various rainfall durations. However, unlike the modified rational method, the runoff volume does not continually increase with the storm duration. In fact, for sustained storms, the entire rainfall volume may be dissipated through infiltration.

7.2.4 Adjusting Detention Storage Volume Estimates

All of the previously described methods of storage volume estimation are based on the assumption that outflow from a detention basin varies linearly with time. However, outflow, even for a single outlet opening, deviates from this assumption. The deviation increases with the number of outlet openings and, particularly when a small, low-flow orifice is incorporated for water quality improvement. A method of estimating the required detention storage for a multistage outlet is presented by Pazwash (1992). For the preliminary design of detention basins, the author suggests that the storage calculated by any one of the methods be increased by 50% to 75%, depending on the type of outlet structure. It is to be noted, however, that the previously described estimation methods are intended for the preliminary sizing of detention basins and that they cannot serve as a substitute for routing computations.

Example 7.2

The post-development peak discharges in Example 7.1, which are calculated based on the rational and modified rational method, are summarized in Table 7.2.

Calculate the required detention storage using

a. Rational hydrograph ($T_d = T_c = 15$ min)
b. Modified rational method

The post-development discharges are as noted in Example 7.1.

Solution

a. Rational Method

The runoff coefficient is calculated at 0.45. As such, use $T_d = 2.3\ T_c$; see Table 3.13 (suggested by the author).

Allowable discharges are calculated based on the existing peak runoff noted in Example 7.1:

$$Q_o = 3.82 \times 50\% = 1.91\,cfs(2\ year)$$

$$Q_o = 5.40 \times 75\% = 4.05\ cfs(10\ year)$$

$$Q_o = 7.63 \times 80\% = 6.10\ cfs(100\ year)$$

Applying Equation 7.23 to 2-, 10-, and 100-year storms:

$$S = \frac{1}{2}(4.74 - 1.91) \times (2.3 \times 15 \times 60) = 2929\ ft^3\ \ (2\ year)$$

$$S = \frac{1}{2}(7.25 - 4.05) \times (2.3 \times 15 \times 60) = 3312\ ft^3\ \ (10\ year)$$

$$S = \frac{1}{2}(10.21 - 6.10) \times (2.3 \times 15 \times 60) = 4255\ ft^3\ (100\ year)$$

b. Modified rational method
Storage is calculated utilizing Equation 7.25:

$$S = \left[Q_i T_d - 0.5\ Q_a\ (T_d + T_e)\right]60$$

Calculations for the peak flows for storms of 2-, 10-, and 100-year frequency are organized in Table 7.2.

The largest calculated storage volume is 4293 ft³ (30 min. storm). Accounting for nonlinear discharge variation, the ponding effect of 3 in. orifice, and free board, add 50% to the calculated required storage. Therefore, the detention basin storage volume should be approximately 7000 ft³.

TABLE 7.2
Peak Runoff for Various Storm Durations

T_d (min)	Q_2	Q_{10}	Q_{100}
$T_d = T_c = 15$	4.74	7.25	10.21
20	4.29	6.07	8.58
30	3.26	4.88	6.96
45	2.66	3.70a	5.33a
60	2.07	2.96[a]	4.44[a]

[a] Peak inflow less than allowable discharge (see Example 7.1). No further reduction is required.

Example 7.3

To attenuate the peak runoff from the site of the development described in Examples 7.1 and 7.2, an aboveground detention basin is provisioned. The storage–stage relation for this detention basin is calculated based on the basin grading plan and is tabulated in Table 7.4. The outlet structure of this basin is described in Example 7.1. Compute the peak discharges for this detention basin.

Solution

Haestad Method's PondPack computer software is utilized to perform the computations. The routing computations for the 100-year, 30-minute storm, according to Example 7.2, requires the largest detention storage, are included in Table 7.5. The computation results for other storm durations and frequencies are summarized in Table 7.6. A comparison of the existing allowable and proposed peak runoff rates from the site is presented in Table 7.7.

In this project, the detention basin is intentionally oversized to reduce downstream flooding caused by an inadequate drainage system. As a result, the detention basin creates peak flow reductions far beyond the requirements. The purpose of a 3 in. orifice at the outlet structure is to create a prolonged retention time in the detention basin to allow coarse and medium-sized suspended sediment to settle. In this project, the 3 in. orifice alone could not meet the applicable water quality standards. Therefore, a water treatment device was incorporated downstream of the outlet structure to supplement the dual-purpose detention basin.

TABLE 7.3
Required Storage Volume Estimation

	Two-Year Storm, Q_a = 1.91 ft³				Ten-Year Storm, Q_a = 4.05		One Hundred-Year Storm, Q_a =6.10	
T_d (min)	20	30	45	60	20	30	20	30
Q_i (cfs)	4.29	3.26	2.66	2.07	6.07	4.88	8.58	6.96
S (ft³)	3143	3290	3744	3155	3032	3317	3891	4293

TABLE 7.4
Storage–Elevation Rating

Elevation (ft)	Area (ft²)	Avg. Area (ft²)	Δ Volume[a] (ft³)	Volume (ft³)
341.5	1360			0
		1530	760	
342.0	1680			760
		2035	2035	
343.0	2390			2795
		2720	2 720	
344.0	3050			5515
		3500	3500	
345.0	3950			9015

Note: The calculated ΔV deviates less than 0.25% from the prismatic calculation, given by the equation: $\Delta_t = [A_1 + A_2 + (A_1A_2)^{1/2}]/3$.

[a] Δ Vol. = average area × ΔH.

TABLE 7.5
100-Year, 30-Minute Storm Routing Computations Summary

Given Pond Data			Intermediate Routing Computations	
Elevation (ft)	Outflow (cfs)	Storage (cfs)	2S/*t* (cfs)	2S/*t* + 0 (cfs)
341.50	0.0	0	0.0	0.0
342.00	0.1	760	5.1	5.2
343.00	0.3	2795	18.6	18.9
343.50	0.3	4145	27.6	28.0
344.00	1.5	5515	36.8	38.3
344.50	3.7	7250	48.3	52.0
344.75	4.6	8130	54.2	58.8
345.00	8.2	9015	\60.1	68.3

Note: Time increment (*t*) = 5.0 min.
Starting pond W.S. elevation = 341.50 ft
Summary of peak outflow and peak elevation
Peak inflow = 6.96 cfs
Peak outflow = 4.46 cfs
Peak elevation = 344.71 ft[a]
Summary of approximate peak storage
Initial storage = 0 cu-ft
Peak storage from storm = 8005 cu-ft
Total storage in pond = 8005 cu-ft
Inflow hydrograph: A: PR100–30.HYD
Rating table file: A: SPAR-DET.PND
Initial Conditions
Elevation = 341.50 ft
Outflow = 0.00 cfs
Storage = 0 cu-ft

[a] The computed peak discharge for a 100-yr, 15-min storm is 4.43 cfs, which is almost the same as for the 100-yr, 30-min storm.

TABLE 7.6
Routing Computation Summary

Storm Duration (min)	2-Year Storm		10-Year Storm		100-Year Storm	
	Inflow	Outflow	Inflow	Outflow	Inflow	Outflow
15	4.74	0.94	7.25	2.58	10.21	4.42
20	4.29	0.68	6.07	2.03	8.58	3.84
30	3.26	1.02	4.88	2.59	6.96	4.46
45	2.66	1.45	3.70	2.76	5.33	4.33
60	2.07	1.39	2.96	2.52	-	-

Note that the required detention storage volume calculations in Example 7.2 yield critical storm durations of 45 minutes for the 2-year storm and 30 minutes for the 10- and 100-year storms (based on the modified rational method). These approximate results agree with the more accurate routing computations. Thus, in practice, the critical storm duration is determined using the graphical solution, and the routing computations are performed for that duration to determine discharge. This procedure shortens the routing computations.

TABLE 7.7
Comparison of Pre- and Post-Development Peak Flows

Storm Freq. (years)	Q_E = Ex. Peak (cfs)	Q_P = Prop. Peak (cfs)	Reduction % $(1 - Q_P/Q_E)$
2	3.82	1.45	62
10	5.40	2.76	49
100	7.63	4.46	42

7.3 EXTENDED DETENTION BASINS

Extended, or dual-purpose, detention basins are basins that not only attenuate the peak rates of runoff but also improve the water quality. The latter objective is achieved by using a small, low-level opening in the outlet structure to prolong the release of runoff from the detention basin. The opening is commonly sized based on the water quality storm, which varies from state to state as discussed earlier. The required storage volume of a detention basin can be preliminarily estimated based on the methods previously discussed in this chapter. As indicated, a multiplier should be applied to the estimated storage volume to account for the nonlinear discharge from the detention basin and particularly the discharge from the low-level opening, which is to impound the water quality storm in whole or in part. The sizes of openings in the outlet structure, which produce an efficient use of detention basins, are commonly determined through an iterative process. A simple procedure for sizing the openings is presented in a paper by the author (Pazwash, 1992). Design procedures for dual-purpose detention basins are presented in case studies in this chapter. To prevent standing water, which creates mosquito breeding, the bottom of detention basins should lie above the water table. A shallow pond also promotes algae growth, which is unsightly.

CASE STUDY 7.1

This study relates to the design of stormwater management elements for a three-story self-storage commercial building on a 4-acre (1.62 hectares) lot that is subdivided from a 29.84-acre parcel of land. The project site is located in the northeast section of Hillsborough Township Hillsborough Township, Somerset County, New Jersey, and is identified as lot 12.01, block 65 on the township tax map (see Figure 7.17) and is occupied by a nursery and garden center. The site of the project is bound by the Conrail to the south, Route 206 to the north. and the undisturbed part of lot 12.01 to the east. Figure 7.18 shows the predevelopment site layout. This plan indicates that the site is largely disturbed, covered by buildings, gravel parking, and storage sheds. The site is relatively flat, sloping from 71 ft (21.6 m) elevation along the railroad to the south to an elevation range of 67 to 69 ft (20.4–21 m) along Route 206 to the north.

Figure 7.18 depicts the post-development site layout and the footprint area of the building, which is 40,110 ft^2 (3726 m^2). Since the project increases imperviousness, an extended detention basin is installed to reduce the peak runoff rates from the site in compliance with the state's stormwater management regulations and to improve the water quality. The detention basin is proposed at the northwesterly corner of the site that is depressed, and currently drains to a pipe to an existing inlet on Route 206.

To supplement the water quality effect of the extended detention basin, manufactured water treatment devices are incorporated in the stormwater management system. In addition, underground infiltration chambers are installed to address the groundwater recharge requirement for the project. Design calculations for the proposed stormwater management elements are presented as follows.

Runoff Calculations

Based on topography, the site is divided into three subareas, designated A1, A2, and A3. As shown in Figure 7.18 and listed in Table 7.8. Area A1, containing 3.18 acres, flows into an existing

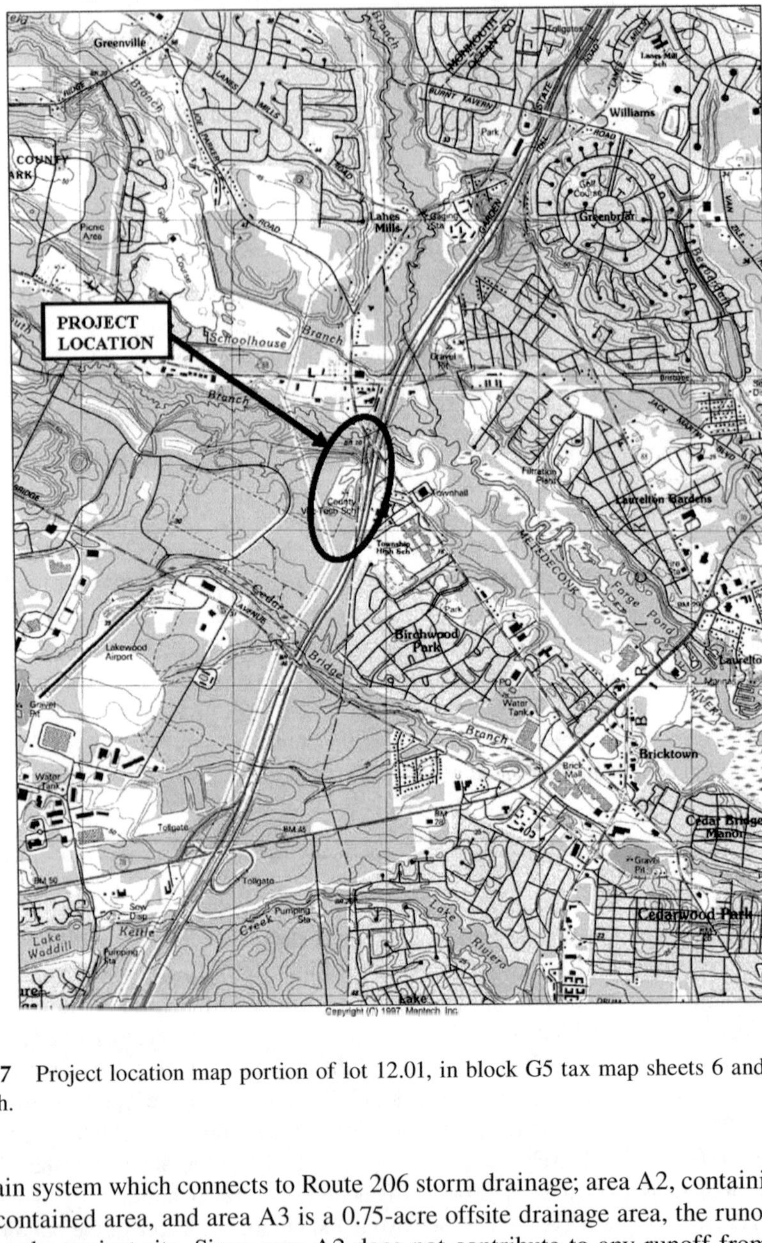

FIGURE 7.17 Project location map portion of lot 12.01, in block G5 tax map sheets 6 and 7, township of Hillsborough.

storm drain system which connects to Route 206 storm drainage; area A2, containing 0.82 acres, is a self-contained area, and area A3 is a 0.75-acre offsite drainage area, the runoff from which flows onto the project site. Since area A2 does not contribute to any runoff from the site, it is excluded from the predevelopment runoff calculations.

The post-development drainage subareas are delineated in Figure 7.19. As seen, the detention basin receives runoff from an offsite area, designated as area A3 in Figure 7.19. Runoff calculations for the project are prepared using rational and modified rational methods. Specifically, the peak rates of runoff are calculated using the former method, and the routing computations are based on the latter method. Runoff coefficients of 0.95, 0.60, 0.45, and 0.35 are selected for the impervious areas, existing gravel pavement, bare soil, and lawn/landscaped areas, respectively.

Consistent with the applicable stormwater management regulations, runoff calculations are prepared for 2-, 10-, and 100-year storms. The calculations also include 25-year storms in compliance with county drainage standards. Table 7.9 summarizes calculated runoff coefficients for areas A1 and A3 and Table 7.10 includes the peak rates of runoff and the allowable discharges from these areas. The allowable discharges are calculated by applying the required reduction

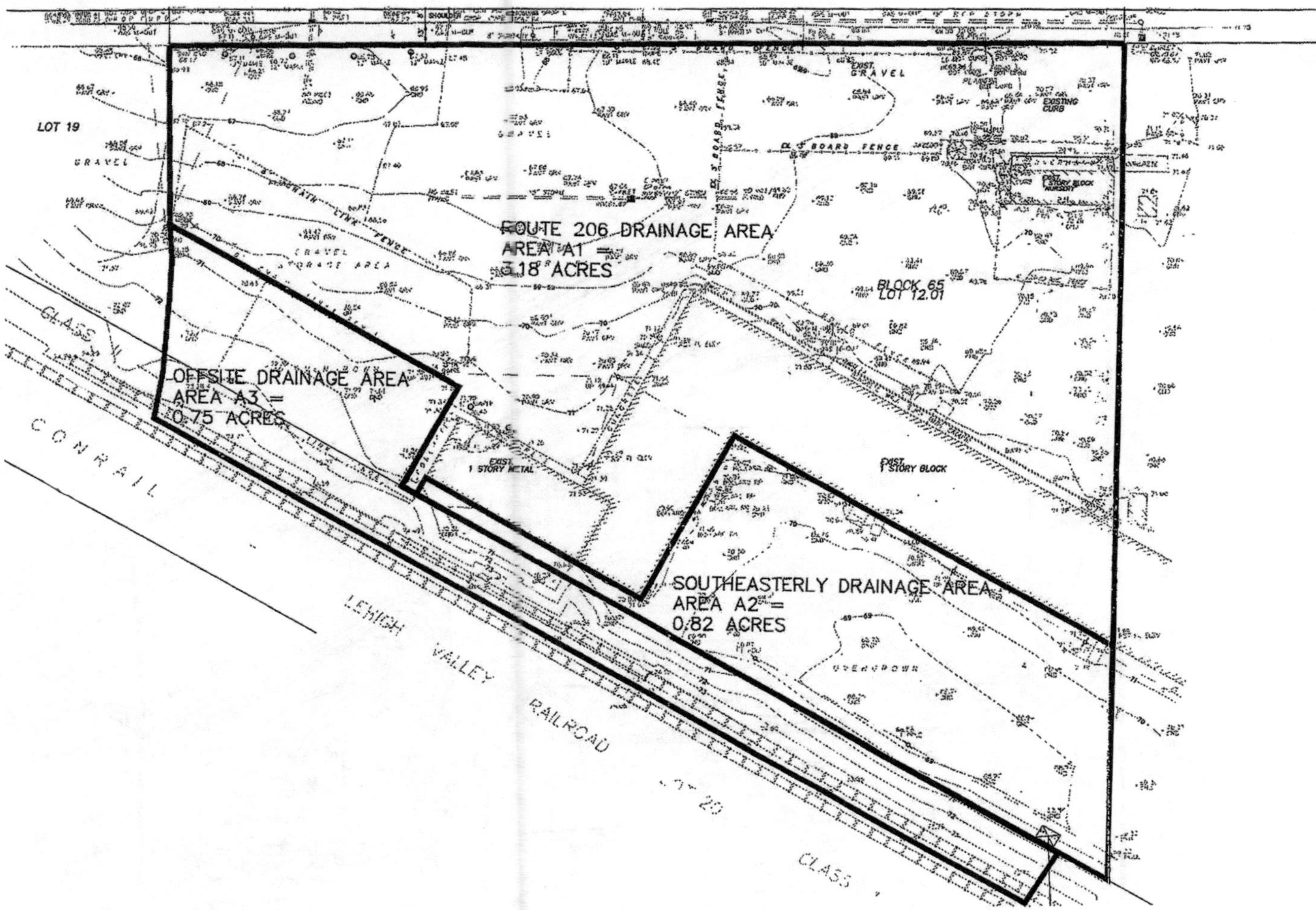

FIGURE 7.18 Pre-development site layout and drainage area map.

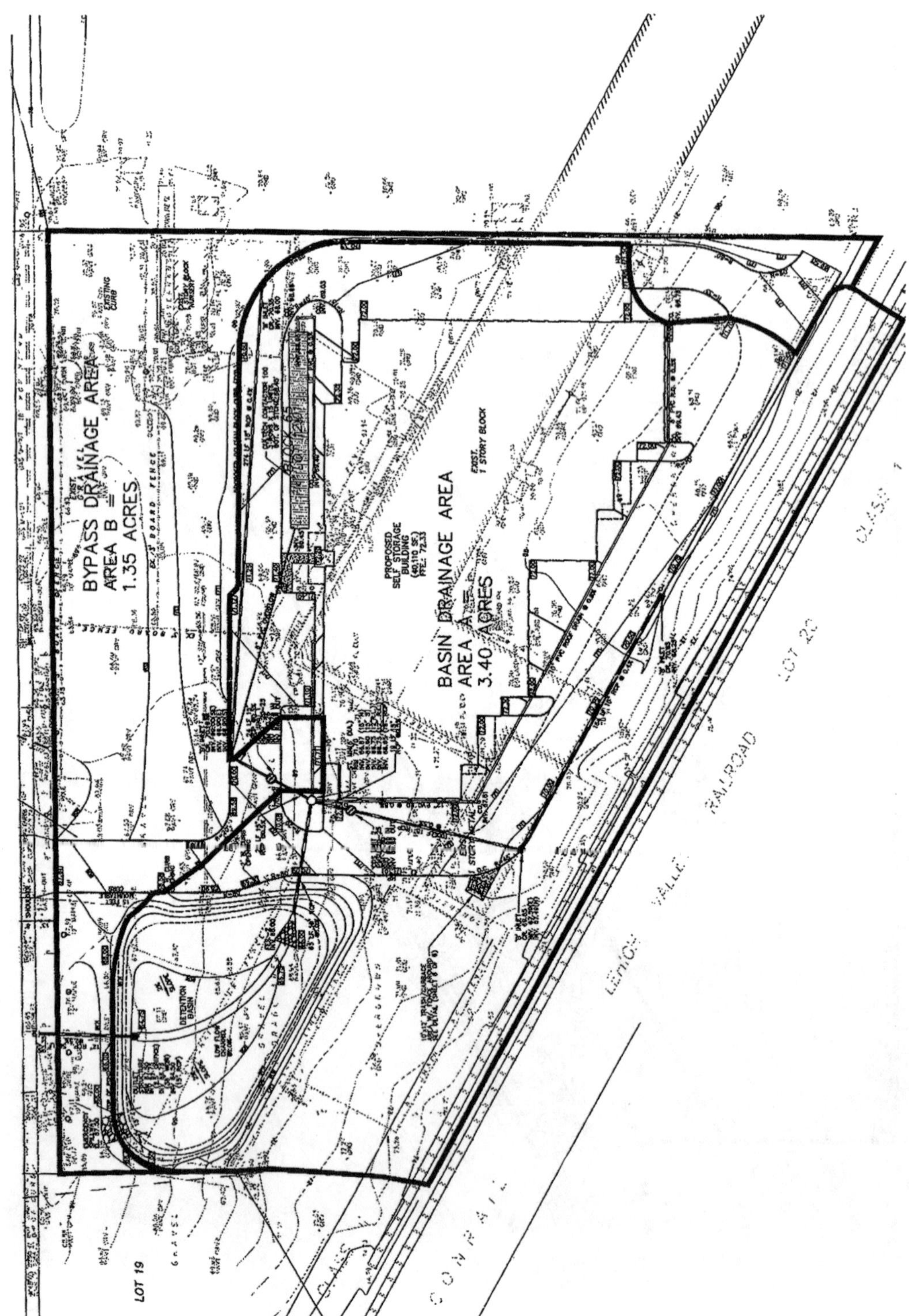

FIGURE 7.19 Post-development site layout and drainage areas.

TABLE 7.8
Predevelopment Runoff Conditions

Total lot area = 4.00 acres
Total area of disturbance (areas A1 and A2) = 4.00 acres
Drainage area A1 to Route 206 drainage system = 3.18 acres
Drainage areas A2 to southeasterly property corner = 0.82 acre
(Area A2 is self-contained on-site and is excluded from the predevelopment calculations.)
Drainage area A3 (offsite tributary to basin) = 0.75 acre

TABLE 7.9
Composite Runoff Coefficients for Areas A1 and A3

	Imp., C = 0.95		Bare Soil, C = 0.45		Gravel, C = 0.60			
	(Acre)	$A \times C$	(Acre)	$A \times C$	(Acre)	$A \times C$	Total Area	$C_w = (\Sigma A^*C)/A$
Area A1	0.73	0.69	1.10	0.50	1.35	0.81	3.18	0.63
Area A3	-	-	0.75	0.34	±	±	0.75	0.45

T_c = 15 min.
$Q_{A1} = A \times C \times I = 2.00 \times I.$
$Q_{A3} = A \times C \times I = 0.34 \times I.$

TABLE 7.10
Peak Rates of Runoff

Storm Frequency (years)	I (in./h)	Q_{A1} (cfs)	Reduction	Q_{A3} (cfs)	Q_S (cfs)
2	3.5	7.00	50%	1.19	4.69
10	4.8	9.60	25%	1.63	8.83
25	5.6	11.20	±	1.90	13.10
100	6.8	13.60	20%	2.31	13.19

Note: Allowable from site, $Q_S = \% (Q_{A1}) + Q_{A3}$.

factors to the peak rates of runoff to the onsite area (A1), noting that offsite runoff is not subject to a reduction.

Runoff calculations for the post-development conditions separate the tributary area to the detention basin, area A, from the bypass area, area B. Calculations are arranged in Table 7.11, which includes the tabulated runoff coefficients for areas A and B, the peak rates of runoff from these areas, and the allowable discharges (Q_O) from the detention basin. The latter discharges are conservatively calculated algebraically by deducting the bypass runoff from the allowable discharge from the site (shown in Table 7.10) for each storm event.

Detention Basin Design

To simplify routing computations, first, the critical storm duration, the one that requires the largest detention storage, is calculated based on the modified rational method. Calculations are organized in Table 7.12 for 2-, 10-, and 100-year storms. This table indicates a critical storm duration

TABLE 7.11
Post Development Runoff Calculations and Allowable Discharges from the Detention Basin

	Imp., $C = 0.95$		Landscape, $C = 0.35$		Offsite, $C = 0.45$			
	(Acre)	A × C	(Acre)	A × C	(Acre)	A × C	Total Area	$C_w = (\Sigma A^*C)/A$
Area A	1.93	1.83	0.72	0.32	0.75	0.34	3.40	0.73
Area B	0.19	0.18	1.16	0.41	-	--	1.35	0.43

$T_c = 15$ *min.*
$Q_A = A \times C \times I = 2.48 \times I$
$Q_B = A \times C \times I = 0.58 \times I$
Drainage area A to basin = 3.40 acres
Drainage area B bypassing basin = 1.35 acres

Storm Frequency (years)	I (in./h)	Q_A (cfs)	Q_B (cfs)	Q_O (cfs)
2	3.5	8.68	2.03	2.66
10	4.8	11.90	2.78	6.05
25	5.6	13.89	3.25	9.86
100	6.8	16.86	3.94	9.25

Note: Allowable from basin, $Q_O = Q_S{}^{a} - Q_B$.
[a] See Table 7.10.

TABLE 7.12
Critical Storm Calculations, Modified Rational Method $V = [T_b \times 60s \times Q_c] - [0.5 \times T_b \times 60s \times Q_O]$; T_b (base time) = $T_d + T_c$, $T_c = 15$ min

Q_O = allowable discharge
Refer to Table 7.11 for calculations of Q_O.

Storm Duration (min)	I (in./h)	Q_A (cfs)	Required Vol. (ft³)
2-Year			$Q_O = 2.66$
15	3.50	8.68	5418
20	3.00	7.44	6135
30	2.40	5.95	7123
45	1.80	4.46	7265
60	1.50	3.72	7407
90	1.10	2.73	6352
10-Year			$Q_O = 6.05$
15	4.80	11.90	5269
20	4.00	9.92	5552
30	3.20	7.94	6117
45	2.50	6.20	5850
60	2.00	4.96	$Q_A < Q_O$
100-Year			$Q_O = 9.25$
15	6.80	16.86	6853
20	5.60	13.89	10,103
30	4.50	11.16	11,651
45	3.60	8.93	$Q_A < Q_O$
60	3.00	7.44	$Q_A < Q_O$

of 60 minutes for a 2-year storm and 30 minutes for the 10- and 100-year storms. The calculations also indicate a required detention storage of approximately 11,650 ft^3. However, considering that the detention basin is dual purpose and, more importantly that there is room available for a larger basin, a detention basin approximately 2.4 times larger than the required volume obtained by preliminary estimates (Table 7.12) was installed. This would lessen surcharges experienced from an inlet on Route 206 to which the outlet pipe from the detention basin will be connected. To prolong the retention time in the detention basin, a 3 in. orifice is proposed for the discharge of the water quality storm. Tables 7.13 and 7.14 provide storage and discharge amounts for the detention basin, respectively.

TABLE 7.13
Elevation, Area, and Volume Relationships

Elevation	Area (ft^2)	Ave. Area	Storage (ft^3)	ΣVolume (ft^3)
64.75	0		0	0
		50		
65.0	100		13	13
		700		
65.5	1300		350	363
		3580		
66.0	5860		1790	2153
		8900		
66.5	11,940		4450	6603
		12,270		
66.75	12,600		3068	9670
		12,950		
67.0	13,300		3238	12,908
		13,980		
67.5	14,660		6990	19,898
		15,450		
68.0	16,240		7725	27,623

TABLE 7.14
Discharge–Elevation Relation

	Q (cfs)				
Elevation	3 in.	36 × 6 in.	Cover	Emergency Spillway	Total
64.75	0				0
65.0	0.06				0.06
65.5	0.12				0.12
66.0	0.17				0.17
66.5	0.20				0.20
66.75	0.21	0.00			0.21
67.0	0.23	1.13			1.35
67.5	0.25	5.11	0	0	5.36
68.0	0.28	7.22	0	15.91	23.41

3 in. inv. 64.75 ft
36 in. × 6 in opening, inv. elevation. 66.75 ft
Top of concrete outlet structure elevation 67.25 ft
Solid cover on outlet structure elevation 67.50 ft
Emergency spillway inv. 67.50 ft

Equations

Orifice flow:

$$Q_o = f\,[CA\,(2gh)^{1/2}],\ C = 0.6$$
$$\text{Clogging factor, } f = 0.66$$

Weir flow:

$$Q_w = CLH^{1.5},\ C = 3.0$$

3 in. Orifice

$$A = 0.049\ \text{ft}^2$$

$$H = \text{elevation, } 64.875\ \text{ft}$$

36 in. × 6 in. opening

$$L = 3\ \text{ft}$$

$$H = \text{elevation, } 66.75$$

Above elevation 67.25 ft use orifice equation

$$A = 1.50\ \text{ft}^2$$
$$H = \text{elevation, } 67.00$$

Emergency spillway, 15 ft. weir:

$$L = 15\ \text{ft}$$
$$H = \text{elevation, } 67.50$$

Velocity over spillway:

$$V = Q/A = 15.91/5.75 = 2.8\ \text{ft/s}$$

Permissible velocity for 6 in. riprap is 6.0 ft/s (NJ, 2017 Standards).

Table 7.15 summarizes routing computations or the 2-, 10-, and 100-year storms along with allowable discharges from the detention basin (refer to Table 7.11). As shown, the proposed extended detention basin reduces the peak flows far in excess of the regulations. Table 7.16 presents a comparison of the pre- and post- development discharges from the project site.

TABLE 7.15
Routing Computations Summary

Storm Frequency (years)	Critical Storm Duration (min.)	Q_i, Inflow	Q_o, Outflow	Allowable from Detention
2	60	3.72	1.01	2.66
10	30	7.94	1.39	6.05
100	30	11.16	3.95	9.25

TABLE 7.16
Predevelopment, Allowable, and Post-Development Peak Rates of Runoff for the Site

Storm Frequency (years)	Q_E[a] (cfs)	Q_A[b] (cfs)	Q_P (cfs)
2	8.19	4.69	3.04
10	11.23	8.83	4.17
100	15.91	13.19	7.89

[a] Predevelopment flow includes offsite area.
[b] See Table 7.10 for allowable peak flow from site, Q_A; $Q_P = Q_o - Q_B$.

Water Quality Provision

The NJDEP stormwater management regulations call for removing 80% TSS from the increased impervious areas and 50% for the pavements to be reconstructed. The required TSS removal rates are listed in Table 7.17. This table reflects a required TSS removal rate equal to 25% for the entire site.

To determine the TSS removal provided by the extended detention basin, the New Jersey water quality storm, namely 1.25 in. of runoff in 2 hours, is routed through the detention basin. It should be noted that the water quality storm has a nonuniform distribution; however, according to the author's experience, routing computations can be simplified based on the average rainfall

TABLE 7.17
Required TSS Removal Rate for the Project Area

Ground Cover	Area (ac.)	Percent TSS Required Removal Rate	A × Percent TSS
Ex. impervious (excludes ex. roof)	0.17	50%	0.09
Proposed roof	0.92	0%	0.00
Increase impervious	1.03	80%	0.82
Grass/landscape (to basin)[a]	0.72	0%	0.00
Offsite area	0.75	0%	0.00
	$\Sigma A = 3.59$		$\Sigma A \times \%$ TSS = 0.91

Required TSS removal rate $= \dfrac{0.91}{3.59} = 25\%$

Provided TSS removal rate for the project area

CDS units = 50%

Detention = 42% (see Figure 7.20)

Composite = 71%

[a] Grass area bypassing detention basin (1.16 acres) is excluded from the calculations. This area will not be treated for water quality and does not require any treatment for water quality.

intensity (0.625 in./h) without a sacrifice in accuracy. The routing computation, thus performed, shows a retention time (defined as the time from the peak discharge from the basin until 90% of the maximum water quality volume stored in the detention basin is drained) of 13 hours.

Figure 7.20 indicates that the proposed detention basin provides 42% TSS removal for the runoff from the pavement area tributary thereof which is 1.01 acres, excluding the roof. However, since the new impervious area (excluding the roof), which is 1.03 acres requires 80% TSS removal, the detention basin alone is insufficient. To address the required TSS removal, two manufactured water treatment devices are placed en route of drainage lines before the detention basin. The locations of these devices, which are CDS (continuous deflective system) units, are shown in Figure 7.19.

The composite TSS removal rate is provided by the detention basin and the water quality units is calculated considering that the area tributary to the CDS units, which includes the proposed four inlets, receives treatment by not only the CDS units but also the detention basin. However, the remainder of the paved area, which is routed directly through the detention basin, is treated by the basin alone. Table 7.18 summarizes the calculations of the TSS removal rate provided by the system. In this table, the composite TSS removal rate of the water quality storm that passes through the CDS units is calculated as follows:

$$TSS = TSS_1 + TSS_2 - \frac{TSS_1 \times TSS_2}{100}$$

$$TSS = 50 + 42 - \frac{50 \times 42}{100} = 71\%$$

where TSS_1 and TSS_2 are the TSS removal rates of each CDS units and the detention basin, respectively. Flow calculations for the CDS units are prepared based on the drainage areas of the proposed inlets. Based on an inlet drainage area map (not included herein), CDS units 1 and 2 receive runoff from 0.52 and 1.04 acres, respectively. Tables 7.19 and 7.20 present flow calculations for the CDS units. Based on the calculated flows. CDS model number PMSU20–25 (is renamed CDS

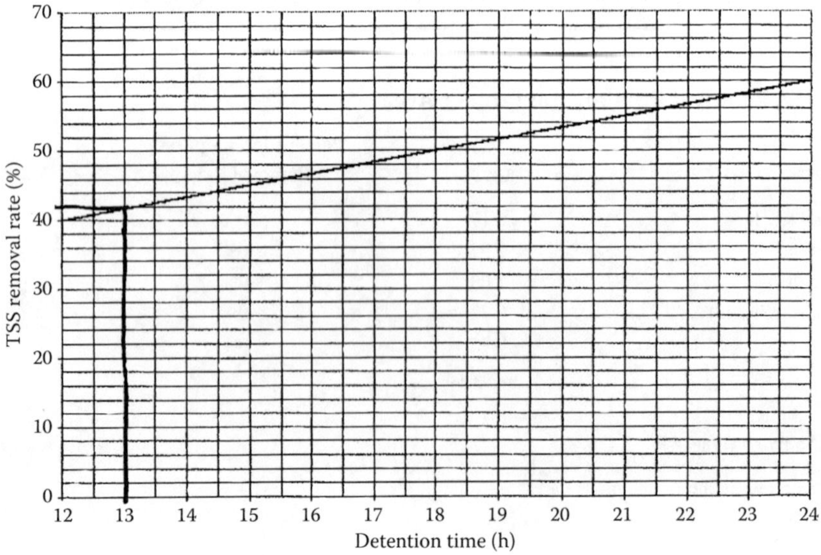

FIGURE 7.20 TSS removal rate versus detention time.

Source: Stormwater BMP Manual, Chapter 9.4: a Standard for Extended Detention Basins, February 2014 (pp. 9.4–7).

TABLE 7.18
Provided TSS Removal Rate for Paved Areas

Ground Cover	Area (ac.)	Percent TSS Required Removal Rate	A × Percent TSS
To CDS units and basin	1.56	71%	1.108
To basin only	0.37[a]	42%	0.155
Bypass	0.10	0%	0.000
	ΣA = 3.59		ΣA × % TSS = 1.263

Note: Provided TSS removal rate = 1.263/3.59 = 0.35%.
[a] 1.93 acres tributary area to the basin; 1.56 acres treated by CDS units.

TABLE 7.19
Composite Runoff Coefficients for CDS Units Sizing

	Imp., C = 0.95		Pervious, C = 0.35			
	(ac)	A × C	(ac)	A × C	Total Area	Cw = (∑× AC)/A
CDS unit 1	0.41	0.39	0.11	0.05	0.52	0.84
CDS unit 2	0.38	0.36	0.66	0.30	1.04	0.63

1.25 in. in 2 h
T_C = 15 min
I = 2.6 in./h

TABLE 7.20
Flow Calculations and CDS Sizing

CDS Unit	Q = CIA	Model No.	Treatment Rate
1	1.14	PMSU 20–25	1.7
2	1.70	PMSU 20–25	1.7

2025-W), capable of treating 1.7 cfs, is selected each for unit 1 and unit 2. (This device is designated as CDS-6 in the NJDEP verification letter and has a treatment capacity of 1.6 cfs).

Groundwater Recharge Calculations

The site is located where the NJDEP groundwater recharge requirement is to be met. The criteria for meeting this requirement were presented in a previous chapter. In this project, the annual groundwater recharge criterion is selected for the design of a groundwater recharge system. The groundwater recharge calculations are prepared using the New Jersey Geological Survey (NJGS) spreadsheet GSR-32 (1993). This spreadsheet calculates both the annual groundwater deficit due to a project and the groundwater recharge provided by a selected infiltration system.

The spreadsheet calculations for the groundwater recharge deficit are presented on sheet 1 of 2 of Figure 7.21. The calculations on Sheet 1 of this Figure, which are based on the 4.0 acre area of

Project Name	Description	Analysis Date	BMP or LID Type
RAIA Properties Corp	Block 65, Lot 12.01	04/02/07	CULTEC Contractor 100HD (3 Rows X 15 Units)

Recharge BMP input parameters

Parameter	Symbol	Value	Unit
BMP Area	ABMP	1265.0	sq.ft
BMP effective depth, this is the design variable	dBMP	10.9	in
Upper level of the BMP surface (negative if above ground)	dBMPu	9.0	in
Depth of lower surface of BMP, must be >=dBMPu	dEXC	27.5	in
Post-development land segment location of BMP, Input Zero if location is distributed or undetermined	SegBMP	2	unitless

Root zone water capacity calculated patameters

Parameter	Symbol	Value	Unit
Empty portion of RWC under post-D natural recharge	ERWC	1.02	in
ERWC modified to consider dEXC	EDRWC	0.40	in
Empty portion of RWC under infilt BMP	RERWC	0.31	in

Recharge design parameters

Parameter	Symbol	Value	Unit
Inches of Runoff to capture	Qdesign	1.22	in
Inches of Rainfull to capture	Pdesign	1.44	in
Recharge provided avg. over imp. area		29.3	in
Runoff captured avg. over imp. area		31.9	in

BMP calculated size parameters

ABMP/Aimp	Aratio	0.11	unitless
BMP volume	VBMP	1,149	cu.ft

Calculation check messages

Volume balance -> Solve problem to satisfy annual recharge

dBMP check -> OK

dEXC check -> OK

BMP location -> OK

Parameters from annual recharge worksheet

Parameter	Symbol	Value	Unit
Post-D deficit recharge (or desired recharge volume)	Vdef	29,024	cu.ft
Post-D impervious area (or target impervious area)	Aimp	12,000	sq.ft
Root zone water capacity	RWC	4.07	in
RWC modified to consider dEXC	DRWC	1.59	in
Climatic factor	C-factor	1.50	no units
Average annual P	Pavg	45.7	in
Recharge requirement over imp. area	dr	3.8	in

System performance calculated parameters

Parameter	Symbol	Value	Unit
Annual BMP recharge volume		29,294	
Avg BMP recharge efficiency		91.8%	Represents % infiltration recharged
% Rainfall became runoff		77.9%	
% Runoff infiltrated		89.6%	
% Runoff recharged		10.7%	
% Rainfall recharged		8.3%	

Other notes

Pdesign is accurate only after BMP dimension are updated to make rech volume = deficit volume. The portion of BMP infiltration prior to filling and the area occupied by BMP are ignored in these calculations. Results are sensitive to dBMP, make sure dBMP selected is small enough for BMP to empty in less than 3 days. For land segment location of BMP if you select "Impervious areas" RWC will be minimal but not zero as determined by the soil type and a shallow root zone for this land cover allowing consideration of lateral flow and other losses.

How to solve for different recharge volumes: By default the spreadsheet assigns the values of total deficit recharge volume "Vdef" and total proposed impervious area "Aimp" from the "Annual recharge" sheet to "Vdef" and "Aimp" on this page. This allows solution for a single BMP to handle the entire recharge requirement assuming the runoff from entire impervious area is available to the BMP. To solve for a smaller BMP or a LID-IMP to recharge only part of the recharge requirement, set Vdef to your target value and Aimp to impervious area directly connected to your infiltration facility and then solve for ABMP or dBMP. To go back to the default configuration click the "Default Vdef and Aimp"button.

FIGURE 7.21 Annual groundwater recharge analysis based on GSR-32 (Sheet 1).

New Jersey Groundwater Recharge Spreadsheet Version 2.0 November 2003

Annual groundwater recharge analysis (based on GSR-32)

Select Township↓	Average Annual P (in)	Climatic factor
Somerset Co., Hillsborough TWP	45.7	1.50

Land Segment	Area (acres)	TR-55 Land cover	Soil	Annual Recharge (in)	Annual Recharge (cu.ft)
1	0.73	Impervious areas	Penn	0.0	–
2	1.35	Gravel, dirt	Penn	5.6	27,198
3	1.92	Open space	Penn	12.6	87,678
4					
5					
6					
7					
8					
9					
10					
11					
12					
13					
14					
15					
Total =	4.0			Total Annual Recharge (in)	Total Annual Recharge (cu.ft)
				7.9	114,875

Project name:	RAIA Properties Corp
Description:	Block 65, Lot 12.01
Analysis state:	04/02/07

Land Segment	Area (acres)	TR-55 land cover	Soil	Annual Recharge (in)	Annual Recharge (cu.ft)
1	2.12	Impervious areas	Penn	0.0	–
2	1.88	Open space	Penn	12.6	85,851
3					
4					
5					
6					
7					
8					
9					
10					
11					
12					
13					
14					
15					
Total =	4.0			Total Annual Recharge (in)	Total Annual Recharge (cu.ft)
				5.9	85,851
				Total Impervious Area (sq.ft)	92,347

Annual recharge requirements calculations		
% of Pre-developed annual recharge to preserve	100%	
Post-development annual recharge deficit =	29,034	(Cubic feet)

Recharge efficiency parameters calculations (area averages)				
RWC = 391	(in)	DRWC =	0.57	(in)
ERWC = 0.98	(in)	EDRWC =	0.14	(in)

Procedure to fill the pre-developemtn and post-development conditions tables

For each land segment, first enter the erea, the select TR-55 Land Cover, then select Soil. Start from the top of the table and proceed downward. Don't leave blank rows (with A-C) in between your segment entries. Rows with A=D will not be displayed or used in calculations. For impervious areas outside of standard lots select "Impervious Areas" as the Land Cover. Soil type for impervious areas are only required if an infiltration facility will be built within these areas.

FIGURE 7.21 (CONTINUED) Annual groundwater recharge analysis based on GSR–32. (Sheet 2)

disturbance, indicate a deficit of 29,034 ft^3 (822 m^3). To offset this deficit, three rows of 15 units each CulTec Contractor 100HD in a 115 ft by 11 ft (35 × 3.35 m) stone trench are provisioned.

The recharge system, which is shown in Figure 7.19, receives roof runoff from a portion of the self-storage building. Sheet 2 in Figure 7.21 shows that by draining 12,000 ft^2 (1115 m^2) of the roof area into the CulTec chambers, 29,294 ft^3 (829.5 m^3) of annual groundwater recharge can be attained. The parameters in this table are

$$ABMP = recharge\ area = 115 \times 11 = 1265\, ft^2$$

$$V = 45x14 + 0.40\left[\left(\frac{18.5}{12}\right)x11 \times 115.0 ft - 45x14\right] = 1158 ft^3$$

Adjusting V for the volume taken by chambers walls VBNP= 1149 ft_3

where:

18.5 in. is the effective depth of recharge system.

$dBMP_u$ = depth of top of infiltration system = 9.0 in.

dEXC = depth of infiltration system bottom = 9 + 18.5 = 27.5 in.

Other parameters in Figure 7.21 Sheet 2 are filled automatically by the spreadsheet when inputting the name of a municipality in New Jersey.

To determine that the bottom of the infiltration chambers and the extended detention basin lie above the ambient water table two test pits were conducted at the site. Soil samples were also sent to a laboratory for gradation tests and soil permeability ratings. Groundwater was encountered at 5 ft, 8 in. (1.7 m) below the existing grade, and mottling (which is an indicator of seasonal high-water table) was found at 3 to 5 ft (1–1.5 m) depth. The soil gradation indicated a soil permeability rating of K3 in one pit and K2 in another. These ratings reflect a permeability range of 2–6 in./h (50–150 mm/h) and 0.6–2 in./h (15–50 mm/h), respectively. The tests verify that the designed bottom of the detention basin and the infiltration chambers lie above the water table. As such, the chambers will function satisfactorily.

CASE STUDY 7.2

Case Description

This case relates to a residential project in the town of North Greenbush in Rensselaer County in New York. The Westview Estates project includes the construction of 38 single-family homes in a 44.5-acre (18.0 ha) parcel of property along County Route 74, also known as Winter Street Extension. The project disturbs approximately 23.7 acres (9.59 ha) while the remainder of the property remains intact and serves as conservation land. Figure 7.22 depicts the predevelopment layout of the vicinity of the development area. This plan, which is prepared based on 2 ft (o.6 m) contour lines of an aerial map of the site, shows that a single building and its driveway, covering 0.5 acres (2024 m^2), are the sole impervious areas on the site. Figure 7.23 shows the post-development layout and two extended detention-water quality ponds. Design calculations for these ponds, which are provisioned to address the stormwater management requirements for the project, are discussed herein.

Description of Stormwater Management Provisions

Figure 7.22 shows the limits of the area to be disturbed by the project, which, as indicated, measures 23.7 acres. Of this, approximately 18.9 acres (7.65 ha) are tributary to northerly properties, and an existing drainage ditch along Winter Street Extension and 4.8 acres (1.94 ha) flow in

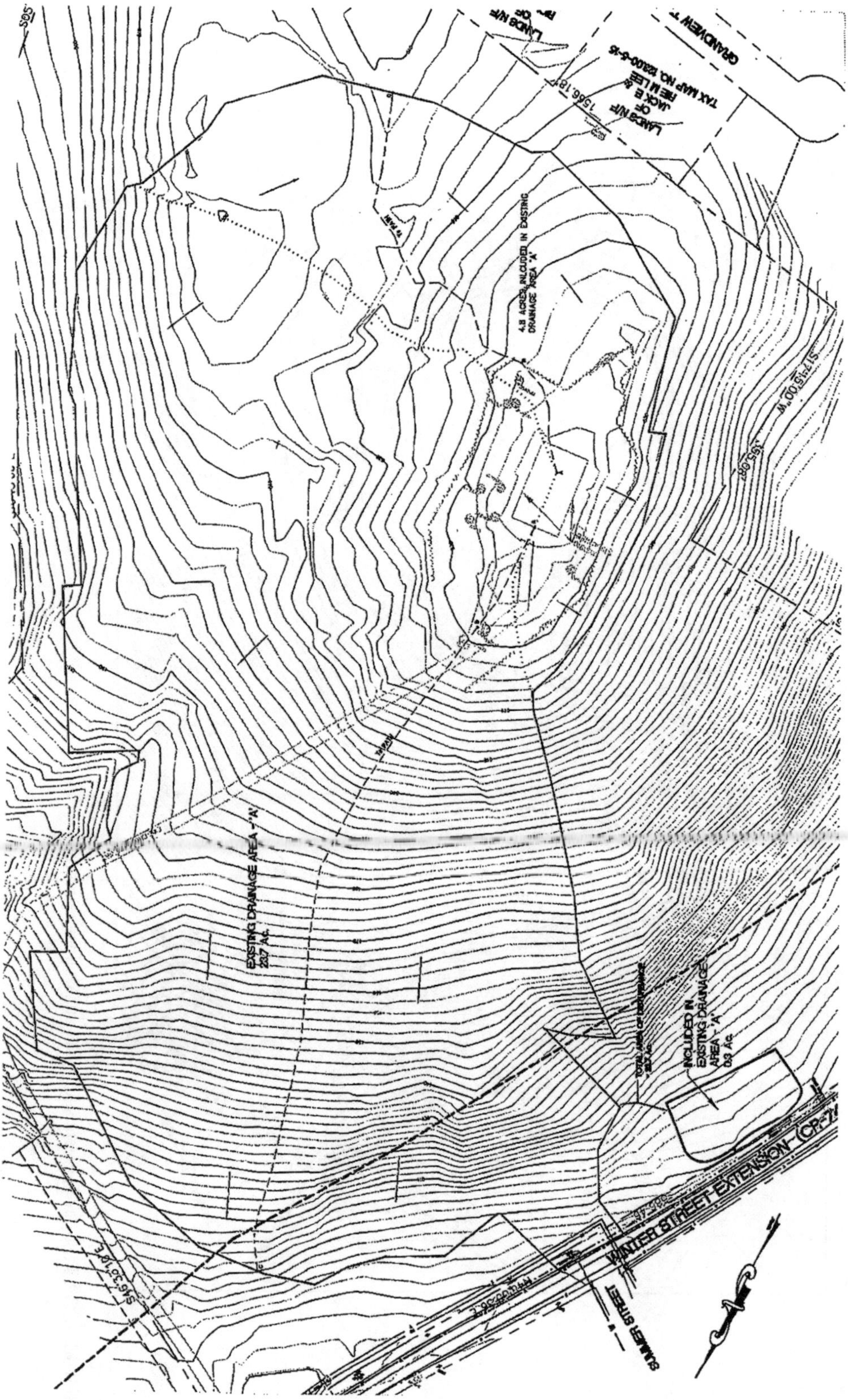

FIGURE 7.22 Predevelopment site layout and drainage.

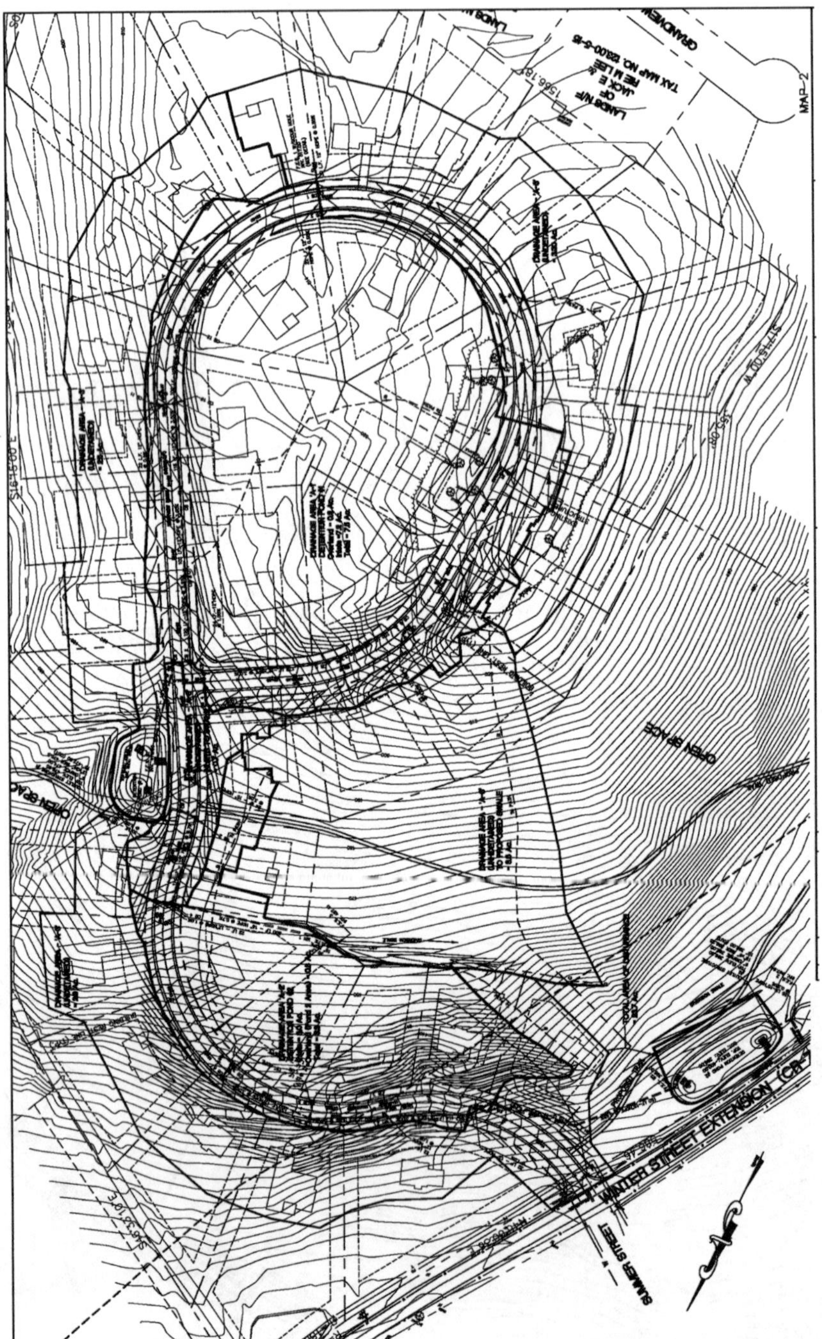

FIGURE 7.23 Post-development site layout and drainage subareas.

southerly and easterly directions to neighboring properties. The remainder of the site area discharges to a northerly property.

Delineated on Figure 7.23 are the drainage areas tributary to the detention ponds and subareas bypassing them. To reduce the runoff to the northerly property, where the soil is reported to be soggy, the discharge from detention pond 1 is redirected by a proposed swale to the southerly side of the site, where it will sheet flow to the existing drainage ditch. A close review of Figures 7.22 and 7.23 indicates that the proposed detention pond 1 collects runoff from a large portion of the area that would otherwise flow to the northerly property, and, as such, it reduces the runoff to that property. The plans also indicate that the project will reduce the tributary area to the easterly and southerly properties from 4.8 acres (1.94 ha) to 3.3 acres (1.34 ha) and thereby reduce the runoff to these properties as well.

Runoff Calculations

Runoff calculations are prepared using the SCS TR-55 method. According to the Rensselaer County soil survey maps, the site soil is mostly hydrologic group C. The predevelopment soil curve number and time of concentration are calculated to be 70 and 0.4 hours, respectively. Consistent with the New York State stormwater management regulations (2015), the calculations are performed for the storms of 1-, 10-, and 100-year frequency. For ease of comparison with the post-development conditions, the entire 23.7 acres of the disturbance area, which is designated as area A in Figure 7.22, is lumped as one area in performing the predevelopment runoff calculations. Runoff calculations for the post-development conditions are performed for the tributary areas to the detention ponds (Pond 1 and Pond 2), designed as areas A-1 and A-2 in Figure 7.23. Also included in the calculations are the un-detained areas, which comprise 12.8 acres (5.18 ha) in total and are designated as area A-3 on this plan. For brevity, runoff calculations for pre- and post-development areas A-1, A-2, and A-3 are not included herein. However, the overall predevelopment peak runoffs from the entire site are compared with those of post-development conditions are summarized in Table 7.22.

Detention Ponds Design

Figure 7.24 shows schematic details of the proposed detention ponds. This type of pond is referred to as a pocket pond P-5 in the NY State Stormwater Management Design Manual (2015). The manual stipulates that each pond consist of a forebay and a permanent pool. The forebay has a minimum capacity equal to 10% of the water quality volume, WQ_v, and the permanent pool has a capacity of at least equal to 50% WQ. For brevity calculations for the storage volume of forebays and permanent pools of the detention ponds, which are calculated from the elevation-area relation for each, are not included herein.

The water quality volume in New York State is the 90% rule, defined in a previous chapter. Calculations for the water quality volumes in ponds 1 and 2 are presented next.

WQ_V CALCULATIONS

Detention Pond 1:

$$WQ_v = PR_v \frac{A}{12}$$

$$A = 7.5 \text{ acres}$$

$$P = 0.95\, in. \left(see\, Figure\, 5.10\; in\, Chapter\, 5 \right)$$

$$I = \text{percent pavement} = 26\%$$

$$R_v = 0.05 + 0.009(I) = 0.284$$

$$WQ_v = 0.95 \times 0.284 \times \frac{7.5}{12} = 0.284\ ac - ft = 7357\ ft^3\left(208.3\ m^3\right)$$

Detention pond 2:

$$A = 3.3 \text{ acres}$$

$$P = 0.95$$

$$I = 0.24$$

$$R_v = 0.05 + 0.009 \times 24 = 0.266$$

$$WQ_v = 0.07\ ac - ft = 3050\ ft^3\left(86.4\ m^3\right)$$

Detention pond 1 has been designed with a forebay and permanent pool, each extending from elevation 469.5 ft to 473.0 ft with a storage volume as follows*:

$$Forebay\, 750\ ft^3 > 10\%\, WQ_v$$

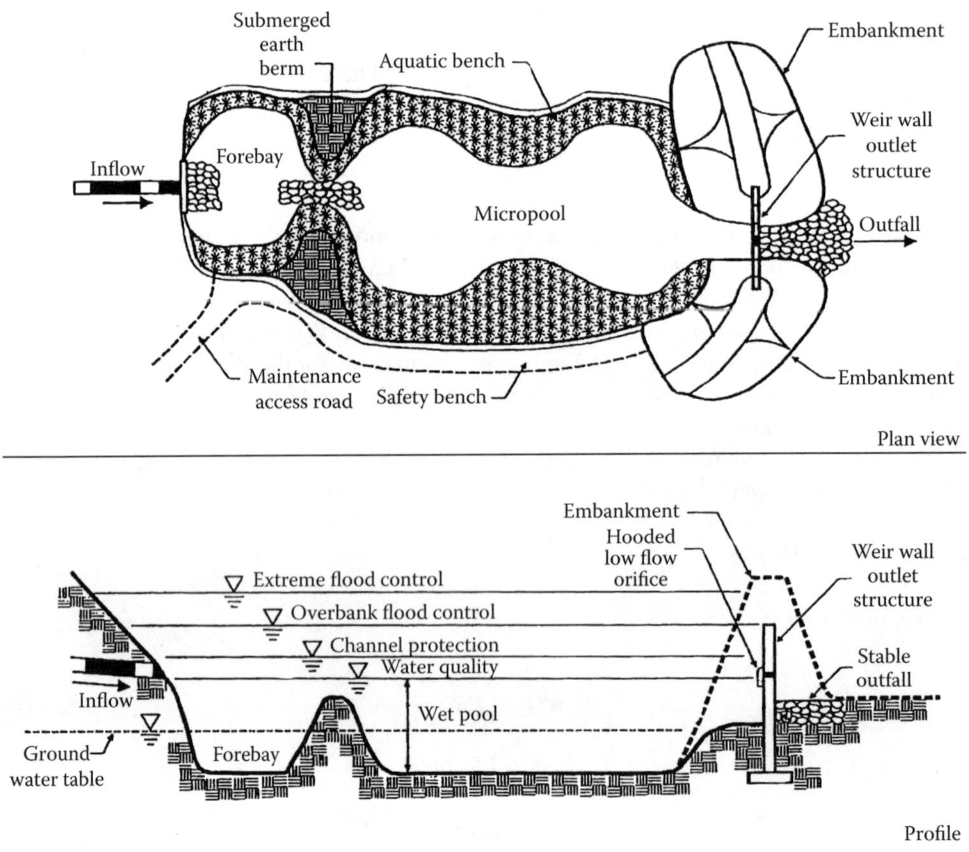

FIGURE 7.24 Pocket- pond (P-5).

$$Permanent\ pool\ 3700\ ft^3 > 50\%\ WQ_v$$

Likewise, the forebay and permanent pool of detention pond 2 are designed as follows:

$$Forebay\ elev.394\ ft\ to\ 397.25\ ft;,\ storage = 310\ ft^3$$

$$Permanent\ pool\ elev.394\ ft\ to\ 397.25\ ft,\ storage = 1550\ ft^3$$

Cp_v, Q_p, and Q_f Design Calculations

Detention ponds 1 and 2 incorporate active storage above their permanent pools, and the lowest opening in the outlet structure of each pond is set just above its permanent pool water level. Figure 7.25 depicts the openings and elevations of outlet structures of detention ponds 1 and 2.

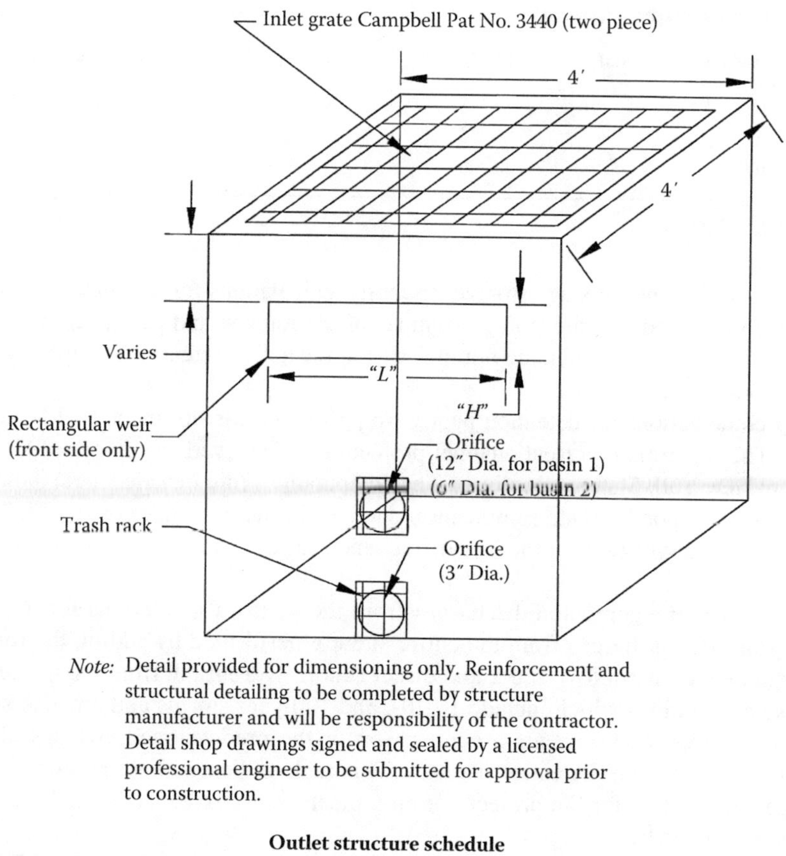

Note: Detail provided for dimensioning only. Reinforcement and structural detailing to be completed by structure manufacturer and will be responsibility of the contractor. Detail shop drawings signed and sealed by a licensed professional engineer to be submitted for approval prior to construction.

Outlet structure schedule

Pond	Bottom of pond	3 in. orifice invert	6 in. orifice invert	15 in. orifice invert	Weir			Top of structure[a]
					L	H	Invert	
1	473.00	473.00	–	475.00	18 in.	12 in.	476.50	478.10
2	397.25	397.25	398.00	–	12 in.	6 in.	399.00	400.00

FIGURE 7.25 Outlet structure detail.

TABLE 7.21
Composite Peak Rates of Runoff Post-Development Conditions

Storm Freq. (years)	Pond 1 Q (cfs)	Pond 2 Q (cfs)	Area A-3 Q (cfs)	Site Total Q^a (cfs)
1	0.35	0.01	6.0	6.0
10	5.75	1.0	\24.0	29.4
100	14.46	3.53	41.0	56.7

[a] Calculated by adding hydrographs.

TABLE 7.22
Existing and Proposed Peak Rates of Runoff Entire Area of Disturbance (Area A)

Storm Freq. (years)	EX. Q (cfs)	PR. Q (cfs)	Reduction (%)
1	8	6	25%
10	34	29.4	14%
100	60	56.7	5%

For brevity, calculations for stage-storage-discharge calculations for the ponds are not included herein. It is to be noted that the storage volumes of the forebay and permanent pool, which are non-effective in flow attenuation, are not included in the active detention storage tabulations and routing computations.

Routing computations for detention ponds are performed for storms of 1-, 10-, and 100-year frequency. These storms represent channel protection (CP_v), overbank flood (Q_p), and extreme flood (Q_f) in New York State, respectively. The computations (not included here) indicate that the proposed detention ponds create significant reductions in the peak rates of runoff. The attenuation effect is most significant for the 1-year frequency storm, which occurs more frequently than others.

Table 7.21 lists the computed discharges from the ponds. Calculations for the post-development composite discharges from the entire site are performed by adding the routed hydrographs from detention ponds 1 and 2 and direct runoff hydrograph from the un-detained area A–3. These calculations which include 1-, 10-, and 100-year storms and are also summarized in Table 7.21. Table 7.22 presents a comparison of the predeveloped and post-development peak rates of runoff from the entire site area. This table indicates that the proposed stormwater management system for the project not only meets, but also exceeds the NYS stormwater management standards.

7.4 UNDERGROUND DETENTION BASINS

Underground detention basins are commonly installed under pavements such as parking lots and driveways. Where open space is sparse, underground detention systems are far more practical than above ground detention basins. In highly developed areas where no open space is available, underground detention systems form the most feasible means of attenuating the peak rates of runoff. However, it is to be noted that many jurisdictional agencies do not accept underground detention basins for water quality.

Solid and perforated pipes, chambers, vaults, and modular structures are employed in constructing underground detention basins. Solid pipes and vaults serve only as storage facilities. However, perforated pipes and chambers in stone trenches also provide storage in stone voids and, in addition, dissipate runoff through infiltration. As such, these systems are more cost-effective than solid pipes. To avoid contaminating groundwater, the bottom of the stone trench should be at least 2 ft (0.6 m) above the water table. The soil should have a minimum permeability of 1 in./h (25 mm/h) to allow the system to fully drain within 3 but preferably 2 days.

7.4.1 Solid and Perforated Pipes

Solid reinforced concrete, high-density polyethylene (HDPE), and corrugated metal pipes are widely employed. Among these, corrugated metal pipes (CMPs) are far more vulnerable to deformation and deterioration than other types of pipes. The author has seen many CMPs that were rusted, deformed, and/or collapsed in less than 20 years of service. Thus, the use of this type of pipe in any underground detention system is discouraged.

Using solid pipes in parallel, a minimum spacing should be kept between pipes to allow rock placement and adequate compaction. For Advanced Drainage System (ADS) N-12 (dual wall pipes), the minimum spacing is 12 in. (30 cm) or one-half of the nominal pipe size, whichever is greater. The same spacing should be allowed for concrete pipes American Concrete Pipe Association (ACPA, 2007). Perforated RCP and HDPE pipes are installed in the stone trench with a minimum 4 in. (10 cm) bedding for 12 to 24 in. (300–600 mm) pipes and 6 in. (15 cm) for 30 to 60 in. (750–1500 mm) pipes. Also, a 6 in. stone cover is commonly placed over the pipes.

Table 7.23 presents ADS's recommended spacing for perforated HDPE pipes, which come in 12 to 60 in. (300–1500 mm) in diameter. This table also lists the storage volume in the pipe, the void volume in the stone trench, and the total retention storage foe each pipe size. The table serves as an aid in design and is applicable for cases where a large number of pipes are laid in parallel in a wide stone trench. Because of having thicker walls, perforated concrete pipes require a slightly larger stone trench than HDPE pipes for a given pipe size. Accurate calculations of storage volume for any perforated pipe in the stone trench can be performed by adding the inside volume of pipes to the void volume in the stone trench. The latter volume is calculated by deducting the outer volume of pipe from the trench volume and multiplying the difference by the stone porosity, the ratio of the void to the overall stone volume. A 40% ratio is used in common practice.

Because of ease of construction and lower initial cost, both solid and perforated HDPE pipes are more widely used than solid or perforated reinforced concrete pipes for underground basins. It is to be noted, however, that the structural integrity of the HDPE pipes is more sensitive to proper installation, including backfill material and compaction than that of reinforced concrete pipes. It is advisable to perform a long-term cost analysis before selecting a make of pipe.

7.4.2 Chambers

In recent decades a variety of chambers, mostly made of plastic material, have been introduced to the market. Plastic chambers, which are arch shaped, can be stacked up to save storage and transportation cost and are rapidly becoming more popular than perforated pipes. StormTech chambers, Storm Chambers and Trinton which are among the widely used retention–infiltration systems in the US are described herein.

TABLE 7.23
Storage Capacities of N-12®, N-12® ST, and N-12® WT Pipes

Nominal Inside Diameter, in. (mm)	Average Outside Diameter, in. (mm)	X Spacing, in. (mm)	C Spacing, in. (mm)	Pipe Volume ft³/ft (m³/m)	Total Volume ft³/ft (m³/m)
12 (300)	14.5 (368)	8 (210)	25.4 (650)	0.81 (0.07)	1.65 (0.15)
15 (375)	18 (457)	8 (210)	28.9 (750)	1.2 (0.11)	2.3 (0.21)
18 (450)	21 (533)	9 (230)	35.3 (900)	1.8 (0.16)	3.2 (0.29)
24 (600)	28 (711)	10 (260)	41.4 (1050)	3.1 (0.29)	5.1 (0.47)
30 (750)	36 (914)	18 (460)	53.1 (1350)	4.9 (0.46)	8.0 (0.74)
36 (900)	42 (1067)	18 (460)	63.0 (1600)	7.1 (0.66)	11.3 (1.05)
42 (1050)	48 (1219)	18 (460)	72 (1830)	9.2 (0.87)	15.0 (1.40)
48 (1200)	54 (1372)	18 (460)	78.5 (2000)	12.4 (1.15)	19.1 (1.77)
60 (1500)	67 (1702)	18 (460)	90 (2290)	19.3 (1.79)	27.8 (2.57)

Note: See figure below for typical cross section used in volume calculations. Bedding depth assumed 4 in. for 12–24 in. pipe and 6 in. for 30–60 in. pipe.
Based on A-profile pipe.
Actual ID values used in calculation.Stone porosity assumed 40%.
Stone height above crown of pipe is not included in void volume calculations.
Calculation is based on the average OD of the pipe.

a. StormTech chambers. In 2009, StormTech was acquired by Advanced Drainage Systems (ADS), which is the largest manufacturer of HDPE pipes and plastic chambers in the U.S.A. StormTech chambers, which are made of propylene (PP), were introduced in two different models, SC-310 and SC-740, decades ago. Now ADS makes seven chambers SC-160 LP, SC-310, MC-780, SC-740, MC-3500, MC-4500, and MC–7200. SC-160 LP, which is the smallest chamber, is the strongest and is employed for retention–infiltration in shallow applications. This chamber is 85.4 in. long by 25 in. wide by 12 in. high (207 × 63.5 × 30.5 cm), weighs 24.0 lbs (10.9 kg), and has 6.85 ft³ (190 L) of bare storage. With 4 in. (10 cm) stone below and 6 in. (15 cm) stone above, this chamber provides 15.0 ft³ (0.42 m³) of installed storage. Figure 7.26 depicts a SC-740 chamber with end cap, and Table 7.24 presents the dimensions, bare and installed storage volumes, and weights of SC-310, DC-780, SC-740, MC-3500, MC-4500, and MC-7200 chambers.
b. ChamberMaxx. In 2008, Cultech Inc. introduced a large open bottom chamber, called ChamberMaxx. which was made of polypropylene resin. This chamber which is now manufactured by Contech of West Chester, OH is approximately 91 in (2.31 m) long, 51 in. (1.3 m) wide and 30 in. (0.76 m) high. A mid chamber with an installed length of 85.4 in. (2.17 m) provides 47 ft³ (1.3 m³) of bare storage volume. For ease of installation Cultec Inc. now makes Cultec chambers to be discussed next.
c. Cultec chambers. Cultec Inc. now located in Brookfield, CT, makes 902HD, chamber and, 330XLHD, 280HD, 180HD and 150XLHD Recharges. Chamber 902HD dimensions are 4.25ft (1.3 m) long by 78 in. (198.1 cm) wide and 48 in. (121.9 cm) high. With an installed length of 3.67 ft (1.18 m), each 902HD chamber provides 27.06 ft³ per foot, (2.53 m³ of storage per meter). All CulTec chambers are made of high molecular weight HDPE.

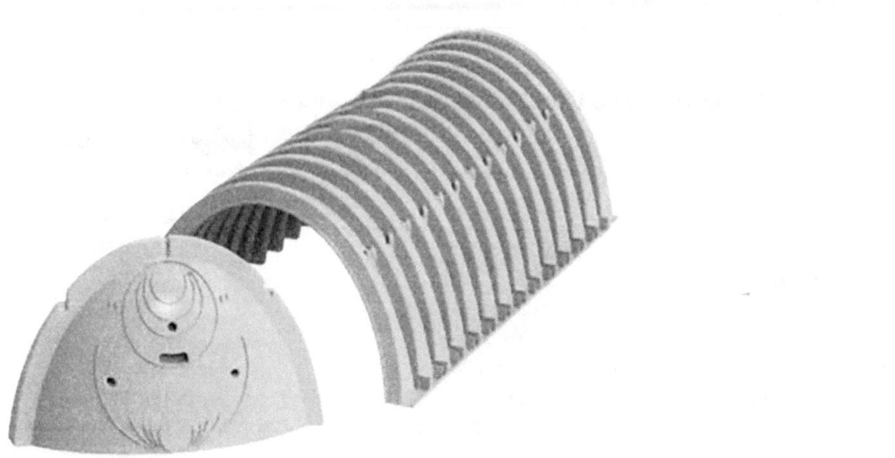

FIGURE 7.26 StormTech SC-740 with end cap

d. StormChambers Hydrologic Solutions Inc. manufactures SC-44 and SC-34 chambers. Both chambers are corrugated, made of polyethylene and have open bottoms to allow for infiltration. SC-44 which is one of the larger chambers currently available on the market, is 90 in. long by 75 in. wide by 44 in. high (28.6 × 198.1 × 111.7 cm) and weighs 120 lbs (54.4 kgs). This chamber provides an internal capacity of 105 ft^3 (2.97 m^3), and with 12 in. (30.5 cm) stone above and 9 in. (22.9 cm) stone below has an installed capacity of 166 ft^3 (4.7 m^3). The SC-34 length, width, and height dimensions are 102 x 60 x 34 inches (259 × 152.4 × 86.4 cm). Each chamber weighs 100 lbs (45.3 kgs) and provides internal and installed capacities of 115 ft^3 (3.25 m^3), respectively.
e. Trinton Chambers. Triton Stormwater Solution of Brighton, Michigan, introduced a chamber made of high-strength resin in 2005. These chambers are soy-oil based and a carbon-neutral product and are nearly 45% lighter than HDPE chambers per cubic foot of storage. Triton chambers come with a side portal feed, which serves as a lateral to which the chambers connect. These chambers are independently tested and approved for double-stack installation. Trinton, chambers like StormChambers are strong enough to carry heavy traffic loads in such configurations. They have also been safely installed at a depth of over 20 ft (6 m). Trinton chambers come in S-22, S-29, C-10, and M-6 models. S-29 which is the largest model, is 89.8 cm long, 149.9 cm wide, 91.4 cm high (35.36 × 59 × 36 in.), and weighs only 14.515 kgs (32 lbs). Each chamber with a lay-up length 33.35 in. (84.7 cm) provides 0.774 m^3 (27.35 ft^3) of storage. These chambers can be stacked up to increase storage per unit area. Table 7.25, prepared by Triton, presents a comparison of Triton S-22 (mid-unit), C-40, and M-6 chambers with some other plastic chambers.

7.4.3 Plastic and Concrete Vaults

In the past two decades, a number of underground stormwater storage vaults and modules have been introduced to the market. These modules are made of either reinforced concrete or plastic material. They can be used as a detention system by using either a concrete slab base or impermeable PVC liner or as a detention/retention-infiltration basin using a geotextile fabric. Depending on their type, these modules provide anywhere from 90% to 97% void space; and as such for a given storage, they

TABLE 7.24
StormTech Chambers Dimensions, Storage, and Weights (by ADS)

Chamber Model	Dimensions LxWxH, in (cm)	Storage Volume Chamber bare, installed ft^3 (m^3)	Weight lb (kg)
SC-310	85.4 xx 34 × 16 (217 × 86.4 × 40.6)	14.7, 31.0 [(1)] (0.42, 0.88)	37 (16.8)
DC-780	78.4 × 51 × 30 (217 × 129.5 × 76.2)	46.2, 78.4 [(2)] (1.30, 2.20)	80 (36.3)
SC-740	85.4 × 51 × 30 (217 × 129.5 × 76.2)	45.9, 74.9 [(1)] (1.30, 2.12)	74.9 (33.6)
MC-3500	90 × 77 × 45 (228.6 × 195.6 × 114.3)	109.9, 175.6 [(2)] (3.11, 4.96)	134 (60.8)
MC-4500	52 × 100 × 60 (132.1 × 254 × 152.4)	106.5, 162.6 [(3)] (3.01, 4.60)	125 (56.7)
MC-7200	83 × 100 × 60 (210.8 × 254 × 152.4)	175.9, 267.3 [(3)] (4.98, 7.57)	202 (91.2)

(1) Includes 6 in. (15 cm) stone below, above, and between chambers.

(2) Includes 9 in. (23 cm) stone below, 12 in. (30 cm) stone above, and 6 in. (15 cm) between chambers.

(3) Includes 9 in. stone below, 12 in. above, and 9 in. between chambers.

TABLE 7.25
Comparison of Triton Chambers with Some Other Chambers

Manufacturer		Triton			Stormtech Rain Arch		Ecoad	Chamber Maxx	Aqua Tunnel
Model	Unit	S-22	C-10	M-6	SC-310	SC-740	EC-1000	Standard	AT300
Width overall	mm	1397	1016	865	863	1245	1160	1310	800
Height overall	mm	887	635	445	406	762	680	770	510
Lay-up length, (excludes lap)	mm	702	754	754	2170	2170	1000	2170	1220
Bare lay-up volume (no stone)	m^3/Lm	0.940	0.371	0.212	0.184	0.594	0.516	0.645	0.245
Lay-up volume (with stone[1])	m^3/Lm	1.296	0.525	0.412	0.414	0.976	0.822	1.0	___
Chamber weight	kg	12.7	6.8	6.8	16.8	33.5	15	34.92	11
SWL/axle load	kN	210[(2)]	210[(2)]	210[(2)]	142[(3)]	142[(3)]	___	142[(3)]	35

Source: Based on information freely available in the public domain.

Comparable chamber systems

(1) Based on 150 mm aggregate below and above chamber, aggregate with 40% void to solid.

(2) Based on 450 mm cover (no pavement) constructed as specified by Triton USA. Independently tested to AASHTO H20/H25 standard with minimal to no deflection. Greater loads (AASHTO H30) can be achieved by employing pavement or geogrid systems or increasing cover to distribute the load.

(3) Based on 450 mm cover with pavement. Tested to AASHTO H20/H/25 standard.

take less room than both solid or perforated pipes or chambers in stone trenche. Some of the widely employed modules are as follows:

a. StormTrap: This trademark module is manufactured by a company of the same name with headquarters in Romeoville, Illinois. The modules are single trap or double traps made of high-strength reinforced concrete exceeding HS-20 loading. The single traps come in 1 ft, 1 in. to 7 ft, 6 in. high (33 to 190.5 cm). The single traps can be stacked up with the bottom tray upside down to double the height of the system, from 2 ft, 2 in. to 15 ft (66 cm to 4.57 m) high. The single traps can be placed either on a concrete slab foundation or stone base to serve as detention basins or retention–infiltration basins, respectively. Figure 7.27 shows single and double StormTrap schemes.
b. Atlantis America, headquartered in New York City, manufactures plastic cells and tanks for various applications. Some of their products include drainage cells, Flo-Tank, Flo-Channel, and Titan. Drainage Cells can be used as a subdrain under walkways and Flo-Tanks as a retention–infiltration system. Flo-Tanks are made in mini and single units. Mini and single units are both 685 mm (29.97 in.) long by 408 mm (16.06 in.) wide, but the former is 240 mm (9.45 in.) high and the latter 450 mm (17.72 in.) high. By staking two through seven single units, double, triple, quad, penta, hexa, and septa models are formed. These models are 880, 1310, 1740, 2170, 2600, and 3030 mm high, respectively, and provide storage volumes of 64 L (16.91 gal) for the mini to 804 L (212.4 gal) for the septa. Figure 7.28 shows an Atlantis single Flo-Tank model.
c. StormTank® is another trademark modular tank manufactured by Brentwood Industries. These modules are constructed of rugged, lightweight, polypropylene sides, top and bottom panels, and rigid PVC columns. StormTanks come in modules 20 and 25 series. Module 20 is HS-20 load holding rated and comes in 18. 24 and 36 in. high, each 21 in. wide by 42 in. long. A 36 in. high tank provides 17.88 ft^3 (0.506 m^3) of storage. Module 25 is rated for HS-25 loading and comes in 12, 18, 24, 30, 33, and 36 in. high, each 18 in. by 36 in. footprint. Figure 7.29 depicts a Module 25 StormTank, and Table 7.26 lists the storage capacity, void ratio, and the weights of all six Module 25 StormTanks. These tanks can be stacked up to 6 ft high and require a minimum of 24 in. (61 cm) cover for HS-25 load holding. StormTank components are shipped separately packed to save shipping costs and assembled onsite without a need for any tools or bonding agents. These modules provide 97% void volume, which is one of the most effective space-saving tanks currently on the market. Similar to other modular storage tanks and cells, StormTanks can be stacked up on each other to reduce the footprint of the detention system. In a turf field project in North

(a) (b)

FIGURE 7.27 StormTrap: (a) single unit, (b) double unit.

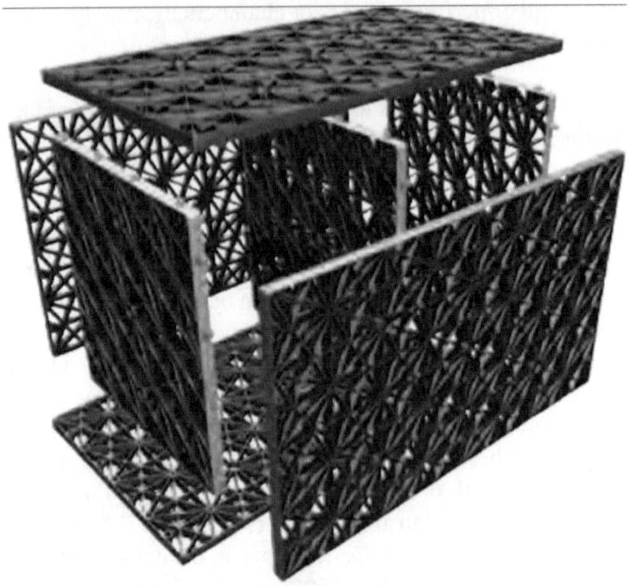

FIGURE 7.28 Atlantis® Matrix® single Flo-Tank module.

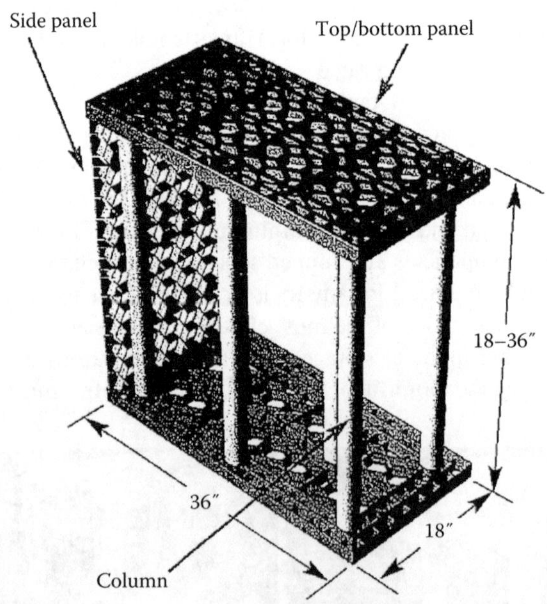

FIGURE 7.29 StormTank module by Brentwood Industries.

Bergen, New Jersey, the author designed 36 in. high StormTanks as an underground detention system. Brentwood also makes a plastic chamber called The Arch.

d. Rainstore3 is another plastic module manufactured by Invisible Structures, Inc. in Aurora Colorado. Each Rainstore3 module is 1 m × 1 m (40 in. × 40 in.) × 10 cm (4 in.) high with 36 columns and weighs 6.3 kg (14 lb). Figure 7.30 depicts the geometry and dimensions of a Rainstore3 unit. Columns are thin-walled cylinders made by injecting molded recycle resins of either high-impact polypropylene (HIPP) or HDPE. Cylinders are 10 cm (4 in.) tall, 10 cm in diameter, 5 mm (0.2 in.) thick, and spaced 16.7 cm (4.6 in.) apart. T-shaped beams connect cylinders and resist external lateral soil/water pressure. A stack of 10 units occupies 1 m^3 (35.3 ft^3) and holds approximately 250 gal (950 L) of water. Rainstore3 units can be placed against each other on geotextile fabrics and can be stacked up to 2.4 m (7.9 ft) high.

TABLE 7.26
Dimensions and Volumes of Storm-Tanks Module 2

Module	Height, in. (mm)	Nominal Void Space (%)	Storage, ft^3 (m^3)	Installed Storage, ft^3 (m^3)	Weight, lb. (kg)
ST-12	12 (3.05)	93.7	4.21 (0.12)	6.91 (0.20)	17.56 (7.97)
ST-18	18 (457)	95.5	6.54 (0.18)	9.15 (0.26)	27.7 (10.3)
ST-24	24 (610)	96.0	8.64 (0.24)	11.34 (0.32)	26.3 (11.94)
ST-30	30 (762)	96.5	10.88 (0.31)	13.56 (0.38)	29.5 (13.38)
ST-33	33 (838)	96.9	11.99 (0.34)	14.69 (0.42)	1.30 (14.20)
ST-36	36 (914)	97.0	13.10 (0.37)	15.80 (0.45)	33.1 (15.0)

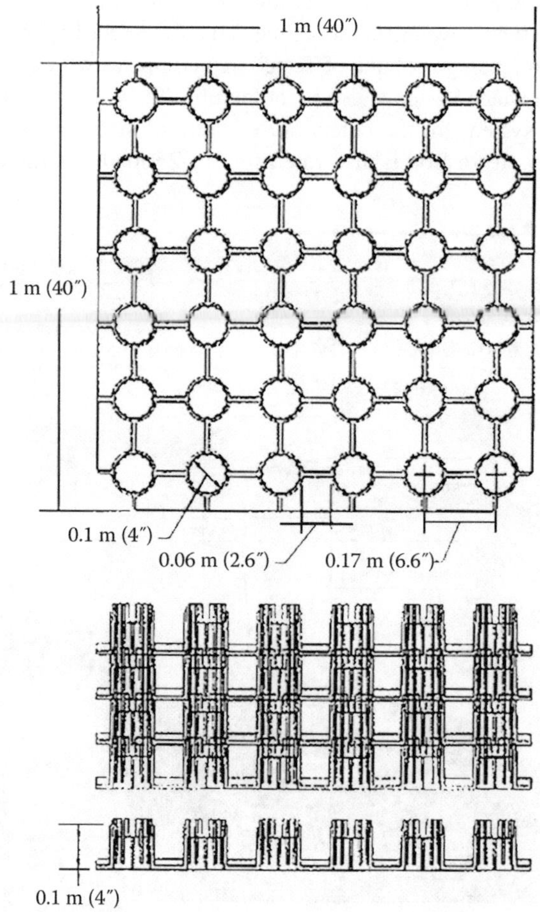

FIGURE 7.30 Rainstore3 module (by Invisible Structures, Inc.; www.invisiblestructures.com).

e. Rain-It Stormwater Management Systems. Located at Arrow Concrete Products Inc. in Granby, Connecticut, makes concrete box structures for detention–retention applications. The boxes (known as Rain-It®) come in three models: end, meddle, and side pieces. Each piece is 8 ft × 8 ft (2.44 × 2.44 m) and is available in 2, 3, 4 and 5 ft (0.61, 0.91, 1.22 and 1.52 m) standard heights. Intermediate heights can be made by special order. The boxes may be placed on a concrete platform to serve as a detention basin or a stone base to form a retention–infiltration basin. A single 3 ft (0.91 m) high model can store approximately 130 ft^3 (3.7 m^3) of water.

f. Terre Arch™/Terre Box™. Terre Arch is a modular multi-chambered precast concrete stormwater storage structure manufactured by Terre Hill Stormwater Systems of Terre Hill Concrete Products in Terre Hill, Pennsylvania. This open-bottom, Roman arch-shaped structure comes in two different sizes: Terre Arch 26 and Terre Arch 48. The former structure is 8 ft wide by 20 ft long by 34 in. high (2.44 m × 6.10 m × 0.86 m) and includes four 52 in. wide by 26 in. high internal openings. The latter is an 8 ft × 20 ft × 55 in. (2.44 × 6.10 × 1.4 m) structure with three 48 in. (1.22 m) high by 72 in. (1.83 m) wide arches. Figure 7.31 shows the Terre Arch 26 detention–infiltration structure. Each Terre Arch 26 section weighs 13,500 lbs (6100 kgs), covers 152 ft^2 (4.30 m^2), and provides 238 ft^3 (6.74 m^3) of net storage. Filling valleys between arches with stone, the Terre Arch 26 and Terre Arch 48 provide 277 ft^3 (7.84 m^3) and 541 ft^3 (15.31 m^3) of store volume, respectively.

Terre Arch fiber-reinforced design is not lightweight but has an HS-25 load-bearing rating. Thus, unlike the HDPE chambers, it can withstand heavy trucks. Terre Arch 26 and Terre Arch 48 structures save installation time and cost. According to the manufacturer, the finished cost per unit storage volume of Terre Arch competes with that of HDPE chambers.

Terre Hill Stormwater Systems also manufactures Terre Box. This is a watertight joint seal concrete box system for the detention/retention of stormwater. Designed for underground installation, Terre Box is also rated for HS-25 load bearing and can be installed

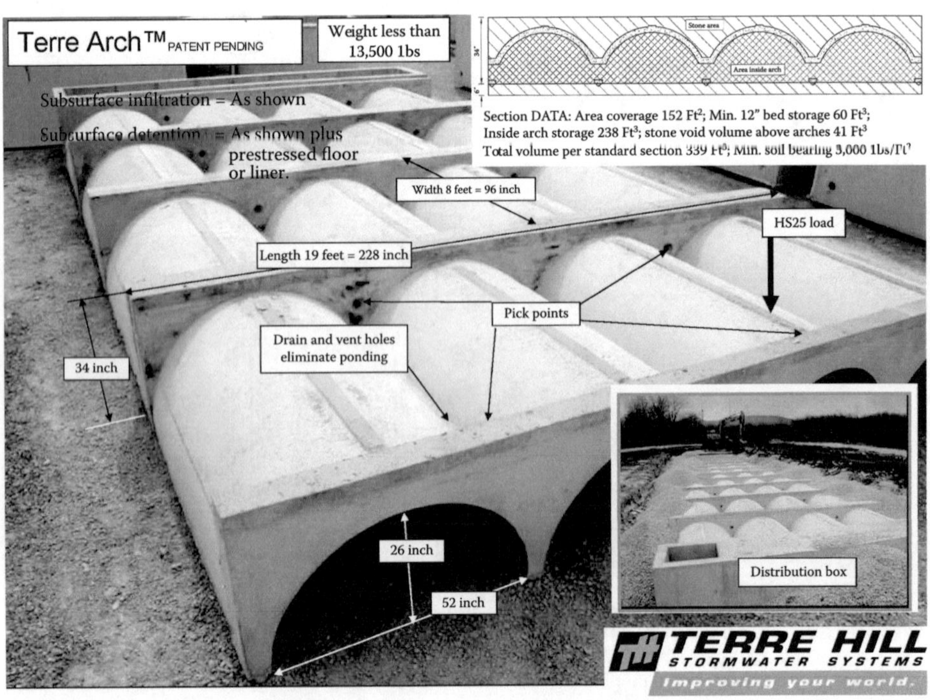

FIGURE 7.31 Terre Arch 26 modular arch structure.

below parking lots or traveled roadways. This system is a preferred choice for below-grade detention systems when recharge is not desired.

g. StormCapture. Oldcastle Infrastructure, one of the largest manufacturers of building products in the United States, makes a concrete vault named StormCapture. This chamber is an open-bottom box with side openings and can be custom-made to suit a specific application. It can be used as an underground detention system, retention–infiltration basin, and rainwater harvesting cistern. StormCapture comes in SC2 Clamshell models. The SC2 model measures 8 ft wide by 16 feet long (7 × 15 feet inside openings) and has customized heights ranging from 2 ft to 14 ft at one-foot intervals. This model provides storage capacities of 210–212 ft^3 for 2 ft high Clamshell to 1470–1608 ft^3 (41.6–45.5 m^3) for the 14 ft high clamshell module. Oldcastle's infrastructure head office is in Atlanta, Georgia. It has offices in New York, New Jersey, Florida, Texas, Minnesota, and California.

h. CUDO: Oldcastle® Infrastructure also makes a cube-like modular hollow box called the CUDO water storage system. This box, which is made of high-strength injection molded polypropylene plastic, is designed to support HS-20 traffic loads. CUDO measures 24 × 24 × 24 in. (61× 61 × 61 cm) and can be installed as a single layer or stacked up to four units to increase storage volume. Each unit provides 95% storage efficiency and comes in two halves and two end caps, which save transportation costs and can be readily assembled in the field. Figure 7.32 shows a single CUDO® configuration. More information on CUDO® can be obtained from its manufacturer's website (http://www.oldcastlestormwater.com).

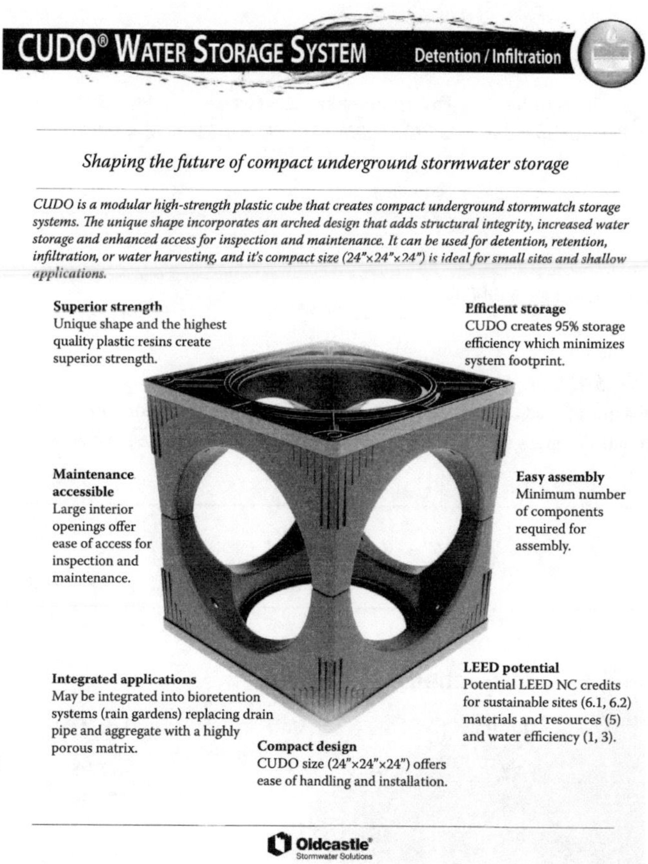

FIGURE 7.32 CUDO® water storage system.

CASE STUDY 7.3

A roadway improvement project in Woodcliff Lake, New Jersey, results in a 0.37-acre increase in impervious area and 1.02 acres (0.413 ha) of land disturbance. To address the applicable stormwater management regulations, a detention basin is provisioned to reduce the predeveloped peak rates of runoff from the disturbed area by 50%, 25%, and 20% for the 2-, 10-, and 100-year frequency storms, respectively. Because of offsite runoff entering the roadway, the overall drainage area at the point of discharge downstream of the detention basin is 9.14 acres (37 ha). Because of limited open space, an underground detention system comprising HDPE pipes is selected in this project. This case study presents step-by-step calculations for the detention basin design and water treatment device.

Runoff Calculations

Runoff calculations in this project are based on the rational method. Since a large percentage of the area is covered by paved surfaces and the roadway is drained by a drainage system, a 10-minute time of concentration is used in calculating the peak rates of runoff. Calculations for the composite runoff coefficient for the area of disturbance, E1, and the undisturbed and/or offsite area, E2 (shown on a drainage area map that is not included here), the peak rates of runoff from these areas, and the allowable discharges, Q_A, from the site are summarized in Tables 7.27 through

TABLE 7.27
Runoff Coefficients Preconstruction

	Reconstruct Ex. Pav. C = 0.95	Grass to be Paved C = 0.35	Pav. to Remain[a] C = 0.95	Grass[b] C = 0.35	Woods[c] C = 0.25	Total Area (ac)	$C_W = \Sigma(A \times C)/A$
E1	0.65	0.37				1.02	0.73
E2			2.59	4.43	1.10	8.12	0.53
Total	0.65	0.37	2.59	4.43	1.10	9.14	0.55

Note: Peak rates of runoff, existing conditions:
$Q_{E1} = 1.02 \times 0.73 \times I = 0.745 \times I.$
$Q_{E2} = 8.12 \times 0.53 \times I = 4.304 \times I.$
$Q_{E\text{-}TOTAL} = 9.14 \times 0.55 \times I = 5.027 \times I.$
[a] This area includes 2.06 acres of road to be milled/repaved and 0.53 offsite impervious area.
[b] This area includes 1.21 acres of grass within project limits and 3.22 acres of offsite grass area.
[c] All woods is offsite area.

TABLE 7.28
Peak Rates of Runoff, Preconstruction

Storm Frequency	I (in./h)	Q_{E1} (cfs)	Q_{E2} (cfs)	$Q_{E\text{-}TOTAL}$ (cfs)
2	4.2	3.13	18.08	21.11
10	5.8	4.32	24.96	29.16
100	8.0	5.96	34.43	40.22

Note: T_c = 10 min.

TABLE 7.29
Allowable Peak Flows from Site

Storm Frequency	Q_{E1} (cfs)	Reduction	Q_{E2} (cfs)	Q_A (cfs)
2	3.13	50%	18.08	19.64
10	4.3.2	25%	24.96	28.20
100	5.96	20%	34.43	39.20

Note: $Q_A = \%Q_{E1} + Q_{E2}$. Calculations for the offsite area runoff coefficient are estimated based on aerial photography.

TABLE 7.30
Runoff Coefficients, Post-Construction

	Imp. $C = 0.95$	Grass $C = 0.35$	Woods $C = 0.25$	Total Area(ac)	$C_W = \Sigma(A \times C)/A$
P_1	1.59	2.53	1.10	5.22	0.51
P_2	2.02	1.90	0.00	3.92	0.66
Total	3.61	4.43	1.10	9.14	0.57

Note: Peak rates of runoff, proposed conditions:
$Q_{P1} = 5.22 \times 0.51 \times I = 2.662 \times I$.
$Q_{P2} = 3.92 \times 0.66 \times I = 2.587 \times I$.
$Q_{P\text{-TOTAL}} = 9.14 \times 0.57 \times I = 5.210 \times I$.

TABLE 7.31
Peak Runoff, Post-Construction

Storm Frequency	I (in./h)	Q_{P1} (cfs)	Q_{P2} (cfs)	$Q_{P\text{-TOTAL}}$ (cfs)
2	4.2	11.18	10.87	22.05
10	5.8	15.44	15.00	30.44
100	8.0	21.30	20.70	41.99

Note: $T_c = 10$ min.

7.29, respectively. It is to be noted that the portion of the roadway to be milled and repaved is not considered an area of disturbance.

Calculations for the post-construction condition are prepared by differentiating the tributary area to the detention basin, P_1, from the area bypassing it, P_2. Composite runoff coefficient and peak runoff calculations for these areas are summarized in Tables 7.30 and 7.31, respectively.

Detention Basin Design

Calculations for the approximate detention storage volume and the critical storm durations are performed using the rational and modified rational method (Equations 7.23 and 7.25).

TABLE 7.32
Required Detention Storage Estimation

Storm Duration (min)	I (in./h)	Q_{P1} (cfs)	Q_{P2} (cfs)	Q_O (cfs)	Required Vol., V_R (ft³)
2-Year					$Q_A = 19.64$
10	4.20	11.18	10.87	8.78	1800[a]
15	3.60	9.58	9.31	10.33	$Q_{P1} < Q_O$
20	3.00	7.99	7.76	11.88	$Q_{P1} < Q_O$
30	2.40	6.39	6.21	13.43	$Q_{P1} < Q_O$
45	1.80	4.79	4.66	14.98	$Q_{P1} < Q_O$
10-Year					$Q_A = 28.20$
10	5.80	15.44	15.00	13.20	1680
15	4.20	11.18	10.87	17.34	$Q_{P1} < Q_O$
20	4.00	10.65	10.35	17.86	$Q_{P1} < Q_O$
100-Year					$Q_A = 39.20$
10	8.00	21.30	20.70	18.50	2100
15	6.90	18.37	17.85	21.35	$Q_{P1} < Q_O$
20	5.80	15.44	15.00	24.20	$Q_{P1} < Q_O$
30	4.60	12.25	11.90	27.30	$Q_{P1} < Q_O$

Modified rational method, critical storm calculations:
$V_R = [T_D \times 60s \times Q_{P1}] - [0.5 \times T_b \times 60s \times Q_O]$; T_b (base time) $= T_D + T_c$, $T_c = 10$ min.
$Q_{P1} = 5.22 \times 0.51 \times I = 2.662 \times I$
$Q_{P2} = 3.92 \times 0.66 \times I = 2.587 \times I$
Q_O = Allowable peak outflow from basin = $Q_A - Q_{P2}$

See Table 7.29 for Q_A calculations.
[a] $V = 60 \times T_b (Q_{P1} - Q_O)/2$; $T_b = 2.5\ T_c$.

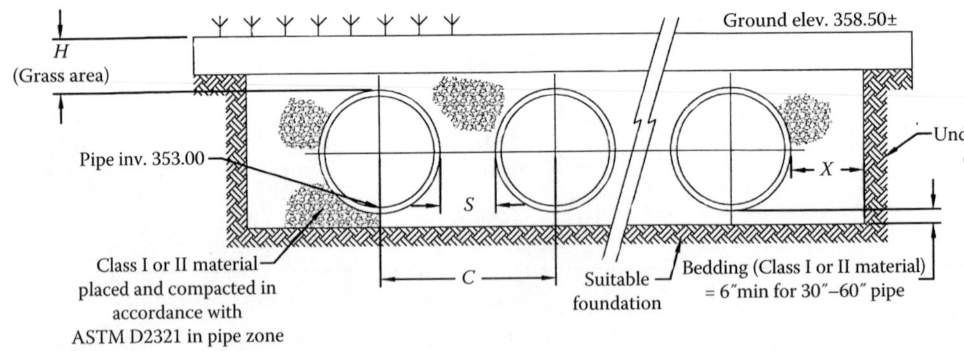

FIGURE 7.33 Typical detention basin cross-section detail.

Calculations for the required detention storage are performed for storms of 2-, 10-, and 100-year frequencies and are organized in Table 7.32. This table indicates a minimum detention basin volume of 2100 ft³. It also shows the critical storm duration of 10 minutes for storms of 2-, 10-, and 100-year frequency. Considering nonlinear variation of discharge and further reduce discharges downstream of the site, the estimated storage volume is increased by nearly 60% in the design of detention pipes. Figures 7.33 and 7.34 show typical cross-section and layout of

Parts summary

Item no.	Item	Quantity
1	48″ Single manifold tee	1 Unit
2	48″ Double manifold tee	1 Unit
3	48″ × 90″ Manifold bend	1 Unit
4	48″ × 24″ Reducer	2 Units
5	48″ HDPE (solid pipe)	240 LF

Detention basin maintenance notes:

1. Detention basin to be maintained by the county of Bergen Department of Public Works.
2. Basin should be inspected for sediment and debris quarterly and after storms exceeding 1 inch of rainfall in 24 hours. Removal of debris and sediment should take place when the basin is thoroughly dry.
3. Concrete outlet structure to be checked annually for cracking, subsidence, spalling, erosion and deterioration.
4. Vortsentry unit to be maintained and inspected as per manufacturer recommendations.

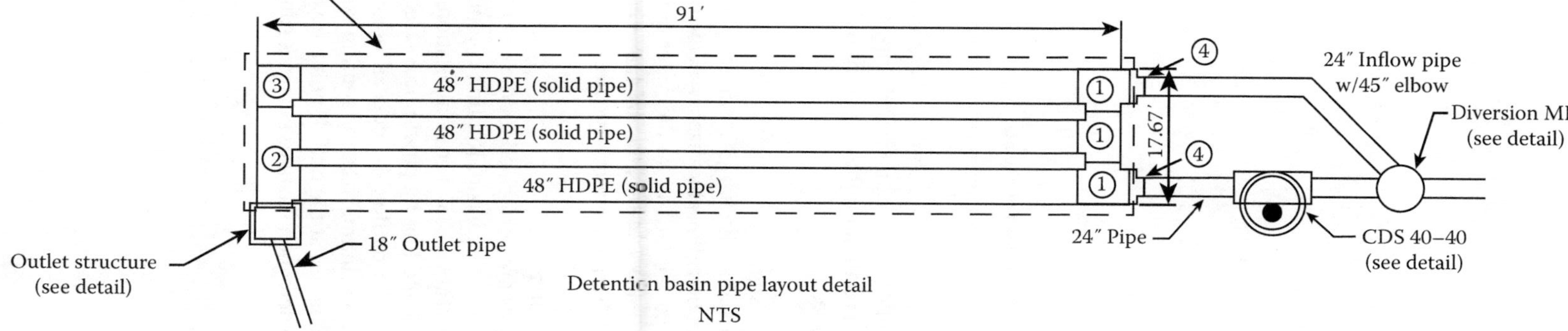

FIGURE 7.34 Detention basin pipe layout detail.

TABLE 7.33
Storage–Elevation Relation

Pipe dia.: 4 ft.
Three rows of 80 L.F. of 48 in. HDPE; headers on either end
(See Figures 7.36 and 7.37.)

Elev. (ft)	Depth (ft)	Depth/Dia.[a]	Ratio to Total Volume[a]	Volume (ft^3)
353	0	0.00	0	0
353.5	0.5	0.13	0.075	255
354	1	0.25	0.2	680
354.25	1.25	0.31	0.27	918
354.5	1.5	0.38	0.33	1122
355	2	0.50	0.5	1700
355.5	2.5	0.63	0.65	2210
356	3	0.75	0.82	2788
356.5	3.5	0.88	0.93	3162
357	4	1	1	3400

[a] Partly full volumes are calculated based on Figure 2.5.

detention pipes, respectively. As shown, the proposed detention system comprises three 80 ft (24.4 m) long rows of 48 in. (1.22 m) HDPE pipes connected with 48 in. headers. Table 7.33 lists the storage-elevation relation for this underground detention basin, which provides 3400 ft^3 (96.2 m^3) of storage volume.

Outlet Structure Design

The maximum allowable discharge from the detention basin is calculated at 18.50 cfs (see Table 7.32). To attain this allowable discharge, the outlet structure includes a 6 in. low-flow orifice designed to discharge rainfalls up to the water quality storm and a 1.25 ft wide weir above the water quality level. Table 7.34 lists the discharge calculations for each of these openings and the composite discharge from the outlet structure.

Routing Computations

Routing computations are performed for the critical storm durations determined in the preliminary sizing of the detention basin. Routing computations for the 100-year, 10-minute duration storm are shown in Table 7.35. The computations indicate that the detention system is effectively (97%) utilized and that by providing a larger detention basin than that calculated by the preliminary design, the outflow is reduced below the allowable discharge of 18.50 cfs (see Table 7.32). Table 7.36 summarizes the routing computations for 2-, 10-, and 100-year storms. The composite discharges at the outfall point are conservatively calculated by adding the peak discharge from the detention basin to the peak runoff from bypass areas. The results are shown in Table 7.37.

TABLE 7.34
Discharge Rating of Outlet Structure

Orifice			Weir		
Inv.: 353			Inv.: 354.25		
Dia. (in.): 6			L (ft): 1.25		
Area (sf): 0.196					
Elev.	**H (ft.)**	**Q (cfs)**	**H (ft.)**	**Q (cfs)**	**Total Q (cfs)**
353	0	0			0.00
353.5	0.25	0.47			0.47
354.0	0.75	0.82			0.82
354.25	1.00	0.95	0.00	0.00	0.95
354.5	1.25	1.06	0.25	0.47	1.53
354.75	1.50	1.16	0.50	1.33	2.48
355.0	1.75	1.25	0.75	2.44	3.69
355.5	2.25	1.42	1.25	5.24	6.66
356.0	2.75	1.57	1.75	8.68	10.25
356.5	3.25	1.70	2.25	12.66	14.36
357.0	3.75	1.83	2.75	17.10	18.93

Equations used:
Orifice flow: $Q_o = [CA\ (2gh)^{1/2}]$, $C = 0.6$
Orifice (partially submerged): $Q_O = (\pi/4)x(CLH^{1.5})$, $C = 3.0$
Weir flow: $Q_w = CLH^{1.5}$, $C = 3.0$
Submerged weir: $Q_{SW} = [C(L \times H)\ (2gh)^{1/2}]$, $C = 0.6$
Head (orifice): H = Elev. – Invert + 0.5 dia.
Head (weir): H = Elev. – Invert

Table 7. 38 presents a comparison of the calculated allowable and post construction peak runoff at the point of discharge. As can be seen, the proposed oversized detention basin system reduces the peak runoff below the allowable values. If composite discharges were calculated by adding hydrographs, the post-development peak runoff would be smaller than those shown in Table 7.37.

The required TSS removal for the project and the TSS removal provided by the CDS water quality device are presented in Tables 7.39 and 7.40 respectively. The TSS removal rate criteria are defined under the former table. The results in the latter table are simply based on the paved area, which is treated by the water quality device before entering the detention system.

The size of the CDS unit provisioned in this project is determined based on the impervious area tributary to the unit and the rainfall intensity of the water quality storm.

TABLE 7.35
Routing Computations for the 100-Year, 10-Minute Storm

Inflow Hydrograph		Routing Computations				
Time (min)	Inflow (cfs)	I1 + I2 (cfs)	2*S*/*t* – *O* (cfs)	2*S*/*t* + *O* (cfs)	Outflow (cfs)	Elevation (ft)
0.0	0.00	-	0.0	0.0	0.00	353.00
5.0	10.65	10.7	6.4	10.7	2.12	354.64
10.0	21.30	32.0	5.3	38.4	16.53	356.74
15.0	10.65	32.0	5.8	37.3	15.71	356.65
20.0	0.00	10.7	7.7	16.5	4.37	355.12
25.0	0.00	0.0	5.4	7.7	1.15	354.34
30.0	0.00	0.0	3.8	5.4	0.83	354.01
35.0	0.00	0.0	2.5	3.8	0.65	353.75
40.0	0.00	0.0	1.5	2.5	0.51	353.55
45.0	0.00	0.0	0.8	1.5	0.32	353.34
50.0	0.00	0.0	0.5	0.8	0.18	353.19
55.0	0.00	0.0	0.3	0.5	0.10	353.11
60.0	0.00	0.0	0.2	0.3	0.06	353.06
65.0	0.00	0.0	0.1	0.2	0.03	353.04
70.0	0.00	0.0	0.0	0.1	0.02	353.02
75.0	0.00	0.0	0.0	0.0	0.01	353.01

Summary of peak outflow and peak elevation:
Peak inflow = 21.30 cfs.
Peak outflow = 16.53 cfs.Peak elevation = 356.74 ft.
Summary of approximate peak storage:
Initial storage = 0 cu-ft.
Peak storage from storm = 3275 cu-ft.
Total storage in pond = 3275 cu-ft.

TABLE 7.36
Routing Computations Summary: 2-, 10-, and 100-Year Storm Events

Storm Freq. (years)	Rainfall Intensity (in./h)	Storm Duration (min)[a]	Q_{IN} (cfs)	Q_{OUT} (cfs)
2	4.20	10	11.18	8.30
10	5.80	10	15.44	11.90
100	8.00	10	21.30	16.53

[a] Critical storm duration; see Table 7.32.

TABLE 7.37
Composite Peak Flows from Site: 2-, 10-, and 100-Year Storms

Storm Freq. (years)	Q_{P2}a (cfs)	Q_{OUT}b (cfs)	Total Q (cfs)
2	10.87	8.30	19.17
10	15.00	11.90	26.90
100	20.70	16.53	37.23

[a] Q_{P2} is peak flow of the bypass area. See Table 7.32.
[b] Q_{OUT} is peak outflow from the detention basin. See Table 7.36.

Water Quality Provision

The project increases impervious coverage by 0.37 acres. Since this is larger than the 0.25 acres threshold, the project is subject to the state of New Jersey water quality requirement. To address this requirement a CDS unit is provisioned for the project. The unit is placed before the detention basin and offline from the inflow pipe such that storms larger than the water quality storm bypass the unit.

TABLE 7.38
Calculated Allowable and Proposed Peak Rates of Runoff from Site

Storm Frequency	Q_A (cfs) [a]	Q_P (cfs) [b]
2	19.64	19.17
10	28.20	26.90
100	39.20	37.23

Note: Area = 9.14 acres.
[a] Q_A = allowable from site (see Table 7.29).
[b] Q_P = total proposed flow (calculated by adding peak flows; see Table 7.37).

TABLE 7.39
Required TSS Removal Rate for the Project Area

Ground Cover	Area (ac.)	% TSS Required Removal Rate	A × % TSS
Increase in impervious area[a]	0.37	80%	0.30
Ex. pavement to be fully re- constructed[b]	0.65	50%	0.33
	ΣA = 1.02		ΣA × % TSS = 0.62

Note: Pavement being milled/repaved does not count as disturbance and does not trigger the stormwater management rules.
[a] New pavement is subject to 80% TSS removal.
[b] Pavement being fully reconstructed is considered disturbed and pursuant to the stormwater management rules requires a TSS removal rate of 50%.

TABLE 7.40
Provided TSS Removal Rate for the Project Area

Ground Cover	Area (ac.)	% TSS Provided Removal Rate	$A \times$ % TSS
Impervious to water quality unit	1.59	50%	0.80
	$\Sigma A = 1.59$		$\Sigma A \times$ % TSS = 0.80

Note: 0.80 > 0.62. Proposed water quality treatment is adequate.

Water Quality Unit Sizing

Water quality peak flow rate to basin:

A (acres) = 1.59[a]

$C_W = 0.90$

$T_C = 10$ min

$I = 3.2$ in./h[b]

Q (cfs) = 4.58 cfs

Note: Use CDS Model CDS 4040, which is rated for 6 cfs treatment rate by the manufacturer.

[a] Excludes grass and woods areas; they retain runoff during WQ storm.

[b] See Chapter 3.

CASE STUDY 7.4

This case exemplifies the application of underground chambers in the stone trench. The project involves widening the New Jersey Garden State Parkway in the vicinity of Interchange 88/89. The project location is shown in Figure 7.35 which is prepared based on the US Geological Survey Lakewood Quadrangle. Figure 7.36 depicts the limits of road widening, which extends nearly one-half mile (800 m) along north- and south-bound lanes of the parkway. The project is located entirely within a CAFRA (Coastal Area Facility Review Act) zone and disturbs approximately 0.77 acre (0.31 ha) of land. Of this, 0.642 acre (0.26 ha) will be new pavements due to the road widening, and the remainder includes clearing some wooded areas associated with grading along the pavements. The existing pavements within the project limits, which will be milled or resurfaced, are not considered as a land disturbance and, therefore not subject to the NJDEP's stormwater treatment requirements.

Stormwater Management Provisions

To control the peak rates of runoff, an underground detention infiltration system is provisioned along the right shoulder of southbound lanes. The system comprises two rows of Cultec Recharger V8 chambers placed in an 11.5 ft (3.77 m) wide by 183.7 ft (60.25 m) long stone trench. This trench is 46 in. (125.6 cm) thick and includes a 6 in. (15 cm) stone base and 6 in. stone cover. The system provides 5186 ft^3 (147 m^3) of storage. Figure 7.37 shows a typical section of the Recharger V8 detention system. Before entering the chambers, the entire runoff is routed through a CDS PMSU 30–30 (redesignated as CDS30–30) water quality device that has 3 cfs treatment capacity. This device removes floatable objects, silt, debris, oil, and grease, thereby preventing premature clogging of the chambers and lessening their maintenance. The CDS unit is approved for 50% total suspended solids (TSS) removal, is easy to maintain, and partially addresses the water quality requirements for the project. To supplement this device, existing woodlands along the northbound shoulder are utilized as a vegetative strip, which provides 80% TSS removal for the runoff from the area tributary thereof.

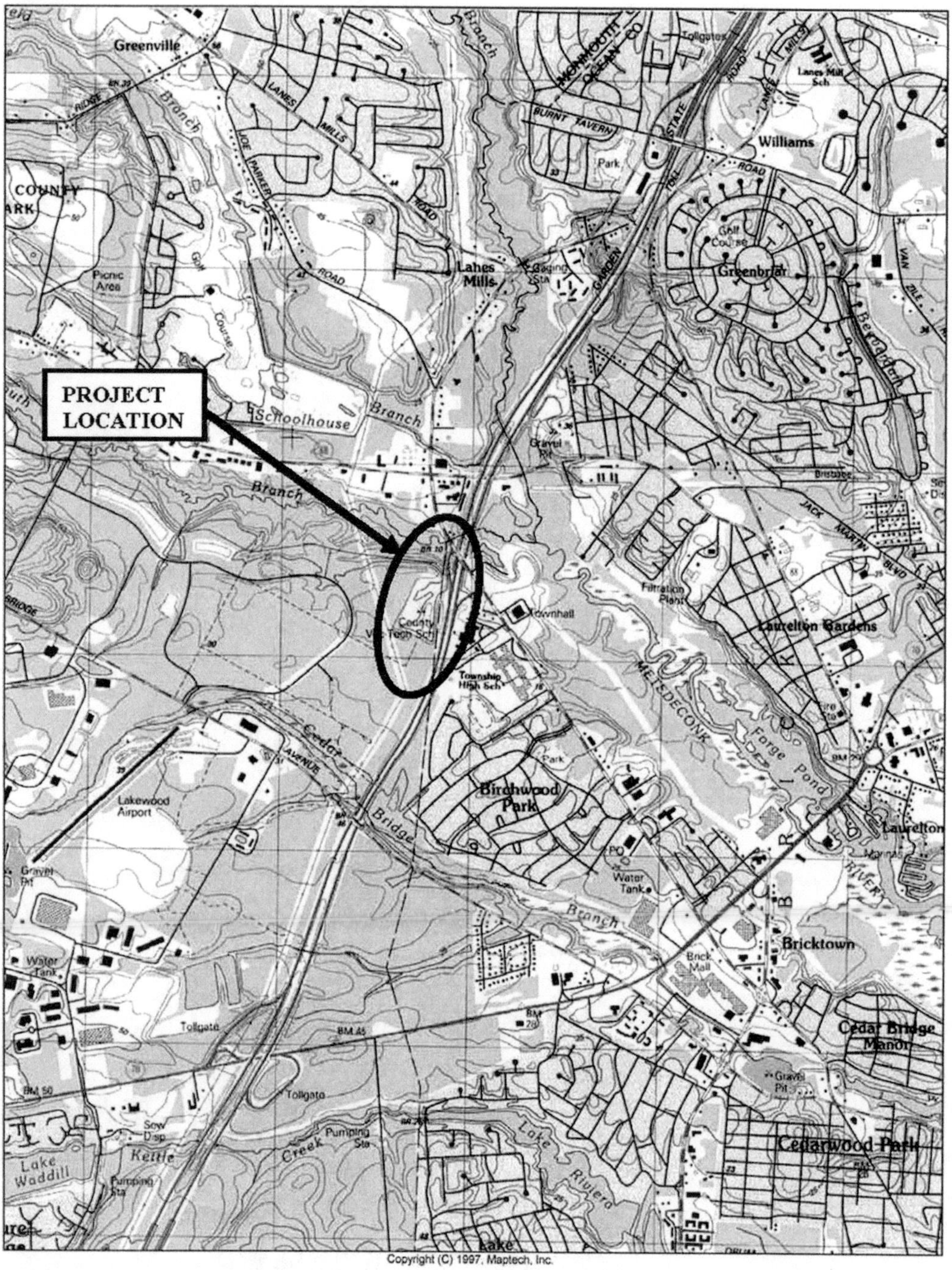

FIGURE 7.35 Location map.

Runoff Calculations

Figure 7.38 shows the preconstruction drainage areas of the north- and south-bound lanes of the parkway. This figure is prepared using the 1 ft contour lines generated from the topographic survey conducted in connection with the design. The drainage areas shown in this figure cover the tributary areas to the project site. Figure 7.39 shows the post-construction drainage areas. As

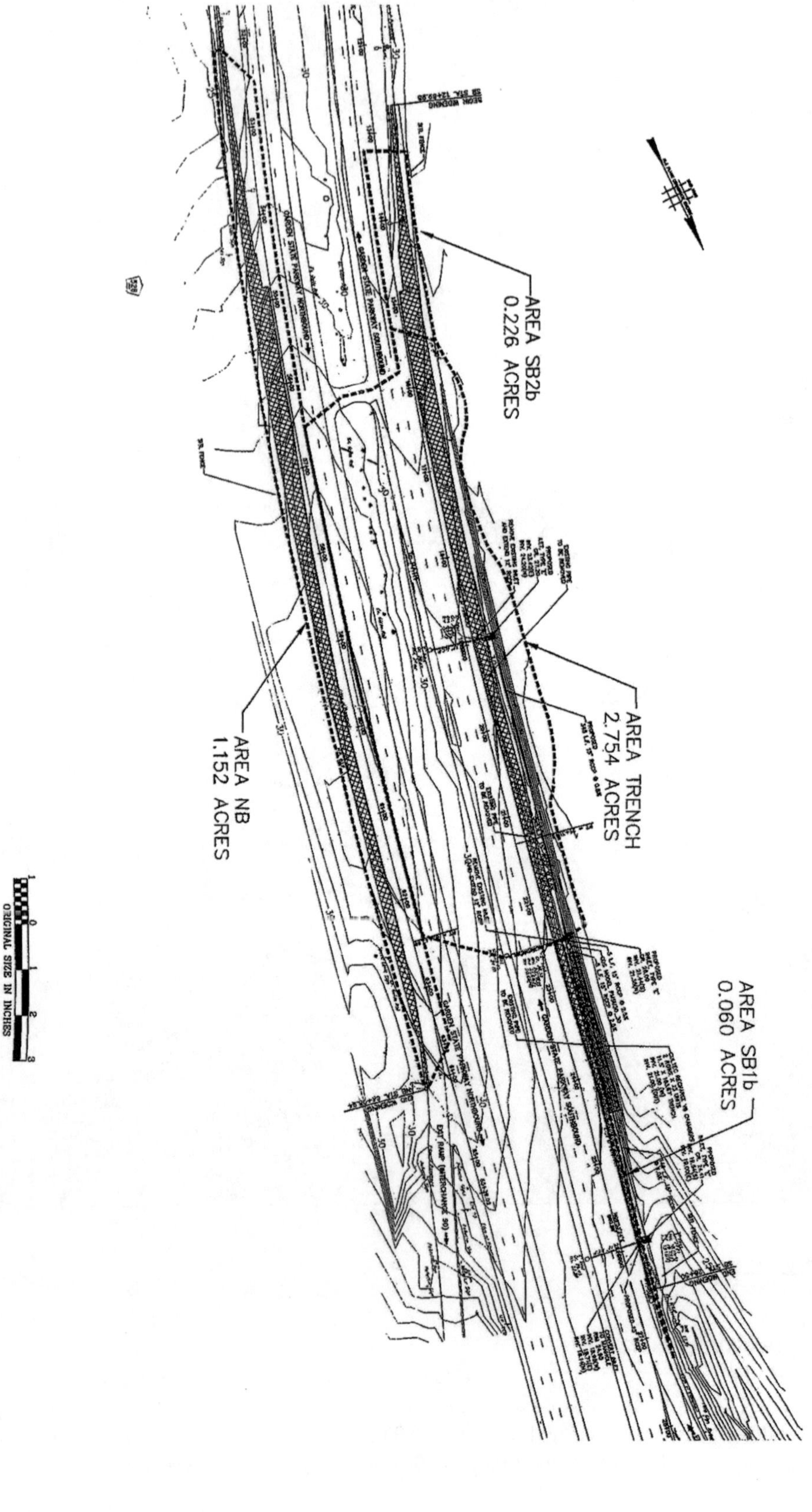

FIGURE 7.36 Overall map of altered area.

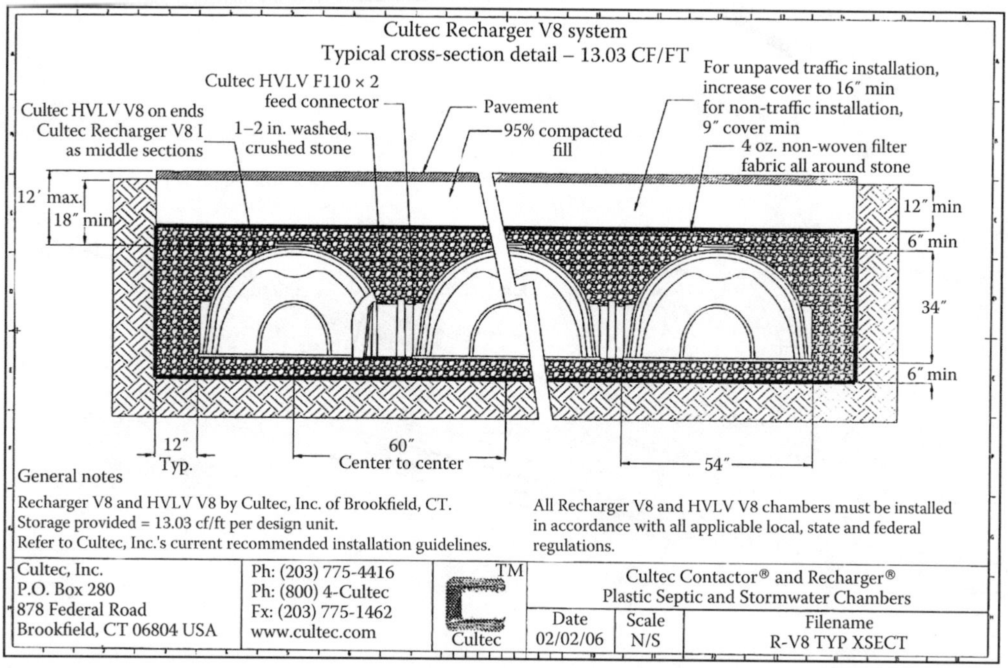

FIGURE 7.37 Cultec recharger V8 system. Cultec now makes Recharger 330HD in lieu of V8.

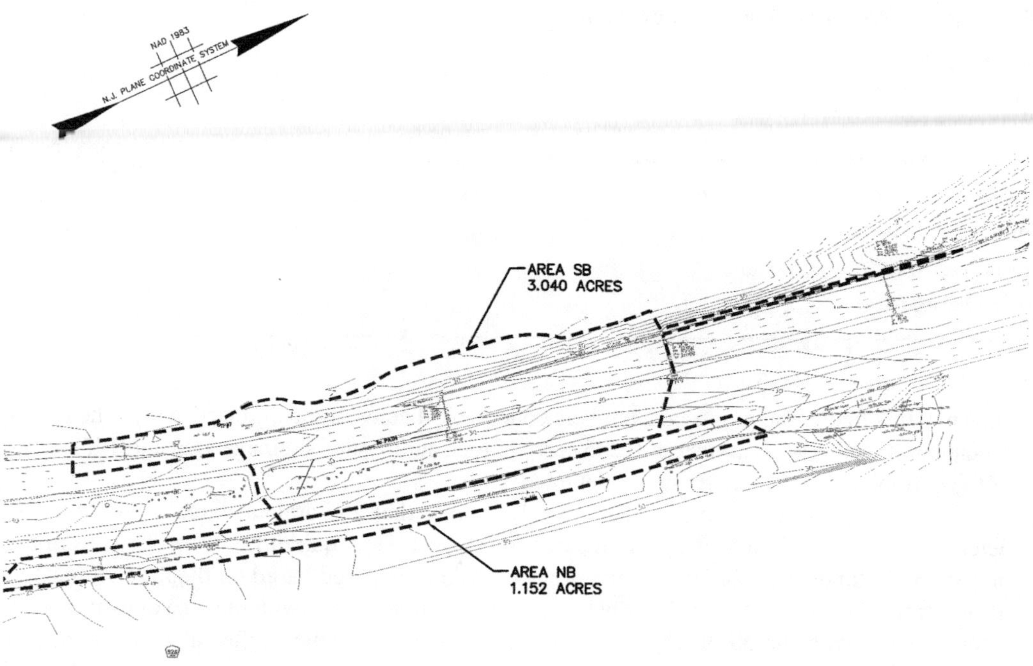

FIGURE 7.38 Preconstruction drainage areas.

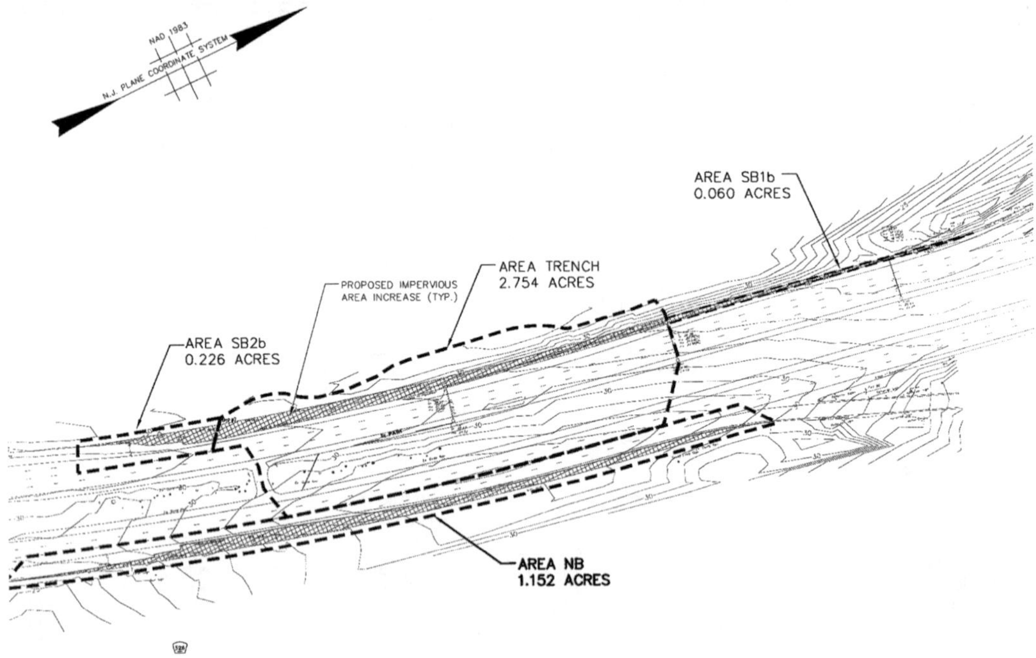

FIGURE 7.39 Post-construction drainage areas.

TABLE 7.41
Existing and Proposed Runoff Coefficients

Area being altered: 0.770 acres
Open space being paved: 0.642 acres
Woods to be open space: 0.128 acres

	Imp. C = 0.95	Grass C = 0.35	Woods C = 0.25	Total Area (ac)	$C_W = \Sigma(A \times C)/A$
Pre	0	0.642	0.128	0.770	0.33
Post	0.642	0.128	0	0.770	0.85

$Q_E = 0.770 \times 0.33 \times I.$
$Q_P = 0.770 \times 0.85 \times I.$

shown in this figure, the proposed chambers collect 2.754 acres of the drainage area in the south-bound lanes, with the remainder of 0.286 acres of the south-bound drainage area and 1.152 acres of the northbound lanes remaining un-detained.

Since the area of disturbance is significantly smaller than the drainage area tributary to the detention basin, calculations are first prepared to establish the required flow reductions from the project area. Runoff calculations for the project area are prepared based on the rational method using a time of concentration of 10 minutes, which is commensurate with the area of disturbance. In compliance with the NJDEP's stormwater management regulations, the calculations include 2-, 10-, and 100-year storms. Tables 7.41 and 7.42 present the pre- and post-construction runoff

TABLE 7.42
Required Flow Reductions

Storm Frequency	I (in./h)	Q_E (cfs)	Reduction	Q_P (cfs)	Q_R (cfs)
2	4.2	1.07	50%	2.75	2.22
10	5.8	1.47	25%	3.80	2.69
100	8	2.03	20%	5.24	4.83

Note: Storm duration and time of concentration = 10 minutes.
Required reduction, Q_R
$Q_R = Q_P - \%Q_E$

TABLE 7.43
Post-Construction Peak Flows for Detained and Bypass Areas and Allowable Discharges from the Detention-Infiltration System

	Imp. $C = 0.95$	Grass $C = 0.35$	Woods $C = 0.25$	Total Area (ac)	$C_W = \Sigma(A \times C)/A$	
Bypass Areas						
SB1b	0.06	0	0	0.060	0.95	
SB2b	0.182	0.044	0	0.226	0.83	
NB	0.905	0.247	0.000	1.152	0.82	
Total	1.147	0.291	0	1.438	0.83	
Detained Areas						
Trench	1.575	0.7	0.479	2.754	0.68	
Storm Frequency	I (in./h)[a]	Q_B (cfs)[b]	Q_D (cfs)[c]	Q_P (cfs)[d]	Q_R (cfs)[e]	Q_A (cfs)[f]
2	4.2	5.01	7.87	12.88	2.22	5.65
10	5.8	6.92	10.86	17.78	2.69	8.17
100	8.0	9.55	14.98	24.53	4.83	10.15

[a] Storm duration and time of concentration = 10 minutes.
[b] Q_B = bypass flow = 1.438 × 0.83 × I.
[c] Q_D = inflow to underground chambers = 2.754 × 0.68 × I.
[d] Q_P = total post-development flow.
[e] See Table 7.43 for calculations of Q_R (required flow reduction).
[f] $Q_A = Q_D - Q_R$, allowable max. discharge from chambers.

coefficient calculations for the disturbed area and the required reductions in the peak rates of runoff, respectively.

Runoff calculations for the entire 4.192 acres (1.697 ha) drainage area are prepared differentiating the tributary area to the proposed underground chambers and the bypass areas. While the calculations show a time concentration of 15 minutes, runoff calculations for the detention basin are conservatively performed using T_c = 10 minutes, which is the same as that of the bypass areas, including the tributary area to the northbound lanes. Table 7.43 includes runoff calculations together with support data for the post-construction peak rates of runoff from the detained and bypass areas. Also listed in this table are the maximum allowable discharges from the proposed detention system. Accounting for the required flow reductions shown in Table 7.42.

Detention–Infiltration System Design

To avoid iterative routing computations, simple calculations are first performed to determine the critical storm duration, the one that will require the largest detention storage volume for each storm event. Table 7.44 presents the critical storm duration calculations performed using the modified rational method. Conservatively, the percolation losses are neglected in these preliminary calculations. The critical storm duration is found to be 15 minutes for the 2- and 100-year storm events and 10 minutes for the 10-year storm event; the required detention storage is estimated to be 2934 ft^3 (83.1 m^3).

Tables 7.45 and 7.46 list elevation- storage and elevation- discharge relations for the detention-infiltration system, respectively. The former table shows that the proposed system is nearly 77% larger than the preliminary calculations of detention basin sizing indicate. The listed discharges on the latter table account for the seepage loss through the bottom of the stone trench housing the chambers. The seepage loss is calculated to be 0.749 cfs (13.9 L/s) based on a soil permeability equal to 10 in./h (250 mm/h), which is one-half of the results from three percolation tests conducted at the site of the chambers; factor 0.5 represents a design safety factor equal to 2.0.

Routing computations are performed using hydrographs for the critical storm durations, which are derived from Table 7.44 and the storage- discharge- elevation rating (Tables 7.45 and 7.46). The routing computations are summarized in Table 7.47. Listed in this table are the outflows from the system, adjusted for the rate of seepage loss. The table also presents a comparison of the outflows and the allowable discharges from the Cultec chambers. The table shows that the oversized detention system creates flow reductions of 16% to 36% more than required.

TABLE 7.44
Critical Storm Duration and Preliminary Sizing of Detention Basins

Modified rational method

$V = [T_D \times 60s \times Q_D] - [0.5 \times T_b \times 60s \times Q_A$[a]

T_b (base time) = $T_D + T_c$, $T_c = 15$ min

Q_D = Detention inflow = $2.754 \times 0.68 \times I$

Q_A = Max. allowable discharge from detention infiltration basin

Storm Duration (min)	*I* (in./h)	*Q*, *cfs*	Required Volume (ft³)
2-Year		$\mathbf{Q_{ALLOWABLE} = 5.65}$	
10	4.20	7.87	1662
15	3.60	6.74	1830
20	3.00	5.62 < 5.65	1657
10-Year		$\mathbf{Q_{ALLOWABLE} = 8.17}$	
10	5.80	10.86	2019
15	4.80	8.99	1963
20	4.00	7.49 < 8.17	1636
100-Year		$\mathbf{Q_{ALLOWABLE} = 10.15}$	
10	8.00	14.98	2709
15	6.80	12.73	2934
20	5.60	10.49	1931

[a] Preliminary required storage volume calculations conservatively neglect infiltration.

TABLE 7.45
Stage–Storage Calculation Results

Elev. (ft)	Depth	Storage (ft³) Incremental	Storage (ft³) Cumulative
23.83	3.83	70.41	5186.1
23.75	3.75	211.22	5115.7
23.50	3.5	213.18	4904.4
23.25	3.25	253.51	4691.3
23.00	3	323.49	4437.8
22.75	2.75	357.93	4114.3
22.50	2.5	381.26	3756.3
22.25	2.25	397.61	3375.1
22.00	2	409.16	2977.5
21.75	1.75	415.92	2568.3
21.50	1.5	421.59	2152.4
21.25	1.25	432.7	1730.8
21.00	1	434.23	1298.1
20.75	0.75	441.42	863.9
20.50	0.5	211.22	422.4
20.25	0.25	211.22	211.2
20.00	0	0	0.0

Cultec recharge V8 chambers

No. of units: 50

2 rows × 25 units

11.5 ft × 183.67 ft stone bed area, 6 in. stone box, 6 in. stone cover

TABLE 7.46
Outlet Structure Elevation–Discharge Relation

15 in. dia. outlet
Invert = bottom of chamber + 6 in.
Inv. = 21.0 ft

Elev.	Depth	H	Q (cfs)	Inf. (cfs)	Total Q
23.83	3.83	2.21	8.78	0.49	9.27
23.75	3.75	2.13	8.62	0.49	9.11
23.50	3.5	1.88	8.09	0.49	8.58
23.25	3.25	1.63	7.53	0.49	8.02
23.00	3	1.38	6.93	0.49	7.42
22.75	2.75	1.13	6.27	0.49	6.76
22.50	2.5	0.88	5.53	0.49	6.02
22.25	2.25	0.63	4.67	0.49	5.16
22.00	2	1.00	2.95	0.49	3.44
21.75	1.75	0.75	1.92	0.49	2.41
21.50	1.5	0.50	1.04	0.49	1.53
21.25	1.25	0.25	0.37	0.49	0.86
21.00	1	0	0	0.49	0.49
20.00	0	0	0	0	0.00

Fully Submerged	Partly Submerged	Infiltration
$Q = \text{CA}\,(2gh)^{1/2}$, dia. = 15 in.	$Q = (\pi/4)(CLH^{1.5})$[a]	$Q = A \times I \times$ (ft/12 in.) (h/3600 s)
$A = 1.227$ ft^2	$L = 1.25$ ft	Trench area = 11.5 × 183.67 ft
$C = 0.6$	$C = 3.0$	$A = 2112$ ft^2
$H =$ Elev.–21.625	$H =$ Elev.–21.0	$I = 10$ in./h
$Q = 5.91(H^{0.5})$	$Q = 2.95(H^{1.5})$	$Q = 0.49$ cfs

[a] This equation is proposed for estimating weir flow discharge from circular openings.

TABLE 7.47
Peak Inflows, Outflows, and Allowable Flows for Detention–Infiltration System

Storm Freq. (years)	Q_{IN}	Q_{DIS}	Q_O	Q_A
2	6.74	4.66	4.17	5.65
10	8.99	6.48	5.99	8.17
100	12.73	9.22	8.73	10.15

Note: Q_A = allowable discharge. See Table 7.43.
Q_{DIS} = peak discharge from chambers.
Q_O = outflow = peak discharge – infiltration rate (0.49 cfs).

7.5 WATER TREATMENT STRUCTURES

Manufactured water treatment devices were discussed in a previous chapter. Vegetative swales and sand filters, which are among the commonly water quality structures, are discussed herein. Filter strips, which are among the non-structural water treatment devices, will also be discussed in the next chapter.

7.5.1 Vegetative Swales

Grassed waterways, also referred to as grass swales, are one of the earliest measures taken to treat stormwater runoff. Swales are used alongside roadways or in medians of divided highways. They are also employed in urban development as a substitute for inlets and pipes. Grass swales can provide moderate improvements in urban runoff quality. To effectively treat the runoff and to be stable, the swale should be properly designed and regularly maintained. The swale should also be a shallow trench no more than 3 ft (1 m) deep and have side slopes less than 3:1 (3H, 1V) to facilitate mowing. Figure 7.40 shows a wet grass swale. in Georgetown, South Carolina. This type of grass swale provides long residence time and natural growth to treat stormwater prior to its discharge to a drainage system or surface water body.

Grass swales are practical in areas of moderate longitudinal slope where the flow velocity will be less than that which would cause erosion during the design storm. This storm is specified as a 10-year event by many jurisdictional agencies. In the Northeast and other parts of the United States, where there is sufficient rain during the growing season, swales may not need any watering. However, when constructed in semiarid climates, swales require irrigation to maintain grass growth.

To be effective, the water depth during the water quality storm should not rise significantly above the grass. Once the flow rises over the grass, the grass begins to bend down in the direction of flow, and the water passes over the grass without receiving much treatment.

Manning's n for grass swales may be as high as 0.3 or 0.4 before submergence begins. However, when the grass becomes deeply submerged, the n value may be dropped by 10-fold to 0.04. The result is a faster velocity, shorter resident time, and low pollutant removal. Swale tests by Yu et al. (2001) with controlled flows at detention times ranging from 5.5 to 18 minutes showed pollutant removal efficiencies of 48% to 86%. Tests by others, referenced in Minton (2005), indicate similar results. However, particles smaller than 15 microns require prolonged detention time. Also, due to erosion, swales lose their effectiveness. The New Jersey Department of Environmental Protection Stormwater Best Management Practices Manual (July

FIGURE 7.40 A wet grass swale in Georgetown, SC.

Source: Photo by the author (October 2023).

17, 2023; last updated October 9, 2024) accept grass swales as a water treatment structure for up to 50% TSS removal.

7.5.2 Sand and Organic Filters

Sand and organic filters are intended primarily for water quality improvement; they provide no peak flow control. They are among the most expensive treatment systems. A review of sand filters is presented by EPA (1999). EPA has also prepared another publication, titled *NPDES Stormwater Best Management Practices, Sand and Organic Filters* (EPA, December 2021). There are efficient types of sand filters that can be installed above ground or underground. Surface sand filters are more efficient and provide far more treatment capacity than all other types of filters. Table 7.48 is a replicate Table 2 in the latter EPA report. According to this report, the initial construction and annual maintenance costs of surface sand filters to treat on acre (0.42 hectares) are $56,000 and $1700, respectively. The costs are about 10% higher for below-ground filters.

A typical sand filter system consists of two or three chambers. The three-chambers system includes a sedimentation, a filtration chamber, and a discharge chamber. Many sand filters have only two chambers: one for sedimentation and the other for filtration and discharge. The sedimentation chamber removes coarse sediment and floating objects, and the filtration chamber contains a sand bed, typically 18 to 24 in. (45 to 60 cm) thick, for removing finer sediments and other pollutants. Figure 7.41 shows a two-chamber sand filter. Water enters the first chamber through grates, overflows a weir that partitions the two chambers, and enters the second chamber filled with sand. The infiltrated water is discharged directly through an outlet pipe or indirectly through a 6 in. perforated pipe, which may terminate to a drainage system or detention basin.

A variety of above- or underground sand filter designs are used in different parts of the country. Regardless of deviations in shape, sand filters are generally designed to treat the water quality storm and sized based on the following equation:

TABLE 7.48
Percent reductions in pollutant Concentrations for sand filters

Parameter	Units	Median Influent EMC	Median Effluent EMC	Percent Reduction
Total suspended solids	mg/L	56	9.0	84%
Fecal coliform	MPN/100 mL	900	400	56%
Total arsenic	μg/L	0.91	0.74	19%
Total cadmium	μg/L	0.30	0.08	73%
Total chromium	μg/L	2.0	1.0	50%
Total copper	μg/L	10	5.5	45%
Total iron	μg/L	642	210	67%
Total lead	μg/L	10	1.7	83%
Total nickel	μg/L	3.3	2	39%
Total zinc	μg/L	63	14	78%
Total phosphorus	mg/L	0.15	0.09	40%
Total Nitrogen	mg/L	1.2	1.1	14%
Nitrate + nitrite (as nitrogen)	mg/L	0.35	0.57	-63%

Source: EPA, NPDES, Sand and Organic Filters, December 2021.
EMC= event mean concentration

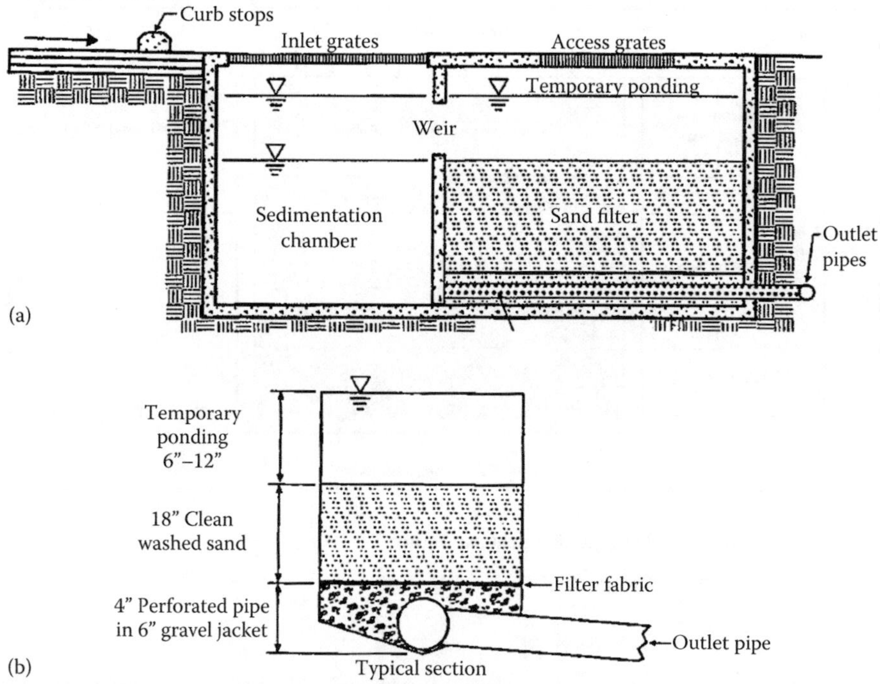

FIGURE 7.41 A two-chamber sand filter: (a) profile, (b) section.

Source: Maryland Stormwater Design Manual, Figure v3.14.

$$Q = A_s K \frac{\left[\left(D + d/2\right)/D\right]}{3600} \tag{7.29}$$

where

Q = average infiltrated flow rate through the sand bed in m³/s (cfs)
D = thickness of sand filter, m (ft)
d = maximum water depth (depth of water on sand bed at the end of storm event, m (ft))
K = sand permeability, m/h (ft/h)
A_s = surface area of sand bed, m² (ft²)

The average discharge from the sand filter, Q, is related to the water quality volume, WQ_v by

$$Q = \frac{WQ_V}{T} \tag{7.30}$$

where T is dewatering time of the sand filter in hours.

The maximum area to be treated by a sand filter is specified at 2.5 acres (1 ha) in New Jersey, where the water quality storm is 1.25 in, (3 cm) in 2 hours. The New York State Stormwater Design Manual (Draft- May 2022) specifies 2 acres as the maximum contributory area to an underground sand filter. By placing sand filters past detention basins, they may be sized for the attenuated runoff from a larger area. In areas of high groundwater, sand filters are preferred over infiltration trenches

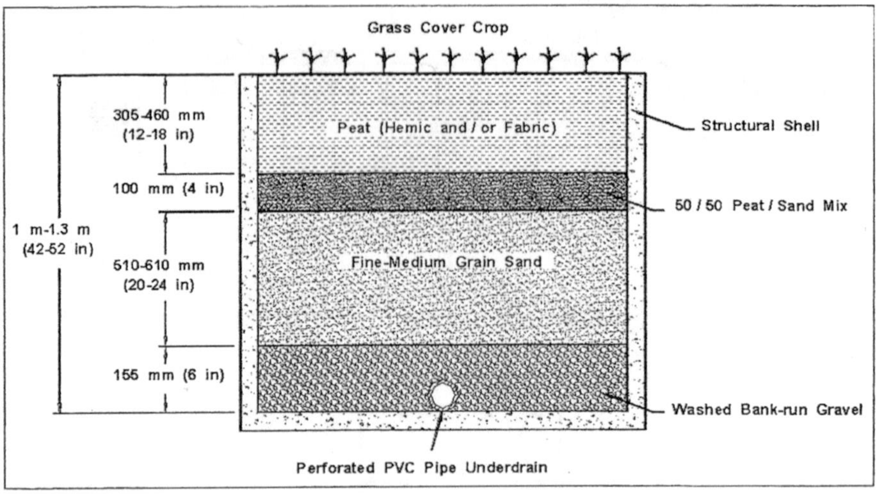

FIGURE 7.42 Cross-section schematic of a peat sand filter.

Source FHWA, HEC-22, 2013, Figure 10.6.

and infiltration basins, where contamination of groundwater with common pollutants such as BOD, suspended solids, or coliform is of concern.

Sand filters are accepted for 80% TSS removal by many jurisdictional agencies. They are also effective in removing some other pollutants if properly maintained. However, they are totally useless if their maintenance is neglected. The author has witnessed sand filters that were fully clogged and discharged the inflow runoff from their overflow weir without any treatment. To prolong their useful life, sand filters should be include a pretreatment stilling basin (see Figure 7.41). Figure 7.42 shows an organic filter in which a peat/sand mixture is substituted for sand bed.

7.6 INFILTRATION BASINS

Infiltration basins are primarily intended for water quality and/or groundwater recharge; however, they may be designed to serve as a multipurpose infiltration–detention system. To improve water quality, infiltration basins are sized to fully contain the runoff from a given storm event. This storm, which is generally known as the water quality storm, is specified by local or state jurisdictional agencies. An infiltration basin may be placed online or offline. In the former case, it receives the entire runoff, but in the latter case, it infiltrates only the water quality storm and diverts the excess runoff through an overflow weir. Discharge from the weir is then routed through an aboveground or underground detention basin. Figure 7.43 depicts an online infiltration basin with a pretreatment stilling basin.

An important criterion in the design of infiltration basins is the water quality storm. The water quality storm definition, as discussed in a previous chapter, varies from state to state. In New Jersey, for example, the water quality storm is specified as a storm of 2-hour duration, having nonuniform temporal distribution and amounting to 1.25 in. (32 mm) of rainfall. As far as the sizing of infiltration basins is concerned, the rainfall intensity has no significance. Preferably, the basin should be sized to retain the entire volume of the design storm and drain it overtime. Drawdown time (also known as dewatering time) and the infiltration bed area are related by the following equation:

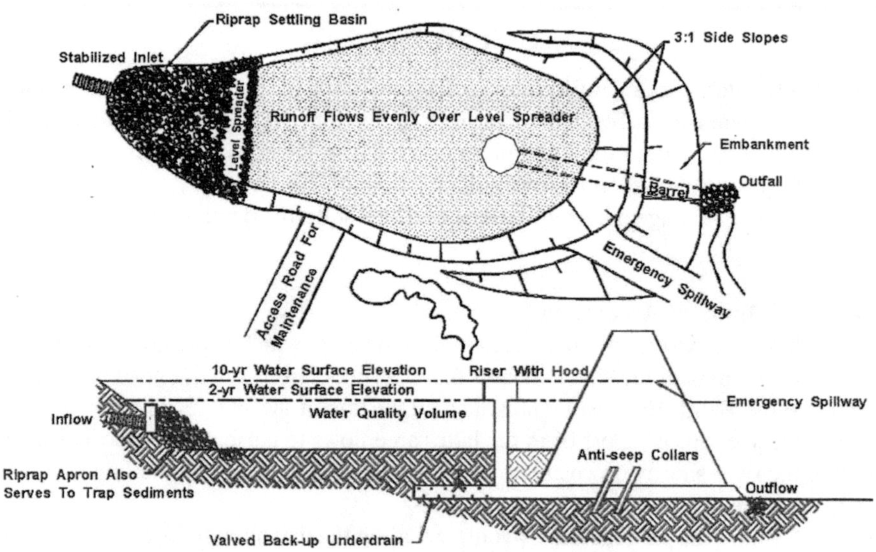

FIGURE 7.43 Online infiltration basin.

Source: HEC-22, Figure 10.4.

$$T = \frac{3600\left(WQ_v / A_f K\right)}{\left[\left(D + d/2\right)/D\right]} \tag{7.31}$$

where

T is drawdown time in hours
A_f is the area of infiltration bed in m^2 (ft^2);
other terms are the same as defined for sand filters

For an infiltration basin, the drawdown time is commonly limited to 24 or 48 hours, depending on whether the basin is vegetated or not. In practice, infiltration basins are commonly sized to store the water quality storm in full or in part and Equation 7.31 is used to calculate the drawdown time.

Note that clean sand has a permeability of 2 to 5 ft per hour (not per day), but the design criterion of 2 to 4 ft/day (0.6 to 1.3 m/day) is used to represent relatively clogged sand, the condition when the infiltration basin (and sand filter as well) should be maintained. Climate plays an important role in drawdown time. In cold climates, the soil media supporting vegetation takes longer to become anaerobic, and a longer drawdown time may be allowed. In humid climates, sand beds of infiltration basins (and sand filters) should have a higher permeability to be completely dry before the next storm. To discharge large storm events, a spillway may be incorporated into the infiltration basin design. Alternatively, the top of infiltration basins may be constructed as a level spreader, or a diversion system is provisioned.

In areas of high soil permeability, the infiltration basin may be expanded to serve as an infiltration–detention basin. In such a case, the basin will include an outlet structure to attenuate the peak rates of runoff beyond the water quality storm. The design of a dual-purpose infiltration–detention basin follows the same procedure as a detention basin: Develop stage–storage–discharge relation and calculate routing. The infiltration loss through the bottom of the basin during the design storm helps reduce the required storage volume. It is to be noted, however, that the infiltration during the storm event is generally small and is conservatively neglected in design. The following case study exemplifies the design of a detention–infiltration basin.

CASE STUDY 7.5

This case study relates to an infiltration–detention basin to address stormwater quality and quantity requirements for a residential project in Clifton, New Jersey. The site of development is situated partly within the flood plain of the Third River, which abuts the eastern property boundary. The project disturbs more than 1 acre (0.4 ha) of land; as such, it is subject to NJDEP stormwater management regulations and the state residential site improvement standards (RSIS) as well.

Preparation of Drainage Area Maps

Figure 7.44 shows an existing drainage area map. This map, which is prepared using 1 ft contour lines, and an aerial map of the site, indicates that the runoff from a major portion of the site travels overland to the Third River, while only a small portion of the site drains to an existing access road west of the site. Since runoff from the latter area flows to a tributary of the Third River that merges with the river shortly downstream of the site, the entire area of disturbance is considered as a single drainage area.

Figure 7.45 shows the proposed site layout, including the detention infiltration basin and drainage conveyance system. Also shown on this map are the main drainage areas, designated as areas A1 and B2. Area A1 includes most of the development site, which is a tributary to the proposed detention basin. This basin, which is located in the southeast corner of the property, discharges to the Third River and is designed as a dual-purpose detention–infiltration basin to address both flow reductions and water treatment requirements. Area B2 is the balance of the site, where the runoff occurs as overland flow. This area includes the stream buffer and a vegetated depression east of the development and south of the detention basin. Also included in B2 is a small portion of the development that due to topography drains to the access road west of the site.

Runoff Calculations

Runoff calculations for the site are performed using the SCS TR-55 method. Based on the soil survey map of the county in which the site is located, the site soil is urban land, Boonton–Wethersfield; the latter two soils are classified as hydrologic group C. In compliance with the RSIS and the NJDEP runoff calculations include the storms of 2-, 10-, and 100-year frequency. Additionally, calculations are performed for the 25-year storm which accords with the local soil conservation district standards. Table 7.49 lists the computed existing peak rates of runoff from the site.

Runoff calculations for the proposed conditions are prepared separately for the area tributary to the detention–infiltration basin, area A1, and the area bypassing it, area B2. Similar to the existing conditions, the runoff calculations for the proposed conditions are prepared for 2-, 10-, 25-, and 100-year- 24 hour storms. Consistent with the Residential Site Improvement Standards (RSIS) a minimum time of concentration of 10 minutes is used for area A1. Haestad Methods software, which is employed for runoff computations, rounds this time to 0.2 hour.

The proposed stormwater management system for the project includes a detention–infiltration basin, also an underground rain cistern for collecting the roof runoff from two of the interior buildings for watering of lawns and landscaped areas. This will not only reduce the volume of runoff discharged to reduce the volume of runoff discharged to detention–infiltration, but also conserve the water demands of the development. To arrive at a conservative detention basin design, however, it is assumed that the roof runoff from these two buildings will drain directly to the detention basin if the storm occurs when the cistern is full.

To meet the NJDEP's 80% TSS removal requirement, the proposed dual-purpose basin has been designed to fully retain and infiltrate the water quality storm. Figure 7.46 shows a schematic detail of the outlet structure. This figure indicates that the crest of the weir is located 2 ft (0.6 m)

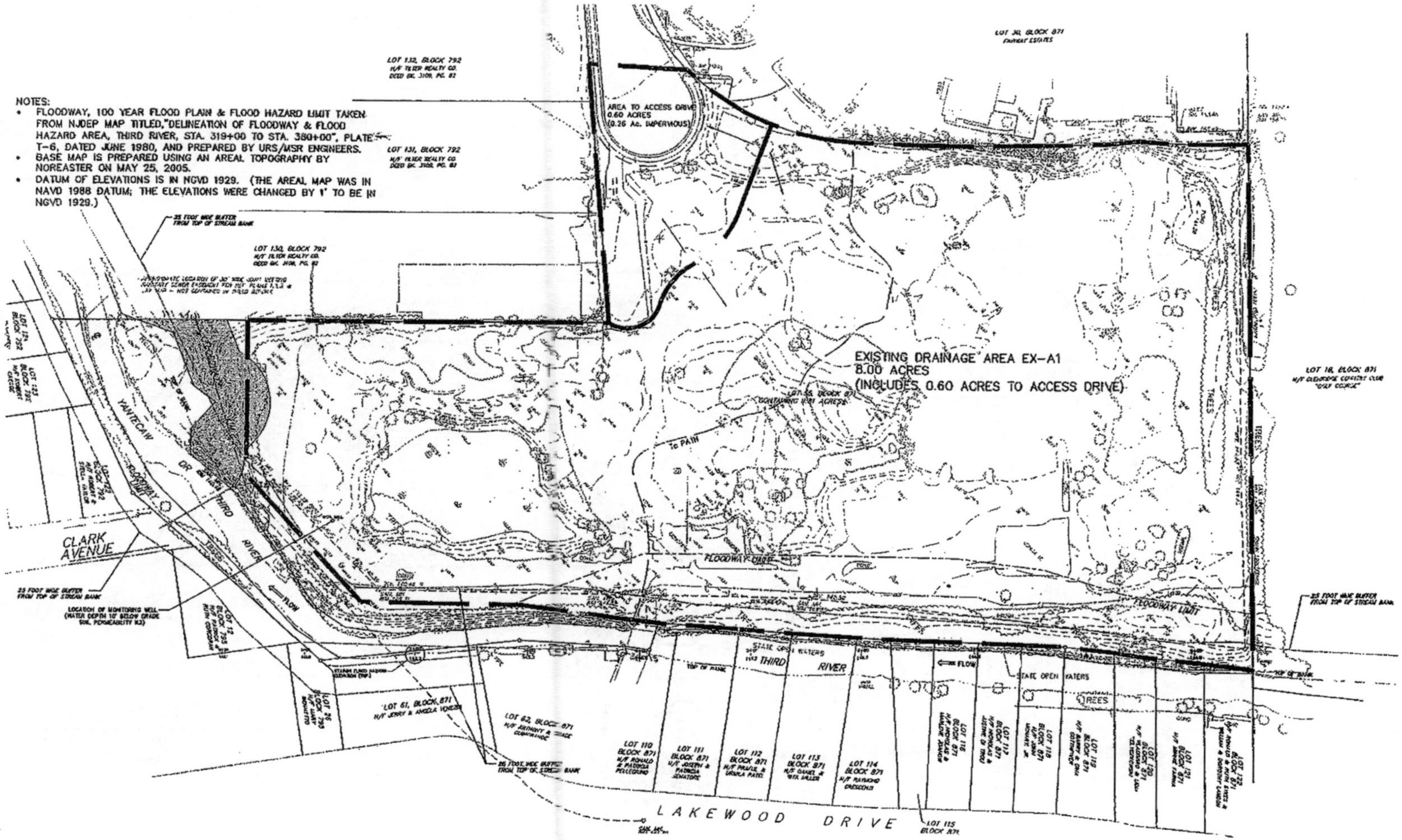

FIGURE 7.44 Existing drainage area map.

FIGURE 7.45 Post-development drainage area map.

TABLE 7.49
Peak Rates of Runoff, Existing Conditions

Storm Frequency (years)	Rainfall Amount (in.)	Q (cfs)
2	2.4	14
10	5.2	24
25	6.4	31
100	8.7	43

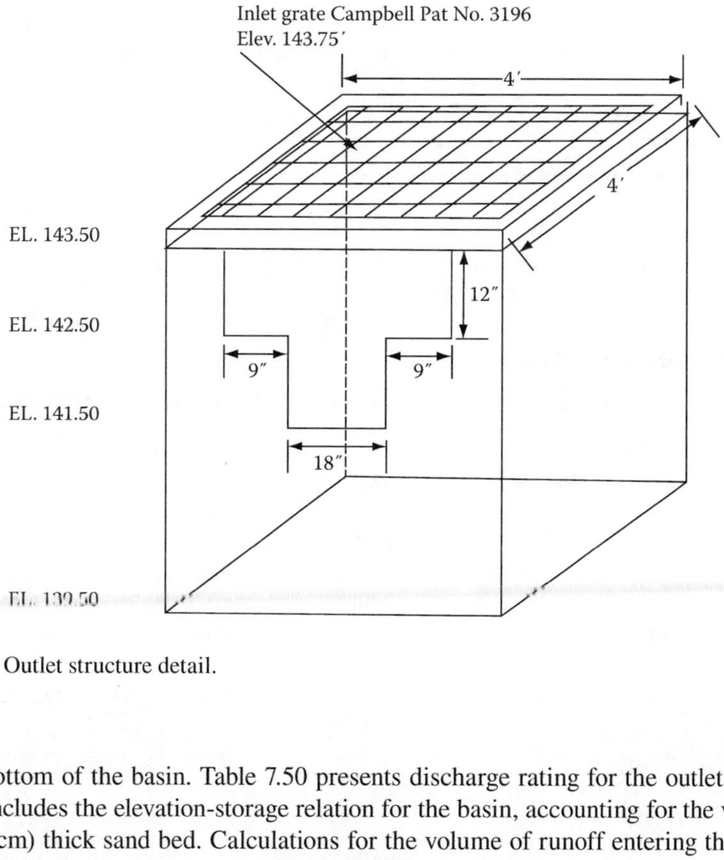

FIGURE 7.46 Outlet structure detail.

above the bottom of the basin. Table 7.50 presents discharge rating for the outlet structure, and Table 7.51 includes the elevation-storage relation for the basin, accounting for the void volume in the 1 ft (30 cm) thick sand bed. Calculations for the volume of runoff entering the basin during the water quality storm, are presented in Table 7.52. As shown, storms up to and including the water quality storm are fully contained below the weir in the outlet structure and infiltrate into the sand bed. Conservatively, the infiltration losses during the water quality storm are neglected in the calculations. The maximum depth of standing water within the infiltration basin is less than 2 ft during the 2-hour water quality storm and, as such, the infiltration basin is in compliance with the NJDEP *Stormwater Best Management Practices Manual.*

Routing computations for the detention–infiltration basin are performed for 2-, 10-, 25-, and 100-year storms Table 7.53 summarizing the computations, show that the proposed basin significantly attenuates the peak runoff rates. The composite discharges from the site are calculated by adding the outflow hydrographs from the basin to the runoff hydrographs for the bypass area. The composite discharges are listed in Table 7.54 and Table 7.55 presents a comparison of the

TABLE 7.50
Discharge–Elevation Relation[a]

Elev.	Q (cfs)				
	18 in. × 24 in.[b]	18 in. × 12 in.	Grate	20 ft spillway	Total
141.5	0.00				0.00
142.0	1.59				1.59
142.5	4.50	0.00			4.50
143.0	8.27	1.59			9.86
143.5	12.73	4.50			17.23
143.75	16.14	6.25	0	0	22.40
144.0	17.69	7.22	4.48	7.00	36.39

[a] Refer to Figure 7.46.
[b] Center weir.

TABLE 7.51
Stage- Storage Relation

Elev.	Area (ft²)	Ave. Area (ft²)	Volume (ft³)	Total Volume (ft³)
138.5	6900		0	0
	(12 in. sand; 30% void)	6900		
139.5	6900		2070	2070
		7240		
140.0	7580		3620	5690
		8240		
141.0	8900		8240	13,930
		9240		
141.5	9580		4620	18,550
		9940		
142.0	10,300		4970	23,520
		10,650		
142.5	11,000		5325	28,845
		11,370		
143.0	11,740		5685	34,530
		12,110		
143.5	12,480		6055	40,585
		12,860		
144.0	13,240		6430	47,015

existing and proposed peak rates of runoff. The latter table indicates that the selected stormwater management system for the project reduces the peak rates of runoff far in excess of the applicable requirements.

TABLE 7.52
Water Quality- Area A1- Detention–Infiltration Basin Drainage Areas

Cover	C	Area (ac.)	$C \times A$
Impervious	0.95	3.5	3.33
Grass	0.35	1.38	0.48
Brick paver	0.6	0.27	0.16
Totals		5.15	*3.97*
			$C_W = 0.77$

Water quality storm = 0.625 in./h for 2 h
$Q = 0.625$ in./h $\times C_W \times A = 2.48$ cfs
$WQ_V = Q \times 2$ h $\times$ 3600 s/h = 17,865 ft^3
Storage provided below outlet = 18,550 ft, at elev. 141.5
See Table 7.58.

TABLE 7.53
Detention Basin Routing Computations Summary, 24 h, SCS Type III Storms

Storm Frequency (years)	Q_{IN} (cfs)	Q_{OUT} (cfs)
2	11.0	2.10
10	17.0	6.94
25	22.0	11.33
100	31.0	19.90

TABLE 7.54
Composite Peak Rates of Runoff Proposed Conditions

Storm Frequency (years)	Q[a] (cfs)
2	3.0
10	9.7
25	15.5
100	29.4

[a] Calculated by adding hydrographs.

100-Year Storm			
Time (h)	Out (cfs)	B2 (cfs)	Total (cfs)
12.2	2.2	12.0	14.1
12.3	8.1	14.0	22.0
12.4	15.3	12.0	27.3
12.5	19.4	10.0	29.4
12.6	19.9	7.0	26.9
12.7	18.4	5.0	23.4

TABLE 7.55
Existing and Proposed Peak Rates of Runoff from the Site

Storm Frequency (years)	Q_E (cfs)	Q_P (cfs)	Reductions (%)
2	14	3.0	79
10	24	9.7	60
25	31	15.5	50
100	43	29.4	32

Design of Drainage System

Consistent with the RSIS, the internal drainage pipes are designed based on a 25-year storm. Flow calculations are based on the rational formula using a runoff coefficient of 0.35 for lawn and landscaped areas and 0.95 for impervious areas. The discharge capacity of the proposed RCP drainage pipes is based on Manning's roughness coefficient of 0.013. An 18 in. RCP is used as the outlet pipe from the detention-infiltration basin. This pipe has ample capacity to convey the 25-year storm discharge based on the slope of the pipe. The pipe can also adequately discharge the 100-year storm under pressure due to the standing water in the basin. To prevent river erosion, a scour hole is provisioned at the terminus of the outlet pipe at the river. The design of the scour hole is conservatively based on the 100-year storm discharge from the basin.

7.7 RETENTION–INFILTRATION BASINS

Retention–infiltration basins, unlike detention basins, have little discharge while filling but discharge without any attenuation once full. This major difference is illustrated in Figure 7.47, which shows that for a given storage volume, a detention basin attenuates the inflow runoff, but a retention–infiltration basin may surcharge at the same rate as the runoff that enters it. Thus, contrary to a common design procedure by many engineers, the required detention storage calculations are not relevant to retention basin design. To prevent frequent surcharge, retention–infiltration basins should be designed to fully retain the runoff they receive during a design storm. This storm should be selected based on the type of basin, nature of development, and site climatic conditions.

Retention–infiltration basins can be constructed aboveground or underground. In urban development projects the latter type of basin is more common than the former. To discharge runoff in excess of the design storm, retention basins can be provided an emergency outlet or spillway. A flat spreader on the top of aboveground detention basins and an overflow pipe in underground basins

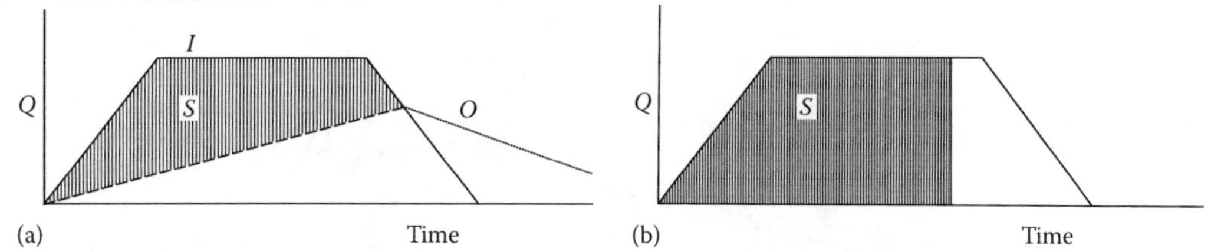

FIGURE 7.47 Flow routing through (a) detention basin and (b) retention basin.

serve this purpose. Perforated pipes or chambers in a stone trench, storage modules, and dry wells are examples of retention–infiltration basins. Systems that comprise chambers and storage modules were described previously. Dry wells (commonly referred to as seepage pits and are widely employed used for small pavement areas and particularly for dewelling roofs of single-family homes) and their design procedure will be discussed in the next chapter.

CASE STUDY 7.6

This case involves a residential–commercial development project in a 5.4-acre (21,850 m^2) parcel of property in the town of Secaucus in Hudson County, New Jersey. The site is situated within the jurisdiction of the New Jersey Meadowlands Commission and is partly located within the tidal floodplain of the Hackensack River. Figure 7.48 shows the predevelopment site layout. The site is covered by a building, asphalt pavement, gravel parking, and strips of grass. The development consists of a retail building, three residential apartment buildings, and associated parking and driveways. The post-development site layout is shown in Figure 7.49.

Stormwater Management Plan

The predevelopment drainage subareas are delineated in Figure 7.48. Shown on this plan are areas El and E2; area El drains to the existing drainage system at Seaview Drive extension and New County Road. The runoff from area E2 drains to the rear properties, lots 9.01 and 10, and mostly enters an inlet in lot 9.01. Because of the inadequacy of the drainage system in these lots and the drainage system downstream thereof, this inlet experiences frequent flooding.

To alleviate flooding in the northerly properties, the stormwater management system is designed to divert the runoff from a portion of area E2 to the drainage system on the Seaview Drive extension. However, to offset the impact of this diversion on the roadway drainage system, an aboveground detention basin is incorporated into the stormwater management system. To further reduce flooding in the rear properties, the runoff from the roofs of residential buildings 1, 2, and 3 is directed to underground retention–infiltration basins and fully detained for a selected design storm.

The layout of the drainage system is shown in Figure 7.49. Also delineated on this plan are the post-development drainage subareas, labeled as area P1, area P1-D, and P2. P1, covering 1.46 acres, drains directly to the roadway drainage system; area P1-D, consisting of 1.75 acres, is routed through the detention basin and is discharged into the roadway drainage system. Area P2 includes 0.9 acre of green area behind the residential buildings and is the only area tributary to the rear properties. An underground retention–infiltration basin is also provisioned for building 4 (office building) to reduce the runoff to Seaview Drive extension.

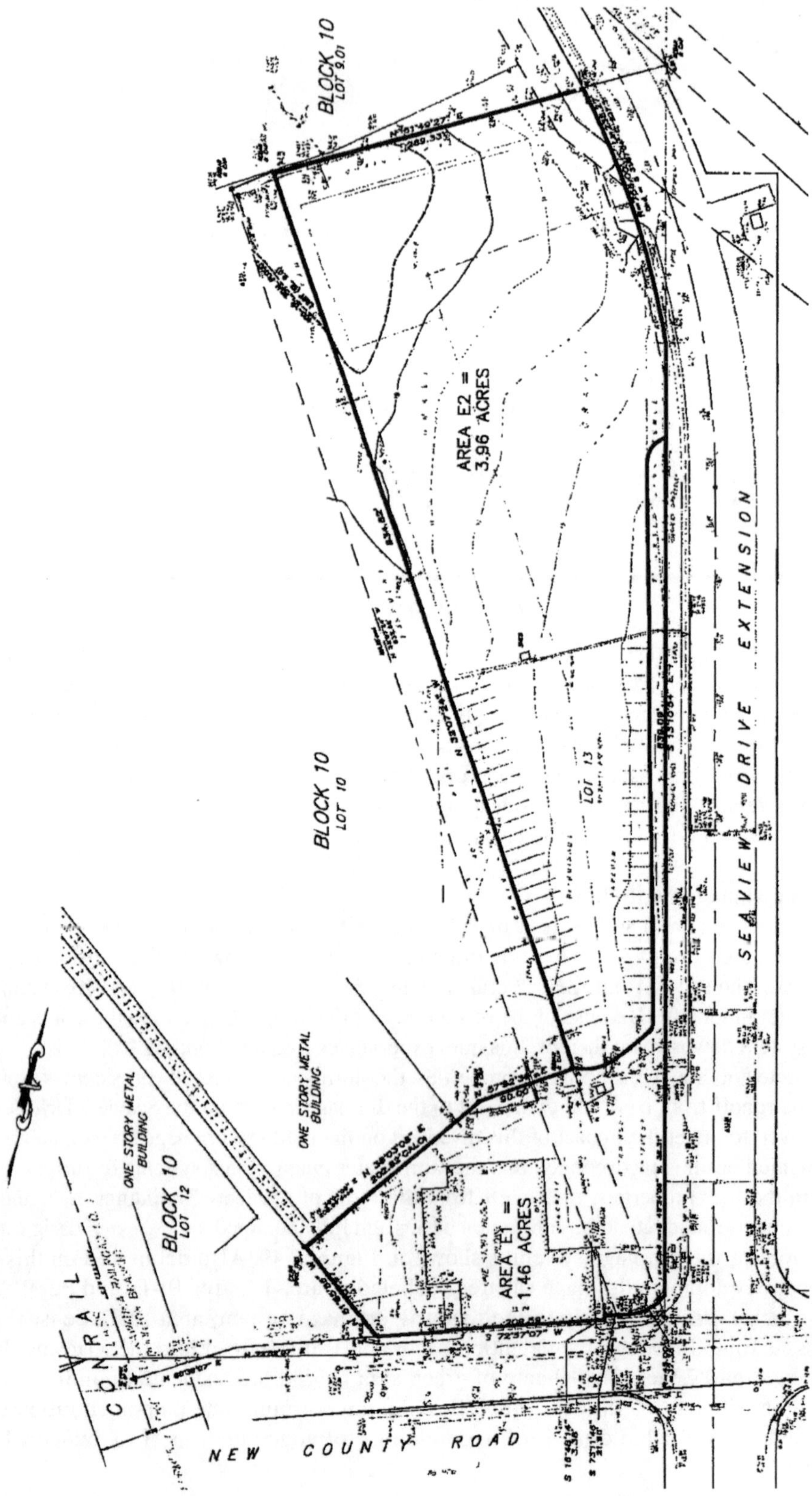

FIGURE 7.48 Predevelopment site layout and drainage areas.

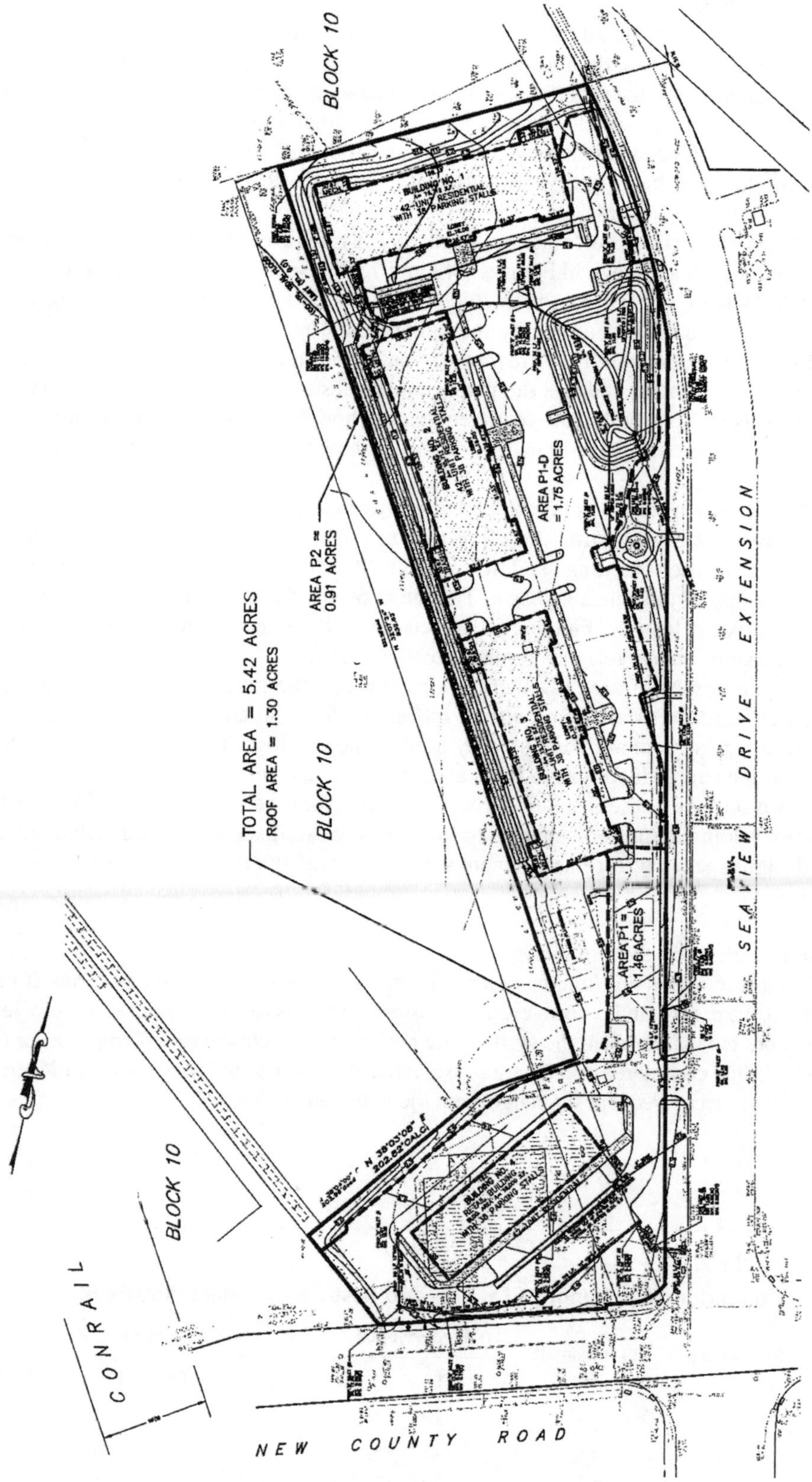

FIGURE 7.49 Post-development site layout and drainage subareas.

Runoff Calculations

Runoff calculations are prepared using SCS TR-55 method. General soils map of Hudson, Essex, and Passaic (HEP) Counties, designate the site soil as hydrologic group C. Considering that the site is mostly impervious, the runoff calculations are based on a time of concentration of 10 minutes for both pre- and post-development conditions. In compliance with the NJDEP stormwater regulations, the runoff calculations include 2-, 10-, and 100-year storms. However, internal drainage pipes are designed for 25-year storms to comply with the RSIS and the N.J. Meadowlands Commission drainage standards.

Table 7.56 summarizes the computed peak runoff rates from the pre- and post-development drainage subareas shown in Figures 7.48 and 7.49, respectively. Also listed in this table are the SCS 24-hour rainfall depths in Hudson County. Note that the roof areas of all four buildings, which are fully detained by the underground retention–infiltration systems, are excluded from the computations. A review of Table 7.56 indicates that the proposed stormwater management system reduces the peak runoff rates to the rear properties by sixfold or more for storms of 2- through 100-year frequency. Though it will increase the runoff to the roadway drainage system. This impact is offset by the proposed detention basin, described next.

The detention basin shown in Figure 7.49 is approximately 4 ft (1.2 m) deep, 10,340 ft^2 (961 m^2) on plan area and provides over 26,700 ft^3 (756 m^3) of storage volume. The basin is designed with an outlet structure consisting of a low-flow 4 in. (100 mm) orifice and an overflow grate just 2 in. below the top of the detention basin. The intent of this design is to attenuate the runoff to the maximum extent feasible. Figure 7.50 depicts the outlet structure, and Table 7.57 presents elevation-storage-discharge relations for the detention system.

Runoff hydrographs of area PD-1 are routed through the proposed detention basin, and the computed discharges are summarized in Table 7.58. The composite discharges to the roadway drainage system are computed by adding the outflow hydrograph from the detention basin to the runoff hydrograph from area P-1 for each storm event. Table 7.59 lists the composite discharges to the Seaview Drive extension from the site. Also listed in this table are the predevelopment peak discharges to the roadway. An inspection of this table indicates that the designed stormwater management system reduces the runoff to the roadway drainage system.

Roof Runoff Retention System Design

The roof runoff retention–infiltration systems are designed to fully retain the roof runoff generated by the 10-year, 60-minute storm event. This storm amounts to 2 in. of rainfall in New Jersey. Overflow pipes emanating from the systems are designed to discharge runoff from longer duration storms. While considered as an extra conservative design criterion by some practitioners, the selected storm prevent frequent surcharge from the retention system.

TABLE 7.56
Computed Peak Runoff, CFS, Pre- and Post-development, Subareas

Storm Freq. (years)	24-h Rainfall (in.)	Predeveloped		Post developed		
		E1	E2	P1	P2	P1-D
2	3.3	4	8	3	1	3
10	5.0	6	14	5	2	6
100	8.3	10	25	9	4	11

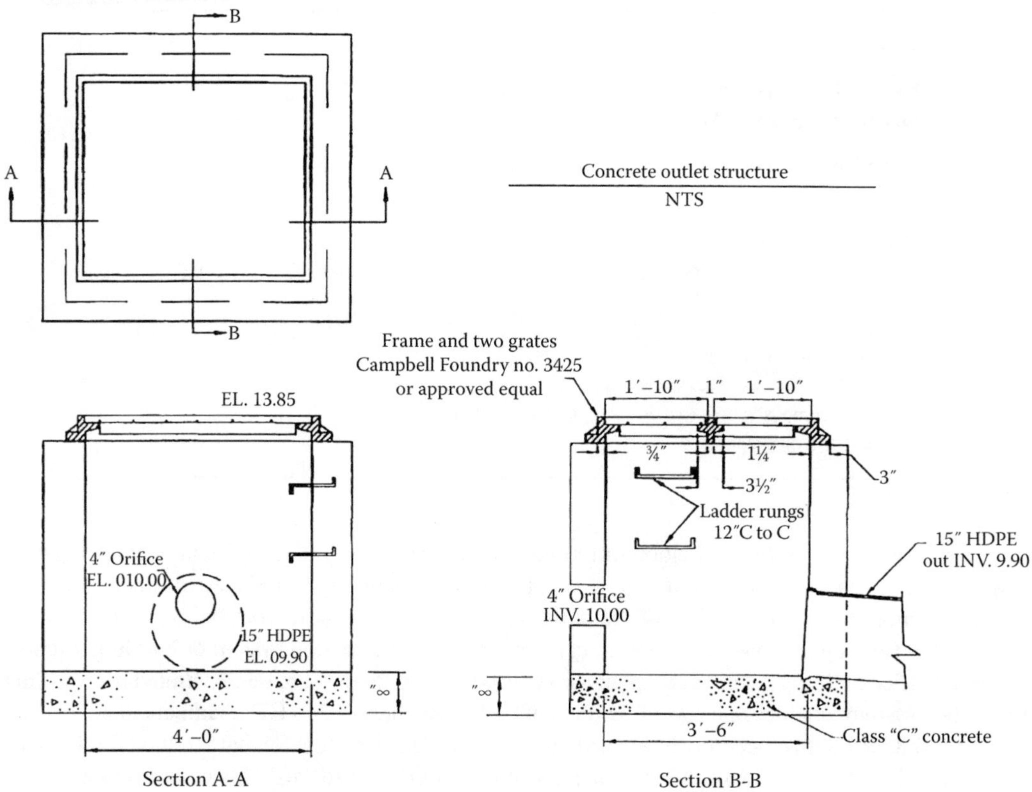

FIGURE 7.50 Outlet structure detail.

TABLE 7.57
Detention Basin Elevation–Storage–Discharge Relation

Elev., ft (m)	Storage, ft³ (m³)	Discharge, cfs (L/s)
10.0 (3.28)	0	0
10.5 (3.44)	625 (17.7)	0.24 (6.8)
11.0 (3.61)	2695 (76.3)	0.38 (10.8)
11.5 (3.77)	5800 (164.3)	0.48 (13.6)
12.0 (3.94)	9315 (263.9)	0.57 (16.1)
12.5 (4.10)	13,140 (372.2)	0.64 (18.1)
13.0 (4.26)	17,200 (487.3)	0.71 (20.1)
13.5 (4.43)	21,760 (616.4)	0.77 (21.8)
14.0 (4.59)	26,730 (757.2)	0.82 (23.2)

TABLE 7.58
Routing Computations Summary

Storm Freq. (years)	Inflow Peak, cfs (L/s)	Outflow Peak, cfs (L/s)
2	3.0 (84.6)	0.49 (13.9)
10	6.0 (169.9)	0.63 (17.8)
100	11.0 (311.5)	0.80 (22.7)

TABLE 7.59
Existing and Proposed Discharges to New County Road and Seaview Drive Extension (Area 1)

Storm Freq. (years)	Q_{E1} (cfs)	Q_{P1} (cfs)	Q_O (cfs)	Total Q_P (cfs)
2	4	3	0349	3.4
10	6	5	0.63	5.5
100	10	9	0.80	9.6

Notes:
Q_{P1} = direct runoff to roadway
Q_O = peak outflow from detention basin
Total Q_P = composite flow to roadway; computed adding Q_{P1} and Q_O hydrographs

In this project, a shallow retention–infiltration system consisting of Cultec chambers is selected to retain roof runoff. As shown in Figure 7.49, the roofs of buildings 2 and 3 are drained to a common retention system in the back of the buildings and the roof runoff from buildings 1 and 4 are drained to two separate systems. Design calculations for the retention system of building 4 (commercial/office building), with an on-plan area of 10,000 ft^2 (929 m^2), are demonstrated herein. This retention system includes one row of 22 units of Cultec Recharger 280 HD chambers in a 157 ft by 5.92 ft (47.85 × 1.8 m) stone trench. According to the manufacturer, the Recharger 280 HD chamber alone provides 6.079 ft^3 of storage volume per lineal foot (0.565 m^3/m). This chamber is 8 ft long by 47 in. wide by 26.5 in. high (2.43 m × 119.4 cm × 67.3 cm). Calculations for the required storage volume and the retention volume of the retention–infiltration system are presented as follows.

Roof Runoff Volume

a. Roof area = 10,000 ft^2
b. Design storm = 2 in. (50 mm) rainfall
c. Required storage volume = (roof area) × (rainfall) = 10,000 × 2 × 0.98/12 = 1633 ft^3 (46.2 m^3)

where the coefficient 0.98 represents the percentage of rainfall that generates runoff; surface retention is estimated at 2%.

Storage Volume of Retention System

Trench dimensions are 157 ft × 5.92 ft × 3.21 ft; calculated as follows:

$$Length = 20\times 7.0\left(inner\ pieces\right)+2\times 7.5\left(end\ pieces\right)+2\times 1\left(stone\ extension\right)=157\ ft$$

where 7.5 ft is the effective length of end pieces

$$Width = 47\ in.\left(chamber\right)+2\times 12\ in.\left(side\ stone\right)=71\ in.=5.92\ ft$$

$$Depth = 6+26.5+6=38.5\ in.=3.21\ ft\left(0.98\ m\right)$$

$$Bed\ area = 157\ x 71/12 = 928.92\ ft^2\left(86.30\ m^2\right)$$

Volume in chambers:

$$\left(157-2\right)\times 6.079\ cf/ft = 942.2\ ft^3\left(26.7\ m^3\right)$$

Void volume in stone trench:

$$\left(928.92\times\frac{38.5}{12}-942.2\right)\times 0.4=815.2\,ft^3\left(23.1\,m^3\right),$$

where 0.4 represents porosity (void volume/trench volume)

$$Total\ volume=942.2+815.2=1757.4\,ft^3\left(49.86\,m^3\right)$$

The above calculations show that the retention–infiltration system can fully retain the roof run-off. It should be noted that the above calculations conservatively neglect percolation through the bottom of stone trench during the storm period; however, the calculations also ignore the volume occupied by the chamber walls.

To assure the proper functioning of the roof runoff retention–infiltration systems, two rest pits were dug to measure the water table. One of the test pits was dug at the location of the detention basin, and the other behind the proposed building 2, where a retention–infiltration system would be placed. The soil samples from the two test pits were sent to Johnson Soils Company (a geotechnical engineering firm in northern New Jersey). A letter report by Johnson Soil company presenting the test results indicated the following:

A permeability rating of K2 at test pit 1 (TP-1, behind building 2) and K3 at test pit 2 (TP-3, at detention basin site). Depths of water table were also measured at these two pits and are noted here:

TP-1: approximate ground elevation 10.2 ft
Water at 3 ft, 6 in. depth
TP-3: approximate ground elevation 12.3 ft
Water at 4 ft, 0 in. depth

These findings indicate that the water table lies approximately 2 ft (0.6 m) below the post -development elevations at the bottom of the retention system and 1.7 ft (0.5 m) below the bottom of the detention basin.

Time to drain system. Based on soil permeability results, the drain time of the infiltration basin behind building 4 is estimated as follows:

$$Average\ depth\ of\ system=\frac{1757.4}{928.92}=1.89\,ft=22.7\ in.$$

$$Soil\ rating=K2, 0.6-2\ in.\ per\ hour.\left(based\ on\ soil\ tests\right)$$

$$Average\ permeability=1.3\ in./h$$

$$Time\ to\ drain=\left[\frac{22.7}{\left(1.3\ \text{in.}/\text{h}\right)}\right]=17.46\ h$$

$$\times 2\ safety\ factor\ =34.9h\ =1\ day, 10.9\ h$$

Water Quality Provisions

The project replaces gravel pavement (which is considered as a pervious surface in terms of water quality) with bituminous pavement. As such, it produces a net increase of 0.15 acre in impervious areas. Since the increase in impervious area is under 0.25 acre, the project is exempt from the state water quality requirement. However, three CDS water quality units were installed to treat the stormwater discharges from the site. One of these units (CDS no. 2) is placed in area P1-D (past

TABLE 7.60
Water Quality Calculations: TSS Removal Rates and CDS Sizing

Required TSS Removal Rate

Cover	Area (ac.)
Existing impervious	1.96
Increase in impervious area (excluding bldgs.)	0.15

Although no water quality measure was required, CDS units were installed to treat most of the site's impervious area.

Proposed TSS Removal Rate

Area	(Acres)	TSS Removal Rate	A × % Removal
Total impervious area to CDS units	1.96	50%	0.98

Weighted *C* for CDS Unit No. 1

Cover	Area (ac.)	*C*	*C* × *A*[a]
Impervious	0.66	0.95	0.63
Grass	0.09	0.35	0.03
Total area and C value	0.75		0.66

Weighted *C* for CDS Unit No. 3

Cover	Area (ac)	*C*	*C* × *A*
Impervious	0.152	0.95	0.14
Grass	0.038	0.35	0.01
Total area and C value	0.190		0.16

CDS Unit Sizing

	C × *A*	*I* (in./h)	*Q* (cfs)	CDS Unit Model No.[b]
CDS unit no. 1	0.66	3.2	2.11	CDS 30–30
CDS unit no. 2			0.41[c]	CDS 20-15-4
CDS unit no. 3	0.16	3.2	0.50	CDS 20–15-G

[a] Rounded to second decimal place.
[b] CDS model number. based on NJDEP and NJCAT treatment rates (approved at the time of design).
[c] Peak flow discharge from detention basin during water quality storm event.

outlet structure of detention basin) and two units (no. 1 and no. 3) are in area P1. Design calculations for these CDS units are provided in Table 7.60.

7.8 STORMWATER WETLANDS

Stormwater wetlands, also known as constructed wetlands, are structural stormwater control practices that can both treat and store runoff. They are similar to wet ponds except that they are generally shallow and incorporate wetland plants in a shallow marshy area.

Natural and constructed wetlands have many beneficial functions, one of which is water filtration. As water flows through a wetland, it slows down, and most of the suspended solids become trapped by vegetation and settle out. Other pollutants are transformed into less soluble forms and are taken up by plants or become ineffective. Wetland plants also create favorable conditions for microorganisms to live and reproduce. Through a series of complex processes, these microorganisms also transform and remove pollutants from water.

Natural wetland systems have often been described as the "earth's kidneys" because they filter pollutants from water that flows through on its way to receiving lakes, streams, and oceans. Because these systems can improve water quality, engineers and scientists construct systems that replicate the functions of natural wetlands. Constructed wetlands use natural processes involving wetland vegetation, soils, and their associated microbial assemblages to improve water quality.

Stormwater wetlands are designed specifically for stormwater treatment and are one of the most effective stormwater practices in that regard. They also provide aesthetic amenity and habitat value. Stormwater wetlands are basically different from natural wetlands in that they have less biodiversity than natural wetlands. Several variations of stormwater wetland design exist, differing in the relative amounts of shallow and deep water and dry storage.

Stormwater wetlands are applicable in many parts of the United States other than arid climates and have limited applicability in highly urbanized settings. If stormwater wetlands are used in high spots where land use generates highly contaminated runoff, they will need significant separation from groundwater. Stormwater wetlands can be used in almost all types of soil and sites with slopes of up to 15%.

Constructed wetlands are generally built on uplands and outside flood plains to prevent damage to natural wetlands and other aquatic systems. Wetlands are commonly constructed by excavation, backfilling, grading, construction of berms, and occasionally installation of flow control structures. In highly permeable soils, an impermeable compacted clay liner is usually installed, and the original soil is placed over the liner. Wetland vegetation is then planted or allowed to establish naturally. Constructed wetlands are treatment systems that use natural processes involving wetland vegetations, soils and their associated microbial assemblages to improv water quality EPA (2025).

Stormwater wetland design varies considerably depending on site constraints. However, some features, including pretreatment, conveyance, maintenance reduction, and landscaping, should be incorporated into most wetland designs. Design variations of stormwater wetlands include shallow marsh, extended detention wetland, pond/wetland design, pocket wetland, and gravel-based wetlands. Figure 7.51 shows a pocket wetland. In the shallow marsh, most of the wetland volume is in a shallow high or low marsh, and the only deep portions are the forebay at the inlet to the wetland and the micro-pool at the end.

An extended detention (ED) wetland is similar to the shallow marsh, with additional storage above the surface of the marsh that provides an extended detention basin. A pond/wetland system, as the name implies, combines a wet pond with a shallow marsh. Stormwater runoff enters the wet pond and then flows through the marsh. Since the pond can be 6–8 ft (1.8–2.4 m) deep, this type of wetland requires less surface area than shallow marsh.

In the pocket pond, the bottom of the wetland intersects the groundwater, which helps in maintaining a permanent pool. Because of groundwater flows, this type of wetland may not be as effective as others and should be used when the drainage area is not large enough to maintain a permanent pool. In gravel-based wetlands, water flows through a rock filter. At the surface, pollutants are removed through uptake by the plants and biological activity on the surface of the rocks. This type of wetland is basically different from other wetland designs; it is more similar to filtering systems.

Table 7.61, which is based on limited data points compares TSS removal efficacy of various types of stormwater wetlands. Stormwater wetlands are relatively inexpensive, but consume 2%–4% of their drainage area, which is high compared with other stormwater management practices. Construction cost information for stormwater wetlands is scarce, but it can be assumed that wetlands cost about 25% more than wet ponds of an equivalent storage volume. The cost of a wetland to store 1000 m^3 (35,000 ft^3) of a 10-year storm runoff may be estimated at $60,000.

Suggested design criteria for constructed wetlands are as follows:

Depending on its depth, the surface area of stormwater wetland should be between 2% and 4% of the contributing drainage area. For gravel-base wetlands, the area also varies with the thickness of the gravel.

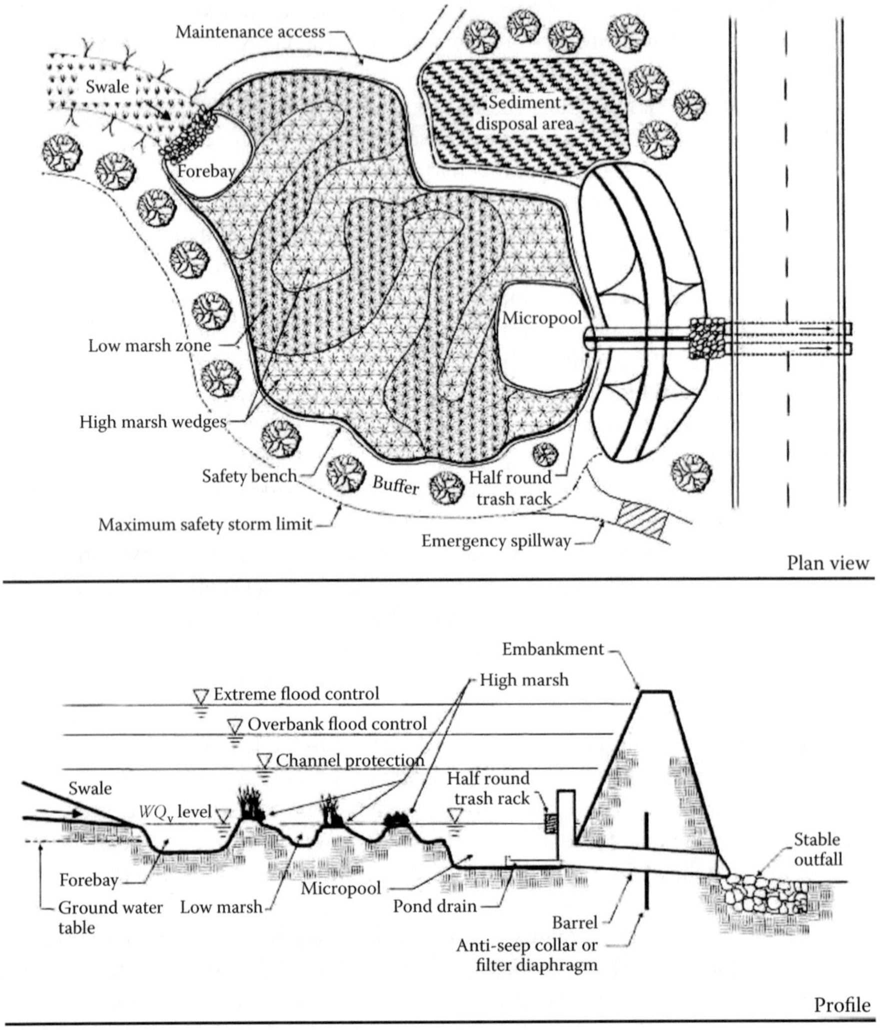

FIGURE 7.51 Pocket wetland.

Source: Maryland Stormwater Management Design Manual (2009).

TABLE 7.61
Typical Removal Rates (%) of Wetlands

	Stormwater Treatment Practice Design Variation			
Pollutant	**Shallow Marsh**	**ED Wetland**[a]	**Pond/Wetland System**	**Submerged Gravel Wetland**[a]
TSS	83–51	69	71–35	83
TP	43–40	39	56–35	64
TN	26–49	56	19–29	19
NOx	73–49	35	40–68	81
Metals	36–85	(80)–63	0–57	21–83
Bacteria	76[a]	NA	NA	78

Source: EPA, NPDES Stormwater Wetlands (December 2021).

[a] Data based on fewer than five data points.

In a constructed wetland, contrary to a wet pond, only a small fraction (less than one-fifth on average) of the surface area is deep open water. At a minimum, one-third of the surface area should be shallower than 15 cm (6 in.) and at least two-thirds of the surface area 50 cm (18 in.) or shallower. A forebay shall be located at the inlet, and a 1.2–1.5 m (4–6 ft) deep micro-pool that stores approximately 10% of WQ_v shall be located at the outlet to prevent clogging of the low-flow opening and sediment resuspension (Maryland Stormwater Management Manual, 2009).

If extended detention is used in a stormwater wetland, provide a minimum of 50% of the WQ_v in the permanent pool; the maximum water surface elevation of WQ_v-ED must not extend more than 1 m (3 ft) above the permanent pool (*New York State Stormwater Management Design Manual*, 2022). Use of stormwater wetlands on trout waters is strongly discouraged as these practices increase stream temperatures.

7.8.1 Subsurface Gravel Wetlands

A subsurface gravel wetland, also referred to as submerged gravel wetland (SGW) functions similarly to natural wetlands except that it is partly underground. This system, which was originally designed at the University of New Hampshire Stormwater Center (UNHSC), is based in part on the idea of multiple-pack reactors commonly used in wastewater treatment.

The original SGW consisted of two treatment cells, preceded by a forebay to retain coarse particles and gross debris. Both cells include a surface wetland and a crushed stone sublayer separated by a minimum 75 mm (3 in.) of graded aggregate filter to prevent wetland soil from entering the gravel sublayer. Figure 7.52 depicts the layout of the UNHSC subsurface gravel wetland. The water leaving the discharge pipe from the forebay is retained in the first cell and slowly moves down to the gravel sub-base through holes in a standpipe. Any water in excess of the capacity of the first cell enters the second cell and is also filtered down through the standpipe. The water in the gravel layer moves under the first cell to the second and is discharged through an outlet with its invert at the top of the gravel subbase. The system is preceded by a hydrodynamic separator device, swale, or forebay.

For permit compliance with the regulations, water quality monitoring was performed from July 2007 through October 2010 at the GSW of Greenland Meadows, a high-use commercial site. The discharge from the GSW enters Pickering Brook, which is an impaired waterway. The results from this project indicated that the treated runoff had a median TSS concentration of 3 mg/L (3 ppm by weight). This figure was better than 5 ppm for preconstruction TSS and far less than the measured 53 ppm TSS concentration at Pickering Brook (Gunderson et al., 2012).

The long-term monitoring data of subsurface gravel wetland at the UNHSC has indicated an average annual nitrate removal of 75–85% during the growing season and 33% during winter.

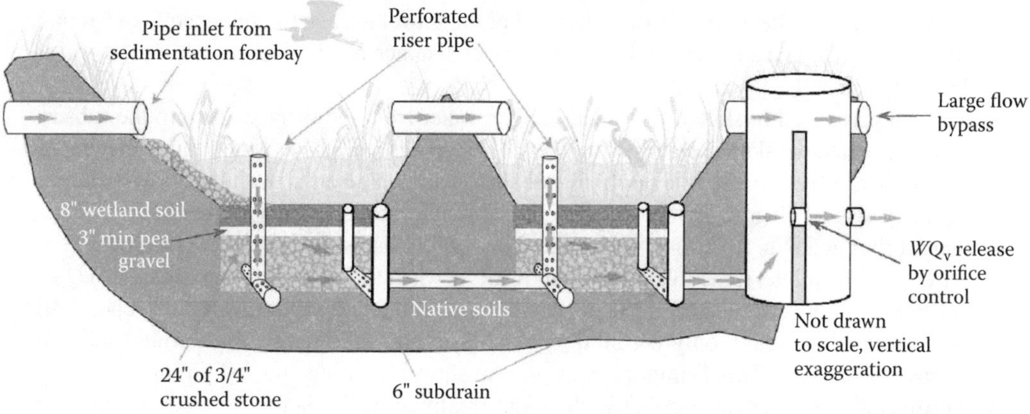

FIGURE 7.52 UNHSC subsurface gravel wetland profile.

Annual average phosphate removal was measured to be 75% and total suspended solid removal was greater than 95%. The UNHSC's specifications for GSW are as follows:

- A subsurface water level is maintained through the design of the outlet invert elevation (invert just below the wetland soil surface).
- Retain and filter the entire water quality volume (WQ_v), 10% in the forebay and 45% above each of the respective treatment cells.
- Option to retain the channel protection volume (Cp_v) for 24–48 h.
- No geotextile or geofabric layers are used within this system, but may be used to line walls.
- If a native low-hydraulic conductivity soil is not present below the desired location of the SGW, a low-permeability liner or soil (hydraulic conductivity less than 10^{-5} cm/s = 0.03 ft/day) below the gravel layer should be used to minimize infiltration, preserve horizontal flow in the gravel, and maintain the wetland plants.
- Gravel length to width ratio (L/W) of 0.5 or greater is needed for each treatment cell with a minimum flow path (L) within the gravel substrate of 15 ft (4.6 m).
- There shall be 8 in. (20 cm) minimum thickness of a leveled wetland soil as the top layer.
- There shall be 3 in. (8 cm) minimum thickness of an intermediate layer of a graded aggregate filter to prevent the wetland soil from moving down into the gravel sublayer. Material compatibility between layers needs to be evaluated.
- There shall be 24 in. (0.6 m) minimum thickness of 3/4 in. (2 cm) crushed stone (gravel) sublayer. This is the active zone where treatment occurs.
- The primary outlet invert shall be located 4 in. (10 cm) below the elevation of the wetland soil surface to control groundwater elevation. Care should be taken to not design a siphon that would drain the wetland. The primary outlet location must be open or vented; the outlet can be a simple pipe.
- An optional high-capacity outlet at equal elevation to or lower than the primary outlet may be installed for maintenance. This outlet would need to be plugged during regular operation. This optional outlet allows for flushing of the treatment cells at higher flow rates. If it is located lower, it can be used to drain the system for maintenance or repairs.
- The bypass outlet (emergency spillway or secondary spillway) is sized to pass designs flows (10-year, 25-year, etc.). This outlet is sized by using conventional routing calculations of the inflow hydrograph through the surface storage provided by the subsurface gravel wetland system. Local criteria for peak flow reductions are then employed to size this outlet to meet those criteria.
- The primary outlet structure and its hydraulic rating curve are based on a calculated release rate by orifice control to drain the WQ_v in 24–48 h.
- The minimum spacing between the subsurface perforated distribution line and the subsurface perforated collection drain at either end of the gravel in each treatment cell is 15 ft (4.6 m). There should be a minimum horizontal travel distance of 15 ft (4.6 m) within the gravel layer in each cell.
- Vertical perforated or slotted riser pipes deliver water from the surface down to the subsurface perforated, or slotted distribution lines. These risers shall have a maximum spacing of 15 ft (4.6 m). Oversizing of the perforated or slotted vertical risers is useful to allow a margin of safety against clogging with a minimum recommended diameter of 12 in. (30 cm) for the central riser and 6 in. (15 cm) for end risers. The vertical risers shall not be capped, but rather covered with an inlet grate to allow for an overflow when the water level exceeds the WQ_v.
- Vertical cleanouts connected to the distribution and collection subdrains, at each end, shall be perforated or slotted only within the gravel layer and solid within the wetland soil and storage area above. This is important to prevent short-circuiting and soil piping.
- Berms and weirs separating the forebay and treatment cells should be constructed with clay or nonconductive soils and/or a fine geotextile, or some combination thereof, to avoid water seepage and soil piping through these earthen dividers.

- The system should be planted to achieve a rigorous root mat with grasses, forbs, and shrubs with obligate and facultative wetland species.

More ore information about SGW, can be found at a UNHSC (2022) publication.

A subsurface gravel wetland does not necessarily have to be two cells like the UNHSC's SGW. Figure 7.53 shows a submerged gravel wetland comprising of only one cell. The figure shows two options for inflow–outflow arrangement. As shown, the inflow can enter the body of water above

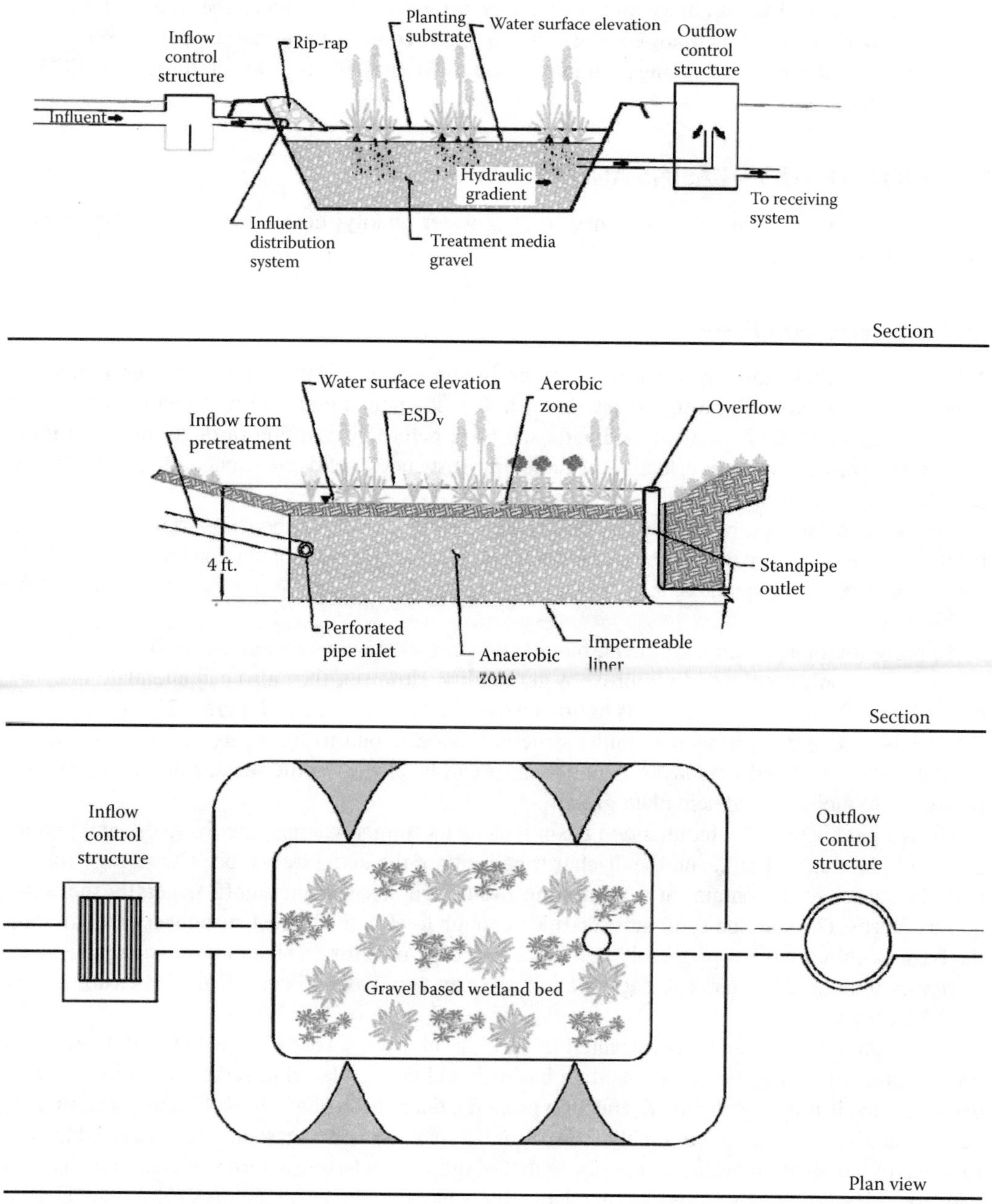

FIGURE 7.53 Submerged gravel wetland alternate plans.

Source: Maryland Stormwater Design Manual, Figure 5.10, 2009.

the treatment media gravel or enter directly into the gravel media. The outlet pipe is issues from the treatment media in the former case and the water body in the latter case.

The Maryland Department of the Environment specifies that subsurface gravel wetlands be used exclusively on HSG C or D soils based on the design guidelines by the same agency indicate the following:

- SGW is best suited for soils of poorly drained or high- water table.
- SGW implementation on a site should not exclude the use of other ESD options. Since SGW does not have drainage area limitations, using one SGW undermines the MEP mandate to mimic local hydrology and distribute runoff controls uniformly across the site.
- Only the aboveground storage volume in the SGW may be used for credit toward ESD treatment.

7.9 BIORETENTION BASINS AND BIOSWALES

These structures are commonly designed to treat water quality; however, they can also provide some runoff reduction.

7.9.1 Bioretention Basins

A bioretention also known as bio-infiltration basin consists of a planted soil bed over an underdrained layer of sand and gravel. Stormwater runoff that enters a bioretention basin is first filtered by vegetation and then by the soil bed and sand layer before discharging through the underdrain. Bioretention basins are very effective in removing suspended solids, nutrients, metals, hydrocarbons, and bacteria from runoff.

Bioretention basins are similar to infiltration basins except for bed material and planting. Infiltration basins are covered with sand and have no plants, whereas bioretention basins are covered with a soil mix and are planted. Also, infiltration basins have no underdrain, but some bioretention basins may have.

To prevent prolonged impoundment, bioretention basins are constructed 2–3 ft (0.67–1 m) deep, and are commonly employed to improve water quality. However, they also can attenuate the peak rates of runoff if they are designed as multipurpose, dual-stage systems. Figure 7.54 shows a multipurpose bioretention basin with an outlet structure. Labeled on this figure are the suggested thickness of the soil bed and sand layer. A mulch layer can be placed on the surface of the planted soil bed to retain moisture and help plant growth.

Design parameters of a bioretention basin include its storage volume, the thickness and permeability of planting soil mix, and the discharge capacity of the underdrain pipe. The storage volume must be sufficient to contain, at a minimum, the design stormwater runoff, generally the water quality storm. The soil and sand must be thick enough to effectively treat stormwater and to drain the basin within three but preferably two days. Also, the underdrain should have enough capacity to convey the filtered water. The depth of soil mix can vary from 45 cm (18 in.) for shrubs to 1 m (±3 ft) for trees.

Bioretention basins can be constructed in parking lot islands, roadway medians, and lawns. To prevent premature clogging, a bioretention basin should not be placed in service until its tributary drainage area is fully stabilized. To function properly, the bottom of a bioretention underdrain must be at least 30 cm (1 ft), but preferably 0.6–0.9 m (2–3 ft) above the seasonal high-water table. The function of bioretention basins is superior to that of infiltration basins in terms of water quality, and they are also aesthetically more pleasing. However, a bioretention basin requires more maintenance than both an infiltration basin and retention–infiltration basin. The fallen leaves should be raked and shrubs should be pruned at least once a year to avoid the return of nitrogen and phosphorus to the soil. If privately owned, a bio-infiltration basin must be protected through easement or deed

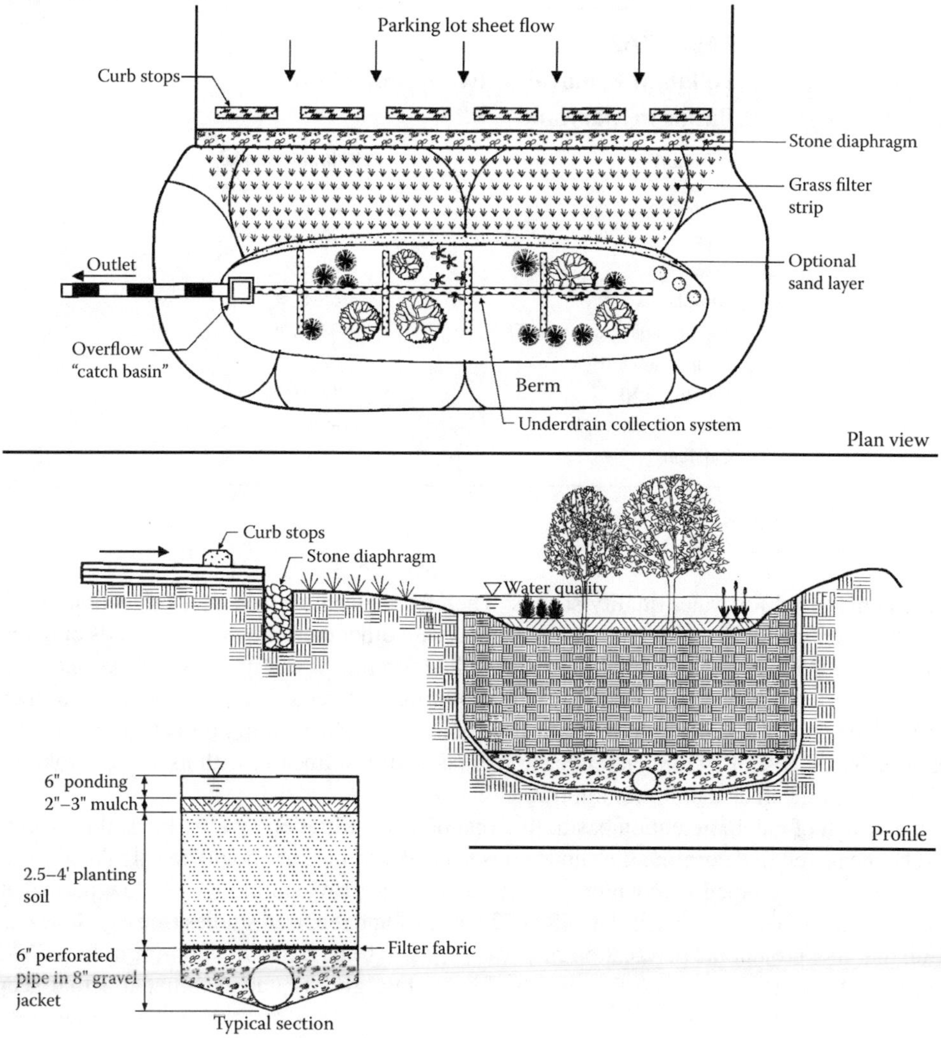

FIGURE 7.54 A bioretention basin variant.

Source: New York State Stormwater Management Design Manual, 2015, Figure 6.19.

restrictions to ensure that they will not be adversely altered. More information on the design of bioretention basins can be found in the *Maryland Stormwater Design Manual (*2009) and NJDEP Stormwater Best Management Practices Manual (2023, 2024).

Specific designs of bioretention basins, similarly to rain gardens and constructed wetlands vary considerably depending on the preferences of designer and site conditions. But, some main features should be incorporated into all designs. These features, similar to those described for constructed wetlands, are pretreatment, treatment, conveyance, maintenance reduction, and landscaping. In arid climates, bioretention basins should be planted with drought-tolerant species.

Bioretention basins have a few limitations: They cannot be used to treat runoff from large drainage areas and cannot be employed where the space is limited. However, they can be used as a stormwater retrofit by modifying existing lawn or landscape areas. As indicated, bioretention basins

TABLE 7.62
Pollutant Removal in Two Bioretention Basins in Maryland

Pollutant	Pollutant Removal
Copper	43–97%
Lead	70–95%
Zinc	64–95%
Phosphorus	65–87%
Total Kjeldahl nitrogen (TKN)	52–67%
Ammonium (NH_4)	92%
Nitrate (NO_3)	15–16%
Total nitrogen (TN)	40%
Calcium	27%

are commonly sized for water quality storms. Therefore, to achieve channel protection and flood control, bioretention basins should be supplemented with other measures such as ponds or detention basins. Table 7.62 presents field test results of pollutant removal effectiveness of two bioretention basins in Maryland. Bioretention basins with subdrains can remove 80% or more of suspended sediment. The cost of bioretention basins varies considerably depending on design details. On an average basis, the cost of a bioretention basin increases almost linearly with its storage volume and may be twice as much as that for an infiltration basin.

In the design of any bioretention basin, like that of any infiltration system, the soil permeability should be measured to determine if a subdrain is needed. Drain time should be calculated based on actual rather than assumed percolation rate. A subdrain is provided in areas of low permeability, where the ponding period is longer than 48 or 72 hours. Some reported studies (see e.g., Jones, 2012) indicate that in a project the percolation rate was assumed to be 0.3 in./h (7.6 mm/h) for HSG A to 0.025 in./h (0.6 mm/h) for group D soil. The former rate grossly underestimated actual seepage losses through the ambient soil.

7.9.2 Bioswales

A bioswale, also referred to as bioretention swale, is similar to the previously described grass swale. Therefore, only their differences are discussed herein. Unlike grass swales, bioretention swales are covered with shrubs and taller grass species and are not mowed. Design parameters for a bioretention swale follow those of a grass swale; however, Manning's n for the former is significantly greater than that for the latter. As a result, bioswales retard the flow and increase the time of concentration far better than grass swales. Also, dense, taller vegetation provides a higher pollutant removal than grass.

An early application of bioswales occurred in Portland, Oregon. In 1996 a total of 2,330 linear feet (710 m) of bioswale were designed and installed in Willamette River Park in Portland to capture pollutants from runoff entering the Willamette River (http://en.wikipedia.org/wiki/Bioswale).

To improve infiltration, the bottom of bioswales may be covered with 0.35–0.9 m (1.2–3.0 ft) thick sandy loam over 15 cm (6 in.) of coarse sand. In areas of low permeability, a perforated subdrain may also be installed in the sand layer. To promote side stability, the side slopes should be no more than 1 by 4 (1V/4H). Bioswale applications include parking lot islands highway medians or roadside swales, and residential roadside channels.

PROBLEMS

7.1 Storage versus outflow relation for a detention basin is given below. Calculate the storage–outflow function, $2S/\Delta t + O$, versus O (outflow) for each of the tabulated values, using $\Delta t = 5$ minutes. Plot a graph of the storage-outflow function versus O.

Storage (10^3 ft^3)	75	81	87.5	100	110.2
Outflow (cfs)	2	6	12.0	22	34.0

7.2 Rework Problem 7.1 for the storage versus outflow discharge relation shown below:

Storage (10^3 m^3)	2	3	3.5	4	5
Outflow (m^3/s)	0.06	0.15	0.35	0.70	1.00

7.3 Consider a 0.5-acre detention basin with vertical walls. The inflow discharge increases linearly from zero to 60 cfs at 30 minutes and then decreases linearly to zero at 75 minutes. The outlet from the detention basin is a 24 in. RCP culvert, located at the bottom of the basin. The basin is 5 ft deep and is initially empty. Use the level pool routing procedure with a 5-minute time interval to calculate the maximum discharge and the maximum water depth in the detention basin. The outlet pipe is 50 ft long at 2% slope and has an unsubmerged outlet.

7.4 Solve Problem 7.3 for the following conditions:

Detention basin size = 2050 m^2, 1.5 m deep;
Inflow peak = 1.75 m^3/s.
Diameter of pipe=60 cm.

Use a time interval of 10 minutes.

7.5 Using the TR-55 method, the pre- and post-development 100-year peak rates of runoff for a 20-acre site are calculated at 25 and 50 cfs, respectively. Estimate the required detention storage to regulate the 100-year discharge from the site to 80% of the predevelopment peak rate of runoff. The soil curve number for the proposed condition is calculated at CN =80 and depth of the 100-year, 24-hour storm is 8.0 in. Use type III storm.

7.6 Solve Problem 7.5 for a 5 hectare site and 100-year peak runoff discharge at 1 m^3/s and 2 m^3/s, respectively. The 100-year, 24-hour storm is 200 mm.

7.7 Using the rational method, the 100-year pre- and post-development peak runoff rates from a 50-acre residential development site are calculated at 20 and 50 cfs, respectively. Estimate the required detention storage to regulate the 100-year discharge from the site to 80% of the predevelopment. Assume that the runoff from the entire site is drained.

7.8 In a 10 hectare residential development, the pre- and post-development peak rates of runoff are calculated at 1 m^3/s and 2 m^3/s, respectively. Estimate the required detention storage to attenuate the post-development peak runoff to 80% of the predevelopment condition. Use the rational method and assume $T_c = 30$ minutes for a post-development hydrograph.

7.9 The post-development conditions of an urban watershed are as follows:

$A = 20$ ha
$T_c = 30$ min
$C = 0.3$ for lawn/landscaped
$C = 0.95$ for impervious areas
Impervious coverage = 25%

Calculate the required detention storage volume to reduce the 2-year and 100-year post-development discharges from the watershed to 50% and 80% of the respective predeveloped conditions where $T_c = 60$ minutes and the watershed is woodland, $C = 0.25$. Base your calculations on the rational/modified rational method and assume a discharge from the detention basin. Use Figure 3.1a for rainfall curves.

7.10 Solve Problem 7.7 for a 50 acre watershed; use Figure 3.1b rainfall curves.

7.11 Repeat Problem 7.9 for a 50 acre watershed using the TR-55 method, assuming hydrologic group B and 100% wood cover for existing conditions and 20% wood cover for post- development. The 2- and 100-year SCS 24 storms are 3.3 and 8.0 in., respectively. The storm is type III.

7.12 In Problem 7.11, design a single outlet using the following geometrics:

a. Circular orifice
b. Rectangular opening
c. V notch weir

The detention basin is 4.5 ft deep. Allow 6 in. freeboard.

7.13 A 75-acre development will consist of 35% pavement, 35% lawn, and 30% wood. Using the rational method, calculate the 100-year peak rate of runoff if the time of concentration is 30 minutes. The entire site runoff is drained to a detention basin. Using the IDF curves of Figure 3.1b (in Chapter 3), estimate the required detention storage to create a 20% reduction in the 100-year peak rate of runoff below that of the existing conditions. The site is entirely wooded under predeveloped conditions and has a time of concentration of 60 minutes.

7.14 Solve Problem 7.13 using the SCS method. The site soil is hydrologic group C. The 100-year, 24-hour storm is 8.3 in. The storm is type III.

7.15 Solve Problem 7.13 for the following conditions:

Area = 30 ha
$T_c = 60$ minutes, predevelopment
$T_c = 30$ minutes, post-development

Use the IDF curves in Figure 3.1a.

7.16 Calculate the required detention storage in Problem 7.13 using the universal runoff model. The soil is silty sand and permeability is measured at 2.5 in./h. The runoff from each soil cover is directly connected to the detention basin.

7.17 Calculate the required detention storage in Problem 7.15 using the universal runoff model and 50 mm/h permeability.

7.18 A 20 ha development site is wooded and has a permeability rate of 35 mm/h. The development creates 5 acres of pavement and turns another 5 acres into lawn, all hydraulically connected. The time of concentration is calculated at 30 minutes and 20 minutes for the pre- and post-development conditions, respectively. Using the universal runoff model, developed by the author, estimate the required detention storage to produce 50% and 20% reductions in the peak rates of runoff for 2- and 100-year storms, respectively. The local IDF curves show the following rainfall intensities:

Storm Frequency (years)	Storm Duration	Rainfall Intensity (mm/h)
2	20	50
	30	40
100	20	120
	30	80

7.19 Calculate the required detention storage in Problem 7.18, if the pavements and pervious areas separately drain to the detention basin.

7.20 A detention basin is designed to attenuate the runoff from 5.16 acres comprising 4.65 acres of roadway/pavement, 0.28 acres of lawn, and 0.23 acres of brick pavers. The allowable discharges from the detention basin are calculated at 7.94, 10.90, and 15.88 cfs for 2-, 10-, and 100-year storms, respectively. Elevation–discharge–storage relations for the detention basin are listed in the following table:

Elevation (ft)	Discharge (cfs)	Storage (ft^3)
869.5	0	0
870.0	0.8	1,150
870.5	2.0	3,400
871.0	2.7	5,670
871.5	4.1	7,950
872.0	6.0	10,200
872.5	8.4	12,600
873.0	11.0	15,200
873.5	13.8	17,950
874.0	17.4	20,980

a. Perform calculations to determine the critical storm duration, the storm that requires the largest detention storage. Use the IDF curves in Figure 3.1b.

b. Perform routing computations for the critical storm.

c. Compare the computed peak outflows from the detention basin with the allowable discharge from it.

7.21 The roof runoff from a 10,000 m^2 commercial building is to be retained using 36 in. perforated HDPE pipes in stone trench. Size this retention system and sketch its layout for a 50 mm rainfall.

7.22 In Problem 7.21, StormTech MC-3500 chambers in stone trench are to be used in lieu of the perforated pipes. Calculate the number of chambers required and define the foot-print dimensions of the stone trench.

7.23 Redo problem 7.22 for a 100, 000 square feet roof and 2 in. rain.

7.24 A 5000 m^2 commercial building roof is drained into Terre Arch 26 structures. Calculate how many Terre Arch units are required to retain 30 mm of rainfall. The structures are to be placed on a 150 mm thick stone base, and their valleys are also filled with stone. If the soil permeability is 50 mm/h, how many hours will it take for the water to drain?

REFERENCES

Advanced Drainage Systems (ADS), Inc., Hillard, OH, info@ads-pipe.com, https://www.adspipr.com

American Concrete Pipe Association, 2007, *Concrete pipe design manual*, 19th printing, American Concrete Pipe Association.

ASCE, 2006, *Standard guidelines for the design of urban stormwater systems*, ASCE/EWRI 45-05, *Standard guidelines for installation of urban stormwater systems*, ASCE/EWRI 46-05, *Standard guidelines for the operations and maintenance of urban stormwater systems*, ASCE/EWRI 47-05.

ASCE/WEF, 1992, *Design and construction of urban stormwater management systems*, American Society of Civil Engineers Manuals and Reports of Engineering Practice No. 77 and Water Environment Federation manual of practice FD-20.

Atlantis America US headquarters Bellerose New York, https://www.atlantis-america.com

Atlantis Corporation-Green City for Life, Head Office: 3/19-21 Gibbes Street, Chatswood, NSW 2067, Australia, https://atlantiscorpation.com.

Brentwood Industries, Global Headquarters, Reading, PA, https://www.brentwoodindustries.com
Campbell Foundry Company, Kearny, NJ, https://campbellfoundry.com
Contech Engineered Solutions, West Chester, OH, https://www.conteches.com
Cultec Inc., 878 Federal Road, Brookfield, CT, https://www.cultec.com
EPA (US Environmental Protection Agency), August 2004, *Constructed treatment wetlands*, EPA 842-F-03-013, Office of Water.
EPA (US Environmental Protection Agency), August 14, 2025, *Constructed wetlands.*
EPA (US Environmental Protection), December 2021, NPDES Stormwater wetland, EPA-832-21-031AA-
EPA (US Environmental Protection Agency), September, 1999, *Stormwater technology fact sheet sand filters*, EPA-832-F-99—007, September.
EPA (US Environmental Protection Agency), December – 2021, *NPDES: Stormwater best management practices, sand and organic filters.*
FHWA, 2013, *HEC-22, urban drainage design manual*, 3rd ed., U.S. Department of Transportation, Federal Highway Administration. FHWA, February 2024, *HEC-22*, 4th ed.
Gunderson, J., Rosen, R.M., Ballestero, T.P., Watts, A., Houle, J. and Farah, K., November/December 2012, Subsurface gravel wetlands for stormwater management, *Stormwater*, pp. 8–17.
Hudson-Essex- Passaic Soil Survey, H-E-P Soil Conservation District, Bloomfield, NJ. https://hepsoilnj.org
Jones, M., May 2012, Design review and simulation for permeable pavement and bioretention, *Stormwater*, pp. 54–58.
Maryland Department of the Environment, Maryland Stormwater Design Manual, May 2009, Vols. 1 and 2
Minton, G.R., 2005, Revisiting design criteria for stormwater treatment systems, *Stormwater*, pp. 28–43.
New Jersey Geological Survey, 1993, *NJGS, GSR-32*, https://www.nj.gov> njgws PDF, 103 pp.
New York State Stormwater Management Design Manual, Draft, 2015. NYDEC (NY Department of Environmental Conservation).
New York State Stormwater Management Design Manual, Draft, May 2022. NYDEC (NY Department of Environmental Conservation).
NJDEP, 2004, revised July 2023, May 2024, *Stormwater best management practices manual*, New Jersey Department of Environmental Protection.
NJDEP, N.J.A.C.7:8, last amended July 17, 2023, *Stormwater management*, NJDEP.
Oldcastle Infrastructure, Atlanta, GA, 30328, https://oldcastleinfrastructure.com.
Oldcastle Infrastructure, 7000 Central Parkway, Atlanta, Georgia, 30328, https://oldcastleinfrastructure.com.
Pazwash, H., August 2–6, 1992, *Simplified design of multi-stage outfalls for urban detention basins*, Proceedings of the ASCE Water Resources Sessions of the Water Forum 92, Baltimore, pp. 861–866.
Pazwash, H., 2007, *Fluid mechanics and hydraulic engineering*, Tehran University Press.
Pazwash, H., 2009, Universal runoff model, paper R13, *StormCon*, Anaheim, CA, August 16–20.
Pazwash, H., 2011, *Urban stormwater management*, CRC Press, Boca Raton, FL.
Residential Site Improvements Standards, 2009, adopted January 6, 1977, last revised 2023, *New Jersey administrative code title 5, chapter 21*,
SCS, 1986, *Urban hydrology for small watersheds, technical release 55 method (TR-55),* Soil Conservation Service (SCS) a part of U.S.DA. TR-55; first issued 1975
StormChamber, P.O. Box 672, Occoquan, VA, 22125, https://wwwstormchambers.com.
StormTrap, 1287 Windham Pkwy Romeoville IL, https://www.stormtrap.com.
Terre Hill Stormwater Systems, Terre Hill, PA 1, https://www.terrehill.com.
Triton Stormwater Solutions, 7600 Grand River Ave. Ste 195, Brighton, MI 48114, https://www.tritonsws.com.
UNHSC, January 2022, *Subsurface gravel wetland design specifications*, University of New Hampshire by TP Ballestero
USDA, January 2009, *WinTR-55, urban hydrology of small watersheds (TR-55) for window*, last updated March 30, 2024.
Yu, S.I., Kuo, J., Fassman, E.A. and Pan, H., 2001, Field test of a grassed swale performance in removing runoff pollution, *Journal of Water Resources Management*, 137 (3): 168–171.

8 Source Reduction and Control; Green Infrastructure

This chapter includes a review of source reduction and control measures for stormwater management. These measures which are a newer trend in stormwater management are intended to reduce the runoff production and/or control the runoff at its source, rather than collecting and conveying it by a drainage system into a detention or retention basin. Source control and reduction practices, are also referred to as green infrastructure, a concept that originated in the White House in 1993, during the Administration of then-President Clinton. Subsequently, the EPA extended the concept to the management of stormwater runoff through waterways, wetlands, or drainage courses to mimic nature in controlling stormwater, containing it at its source.

8.1 SOURCE REDUCTION AND CONTROL MEASURES

8.1.1 Introduction

Source reduction and control include measures that reduce the volume of runoff and its peak flow before the runoff is discharged from its source into storm drains or a combined sewer system. Effectiveness of structural BMPs in general and stormwater treatment devices in particular in removing pollutants was discussed in previous chapters. As noted, structural BMPs and manufactured devices may not remove all pollutants effectively. Some pollutants, such as salt, cannot be removed by any structural BMP. Apart from inefficacy, structural BMPs are expensive. Their cost includes not only the construction and maintenance, but also the land in the case of open detention, retention, or infiltration basins. Although structural BMPs may sometimes be necessary, they should be considered a supplement rather than the prime water quality measure in any stormwater management design. A sound practice should be based on prevention, realizing that source reduction and/or control is a far better option than treatment. In fact, source reduction is an Environmental Protection Agency (EPA)-mandated component of the National Pollutant Discharge Elimination System (NPDES).

8.1.2 Source Reduction

Source reduction refers to measures that reduce the generation of runoff at its source. These measures, which include low-impact developments, can be effectively achieved by reducing impervious surface pavements, including narrower streets, sidewalks, and driveways; small cul-de-sacs; and shared driveways. Taller residential and office buildings and garage structures also reduce building footprints and impervious surfaces. Reducing impervious surfaces not only generates less runoff and pollutants but also lessens both construction and maintenance costs.

While local codes may not permit narrower streets or smaller cul-de-sacs, waivers or alternatives such as "tee" turn-a-rounds or looping lanes can be discussed with local officials. Although the public may view narrower streets as unsafe, narrower roads, in fact, reduce the speeds at which vehicles can travel and make the roads safer. Iceland, Sweden, and Norway, which have narrow streets and reduced city speed limits, have far lower number of pedestrian fatalities than other countries including the United States.

A genuine source reduction, as will be discussed later in this chapter, should be based not on the size of a development project but rather on a per capita impervious area. On this basis, multifamily residential developments, high-rise buildings, and, above all, city living reduce the per capita impervious coverage and the rate and volume of runoff as well.

DOI: 10.1201/9781003460121-8

Using pavers or pervious concrete in lieu of concrete walkways, and driveways and patios and grass pavers for driveways and emergency and maintenance vehicles access also provides source reduction. For an in-depth discussion on source reduction, the reader is referred to Chapter 8 in the New York State Stormwater Management Design Manual (2022) and an EPA publication titled National Menu of *Best Management Practices (BMPs) for Stormwater*, last updated on July 9, 2025.

8.1.3 Source Control

Source control involves measures that capture the runoff after it is generated at a site and before it is discharged offsite. A variety of measures can be employed to achieve source control. These measures, referred to as green infrastructure, include rain gardens, bioretention cells, roadside drainage swales, disconnection of roof leaders from pavements and/or drainage systems, pervious pavements, green roofs, blue roofs, rain gardens (planted depressions) rainwater harvesting, and constructing a grass strip or planting tree pods along sidewalks.

If these measures are properly implemented, it is possible to avoid any increase in discharge of runoff from the site of developments during moderate 1-, 2-, and even 5-year storms and to significantly reduce both volume and peak rate of runoff for 10-, 25-, and even 100-year storms. Runoff control and reduction techniques also utilize the natural features of the site to reduce runoff. Figure 8.1 presents the layout of a cluster subdivision site plan in which natural conservation areas are preserved.

The Federal Clean Water Needs Survey estimates that over $100 billion of infrastructure investment is needed in the next twenty years to address stormwater and sewer overflows alone. Many jurisdictional agencies in the United States recommend the use of green infrastructure measures as a primary method of stormwater management. Cities are increasingly considering green infrastructure as a cost-effective solution to address stormwater and sewer overflow problems. A number of cities in the United States are also experimenting with or implementing some of these measures. Portland, Oregon, and New York City are exemplified here.

With a population of 2.2 million, Portland Metro area is the third most populated city in the northwestern United States. The city implemented a Green Streets Program starting in 2003 that combines rain gardens and such LID features as permeable pavement, green roofs, and roadside/sidewalk swales and planters that allow water to infiltrate (Rogers and Faha, 2007). With this program, the city estimates to reduce the peak flows by as much as 85%, stormwater volume by 60%, and water pollution by up to 90%.

To further reduce the combined sewer overflows (CSOs) during heavy storm events that pollute the Willamette River and Columbia Slough Watershed, the city constructed two massive tunnels along the east and west banks of the river. The east tunnel is 22 ft (6.7 m) in diameter and 6 miles (9.7 km) long, and the west tunnel is 14 ft (4.27 m) in diameter and 3.5 miles (5.6 km) long. The latter tunnel crosses under the river and merges with the former. The tunnels were bored using a microtunneling process that had also been used in the Big Dig project in Boston.

New York City, like many other older cities, is served by combined sewer systems that convey both stormwater and sewer to treatment plants. There are approximately 460 CSOs in the five boroughs of New York City, and during sustained or heavy rainfalls, overflows occur to rivers and bays. To reduce CSOs, the city planned to manage runoff from 10% of impervious surfaces through green infrastructure by 2020. To evaluate the effectiveness of various measures, the city began pilot studies in a number of sites throughout the borough of Queens in 2011. Included in the measures were enhanced tree pots in 20 ft long by 5 ft wide (6 m × 1.5 m) and 2 ft (0.6 m) deep sandy soils, each receiving runoff from 2000 to 6000 ft^2 (200–600 m^2) of paved area; bioswales along open curb cuts; and installation of 6400 ft^2 (595 m^2) of porous asphalt and 4200 ft^2 (390 m^2) of crushed glass in a section of Queens. The pilot study indicated that green infrastructure can be effective in one of the most densely developed urban areas in the United States (McLaughlin et al., 2012).

8.1.4 Source Reduction and Control Benefits

Source reduction/control, which may also be termed a decentralized solution, has the following advantages compared to structural BMPs:

- *Effectiveness*: Source reduction and control can be more effective than structural BMPs for a number of pollutants, including salt, phosphorus, and metal. Some pollutants, such as salt and soluble phosphorus, cannot be removed by detention basins and ponds.
- *Reducing the size of required BMPs*: By controlling runoff quantity and water quality, source reduction lowers the overall cost of stormwater management provisions.
- *Partial, if not total, elimination of water quality devices*: Source reduction may eliminate the need for any water quality device. Pursuant to the author's recommendation, the need for any water quality devices was eliminated in some large single-family homes in Franklin Lakes, New Jersey, simply by using pavers in lieu of paved driveways.

Source control can also prevent flood and flood-related damage to downstream properties. This is particularly true during construction, when a large amount of sediment is produced on site. Soil erosion control measures greatly reduce sediment discharge to drainage systems and detention basins. Serving as an expert witness, the author found that due to inadequate soil erosion control measures and lack of maintenance, the runoff from a residential project which was under construction caused millions of dollars of damage to a car dealership downstream thereof. Specifically, during a heavy storm, the runoff from the development surcharged from the drainage system, which was half filled with silt, and the muddy water flooded the dealership lot, ruining many new cars. The legal and punitive costs to the developer were much higher than the cost of source control, in essence installing and maintaining adequate soil erosion measures and cleaning the storm drains.

Source boundaries may be defined up to the stormwater conveyance system (Baker, 2007). However, accounting for pollutants entering catch basins due to road deicing and oil leaks from cars and trucks, the conveyance system may be included in the source boundary. It should also be noted that contrary to a common perception, lawns produce more pollutants than roads. Lawn runoff is laden with nutrients and phosphorus. In fact, the total concentration of nitrogen and phosphorus in lawns is higher than effluent from an advanced secondary treatment plant. Therefore, it is prudent to begin source control from lawns. To reduce pollution, soil should not be compacted during construction which reduces infiltration. Also, fertilizer and pesticides should be used at the minimum amount necessary and not be applied before rainfall.

To reduce lawn pollutants, some states have passed laws that restrict the use of lawn fertilizers above a certain limit of phosphorus. A widespread enactment of such laws will eventually lower the concentration of phosphorus in lawn runoff due to fertilizer, which is the largest source of this urban pollution. Homeowners are generally unaware of the ill effects of lawn chemicals on the environment. They may participate in any lawn pollution reduction plan if they are informed. Public outreach is the key in educating homeowners and in the success of any public planning. A number of measures for source reduction is are discussed later in this chapter.

8.2 LOW-IMPACT DEVELOPMENT

Urban development severely disrupts natural hydrology. Low-impact development (LID) is an environmentally friendly development approach with minimal impacts on the hydrologic regime and water quality. The afore described Figure 8.1 shows a low-impact residential subdivision. Specifically, the objectives of LID, which is a site-based process, are:

- Minimizing the increase in the amount of runoff by reducing impervious covers, such as roads, driveways, and parking areas

- Maximizing on-site infiltration and retention by conserving natural vegetation, maintaining natural drainage courses, and minimal use of pipes and inlets
- Reducing clearing and grading to minimize erosion and sedimentation
- Providing measures to store runoff in small depressions and bio-retention and infiltration basins spread out throughout the site
- Maintaining predevelopment time of concentration by routing runoff and disconnecting impervious surfaces such as roofs and driveways from roads
- Minimizing the use of treatment systems, which require high maintenance and fail when neglected

Promoting public awareness of, knowledge of, and participation in pollution prevention measures will help protect the environment. The public is not fully aware of the extent to which traditional developments involving clear-cutting large strips of land harms the environment. Compacted lawns create more runoff than natural open space and woodland. LID is intended to eliminate or reduce these problems through an alternate, comprehensive approach to stormwater management, including retention and storage of rainfall. The LID storage techniques include blue roofs, green roofs, rain gardens, rain barrels, porous pavements, pavers, dry wells, constructed wetlands, conservation of existing woodland and open space, landscape bioretention cells and basins, lawn and landscape depressions, grass swales, infiltration trenches at or below grade, stone basins, and landscaping. An October 2000 EPA publication, presents effectiveness of LID. The following case exemplifies one of the early applications

The Ipswich River in northeastern Massachusetts, which is the source of drinking water for more than 330,000 residents and businesses, had been classified as a highly stressed stream because of low or no base flow along its upper reach and chronic fecal coliform bacteria in the early 2000s. To improve water quality and increase base flows of the river, the town of Wilmington implemented a LID stormwater management project within the Silver Lake Watershed. Silver Lake is a 28.5-acre (11.5 ha) kettle lake, has a drainage area of 132 acres (53 ha), and is situated in the headwaters of the Ipswich River. The project included the installation of a number of LID stormwater management measures in a parking lot in the vicinity of the Silver Lake town beach and a residential area on the opposite side of the lake, as follows:

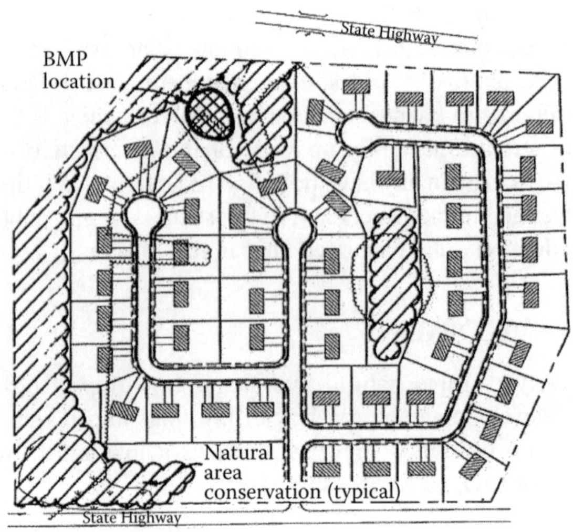

FIGURE 8.1 Schematic layout plan of a residential subdivision with natural conservation areas preserved.

Source: New York State Stormwater Management Design Manual, Figure 5.32, 2022.

- Concrete pavers to replace asphalt pavement along one side of a street
- Porous asphalt and interlocking concrete pavers on a parking lot
- Residential roof rain harvesting systems, 200 to 850 gallons (0.76 to 3.2 m^3) capacity
- An underground 8000 gal. (30 m^3) tank to collect roof runoff from an elementary school
- Block pavers in a parking lot
- Gravel Pave (Geoweb cells filled with stone) in an overflow parking lot
- Flexi-Pave (poured-in-place rubber paving surface) in another overflow parking lot
- Rain gardens (12 in all) along streets to collect runoff
- Bioretention basins (depressed landscape with engineered soil mixture) along the perimeter and a median island in the paver parking area
- Two vegetated swales to replace drainage pipes terminating at the lake

In situ permeability tests were conducted using single- and/or double-ring infiltrometers at rain gardens, bioretention basins, porous asphalt, concrete pavers, and Gravel Pave. The measured permeability ranged from 12 in./h (300 mm/h) in a rain garden to 22 in./h (560 mm/h) in bioretention basins to 57 in./h (1450 mm/h) in pavers, 78 in./h (1980 mm/h) in porous asphalt, and over 9 ft/min (2.75 m/min) for Gravel Pave. Overall, the porous surfaces performed better than the design infiltration rates. To maintain its permeability, the porous asphalt has been vacuum-swept twice a year. The project was successfully completed in the summer of 2006, and it had a total construction cost of $340,000 (Roy and Braga, 2009).

8.3 POROUS PAVEMENTS

A porous pavement is a material through which water can seep. The words pervious and permeable are interchangeably applied to porous pavements. Pervious means the ability of a material to let water penetrate at its surface; permeable represents a material allowing water to infiltrate through it. Porous pavements can be divided into a number of categories in terms of material and construction. They include open decks, open-celled paving grids, gravel pavement, porous asphalt, pervious concrete, and open-joint paving blocks (known as pavers). A review of design and simulation for permeable pavement and bioretention systems was presented in a paper by Jones (2012). A book by Ferguson (2005) provides detailed information on porous pavements, and an EPA publication titled "Soak Up the Rain: Permeable Pavement," dated February 3, 2025, includes sources of information about permeable pavements.

Decks are level or elevated wooden structures to bear foot traffic and are suitable for where they can be built around the existing environment, such as wetlands and boardwalks in coastal areas and of course in single family homes and residential developments. Open-graded aggregate is one of the most permeable and the least costly materials. To avoid creating dust, aggregate is made of single-sized angular material. This type of pavement has 30–40% void and tens of inches per hour permeability.

Because of reducing runoff through infiltration, porous pavements, and particularly porous asphalt, pervious concrete and pavers are gaining great interest from site planners and public work departments. Regardless of terminology, some porous pavements, including porous asphalt, pervious concrete, and block pavers, have a common feature: a permeable surface and an underlying stone reservoir to temporarily retain surface runoff and infiltrate it into the subsoil over time. Also, these pavements have the same design considerations: Soils should have permeability greater than 0.5 in./h (13 mm/h), but preferably 1 in. (25 mm) in one hour; the bottom of porous pavement should be at least 2 ft (61 cm) above the seasonable high-water table; and the bottom of the stone reservoir should be completely flat so that water can infiltrate the entire surface.

8.3.1 Open-Cell Paving Grids

A variety of open-cell paving systems are available on the market. These paving systems are made of concrete or plastic materials with open cells for gravel fill or grass to grow. AgTec heavy duty plastic grid and Turfstone concrete pavers are described herein.

FIGURE 8.2 Turfstone pavers on a driveway.

Turfstone grid pavers by Belgard are open cells with 40% surface openings. Each lattice grid measures 60 cm by 40 cm, is 8 cm thick (23–5/8 × 15.75 × 3–1/8 in), includes 8 cells, and covers 0.24 m^2 (2.56 ft^2). The grids are installed on a minimum of 6 in (15 cm) uncompacted stone base, and the cells are filled with gravel or grass. These pavers are durable and can be used for driveways and parking lots. In areas with low rain, the use of gravel fill is recommended. These pavers can support up to 5000 psi (350 kg/cm^2) of load. Figure 8.2 shows Turfstone pavers filled with grass on a driveway. For driveways, these pavers can be installed over 1.5 in. (3.75 cm) of sand and at least 4 in. (10 cm) of stone base.

AgTec pavers are made of recycled plastic. These pavers, similarly to Turfstone pavers measure 60 × 40 × 8 cm; are heavy duty, frost free and UV resistant; and last 20 years. Each lattice weighs roughly 20 lbs. (8.2 kg) and provides 95 kg/cm^2 (1350 psi) load capacity. Because of the light weight, 135 grids fit on a pallet compared with 50 grids of concrete pavers. As such, they cost less to transport and to install. The paver cells can be filled with gravel or grass and used on driveways and parking lots.

AgTec also makes geo cell grid paver mats made of black, UV-stabilizing HDPE. Each mat is 2 in. (5 cm) thick and covers 8.4 ft by 27.4 ft (2.56 m by 8.35 m). The plastic webs of these mats are 1.5 ± 0.1 mm thick, providing 32 by 28.7 cm open cells for fill by grass. The shallow 2 in (5 cm) thick paver is ideal for light traffic areas, service roads, golf course paths, and dog parks. For more information about these products one can visit www.belgard.com and www.agtech.com. There are a number of other types of pavers such as Grass Pave, Glass Foam Gravel, and Flexi-Pave the information for which can be obtained by visiting their manufacturers website.

8.3.2 Porous Asphalt

Porous asphalt was first developed in the 1970s at the Franklin Institute in Philadelphia, Pennsylvania (Adams, 2003). This pavement, also referred to as "open-graded mix," "gap-graded mix," and "popcorn mix," contains the same elements as conventional asphalt; however, it has a different recipe. Porous asphalt is basically bituminous asphalt except that fine aggregate, smaller than 600 μm (no. 30 sieve), has been screened and reduced. A standardized mix of porous pavement that was found to perform well is listed in Table 8.1. Porous asphalt is underlain by a choke course on an open-graded sub-base reservoir. A choke course is a permeable layer, typically 1.5–2.0 in. (38–50 mm) thick, of small, open-graded aggregate that provides a level, stabilized bed surface for the porous asphalt.

The open-graded sub-base reservoir is a thick layer of uniformly grade 1.0 to 2.5 in. (25–65 mm) clean stone to retain the design storm. However, based on availability, larger or smaller stones can be used. Depending on the storm design frequency and the soil permeability, the sub-base reservoir

TABLE 8.1
Standard Mix of Porous Asphalt[a]

US Standard Sieve Size, in. (mm)	Percent Passing
1/2 (12.5)	100
3/8 (10)	95
#4	35
#8	15
#16	10
#30	2

Note: 5.75%–6% bituminous by weight.

[a] See Appendix B at the end of the book for the Unified Soil Classification System, US standard sieve sizes, and coarse aggregates.

depth can vary from 12 in. (30 cm) to 36 in. (90 cm). The thickness of the stone reservoir should be selected based on climate. In cold regions, the bottom of the stone reservoir should be no lower than the frost line. The soil beneath the stone sub-base should be minimally compacted to retain porosity and permeability. A 3–4 in. (75–100 mm) thick layer consisting of 3/8 to 3/4 in. (10–19 mm) stone may be placed between the chock layer and the subbase. This high infiltration stone layer, which is intended to prevent the migration of stones from the choke layer into the sub-base reservoir, also stores water.

In instances where the subgrade soil has very low permeability, an underdrain may be installed in the sublayer to facilitate water removal. An underdrain is a small-diameter, perforated, typically PVC or HDPE pipe. The use of a filter fabric between the sub-base reservoir and subgrade uncompacted soil is optional. Stormwater drains through the asphalt and is held in the stone base, which has 35% to 40% void space and percolates slowly into the underlying soil. A layer of geotextile filter fabric is placed below the stone to prevent the movement of soil into the stone bed. The University of New Hampshire Stormwater Center (UNHSC) has been experimenting with porous asphalt for a number of years. The asphalt being studied consisted of porous asphalt at the surface and a stone choker course (base course) on an underlying layer of finer material consisting of sand and gravel (Roseen and Ballestero, 2008).

A general misperception exists in the industry about the functionality of porous asphalt (and pervious concrete, as well) in cold-weather environments. Based on field observations, as well as the UNHSC experiments, freeze-thaw action and freezing of filter material is not an issue in the porous asphalt. By allowing the runoff to infiltrate, porous asphalt has fewer icing conditions, and there is less need for snow plowing. While the filter may freeze, it does not freeze solidly but still allows water to percolate.

Research findings at the UNHSC show that with porous asphalt, salt application could be reduced by up to 75% based on snow and ice cover. According to Dr. Robert Roseen, the director of UNHSC from 2004 to 2012, "With only 25% of the salt, the snow and ice cover on the porous asphalt was the same as on the dense-mix asphalt parking lot." He indicates that "even with no salt, porous asphalt has higher frictional resistance than dense-mix asphalt with the 100% of normal salt application." For more information about UNHSC pavers asphalt, one is referred to Design Specifications for Porous Asphalt, 2016 by J.F. Briggs (https://scholars.unh.edu).

Porous asphalt is well suited for parking lots, sidewalks, playgrounds, and low- to moderate-traffic areas. It has been used in hundreds of sites in the United States. One of the first porous asphalts, which was constructed in a parking lot in Walden Pond State Reservation in Massachusetts in 1972, is good example of porous asphalt performance. This parking lot did not need any repair for over two decades in spite of the freeze–thaw cycles of New England winters. Among the different mixes that were used in this lot, the one that performed best consisted of 2.5 in. (63 mm) porous asphalt on 1.5 in. (38 mm) type A stone on 10.5 in. (27 cm) type B stone on a gravel fill. Type A consisted of 3/8 and 1/2 in. (10–13 mm) stone, and type B was a mix of stone 3/4 to 2.5 in. (19–64 mm) at a certain percentage by weight. The porous asphalt included a stone mix ranging from 1/2 in. to sieve no. 200 and AC-20 viscosity grade asphalt (an acrylic latex sealant) 4.5% to 5.5% by weight.

Because of having less fine material, porous asphalt has lower shear stress capacity than conventional pavements; therefore, it is not recommended for airport runways or slopes greater than 6%. Likewise, porous pavements are not suitable where sand is applied for snow removal. Porous asphalt also is not recommended for truck stops or heavy industrial areas where there is a threat of groundwater contamination because of spills or leakage.

Different states have tried a variety of recipes. In a 1980 report titled *Porous Pavement Phase I Design and Operational Criteria*, the EPA presented a standard pavement design. This report notes that the initial cost for a porous pavement can be up to 35%–50% higher than the cost of conventional paving but that this extra cost is more than offset by the elimination or downsizing of conventional stormwater management systems. The cost saving is even greater in urban settings where the land to install a detention basin or other stormwater management facility is expensive.

Additionally, based on life-cycle cost analysis, pervious pavement systems appear to be more economical in the long run. Common parking lots made from impervious pavement typically last 12 to 15 years in northern climates where freeze-thaw is prevalent, while pervious pavement lots can last more than 30 years, Roseen says. Considering performance and stormwater management benefits, the use of porous asphalt and other pervious pavements has been increasing rapidly in recent decades. This trend is expected to grow even faster in the future.

In Portland, Oregon, one of the largest porous asphalt parking lots to store Hyundai cars was installed along the Port of Columbia River in 2006. This lot includes 35 acres (14.2 ha) of porous asphalt and 11 acres (4.45 ha) of standard dense graded asphalt that suffer heavy traffic from delivery trucks. The pervious pavement consists of a 3 in. (7.5 cm) course of open-graded porous asphalt made of native river sand and a 10 in. (25 cm) open-graded rock base. A reason for selecting porous asphalt was the availability of native river sand along the Columbia River. Although this porous pavement was somewhat more expensive than standard pavement, it eliminated the need for any drainage systems and stormwater treatment devices, resulting in large cost savings. The 46-acre (18.6 ha) pavement cost \$6.4 million (\$29/square yard (\$34/m^2)) (Brown, 2007). The pavement is swept two to three times a year to keep leaves and cottonwood seeds from clogging the voids and has been functioning well in infiltrating the runoff.

Figure 8.3 shows experimental porous asphalt at the EPA laboratory in Edison, New Jersey. As the figure indicates, this porous asphalt fully infiltrates the sprinkler water. In Clark, New Jersey, porous asphalt was used at parking stalls on both sides of the parking lot of the municipal building

FIGURE 8.3 Experimental porous asphalt, EPA laboratory in Edison, New Jersey.

FIGURE 8.4 Porous asphalt at parking stalls, Municipal Building, Clark, New Jersey.

to absorb the runoff from the lot. Figure 8.4 shows the porous asphalt at parking stalls abutting conventional asphalt driveways at the parking lot of the Clark municipal building in New Jersey.

8.3.3 Pervious Concrete

Pervious concrete is a mixture of cement, coarse-graded aggregate, and water. The main difference between pervious concrete, which was developed in 1970s, and conventional concrete is that little or no sand is used in pervious concrete. Fibers and admixtures may be added to improve strength or other properties. Still, the common composition is the amount of water, the cementitious content, and the absence of fine material. The result is a low-slump paste that forms a coating that bonds the aggregate without flowing during mixing and placing. Using just enough paste produces a medium with interconnected voids.

Because of high void percentage, typically 15–25%, pervious concrete is relatively lightweight, 100–120 lb/cf (1600–1920 kg/m^3). Upon placement, pervious concrete looks like a sponge and provides an ultimate compression strength of 500–4000 psi (3400–27,600 kPa). Cured pervious concrete can pass 3–8 gallons of water through each square foot (125–330 L/m^2) per minute. This reflects an infiltration rate of 4.8–13 in./min (12–33 cm/min). At this rate, pervious concrete can capture the most intense rainfalls. Thus, pervious concrete, similar to, yet better than, porous asphalt, reduces or eliminates the need for detention ponds. Likewise, it can downsize drainage systems and their costs.

Pervious concrete has been used since the late 1970s in various applications, including driveways, sidewalks, and parking lots. In recent years, the use of pervious concrete has grown rapidly to address EPA and local stormwater management regulations. In fact, the use of pervious concrete is recommended by the EPA as one of the best stormwater management practices.

A benefit of pervious concrete its light gray color, which reflects more solar radiation and does not become as hot as the darker asphalt. In an experiment involving thermographic imaging, an asphalt roadway in Rio Verde, Arizona, was found to be 30 °F warmer than an adjacent regular concrete parking lot at the same time of day. The lower temperature helps the healthy growth of trees and plants, which in turn can reduce the temperature in an urban setting.

Pervious concrete also offers a safety benefit in wet weather. A study by the University of Illinois showed a stopping distance of 49 m (162 ft) on dry concrete and 58 m (190 ft) on dry asphalt for a given speed. The effect was even more significant in wet conditions, with 96 m (316 ft) of stopping distance for pervious concrete versus 134 m (440 ft) for porous asphalt. However, according to Dr.

Roseen, porous asphalt functions better than pervious concrete in winter months. By absorbing more heat, porous asphalt will be warmer in winter, promoting deicing. This author observed similar behavior at his former residence, where the asphalt driveway abutted a concrete paver sidewalk.

UNHSC in conjunction with the Northern New England Concrete Promotion Association experimented with pervious concrete. At the UNHSC, the makeup of the sub-base in an installed pervious concrete pavement is considerably different from most concrete systems. The UNHSC pervious concrete consisted of a thick stone base made of 1–3 in. (2.5–7.5 cm) stone, which provided considerable storage capacity and a sand layer that functioned as a filter to improve water quality (Gunderson, 2008).

The sub-base should be thick enough to store an extreme storm event, and the subsoil should be able to drain the water within three days to keep water from standing too long in a freeze–thaw environment. In soils of poor permeability, a subdrain pipe can be installed in the sub-base. Figure 8.5 shows a schematic cross-sectional view of typical pervious concrete pavement and other pervious covers including porous asphalt, open cell grids, and concrete pavers, to be discussed later in this section. The thickness of the stone reservoir varies based on the application and the required retention storage.

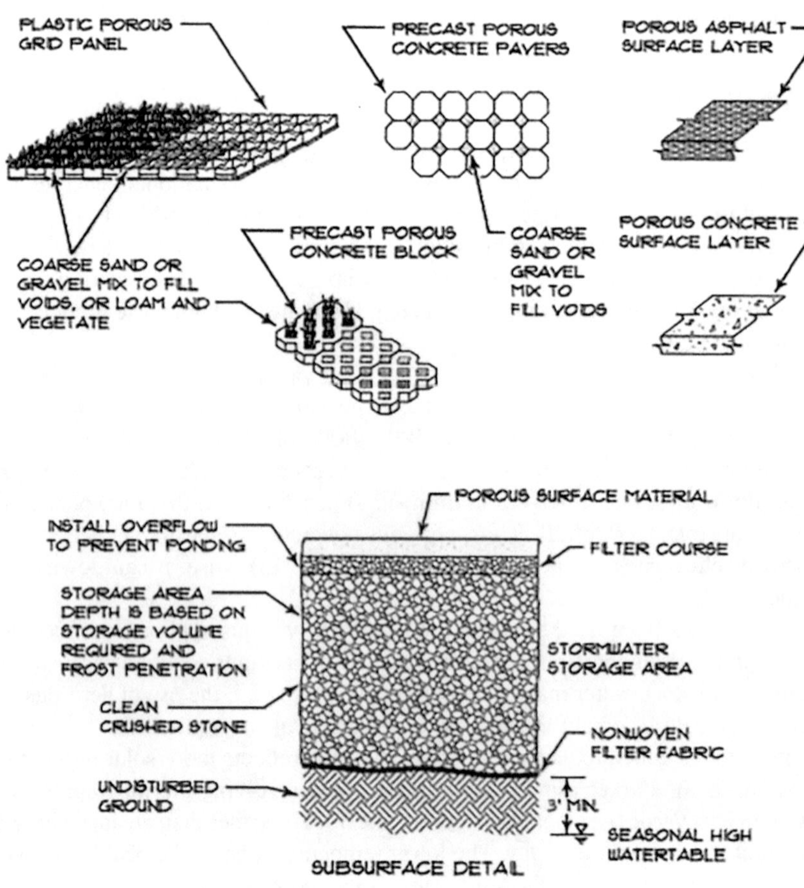

FIGURE 8.5 Cross-sectional view of typical pervious/porous pavements.

Source: ***Maine Stormwater Design Manual*** **(2016), Figure 7.7.1.**

In terms of initial cost, the concrete industry claims that pervious concrete is comparable with porous asphalt, but it requires fewer repairs and has lower ownership costs in the long run. However, in a parking lot project at the University of North Carolina in Chapel Hill, where some sections were constructed using pervious concrete and other porous asphalt, pervious concrete pavement cost four times as much as the porous asphalt. A major problem with pervious concrete is the difficulty of installation. Unlike porous asphalt, which can be handled by any qualified installer, pervious concrete generally requires a certified installer. This difficulty in the preparation of a proper paste and installation was experienced in a 2003 pilot project on the Villanova University campus in Pennsylvania by the concrete manufacturer for the selection of a contractor to place pervious concrete (Traver et al., 2004, 2005). New Jersey Concrete and Aggregate Association is one of the companies that makes pervious concrete and can be visited at http://www.njconcrete.com for information on pervious concrete installers.

Pervious concrete has long been used for sidewalks, driveways, and parking areas. However, there is a growing interest in using it in low- to moderate-traffic roadways, and this trend is expected to increase with more stringent stormwater management regulations. In a project reviewed by this author, pervious concrete was used to pave the parking lot of Unilever Research headquarters in Englewood Cliffs, New Jersey. The pavement, which consisted of 4 in. (10 cm) thick pervious concrete over a 24 in. (60 cm) stone reservoir, was installed at a cost of approximately $10/ft^2 ($108/m^2) in 2012. This project eliminated the need for any detention structures and water treatment devices that would otherwise be required to address the NJDEP stormwater rules.

8.3.4 Concrete Pavers

Concrete pavers were first developed in Europe over 60 years ago. In the United States, the use of pavers for sidewalks, patios, driveways, and parking lots began nearly five decades ago. Pavers are a modern version of stone block pavements that were used by Romans and other old civilizations to construct roads for thousands of years. Figure 8.6 shows cobblestone pavement in Modena town center in Italy.

FIGURE 8.6 Cobblestone pavement in Modena, Italy.

Source: Photo by the author (September 2023).

In Chicago alone, over several million square feet of permeable pavers with open-graded aggregate have been installed over the past three decades; this city is at the forefront among large American cities in a green program to replace paved alleyways with permeable pavement to rectify vast flooding problems (Buranen, 2008). A program adopted in 2007 covered 20 alleys and incorporated three types of pervious pavements: pervious concrete, porous asphalt, and pavers. Depending on the circumstances, an alley was partly or fully covered with permeable pavement. For example, permeable concrete was placed at the alley center with regular concrete on either side.

Concrete pavers are installed by placing a layer of an open-graded aggregate and compacting it to provide a stable surface, placing a setting bed on top of a base course, and setting pavers either by hand or machine. The joints between the blocks are filled with finer material. No filter fabric is generally used to separate aggregate layers. Experience shows that separating aggregate layers introduces clogging and causes more rapid degradation of the system. However, depending on the subsurface condition, a nonwoven filter fabric may be used below the base coarse aggregate to increase stability (see Figure 8.5).

Depending on the application and the intended use, the aggregate layers may be from 6 to 12 in. (15–30 cm) thick. A 6 in. base is commonly used for patios and sidewalks, 8 in. (20 cm) in driveways, and 12 in. (30 cm) for significant exfiltration, light-traffic roads such as highway ramps, and parking lots. The thicker base also provides a larger retention capacity. The layers from bottom to top are as follows:

Subbase: Six to twelve inches (15–30 cm) of 1.5 in. (38 mm) crushed stone aggregate. This stone, which is called #4 aggregate, is compacted in 4–5 in. (10–13 cm) lifts. Railroad ballast, which is called #2 aggregate, may be used (DeLaria, 2008). Recycled concrete is also used, however, as a mix with dense graded aggregate (dga) or quarry-processed stone (known as QP).

Base course: Four inches (10 cm) of 0.75 in. (18 mm) all-fracture-face aggregate, called #67. If #2 aggregate is used as the sub-base, #57 aggregate, which is 1 in. (25 mm) or smaller stone should be used as the base course in order not to fill pore space in the sub-base. The base course is also compacted.

Setting bed: One to 1.5 in. (2.5–4 cm) of coarse sand, also called concrete sand or stone dust. It is important that the sand has no more than 10% fines smaller than 0.075 mm (no. 200 sieve) to avoid slowing down the drainage of the setting bed, also called bedding sand. This bed is not compacted.

In sidewalks and residential patios which are not subjected to any traffic, 3/4 in. stone may be used for sub-base and the base course is eliminated. Figure 8.7 shows concrete block pavers on 15 cm (6 in.) stone base at the author's driveway in Paramus, New Jersey.

Like pervious concrete and porous asphalt pavements, pavers need maintenance, which basically involves sweeping or vacuuming annually to remove surface grit. Over a long period, the subsurface may accumulate enough silt to lose infiltration and detention capacity in any porous pavement. When this happens, the asphalt and concrete have to be removed completely and discarded; however, pavers can be deconstructed, the aggregate layers cleaned, and all materials reinstalled with minimal waste. Thus, pavers, while more expensive to construct than both pervious concrete and porous asphalt pavements, cost far less in the long run.

Another advantage of pavers compared with concrete and asphalt pavement is local repair, such as for the installation of a utility line. If pavers are damaged and have to be removed, they can be replaced without the repaired area being noticed. Also, pavers come in different sizes, colors, and

FIGURE. 8.7 Pavers at author's driveway.

shapes and can be arranged in different patterns. A variety of concrete pavers are available on the market, Belgard, Cambridge, and PaveDrain concrete block among them.

The EPA accepts the use of pervious pavement as an alternative to traditional SWPP BMPs, such as grassy swales and trench infiltration systems. At the Chesapeake Bay visitor center in Annapolis, Maryland, the worn-out asphalt pavement was retrofitted with Pine Hill brick (made of clay) pavers. The pavers infiltrated most of the runoff during Tropical Storm Nicole, which dropped slightly over 9 in. (23 cm) of rainfall in a 24-hour period in 2010. The pavers also functioned well during both Hurricane Irene and Tropical Storm Lee in 2011 and Ophelia in September 2023.

8.4 DISCONNECTION OF ROOFS AND PAVEMENTS

Disconnecting roof leaders from pavements and drainage systems and instead terminating downspouts to lawn and landscape areas create significant reduction in runoff from residential and commercial buildings. Figure 8.8 demonstrates how this may simply involve turning the downspout from driveway to lawn.

Likewise, terminating downspouts to landscape areas can significantly reduce the roof runoff discharge to drainage systems. Three of the roof leaders from this author's dwelling terminate at the landscape area, and the fourth one discharges to a rain garden in his backyard. Figure 8.9 depicts the planted area at the downspout of one of the roof leaders. The landscape area fully retains the roof runoff, and thus no runoff leaves the property.

Micro-bioretention basins along sheets and grass strips next to sidewalks also reduce the discharge of runoff from urban areas. This will also reduce the size and costs of drainage infrastructure. Figure 8.10 shows a strip of grass along the sidewalk of the author's property. The grass has been effectively retaining the runoff from the sidewalk.

The following example illustrates the reduction effect of discharging roof runoff to lawn and landscape areas.

FIGURE 8.8 Downspout location at a residential dwelling.

FIGURE 8.9 A downspout at the author's roof terminating at the planted area.

FIGURE 8.10 A strip of grass along the sidewalk by the author's property.

***Source*: Photo by the author (May 2024).**

Example 8.1

A roof leader spreads runoff from a 40 m²/400 ft² area of a roof through a splash guard onto a 50 m²/500 ft² lawn. For a low to moderate soil permeability of 100 mm/h (4 in./h). Calculate the runoff discharged from the site during a 50 mm/h (2 in./h) rainfall. This 50 mm/h (2 in./h), calculate the runoff discharged from the site. This storm represents a 10-year, 60-minute storm in New Jersey.

First, calculate the volume of rain on the roof and lawn in both SI and CU:

$$V_r = (40 + 50) \times 50 / 1000 = 4.5\ m^3$$

$$V_r = (400 + 500 \times 2 / 12) = 150\ ft^3$$

Then, calculate the infiltration capacity of the soil

$$I_v = 50 \times 100 / 1000 = 5.0\ m^3$$

$$I_v = (500 \times 4 / 12) = 166.7\ ft^3$$

$$I_v > V_r$$

These simple calculations conservatively neglect the initial abstractions. Estimating the initial losses at 2 mm for the roof and 10 mm for the lawn, more accurate calculations are performed as follows:

- Duration to satisfy initial abstraction on the roof

$$tr = (2mm / 50mm) \times 60 = 2.4\ min$$

After that time, the runoff from the roof discharges onto the lawn. Within the same period, 2 mm of the 10 mm depression storage of the lawn will be used. As shown below, within the next 5.33 minutes, the rainfall on the lawn and the runoff from the roof satisfy the remaining 8 mm initial abstraction of the lawn:

$$Water\ entering\ the\ lawn = (40+50)\times 50mm/60 = 75\ mm/h$$

$$t_l = 8\times 50\ \mathrm{m}^2/75 = 5.33\ min.$$

As such, during the first 7.73 (2.4+5.33) minutes of the rainfall, the initial abstractions retain the rain, conservatively neglecting any infiltration. Then, the infiltration fully absorbs the rainfall. The following calculations display the process.

- During 0 to 7.73 minutes:

$$\text{Rainfall amount} = (7.73\times 50/60)\times(40+50)/1000 = 0.58\ m^3$$

$$\text{Initial abstractions} = (40\times 2 + 50\times 10)/1000 = 0.58\ m^3$$

- From 7.73 min to 60 minutes:

$$\text{Rainfall amount} = (60-7.73)\times 90\times(50/60)/1000 = 3.92\ m^3$$

$$\text{Infiltration rate} = \frac{3.92\ m^3}{(52.27\ min\times 50)} = 0.0015\ \text{m/min} = 1.5\ \text{mm/min}$$

$$\text{Infiltration capacity} = 100/60 = 1.67\ \text{mm/min}$$

The lawn infiltrates the entire rainfall volume. These calculations which are based on the author's universal method show the effect of disconnecting pavements. Note that the composite CN method in the TR-55 cannot be applied to this case where the impervious area exceeds 30% of the total area, and the rainfall duration is one hour. Likewise, the NJDEP's RCN is not applicable to this storm. However, applying the TR-55 method to a 24-hour storm of 10-year frequency, which amounts to 5.3 inches of rainfall, and using CN-39, which is the smallest soil curve number for pervious areas, yield a discharge of 39.051 cu ft (1.12 m^3) of runoff from the lot, which is unreal.

Even assuming that the roof runoff is separated from the lawn, the TR-55 method shows a runoff depth of 0.26 in. (0.6 mm) or 10.83 ft^3 (0.3 m^3)from the lawn alone during the above-indicated 24-hour storm. As such, the TR-55 method, though specified as the sole method of calculations by a large number of regulatory agencies, including the NJDEP, is improper for calculating the run-off from impervious surfaces connected to pervious areas, This method yields erroneous results, grossly exaggerates the runoff conditions, and does not agree with observations.

As was indicated, three of the downspouts at the author's property terminate at a planted area adjacent to the lawn. Though the topsoil is underclaimed by clay less than 10 inches (25 cm) below grade, the author affirms never having seen a drop of runoff from the lawn flowing onto the side-walks along the property during the past 16 years. It is this author's opinion is that disconnecting pavements is more effective than green roofs and blue roofs. It is also inexpensive and free of main-tenance. The author highly recommends that site planners and stormwater management profession-als give serious considerations to its use.

8.5 GREEN ROOFS

Green roofs originated in Europe to reduce runoff from cities where nearly one-quarter of the total urban land is covered with roofs. The concept moved to the United States in the 1990s, and green roofs have since become popular. Now all across the country, college and university administrators are raising green roofs on dormitories and department buildings. An example is Pennsylvania State

FIGURE 8.11 Undulating green roof of California Academy of Sciences in San Francisco.

Source: Photo by the author (2010).

University, which has at least thirty green roofs. Figure 8.11 shows the undulating green roof of the California Academy of Sciences in San Francisco, a showcase that serves as a wildlife habitat, among other environmental and educational objectives. Nearly 87% of the 2.5 acres (1.01 ha) of this roof is covered by numerous native plants in biodegradable trays.

8.5.1 Green Roof Structure

The main components of a green roof typically include a lightweight growing medium with a high percentage of porous inorganic material, a root-repellent system, filter cloths, a drainage system, and a variety of native plants. The objective of the filter fabric is to prevent growing medium from clogging the drainage system that is intended to drain excess water from the roof and save plants from drowning Figure 8.12 shows a layout of a green roof. Green roof infrastructure can be a loose lid, with each of the layers placed on another, or modular, with a number of layers placed in a prefabricated tray. Because of ease of installation and lower cost, modular green roofs are used more often in practice.

In terms of construction, green roofs are divided into three types: extensive, semi-intensive, and intensive roofs. Extensive and intensive green roofs are also referred to as above-grade and below-grade, respectively. Extensive green roofs require soil or lightweight medium no more than 6 in. (15 cm) thick that are planted with some of the many varieties of the sedum family. The plant selection is critical in the performance of green roofs and is dependent on the local climate. In addition to being attractive, low-growing sedum is both drought tolerant and a good water absorbent.

Intensive green roofs, as their name implies, require more effort to establish and maintain. They are also thicker, at least 1 ft (30 cm) thick. This type of green roof can include small trees and shrubs, which can be referred to as rooftop gardens, perennial gardens, and/or annuals. The services of gardeners or maintenance crew for regular weeding, mowing, pruning, and fertilizing and even a landscape architect are needed just as for a regular garden. Irrigation and drainage systems

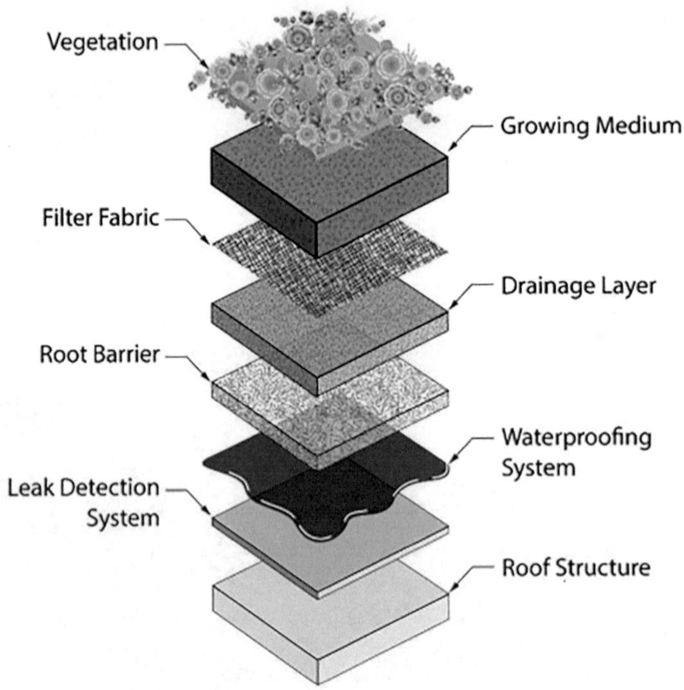

FIGURE 8.12 Cross-section of green roof layers.

Source: ***New Jersey Stormwater Best Management Practices Manual*** **(July 2023).**

for intensive green roofs are more complex, and the amount of fertilizer must be monitored so that pollution in runoff does not increase. One such green roof was constructed on portions of a new parking garage of the Stevens Institute of Technology in Hoboken, New Jersey; the author served as a drainage consultant on this project.

Semi-intensive green roofs consist of varying levels of growing plants with more plant variety than extensive systems. This type of roof is used when the roof structure does not have the capacity to bear the weight of a fully intensive green roof. All types of green roofs require waterproofing membranes to avoid leakage into buildings and an additional barrier layer to protect the membrane from root penetration and accidental damage by workers. Apart from requiring more materials and labor than extensive roofs, intensive green roofs also require far more structural reinforcement of the roof to bear their extra weight.

The extensive green roof adds, on average, 7–8 lb/ft^2/in. (130–150 N/m^2/cm) of soil. Thus, an extensive green roof adds between 30 and 45 lb/ft^2 of dead load to a standard roof, which weighs 25–35 pounds per square foot (1200–1700 N/m^2). According to the EPA, an intensive green roof adds 80 to 150 lb/ft^2 (3800–7200 N/m^2:390–730 kg/m^2) of load to the roof structure (http://www.epa.gov/neatisland/strategies/greenroofs.html). The design of intensive and semi-intensive green roofs commonly includes a storage reservoir, which may be filled with stone. This reservoir also serves as a root barrier and improves drainage and aeration.

Because of city living, green roofs are far more popular and practical in Europe than in the United States. In Germany, 10%–20% of buildings are covered with green roofs compared with less than 1% in the United States. One reason is that most people in Europe live in multistory apartment buildings with flat roofs, while the majority of dwellings in the U.S. are single family homes with steep roofs. Figure 8.13 shows green roofs of an apartment complex in Munich, Germany, where the roofs are partly covered with potted plants. The excess roof runoff from the buildings enters a vegetable garden on the ground surface (at the bottom of the picture). Figure 8.14 depicts the Bosco Verticals, a green building in Milan, Italy. The entire living façade of this building, which may be the greenest building in the world, incorporates numerous trees and over 90 species of plants.

FIGURE 8.13 Green roofs in an apartment complex in Munich, Germany. Photo by the author (2010)

FIGURE 8.14 Bosco Verticale in Milan, Italy.

Source: Photo by author (September 2023).

8.5.2 Cost of Green Roofs

By dissipating heat, green roofs are supposed to reduce the cost of heating and cooling and effectively mitigate the heat island effect of buildings in cities, but this effect is highly dependent on the climate. Researchers in Canada modeled the heating and cooling energy savings from a 32,000 ft^2 (2970 m^2) green roof of a one-story commercial building in Toronto and estimated that the green roof could save approximately 6% of total cooling energy and 10% in total heating, amounting to 21,000 kWh annually (https://www.epa.gov/heatisland/mitigation/greenroofs.htm). Estimating the annual maintenance cost of a green roof at just \$1.00/ft^2 and the energy cost on the average at \$0.20/kWh (ranges from 10 to 40 Canadian cents), the annual costs of maintenance and energy savings for green roofs would be \$32.000 and \$4,200, respectively. Thus, the savings do not even cover 15% of the annual maintenance cost of the roof.

It is also claimed that green roofs have long useful lives. In a 2007 paper, Stephen Peck, the founder and president of Green Roofs for Healthy Cities, noted that the standard expectation in Germany is that the membrane with a green roof above it will last forty years; this may be an overly optimistic expectation. Relating to water quality, green roofs may have the ability to remove pollutants such as dust from runoff. However, this effect is offset by the fertilizer and compost used for the grass and plants.

An EPA website, updated February 9, 2009, reports estimated costs of installing a green roof starting from \$10/ft^2 for simple extensive roofing to over \$25/ft^2 for intensive roofs. The annual maintenance cost of either type ranges from \$0.75 to \$1.50/ft^2 (https://www.epa.gov/heatisland/mitigation/greenroofs.htm).

Today, the green roof installation cost is estimated \$150/m^2 (\$14/ft^2) for simple extensive roofs and up to \$375/m^2 (\$35/ft^2) for intensive ones. The annual maintenance cost of extensive and intensive green roofs is estimated at \$10 / m^2 (\$1/ft^2) for the former and \$30/m^2 $\left(\$3 / ft^2\right)$ for the latter. In comparison, traditional shingle roofs cost about \$70/m^2 (\$7/ft^2), have low maintenance costs, and last twenty to twenty-five years. Additionally, in arid regions such as the US Southwest, green roofs are hardly practicable.

Apart from the initial and maintenance costs, according to the author's estimates, the green roof cost of retaining 1 m^3 of rainwater ranges from \$5,000 to \$7,500 (\$140–\$210/ft^3). Therefore, if reducing runoff is the primary concern, green roofs are not the solution. Other, far more cost-effective alternatives are available. These include dry wells, porous pavements, rain gardens, and rain tanks. In addition to being far less expensive than green roofs, these other options are adaptable to single-family dwellings, which commonly have steep roofs that inhibit the installation of green rooftops. In lieu of green roofs, one should consider rain tanks, which are not only far less expensive but also harvest roof runoff water for reuse.

Alternatively, herbs and vegetable planters and potted shrubs can be placed on flat roofs and balconies. Figure 8.15 shows planted pots on the roof of the Hilton Hotel in Cincinnati, Ohio. Apart from reducing runoff, such arrangements leave room on the roof for placing tables to serve food and beverages. Estimating the retainage capacity of each pot at 4 in (10 cm), covering only one-fourth of the roof area by planted pots, provides more reduction of roof runoff than a 6 in. (15 cm) extensive green roof covering the entire roof area.

8.5.3 Green Roof Stormwater Management Analysis

The descriptions of green roofs so far have not addressed their water retention capability; this section presents a proper method of estimating the volume and the peak runoff from green roofs. A number of jurisdictional agencies recommend SCS TR-55 with a reduced soil CN for estimating the peak and/or volume of runoff discharged from green roofs. New York, New Jersey, Maryland, and Maine are discussed. The NY State Stormwater Management Design Manual (2022) uses the following modified CN for the green room retention roof.

$$CN = 1000 / \left(B + 10\right) \tag{8.1}$$

FIGURE 8.15 Herb pots on a partial roof of the Hilton Hotel in Cincinnati, Ohio.

Source: Photo by the author (2013).

TABLE 8.2
Reduced CNs for Green Roofs, Maryland

Retention Depth of Green Roof in in. (cm)	Effective RCN*
0.6 (1.5 cm)	94
1.0 (2.5 cm)	92
1.2 (3.0 cm)	90
1.4 (3.5 cm)	88
1.6 (4.0 cm)	86
1.8 (4.5 cm)	84
2.0 (5 cm)	82
2.2 (5.5 cm)	81
2.4 (6 cm)	77

where the parameter B the cumulative product of thickness of each larger by its porosity. For an extensive green roof consisting of 6 in. (15 cm) soil media at 30% porosity and 2 in. (5 cm) drainage layer at 0.02 porosity, the above equation gives CN = 84.

In the Maryland Stormwater Design Manual (Chapter 5, 2009), the green roof's contribution to ESD was based on the RCN, which depended on the green roof thickness. Because of discrepancies between observations and the calculations, MDE (Maryland Department of the Environment) has come to conclusion that a different procedure for analyzing the performance of green rooftops needs to be developed. In the interim, MDE recommends that designers and plan reviews use the reduced CNs based on retention depth of the green roof tops (Table 8.2).

A roof with a retention capacity of 25 mm (1 in.), which closely relates to a 6 in, extensive green roof for example, is assigned a reduced CN of 92.

The green roof design in New Jersey is also based on soil CNs. The design procedure of a green roof described in Section 4 of Chapter 9 in the NJDEP Stormwater Best Management Practices Manual (July 2023) includes runoff volume reduction and peak flow attenuation. The manual treats green roof as an impervious surface and applies CN = 98 to calculate the rainfall depth—which it calls the runoff—and then use unsupported assumptions to calculate the peak flow reductions for different storm events. As such, the analysis is totally erroneous.

First of all, using CN = 98 underestimates the amount of rain that falls on the green roof. For the 10-year, 24-hour storm, which amounts to 5.3 in. of rain for example, such calculations suggest a rainfall depth of 4.9 in., which is 7.5% less than its actual amount. The reduced rainfall depth is even 9.3% for 2-year storms (about 3.5 in. in New Jersey) and even higher for smaller storms. The green roof design in that manual goes one step further and based on unsupported assumptions includes a procedure to manipulate the peak roof runoff reductions. It is evident that contrary to the calculations, the green roof discharges without any attenuation once its retention capacity is exhausted.

8.5.4 Proper Procedure for Green Roof Analysis

The author's suggested procedure for green roof analysis is as follows.

8.5.4.1 Retention Capacity Calculations

The maximum retention capacity of a green roof is the difference between the field capacity and the wilting point of its soil medium. The former represents the amount of water that a soil can hold in its pores against the force of gravity, and the latter is the threshold moisture content in the soil that the vegetation needs to survive. These parameters depend on the types of soil; typical field capacity and wilting point are 30% and 10% of the soil volume, respectively. Therefore, the maximum retention capacity of a green roof is 2 mm for every 1 cm of soil depth. Considering that soil moisture has to be kept above the wilting point for vegetation survival, the retention capacity of soil may be taken as 1.5 mm per every 1 cm of soil. Therefore, the retention capacity per unit area of a green roof can be calculated using the following equation:

$$R_{\mathrm{d}} = 0.15 d_{\mathrm{s}} + 0.02 d_{\mathrm{d}} \tag{8.2}$$

where

R_{d} = retention capacity (depth) per unit area, cm in SI; inches in CU

d_{s} = thickness of soil media, cm S.I; inches in *CU*

d_{d} = thickness of drainage layer, cm S.I; inches in *CU*

and 0.15 and 0.02 are the retainage capacity per unit depth of soil media and drainage layer, respectively. Table 8.3 shows typical retention depths of green roofs with 2 in (5 cm) stone drainage base. In practice, the retention in the drainage layer, which is due to wetting stone particles, may be neglected. Note that the water adhesion to the surface of stones in the drainage layer amounts to only 1 mm (0.04 in.) of the total retainage in this table.

TABLE 8.3
Typical Retention Capacity of Green Roofs

Roof Thickness, in./cm	Retention Depth, in./cm
4/10	0.64/1.63
6/15	0.94/2.29
8/20	1.24/3.05
10/25	1.54/3.81
12/30	1.84/4.57

Based on the earlier equations, the volume of runoff discharged from a green roof during rainfalls exceeding the retention capacity ($P > R_d$) is given by the equations:

$$V_o = A(P - R_d)/1000\ SI \quad (8.3)$$

$$V_o = A(P - R_d)/12\ \text{CU} \quad (8.4)$$

where
A = green roof area m² (ft²)
P = rainfall depth mm (in.)
V_o = discharge volume m³ (ft³)

8.5.4.2 Peak Flow Calculations

The previous equations indicate that for small rainfalls, green roofs can retain the entire rainfall but that during sustained rainfalls, once the retention storage in the soil is exhausted, the green roof functions the same as a conventional roof. Therefore, the peak runoff from green roofs during heavy storms can be calculated using the following equations, which are based on the universal runoff model using zero infiltration and considering that the initial losses have been satisfied:

$$Q = IxA/3600 \quad SI \quad (8.5)$$

$$Q = Ix\ \text{A}/43200 \quad CU \quad (8.6)$$

I = rainfall intensity, mm/h (in./h)
A = green roof area, m² (ft²)
Q = peak discharge, L/s (cfs)

Depending on the type of green roof and the depth of soil media, green roofs can retain from 40% to over 60% of rainfalls, annually. The average annual retainage in %, as listed in Table 3.6.2 in the *Maine Stormwater Management Design Manual (2016)*, ranges from 40% to 60% for extensive green roofs of 10 to 20 cm (4–8 in.) thick and from 60 to 75% for intensive green roofs with 20 to over 50 cm (8–15 in) thick media. These actual measurements as reported in the Maine manual support the author's analysis.

Example 8.2

An extensive green roof consists of 10 cm thick soil media and a 3 cm drainage layer, covering 1000 m² of a 1200 m² roof. Calculate the retention capacity of the green roof and the cost of roof structures. Base your cost calculations on \$120/m² of roof area.

Solution

Estimating the effective retention (field capacity minus wilting point) at 15% and the retention of drainage media at 2%.

$$\textit{Retention volume} = 1000 \times (10\ cm \times 0.15 + 3 \times 0.02)/100 = 15.6\ m^3$$

$$\textit{Added roof cost} = 1200 \times 120 = \$144{,}000$$

Example 8.3

Estimating the annual maintenance cost of the green roof in Example 8.2 at \$10/m² and assuming that one half of the annual rainfalls are retained by the roof, calculate the maintenance cost for every 1 m^3 of the rainwater the roof retains. The average annual precipitation at the locality is 1000 mm (40 in). Also, calculate the peak discharge for the roof during a 5-year, 30-minute storm with intensity of 70 mm/h.

Solution

a. *Annual maintenance cost* $= 1000 \times 10 = \$10{,}000$

$$\text{Volume of annul runoff reduction} = 1000\,\text{m}^2 \times 1000\,\text{mm} \times 50\% / 1000 = 500\ m^3$$

$$\textit{Maintenance cost of runoff reduction} = \frac{10{,}000}{500} = \$20 / m^3$$

$$\textit{Runoff volume during the storm} = 1000\,\text{m}^2 x \left(70\ mm \times \frac{30}{60}\right) / 1000 = 35\ m^3\ 35 > 15.6\ m^3$$

Since the rainfall volume exceeds the retention capacity, the roof discharge occurs. The time that the roof begins to discharge after the beginning of the rain is calculated as follows:

$$(15.6/35) \times 30 = 13.7\ min$$

From that time on, the green roof discharges with zero attenuation and the peak runoff from the green roof and the entire roof area are calculated as follows

$$Q = 1000x(70/3600) = 19.4\ L/s \text{ from the green roof area}$$

$$Q = 1200x(70/3600) = 23,.3\ \text{L/s from the entire roof area}$$

These examples illustrate that green roofs are not only very costly means of reducing runoff but also are totally ineffective for peak flow attenuation.

8.6 BLUE ROOFS

A blue roof is a non-vegetated stormwater management source control practice. It includes impermeable structures such as trays that store rain temporarily. The stored water is dissipated partly by evaporation. By releasing the stored water after the storm has ended, blue roofs also mitigate the peak flows to storm drains or combined sewer systems. Alternatively, the stored water can be reused for irrigation. Keeping the water on the roof during hot days will also reduce the load on HVAC equipment.

A variety of flow controls are available including modified inlets, check dams, and trays. Blue roofs were conceived in 2008, and its application is rapidly expanding in the United States and around the world. The installation of blue roofs has proved to be an emerging technology in densely populated urban areas such as New York City, Philadelphia, and San Francisco. Depending on the types of control devices used to control the discharge of water that is stored temporarily, blue roofs are classified as active or passive. All types of flow control have some detention effects. However, the tray and check dam systems are more effective than others. Trays, which may be made of metal or recycled plastic, provide a simple means of storing rain on the roof. They are partly filled with blasts, such as rocks, to be held in place.

A number of blue roof pilot projects have been conducted across the United States A significant blue roof pilot project was undertaken by the New York City Department of Protection (NYDEP) between 2010 and 2012. This project was the first to utilize a novel passive blue roof tray design developed by Geosyntec Consultants. The design relies on the lateral transitivity of nonwoven filter fabric for drawdown control in a full-scale pilot project. Monitoring of these systems has demonstrated their performance as an effective means for peak flow mitigation and lagging the peak flow in combined sewer systems. Coupled with light-color roofing material, the blue roofs provide energy cost savings through rooftop cooling.

The New York City Department of Environmental Protection (NYCDEP) has led the advancement of blue roofs through widespread technology by conducting tests. In 2015, NYCDEP renewed its commitment to sustainability. Now, blue roofs account for 10% to 20% of hard rooftop areas. In some cases, recycled plastic trays held in place by ballast, such as rocks, are employed to store rainwater and to reduce the annual runoff from the roof.

NYCDEP has experimented with trays and check dams. The tests indicate that both types of blue roofs provide runoff reduction benefits. In particular, the tests showed that tray systems retain 75% to 80% of rainfalls of up to 50 mm (2 in.). This retention depth is about twice as much rainfall as a 15 cm (6 in.) extensive green roof holds: For the same 50 mm retention of rain, blue roofs add less than one-tenth of the load on the roof. As such, blue roofs significantly reduce the added cost of roof structure and eliminate the need for an impermeable membrane. Also, because of the absence of plants, the maintenance costs for blue roofs are significantly lower than those for green roofs. In addition, blue roofs do not require complex repairs as green roofs do. Thus, the author recommends using blue rather than green roofs on flat top buildings.

8.7 DRY WELLS

A dry well (commonly referred to as seepage pit) is a perforated hollow cylinder installed in a stone trench. Traditionally, cylinders are made of concrete; more recently, they have also been fabricated of plastic materials. Figure 8.16 shows a typical 7.0 ft (2.1 m) O.D., 5.0 ft (1.5 m) tall concrete seepage pit manufactured by Peerless Concrete Products Co. (Butler, NJ, USA). This company makes seepage pits ranging from 6 ft, 6 in. (2 m) O.D., 6.0 ft (1.83 m) I.D. to 10 ft (3.05 m) O.D., and 3 ft to 9 ft, 10 in. (0.915 to 3.0 m) tall. The top slab thickness of pits varies from 4 in. (10 cm) in nontraffic areas such as backyards to 8.0 in. (20 cm) for HS-20 load rating. Soil Erosion and DEP/DEC, Department of Environmental Protection/Conservation manuals
in many states include details of seepage pits filled with rock (see Figure 9.21). This arrangement, however, reduces the effective retention volume of a dry well cylinder by over 60%.

Dry wells not only serve as a cost-effective system for runoff reduction but also provide groundwater recharge and water quality enhancement. They are one of the best means of managing presidential roof runoff. As will be shown, when sized properly, they can maintain the preexisting runoff conditions for a new dwelling. Many states, including New York, accept dry wells for water quality, runoff reduction, and extreme flood protection, and Maryland has approved dry wells for ESD and Rev (runoff reduction). NJDEP (2023), however, does not accept dry wells for either water quality or runoff quantity, which is absurd.

8.7.1 Dry Well Analysis

Dry wells function as retention basins. Therefore, to avoid premature clogging from grass clippings, leaves, silt, and debris, seepage pits are not recommended for retainage of runoff from lawn areas and supermarket parking lots. They are best suited for roof runoff, where few impurities are present. Even for this application, roof gutters should have a screen to remove leaves and roofing shingle grits.

In New Jersey, where the author practiced, and many other parts of the country as well, seepage pits are widely used for the retention of runoff from dwelling roofs to maintain/reduce runoff from residential lots. As indicated, seepage pits function as retention basins in that they fully retain the rainfall until filled completely but discharge with zero attenuation once full (Figure 8.17). To prevent frequent surcharges, the author recommends that designing seepage pits for at minimum a 10-year, 60-minute storm. In New Jersey, this storm has approximately 2 in./h (50 mm/h) intensity

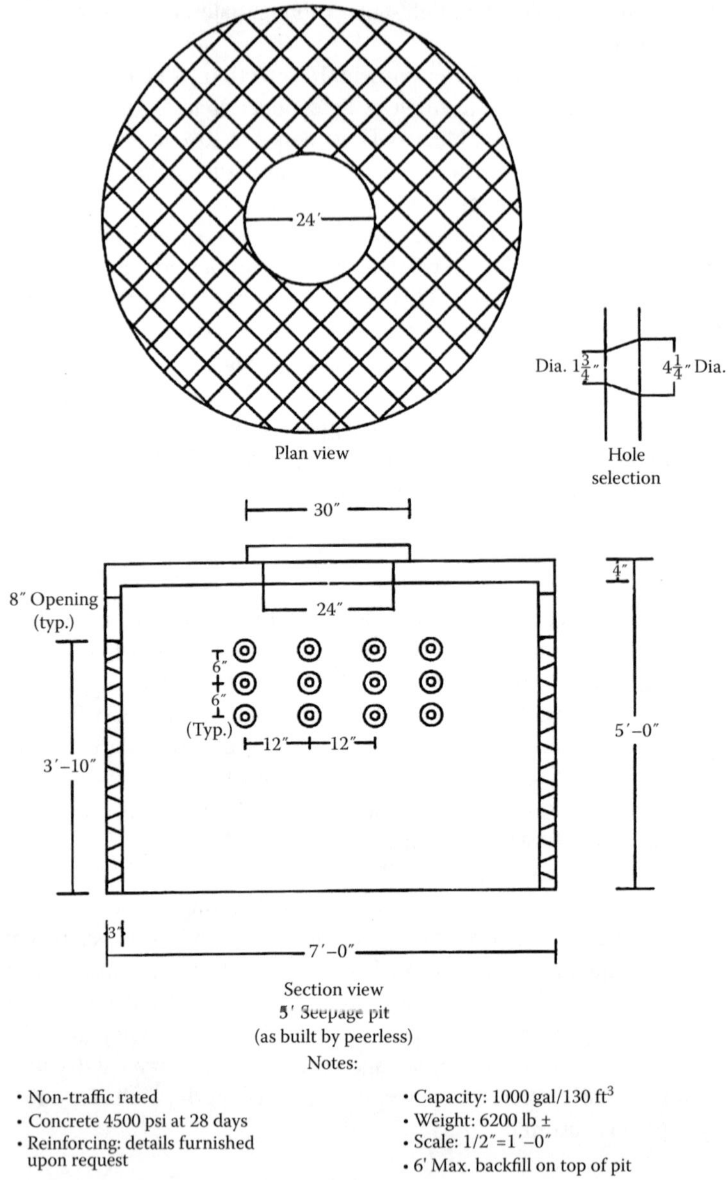

FIGURE 8.16 A typical 7 ft O.D., 5.0 ft tall seepage pit by Peerless Concrete.

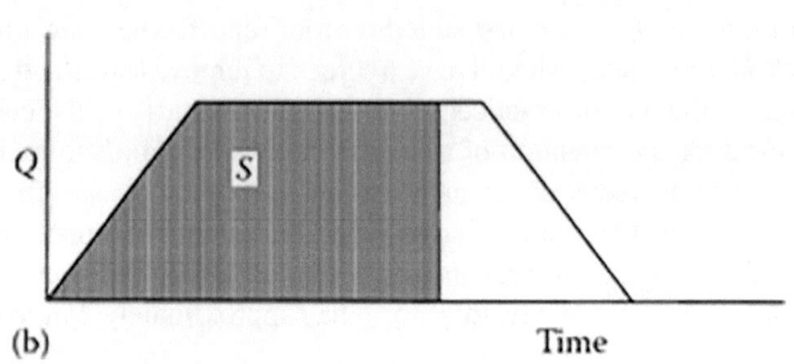

FIGURE 8.17 Flow routing through a dry well.

and amounts to 2 in. (50 mm) of rainfall. Universally, the required storage volume of a roof runoff seepage pit can be calculated as follows:

$$V = A \times P \; SI \tag{8.7}$$

$$V = 0.083A \times P \; CU \tag{8.8}$$

where
V = rainfall volume = retention storage, L (ft^3)
A = roof area, m^2 (ft^2)
P = rainfall depth, mm (in.)

Alternatively, the required retention volume may be calculated using the following equations based on rainfall intensity. Note that these equations are based on a runoff coefficient of 1.0, which is based on the author's universal method and which properly reflects the runoff volume from the roof.

$$V = A \times I \times T_d \tag{8.9}$$

$$V = 0.083A \times I \times T_d CU \tag{8.10}$$

where
I = rainfall intensity, mm/h (in./h)
T_d = storm duration, hours
A and V are as defined before

The preceding equations conservatively neglect percolation losses through the bottom and sides of the seepage pit during the design storm. For storms longer than the design (suggested one hour), the percolation losses through the bottom can be included in the seepage pit design. This is illustrated by Example 8.4. This example shows that for soils of high and moderate permeability, the suggested 10-year, 60-minute storm provides a satisfactory design. To discharge heavy storm events, roof drains can be equipped with an overflow pipe (see Figure 8.18).

Figure 8.19 depicts the increase in impervious areas from a vacant lot due to the construction of a single-family home. In this figure, the lot size, dwelling roof, and driveway areas are denoted as A, R, and D, respectively. For simplicity, the changes in runoff are expressed in terms of the product of runoff coefficient and area, CA. Also, for numerical illustration, runoff coefficients of 0.3 and

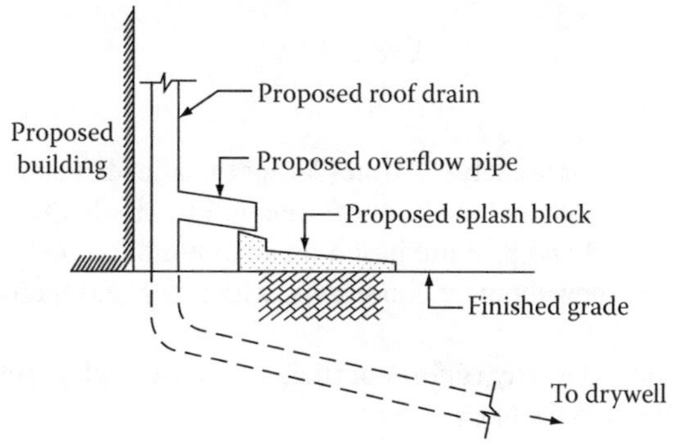

FIGURE 8.18 Drywell overflow

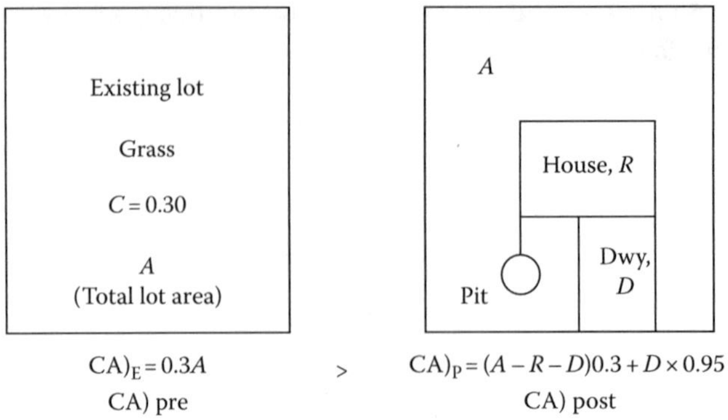

FIGURE 8.19 Schematic changes in impervious areas from a residential lot.

0.95 are employed for the pervious and paved areas, respectively; pervious areas commonly consist of grass–woods combination for open lots and lawn/landscape areas for the improved sites. For the runoff from the roof shown on the above figure is to be retained by a seepage pit (or an underground chamber) in the stone trench, the pre-and post-development peak rates of runoff from the lot can be expressed by the following proportionality equations, respectively (Pazwash, 2012):

$$Q\_Pre \approx 0.3A \tag{8.11}$$

$$Q\_Post \approx 0.3(A - R - D) + 0.95D \tag{8.12}$$

where
A = lot area, m^2 (ft^2)
R = roof area, m^2 (ft^2)
D = driveway, m^2 (ft^2)

The rainfall intensity is the proportionality factor in these equations. The condition that the post-development peak runoff not exceed that of predevelopment is determined from the following equation:

$$0.3(A - R - D) + 0.95D \le 0.3A \tag{8.13}$$

This equation simplifies as

$$0.3R \ge 0.65D \tag{8.14}$$
$$\text{or } R \ge 2.2\text{ D}$$

Thus, if the roof area is approximately 2.2 times larger than the driveway area and the roof runoff is fully retained, the post-development peak runoff rate will be smaller than that of the predevelopment condition. Equations 8.13 and 8.14 are based on an assumption that the time of concentration is the same for pre- and post-development conditions, which is a reasonable assumption for single-family homes.

This analysis can be applied to any runoff coefficient. For example, using a runoff coefficient of 0.25 for pervious areas, Equation 8.14 changes to

$$R \ge 2.8D \tag{8.15}$$

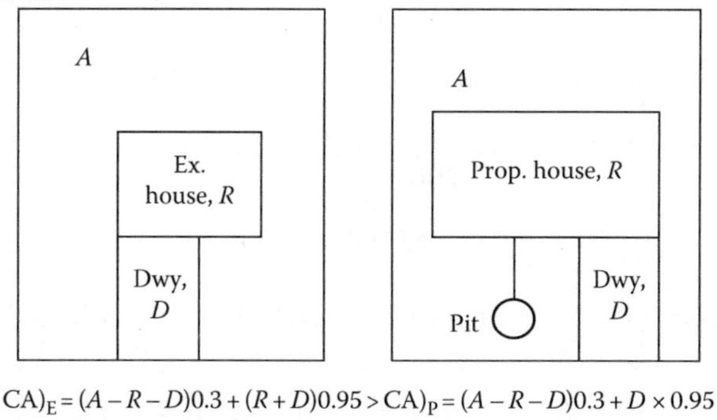

FIGURE 8.20 Changes in imperviousness

If the preceding condition (the driveway being 2.8 times smaller than the roof area) cannot be met, then the driveway may be covered with paver blocks rather than asphalt or concrete pavement. This will reduce the runoff coefficient for the driveway to 0.5 or less, and therefore, the condition of no increase in peak runoff rate (Equation 8.15) becomes

$$R \geq 1.67D \tag{8.16}$$

Figure 8.20 shows the changes in the peak runoff from a reconstructed single-family home in which the runoff from the new dwelling roof is retained by a seepage pit. In this case, the condition of no increase in peak runoff from the lot is expressed by the following equation:

$$\left[C_\mathrm{p}(A-R-D)+C_\mathrm{I}D\right]_\mathrm{Post} \leq \left[C_\mathrm{p}(A-R-D)+C_\mathrm{I}(R+D)\right]_\mathrm{pre} \tag{8.17}$$

In this equation, C_p and C_I are the runoff coefficients for pervious and paved areas, respectively, and the other parameters are as defined previously. Except for unusual circumstances, the preceding equation holds true; therefore, retaining the roof runoff will more than satisfy the condition of no increase in the peak runoff from the reconstructed dwelling.

Example 8.4

Design a seepage pit to retain the runoff from a 200 m^2 (2150 ft^2) dwelling roof for a 10-year, 60-min, 50 mm/h (2 in./h) storm. Also, determine if a surcharge occurs during 25-year storms (see Table 8.4 for the 25- year rainfall data). The soil permeability is 100 mm/h (4 in./h).

TABLE 8.4
Twenty-Five-Year Storm Durations and Intensities

Duration (h)	Intensity mm/h (in./h)
1.5	38 (1.50)
2.0	30 (1.20)
3.0	23 (0.90)
4.0	18 (0.73)
6.0	13 (0.53)

Intensities in mm/h are rounded to full digits.

Solution

Required storage volume is calculated using Equations 8.7 and 8.8:

$$V = 200 \times \frac{50}{1000} = 10\ m^3$$
$$V = 0.083 \times 2150 \times 2 = 356.9\ ft^3$$

Select a 7.0 ft O.D., 6.5 ft I.D. 5.0 ft deep pit (Figure 8.16) in 11 ft × 11 ft × 6.5 ft stone trench

$$Pit\ Vol. = \left(\frac{\pi}{4}\right) \times (6.5^2 \times 5) = 165.9\ ft^3 = 4.70\ m^3$$
$$Void\ vol. = \left[11 \times 11 \times 6.5 - \left(\frac{\pi}{4}\right) \times 7^2 \times 5\right] 0.4 = 237.6\ ft^3 = 6.73\ m^3$$

$$Totalval. = 165.9 + 237.6 = 403.5\ ft^3 > 356.9\ ft^3$$

$$Total\ vol. = 4.70 + 6.73 = 11.43\ m^3 > 10\ m^3$$

During the 25-year storms, the rainfall amount is compared with the seepage pit volume (11.43 m^3) plus the percolation from the bottom of the seepage pit, which is calculated as follows:

$$V_r = 200 \times I \times \frac{T_d}{1000} = 0.2 \times I \times T_d\ m^3$$
$$V_{per} = (11 \times 0.305)^2 \times \frac{100}{1000} \times T_d = 1.125 T_d$$

where
V_r = Rain volume, m^3
V_{per} = percolation volume, m^3
T_d = rainfall duration, h

In these equations, seepage losses from the sides of stone trench are conservatively ignored. Table 8.5 lists the calculated rainfall volume, the percolation volume, the sum of percolation and seepage pit volumes, and the percent of the seepage pit volume utilized during each storm. In the table, all volumes are expressed in m^3.

Calculations in CU are left as an exercise by students.

TABLE 8.5
Seepage Pit Performance during 25-Year Storms

T_d (h)	I (mm/h)	V_r (m^3)	V_{per} (m^3)	V_{pit}+V_{per} (m^3)	Seepage Pit Storage (%) Used
1.5	38	11.4	1.69	13.09	87
2.0	30	12.0	2.25	13.68	88
3.0	23	13.8	3.38	14.81	93
4.0	18	14.4	4.50	15.93	90
6.0	13	15.6	6.75	18.18	86

8.8 RAIN GARDENS

A rain garden is a shallow planted bed on topsoil, underlain by a sand base to retain the runoff from a paved area and infiltrate it into the ground. Though the concept of rain gardens is not new, the term rain garden has been introduced to stormwater management practices in the recent past. Ancient civilizations practiced directing runoff into vegetable gardens, particularly rice fields, which require ponding water. In this country, it appears that rain gardens originated in Maryland, developed in Minnesota, and quickly spread to other states.

In New Jersey, with an average monthly rainfall of approximately 4 in. (100 mm) during the growing season, rain gardens are becoming increasingly popular in retaining rainwater from dwelling roofs and driveways. The Borough of Englewood Cliffs in New Jersey formed a committee to develop and publicize rain gardens as well as other green stormwater management measures in the borough. This author, who served as a member of this committee, reviewed and made site-specific recommendations for reducing runoff from developments.

A regular flower garden, a vegetable garden, and even a simple lawn depression all function like a rain garden. However, a rain garden is commonly a planted shallow depression or series of shallow depressions on a sand bed to retain the runoff it receives and infiltrate it over time. To discharge water in excess of its capacity, a rain garden can incorporate a riser and trash rack. Figure 8.21 shows a sectional view of a typical rain garden with an overflow system.

Some design considerations for rain gardens and their landscaping are presented at a publication by Penn State Extension (2022). An EPA (2025) publication titled, "Soak up the Rain; Rain Gardens," includes information about rain garden and references local resources of New England States. Rutgers University and the Native Plant Society of New Jersey have prepared a rain garden manual titled Rain Garden Manual of New Jersey (2010). This manual was updated by the Native Plant Society of New Jersey in 2023/24. Both of these manuals provide useful information on the design and construction of rain gardens and the selection of proper plants. See (https://www.npsnj.org).

To improve infiltration and aesthetics, rain gardens are partly covered with a variety of native plants, herbs, and woody plants that are adapted to the soil and site climate. Native plants that are drought tolerant and have deeper root systems that help increase the recharge of groundwater and that require no fertilizer are ideal. The root systems of plants in rain gardens filter pollutants from rain that infiltrates the soil. Bushy flowers may be planted in rain gardens, as well. The plants should be maintained twice annually to keep them healthy and aesthetically pleasing. Rain gardens are

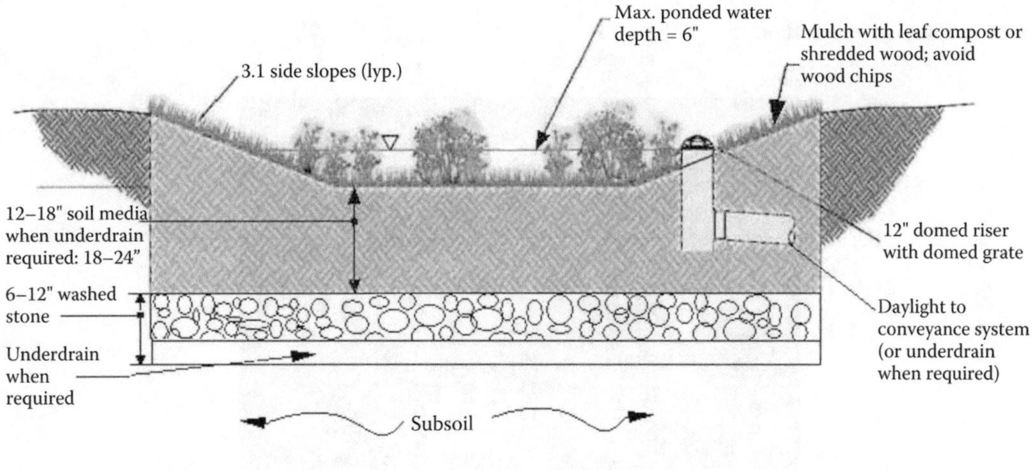

FIGURE 8.21 Typical rain garden section. (From New York State Stormwater Management Design Manual, Figure 5.41, 2015)

effective in reducing flooding and improving water quality and are becoming popular as a green solution to flooding and pollution problems.

The city of Aurora, Illinois, performed a model study of pollutant removal effectiveness of rain gardens using the Source Loading and Management Model (SLAMM). Five hydraulically connected rain gardens, each 11 ft (3.35 m) wide, were placed between the curb line and sidewalks along Spring Street. One of the rain gardens was 60 ft (18.29 m) long, and the remaining four were 50 ft (15.24 m) long. All rain gardens included a substrate and plants that consisted of native shrubs and flowering plants (dogwood and nannyberry). Three of the gardens had a 0.49 acre (1983 m^2) drainage area. SLAMM modeling for these three gardens showed that they received 29,085 ft^3 (824 m^3) of runoff during the two-year modeling period. Of this, 82% was infiltrated, and the rest overflowed to a catch basin in the third rain garden. The modeling study also showed that the three garden systems captured 85% of total solids, 89% of particulate phosphorus, 84% of total Kjeldahl nitrogen, 85% of heavy metals (copper, lead, and zinc), and 83% of fecal coliform bacteria (Seth, 2011). However, no field measurements were taken to confirm these results.

Figures 8.22 and 8.23 depict a lawn depression, 6–12 in. (15–30 cm) deep, that the author designed to retain the runoff from the roof and parking lot of the Fire Station of Hawthorne Township in

FIGURE 8.22 Rain basin retains runoff from roof and driveway of Hawthorne, New Jersey, municipal fire station. The rocks at right protect the basin from erosion at the roof drain.

Source: Photo by the author.

FIGURE 8.23 Another view of s lawn depression (a simple, inexpensive rain garden).

FIGURE 8.24 Rain basin in the author's backyard.

Source: Photo by the author (July 2024).

northern New Jersey. This lawn depression (which is basically the simplest, least expensive rain garden) measures approximately 450 m^2 (4840 ft^2) and receives runoff from 2980 m^2 (32,080 ft^2) of impervious area consisting of 430 m^2 (4630 ft^2) of roof and 2550 m^2 (27,450 ft^2) of parking lot. This lawn depression has been functioning satisfactorily without any maintenance other than regular mowing since its construction in 2005.

Figure 8.24 shows a rain garden in the backyard of the author's residence. This garden measures approximately 25 m^2 (270 ft^2) on plan area and is up to 15 cm (6 in.) deep. It is planted with Hosta, Pachysandra, a dogwood, and six giant green arborvitaes. This garden receives runoff from approximately 45 m^2 (485 ft^2) of roof area (one of the four roof leaders of the dwelling) plus a gazebo, cabana, and 35 m^2 (375 ft^2) of concrete pavement. The planted depression has no stone/sand subbase and, in fact, is underlain by a clay layer 25–30 cm below the surface. It fills up 10–15 cm (4–6 in.) after heavy rains and drains out within 24 to 36 hours. The depression was filled approximately 15 cm (6 in.) deep after two consecutive thunderstorms, which dropped nearly 13 cm (5 in.) of rain on May 22 and 23, 2014, but drained completely a day and a half later. This garden has also fully retained every heavy rainfall ever since and has required no watering whatsoever.

Based on the author's review and recommendation, a developer used a rain garden to manage the runoff from the building roof and the parking lot of a new apartment building in the Township of S. Hackensack, New Jersey. This option eliminated the need for an underground detention basin and water treatment devices and eased the acquisition of the required stormwater permit from the NJDEP. A rain garden does not have to be sunken. It may be constructed as a nearly flat landscaped garden when it is downslope from its tributary area. Figure 8.25 shows a rain garden the author designed to capture the runoff that was entering the yard of a property from an uphill neighbor in a northern New Jersey town.

Despite the benefits of rain gardens, many people consider them unattractive and shy away from them. On the contrary, a rain garden, if properly constructed and decorated with ornamental plants, can be quite appealing. A case in point was a 200-acre mixed-use project in Lynnfield, Massachusetts. Another example worth visiting is a demonstration rain garden that was constructed in the spring of 2007 at the Nature Center of the Stony Brook-Millstone Watershed Association in New Jersey.

Rain garden plants may be selected based on the USDA Plant Hardiness Zone Map (US National Arboretum), which divides the United States and Puerto Rico into 13 zones that range from below −50 °F for zone 1 to above 60 °F in 10° increments, during cold season. Each zone is further divided

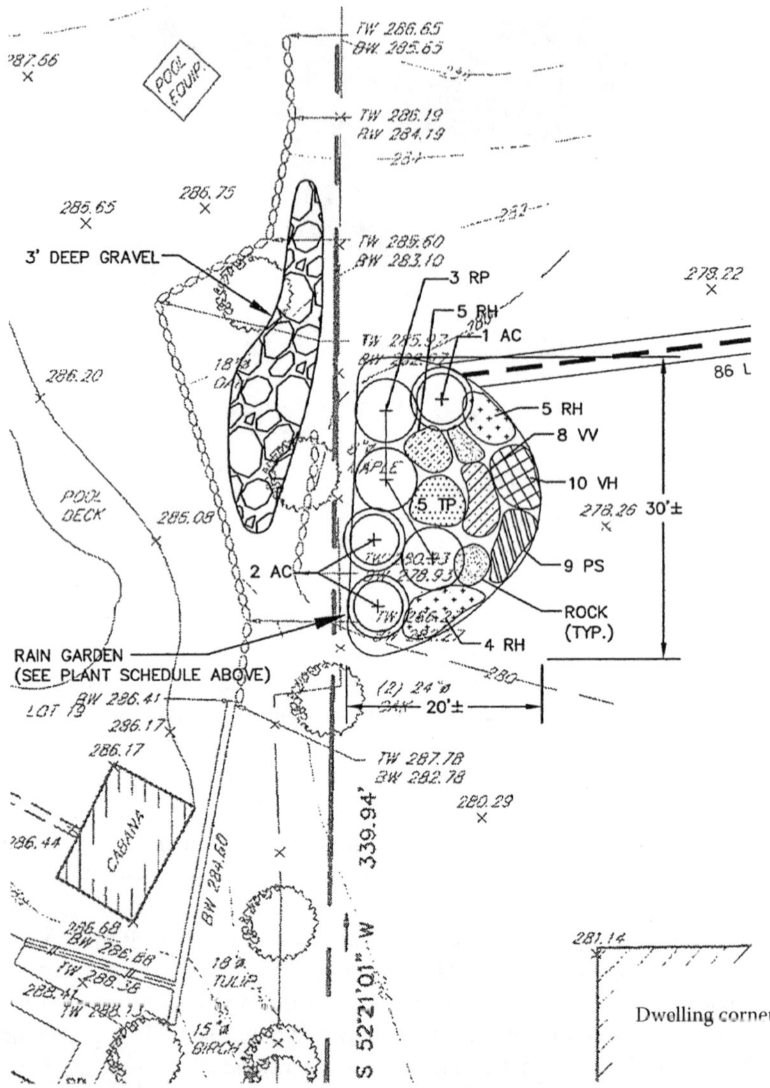

FIGURE 8.25 Rain-garden plan for retainage of runoff entering a residential yard from an uphill property. Plant selection listed below.

into a and b in 5° increments; New Jersey for example includes planting zones 6b, 7a, 7b, and 8a. In New England states and the eastern United States down to North Florida, designated as zones 3a to 9b, the following plants may be selected for rain gardens: Bur Oak, Switchgrass, Hosta, Little Bluestem, and Butterfly Milkwood for full sun and wild bergamot and Hydrangeas for sun and shade. Figure 8.26 shows Aphrodite plantain lily, a Hosta family plant. To select proper plants, consult with state plant societies or local nurseries.

The site of the rain garden should be tested to determine if a permeable soil is present. Low permeability soils need to be removed and replaced with a good soil mix comprising of 50% sand, 20% topsoil, and 30% compost. If the soil contains less than 10% clay, it can be used in place of imported topsoil in the mix. Existing soil can also be improved by adding amendments such as lime, gypsum, and certain nutrients. Alternatively, the rain garden can incorporate an underdrain.

FIGURE 8.26 Aphrodite plantain lily, a Hosta family plant, suitable for plant hardiness zone 7 (New Jersey included).

Rain gardens can be almost any size and shape. A typical residential rain garden ranges in size from 10 to 40 m^2 (110–440 ft^2). Rain gardens smaller than 10 m^2 (110 ft^2) cannot have desirable plant variety, and gardens larger than 40 m^2 are not easy to maintain. As a rule of thumb, a rain garden should be twice as long (perpendicular to the slope) as it is wide. The depth of rain gardens is determined based on the ground slope. For a flat garden, the depth can vary from 7.5 to 15 cm (3–6 in.) for slopes less than 4% to 20 cm (8 in.) for slopes 8% to 12%.

The size of a rain garden is a function of its depth and its distance from downspouts. Because of infiltration losses along the flow path, the farther the rain garden is away from a downspout, the smaller it becomes. For soils of moderate permeability, the author suggests using an area factor ranging from 0.25 for a rain garden 3 m (10 ft) from a downspout to 0.05 for a rain garden more than 10 m (30 ft) away from it; the area factor is the ratio of rain garden area to that of the roof or pavement area being treated.

Rain gardens cost between \$6 and \$30 per square foot (\$65–325/m^2). The latter figure is associated with cases where there is a need for other control structures, such as storm drains, underdrains, and curbing. In the absence of any drainage structure, the cost of plants forms the majority of the overall cost of a rain garden. More detailed information on rain gardens can be found in Rain Garden Manual 2025, prepared by the Native Plant Society of New Jersey and Rutgers Water Resources Program and an EPA February 3, 2025 publication.

Plant Schedule

Key	Quantity	Botanical Name	Common Name	Type
AC	3	*Viburnum trilobum*	American cranberry bush	Shrub
AV	6	*Andropogon virginicus*	Busy broom sedge	Grass
PS	9	*Phlox stolonifera*	Creeping phlox	Herb
RH	9	*Rudbeckia Hirta*	Black-eyed Susan	Herb
RP	3	*Rosa palustris*	Swamp rose	Shrub
TP	5	*Thelypteris palustris*	March fern	Fern
WV	10	*Verbena hastata*	Blue vervain	Herb
VV	8	*Veronicastrum virginicum*	Culver's root	Herb

Example 8.6

A 15 m^2/150 ft^2 rain garden receives runoff from a 100 m^2/1000 ft^2 roof area and 15 m^2/150 ft^2 of driveway. The rain garden consists of a 15 cm/6 in. stone base, 20 cm/8 in. soil mix and is 15 cm/6 in. deep. Calculate the rainfall depth that the rain garden can retain.

Solution

Calculations are performed separately in SI and CU. A 40% void in stone and 15% effective porosity (porosity less moisture content for soil mix) is assumed.

- Storage volume in stone and soil mix:

$$V_s = 15\times\left(\frac{15}{100}\right)\times 0.4 = 0.9 m^3; 150\times\left(\frac{6}{12}\right)\times 0.4 = 30\ ft^3$$

$$V_{sm} = 15\times\left(\frac{20}{100}\right)\times 0.15 = 0.45\ m^3; 150\times\left(\frac{8}{12}\right)\times 0.15 = 15\ ft^3$$

- Dry storage above bed:

$$V_d = 15\times(15/100) = 2.25 m^3; 150 x\left(\frac{6}{12}\right) = 75 ft^3$$

$$Total\ storage = 0.9 + 0.45 + 2.25 = 3.6\ m^3; 30 + 15 + 75 = 120\ ft^3$$

$$Retainage\ depth = \frac{V}{\left(A_{roof} + A_{gar}\right)}$$

$$Rainfall\ depth = 3.6/(100+15) = 0.031\ m = 31\ mm;$$

(120/1250) × 12 =1.25 in.

8.9 COST-EFFECTIVENESS OF BMPs

To arrive at a meaningful comparison of various BMPs, their costs should be expressed per unit volume of runoff being retained. While the unit volume is commonly taken as gallons, the costs are expressed in cubic meters and cubic feet herein. The average cost of the stored water is \$175/m^3 (\$5/ft^3) for infiltration basins. Bioretention basins' cost ranges from \$250 to \$525/m^3 (\$7-\$15/ft^3). Cost of retainage storage for Rain gardens ranges from \$270 to \$600/m^3 (\$8-\$17/ft^3), depending on their depth and whether they need to be equipped with an overflow and/or discharge system. The cost of stormwater wetlands (discussed in Chapter 7) ranges from \$75 to \$210/m^3 (\$2–6/ft^3) depending on their size; the larger the size is, the smaller the unit cost.

Lawn depressions cost less than \$10/m^2 (\$0.90/ft^2) of their surface area. This reflects an average unit cost of \$65/m^3 (\$1.80/ft^3) of stored runoff for a 15 cm (6 in.) depression. The cost of blue roofs is in the range of \$1000-\$2000/m^3 (\$30-\$60/ft^3) of the stored water. The costliest system is a green roof, which costs anywhere from \$125 to \$350/m^2 (\$12-\$30/ft^2) of roof area in the United States. In terms of stored water, this reflects a cost of \$5000 to over \$10,000/m^3 (\$140-\$275/ft^3). In fact, the latter estimates are less than some reported costs. For example, Grey et al. (2013) note the cost of green roofs at \$10 to \$325/ft^2 and \$160 to \$522 per gallon of water managed. The latter figures represent \$42,000 to nearly \$138,000 per cubic meter of stored water for green roofs.

The cost of porous pavements depends on the thickness of the stone reservoir; the thicker the stone base is, the smaller the unit cost of stored water. Pavers cost in the range of \$110–\$160/m^2

(\$10–15/ft^2) and provide up to 6–8 cm (2.4–3.2 in.) of storage volume. As such, the cost of storing runoff is estimated at \$1750/m^3 (\$50/ft^3) on average. Both the construction cost and the retention cost of porous asphalt and pervious concrete are lower than pavers. The average construction costs of porous asphalt and pervious concretes are \$100/m^2 (\$9/ft^2) and \$150/m^2 (\$14/ft^2), respectively. These permeable pavements, with 75 cm (30 in.) of stone reservoir, can hold up to 25 cm (10 in.) of water per unit area. Therefore, the cost of storing water is approximately \$400/m^3 (\$11/ft^3) for porous asphalt and \$600/m^3 (\$17/ft^3) for pervious concrete.

8.10 BIORETENTION CELLS

Bioretention cells are compact basins for application at curbside or parking areas. These cells typically consist of a soil medium placed in a bottomless box and planted with a tree. A cell may have a grate or curb opening to capture runoff. Some bioretention cells, such as Urban Green and Ultra-Urban and were discussed in Chapter 6.

Upon acquiring Filterra from AmeriCast, Contech stopped making Urban Green and now makes a biofilter called Filterra. Filterra comes in Filterra Tree Pod, which is somewhat similar to Ultra-Urban except that it receives runoff from curb openings. They also produce Filterra® Boxless™, which, as the name implies, has no box. In this type of filter, soil media are placed in an excavated area and are planted with trees. This type of filter is particularly suited for parking areas where runoff from the parking may be filtered through a short gravel berm before entering Filterra. This results in considerable cost savings and, for larger applications, may cost about the same as or less than bioretention basins. Table 8.6 presents the pollutant removal efficiency of Filterra Boxless, and Table 8.7 lists the available Filterra Tree Pod dimensions for contributory drainage areas of 1/6 to 1.21 acres (675 to 4900 m^2).

Fabco Industries (Hauppauge, NY, USA) recently introduced a bio-infiltration cell called FocalPoint that unlike UrbanGreen and Filterra Tree Pod has no soil medium. Instead, it includes four elements that filter stormwater. The top layer is made of clean, shredded hardwood mulch; the second, thicker layer is a performance biofiltration medium; and the third layer is clean stone on an open mesh separation fabric. These are placed on a hollow modular underdrain storage. Runoff entering the unit is filtered through the layered elements, which remove pollutants, and clean water enters the storage module. Fabco claims that the system removes up to 73% of phosphorus, 43% of total nitrogen, 85% of total suspended solids, 90% of oil and grease, and 87% of bacteria.

TABLE 8.6
Filterra Boxless Expected Pollutant Removal[a]

TSS	85%
Phosphorus	60%–70%
Nitrogen	43%
Total copper	>58%
Dissolved copper	46%
Total zinc	>66%
Dissolved zinc	58%
Oil and grease	>93%

Note: Information on the pollutant removal efficiency of the filter soil/plant media is based on third-party lab and field studies.

TABLE 8.7
Filterra Quick Sizing for Type I and Type II Street Trees

Available Filterra Street/Shade Tree Box Sizes (ft)	Maximum Contributing Drainage Area (acres) C = 0.85
4 × 6 or 6 × 4	0.16
4 × 8 or 8 × 4	0.23
Standard 6 × 6	0.26
8 × 6 or 6 × 8	0.36
4 × 12 or 12 × 4	036
10 × 6 or 6 × 10	0.46
12 × 6 or 6 × 12	0.56
13 × 7 or 7 × 13	0.71
Available Filterra Street/Shade Tree Box Sizes (ft)	**Maximum Contributing Drainage Area (acres) C = 0.50**
4 × 6 or 6 × 4	0.28
4 × 8 or 8 × 4	0.39
Standard 6 × 6	0.45
8 × 6 or 6 × 8	0.61
4 × 12 or 12 × 4	0,61
10 × 6 or 6 × 10	0.78
12 × 6 or 6 × 12	0.95
13 × 7 or 7 × 13	1.21

Note: Typical street tree standards recommend a 1.5 in. to 2.5 in. caliper. To accommodate these size requirements, Filterra appropriately sized each unit at a 5 ft, 2 in. depth (INV to TC) for type I and type II trees. (Three inches or greater caliper trees will require a 6 ft, 2 in. depth unit.) A standard schedule 40 pipe coupling is cast into the wall for easy connection to the discharge drain.

Source: Filterra Stormwater Bioretention Street/Shade Tree System.

8.11 FILTER STRIPS

8.11.1 Application and Treatment Rates

A filter strip is a vegetated surface that is designed to treat sheet flow from adjacent pavements. Filter strips function by slowing runoff velocities and removing sediment and other pollutants via filtration, absorption, and uptake by plants and infiltration. Although filter strips also considerably reduce runoff volumes, this effect is commonly ignored in practice. To be effective and stable, filter strips should receive runoff as sheet flow and have a small or moderate slope. NJDEP specifies the maximum slope of a filter strip not to exceed 8% and the length of a strip no less than 25 ft (8 m) and no more than 100 ft (30 m).

The treatment rate of a filter strip depends on its length, slope, soil cover, and hydrologic soil group. The maximum accepted treatment rate of filter strip varies from 60% to 80%. To receive the credit for these TSS removal rates, the filter strips must be designed to treat the water quality design storm, which, as has been indicated throughout this book, is 1.25 inches (32 mm) in 2 hours in New Jersey, 1.0 in. (25 mm) in New York, and 0.9–1.0 in. (23–25 mm) in Maryland. Table 8.8 shows the NJDEP maximum allowable TSS removal rate for different categories of vegetation. Figure 8.27 shows the TSS removal rate of filtration for various types of soil cover in HSG B.

TABLE 8.8
Treatment Rates of Filter Strips

Type of Vegetation	TSS Removal Rate
Turf Grass	60%
Meadow Cover	70%
Planted woods	70%
Existing Forest Areas	80%

Source: NJDEP (May 2024), Filter Strips March 2021.

Chart C. Vegetative Filter Strip Length:
Contributory Drainage Area – Hydrologic Soil Group B & Soil Type = Loam, Silt Loam

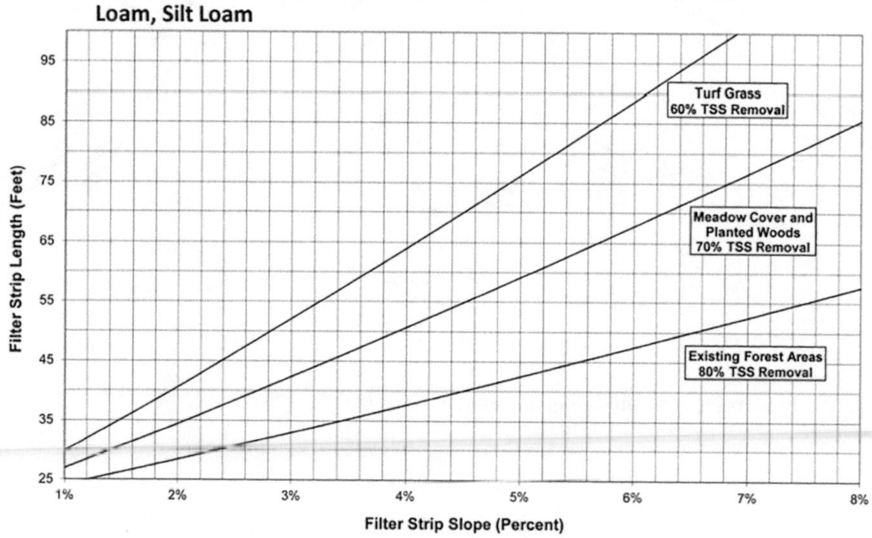

FIGURE 8.27 TSS removal for vegetated filter strips drainage area soil: loam, silt loam HSG B. ***Source***: NJDEP, Chapter 9, May 2024.

The NRCS of Illinois (Code 853, 1999) calls for 80% TSS removal for a filter strip length (along the flow path) of at least one-half of that of the paved area to be treated. This is significantly larger than the TSS removals adopted in New Jersey. Filter strips also remove other pollutants such as nitrate, total phosphorus, and heavy metals. However, removal of these other pollutants is not regulated by many jurisdictional agencies as yet.

Filter strips can be natural or manmade, covered with a variety of plants and trees. Some typical locations for filter strip applications are adjacent to roadways, parking lots, and other impervious surfaces. However, they are not feasible for steep slopes (>10%). Filter strips, similar to roof runoff disconnection, reduce flow; they also provide groundwater recharge. However, as was indicated, the NJDEP stormwater regulations totally disregard these beneficial effects of filter strips. Case Study 7.4 included the application of a filter strip for addressing water quality requirements of a New Jersey Garden State Parkway project. The following case study discusses an application of filter strips for a municipal parking lot expansion.

CASE STUDY 8.1

This study represents the combined use of a detention basin and a vegetative filter strip to treat the runoff from a parking lot of the municipal building in the Borough of Franklin Lakes, New Jersey. Figure 8.28 depicts the existing parking lot layout and drainage pipes that traverse this lot. Figure 8.29 shows the layout of the parking lot expansion. Also shown on the latter plan is a vegetative filter strip and a shallow detention basin that were provided to address the water quality and peak flow reduction requirements for the project.

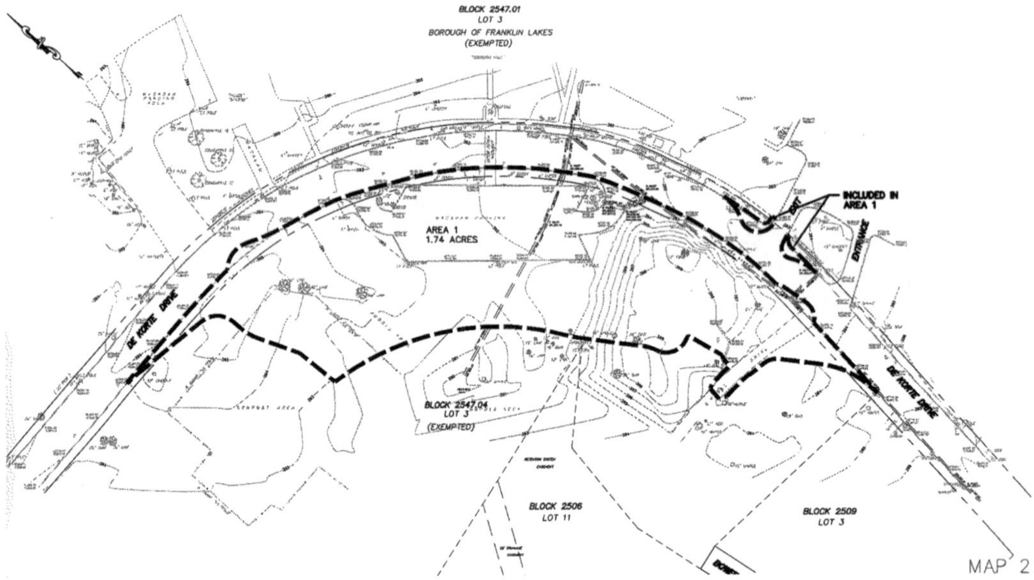

FIGURE 8.28 Predevelopment site layout and area of disturbance.

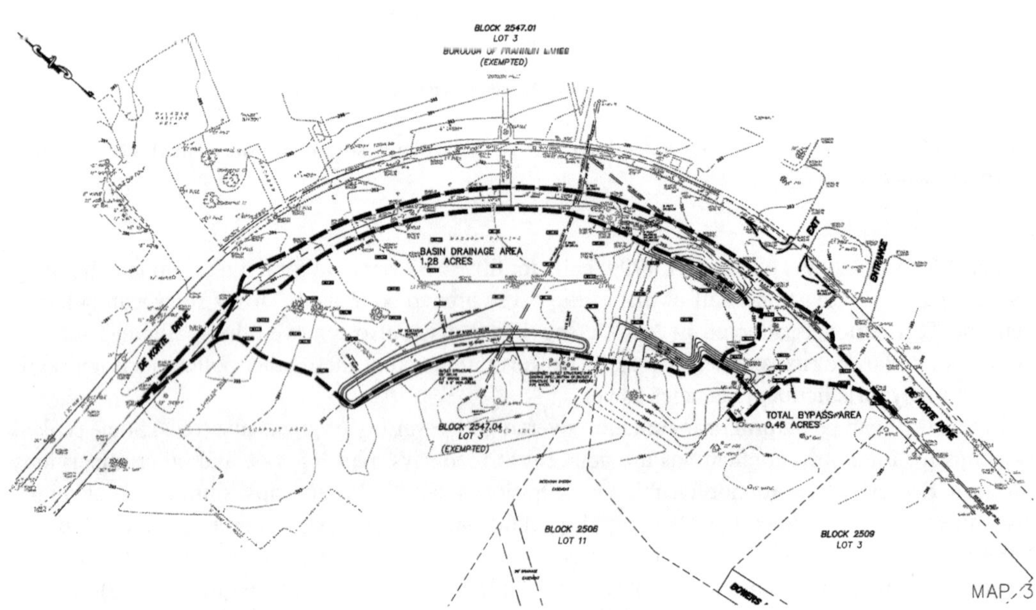

FIGURE 8.29 Post-development site layout and drainage subareas.

Detention Basin Design

Delineated in Figure 8.29 are the limits of the drainage area tributary to the detention basin and the area where the runoff bypasses the basin. To allow for a meaningful comparison between the pre- and post-development peak rates of runoff, the same overall drainage area as that of post-development conditions is drawn on the predevelopment plan.

Runoff calculations for the project site are based on the rational method using a runoff coefficient of 0.95 for pavements and 0.35 for lawns. Tables 8.9 and 8.10 summarize pre- and post-development runoff coefficients and peak runoff rates for the subareas shown in Figures 8.28 and 8.29, respectively. Table 8.10 also lists the maximum allowable discharges from the site for 2-, 10-, and 100-yr storms, all calculated based on a 10-min storm duration.

In designing a detention basin–outlet system, the critical storm duration is first determined. Table 8.11 presents calculations for the critical storm duration and the required detention storage estimates for the above indicated storms. The table shows that a detention basin with minimum storage volume of 2550 ft^3 is required and that critical storm durations are 30 minutes for the 2-year storm and 15 minutes for the 10- and 100-year storm events. Considering that the preceding calculations are based on a linear outflow hydrograph assumption and accounting for freeboard above the maximum water surface elevation, an aboveground detention basin is designed with nearly double the capacity estimated in the preliminary calculations.

TABLE 8.9
Runoff Coefficient Calculations

a. Predeveloped Conditions

	Area (ft^2)	C	$A \times C$
Gravel	1565	0.65	1017
Paved	16,630	0.95	15,799
Grass	57,575	0.35	20,151
Total	75,770		36,967

Area 1: area of disturbance = 1.74 acres (75,770 ft^2)
$C_W = (\Sigma A \times C)/\text{Area} = 0.49$

b. Post- development Conditions

	Paved $C = 0.95$	Grass $C = 0.35$	Total Area (ft^2)	Total ($A \times C$)	C_w
Basin	30,575	24,975	55,550	1.28	0.68
Bypass	5820	14,400	20,220	0.46	0.52

TABLE 8.10
Calculated Peak Rates of Runoff Pre- and Post-development Conditions

Storm Freq. (years)	I (in./h)	Q_E[a]	Q_{DET}[b]	Q_{BYPASS}[c]	Q_A[d]
2	4.2	3.58	3.66	1.00	1.79
10	5.8	4.95	5.05	1.39	3.71
100	8.0	6.82	6.96	1.91	5.46

[a] $Q_E = 1.74 \times 0.49 \times I$.
[b] $Q_{DET} = 1.28 \times 0.68 \times I = 0.87 \times I$.
[c] $Q_{BYPASS} = 0.46 \times 0.52 \times I = 0.24 \times I$.
[d] $Q_A = Q_E \times \%$ reduction = allowable discharge from site.

TABLE 8.11
Required Detention Storage Calculations

Storm Duration (min)	I (in./h)	Q_{DET} (cfs)	Q_{BYPASS} (cfs)	Q_o (cfs)	V (ft³)
2-Year Storm					
10	4.20	3.66	1.01	0.78	1724
15	3.60	3.13	0.86	0.93	2126
20	3.00	2.61	0.72	1.07	2170
30	2.40	2.09	0.58	1.21	2303
45	1.80	1.57	0.43	1.36	1989
60	1.45	1.26	0.35	1.44	n/a[a]
10-Year Storm					
10	5.80	5.05	1.39	2.32	1638
15	4.90	4.26	1.18	2.53	1938
20	4.00	3.48	0.96	2.75	1703
30	3.30	2.87	0.79	2.92	n/a
100-Year Storm					
10	8.00	6.96	1.92	3.54	2054
15	6.90	6.01	1.66	3.80	2552
20	5.80	5.05	1.39	4.07	2397
30	4.70	4.09	1.13	4.33	n/a

[a] Q_o>/Q_{DET}.
$T_c = 10$ min
$V = [T_D \times 60s \times Q_{DET}] - [0.5 \times T_b \times 60s \times Q_o]$
T_b (base time) $= T_D + T_c$
Q_o = max. allowable outflow from basin, Q_A -$Q_{BYPASS}Q_{BYPASS} = 0.24 \times I$ (see Table 8.10)
$Q_{DET} = 0.87 \times I$ (see Table 8.10)
Q_A = Allowable from site (see Table 8.10)

To prolong the retention time for the water quality storm, the outlet structure is equipped with a 2.5 in. low-flow orifice, which is the smallest permissible orifice. Also, an 18 in. wide by 6 in. high opening is included to release storms of 2- through 100-year frequency. Figure 8.30 depicts the proposed outlet structure, and Table 8.12 summarizes outputs for storage and discharge relation with stage. Routing computations are prepared for 2-, 10-, and 100-yr storms based on the previously indicated critical storm durations and is summarized in Table 8.13.

Composite peak runoff rates for the post-development conditions are calculated by adding the hydrographs of direct runoff and the routed flow from the detention basin. Table 8.14 lists composite peak runoff rates for the post-development conditions and presents a comparison of these discharges with the predevelopment peak runoff rates from the project site, the area of disturbance. Since the detention basin is oversized, it creates greater reductions in the peak rates of runoff than the regulations require.

Water Quality Calculations

A 30 ft (9.15 m) wide vegetated filter strip varying in slope from 2 to 2.5% was installed to supplement the extended detention basin to address the water quality requirements for the parking lot expansion. The strip was constructed alongside the parking area and was planted with native vegetation and small bushes. A review of the Bergen County soil survey maps shows that Dunellen soil, which is classified as HSG A, is present at the site. For an average slope of 2.25%, Figure 8.31 indicates a minimum 70% TSS removal for this filter strip.

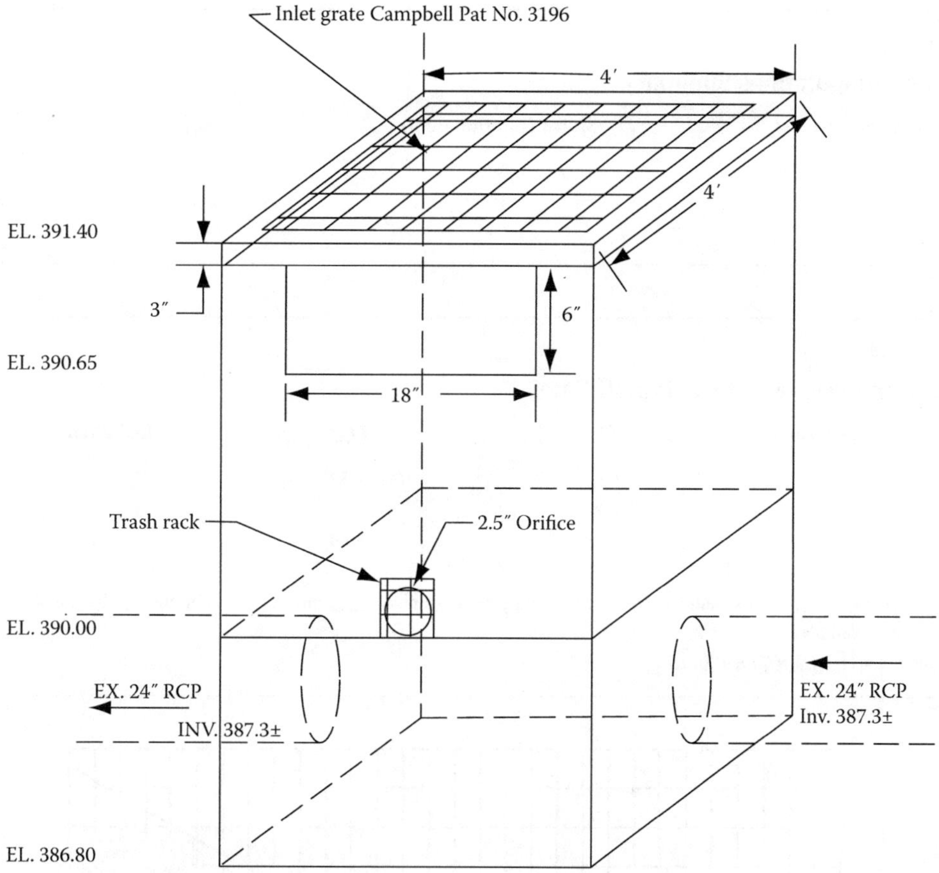

Notes:
1. Detail provided for dimensioning only. Reinforcement and structural detailing to be completed by structure manufacturer and will be responsibility of the contractor. Detail shop drawings signed and sealed by a licensed professional engineer to be submitted for approval prior to construction.
2. Outlet structure to be constructed over existing 24″ RCP to be utilized as an outlet pipe from the basin.

FIGURE 8.30 Pond outlet structure detail (N.T.S.).

TABLE 8.12
Elevation–Storage–Discharge Relation

Elevation (ft)	Storage (cf)	Discharge (cfs)
390.00	0	0
390.65	1795	0.12
391.00	2995	1.09
391.15	3580	1.76
391.40	4630	2.74
391.50	5050	5.78

TABLE 8.13
Routing Computations Summary

Storm Frequency (years)	Storm Duration (min)	Q_{DET}	Q_{OUT}
2	30	2.09	0.99
10	15	4.26	1.18
100	15	6.01	2.23

TABLE 8.14
Existing and Proposed Peak Runoff Rates

Storm Frequency (years)	Q_E	Q_{PR}[a]	Reduction %[b]
2	3.58	1.57	56
10	4.95	2.36	52
100	6.82	3.89	43

[a] Q_{PR} = Post-development discharge, calculated by adding hydrographs of Q_{out} and Q_{BYPASS}. (For brevity, these hydrographs are not included here).

[b] Reduction = $(1 - Q_{PR}/Q_E) \times 100$.

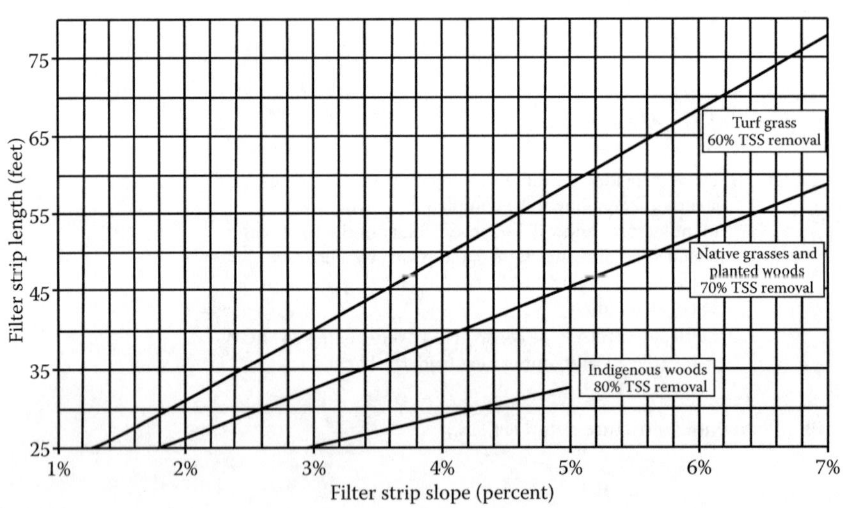

FIGURE 8.31 TSS removal of vegetated filter strip, Sandy Soil, HSG A (Source: NJDEP, 2014, 2023).

After passing through the filter strip, runoff from the parking area enters the detention basin, which according to routing computations in the previous case study provides 12 hours of retention time. For this retention time, Figure 8.32 indicates 40% TSS removal. The composite TSS removal of the filter strip plus detention basin is calculated as follows:

$$TSS\ removal = 70 + 40 - \left(70 \times 40 / 100\right) = 82\%$$

Therefore, the system removes more TSS than required. Since this site is situated in the PA1 (metropolitan planning area) and was previously disturbed, it is exempt from groundwater recharge.

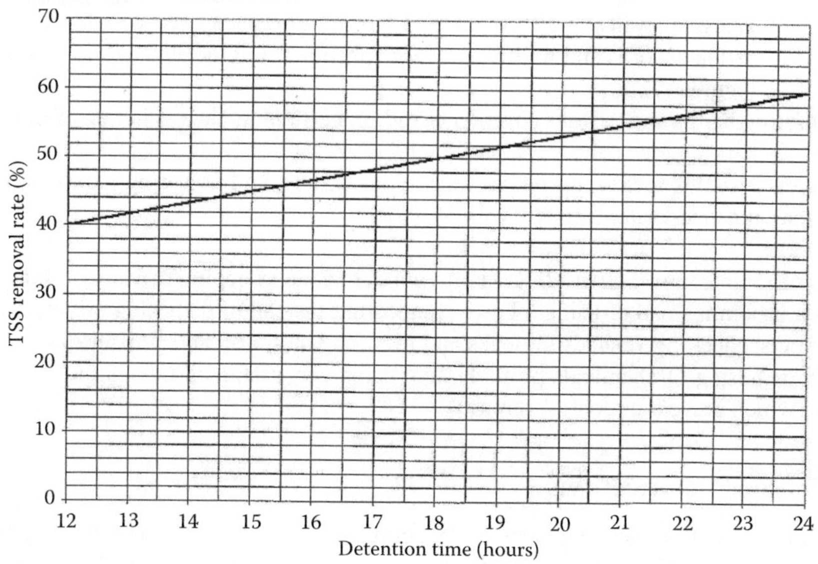

FIGURE 8.32 TSS removal rate of detention basin (40% for 12 hour of detention).

Source: NJDEP.

8.12 OTHER NONSTRUCTURAL MEASURES

As indicated previously, source reduction is more effective and a far better solution than structural BMPs for controlling both quantity and quality of runoff. A number of measures are available for source reduction, some of which are easy to implement. Measures can be grouped into common and specific, as follows.

8.12.1 Common (General) Measures

Common measures, which can be applied to any type of development, include:

a. Separating roof drains and downspouts from driveways and drainage systems, instead directing them to lawn/landscape areas and rain gardens (see Figure 8. 8).
b. Pitching driveways slightly toward lawn and vegetated areas, where practicable.
c. Minimizing land disturbance and grading.
d. Collecting roof runoff in rain barrels/tanks and cisterns for reuse (see Chapter 10), or retaining roof runoff in dry wells, infiltration chambers or the like.
e. Creating lawn/landscape depressions and rain gardens to retain the runoff from roof/paved areas and infiltrating it over time.
f. Constructing a raised planter or planted bed at the foot of a steep slope to retain/reduce runoff flowing onto downhill areas.
g. Appling the lawn chemicals no more than the minimum needed rate.
h. Using water-based, rather than oil-based, paints and household cleaners.
i. Conserving outdoor water use. Overwatering does not make a healthier lawn; It is only a waste.
j. Minimizing the width of access roads, cul-de-sacs, and driveways. Narrowing the roadway width from 22 to 18 ft reduces the peak rate and volume of road runoff by over 20%.

k. Installing roadside swales and/or vegetation buffers to reduce overflows from inlets and pipes; vegetation absorbs and infiltrates a large portion if not the entire runoff from paved roads (see Figure 8.9).
l. Substituting hardscape (porous pavements and, in particular, pavers) in lieu of impervious covers for parking lots, driveways, patios, and sidewalks.
m. Covering pervious areas with native vegetation, shrubs, and flowering plants rather than high-maintenance, water-thirsty lawns.

In terms of maintenance, other effective measures include sweeping driveways, sidewalks, and patios with a broom instead of using a hose; composting instead of disposing of yard waste; picking up after pets; checking vehicles for leaks and recycling motor oil and antifreeze when changing these fluids; disposing of chemicals properly; and using car wash facilities that recycle water, the latter two of which prevent or control runoff.

8.12.2 Clustered Developments

Clustered development is a specific measure for source reduction aimed at maximizing the conservation of trees and natural vegetation and minimizing impervious cover in residential subdivisions. The objectives of clustered developments are:

a. To maximize preservation of open space for recreational and scenic purposes
b. To provide a development pattern harmonious with natural land features
c. To offer a variety of choices and more economical housing
d. To use land efficiently with smaller networks of paved streets and utilities

The common green spaces in clustered developments are maintained by property management associations, although before the association is formed, the developer is responsible for all maintenance. Common green spaces on development tracts include stream buffers, significant stands of trees, individual trees of significance, steep slopes, rock outcrops, open fields and meadows, any historic and archeological sites, and wetlands and their buffers. Submissions for a preliminary cluster residential development review by a municipality (as with any type of development project) generally include:

- Topographic maps
- Wetlands plan
- Preliminary plot plan showing the maximum number of lots allowed under regulating subdivision ordinance
- Preliminary site plan showing the subdivision layout
- Calculations associated with the net residential acreage and density
- A written narrative explaining adherence of the proposed cluster subdivision with the ordinance

8.13 MINIMAL IMPACT DEVELOPMENTS

It is a common misperception that single-family homes with large lots have less adverse impacts on stormwater runoff than multifamily residential developments such as townhouses and condominiums. Basing the environmental impact on the amount of pavement and disturbance on a per parcel of land or a development site is misleading.

It should be realized that people need housing and therefore the impact criteria should be based not on how much impervious area is placed on a parcel of land but rather on how much pavement is created to provide housing for one person (or one family). On this basis, as indicated in Chapter 1, large single-family homes actually disturb much more land, create more stormwater runoff, and have far more adverse environmental impacts than condensed residential developments per person.

Table 8.15 presents typical amounts of pavements for single-family homes on lots ranging from 675 m² (1/6 acre) to 8100 m² (2 acres) in the United States. This table indicates that larger homes, while having a smaller percentage of imperviousness, create far more impervious coverage than smaller homes per capita. The access roads and streets are excluded from the listed impervious covers in this table.

Table 8.16 shows impervious coverage for typical single-family homes, accounting for paved streets along each lot. This table also lists the per capita impervious coverage for lot sizes ranging from 1/6 of an acre to two acres. Also listed in the latter table are the calculated annual volume of runoff from impervious surfaces in each lot. The calculations are based on rainfall of 760 mm (30 in.), the average annual precipitation in the United States and some European countries as well.

The impact of multifamily residential developments will be nil if the project is a brownfield development, that is, developed on an abandoned industrial or commercial site. The following residential project exemplifies this case.

This author was in charge of stormwater management design and permit acquisition for this project, a residential development project called Cambridge Crossings in Clifton, New Jersey. The project included the construction of 47 three-story multifamily buildings, two clubhouses, swimming pools, and tennis courts on a 42.5 acre (17.2 ha) parcel of land. The site had been formerly occupied by American Cyanamid, a chemical industry, and had been left abandoned for some time.

The development reduced the impervious surfaces, inclusive of roads, buildings, driveways, and parking lots, from 20.3 acres (8.22 ha) to 15.6 acres (6.32 ha). The development created 640 residential units: 210 townhouses, 160 flats, and 270 condominiums. All units had two bedrooms and were occupied by two or three people except for the condominiums, which were age-restricted and

TABLE 8.15
Per Capita Lot Impervious Coverage: Single-Family Homes[a]

Lot Size, m² (ft²)	Roof, m² (ft²)	Driveway, m² (ft²)	Patio/Walks, m² (ft²)	Total Imp., m² (ft²)	% Cover
675 (7260)	130 (1400)	74 (800)	84 (900)	288 (3100)	42.7 %
1010 (10,870)	149 (1600)	102 (1100)	111 (1200)	362 (3900)	35.9 %
2025 (21,800)	214 (2300)	195 (2100)	204 (2200)	613 (6600)	30.3%
4050 (43,560)	279 (3000)	372 (4000)	372 (4000)	1022 (11,000)	25.2%
8100 (87,120)	333 (3800)	669 (7200)	465 (5000)	1486 (16,000)	18.4%

TABLE 8.16
Overall Per Capita Impervious Coverage and Annual Runoff: Single-Family Homes

Lot Size, m² (ft²)	Total Imp. m² (ft²)	% Cover[a]	Per Capita[b], m² (ft²)	Per Capita Annual Runoff, m³ (10³ cf)
675 (7260)	446 (4800)	66%	111 (1200)	75.9 (2.70)
1010 (10,870)	595 (6400)	59%	149 (1600)	101.9 (3.60)
2025 (21,800)	929 (10,000)	46%	232 (2500)	158.7 (5.63)
4050 (43,600)	1347 (14,500)	33%	269 (2900)	184.0 (6.53)
8100 (87,200)	1951 (21,000)	24%	390 (4200)	266.8 (9.45)

[a] Based on family of four for smaller than 4050 m² (1 acre) and family of five for 4050 m² and 8100 m² (1 and 2 acre) lots.
[b] Based on 760 mm (30 in.) annual rainfall and 0.90 runoff/rainfall ratio from pavements.

had two residents. As such, the development potentially provided housing for 1465 persons. Since two streams were traversing through the site, 14.1 acres (5.7 ha) of land remained intact to serve as a stream buffer and conservation area.

Of the remaining 28.4 acres (11.5 ha), 12.8 acres (5.2 ha) comprised green space, including landscape islands and vegetated detention basins. Thus, the overall amount of land use and impervious coverage on a per capita basis (assuming that the land was undisturbed prior to the development) were approximately 844 square feet and 464 square feet (78 and 43 m^2), respectively. Table 8.17 provides a summary of pre- and post-development pavement areas and the per capita impervious coverage. A sample photo of townhomes in this residential development is in Figure 8.33.

The amount of impervious coverage per capita for midrise or high-rise apartment buildings would be even smaller than that the above-described development. Table 8.18 describes a six-story apartment building in Laval, Quebec, Canada. Figure 8.34 depicts the apartment buildings (two in all) that were under construction in July 2012. As Table 8.18 indicates, the per capita impervious coverage for this midrise apartment building is estimated at 16.3 m^2 (175 ft^2). For high-rise apartment buildings, the per capita impervious coverage drops to under 100 ft^2 (9.3 m^2) or even less.

In comparison, a typical midsized single-family home in a one-half acre (2025 m^2) suburban subdivision, as shown in Table 8.16, creates over 10,000 square feet (929 m^2) of impervious surface, accounting for streets, sidewalks, driveway, patio, and roof. For a family of four, the suburban

TABLE 8.17
Per Capita Impervious Coverage Multifamily Residential Developments

Example: Cambridge Crossings, Clifton, NJ
Site area = 42.5 acres
Development area = 28.4 acres
Impervious coverage = 15.6 acres[a] (including streets)
No. of units = 640 (one/two bedrooms)
Households (2.2 per unit) = 1465[b]
Per capita impervious area = 464 ft^2

[a] Preexisting impervious coverage = 20.3 acres
[b] 2 per condo; 2.3 persons on average per townhome or flat

FIGURE 8.33 A typical depiction of townhomes in the Cambridge Crossings development.

Source: Photo by the author (2012).

TABLE 8.18
Per Capita Impervious Coverage for a Six-Story Apartment Building

Estimate: 1500 ft^2 apt. (50 ft × 30 ft)
Two household/apt.
Households: 12/1500 ft^2 footprint
Street: 12 × 50 = 600 ft^2
Total imp. area = 2100 ft^2
Per capita impervious area = 2100/12 = 175 ft^2/(16. 3 m^2)

FIGURE 8.34 Two 6-story apartment buildings in Laval, Canada.

Source: Photo by the author (2013).

dwelling creates approximately 2500 square feet (232 m^2) of impervious coverage per capita. Thus, the exemplified residential development and midrise apartment building create less than 19% and 7% impervious coverage per capita of that of the previously indicated 0.5 acres single-family home. Residential towers all around the world disturb even less land and result in even less per capita impervious cover. Thus, city living has a far smaller adverse impact on the environment in general and stormwater runoff in particular than suburbs.

As indicated in a previous chapter, current stormwater management regulations exclude small sites and in particular single-family homes. It is to be noted, however, that single-family homes, which form the majority of residential dwellings in the United States, are the culprit of stormwater runoff and flooding problems. Though each single-family home has a small effect, collectively they create a significant increase in the peak and volume of runoff. To reduce these adverse impacts, the stormwater regulations should therefore cover single family homes. Also as indicated earlier, roof runoff should be directed to landscaped areas or rain gardens or drained into rain tanks/barrels or retention-infiltration systems such as dry wells (seepage pits), chambers in stone trenches and roof downspouts should be terminated at lawn and landscape areas. Figure 8.10 showed a downspout terminating at a front yard landscape area of the author's residence. This landscape area fully retains the roof runoff, resulting in no runoff from the roof and the yard discharging to the municipal storm drain system.

8.14 STORMWATER FEES

There are approximately 3.5 million miles (5.6 million kilometers) of storm sewers and 270 million inlets in the United States including the combined sewer systems in the country's old cities, such as in parts of New York City and the borough of Fort Lee and the city of Hoboken in New Jersey. Because of aging and inadequacy, sewers overflow during heavy storms in combined sewers, and it is estimated that over $100 billion of infrastructure investment is needed in the next 20 years to address stormwater and sewer overflows alone. As was indicated in chapter 5, EPA estimates that more than $630 billion is needed to address CWA of 1972 and replace aging stormwater and wastewater infrastructure, over the next 20 years.

To provide a budget, a number of alternate solutions, such as real estate transfer fees, real estate tax adjustments, sales tax, state revolving funds, voluntary offset programs, and stormwater utility fees, have been sought. Among these, the stormwater utility fee is the most practical. In the 1970s and 1980s, stormwater utility fees were first established in Washington State, and since then, many municipalities have established stormwater fees just like the water, sewer, and energy fees. The use of stormwater utility fees for financing urban stormwater programs is growing.

To arrive at a more informed decision on this matter, the Center for Urban Water Resources at the University of Washington was formed. Communities saw a utility fee as the steadiest and most manageable way to pay for stormwater system maintenance and upgrades, but getting started was a hard battle for municipal officials. Unlike other utilities such as water, electricity, and gas, which are metered and consumers pay for their usage, stormwater fees are not related to any measured quantity. Because of the lack of metering, it was hard for people to pay for something that they could neither see nor measure.

Establishing a rate structure for stormwater utilities was complex. The complexity was partly due to insufficient data on the long-term stormwater management maintenance costs to address the increasingly stringent water quality standards. Thus, for simplicity, many communities charged the stormwater fees based on a flat rate for residential and commercial properties. A stormwater management system designed by the author for a roadway improvement program in a northern New Jersey municipality in Bergen County, New Jersey. constructed in 2010, was the first such system to be maintained by the county.

An April 2009 EPA publication No.833-F-07-012, titled "Funding Stormwater Programs" indicates that over 800 communities and districts across the United States had adopted a stormwater utility by publication. It also exemplifies the utilities in South Burlington, Vermont, and Newton, Massachusetts. Both of these were the first in kind to establish stormwater utility. The latter community established a stormwater drain fee to upgrade infrastructures in nearby streams. properties. The fees were based on a flat rate of $6.25 per quarter for residential properties at the time.

Now there are over 2000 stormwater utilities in the U.S., and the fee rates have risen since their inception. In Newton, Massachusetts, for example, a residential home owner pays a flat rate fee of $140 per year; all other property owners pay based on the amount of impervious area on their property. The stormwater fees are included in the quarterly water–sewer bill. In many US communities, the stormwater fees are still based on a flat rate for all single-family homes regardless of the size of the property and the improvements thereon.

To account for the differences in the amount of runoff discharged from properties, some municipalities have established other rate structures for stormwater fees. One such structure is an equivalent residential unit (ERU). This rate is based on the amount of impervious surface from which water runs off and enters the stormwater systems. One ERU equals 2139 square feet (199 m^2) of non-porous surfaces. Another rate structure is the equivalent hydraulic area (EHA), which entails billing customers based on the combined impacts of pervious and impervious areas on stormwater runoff; the city of Ithaca, New York, established stormwater fees at $4.60 per ERU in the mid-2000s (Zolezi, 2009). In the second edition of this book, the author wrote that the future will tell that these rates will be inadequate. Now, in the same town, single-family homes that contain less than 0.25 ERU (575 ft^2, 53 m^2) of impervious area are exempt from stormwater fees, but the single, two and three-family residentials are assumed to have 1 ERU and to pay a flat rate of $14.25 per quarter in 2023.

After 15 years of planning, Philadelphia adopted a new approach for assessing nonresidential property owners' stormwater fees on July 1, 2010. This fee structure was a parcel-based stormwater billing system that charged owners based on the property's gross area and impervious surface coverage. The Philadelphia Water Department acknowledged that a change from a meter-based rate would be a challenging concept to many customers. However, the change was intended to make the cost of service more equitable to customers (Cunningham, 2011). This rate structure is still followed in that city.

The ERU and EHA, though seemingly equitable, are not fair. Two nearly identical properties in a given municipality may generate different amounts of runoff, and not all pervious or impervious surfaces have the same impact on stormwater runoff. The impact is affected by the type and growth of vegetation, roughness of impervious surfaces, land slope, and the implementation of stormwater provisions on a property.

To account for this difference, credit may be given to onsite measures that reduce the impact on the downstream drainage system. One such measure should be swimming pools, which act as small retention basins. However, many municipalities include swimming pools as impervious coverage in their municipal ordinances, which is not fair. Swimming pools, although do not provide any percolation, act as a retention basin. On average, a swimming pool can retain 6 inches (15 cm) of rain before spilling. As such, they not only reduce runoff volume far more than lawns but also create zero runoff during sustained rainfalls.

In addition to the quantitative aspects of stormwater runoff, the impact on water quality is to be considered in the fee calculations and the appropriation of any credit. However, the latter impact is difficult to assess as it involves evaluating the sources of pollutants of concern and the costs of their treatment. Apart from all these difficulties, this impact does not relate to the costs of maintenance of public drainage and stormwater management facilities.

Stormwater fees vary from municipality to municipality and state to state. For instance, the fees range from \$5 to \$11 per month in various towns in Pennsylvania. And in Hampton, Virginia, the stormwater fee was established at \$141.96 per year, effective July 1, 2023, and is collected along with real estate property tax. Similarly, many municipalities bill single-family residential parcels at a flat rate regardless of impervious coverage, but some base the fee on the amount of impervious area. Los Angeles County for instance set a rate of 2.5 cents per square feet of impervious surfaces in July 2019.

The above-cited 2009 EPA publication titled "Funding Stormwater Programs" also discusses stormwater funding alternatives, including a flat rate ERU, also known as equivalent service unit (ESU) and equivalent hydraulic area (EHA), and advantages and drawbacks of each one. It also presents the implementation of stormwater utility fees in EPA Region III: Pennsylvania, Maryland, Delaware, the District of Columbia, Virginia, and West Virginia (https://www.3epa.gov/npdes/pubs/region3_factsheet_funding.pdf).

An October 11, 2023, EPA pamphlet titled "Paying for Stormwater: The Benefits of a Utility" includes talks by three stormwater professionals on the subject. This pamphlet, last updated February 17, 2025, exemplifies the benefits of stormwater utility in the town of Wilmington, Vermont, which by establishing a stormwater utility addressed its stormwater issues by equitably distributing the cost of compliance with regulations and watershed protection.

PROBLEMS

8.1 A roof leader spreads runoff from a 50 m^2 area of a roof through a splash guard to a 50 m^2 lawn. For a low to moderate soil permeability of 100 mm/h, calculate runoff from the site during a 50 mm/h rainfall which represents a 10-year, 60-minute storm in New Jersey.

8.2 Rework problem 8.1 for a 500 ft^2 of roof area spreading over a 500 ft^2 lawn. For a soil permeability of 4 in./h, calculate runoff from the site during a 2 in./h rainfall.

8.3 A 100 m^2 extensive green roof includes 10 cm of soil media and a 2.5 cm drainage layer. Calculate the retention storage and peak discharge from this roof for a 25 mm/h storm of 2-hour duration. The effective porosity of soil media and the drainage layer are estimated at 15% and 2%, respectively.

8.4 Solve problem 3 above for a 1000 ft^2 green roof consisting of 5 in. soil medium and 1.5 in the stone base for a 2-hr, storm of 1 in/h intensity.

8.5 Design a seepage pit to fully retain the runoff from the 150 m^2 roof of a residential dwelling for rainfalls up to 50 mm. The pit will be placed in a 50 cm thick stone envelope and a 25 cm stone base.

8.6 A seepage pit is to be used to fully retain the runoff from a 1500 square foot roof of a residential dwelling during the 10-year, 60-minute storm. This storm has a 2 in./h intensity and amounts to 2 in. of rainfall. Calculate the required storage volume of the seepage pit.

8.7 A 6.5 ft O.D., 6 ft I.D., 4 ft high seepage pit, enveloped in a 2 ft thick stone and one (1) ft thick stone base, is proposed for the dwelling of Problem 8.6. Is this pit adequate? Use a void ratio of 40%. for stone

8.8 The construction of the Ambulance Corps in Hawthorne, Passaic County, New Jersey, affects 1.073 acres and increases the impervious coverage from 1400 square feet to 32,080 square feet. A rain basin in the form of a lawn depression (see Figures 8.22 and 8.23) is designed to fully retain the entire runoff from the paved area during the 10-year, 60-minute storm, having 2 in./h intensity. Calculate the required storage volume of the rain basin.

8.9 A 20 m^2 rain garden receives runoff from a 150 m^2 roof area. The rain garden includes a 15 cm deep depression, 30 cm thick topsoil and a 15 cm stone sublayer. Calculate the maximum rainfall depth that the rain garden can retain.

8.10 Solve Problem 8.9 for a 200 ft^2 rain garden receiving runoff from a 1500 ft^2 roof area. The garden consists of 6 in. depression, 1 ft thick topsoil, and 6 in. stone base.

8.11 Design a rain garden to retain the runoff from a 1200 ft^2 residential dwelling roof during a New Jersey water quality storm that is 1.25 in. of rainfall in 2 hours. The rain garden includes a 6 in. depression and 1 ft-thick soil medium over a 6 in. layer of sand. Estimate porosity of the soil and sand at 0.2 and 0.35, respectively.

8.12 Solve Problem 8.11 for a 120 m^2 roof area, 15 cm depression, 30 cm thick soil, 15 cm thick sand, and 30 mm rainfall.

8.13 A 1500 ft^2 roof area drains to a 200 ft^2 rain garden that includes 12 in. thick soil, 6 in. thick stone, and allowable ponding of 3 in. Evaluate if this rain garden satisfies the WQ_v requirements of New York State (refer to Chapter 5). The 90% rainfall event in the area is 1.0 in. State any assumptions made.

8.14 Solve Problem 8.13 for a 150 m^2 roof area and 20 m^2 rain garden. The thicknesses of soil and drainage layers are 30 cm and 15 cm, respectively. The 90% rainfall event is 25 mm, and the allowable ponding is 7.5 cm.

8.15 Size a bioretention basin for a 1300 ft^2 pavement and 2.0 in. of rainfall. The bioretention basin includes 2.5 ft thick soil mix, 10 in. thick sand, 1.0 ft deep depression, and 3 in. thick mulch. Estimate the porosity of soil, sand, and mulch at 15%, 35%, and 40%, respectively. The basin abuts the pavement.

8.16 Solve Problem 8.15 for 120 m^2 pavement and 50 mm rainfall. The bioretention basin is 30 cm deep, and the thickness of the soil mix, sand base, and mulch is 75, 25, and 7.5 cm, respectively.

8.17 Solve Problem 8.15 for the case that the bioretention basin is 10 ft away from the pavement. Assume that 35% of the runoff from the pavement is captured by lawn enroute to the basin.

8.18 Solve Problem 8.16, where the bioretention basin is 3 m away from the pavement, and 35% of pavement runoff is absorbed by the lawn between the bioretention basin and the depression.

8.19 A 25-acre single-family residential development comprises 1/3 acre lots and includes 35% impervious coverage. Calculate:
   a. The impervious coverage per dwelling and per capita, in square feet
   b. The annual runoff volume per dwelling and per capita
   Base your calculations on a household of four persons per dwelling and 30 in. annual precipitation. State any assumptions made.

8.20 A 25-acre multi-residential development includes 600 condominium units and is 50% paved. Calculate:
   a. The impervious coverage per condominium unit and per capita in square feet
   b. The annual runoff volume per capita
   Base your calculations on two persons per condo and 30 in. annual precipitation. State any assumptions you need to make.

8.21 A 10 ha residential development comprises 75 single-family homes and is 30% covered with impervious surfaces. Calculate:
   a. The average impervious coverage per dwelling and per capita in square meters.
   b. The average annual runoff volume per dwelling and per capita
   Base your calculations on a household of four persons per dwelling and 750 mm of annual precipitation. State your assumptions.

8.22 A multi-residential development includes 750 apartments on a 10-ha parcel of land that is 50% covered with pavements and roofs. Calculate:
   a. The average impervious surface per apartment and per capita
   b. The annual runoff volume per capita
   Base your calculations on two people per apartment and 750 mm annual precipitation. State any assumptions made.

REFERENCES

Adams, M.C., May/June 2003, Porous asphalt pavement with recharge beds 20 years and still working, *Stormwater*.

Baker, L.A., November/December, 2007, Stormwater pollution, getting at the source, *Stormwater*, pp. 16–42.

Brown, D., 2007, Porous in Portland, parking lot incorporate porous asphalt for optimal stormwater management, *Stormwater Solutions*, January/February, pp. 16–19.

Buranen, M., October 2008, Chicago's green alleys, *Stormwater*, pp. 50–57.

Cunningham, C., May/June 2011, GIS for green, Philadelphia introduces parcel-based stormwater billing and corresponding online application, *Stormwater Solutions*, pp. 12–13.

DeLaria, M., 2008, What I learned in paver school, the role of permeable paver systems as a stormwater management technique. *Stormwater*, pp. 54–58.

EPA, 1980, *Porous pavement, phase I design and operational criteria*, Report No. EPA-600/2-80-135, NTIS #PB8110-4796.

EPA, March 2012, *Benefits of low impact development (LID): A literature review*, EPA-841-N-12-003A.

EPA, October 2000, *Low Impact Development, a Literature Review*, EPA-841-B-00-005.

EPA, 2023, *Paying for stormwater – The benefits of utility*, last updated February 17, 2025.

EPA, 2024, *Soak up the rain: Permeable pavement*, https://www.epa.gov/soakuptherain/soak-rain-permeable pavement.

EPA, 2025, *Soak up the rain: Rain gardens*, February 3.

EPA, 2025, *National Menu of Best Management Practices (BMPs) for Stormwater*, July 9.

Ferguson, B.K. (2005), *Porous pavements*, 1st. ed. Taylor and Francis, Boca Raton, FL.

Grey, M., Sorem, D., Alexander, C. and Boon, R., March/April 2013, Low-impact development BMP installation, operation and maintenance cost in Orange County, CA, *Stormwater*, pp. 26–35.

Gunderson, J., September 2008, Pervious pavements, *Stormwater*, pp. 62–71.

Jones, M., May 2012, Design review and simulation for permeable pavement and bioretention, *Stormwater*, pp. 54–58.

Maine Stormwater Management Design Manual, 2016, Maine Department of the Environment, 114 pp, https://www.maine.gov > vol 3pdf

Maryland Stormwater Design Manual, May 2009, Vols. 1 and 2. Maryland Department of the Environment.

McLaughlin, J., Stein, J., Mehrotra, S., Leo, W. and Jones, M., September 2012, Stormwater source control in New York City, implementing green infrastructure within the nation's largest urban landscape, *Stormwater*, pp. 40–47.

Native Plant Society of New Jersey, 2010, *Rain garden manual for New Jersey*, Office of Continuing Professional Education, Cook College, 102 Ryders Lane, New Brunswick, NJ.

Native Plant Society of New Jersey, 2025, *Rain garden manual, in cooperation with Rutgers Cooperative Extension Water Resources Program.*

New Jersey Concrete and Aggregate Association, *Pervious concrete*, https://www.njconcrete.com.

New Jersey Stormwater Best Management Practices Manual, first adopted 2004, July 2023, last updated October, 2024.

New York State Stormwater Management Design Manual, January 2015, Stormwater Management Design Manual.

New York State Stormwater Management Design Manual, Draft-May 2022, New York State Department of Conservation.

Pazwash, H., October 2–3, 2012, *Stormwater management techniques for single family homes*, presented at the 8th Annual NJAFM Conference, Somerset, NJ.

Peck, S., 2007, In the green, green-roof infrastructure gains popularity, *Stormwater*.

Peerless Concrete Products Co., 246 Main Street, Butler, NJ 07405, www.peerlessconcrete.com.

Penn State Extension, 2022, *Rain gardens (bioretention cells), a stormwater BMP*, uploaded August 24.

Rogers, W.M. and Faha, M., September 2007, Pulling together the pieces of the stormwater puzzle, porous pavement and other techniques in Portland, *Stormwater*, pp. 92–100.

Roseen, R.M. and Ballestero, T.P., May/June 2008, Porous asphalt pavements for stormwater management in cold climates, *Hot Mix Asphalt Technology*.

Roy, S.P. and Braga, A.M., February 2009, Saving Silver Lake, *ASCE, Civil Engineering, Magazine*, pp. 72–79.

Seth, I., July/August 2011, Runoff modeling for rain gardens, determining volume reduction and pollutant removal rates, *Stormwater*, pp. 36–39.

Traver, R. et al., July/August 2004, Porous concrete, *Stormwater*, pp. 30–45.

Traver, R. et al., July/August 2005, Lessons in porous concrete, an update on the Villanova urban stormwater partnership's demonstration site, *Stormwater*, pp. 130–136.

Turfstone Open Cell Pavers, Belgard, https://www.belgard.com.

US National Arboretum, *USDA plant hardiness zone map*, https://www.usna.usda.gov/Hardzone/ushzmap.html.

Zolezi, C., April 2009, Study examines feasibility of city's stormwater utility user fee, *Urban Water Management*, pp. 17–19.

9 Installation, Inspection, and Maintenance of Stormwater Management Systems

This chapter presents an overview of installation, inspection, and maintenance of stormwater conveyance systems, and structural and non-structural stormwater management facilities. Also presented herein is a brief description of soil erosion and sediment control measures, which are employed to control discharge of stormwater pollutants during construction.

9.1 SOIL EROSION AND SEDIMENT CONTROL MEASURES

During construction, however short, a large quantity of sediment is created due to erosion of loose, unvegetated soil. Figure 9.1 shows a loose, disturbed soil that was highly vulnerable to erosion and water pollution at a construction site. To reduce the soil migration from a site to paved roads and drainage structures offsite by construction vehicles, a stone blanket is installed at the entrance to the site; a plastic silt fence is also installed downhill from the site and hay bales placed around catch basins or geofabric materials installed beneath inlet grates. More recently, filter socks made of leaf compost have been employed in lieu of plastic silt fences, filter pads are used to cover inlet grates. Figure 9.2 shows a commercially available filter mat, for roadway projects known as Inlet Filter pad, manufactured by Blocksom & Co. (Michigan City, IN, USA).

To trap sediment onsite, siltation basins are commonly built within the construction site. Often, the sites of permanent stormwater management ponds/detention basins or infiltration basins are used as temporary siltation basins. However, it is imperative that these basins are cleaned of any silt and that they are fully restored following the completion of site work and prior to the installation of inlet–outlet structures.

On a large construction site of a water supply reservoir in west central New Jersey, natural depressions were utilized as small sedimentation basins pursuant to the author's recommendation in 1986 and served as an effective sediment control measure. On that site, the stored water in depressions was used for sprinkling disturbed areas and for watering temporary vegetation after the silt was allowed to settle between storms.

9.2 TYPES AND INSTALLATION OF DRAINAGE PIPES

Presented in this section are installation of RCPs and HDPE pipes, which are employed more than other makes of pipes for conveyance of stormwater runoff; CMPs are also used nationwide. However, these pipes are less durable and much more vulnerable to deformation, deterioration, and collapse than either RCP or HDPE pipes. The author has seen numerous cases of deformation, damage, and even total collapse of CMP conveyance systems or underground detention basins in the states of New Jersey and New York (particularly in Albany County). Considering longevity, durability, and strength, both RCP and HDPE pipes are far superior to CMPs and are recommended for application in drainage and detention systems

A number of US concrete plants manufacture RCPs including Oldcastle Infrastructure (Atlanta, GA, USA) and AmeriTex Pipe & Products (Seguin, TX, USA). HDPE pipes are made by several companies including Advanced Drainage Systems (ADS; Hilliard, OH, USA). Contech manufactures both PVC

DOI: 10.1201/9781003460121-9

FIGURE 9.1 Loose, highly erodible soil at a construction site, a potential source of water pollution.

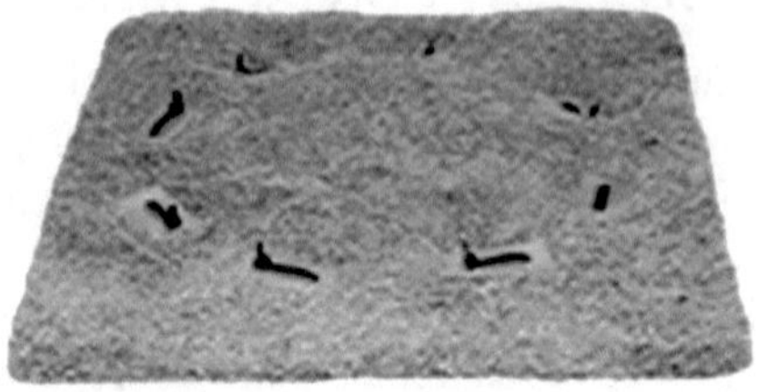

FIGURE 9.2 The Blocksom & Co. Inlet Filter commercially available filter mat.

and CMP and also makes DuroMaxx. This trademarked pipe is extruded of HDPE resin body embedded in ribs of high-yield steel for added strength. It has a smooth internal surface and rib profile outer wall, resulting in exceptional strength and superior hydraulics (n = 0.011–0.013), giving it an advantage over HDPE pipe, which loses its stiffness at high temperatures. DuroMaxx pipes are currently manufactured in 14 and 24 ft long sections of 30 to 120 in. diameter at 12 in. intervals. Because of the lower cost per unit of storage volume, the larger sizes are more suitable for underground detention basins.

9.2.1 Reinforced Concrete Pipes

RCPs have been used for over a century in the United States and many other countries. These pipes come in round, elliptical, and arch shapes. Reinforced arched concrete pipes are employed primarily as culverts under roads and railways and as such have a much smaller market than round and elliptical pipes; they are also only manufactured by a fraction of concrete pipe companies in the United States. Because of transportation cost and delay in delivery, this type of pipe has little application in many parts of the country. Oldcastle and AmeriTex, however, continue to make arched concrete pipes. Hancock Concrete Products, LLC (Hancock, MD, USA) was also a manufacturer of arch concrete pipes that were marketed the northeastern United States. This company was acquired by Oldcastle Infrastructure in 2021.

9.2.1.1 Round Concrete Pipes

Round RCPs are available in three ASTM classes: III, IV, and V. Class V pipes, the strongest, are used only in unusual cases where the pipe does not have more than 6 in. (15 cm) cover. Table 9.1 lists the sizes and weights of round RCPs from 12 to 108 in., which are more widely used than 120 through 180 in. sizes. The minimum required cover for class III pipes is 15 in. (38 cm) for 12 and 15 in pipes, and no more than 12 in. (30 cm) for 18 and 24 in (30–60 cm) pipes and only 6 in.

TABLE 9.1
Dimensions and Approximate Weights of Reinforced Concrete Pipes

	Class III		Class IV[a]		Class V	
Internal Diameter, in.	Minimum Wall Thickness, in.	Approximate Weight, lb/ft	Minimum Wall Thickness, in.	Approximate Weight, lb/ft	Minimum Wall Thickness, in.	Approximate Weight, lb/ft
12	1 3/4	79	2	93	±	±
15	1 7/8	103	2 1/4	127	±	±
18	2	131	2 1/2	168	±	±
21	2 1/4	171	2 3/4	214	±	±
24	2 1/2	217	3	264	3 3/4	366
27	2 5/8	255	3 1/4	322	4	420
30	2 3/4	295	3 1/2	384	4 1/4	476
33	2 7/8	336	3 3/4	451	4 1/2	552
36	3	383	4	524	4 3/4	654
42	3 1/2	520	4 1/2	686	5 1/4	811
48	4	683	5	867	5 3/4	1011
54	4 1/2	864	5 1/2	1068	6 1/4	1208
60	5	1064	6	1295	6 3/4	1473
66	5 1/2	1287	6 1/2	1542	7 1/4	1735
72	6	1532	7	1811	7 3/4	2015
78	6 1/2	1797	7 1/2	2100	8 1/4	2410
84	7	2085	8	2409	8 3/4	2660
90	7 1/2	2398	8 1/2	2740	9 1/4	3020
96	8	2710	9	3090	9 3/4	3355
102	8 1/2	3078	9 1/2	3480	10 1/4	3760
108	9	3446	10	3865	10 3/4	4160

[a] ASTM C76, AASHTO M170.

(15 cm) for pipes 36 in. and larger. The thickness of class III pipes ranges from 1.75 in. (4.5 cm) for 12 in. (30 cm) pipes to 15 in. (38 cm) for 180 in. pipes. All reinforced concrete pipes come in 8 ft (2.4 m) long sections and the weights of class III pipes which have thinner walls than other classes vary from approximately 632 lb. (287 kgs) for 12 in pipe to 76,580 lb. (34.8 metric tons) for 180 in. pipes. Table 9.2 shows the maximum cover of reinforced concrete pipes.

The American Concrete Pipe Association (ACPA, 2007) has specified four standard installations identifying principal zones within the lower half of the pipe: middle bedding, outer bedding, haunch, and lower side. The type of material and level of compaction of these zones vary with the installation type (1, 2, 3, and 4). In all these types (except class V which requires no bedding), the minimum bedding depth is specified as the outer pipe diameter divided by 24 ($D_o/24$) or 3 in. (7.5 cm), whichever is greater. In rock foundation, the minimum bedding for all types is $D_o/12$ or 6 in. (15 cm). The level of compaction in the haunch and the lower side varies for soil categories I, II, and III. Soil category I is gravelly sand and includes SW, SP, GW, and GP. category II is sandy silt including GM, SM, and ML as well as GC and SC with less than 20% passing #200 sieve; category III is silty clay, which includes CL, MH, GC, and SC (for soil classification, see Appendix B).

Figure 9.3 shows a typical pipe trench for round concrete pipes as specified by NJDOT (2015) for category III construction, which is more widely used than other types. The specified trench width

TABLE 9.2
Maximum Permissible Depth of Cover over Reinforced Concrete Pipes

Pipe Diameter	C-76 Class III	C-76 Class IV	C-76 Class V
12 in.	9 ft	18 ft	50 ft
15 in.	10 ft	21 ft	50 ft
18 in.	11 ft	24 ft	50 ft
21 in.	12 ft	25 ft	50 ft
24 in.	10 ft	18 ft	50 ft
30 in.	11 ft	19 ft	50 ft
33 in.	11 ft	19 ft	49 ft
36 in.	12 ft	20 ft	49 ft
42 in.	12 ft	21 ft	50 ft
48 in.	12 ft	21 ft	47 ft
54 in.	11 ft	17 ft	31 ft
60 in.	11 ft	18 ft	32 ft
66 in.	11 ft	18 ft	32 ft
72 in.	12 ft	19 ft	33 ft
78 in.	12 ft	19 ft	33 ft
84 in.	12 ft	19 ft	33 ft

is equal to pipe OD plus 18 in. (46 cm) for 18 in. (45 cm) and smaller pipes and pipe OD plus 24 in. (60 cm) for 24–48 in. (60–120 cm) pipes. Table 9.2 presents the maximum permissible depth of cover over RCPs. The depths listed in this table were calculated assuming backfill material having a unit weight of 120 lb/ft^3 (1920 kg/m^3, 19 kN/m^3). It is to be noted that reinforced concrete pipes have been installed at depths greater than 100 ft (30 m) with engineered backfill and compaction.

Elliptical Concrete pipes come in more than twenty sizes ranging from 23 × 14 in. (58 × 36 cm) through 180 × 116 in (457 × 295 cm). These pipes are approximately equivalent to 18 in. to 144 in. round RCPs in terms of discharge capacity, respectively. However, only a few concrete plants still make pipes larger than 76 × 48 in. (193 × 122 cm), equivalent to 60 in. round pipe. The minimum required cover on 23 × 14 class III pipe is 12 in. (30 cm) and only 6 in. (15 cm) on 30 × 19 in. (76 × 48 cm) and larger pipes. The weight of elliptical concrete pipes ranges from 1560 lb. (700 kg) for an 8 ft long section of 23 × 14 in. pipe to 56,000 lb. (254 metric tons) for 180 × 116 in. pipe. The ACPA (2007, 2011) presents tables of trench widths based on 1.25 times the pipe outer diameter plus 1 ft. The latter specified trench width is not significantly different from that of the NJDOT. For elliptical concrete pipes, the middle bedding width should be at least equal to one-third that of the outer diameter pipe, $D_o/3$, where D_o is the diameter along the bedding (longer diameter for horizontal elliptical or shorter for vertical elliptical). The level of compaction for various soil types is the same as that of round pipes.

Both round and elliptical RCPs come in solid or perforated walls. Solid pipes are used for either conveyance or detention, and perforated pipes are employed only for underground retention–infiltration basins. The minimum spacing between round and elliptical pipes is commonly specified as 12 in. (0.3 m) or one-half of the pipe diameter (longer axis for horizontal elliptical pipes), whichever is greater. To ensure that pipes are installed in accordance with plans, they should be inspected during the entire construction period. Appendix 9A presents an overview of installation and inspection of any make of pipe.

9.2.1.2 Prestressed Concrete Pipes

Concrete pipes are also made in prestressed for pressure flow applications. Vianini Pipe Inc. (Whitehouse, NJ, USA; www.viananipipe.com) and Thompson Pipe Group (Rialto, CA;

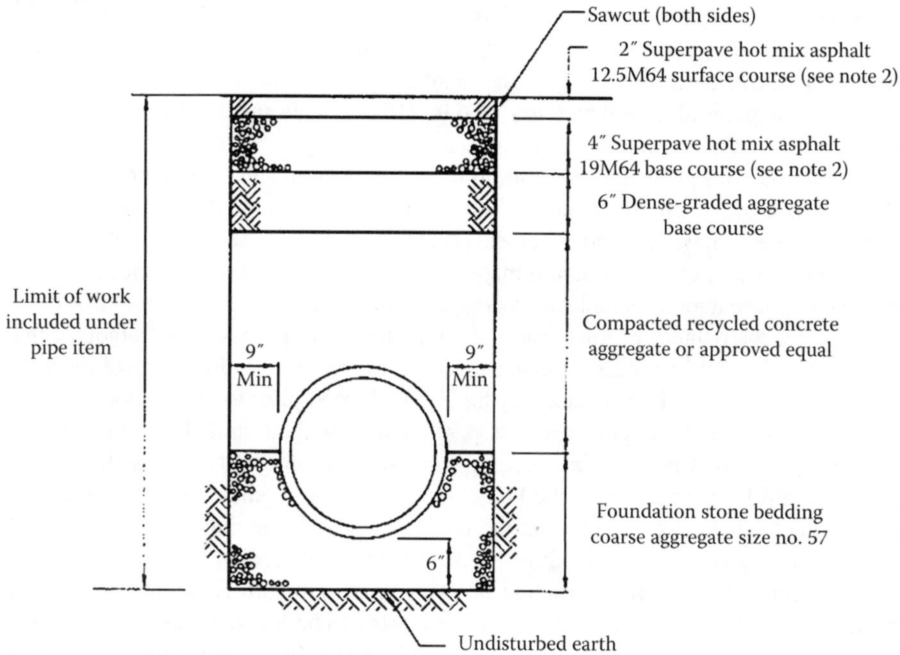

Pipe trench and pavement repair

NTS

Note: 1. The following items of work are included under the pipe item:

- Recycled concrete aggregate
- Foundation stone bedding
- Sawcut
- Superpave hot mix ashpalt 12.5M64 surface course
- Superpave hot mix asphalt 19M64 base course
- Dense-graded aggregate base course
- Dewatering (if necessary)
- Bypass pumping
- Sheeting (if necessary)

2. This detail shows the final condition. Pipe trenches in any areas of milling and/or resurfacing shall be constructed with superpave hot mix asphalt 19M64 base course up to the level of existing pavement. Thickness of superpave hot mix ashpalt 12.5M64 surface course shall be adequate to produce the above shown final condition after milling and/or resurfacing is complete.

FIGURE 9.3 Typical concrete pipe trench under pavement.

www.thompsonpipegroup.com) are among companies that manufacture prestressed concrete pipes. Prestressed concrete pipes are employed for domestic water projects rather than storm sewers and as such they are not discussed further in this book.

9.2.1.3 Concrete Box Culverts

Box culverts have a broad, flat surface; as such, they normally do not settle, and they cannot be forced down to be level. Therefore, it is imperative to have a perfect level grade established before the box is placed. Base material should be fine to medium granular stone. It is best to place and compact a 6 in. (15 cm) thick medium granular material first and cover it with 2 in. (5 cm) thick fine granular material as a leveling course. The boxes should be aligned accurately prior to pulling in place. This is especially important for the first few sections as it will impact the line and grade of the sections that follow. To prevent the granular material from moving into the joint as the boxes are pulled together, the material should be removed 6 in. (15 cm) wide and 2–3 in. (5 to 7.5 cm) deep in front of the groove of the last installed section.

Any unstable material present at the bottom of the excavation should be removed to the depth and width specified by the engineer and replaced with granular material. Also, rock and boulders encountered at the bed must be removed at least 6 in. (15 cm) below the bottom of the box section and replaced with medium to fine granular material. When putting two sections together, their joint should be sealed. Place a joint sealant such as butyl on the bottom half of the groove of the last section first and place the balance of joint material on the top half of the tongue (spigot) of the box to be set. The material should be placed about 1 in. (2.5 cm) from the leading edge of the groove and tongue. In cold weather, the joint sealant may have to be heated prior to application.

For waterproofing, an expanding water-stop secondary sealant should be applied after butyl sealant has been placed. Upon placing sections and waterproofing, lift inserts (holes) should be filled with grout flush with the top of the box culvert and backfill should be placed uniformly on each side and compacted, taking care that the box's alignment is maintained. The backfill material, the compaction level, and the depth of layers should follow contract specifications. Northern Pipe Company and Oldcastle Infrastructure are among manufacturers of concrete box culverts. More information on installation of concrete box culverts can be found at the websites of Oldcastle Infrastructure (www.Oldcastleinfrastructure.com), and Geneva Pipe and Precast, a Northwest Pipe Company (www.nwpipe.com) and AmreiTex Pipe and Concrete Products (www.ameritexpipe.com).

9.2.2 HDPE Pipes

HDPE pipes are manufactured by a number of companies, including ADS in the United States. After acquiring its competitor, Hancor Inc., in August 2005, ADS is now the largest HDPE pipe company in the world. HDPE pipes come in single and double walls. Single-wall pipes are available in 3 to 24 in. (75–600 mm) for highway drainage application; double-wall pipes have smooth interiors for hydraulic efficiency and corrugated exteriors for added strength. ADS dual-wall pipes are labeled N–12. The N-12 pipes come in standard N-12 IB, N-12 WT (watertight) and N-12 ST (soil tight). These pipes provide superior corrosion and abrasion resistance. As such these pipes serve as an excellent alternative for use in drainage and stormwater management projects. HDPE double-wall pipes are available in solid and perforated walls, varying from 4 to 60 in. (100–1500 mm) in diameter.

ADS also makes HP Storm pipes. These pipes are dual wall high performance polypropylene, come in 12 in. to 60 in. (300–1500 mm), in diameter, 20 ft (6 m) long and are and are used as gravity flow-storm sewers. In addition, ADS makes triple-walled pipes labeled ADS 3000. These pipes come in 10 ft (3 m) length, and 3 and 4 inch in diameter and three different hole patterns, The ADS 3000 pipes have high strength and are corrosion resistant to acid, alkalis, and salt and as such are superior in stiffness than PVC pipes.

HDPE N-12 pipes come in three classes based on stiffness and strength. Class 1 is the strongest and can be installed in a trench with or without complication of backfill. With dumped backfill, a 48 in. (100 cm) N-12 class 2 requires 90% compaction but a 48 in. class 3 needs 95% compaction for the same maximum cover. N-12 pipes come in 20 ft (6 m) long sections. HDPE pipes are lightweight and can be easily handled by two-man crews for sizes up to and including 18 in. Pipes of 24 and 30 in. are lifted by one sling and larger pipes by a two-point sling.

HDPE pipe sections are 25–30 times lighter than equivalent size concrete pipes. A 20 ft (6 m) section of 18 in. ADS double-walled pipe, for example, weighs approximately 128 lb. (57 kg). In comparison, an 8 ft (2.4 m) section of 18 in. RCP class IV weighs over 1350 lb. Table 9.3 presents nominal sizes and weights of double-wall HDPE pipes from 4 to 60 in. (100 to 1500 mm) in diameter. This table also lists approximate outside diameters of ADS, HDPE pipes, rounded to the nearest whole mm.

TABLE 9.3
Sizes and Approximate Weights of Dual-Wall (N-12) HDPE Pipes

Inside Diameter		Outside Diameter		Weight	
in.	**mm**	**in.**	**mm**	**lb/ft**	**kg/m**
4	100	4.6	117	0.44	0.65
6	150	7.0	178	0.85	1.30
8	200	9.5	241	1.50	2.20
10	250	12.0	305	2.10	3.10
12	300	14.5	368	3.20	4.70
15	375	18.0	457	4.60	6.80
18	450	22.0	559	6.40	9.50
24	600	28.0	711	11.00	16.40
30	750	36.0	914	15.40	22.90
36	900	42.0	1067	19.80	29.40
42	1050	48.0	1219	26.40	39.30
48	1200	54.0	1372	31.30	46.60
54	1350	61.0	1549	34.60	51.50
60	1500	67.0	1702	45.20	67.30

Source: Advanced Drainage Systems.

For parallel pipe installations, minimum spacing should be allowed between pipes. This minimum is 12 in. (0.3 m) or one-half of nominal pipe size (inside diameter), whichever is greater. It should be noted that the said minimum spacing applies to solid pipes only. For perforated pipes in stone trench, the minimum spacing is generally wider. The spacings for underground perforated HDPE pipes were listed in Table 7.23.

HDPE pipes are designed for use under H-25 and E-80 live loads. They are also highly resistant to corrosion. They have been in use since the mid-1950s. Contractors have had 70 years of satisfactory experience with HDPE pipes. During the past three decades, some governmental agencies, such as the FHWA, have approved the use of HDPE pipes for federal highway drainage projects. Consequently, the use of HDPE pipes has grown rapidly, and this trend is expected to continue in the future. Initially, the useful life of HDPE pipes was estimated at 50 years; however, based on experience of decades of performance, the useful life of HDPE pipes is found to be much longer.

Storage and handling should follow the manufacturer's specifications. ADS recommends the following precautionary measures in handling and storage of HDPE pipes:

- Stack pipes less than 6 ft (1.8 m) high.
- Alternate bells for each row of pipe.
- Do not drop the pipe from delivery trucks into an open trench or onto uneven surfaces.
- Avoid dragging the pipe across ground or striking the pipe against another pipe or object.
- Inspect the pipe and joining system before installation.

Ambient temperature extremes have little short-term effect on the strength of polyethylene pipes. Depending on the product, either carbon black or titanium dioxide is added to the polyethylene to protect against ultraviolet light. However, it is advisable to avoid long-term storage under direct sunlight. At 80 °F (27 °C), the pipe wall temperature can reach 110 °F (43 °C) if the pipe is left in full sunlight.

HDPE pipes are installed in trenches in the same way as RCPs, although the installation requires more care. The pipes are aligned by field survey, known as construction stakeout. The width of the trench depends on the pipe diameter, backfill material, and the method of compaction. Generally, 6–8 in. (15–20 cm) on either side of the pipe is the minimum allowable trench width when compaction equipment is not needed. In poor native soils such as peat, muck, or highly expansive soils, a wider trench than these minimums will be required.

Trench widths for small diameter pipes of all makes are dictated by the bucket size available for the excavation and, in many cases, exceed twice the pipe diameter. Pipe trench should be excavated with sidewalls nearly vertical. In deep excavation or poor soil conditions, it may be necessary to excavate with sidewalls sloped adequately. If this results in an excessively wide cut, trench boxes are used. The length of the box should be suitable for the pipe length. As indicated, the standard RCP length is 8 ft (2.44 m), and ADS N-12 pipe is 6.1 m (20 ft); however, 14 ft (4.27 m) length can be ordered for shorter trench boxes.

To prevent disruptions of backfill envelope, the bottom of the trench box should be placed no higher than 24 in. (0.6 m) from the bottom of the trench. This may require raising the trench box during pipe installation to conform with the previously indicated spacing, which is an OSHA regulation. To properly backfill and compact the soil around the pipe, the trench box may be dragged along the trench but only if it does not damage the pipe or disrupt the backfill. In some cases, it may be necessary to move the trench box two or three times to achieve the required compaction of the soil envelope. Figure 9.4 shows a typical subtrench installation. For more information on the use of trench boxes, the reader is referred to ADS pipes Installation Instructions by (Videos).

When installing pipe in rocky trench, a minimum of 12 in. (0.3 m) of acceptable backfill material should be placed below the bottom of the pipe to provide a cushion between the pipe and the rock. In soft, unstable soils, the bedding should be at least 2 ft (0.6 m) thick. The minimum width of trench for installation in loose soil is pipe diameter plus 4 ft (1.2 m).

In suitable soil conditions, a minimum of 4–6 in. (10–15 cm) of bedding should be placed and compacted on the trench to equalize load distribution along the bottom of the pipe. Upon the placement of the pipe, backfill is placed in the trench. Acceptable backfill materials and compaction are very similar and, in many cases identical, to those for other types of pipes. Figure 9.5 shows a typical backfill structure. Appendix 9A provides a description of backfill material, placement, and cover compaction of HDPE pipes.

ADS recommends a minimum of 1 ft (30 cm) cover on 4–48 in. (100–1200 mm) N-12 pipes and 2 ft (60 cm) for 54 and 60 in. pipes for H-25 live loads. Considering any deviation from the manufacturer's installation specification, a 1.5 ft (45 cm) minimum cover is recommended for pipes smaller than 48 in.

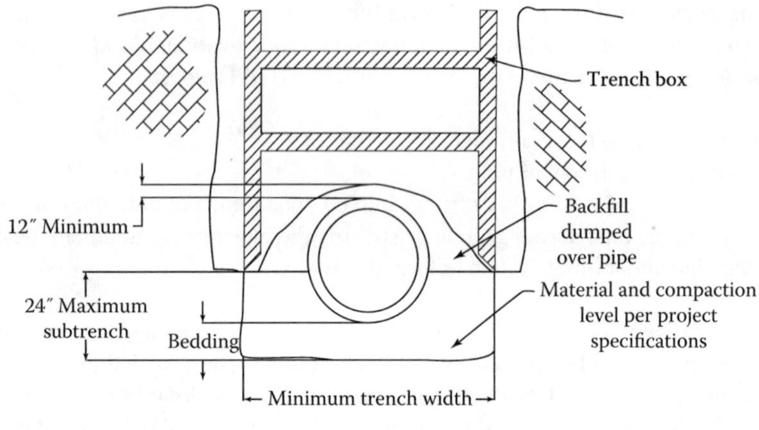

FIGURE 9.4 Subtrench installation.

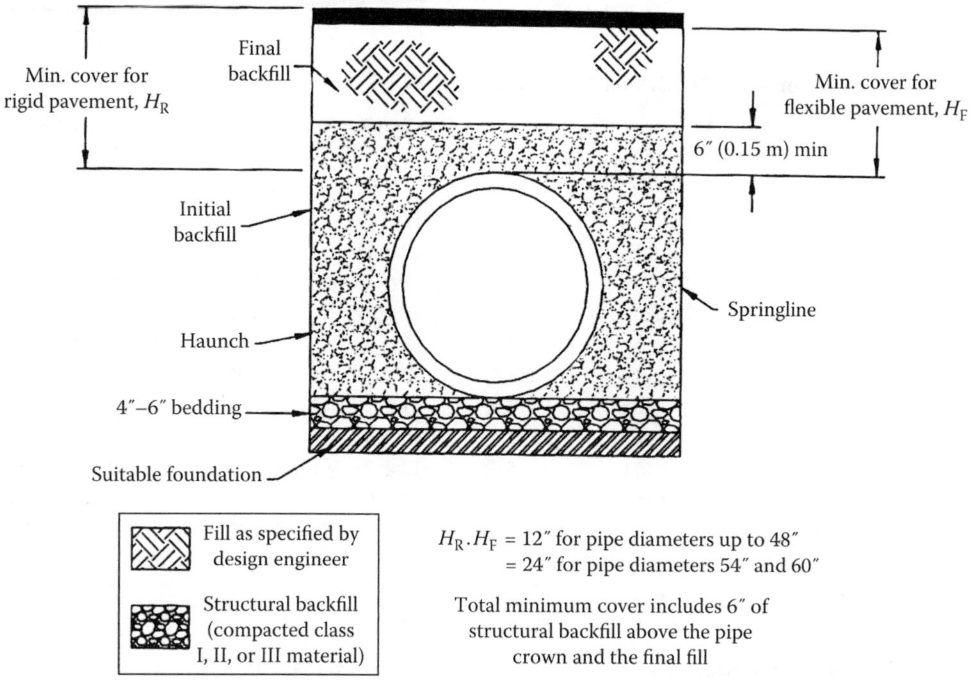

FIGURE 9.5 Typical backfill detail.

The maximum cover on N-12 pipes is highly dependent on the type of backfill material and the level of compaction. Table 9.4 shows the maximum cover on N-12 classes 1, 2, and 3. This table shows that an 18 in. (40 cm) class 1 pipe can be buried under 17 ft (5 m) of cover with no compaction. However, for the same permissible maximum cover, an 18 in. class 2 and 3 pipe requires 90% and 95% compactions, respectively.

When groundwater is present in the trench, it is necessary to dewater in order to maintain stability of the in situ and impacted bedding and backfill materials. The water level in the trench should be kept below the bedding during the pipe installation. A concern with HDPE pipes (and other lightweight material) in high water table areas is floatation due to uplift forces. The condition of stability should be considered when the pipe is empty. Assuming a soil friction angle of 36.8° and specific weight of 120 pcf (1920 kg/m^3, 19 kN/m^3) for backfill material, Hancor calculated minimum covers as shown in Table 9.5.

More information on technical specification of ADS HDPE pipes can be found in ADS (2022) Release Notes: https://www.ads-pipe.com, Appendix 9A describes installation of HDPE pipes in special cases.

9.2.3 Dewatering

In areas of high water table, dewatering may be required to maintain a dry trench during pipe installation. Groundwater conditions should be investigated prior to excavation. Test borings may be performed to determine the depth of water table and the rate, and the direction of groundwater flow. Groundwater is usually controlled by one of or a combination of the following methods:

Drains
Tight sheeting
Pumping
Well points

TABLE 9.4
Maximum Cover for ADS N-12, N-12 ST, and N-12 WT Pipes, ft (m)

Diameter in. (mm)	Class 1		Class 2			Class 3	
	Compacted	Dumped	95%	90%	85%	95%	90%
4 (100)	37 (11.3)	18 (5.5)	25 (7.6)	18 (5.5)	12 (3.7)	18 (5.5)	13 (4.0)
6 (150)	44 (13.4)	20 (6.1)	29 (8.8)	20 (6.1)	14 (4.3)	21 (6.4)	15 (4.6)
8 (200)	32 (9.8)	15 (4.6)	22 (6.7)	15 (4.6)	10 (3.0)	16 (4.9)	11 (3.4)
10 (250)	38 (11.6)	18 (5.5)	26 (7.9)	18 (5.5)	12 (3.7)	18 (5.5)	13 (4.0)
12 (300)	38 (11.6)	18 (5.5)	26 (7.9)	18 (5.5)	13 (4.0)	19 (5.8)	14 (4.3)
15 (375)	42 (12.8)	20 (6.1)	28 (8.5)	20 (6.1)	14 (4.3)	20 (6.1)	15 (4.6)
18 (450)	35 (10.7)	17 (5.2)	24 (7.3)	17 (5.2)	12 (3.7)	17 (5.2)	12 (3.7)
24 (600)	30 (9.1)	15 (4.6)	21 (6.4)	15 (4.6)	10 (3.0)	15 (4.6)	11 (3.4)
30 (750)	25 (7.6)	12 (3.7)	18 (5.5)	12 (3.7)	8 (2.4)	13 (4.0)	9 (2.7)
36 (900)	29 (8.8)	13 (4.0)	20 (6.1)	13 (4.0)	9 (2.7)	14 (4.3)	9 (2.7)
42 (1050)	27 (8.2)	13 (4.0)	19 (5.8)	13 (4.0)	8 (2.4)	13 (4.0)	9 (2.7)
48 (1200)	25 (7.6)	12 (3.7)	17 (5.2)	12 (3.7)	7 (2.1)	12 (3.7)	8 (2.4)
54 (1350)	26 (7.9)	12 (3.7)	18 (5.5)	12 (3.7)	7 (2.1)	12 (3.7)	8 (2.4)
60 (1500)	29 (8.8)	13 (4.0)	20 (6.1)	13 (4.0)	8 (2.4)	14 (4.3)	9 (2.7)

Notes: Results based on calculations shown in the Structures section of the ADS Drainage Handbook (v20.2). Calculations assume no hydrostatic pressure and a density of 120 PCF (1926 kg/m^3) for overburden material.

1. Installation assumed to be in accordance with ASTM D2321 and the installation section of the Drainage Handbook.
2. For installations using lower quality backfill materials or lower compaction efforts, pipe deflection may exceed the 5% design limit; however, controlled deflection may not be a structurally limiting factor for the pipe. For installations where deflection is critical, pipe placement techniques or periodic deflection measurements may be required to ensure satisfactory pipe installation.
3. Backfill materials and compaction levels not shown in the table may also be acceptable. Contact ADS for further details.
4. Material must be adequately "knifed" into haunch and in between corrugations. Compaction and backfill material is assumed uniform throughout the entire backfill zone.
5. Compaction levels shown are for standard Proctor density.
6. For projects where cover exceeds the maximum values listed here, contact ADS for specific design considerations.
7. Calculations assume no hydrostatic pressure. Hydrostatic pressure will result in a reduction in allowable fill height. Reduction in allowable fill height must be assessed by the design engineer for the specific field conditions.
8. Fill height for dumped class I material incorporates an additional degree of conservatism that is difficult to assess due to the large degree of variation in the consolidation of this material as it is dumped. There are limited analytical data on its performance. For this reason, values as shown are estimated to be conservatively equivalent to class 2, 90% SPD.

Source: ADS (TN 2.01, September 2014).

Drains and tight sheeting are practical when the water table is not significantly above the bottom of the excavation. Pumping or well point should be used when the bottom of the excavation is deeply submerged. Figure 9.6 shows a direct pumping from a sump. In this dewatering method, the pump should be submersible or self-priming so that intermittent flows can be discharged. Centrifugal pumps are best suited for larger quantities of water. In mud slurry conditions, diaphragm pumps should be used. Regardless of the specific type of pump, a standby pump should be available in case the operating pump clogs.

Sump pumps lower the water table in a localized area (see Figure 9.7). For lowering the water table along the pipe alignment, a well point system is used (see Figure 9.8). This dewatering method provides an effective means of controlling groundwater in permeable soils and can eliminate the

TABLE 9.5
Minimum Recommended Cover to Prevent Flotation

Nominal Diameter, in. (cm)	Minimum Cover in.(cm)[a]	Nominal Diameter, in. (cm)	Minimum Cover, in. (cm)
6 (15)	4 (10)	24 (61)	17 (43)
8 (20)	5 (13)	30 (76)	22 (56)
10 (25)	7 (18)	36 (91)	25 (64)
12 (30)	9 (23)	42 (107)	29 (74)
15 (38)	11 (28)	48 (122)	33 (84)
18 (46)	13 (33)	60 (152)	40 (101)

Note: For structural purposes, a minimum cover of 12 in. (30 cm) is required for 4-48 in. pipe, and 24 in. (60 cm) for 54–60 in. pipe.

[a] Rounded to nearest full centimeter.

FIGURE 9.6 Sump pump dewatering pipe installation in a high-water table site.

Source: Photo by the author.

need for sheeting and shoring. A well point system consists of 1 or 2 in. pipes (well points) that are driven vertically into the wet soil.

The well points are connected to a leader pipe by swing joints that facilitate their placement. The spacing and number of well points are calculated based on the soil permeability and the amount of water to be pumped. Spacing of 3 ft (1 m) of well points is common for a 6 in. (15 cm) diameter header pipe and 2 ft for 8 in. (20 cm) header pipe. In long header pipes, gate valves are installed at 100–200 ft (30–60 m) intervals. Significant lowering of the water table may result in subsidence of the ground in the surrounding area and structures in close proximity of the dewatered area. In such applications, a geotechnical investigation should be performed and adequate precautions taken.

9.3 WATERTIGHT JOINTS

To prevent leakage, pipe joints should be watertight. Also, the connection of pipes to manholes/inlets should be made leak free.

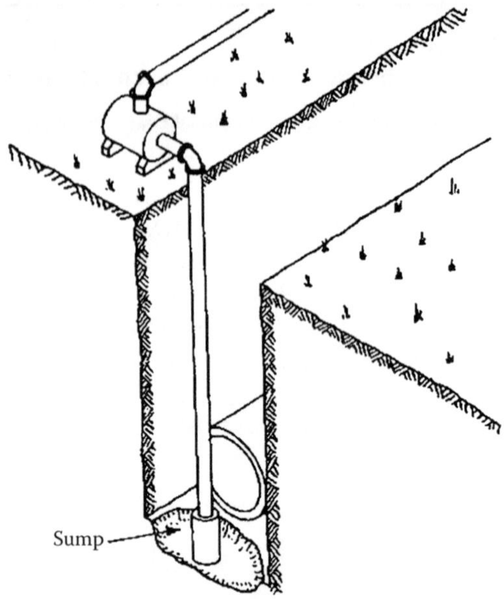

FIGURE 9.7 Dewatering by sump pump.

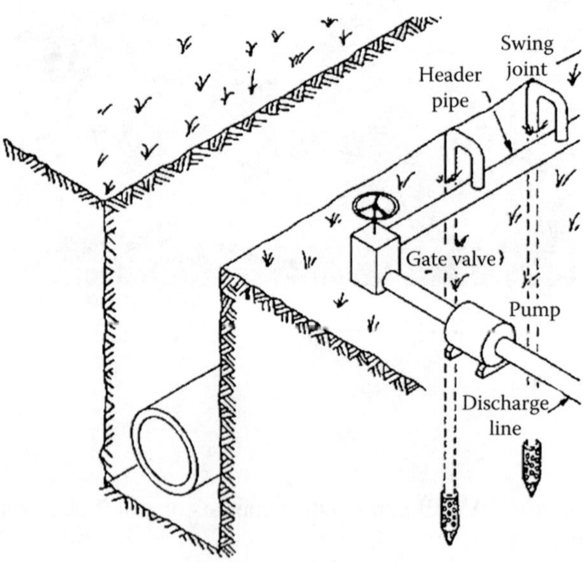

FIGURE 9.8 Well point system.

9.3.1 Pipe Joints

A variety of HDPE pipes come standard with watertight joints. HDPE double-walled smooth interior (N-12) pipes come in N-12 WT 1B and N-12 ST, watertight and soil-tight joints, respectively. These joints are shown in Figure 9.9. A soil-tight joint is adequate for storm drains and watertight joints are used for both storm drains and sanitary sewers.

O-ring and bell and spigot are used to make a watertight joint in concrete pipes. Figure 9.10 shows two variations of O-ring joints. In practice, mortar is also used at joints. Joints employing mortar sealants are rigid, and any movement or deflection after installation may cause cracks and

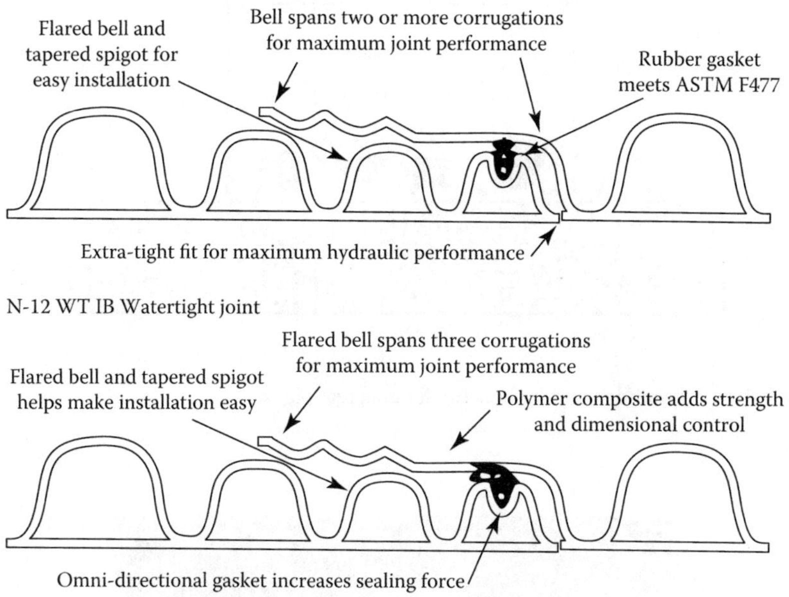

FIGURE 9.9 ADS watertight and soil-tight joints.

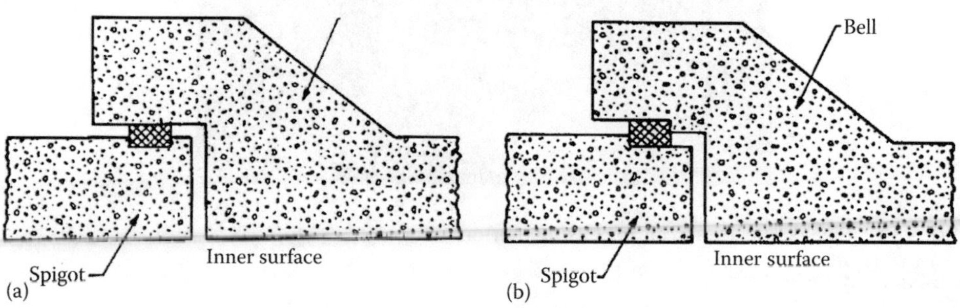

FIGURE 9.10 O-ring joints in concrete pipe: (a) opposing shoulder with O-ring and (b) spigot groove with confined O-ring.

Source: ACPA, *Concrete Pipe Handbook*.

leakage. An external jointing system for concrete pipes utilizes external rubber and mastic sealing band conforming to ASTM C990M. Figure 9.11 shows an external joint. Generally limited to a noncircular pipe with tongue and groove configurations, external sealing bands provide resistance to external loads normally encountered in drainage pipes.

Historically, watertight concrete pipes and structures have been associated only with sanitary sewer systems. Over the past four decades, environmental regulations have resulted in the need for watertight storm drains. However, mortar is still employed widely at concrete pipe joints.

9.3.2 Pipe Connection to Manhole/Inlet

Generally, there is a much larger gap at the connections of pipes to manholes or inlets than the pipe diameters themselves. Therefore, it is important to provide a watertight connection between pipes and manholes or inlets.

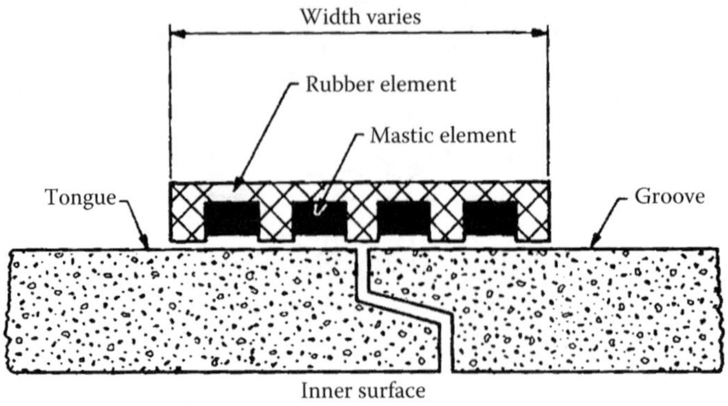

FIGURE 9.11 External sealing band for noncircular concrete pipes.

Source: ACPA.

FIGURE 9.12 Pipe connection to an inlet using mortar.

Source: Photo by the author.

In practice, the joints are filled or covered by mortar. Figure 9.12 shows a typical pipe-to-inlet connection in a shallow installation. Due to settlement, the mortar becomes cracked or broken, and water begins to leak from the joint. Over time, the leakage washes the soil away and creates a cavity at the connection. The result is a collapse of pavement and sinking of the ground around the manhole inlet, especially in steep areas where the water flows at a high velocity in the pipe. Figure 9.13 depicts this problem on a fairly steep street in a suburban town in northern New Jersey. To avoid this common problem, a flexible gasket providing a watertight seal should be employed between concrete manhole/inlet and pipes of any make.

Press-Seal Gasket Corporation (Fort Wayne, IN, USA) is one of the larger manufacturers of flexible seal for both reinforced concrete and HDPE pipes. Cast-A-Seal 802 (Figure 9.14) is a cast-in connector for straight-wall application for any pipe 18 in. (375 mm) and larger. A-LOK Products, Inc. (Tullytown Borough, PA, USA) also manufactures all sizes and styles of watertight connections of pipes to flat and curved walls. Figure 9.15 shows a curved wall connector. Trelleborg Pipe Seals Inc. (Milford, NH, USA) manufactures two types of rubber connections formerly made by NPC, Inc., one of which, the NPC Kor-n-seal I, is employed more often in sewer pipes than storm drains.

A seal called ADS Pipe Adapter is specifically made for ADS N-12 drainage (and sanitary) HDPE pipes up to 30 in. size. Press-Seal makes seals for N-12 pipes. The gasket consists of a rubber

FIGURE 9.13 Pavement damage due to joint leak.

Source: Photo by the author.

FIGURE 9.14 Press-Seal Gasket Corp. Cast-A-Seal boot-type pipe-to-structure connector.

ring with corrugated interior that fits into the HDPE pipe corrugation and a smooth exterior that connects the pipe to a manhole or other structures. Figure 9.16 shows this Press Pipe Gasket.

Connectors cost well under 5% of total job cost but save significant repair costs later. Stopping a leak can be inexpensive, but repairing a pavement failure is costly. A single pavement damage due to subsidence can cost several thousand dollars to repair, plus the danger and disruption of traffic. A flexible joint is advisable, not only at structures but also in pipes,. Occasionally the soil material around a joint settle, and with a grouted connection there is a risk of disconnection or crack of the pipe. To ensure a watertight storm drainage system, an infiltration/exfiltration test may be conducted as described next.

9.3.3 Infiltration/Exfiltration Testing

For pipes in watertight applications without specifying any ASTM specification for testing, an infiltration/exfiltration test is a simple and easy method of ensuring proper joint performance. In an

FIGURE 9.15 A-LOK's curved wall connector.

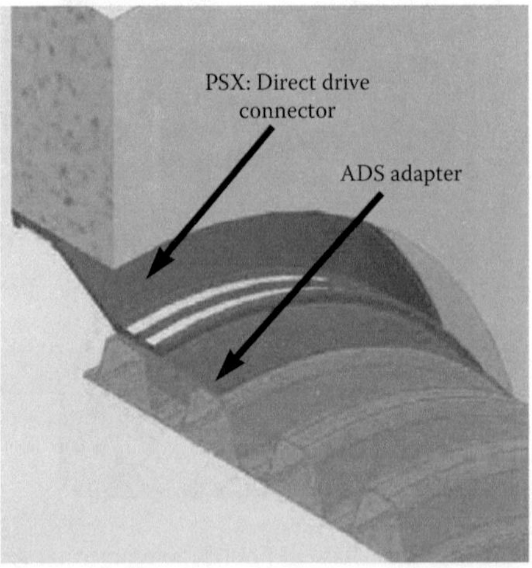

FIGURE 9.16 Press-Seal rubber adapter for ADS N-12 drainage (and sanitary) corrugated pipes.

infiltration/exfiltration test, a reach of pipe is tested by filling the system with water from a structure to the next structure (manhole or inlet), allowing the system to stabilize for 24 hours, measuring the water level, and then measuring the water level again after a time period. The drop of water level can then be converted to gallons of leakage per inch of pipe diameter for a one-mile length of pipe/day and compared to the permissible level established for the project. In the absence of a specified level, 200 gal/in. dia./mi/day (18 L/mm dia./km/day) is commonly considered watertight for storm sewer applications.

9.4 CONSTRUCTION OF STORMWATER STRUCTURES

This section discusses the construction of detention basins, ponds, infiltration basins, and grass swales, which are among the most common structural stormwater BMPs.

FIGURE 9.17 A detention basin covered with overgrown vegetation.

Source: Photo by the author.

9.4.1 Detention Basins

Dry detention basins are commonly constructed through excavation below grade. Due to topography, an embankment may have to be erected on the downhill side of the detention basin. To avoid unnecessary jurisdictional compliance, the height of the embankment should be kept below the limit above which the embankment is classified as a dam.

The side slopes of a detention basin should be selected considering stability and maintenance. To facilitate mowing, grass-lined banks should be at no steeper than 3:1 (3 H, 1 V). In stone lined or riprap-lined detention basins, the banks should be no steeper than the angle of response of granular material, but no more than 2:1. The bottom of detention basins should preferably be covered with vegetation to improve water quality and to enhance infiltration. (See Figure 7.1.)

Detention basins should include a low-flow channel to convey low flows from the terminus of inflow pipe to outlet structure. Riprap stone is preferable to concrete for lining low-flow channels. By draining the bottom of the basin after a storm, a riprap-lined low-flow channel prevents soggy conditions, which facilitates mowing the bottom and banks of the basin, which in turn prevents overgrowth of vegetation. Figure 9.17 depicts overgrowth in a detention basin without a low-flow channel and neglected maintenance just five years after construction. In soils of poor permeability, a subdrain may be installed below the bottom of the basin to help drain the basin.

All detention basins and ponds should include an emergency outlet or spillway to safely release storms that exceed design frequency. Additionally, an access driveway should be provided atop all basins to maintain outlet structure and forebays, if any. In addition, maintenance right-of-way (ROW) or easement should be provided from a public or private road. Maintenance access should be at least 12 ft (4 m) wide and at less than 15% slope.

9.4.2 Infiltration Basins

Infiltration basins, as with detention basins, are constructed by excavation and erecting a low embankment, if necessary. Natural depressions can serve as infiltration basins. These basins may be placed online or offline relative to the drainage route. In an online arrangement, the basin is designed to fully capture a selected storm, typically the water quality storm and discharge through an outlet for larger storm events. Offline infiltration basins are designed to receive only the water quality storm with the larger storms bypassed either downstream or into a detention basin. In either case, the basin is relatively shallow, commonly 2–3 ft (0.6–0.9 m).

Infiltration basins may be also designed to serve as a dual-purpose infiltration–detention basin. In this type of basin, the water quality storm is retained below the lowest opening in the outlet structure so that it is infiltrated slowly into the ground. The stored volume above the water quality storm level is discharged in a regulated manner through an outlet structure. Case Study 7.5 presented design calculations for this type of infiltration–detention basin.

Most failures of infiltration basins stem from lack of prior soil testing, poor construction practices resulting in over compaction, and premature silting. To avoid any loss of permeability due to compaction, excavation should be performed from the perimeter of the basin, avoiding heavy machinery entering the excavation site. Thus, the bottom width of above ground infiltration basins should be limited to 30 ft (10 m) when a backhoe is used for excavation. For wider basins, draglines or other longer-reaching arm machinery or, alternatively light machinery, should be employed. To prevent premature siltation, the use of infiltration basins as temporary siltation basins should be avoided. If due to a lack of space, an infiltration basin has to be used during construction, it must be thoroughly cleaned of silt and sediment before it is lined with a sand layer.

To be functional, infiltration basins should only be considered and constructed in areas where soil permeability is at least 1 in./h (25 mm/h). Also, the bottom of infiltration basins should be at least 2 ft, but preferably 3 ft (1 m) above the seasonal high -water table. The bottom of infiltration basins may be covered with vegetation or sand. In the latter case, the sand layer should be at the minimum 6 in. (15 cm) but preferably 1 ft (30 cm) thick to enhance infiltration. A soil log and permeability test should be conducted at the location of any infiltration basin to ensure that soil is sufficiently permeable, and the water table will not be encountered at least 2 feet (preferably 3 feet) below the bottom of the basin.

Typical problems associated with infiltration basins are standing water, soggy surfaces, sedimentation, and inadequate access. The first two problems usually arise from either insufficient soil permeability or the loss of infiltration capacity due to deposition of silt on the bottom of the basin. In the latter case, the soil infiltration capacity may be restored by scraping the soil from the bed or removing and replacing the sand layer by light machinery. To access the basin for maintenance, ramps sloping 15% or less should be provided to the bottom of the basin. To prolong the longevity of an infiltration basin, it is recommended that runoff receives a pretreatment before entering the basin. The use of inlet filters, particularly in roadways and parking lots, offers a cost-effective means of pretreatment.

9.4.3 Wet Ponds

Wet ponds are suitable in areas of very low permeability soils and/or high-water table. In fact, in high water table areas and costal zones, a wet pond can be the most feasible structural BMP for attenuating the peak rate of runoff. In addition to peak flow reductions, wet ponds improve water quality by slowing down the inflow runoff through the stagnant body of water in the pond. Thus, a wet pond can be more effective than an extended detention basin in terms of removing suspended sediment matter and improving water quality. Where the water table is not shallow and soil is permeable, the bottom and sides of a wet pond should be lined with an impermeable geofabric material, or a clay liner. Wet ponds are commonly constructed through excavation with little or no embankment. Since the water level fluctuates due to filling during a storm and drawdown following a storm, the banks of a wet pond should be constructed at a slope significantly less than the angle of repose of native soil. To stabilize soil, banks may be covered with stone lining. Figure 9.18 shows a pond with rock-lined banks designed by the author.

9.4.4 Grass Swales

A grass swale can be an effective means of removing sediment and pollutants if it is properly designed and maintained. To facilitate mowing, grass swales should be constructed having a parabolic section with sides at 3:1 (3 H, 1 V) slope or flatter. To prevent overgrowth of vegetation, grass

FIGURE 9.18 A rock-lined pond (designed by the author).

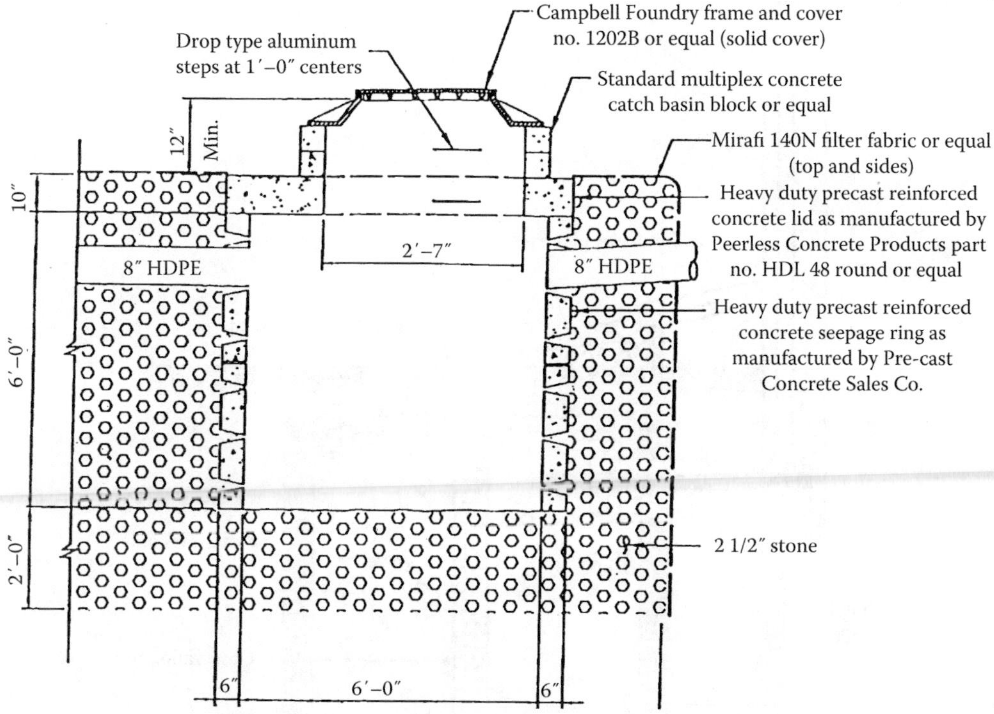

FIGURE 9.19 Seepage pit detail.

swales should be planted with a grass that is suited for mowing. However, the grass should be left long enough to filter the water quality storm. Grass swales are more practical where there is sufficient rain during the growing season. In dry weather conditions, periodic irrigation is necessary for healthy grass growth. In the southwest and other arid or semiarid parts of the country, stone is far more practical than grass as a cover for roadside swales.

9.4.5 Dry Wells and Infiltration Chambers

A dry well, also known as seepage pit, generally consists of a hollow concrete cylinder installed in stone trenches (see Figure 9.19). Figure 9.20 shows two seepage pits during installation in a stone

FIGURE 9.20 Seepage pits being installed in stone trench.

Source: Photo by the author.

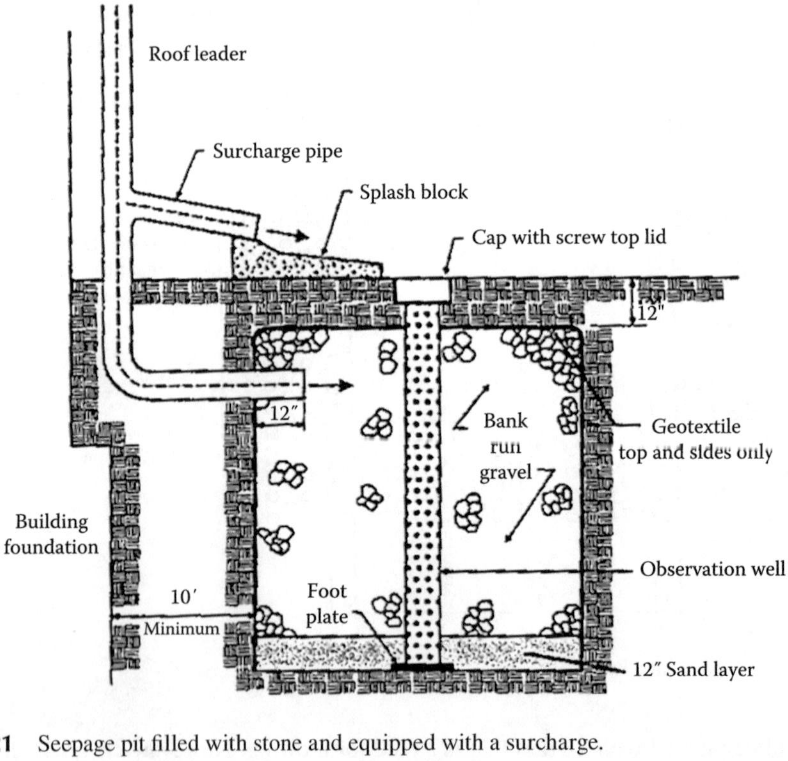

FIGURE 9.21 Seepage pit filled with stone and equipped with a surcharge.

trench. Dry wells filled with stone are also used in practice. Figure 9.21 depicts this arrangement. However, filling a dry well with rocks results in a loss of its internal storage by nearly 40%.

Perforated pipes or chambers in stone trenches also serve the same purpose. Some of the most widely employed chambers include StormTech, StormChamber, Cultec, and Triton, which were discussed in Chapter 7. Dry wells are commonly designed to retain the stormwater runoff from roofs and small impervious areas. To prevent their premature clogging due to leaves, grass clippings, and debris, seepage pits should preferably be used for roof runoff with roof leaders

FIGURE 9.22 Trash rack with parallel bars on an outlet structure (designed by the author).

or gutters equipped with a screen. In addition, seepage pits should include a solid cover or rim to prevent runoff from lawn and paved areas entering them. Furthermore, to allow for inspection and maintenance, seepage pits should be installed where they are readily accessible, and their rim should be placed at grade. If dry wells or infiltration chambers are used in a parking area, inlets should be equipped with filters such as FloGard +Plus (see Chapter 6) or the like to remove silt, debris, oil, and grease from runoff. Design criteria for seepage pits and retention–infiltration basins were discussed in Chapter 8.

A soil log and percolation test should be performed at the location of a dry well or any underground retention chamber to confirm the presence of suitable soil and absence of water table. Following the completion of construction, dry wells should be vacuum cleaned; if deemed necessary and tested to ascertain their actual performance. If tests show that seepage pits fail to completely drain within 48 hours, other measures should be selected for the control of stormwater runoff.

9.4.6 Outlet Structures

Outlet structures should be readily accessible for maintenance. They should have a minimum on plain dimensions of 4 × 4 ft (1.25 × 1.25 m) to enable a maintenance crew to enter. Upstream from an outlet, the basin should be lined with riprap, gabion mattress, or concrete to provide a firm footing for maintenance operations.

Outlets for all types of stormwater management facilities should be equipped with trash racks of properly sized openings. The openings should be small enough to collect brush and large debris that may block the outlet pipe but must be large enough to permit passage of leaves and small debris, which are not likely to clog the outlet pipe. This will minimize the frequency of cleaning and maintenance. If trash rack openings are too small, they become clogged by leaves, and small debris and cause frequent flooding. To facilitate cleaning with a rake, trash racks should be designed with vertical bars rather than square openings (see Figure 9.22). A review of outlet structures of detention basins is presented in an ASCE (1985) publication.

In a development project that included multiple detention basins and ponds, the contractor installed aluminum grids on the outlet structures pursuant to a request by the developer. After a heavy midfall rainstorm, the tree limbs full of leaves completely blocked the trash racks, which had 2.5 in. (65 mm) square openings, and as a result, the basins were surcharged. With the buildup of pressure against the outlets, some of the racks were snapped off from the structures and jammed inside of the outlet pipes downstream. Following this incident, typical trash racks of proper sizing were designed and installed on all the outlet structures. Figure 9.22 shows a typical trash rack designed by the author. Prefabricated plastic trash racks are available for a variety of applications.

Plastic Solutions, Inc., makes various HDPE trash racks, commercially known as StormRax™ (http://www.plastic-solutions.com). StormRax is also available at Contech.

Moreover, to avoid frequent maintenance, the openings in outlet structures should be designed with orifices large enough to prevent frequent clogging. Regulations in many states neglect this important maintenance consideration, allowing the use of small orifices in detention basins.

In New Jersey, for example, orifices as small as 2.5 in. are permitted not only in aboveground detention basins but also underground. Such small orifices are highly vulnerable to clogging. In underground detention systems where the clogging can be left undetected, the detention basins can be partly if not fully filled with water when a storm occurs. Considering proper functioning and ease of maintenance, the author recommends using a 6 in. (150 mm) orifice as the smallest opening in any underground detention system. Following this recommendation, 6 in. orifices were substituted for 2.5 and 3 in. orifices in hundreds of underground detention basins in New Jersey.

9.4.7 Slope Stabilization

In steep slopes and sandy soil, the land is vulnerable to erosion. As indicated in a previous chapter, erosion control blankets (ECBs) and turf reinforcement mats (TRMs) are employed to establish vegetation and control erosion in such cases. ArmorMax, manufactured by Propex Global is one of the high-performance TRM with a life expectancy of 50 years or longer. This TRM and those by other manufacturers can also serve as an effective erosion control measure and stabilization of bed and banks of channels.

9.5 INSPECTION AND MAINTENANCE

Addressing current stormwater management regulations requires maintenance measures for stormwater management facilities in urban development projects in all parts of the country. Even single-family homes that exceed a threshold limit of disturbance (LOD) or imperviousness are subject to these regulations in some states. Regardless of their size, type, and physical features, all stormwater management facilities require regular cleaning and proper maintenance to function effectively. Installed stormwater management systems should be inspected to ensure that they are properly constructed. Appendix 9B includes a construction inspection checklist for drainage systems.

9.5.1 Objectives of Inspection and Maintenance

The primary objective of regular inspection is to ensure that stormwater management facilities operate properly and safely. Stormwater management facilities should be inspected regularly by qualified personnel. The inspection should include all stormwater management elements, including conveyance system, detention and infiltration basins of any type, outlet structures, and water treatment devices. Nonstructural systems, such as pervious pavers, rain gardens, and blue and green roofs, need to be inspected as well. Inspections provide information on the effectiveness of scheduled maintenance procedures and changes in scope and scheduling that are warranted. In addition to stormwater management facilities, vegetative areas tributary to drainage systems and detention and retention basins should be inspected and maintained regularly.

Conveyance systems generally include inlets, manholes, pipes, and grass swales or stone channels. All of these elements should be inspected on a regular basis and cleaned off silt and floating debris such as leaves, paper, bottles, and cans, as necessary. The frequency of inspection depends on where the system is installed and the amount and type of impurities likely to be carried by runoff. Inspection should be conducted at least twice a year; however, in public places such as shopping centers and supermarket parking lots, drainage systems should be inspected at least four times a year.

As indicated, the purpose of maintenance is to keep the stormwater management facilities effective, operational, and safe. Regular maintenance of stormwater management facilities will also minimize the need for major costly repairs. Maintenance is divided into preventative (or regular) maintenance and repairs. Regular maintenance is intended to prevent a breakdown or failure and includes vegetative maintenance and facilities maintenance. There is a general neglect in maintenance of stormwater management systems; see Pazwash (1991) and Bryant (2004).

Because of the lack of common understanding of the impact of urban developments on water quality, the public does not consider maintenance of stormwater management facilities to be important. Also, because many of the stormwater management facilities are hidden underground and out of sight, they remain unattended. As a result, many private homeowners, homeowner associations, and even some municipal officials neglect to maintain drainage and stormwater management systems. Figure 9.23 shows an inlet grate in a park almost completely clogged by leaves and debris and Figure 9.24 depicts a municipal drainage pipe at a headwall, over two-thirds filled with silt.

FIGURE 9.23 Inlet grate clogged by leaves

Source: Photo by the author.

FIGURE 9.24 Drainage pipe, over two-thirds filled with silt.

Source: Photo by the author.

FIGURE 9.25 Grates on a clogged sand filter.

Source: Photo by the author.

Figure 9.25 depicts an unattended, totally ineffective sand filter in a residential development in New Jersey. This filter was installed to address the state's water quality requirements, because of neglect, the filter became fully clogged, bypassing the runoff through an internal overflow weir without getting any treatment.

Delaware DOT's Nonpoint Discharge Elimination System (DelDOT's NPDES) Group quantifies the benefits of good housekeeping practices and nonstructural BMPs. These practices include cleaning inlets, pipes, and detention/retention basins; street sweeping; and inspection of infrastructures. In addition to maintenance, the NPDES Group presents public education programs to convince Delaware citizens not to overwater their lawns or dump their used oil in storm drain systems. These measures help protect such important watersheds as the Delaware River and Chesapeake Bay. See DNREC.Gov. https://dnrec.delaware.org>npdes

The New Jersey Stormwater Management Regulations adopted on February 2, 2004 included a Municipal Stormwater Program that required all required all MS4s (564 in all) to clean and maintain their drainage systems within five to six years depending on the size of their system. By June 2009, over 675,000 inlets had been cleaned and several thousand tons of trash, debris, and silt removed from storm sewers. These figures demonstrate the need for and importance of good housekeeping and maintenance of drainage and stormwater management facilities. A brief review of maintenance measures for stormwater management facilities is presented in the following sections. For more information, the reader is referred to Pazwash (1991), ASCE/WEF (1992), ASCE standards (2006), ANSI/ASCE/EWRI (2017), Brzozowski (2004), and Barron and Lankford (2005). Also visit U.S EPA Stormwater Maintenance https://www.epa.gov>npdes>sto. last updated July 24, 2023.

9.5.2 Vegetative and Paved Areas

9.5.2.1 Lawns/Landscapes

Lawns, landscapes, trees, and shrubs that are located within the drainage area of a detention, retention, or infiltration basin or pond should be maintained to minimize soil erosion and production of vegetative wastes. This will decrease deposition of silt in inlets, pipes, and detention basins. Lawns should be mowed regularly during the growing season to maintain the grass at 2 in. (5 cm) high. Fertilizers should be applied at a minimum rate required and in no case more than 6 lb per 1000 square feet (2.9 kg/100 m^2). Damaged and dead lawn should be repaired and bare areas vegetated. If the season prevents the reestablishment of grass, exposed areas should be covered with salt-hay, mulch, or straw. Where reseeding is not effective in establishing a nonerosive vegetative cover, soil should be protected by other materials including sod, erosion control blankets, riprap, and gravel.

The plant beds should be mulched with hardwood every two years in order to retain moisture around the root zones and to provide a growth medium for shrubbery. Shrubs and trees should be pruned to maintain shape and appearance: trees once a year (preferably in early spring) and shrubs regularly, at least twice a year or as needed. For more information on maintenance frequency, the reader is referred to Kang et al. (2008) and Better Homes and Gardens (2022).

9.5.2.2 Pavements

Maintenance measures for parking lots, streets, and access driveways include the removal of sediment, debris and other pollutants. Paved areas should be swept regularly (at least once every three months), and the swept material must be disposed of properly. Figure 9.26 shows dirt and debris that the author swept from approximately 30 m (100 ft) of curb along his property. This dirt, weighing nearly 11 kgs (25 lb), had accumulated within six months.

Porous asphalt and pervious concrete and permeable interlocking concrete pavers can become clogged with sediment over time and this reduces their infiltration rate which in turn decreases their storage potential. To restore their infiltration capacity, the sediment should be removed from the surface of permeable pavers.

Pervious pavement of any type must be inspected for clogging and excessive sediment accumulation at least twice a year as well as after every intense rainstorm (more than 1 in. i.e. 25 mm of rainfall in one hour in New Jersey). It is best to inspect porous asphalt, pervious concrete, or permeable interlocking concrete pavements (PICPs) immediately after a heavy rainfall.

Any standing water indicates the need for desilting the pavement. The permeability can also be measured by conducting an infiltration test using the ASTM C1701 standard test method for infiltration rate of in-place pervious concrete. The Interlocking Concrete Pavement Institute (ICPI) recommends cleaning if the tested surface infiltration rate falls below 10 in./h (250 mm/h).

Sediments are most effectively removed by vacuum type street cleaning equipment without brooms and water sprayer. This equipment can remove the top 1 in. (25 mm) of sediment. Regenerative air sweepers (i.e., those that blow air across the pavement surface) are not recommended. Vacuuming should be performed twice a year, preferably once in the spring and again in the fall during dry weather. Disposal of sediment should follow all applicable local and state waste regulations. The author uses light duty pressure washing to clean pavers and to remove dirt from their gaps in his backyard and driveway once a year.

FIGURE 9.26 Street dirt and debris from 100 ft (30 m) of curbside.

Source: Photo by the author.

9.5.3 Stormwater Drainage Systems

The following subsections present a brief description of maintenance measures for various drainage elements. More information on this subject can be reviewed in ASCE "Inspection and maintenance of stormwater control measures."

9.5.3.1 Restoration of Grass and Riprap-Lined Swales

Grass and riprap swales should be inspected for erosion at least once a year and maintained as necessary. Maintenance of grass includes mowing on a regular basis and seeding or sodding eroded areas as needed. Regular mowing ensures that large storms do not create a soggy condition due to vegetation overgrowth. The stone-lined swales maintenance includes regrading banks and replenishing eroded areas with new stone.

9.5.3.2 Snow and Ice Removal from Pavements

Accumulation of snow and ice hinders the functioning of inlets, conveyance systems, and outlet structures. Snow should be removed from impervious areas to ensure the drainage facilities will be functional during the winter season.

9.5.3.3 Removal of Sediment and Floatable Objects from Drainage Systems

Stormwater management elements are expected to receive and trap sediment and trash. Inlets, pipes, drainage swales, outlet structures, and water quality filter inserts should be inspected at least twice a year, as well as after every major or intense storm (exceeding one inch in one hour in the East, South, Northwest, and Midwest and flash floods in the arid Southwest). The sediment and trash should be removed and disposed of in accordance with local regulations.

Removal of silt and debris from inlets and manholes may be performed either manually using a shovel or by a vacuum truck. Pipes and culverts can be cleaned by flushing, vacuuming, or a combination thereof. New designs of curb pieces in combination inlets trap floatable objects such as bottles, cans, paper cups, and the like that would otherwise enter the inlets (see Figure 9.27).

9.5.3.4 Control of Potential Mosquito Breading Habitat

Stagnant bodies of water such as ponds and wetlands are potential habitats for mosquito breeding. Standing water in inlets and outlet structures is also a source of mosquito culture. Removal of all obstructions in inlets and drainage conveyance systems helps avoid creation of mosquito

FIGURE 9.27 Can and plastic bottle trapped by curb piece.

Source: Photo by the author.

breeding habitats. Eliminating potential breeding sites is preferable to application of chemicals to control mosquitoes.

9.5.4 Ponds/Detention Basins

9.5.4.1 Algae and Weed Control

Shallow ponds and detention basins with prolonged standing water are vulnerable to weed and algae growth. Excessive algae growth results in oxygen depletion, causing the development of anaerobic conditions. Low oxygen causes the emission of foul odors and an unpleasant scene. Figure 9.28 shows a stagnant irrigation pond fully covered with algae. Weeds associated with ponds and detention basins may be submergent, floating, or emergent. Submergent vegetation is the most difficult to detect and to control.

Ponds and detention basins should be inspected at least twice annually for algae and plant growth. Algae growth can often result from the misuse of fertilizers on lawns. Proper application of fertilizers lessens the algae problem. Aeration offers a practical solution to algae growth in ponds. Weeds that become a problem may be cleared by pond maintenance professionals.

9.5.4.2 Underground Detention Basins

Underground detention basins should be inspected for excessive deposition of sediment, grit, and debris at manholes, inspection ports thereon, and outlet structures, and if feasible within detention chambers/pipes themselves through videoing. Clogging of perforations in pipes and chambers prevents the full utilization of the enveloping stone. The ground above and adjacent to underground detention basins should also be inspected for excessive settlement.

Sediment, debris, and trash should be removed from underground detention basins and disposed of properly. It is recommended that such basins be dewatered before removing the sediment and debris from them. Entering underground detention basins, if necessary, and viable should be performed by a person who is trained for a confined space entry and in compliance with the latest OSHA confined space regulations. In addition, specific manufacturers' maintenance procedures should be followed when cleaning underground pipes and chambers.

9.5.4.3 Wet Ponds

A wet pond functions like a permanent sedimentation basin. As more sediment accumulates in a pond, its effectiveness in removal of pollutants is diminished. Pollutants such as phosphorus, which

FIGURE 9.28 Stagnant irrigation pond fully covered with algae.

Source: Photo by the author.

attaches to sediment, can be chemically released under anaerobic conditions. Therefore, to maintain the effectiveness of this type of BMP, the sediment should be removed on a regular basis. Although guidelines generally call for removing sediment when 25% of the permanent pool volume has been lost, the author suggests removing the silt when it is accumulated to 15 cm thick, or 15% storage, whichever is smaller.

9.5.4.4 Outlet Structures

Outlet structures are the most important flow control element in ponds, and any detention basins, whether aboveground or underground. These structures and their trash racks should be inspected at least four times annually and after every intense storm (suggestively 1 in., 25 mm, or more in one hour) in the northeastern United States and many parts of the country and flash floods in the arid Southwest). Any floating debris, such as brush, leaves, tree limbs, paper, tumbleweeds, and plastic cans, should be removed from trash racks. Also, sediment and other obstructions should be removed from outlet structures and in particular from low-flow openings.

9.5.5 Water Treatment Devices

9.5.5.1 Catch Basin Insets

Catch basin inserts should be regularly inspected for clogging and cleaned as necessary. It is advisable to perform the inspection initially after installation and two to four times annually during the first year as well as after any major storm event. In Encinitas, California, where 32 KriStar catch-basin inserts were used, two decades ago, semi-annual maintenance was sufficient, and the annual maintenance cost of each insert was approximately $40 (Brozowski, 2004). KriStar of Santa Rose, California, was acquired by Oldcastle in January 2014. Oldcastle's FloGard +Plus catch basin inserts are now one of the most widely used inlet filters in the United States. Presently KriStar OEM supplies and installs KriStar catch basin filter parts.

Cleaning of catch basin inserts is relatively simple and can be performed by one person. Maintenance basically involves removing the filter insert, emptying its content, replacing oil absorbent, and placing everything back in the inlet. Alternatively, the insert may be cleaned using a vacuum truck. Specific manufacturers' specifications for inspection and maintenance should also be followed. The frequency of cleaning is highly dependent on the location. While in residential developments, it would be sufficient to clean catch basin inserts once or at the most twice a year, the inserts may have to be cleaned at least three or four times annually in shopping center and supermarket parking lots. Appendix 9C includes specifications for cleaning FloGard +Plus filters.

9.5.5.2 Manufactured Water Treatment Devices

Many structural stormwater management facilities, such as detention basins and ponds, do not provide adequate water quality provisions to address the applicable stormwater management requirements for a project. This is particularly the case for underground detention basins, which commonly are neither intended nor are they accepted by some jurisdictional agencies for water quality. As indicated in a previous chapter, a large number water treatment devices are available and more are introduced to the market.

Water treatment devices, other than inlet inserts, include filter media and hydrodynamic devices. Most hydrodynamic (nonmedia) filter devices, such as CDS, Vortechs, BaySaver, among others, are not approved for 80% total suspended sediment removal in New Jersey and some other states. Many of these devices have to be placed offline, bypassing flows beyond the water quality storm. Filter media devices, which are accepted for 80% TSS removal by many jurisdictional agencies, are commonly employed in connection with underground detention basin and is placed past detention basins. StormFilter, BayFiller, Up-Flo Filler, Jellyfish, and Aqua-Filter are among filter media devices.

The filter media devices are generally employed for treatment of small rates of flow, are far more expensive than nonfilter media devices, and require more maintenance. Manufacturers' recommended inspection and maintenance procedures should be followed for their water treatment devices.

9.5.6 Repair of Stormwater Management Facilities

Inlets, pipes, detention basins, water quality devices, and outlet structures that get damaged because of floods, settlement, vandalism, or other causes must be repaired promptly. The urgency of the repair depends on the effect of damage on functionality and safety of the element. Major structural repairs should be designed by a professional engineer. For deteriorated concrete and corrugated metal pipes, CentriPipe provides a cost-effective rehabilitation solution. CentriPipe is a trenchless method of slip lining pipes up to 144 in. (366 cm) in diameter. The method involves centrifugally casting a specialty mortar and admixtures inside the pipe walls. The method can also be applied to elliptical concrete pipes and old brick culverts

9.5.7 Neglects in Maintenance

While the public can visualize the importance of controlling the runoff to prevent flooding, there had been and still is a general lack of understanding of the need for improving water quality. Some public and many private entities do not consider inspection and maintenance of water quality devices to be essential. An inspection of twenty-nine stormwater treatment practices revealed that approximately 70% of them were improperly installed and nearly the same percentage of these practices needed maintenance (Bryant, 2004).

Likewise, drainage structures and stormwater management systems are not sufficiently inspected and cleaned. A statewide survey of various stormwater management practices, conducted by the University of Minnesota, indicated that over 84% of cities conduct routine maintenance once a year or less (Kang et al., 2008). These and other cases indicate a vast negligence in inspection and upkeep of stormwater management facilities in the past and to a much smaller extent at present. The following recommendation would help rectify neglects in maintenance (Pazwash, 1991):

- Designers and planners should give a more serious consideration to ease of maintenance of stormwater management facilities in general and outlet structures in particular.
- Public and private owners of stormwater management systems should pay more attention to the maintenance of their systems. People generally tend to ignore that these facilities need to be maintained. While maintaining our cars and homes, we pay no or little attention to drainage facilities we own.
- Homeowners should be educated about the need for assuming the maintenance responsibilities of their own drainage and stormwater facilities within their properties. A majority of homeowners even fail to consider that such responsibilities rest with them.
- The responsible party (private or public) should be clearly defined and the required maintenance measures should be specified in the design phase and/or upon the completion of a project.
- Practical means of enforcing the maintenance of stormwater management infrastructure should be identified.

As was indicated in Chapter 8, a large number of municipalities have established stormwater fees for undertaking necessary maintenance measures; others have also prepared stormwater management agreements to ensure that privately owned stormwater management systems are properly maintained, an example of the latter is Douglas County, Georgia, where the county water and sewer authority has created a formal development form to be signed by the owner and developer of every project before a land disturbance permit is issued (Barron and Lankford, 2005). Other means of developing and implementing maintenance programs are presented in some other publications (see, for example; Kang et al., 2008). More recently, EPA has published pamphlets and handbooks for stormwater fees such as "Paying for Stormwater-the Benefits of a Utility," issued October 11, 2023, updated February 17 2025.

Now many states, such as Pennsylvania, Florida, and Virginia have established fees based on the equivalent residential units (ERU). In some states, Pennsylvania included, a memorandum of understanding (MOU) is issued between the municipality and the county conservation district to serve as an additional layer of review of stormwater management maintenance plans. An opinion held by some practitioners is that construction and maintenance of all detention and retention basins in general and infiltration basins in particular should be undertaken by the municipality to ensure proper functioning. In the majority of cases, stormwater management facilities within the ROW of roadway are maintained by the department of public works of the municipality or county or by the state and federal highway agencies. But the maintenance of a privately owned system rests with its owner.

9.5.8 Inspection, Operation, and Maintenance Manual

Depending on the nature of the project, a specific stormwater management inspection, operation, and maintenance manual should be prepared. The manual should include the name and address of the project, the party responsible for maintenance and his or her name, phone and fax number, postal address, and, of course, his or her e-mail. Also included in the manual should be a brief description of the installed facilities, and each manual should include a separate inspection and maintenance checklist for record keeping. An annual cost of estimate maintenance is also helpful. The following case study serves as an example of inspection and maintenance manual.

CASE STUDY 9.1 Stormwater Management Inspection and Maintenance, New Jersey Garden State Parkway Interchange 98

INTRODUCTION

The stormwater management system (designed by the author) for the Garden State Parkway roadway improvement project at Interchange 98 included three infiltration basins. The project and its stormwater management requirements were presented in Chapter 7. In short, these basins were designed to address NJDEP's regulations relating to peak runoff rates. The infiltration basins were also designed to provide water treatment in compliance with the regulations. The location of the infiltration basins was selected considering existing depressions, topographic features, and accessibility. This was intended to minimize the number of structures needed and ease maintenance and inspection of the stormwater management systems, as well. The site location is shown in Figure 9.29.

Soils in the area are predominately Downer, which is a sandy soil of high permeability. This type of soil is ideal for infiltration basins. The infiltration basins are located in highly visible sites to facilitate inspection. All three basins are also readily accessible from the southbound lanes to perform any needed maintenance.

This Operation and Maintenance Manual (O&M manual) is intended to provide guidance for the inspection and maintenance of the infiltration basins by the New Jersey Turnpike Authority. The manual can also serve as a support document in connection with the permit application to the NJDEP.

DESCRIPTION

The infiltration basins are depicted on the grading and drainage plans. For brevity, only a location plan of the basins is enclosed herein. All three infiltration basins have sand bed and grass lined side slopes. The sand beds are specified to have K-5 permeability rating. The sand layer is 6 in. (15 cm) thick in basins 1 and 2, and 12 in. (30 cm) thick in basin 3. In accordance with the

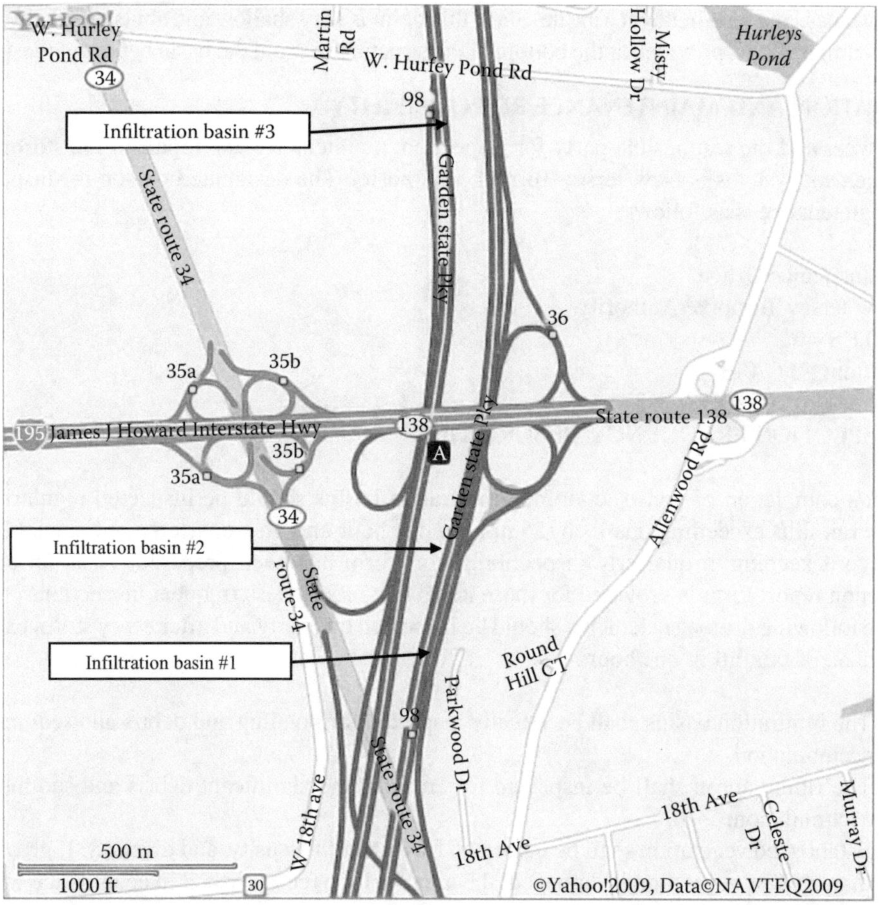

FIGURE 9.29 Location map of infiltration basins.

New Jersey Stormwater Best Management Practices Manual, all infiltration basins are designed to be no more than 2 ft (0.6 m) deep.

Located between the north and south bound lanes, infiltration basin 1 is the most southerly basin. The runoff from the roadway enters this basin as sheet flow, after passing through grass and woods. Only an emergency spillway is incorporated in this basin to allow for discharge of runoff in excess of the design storm. Overflow from the spillway flows downhill overland toward existing wetlands.

Infiltration basin 2 is located about 600 ft (180 m) north of infiltration basin 1. Unlike infiltration basin 1, a conveyance system is included within this infiltration basin. Much of the runoff to the basin does occur as sheet flow. Additionally, a 15 in. (38 cm) RCP carries runoff from two existing inlets along the northbound lanes of the Garden State Parkway to this basin. A riprap apron is provided at the outlet of the 15 in. RCP to prevent erosion and scouring in the basin bottom. A concrete box structure with a flat grate is provided to serve as an overflow structure and emergency spillway for this basin. This outlet structure is constructed above an existing 24 in. (61 cm) RCP drainage pipe that traverses the southbound lanes in a westerly direction.

Infiltration basin 3 is located on the southbound side of the Garden State Parkway between the exit ramp for Interchange 98 and the southbound lanes. Runoff entries this basin overland, similar to infiltration basin 1. The grate atop an existing inlet, located within the basin, is raised to serve as an emergency spillway. An existing 15 in. pipe emanates from this inlet, extends westerly below the exit lanes, and terminates at a swale. Since this basin is very shallow and abuts the roadway, a 3 in. (76 mm) orifice is provided at the bottom of the structure to drain the basin between rainstorms.

OPERATION AND MAINTENANCE RESPONSIBILITY

The owner and the responsible party for inspection, maintenance and repair of the stormwater management systems is New Jersey Turnpike Authority. The designated person for inspection and maintenance is as follows:

Maintenance Office
New Jersey Turnpike Authority
740 US-46
Clifton, NJ 07013
Phone No.: (973) 478–8337
INSPECTION FREQUENCY DESCRIPTION

Upon completion of construction, the infiltration basins should be inspected regularly following rainfalls exceeding one inch (25 mm) in one hour and on a quarterly and annual bases. For record keeping, a quarterly inspection report form has been prepared. Also, an annual inspection report form is provided for those items that require less frequent inspection.

The following drainage elements should be inspected quarterly and after every storm exceeding one in. of rainfall in one hour:

- The infiltration basins shall be visually inspected for ponding and debris and sediment accumulation.
- The riprap apron shall be inspected for erosion, and significant debris and sediment accumulation.
- Established vegetation shall be inspected for its health, density, and diversity. If greater than 25% of vegetation is damaged, the area shall be reestablished in accordance with the original specifications.
- Infiltration basins that do not infiltrate within 72 hours following a storm require immediate corrective measures.
- The conveyance system, inclusive of two inlets and drainage pipes, shall be inspected for sediment and debris accumulation.

The annual inspections include the following:

- Permeability tests of the soil at the bottom of the basins shall be taken.
- Outlet and inlets structures shall be inspected for cracking, subsidence, spalling, erosion, and deterioration.
- Riprap aprons and emergency spillways shall be inspected for erosion.
- The basin's vegetation shall be inspected for unwanted growth and erosion/scouring.

The inspection forms are completed by checking "yes" or "no." Any item that is checked as "yes" requires maintenance and must be corrected as outlined in the next section.

INFILTRATION BASIN MAINTENANCE INSTRUCTIONS

The maintenance shall include the following:

- Removal of sediment and debris. This should take place when the infiltration basins are thoroughly dry. The sediment should be disposed of in compliance with all applicable local, state, and federal regulations.
- Grass should be mowed at least bi-weekly during the growing season.

- Unwanted vegetation should be removed with minimum disruption to surrounding vegetation and basin subsoil.
- Reestablished vegetation should be inspected biweekly during the first growing season until the vegetation is established.
- Routine tilling should be performed with light equipment to maintain the infiltration capacity and break up clogged surfaces when necessary.
- Vegetation should be reestablished in eroded areas and riprap stone placed when erosion or scouring occurs at the emergency spillway.
- The outlet structure should be repaired or replaced, if necessary.
- Sediment and debris should be removed from inlets and conveyance systems and disposed of, in compliance with all applicable local, state, and federal regulations.

INFILTRATION BASINS
Garden State Parkway Interchange 98
Township of Wall
Monmouth County, New Jersey
QUARTERLY INSPECTION REPORT

Date: __________ Time: ______________ Weather Conditions: ________________________

	No	Yes
1. *Infiltration basin #1*		
I. Basin bottom:		
a. Sediment/debris accumulation	________	________
b. Clogging of sand/standing water	________	________
II. Vegetarian/grass:		
a. Areas of vegetation damage (>50%)	________	________
2. *Infiltration basin #2*		
I. Basin bottom:		
a. Sediment/debris accumulation	________	________
b. Clogging of sand/standing water	________	________
II. Vegetarian/grass:		
a. Areas of vegetation damage (>50%)	________	________
III. Riprap apron:		
a. Sediment/debris accumulation	________	________
b. Erosion/scouring	________	________
IV. Conveyance system:		
a. Sediment/debris accumulation	________	________
3. *Infiltration basin #3*		
I. Basin bottom:		
a. Sediment/debris accumulation	________	________
b. Clogging of sand/standing water	________	________
II. Vegetarian/grass:		
a. Areas of vegetation damage (>50%)	________	________
III. Outlet structure:		
a. Clogging of 3-in. orifice	________	________

Additional Notes:
____
____

This report represents the conditions observed by: ______________________________
____

Signature

INFILTRATION BASINS
Garden State Parkway Interchange 98
Township of Wall
Monmouth County, New Jersey
ANNUAL INSPECTION REPORT

Date: __________ Time: __________ Weather conditions: __________

	No	Yes
1. *Infiltration basin #1*		
I. Basin bottom:		
a. Permeability rate of sub soils (<4 in./h)/100 m/h)	____	____
II. Vegetation/grass:		
a. Unwanted growth (i.e., trees)	____	____
b. Erosion/scouring	____	____
III. Emergency spillway:		
a. Erosion/scouring	____	____
2. *Infiltration basin #2*		
I. Basin bottom:		
a. Permeability rate of sub soils (<4″/h)	____	____
II. Vegetation grass:		
a. Unwanted growth (i.e., trees)	____	____
b. Erosion/scouring	____	____
III. Outlet structure:		
a. Cracking/subsidence/spalling/erosion/deterioration	____	____
3. *Infiltration basin #3*		
I. Basin bottom:		
a. Permeability rate of sub soils (<4″/h)	____	____
II. Vegetation/grass:		
a. Unwanted growth (i.e., trees)	____	____
b. Erosion/scouring	____	____
III. Outlet structure:		
a. Cracking/subsidence/spalling/erosion/deterioration	____	____

Additional Notes:

____

____

This report represents the conditions observed by: __________

____
Signature

INFILTRATION BASINS
Garden State Parkway Interchange 98
Township of Wall
Monmouth County, New Jersey
MAINTENANCE LOG SHEET

Date	Location	Maintenance Performed	Performed By

PROBLEMS

9.1 Calculate the volume of a standard trench excavation per lineal feet of a 24 in. RCP. The average cover over the pipe is 18 in.

9.2 Solve Problem 9.1 for a 600 mm RCP and 45 cm average cover.

9.3 Calculate the excavation volume of the trench for a 30 in. HDPE pipe and 24 in. average cover.

9.4 Rework Problem 9.3 for a 750 mm HDPE pipe and 600 mm cover.

9.5 An underground detention system consists of four rows of 80 ft long, 42 in. solid RCP pipes under 18 in. cover. The pipes are capped at one end and terminate to a 4 ft wide by 6 ft deep chamber. Calculate:

a. The storage volume of detention pipes

b. The volume of excavation, assuming that the trench sides are at a slope of 1/5 (1 H, 5 V).

9.6 Solve Problem 9.5 for 1000 mm concrete pipes, each 24 m long and 45 cm cover; the chamber width and height are 1.2 and 1.8 m, respectively.

9.7 An underground detention system includes five rows of 750 mm solid HDPE pipes, 60 m long and 750 mm headers pipes. Calculate:

a. The detention storage volume

b. The amount of excavation; the pipes are under 60 cm of cover. The trench is at 1:5 side slope.

9.8 Solve Problem 9.7 for 180 ft long, 30 in. pipes under 2 ft of cover.

9.9 Estimate the annual cost of maintenance of treating Stormwater runoff at a flow rate of 0.6 cfs for 80% TSS removal by an MTD of your choice.

9.10 Estimate the first year's and 3 years' maintenance of an MTD of your choice that treats 30 L/s for 80% TSS removal. State any assumption made.

9.11 In Problem 9.10, perform an estimate for 20 L/s.

APPENDIX 9A: INSTALLATIONS OF HDPE PIPES BY ADS, SPECIAL CASES

9A.1 Joining Different Pipe Types or Sizes

Drainage systems may involve connecting pipe of different materials or sizes. Options to make these transitions are often limited by the joint quality required. A common method of connecting different types of pipes of the same size, and in some cases different sizes, is through the use of a concrete collar. Another option may be using fittings or adapters specifically designed for this application. ADS offers a selection of fittings designed to make the transition from one material directly to another. Fittings may be more watertight than a concrete collar.

9A.2 Curvilinear Installations

HDPE pipe can be laid on a curved alignment as a series of straight sections deflected horizontally at each joint. Typically, ADS N-12 ST IB and N-12 WT IB bell and spigot pipe joints can only accommodate small deflection angles (<1°). Split couplers and bell-bell couplers used to couple plain end N-12 pipes will also permit small deflection angles (approximately 1° to 3°). For larger angles, custom bends, manholes, or inlets are used.

9A.3 Vertical Installations

N-12 pipe is sometimes installed vertically for use as catch basins or manholes, meter pits, and similar applications. Vertical risers do not behave the same as pipe that is installed horizontally because the pipe/soil interaction is different. Installation requirements are especially important for vertical installations. Backfill shall extend a minimum of 1 ft (0.3 m) completely around the vertical structure. Backfill material recommendations are identical to those for a horizontal installation; compaction levels and maximum lift requirements must be strictly adhered to (refer to Table 9A.1 in the ADS Pipe Installation Instructions.

Additional general applications limits include the following:

- Height of the vertical riser must not exceed 8 ft (2.4 m), unless the design is reviewed by the ADS Application Engineering Department.
- In traffic areas, a concrete collar or similar structure designed to transmit the load into the ground (away from the pipe) must be used at the surface.
- Cast iron frames, holding grates, or lids must be seated on a concrete collar or similar structure so that the weight of the frame and grate or lid is transferred into the ground, not to the vertical pipe.

Vertical installations of any fitting by ADS or its manufacturer should first be reviewed for suitability with ADS Application Engineering or its manufacturer. This includes, but is not limited to, tees, elbows, and reducers of any combination.

9A.4 Steep Slope Installations

Where pipe slope is equal to or greater than 12%, precaution must be taken to ensure that application conditions will not adversely affect the pipe structure or flow characteristics. One design consideration is proper venting, to ensure negative pressure does not form inside the pipe. Venting can be provided along the pipe slope, at the head of the slope, or by designing the flow in the slope not to exceed 75% full in peak design flow conditions. Next, thrust blocks must be designed and constructed at all fittings and grade changes and, above all, a change in flow direction, which can cause excessive force against the pipe wall. Finally, consideration must be given to pipe slippage along the slope, which can result in slope failure of the surrounding soil, structural damage to the pipe wall, or compromising of joint quality for the overall system. Pipe should be restrained through the use of concrete blocks or pipe anchors.

Note: A grade of 12% is listed for reference purposes only; additional design consideration may be necessary for slopes less than 12%, shear slope stabilization, negative pressure, or water hammer may be of concern.

9A.5 Cambered Installations

Pipe installation under high embankments may need to design for uneven settlement regardless of the backfill envelope quality and construction. In order to eliminate low pockets under the embankment, the pipe should be cambered. Cambering is the process of installing the pipe so that the expected settlement will create the design slope. It can be achieved by installing the upstream half of the pipe on a flat grade and the downstream half on a grade that is larger than design. A qualified soils engineer should be consulted for this specialized situation.

9A.6 Sliplining

Due to abrasive or corrosive environments, premature deterioration of some types of pipes may occur. In lieu of a total replacement, slip lining the existing pipe with HDPE pipe or PVC pipe is often an economical and efficient way to prolong a culvert's service life. CentriPipe (centrifugally cast-in-place lining), which was discussed earlier, provides a very effective solution. Typically, HDPE pipe can only be used for open-ended applications where the pipe does not need to be bent for installation. Other considerations during design and preconstruction should include the inside and outside diameter of the carrier pipe and HDPE pipe, length of installation, and grout installation.

APPENDIX 9B: CONSTRUCTION INSPECTION CHECKLIST FOR DRAINAGE SYSTEMS: AN OVERVIEW

9B.1 Soil Erosion and Sediment Control Measures

- Check stone blanket at the entrance to construction site.
- Inspect silt fence/filter socks downhill of disturbed area.
- Regularly check the performance of siltation basins.
- Inspect hay bales/geofabric liner in inlets.
- Inform design engineer and contractor of any erosion control problems/deficiencies.
- Make sure corrective measures are implemented.

9B.2 Excavation

- Frequently monitor the line and grade of the excavation for pipes.
- Check if excavated materials are placed at a minimum distance from the excavation edge.
- Make sure that unsuitable subgrade is over excavated and stabilized.
- Classify rock excavation with contractor's representative.
- Keep a record of the excavated rock quantity.

9B.3 Pipe Installation

9B.3.1 Trenching

- Check if shoring is required and installed for trenches over 5 ft (1.5 m) deep.
- Check, if necessary, fences and barricades are installed to prevent accidental entry of persons or equipment into the trench.
- Check if escape ladders are placed per construction safety standards manual.
- Record soil types that deviate from the plans/specifications.

9B.3.2 Pipe Laying

- Check for proper size and class of pipe.
- Verify that pipe grade is checked frequently.
- Check if the joints (seals, couplings, gasket, bell, and groove) are installed per manufacturer's specifications.
- For concrete pipes, check that pulled holes are grouted and placed on the bottom.
- Verify that bedding material, backfill, and compaction follow specification.

9B.3.3 Manholes. Inlets Connections

- Check for correct pipe grade and alignment at manholes/inlets.
- Check for proper location and size of inlet/manhole.
- Inspect the proper placement of filter gravel/bedding under inlet and outlet.
- Inspect for proper connection of pipes into inlet/manhole (intrusion length, outside joint seal, or mortar).
- Inspect backfilling and placement of the manhole cover/inlet grate.
- Ensure silt/debris are removed from inlets/manholes.
- Check if the outlet end of pipe is at proper location and grade.
- Check stone size and dimensions of riprap apron/scour hole at detention basin(s) or outfall to streams.

9B.3.4 Bachfilling

- Ensure that proper measures are taken to prevent displacement of the pipe.
- Check that no organic matter, large rocks, or ice is placed within the specified distance from the pipe.
- Make sure that backfill is compacted as specified.
- Be sure that site restoration is done properly and on time.
- Record any site damage and its amount.
- Inspect any repairs to verify satisfactory condition.

9B.3.5 Repairs

- Make sure safety procedures are followed.
- Inspect resetting of grade and alignment of pipes.
- Check replacement of gravel envelope and backfill compaction.

9B.4 SITE ESTORATION

- Check debris and rock removal.
- Inspect inlets/manholes for cleanliness; make sure debris and silt are removed.
- Check site restoration.

APPENDIX 9C: GENERAL SPECIFICATIONS FOR MAINTENANCE OF FLOGARD +PLUS® CATCH BASIN INSERT FILTERS

Oldcastle® Stormwater Solutions

DRAINAGE PROTECTION SYSTEMS DPS

GENERAL SPECIFICATIONS FOR MAINTENANCE OF *FLOGARD +PLUS*® CATCH BASIN INSERT FILTERS

Scope

Federal, state, and local Clean Water Act regulations and those of insurance carriers require that stormwater filtration systems be maintained and serviced on a recurring basis. The intent of the regulations is to ensure that the systems, on a continuing basis, efficiently remove pollutants from stormwater runoff thereby preventing pollution of the nation's water resources. These specifications apply to the FloGard +Plus Catch Basin Insert Filter.

Recommended Frequency of Service

ADS drainage protection systems (DPS) recommends that installed FloGard +Plus Catch Basin Insert Filters be serviced on a recurring basis. Ultimately, the frequency depends on the amount of runoff, pollutant loading, and interference from debris (leaves, vegetation, cans, paper, etc.); however, it is recommended that each installation be serviced a minimum of three times a year, with the filter medium changed once per year. DPS technicians are available to do an on-site evaluation, upon request.

Recommended Timing of Service

DPS guidelines for the timing of service are as follows:

1. For areas with a definite rainy season: Prior to, during and following the rainy season.
2. For areas subject to year-round rainfall: On a recurring basis (at least three times per year).
3. For areas with winter snow and summer rain: Prior to and just after the snow season and during the summer rain season.
4. For installed devices not subject to the elements (washracks, parking garages, etc.): On a recurring basis (no less than three times per year).

Service Procedures

1. The catch basin grate shall be removed and set to one side. The catch basin shall be visually inspected for defects and possible illegal dumping. If illegal dumping has occurred, the proper authorities and property owner representative shall be notified as soon as practicable.
2. Using an industrial vacuum, the collected materials shall be removed from the liner. (Note: DPS uses a truck-mounted vacuum for servicing FloGard +Plus Catch Basin Insert Filters.)
3. When all of the collected materials have been removed, the filter medium pouches shall be removed by unsnapping the tether from the D-ring and set to one side. The filter liner, gaskets, stainless steel frame and mounting brackets, etc. shall be inspected for continued

serviceability. Minor damage or defects found shall be corrected on the spot and a notation made on the Maintenance Record. More extensive deficiencies that affect the efficiency of the filter (torn liner, etc.), if approved by the customer representative, will be corrected and an invoice submitted to the representative along with the Maintenance Record.

4. The filter medium pouches shall be inspected for defects and continued serviceability and replaced as necessary and the pouch tethers re-attached to the liner's D-ring. See below.
5. The grate shall be replaced.

EXCHANGE AND DISPOSAL OF EXPOSED FILTER MEDIUM AND COLLECTED DEBRIS

The frequency of filter medium pouch exchange will be in accordance with the existing DPS-Customer Maintenance Contract. DPS recommends that the medium be changed at least once per year. During the appropriate service, or if so determined by the service technician during a non-scheduled service, the filter medium pouches will be replaced with new pouches and the exposed pouches placed in the DOT approved container, along with the exposed debris. Once the exposed pouches and debris have been placed in the container, DPS has possession and must dispose of it in accordance with local, state and federal agency requirements.

DPS also has the capability of servicing all manner of catch basin inserts and catch basins without inserts, underground oil/water separators, Stormwater interceptors and other such devices. All DPS personnel are highly qualified technicians and are confined space trained and certified. Call us at (888) 950–8826 for further information and assistance.

REFERENCES

ADS, 2014, *Water management drainage handbook, specification, section 1, and installation*, ADS.

ADS, May 2022, Technical Notes TN-2.01 Minimum and Maximum Burial Depth for Corrugated HDPE Pipe (per AASHTO).

A-LOK Products Inc., P.O. Box 1647, 697 Main Street, Tullytown, PA 19007, Ph. (215) 547-3366, https://www.a-lok.com.

American Concrete Pipe Association, 19th printing, 2007, *Concrete pipe design manual*, revised October 2011.

American Concrete Pipe Association, 2007–2014, *Concrete pipe and box culvert installation*, Resource No. 01-103, https://info@concrete-pipe.org.

AmeriTex Pipe & Products, Seguin, Texas, https://ameritexpipe.com.

ASCE (American Society of Civil Engineers), 1985, *Stormwater detention-outlet control structures*, ASCE, ISBN0-87262-480-3.

ASCE/WEF, 1992, *Manuals and reports of engineering practice no. 77, WEF manual of practice FD-20, Design and construction of urban stormwater management systems*, WFE and ASCE, Alexandria, VA.

ASCE, 2006, *Standard guidelines for the design. Installation and operation of urban stormwater systems*, Publications No. 46-15, 46-16 and 47-16, 148 pp.

ASCE, 2023, *Stormwater ponds; inflows, routing and outlet structures*, ASCE.

ANSI/ASCE/EWRI, March 2017, *Standard guidelines for the design. Installation, maintenance and operation of urban stormwater systems.*

Barron J. and Lankford, M., September/October 2005, Maintenance of privately owned stormwater infrastructure: One approach to enforcement, *Stormwater*, pp. 100–102.

Better Homes and Gardens, August 23, 2022, Trees and shrubs care guide.

Blocksom & Co., 450 E 5th Street. Michigan City, Indiana 46360, Ph. 1 (219) 878-4455, http://www.blocksom.com/sediment-and-eroson-control/.

Bryant, G., May/June 2004, Stormwater inspection and maintenance, the sleeping giant, *Stormwater*, pp. 8–10.

Brzozowski, C., 2004, Maintaining stormwater BMPs, *Stormwater*, May/June, pp. 36–51.

EPA, July 2023, Stormwater Maintenance, last updated February 24,2025

EPA February17, 2025, Paying for Stormwater-Benefits of a utility

FloGard+Plus Catch Basin Insert Filter, *Olb castle infrastructure*, https://Oldcastleinfrastructure.com.

Kang, J.-H., Weiss, P.T., Wilson, C.B. and Gulliver, J.S., November/December 2008, Maintenance of stormwater BMPs frequency, effort and cost, *Stormwater*, pp. 18–29.

KriStar OEM, *Replacement parts, Anaheim, CA*, https://www.kristar.com.
Oldcastle Infrastructure, 7000 Central Parkway, Atlanta, Georgia, 30328, https://Oldcastleinfrastructure.com.
Pazwash, H., July 29–August 2, 1991, *Maintenance of stormwater management facilities, neglects in practice*, Proceedings of the ASCE National Conference on Hydraulic Engineering and International Symposium on Groundwater, Nashville, TN, pp. 1072–1077.
Press-Seal Corporation, 2424 W. State Blvd., Fort Wayne, IN 46808, https://www.press-seal.com.
StormRax by Plastic Solutions, Winchester, VA, https://stormrax.com; info@stormrax.com.
Trelleborg Pipe Seals 250 Elm Street, P.O. Box 301, Milford, NH 03055. Ph. 1 (800) 626-2180, https://www.trelleborg.com

10 Water Conservation and Reuse

The world population is continually growing while our freshwater supplies are shrinking. This trend is placing a constraint on the use of water, not only on agriculture but also on domestic needs. The southwestern United States and many countries around the globe already experience a water shortage. As this trend continues, the need for conservation and reuse of water becomes a challenging reality. This chapter discusses measures for conservation and reuse of water.

10.1 TRENDS IN WATER SUPPLY AND DEMAND

Approximately 3% of the earth's water is fresh, and about 70% of that is confined in glaciers and polar ice caps; thus, less than 1% of the water on the earth is suitable for drinking, and only 0.08% of that water is accessible to humans. Because of the overdraft, freshwater supplies are being depleted, while because of population growth, the demand is rising. Therefore, if the current supply and demand process is continued, soon we will be running out of water. To prolong the useful life of available supplies, we should reduce our water usage through conservation measures and reuse.

The world population has grown from approximately 6.14 billion in 2000 to 8.2 billion in 2024. This reflects an average growth of 1.124%. The US population has risen from roughly 190 million in 1920 to 249 million in 1990 to 281.4 in 2000, and according to the US Census Bureau population clock, 337.6 million by the end of 2024 (https://www.census.gov). The rise in the US population during the past 24 years has averaged 0.76%. At this rate, the US population is expected to increase by nearly 21.7% to 411 million by the year 2050. The Congressional Budget Office, however, projects the US population at 373 million by 2053, which appears to be underestimated. According to a study at the EPA National Risk Management Research Laboratory in Edison, New Jersey, the total gross water use in the United States already exceeds the total available freshwater supply, particularly in Florida and the Southwest. On a global basis, the water shortage is even gloomier. According to a 2016 report by the UNEP (United Nations Environment Programme), global demand for water is expected to exceed supply by 40% in 2030 if the current trend continues. This water deficiency has detrimental consequences, including on food crises, poverty, and sustainable development (https://www.waterforum.org). UNICEF estimates that 700 million people could be displaced due to intense water scarcity by 2030 (https://www.unicef.org/wash/water-scarcity).

The USGS (2018) prepared a graph showing the trends in total water withdrawals in the United States from 1950 to 2015; According to this figure, in 2010, the US used an estimated 355 billion gallons per day. This was 13% less than the previous five-year estimate in 2005. The 2015 USGS estimate showed an overall withdrawal of 322 billion gallons per day, which was approximately 9% less than that of 2010. Figure 10.1 shows the daily water withdrawal by category. This figure indicates that thermoelectric power, with 41% of the total demands, was the largest water usage in the United States, followed by irrigation and agriculture at 39%. (see U.S.G.S. Circular1441, 2018). The USGS updated water use estimates are available at https://www.usgs.gov/mission-areas/water-resources/science/water-use-united-states.

According to USGS estimates, the average daily residential water demand in the United States was 26.9 billion gallons (101.8 million m^3) in 1995 (Solley et al., 1998). This amounts to 101 GPCD (380 LPCD). The USGS's *Public Supply and Domestic Water Use in the United States, 2015* indicates that total domestic deliveries by public water suppliers averaged 23,300 MGD per day (88.2 million cubic meters) to 283 million people (about 87% of the total US population) in 2015. This reflects an average per capita demand of 82.3 gal (311 L) daily. The EPA estimates that each person

DOI: 10.1201/9781003460121-10

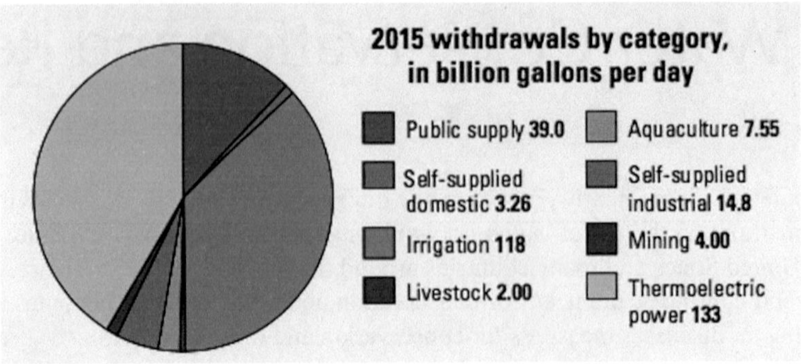

FIGURE 10.1 Water withdrawals for various uses.

uses an average of 82 gal (310 L) per day at home (see https://www.epa.gov>watersense). However, as will be indicated, the average daily per capita use varies from less than 20 gal to over 200 gal (75 to 760 L). Based on the author's records, the average daily water use during the past six years in his home has amounted to 205 gal (755 L), of which 69% was consumed outdoors during the growing season (six months from May 1 to October 31). This reflects an average daily indoor water use of 64 gallons for a household of two or 32 gallons (120 L) per person.

10.2 WATER CONSERVATION

Water conservation meaning reduced use, lowers water bills, and prolongs the useful life of our water supplies. It also lessens the cost of treatment and distribution of domestic water. A brief review of measures to conserve water in urban stormwater management practices was presented in a paper by the author in 1994.

In the past, water purveyors did not promote the idea of water conservation but rather were against it in order to elevate their revenues. Some people were not familiar with measures to save water, and many others were not concerned at all about water conservation. One reason for the lack of concern was that water bills were many times smaller than electric and gas bills in the United States and many other countries, as well. However, to delay the need for upgrades and to provide funding for water supply repairs, water purveyors across the United States have been embracing conservation during the past three decades and raising user rates at the same time. A 2004 survey by the American Water Works Association (AWWA, 2005), indicated the cost of water was \$1.91 per CCF (one CCF =100 CF = 748 gal.) or \$0.70 per 1000 L. Since then, the prices have gone up. A recent survey by the AWWA (2023), indicates the average cost of water in 2022 to US residential customers was \$4.51 per CCF or \$1.59 per 1000 L. The water rate now varies from \$3 to over \$8 per CCF (748 gal or 2830 L) in the United States. The cost of domestic water in Monterey, California, is nearly \$15 per CCF, which is higher than any other city in the country. Contrary to intuition, the rates are higher in the northeastern parts of the country than in the arid western states, where there is a water shortage. In fact, West Virginia has the highest water fee in the country. In Paramus, New Jersey, where the author resides, Veolia water company currently (October 2024) charges \$ 5.60 per CCF (\$1.98/1000 L) plus a FAC charge of 69 cents per day, The average cost of water in Canada is nearly the same as that in the United States but about 1.5 times as much in Germany and more than twice as much in France.

Water conservation as a means of water resources management is on the rise in the United States, especially in California and Florida, where the growth in population is higher than the national average. Many water districts are also considering water recycling as a water management technique. Water conservation, which was initiated in this country nearly three decades ago, is yet to be matured. The public is more aware of indoor conservation measures than outdoor ones. Also, means of reducing indoor use were developed many years before outdoor conservation was given consideration. Water-efficient appliances and fixtures were introduced to the market nearly 40 years ago. However, conservation of outdoor water uses, though more important than indoor water conservation, is still in the development stage. To conserve and reuse water, urban water agencies and environmental groups have formed water conservation councils around the country.

Water conservation and reuse are growing nationwide, even in the eastern states which receive from 40 to 46 in. (1000–1170 mm) of rainfall annually. In Florida, Georgia, and the Carolinas, water reuse has been established for some time. With stringent environmental regulations, it is difficult, if not unfeasible, to build new dams. Thus, with increases in population, it becomes necessary to conserve and reuse water. Water efficiency and reuse are becoming popular recently and this trend is expected to continue in the future. Since 1999, two magazines on the subject of stormwater management and recycling and reuse have begun publications. *Stormwater* magazine began its publication in June 1999, and the first issue of *Water Efficiency* magazine was published in September/October 2006. These magazines are available online (https://www.stormwater.com) and (https://www.waterefficiency.net), respectively. A review of measures to conserve water in urban stormwater management practices was presented in a paper by the author in 2002. An EPA Water Sense publication titled "Start Saving," dated May 28, 2025, discusses various measures for indoor and outdoor water savings (https://www.epa.gov/watersense-current).

10.2.1 Indoor Conservation

Indoor conservation is achieved with water-efficient fixtures and appliances, including shower heads, toilet flushers, faucets, dish washers, and clothes washers. Among these, low-flush toilets and low-flow shower heads result in a larger conservation than others. A report by the Water Research Foundation, titled "Residential End Use of Water, Version 2," dated April 1, 2016 (which was updated to Version 3 on June 18, 2024), indicates that based on records of water purveyors (most of which were in the Western states), the per capita water demand had been reduced by 15% from 69.3 gal in 1999 to 58.6 gal in 2016. The same report also indicates that the percentages of homes that used water-efficient clothes washers, toilet flushers, and shower heads in 1999 were 6%, 5%, and 75%, respectively, and that for the same fixtures, the per centages rose to 46 %, 37% and 80 % in 2016.

According to a report titled "Residential Toilets," by the EPA, dated June 3, 2024, toilets are by far the main source of water use in the home, accounting for nearly 30% of an average home's indoor water consumption. This report also indicates that by using Water Sense labeled toiles (1.28 gallons per flush), each home can save up to 13,0000 gallons (49,000 L) of water annually. In May 2024, EPA released a revised version of Water Sense Specifications for Toilets (Version 2). which took effect on July 1, 2025.

Table 10.1 presents average daily indoor water uses for various fixtures in water-conserving and non-conserving homes in the United States. This table indicates that the average indoor use in a non-conserving single-family home is estimated at 264 L per capita per day (69.5 GPCD), 25% of which is consumed by toilets. In comparison, the average demand in a water-conserving home is 114 LPCD (30.4 GPCD). Thus, using water-conserving fixtures reduces the indoor use by nearly 55%.

This table indicates that the use of water-saver toilets alone can save about 61.000 L (16.200 gal) of water a year in a household of four. By replacing just low-flow shower heads and installing faucet aerators in his former residence, the author had found the indoor water use to drop below 170 LPCD (45 GPCD). Using low flow (1.6 GPM) toilets and side loading (HE) clothes washers in his current residence, the per capita water use has dropped to 32 gallons (120 L) daily.

While indoor conservation measures began nearly 40 years ago in the United States, inefficient shower heads and toilet fixtures are still in use in many older homes. This is also the case in some

TABLE 10.1
Average Indoor Water Use in Liters (Gallons) Per Capita per Day Fixture

	Non-Conserving Home		Conserving Fixtures	
	LPCD	GPCD	LPCD	GPCD
Toilet	66	17.5	24	6.4
Shower	57	15	28	7.5
Faucet	42	11	28	7.5
Clothes washers	61	16	15	4
Leak	23	6	11	3
Miscellaneous	15	4	8	2
Total	264	69.5	114	30.4

Note: These figures are based on 3.5 gallons per flush (gpf) for late 1990s toilets, 3 GPM (gallons per minute) for shower heads, 2.2 GPM for kitchen faucets for non-conserving fixtures, and 1.28 gpf for flushometer toilet tanks, 1.5 GPM for water-saver showerheads. and 1.5 GPM for low-flow faucets. The figures are also based on 40 gal per cycle for top-loading washing machines and 10 gal. for HE, side-loading ones.

other countries. One of the most wasteful fixtures in public and commercial buildings is automatic flush urinals, which waste from 0.5 to 1 gal (3.8 L) of water per flush. Using waterless or low-volume urinals in offices, commercial buildings, and malls will result in significant water savings. Replacing each flusher urinal with water-free urinals such as the Sloan valve or the like can save over 30,000 gallons (115,000 L) of water per year. This would amount to several billion gallons of water saving annually. A review of measures to conserve water in urban stormwater management practices was presented in a paper by the author in 2002. For in-depth information about indoor and outdoor water uses and conservation measures one is referred to an excellent book by Amy Vickers (2001). An easy-reading paperback book by Mark Obmascik (1993) presents easy reader suggestions for water savings for homeowners.

According to a January 19, 2017 EPA publication, titled "US Indoor water use/Water Sense", an American family of four uses an average of approximately 400 gallons (1,500 L) of water daily, about 70% of that water is used indoor of which nearly 27% is consumed by toilets alone. An EPA Water Sense publication titled "Start Savingl U.S. EPA," dated May 28, 2025, presents a brief description of various measures for indoor and outdoor water savings (https://www.epa.gov>wa). By retrofitting indoor fixtures and toilet flushers in the city of Seattle, Washington, the total per capita water consumption (indoor and outdoor) dropped from 150 GPCD in 1990 to less than 100 GPCD in 2008 (Brzozowski, 2009). As a result, while the population grew by 16%, the water consumption in the city declined by nearly 26% from 1990 to 2008.

In spite of the benefits of water saving, still some people are not concerned about water conservation and do not use water saver fixtures. One reason for the lack of concern is that water bills are many times smaller than electric and gas bills in the United States and many other countries. As was indicated the average cost of water to American customers has risen from at $ 1.91 per CCF in 2004 to $4.51 per CCF in 2022. But the water bills are still lower than gas and electric. The average price of water in Canada is nearly the same as that in the United States and over twice as much in France and 1.5 times as much in Germany.

10.2.2 Outdoor Conservation

Outdoor demands for water are far greater than indoor uses. As such, more water can be saved through outdoor conservation measures. Until decades ago, however, little consideration was given to conserving outdoor uses. Irrigation is by far the largest outdoor water user globally and the second-largest water user in the United States and many other countries. In 2008, the USGS reported that over 65% of all the world's freshwater withdrawals were used for irrigation. In 2000, withdrawals for irrigation in the United States were estimated at 137,000 MGD (million gallons per day; 518 million m^3/day. This use represented 40% of total freshwater withdrawals, which was the second largest water user after the thermoelectric power industry. Over 85% of the water used for irrigation was in the 17 contiguous western states, where the average annual precipitation ranges from less than 4 in. to about 20 in., which is insufficient to support crops. California was the largest water user, consuming 22% of the total withdrawals.

Figure 10.2 shows the variations in irrigation water use between 1950 and 2015. This figure indicates that the withdrawals for irrigation had increased to approximately 150 billion gal per day (BGD; 567 million m^3/day) by 1980 but since have been declining. It reached its lowest level at 115 BGD in 2010. The irrigation water use in 2015 was estimated at 7.55 BGD (447 million m^3), and 63.5 million acres (25.7 million hectares) were irrigated, 55% of which was through sprinklers. About 84% of the 447 million m^3 of water was consumed by the 74% of the irrigated land in 17 conterminous western states. The 2015 data as shown by the following calculations indicate an average annual irrigation use of 2.08 acre-ft/acre (634 m^2/ha), that is 25 in./63 cm of water use per year.

$$(7.55 \times 10^9 \times 365/7.48)/(63.5 \times 10^6 \times 43{,}560) = 2.08 \text{ ft} = 25 \text{ in.}$$
$$(447 \times 10^6 \times 365)/(25.7 \times 10^6 \times 10^4) = 0.63 \text{ m} = 63 \text{ cm}$$

Outdoor water uses in a single-family home include lawn and landscape irrigation, car washing, decks and driveway cleaning, and filling swimming pools. Of these, over 85% goes to watering lawns and landscaping. The AWWA estimates that between 50% and 70% of our tap water still goes to outdoor yards. An EPA Water Sense January 2017 report, titled "Outdoor Water Use in the United States" indicates that residential outdoor water consumption in the U.S. can reach nearly 9 billion gallons (34 billion Liters) per day. More than half of that outdoor water is used for sprinkling lawns and gardens. a large portion of which is wasted due to inefficient watering methods and systems.

Outdoor water use varies greatly with geographic location. It is evident that outdoor water consumption is much higher than the above-indicated 30% of the average residential water use in large suburban single- family homes and especially affluent ones. Local climatic conditions and landscape

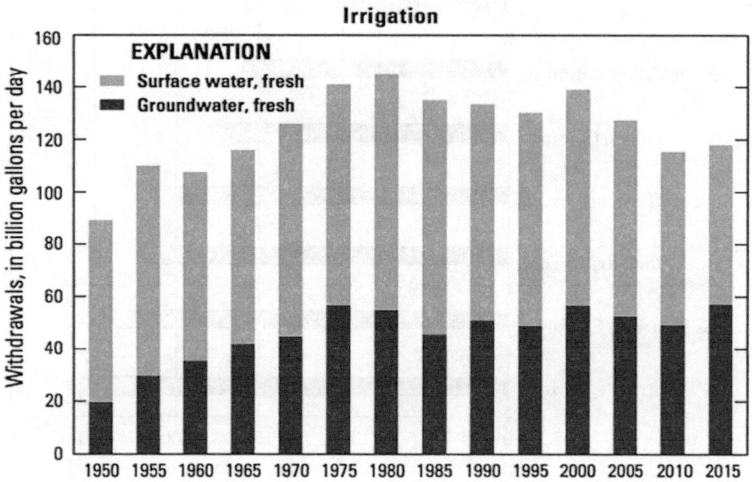

FIGURE 10.2 Irrigation uses, 1950–2015.

Source: USGS (2019).

design also result in outdoor residential water demand that differs significantly from the national average. In fact, actual outdoor water use can range from 1 to 2 GPCD in apartment buildings to more than 200 GPCD in a large single-family home. Figure 10.3 shows a survey of the average indoor and outdoor single-family residential water use in 14 cities in the United States and Canada. As this figure shows, the average daily outdoor water use in the United States varies from less than 25 GPCD in Seattle, Washington, to 190 GPCD in Las Virgenes, California (Mayer et al., 1999). Though these data are not current, they provide a clear picture of ratio of outdoor to indoor water uses in various parts of the United States. Because of conservation measures, per capita and outdoor and indoor water uses have dropped since 1999. According to the Water Research Foundation, per capita water use dropped 15% between 1999 and 2016. During the same period, indoor water use was reduced by 22%.

A January 19, 2017, report in the EPA Water Sense snapshot also includes the current water uses and projected water uses in 2030. Figure 10.4 shows that the projected daily per capita domestic water use will range from less than 50 gal (190 L) to 200 gal (760 L) in the country. The figure also shows the population growth in each state.

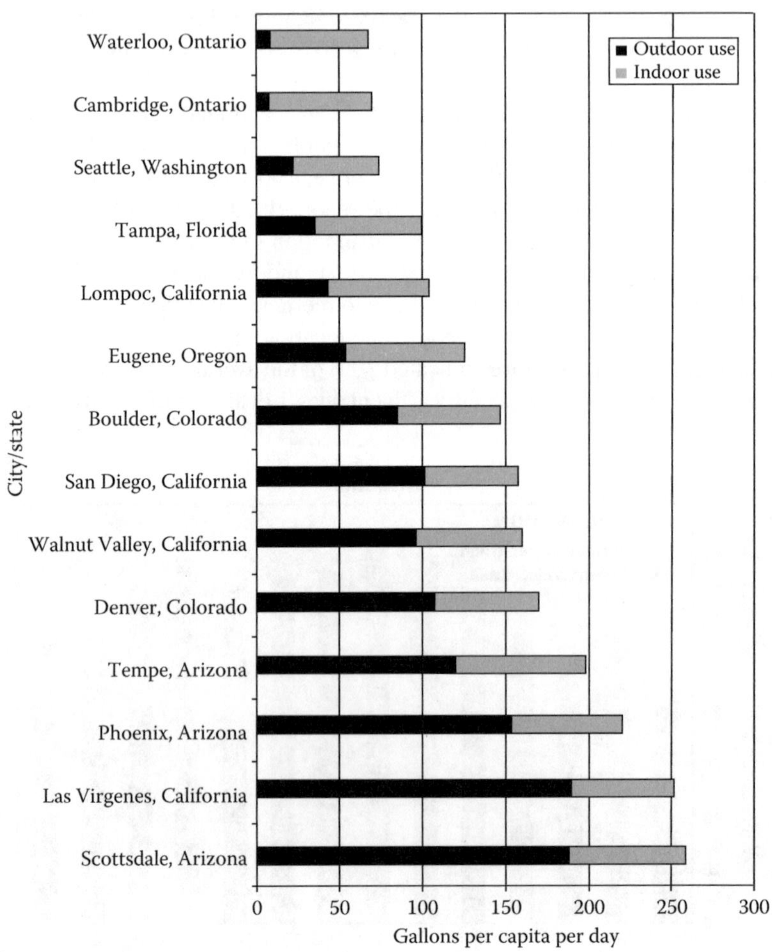

FIGURE 10.3 Average per capita indoor-outdoor water use in single-family homes for certain North American cities.

Source: Mayer et al. (1999.)

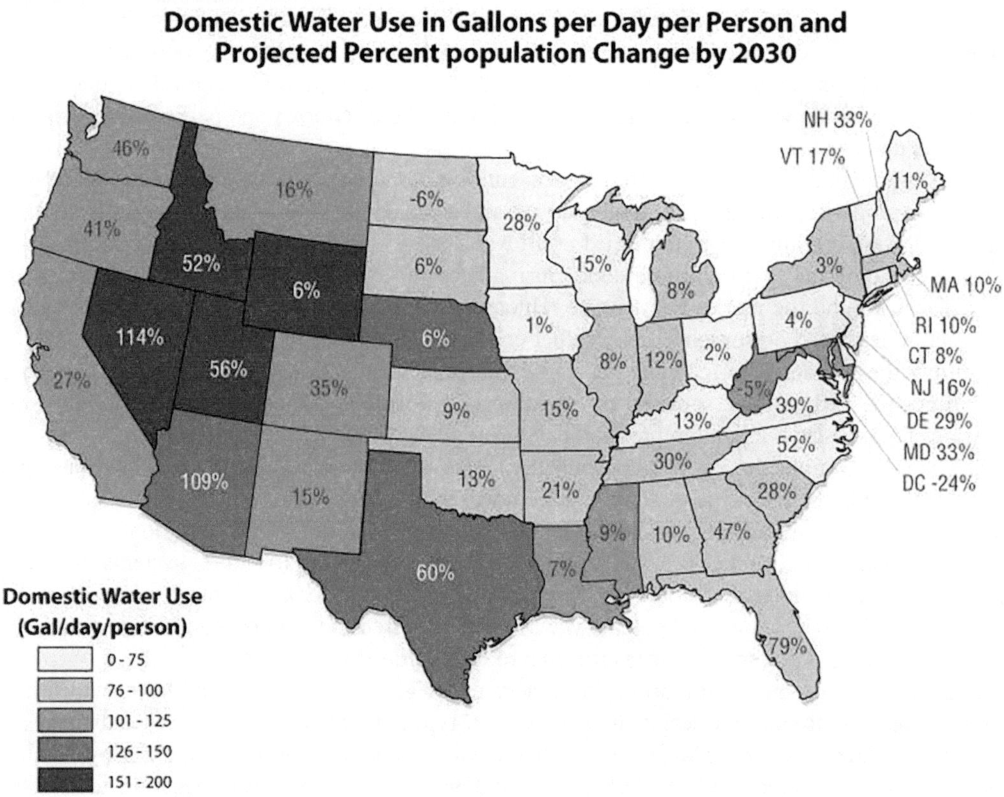

FIGURE 10.4 Domestic water use in gallons per day per person and projected percent change by 2030.

Note: Water data from USGS, Estimated Use of Water in the United States (2005), p. 20, Table 6; population data from US Census Bureau, State Interim Population Projections by Age and Sex: 2004–2030.

As was indicated, the average domestic water use in a household of four is estimated at 320 gallons. However, outdoor water use depends highly on geographic location and climate. During summer months, the daily use in an effluent single- family home can be as high as 1000 to 3000 gal (3800–11,400 L) on a hot day in arid areas. At least 50% of the water used outdoors goes to waste as evaporation, deep percolation, inefficient irrigation systems and methods, and more importantly, over-watering. Overwatering is the largest source of water waste in this country. Most water agencies are aware of this situation, and many are working with municipal officials to offer rebates for efficient irrigation systems such as rain sensors, weather-smart controllers, and drip irrigation. Some communities also offer rebates to property owners for replacing their conventional spray nozzles with low-flow nozzles.

Irrigation technology has changed significantly during the past 30 or so years. New sprinkler nozzles are more efficient than ever before. Installing rain sensors and adjusting the sprinkler timer based on climatic conditions can reduce water use significantly. This author recalls a former neighbor who ran his sprinklers for over one hour five times a week. Lacking any rain sensor, the sprinklers kept running even when it was raining, wasting water and adding to runoff on the street. To conserve outdoor water use, the author has installed a rain sensor and low-flow Hunter sprinkler heads at his current residence, scheduled watering three times weekly, in dry weather conditions and adjusting watering time from 10 to 30 minutes based on outdoor temperature and humidity, and also turning off the timer during rainy days.

To improve the efficiency of outdoor water use, smart box controllers can be installed on sprinkling systems. Controllers can be linked to PCs and smartphones. Smart controllers come in weather-based and soil-based versions. The former creates an irrigation schedule based on evapotranspiration, rainfall, and radiation data taken from local weather stations; the latter controllers operate directly according to soil moisture readings of the sensor buried in the ground. Therefore, the latter can be more accurate than the former. While smart box controllers may be difficult for some homeowners to use, they are well suited for golf courses and parks where irrigation specialists manage lawn watering.

Efficient irrigation is not only suited to residential landscaping but even more suited for public and private properties such as parks, golf courses, and commercial and industrial properties. As was noted, irrigation consumes over one-third of all water used in this country. In fact over two-thirds of water consumption takes place outdoors during summer months in the United States and some other places around the world. Easy-to-use remote-controlled devices can regulate the amount of irrigation water based on seasonal and weather conditions and types of plants. Studies indicate that combining these innovative technologies alone can reduce irrigation water usage by as much as 50%. Since the early 2000s, some states, including California and Florida, have had certification programs for landscape irrigation professionals (Poremba, 2009).

It is worthy to note that more important than innovative technologies to conserve water outdoors is changing the public perception of water availability. Many of us are still under the illusion that existing water supplies are limitless. The best long-term solution to conservation is to educate people, especially at early ages. Apart from using water-conserving irrigation systems, measures that reduce outdoor water use should be sought. Also, recycled water and rainwater from roofs and pavements, rather than municipal potable water, should be used to the maximum extent practicable for irrigation, in addition to other measures that reduce outdoor water uses.

The selection of plants is important in conserving irrigation water. Turf and lawn, which have shallow roots, require more watering than many other types of plants. Native grasses and bushy and deep-rooted plants require far less watering than lawns. Landscape alternatives, including reduced lawn area, applying mulch and weed control around plants, will also reduce irrigation needs. On a steep front yard of his former residence in New Jersey, where the author planted junipers and placed mulch, the plants stabilized the slope and grew naturally. Without any supplemental irrigation other than rain, the plants formed a thick cover.

Using synthetic turf instead of lawn at homes and particularly sports fields, installing rock gardens ornamented with potted plants or partly covered with water tolerant shrubs will save a lot of water. Likewise designating a beautiful and efficient water smart landscape provides a holistic approach to landscaping. Figures 10.5 and 10.6 show parts of the author's front yard, which in addition to the lawn include two 4 m (12 ft) wide planted flowering bushes and small trees adjacent to

FIGURE 10.5 Landscaped areas in the front yard of the author's residence.

Source: Photo taken by the author (May 25, 2024).

FIGURE 10.6 Another view of the landscaped front yard.

Source: Photo taken by the author (May 25, 2024).

the building and three landscaped islands bounded by lawn and sidewalk and covered with large bushes, and flowering trees such as Rhododendrons. The planted areas along the dwelling receive and fully infiltrate runoff from three fourths of the roof area, and the landscape islands retain any runoff from the lawn all of the rainfall they receive. Apart from being aesthetically pleasing, landscaped areas eliminate the use of lawn chemicals and require far less watering than grass.

During the 1990s, a water-wise landscaping called xeriscaping was introduced in this country (Weinstein, 1999). This type of landscaping can reduce the irrigation water needs of traditional landscaping by one-half. Xeriscaping is a water-wise landscaping. Not only does it use significantly less water than grass, it requires less maintenance and is more aesthetically pleasing. Americans are traditionally obsessed with lush green lawns. To gain acceptance, it is important for people to see good examples of xeriscaping. Unless you need to play or walk on it, there is no reason to have a lawn. Xeriscaping does not have to use a plant palette; even a traditional landscape incorporating lawn can be a xeriscape if properly arranged.

Repair, replacement, and retrofitting of outdoor watering systems is another source of conservation. The Denver Zoo, for example, implemented a plan in 1999 to reduce the water usage of the flamingo pond that decreased the water usage from 300 million gal. to 75 million gal in 2005 (Ramos, 2006).

Mandating conservation is another measure for reducing outdoor water consumption. The city of Arroyo Grande in California lowered the water uses from 155 GPCD (587 LPCD) to 115 GPCD (435 LPCD) since 2001 through the prohibition of certain uses. Cleaning driveways, patios, parking lots, sidewalks, and streets, except where necessary for public health and private turf safety, were among them. Other required measures included disallowing irrigation of public and landscaping on Wednesdays, emptying and refilling swimming pools, and use of potable water for dust control and soil compaction (https://www.aroyo-grande.org/553/Mandatory-WaterConservation-Requirement).

A highly water-conserving method of irrigation called jar irrigation was practiced in Persia centuries ago. In this practice, a hollow clay jar was placed near the roots of vegetable and fruit plants, such as melons, and filled with water as needed. As the soil moisture is taken up by plant roots, the soil becomes dry and extracts water from the clay jar through capillary suction. So, little water was lost through evaporation, and none was wasted by runoff. Thus, the Persians could cultivate in arid parts of the country where there was no qanats (ghanats) and the rain was scarce and insufficient to grow vegetables and fresh produce.

10.2.3 Reducing Outdoor Water Use: A Summary

Apart from the above indicated measures, outdoor water use for irrigation can be reduced in a number of ways (Pazwash, 2002):

- Water when needed. Often water is applied when the grass does not need it. Automatic sprinklers without rain sensors are a good example of water waste. They turn on even when it rains or the ground is still wet from a prior rain.
- Water in the early morning or late afternoon to reduce losses. Watering under the hot sun, as many do, results in the loss of over one-third of water to evaporation. Water before 9:00 a.m. in the east and after 6 p.m. in the west, when fungus is not a problem.
- Water at a low rate. Many sprinklers water at a rate beyond which the soil can absorb, and as a result, the water flows onto sidewalks, streets, and storm drains. This is especially the case for steep slopes. To avoid this loss, water should be applied at a slow rate and preferably at several short cycles rather than a single, long burst.
- Let the grass grow longer. Taller grass requires less water than shorter one. By getting a better soil cover, taller grass reduces sunrays and evaporation.
- Use mulch around plants. Mulch traps moisture beneath it, giving plants a steadier, longer water supply.
- Select landscape alternatives using smaller lawn area. Perennials, wildflowers, bushes, and native gasses use a lot less water than water-thirsty lawn such as Kentucky blue grass; better yet, use landscape islands
- Use drought-tolerant, low-maintenance plants in lieu of water-thirsty lawns.
- Avoid automatic timer sprinklers. These are one step backward for conservation. Many homeowners set the timer on a fixed schedule and forget to change the timing with season or to turn them off when it rains or watering is otherwise not needed. Install rain shut-off sensors on automatic sprinklers and adjust timer for seasonal changes in weather.
- Use a broom, rather than a hose, to clean driveways, patios, sidewalks, and loading docks.
- In arid areas, use rock gardens and water-tolerant plants such as cactus in lieu of lawn and landscaping, and stabilize slopes along roadway shoulders with rocks rather than vegetation.
- Harvest rainwater. The roof rain collected in rain tanks and/or barrels provides ample water for sprinkling lawns and irrigating plants and bushes in many parts of this country and many others.

10.3 URBAN RUNOFF VOLUME

In the United States, with an average annual precipitation of 30 in. (760 mm), 1600 ac-ft of water falls on every square mile. Conservatively estimating the average annual runoff at 30% of the rainfall, more than 156 million gallons of stormwater runoff is discharged from every square mile of a typical suburban development. Based on a daily per-capita demand of 100 gallons, this runoff amounts to annual water needs of approximately 4274 persons. For a suburban community of a typical 0.25 acre lot, this runoff is nearly 1.8 times as much as the outdoor water needs of the community. This estimate, as shown in the following calculations, is based on a 35 GPCD outdoor water need; an average of 0.12 acre of streets, sidewalks, and open space per lot; and an average of four households per dwelling.

$$\text{Rainfall volume} = 640 \times 30/12 = 1600 \text{ ac-ft} = 1600 \times 43{,}560 \times 7.48 = 521.3 \times 10^6$$

$$\text{Runoff volume} = 521.3 \times 10^6 \times 0.3 = 156.4 \times 10^6 \text{ gal}$$

$$\text{Number of lots} = 640/(0.25 + 0.12) = 1730$$

$$\text{Annual outdoor water needs per home} = 4 \times 35 \times 365 = 51{,}100 \text{ gal}$$

$$\text{Annual outdoor water demand per mi}^2 = 51{,}100 \times 1730 = 88.4 \times 10^6 \text{ gal}$$

In SI, the runoff volume and the outdoor water needs in every 1 km^2 of a single-family home residential district with an average lot size equal to 1000 m^2 are calculated as follows:

$$\text{Rainfall volume} = 1{,}000{,}000 \text{ m}^2 \times 0.76 \text{ m} = 760{,}000 \text{ m}^3$$

$$\text{Runoff volume} = 760{,}000 \times 0.3 = 228{,}000 \text{ m}^3$$

$$\text{Number of lots} = 1{,}000{,}000 \text{ m}^2/(1000 + 500) = 667$$

$$\text{Daily demand water demand} \approx 130 \text{ L/d per capita}$$

$$\text{Annual outdoor water usage per home} = 4 \times 130 \times 365 = 189{,}800 \text{ L} = 189.8 \text{ m}^3$$

$$\text{Annual outdoor water demand per km}^2 = 189.8 \text{ x } 667 = 126{,}600 \text{ m}^3$$

These calculations represent the average national condition on a conservative basis. In parts of the country where the amount of rainfall is greater than 30 in. the outdoor demand may be met without any or little irrigation., Some other parts, such as southwestern states, receive significantly less rainfall than the national average; have a hot, dry climate; and have a growing season of nearly 12 months a year. In these locations, a combination of using outdoor water conservation programs, drought-tolerant landscaping, and aggressive collection of rainfall will help meet outdoor water needs, in full or in part. Calculations tailored to the local condition will identify water availability and demands. In New Jersey, where the author resides, the calculations are demonstrated as follows.

In New Jersey, where the average annual precipitation is 46.6 in (1180 mm), approximately 243 million gallons of runoff occurs per square mile each year. For the suburban community illustrated earlier, this runoff amounts to over 284,800 gallons per lot. Using the same outdoor demands as before, each dwelling needs 51,100 gallons annually. Thus, the runoff from the said residential area is nearly 2.8 times as much as the outdoor demands of the community. In SI, the average annual runoff and the ratio of annual runoff to the above-calculated annual outdoor demands per square kilometer of the development are calculated as follows:

$$\text{Average annual runoff} = 1{,}000{,}000 \text{ m}^2 \text{ } (1180/1000) \times 0.3 = 354{,}000 \text{ m}^3$$

$$\text{Average annual runoff per lot} = 354{,}100/667 = 530.8 \text{ m}^3$$

$$\text{Annual outdoor demand per square kilometer} = 667 \times 189.8 = 126{,}600 \text{ m}^3$$

$$\text{Annual runoff volume to annual outdoor demands ratio} = 354{,}000/126.600 = 2.8$$

Evidently, the rainfall and outdoor water needs vary with climatic conditions, the type of developments, and the nature of outdoor water demands. Also, the rainfall has temporal variation which inhibits its full utilization. The above calculations, however, indicate that outdoor demands in many parts of the country can be met by using urban runoff alone, if fully captured,

According to the author's records, the annual outdoor water usage in his residence in Paramus, New Jersey, averaged 154,750 L (40,900 gal) from 2019 through 2024. Accounting for an estimated 11,750 L (3,100 gal) of annual water use to maintain the normal water level at the pool, the average annual water consumption by lawn and plants during these six years amounts to 143,000 L

(37,800 gal) or 390 L/day (103 gpd), which is less than the average national daily water use of 140 gallons (530 L) per dwelling. Per unit area of greenery, which is a meaningful measure for outdoor use, 143,000 L reflects 12 mm (0.48 in.) of weekly outdoor water use during the growing season at the author's residence. These figures are calculated based on 450 m^2 (5000 sq ft) of lawn and landscaping that cover about 45% of the author's 1318 m^2 (0.28 acres) and six months of growing season per year. It should be noted that this water use was to compensate for the nonuniform rainfall in New Jersey, which averages 23 inches (580 mm) during the growing season. If the rain had fallen on a regular basis, the lawn would not have required any watering.

10.4 REUSE OF RAINWATER AND STORMWATER RUNOFF

Although conservation can help extend existing water supplies, it will not be enough to cope with population growth. As the water supplies become shorter, the greater will be the need for water reuse. In fact, to ensure sufficient water supplies, water reuse is a must. A cost-effective and environmentally friendly solution is the reuse of stormwater runoff, in general, and rainwater from roofs, in particular. As was indicated in a previous section stormwater is a vast water resource that can be, and should be, captured and reused. It is also more plentiful and far more economical than treating wastewater here in this country and, indeed in many countries around the world.

Wastewater has been reused for nearly fifty years in this country and even longer in Europe, but, the reuse of stormwater runoff has been ignored. One reason for the reuse of wastewater is of course its steady supply. Traditionally, stormwater management practitioners viewed stormwater runoff as waste to be disposed of in a regulated manner. While this view is gradually changing, many practitioners have yet to incorporate measures for retention and reuse in their stormwater management plans. The following case studies exemplify the effective reuse of stormwater runoff.

In the city of Austin, Texas, collecting only 30% of runoff from paved areas provided sufficient water for 330 days of demand in one study (Hall, 2005), and in Los Angeles, six projects captured 1.25 million gallons of water for every inch of rain (1.86×10^6 L/cm). Tucson, Arizona, with 12 in. of annual rainfall, enacted the nation's first municipal rainfall harvesting ordinance for commercial projects, effective June 1, 2010. Tucson officials expected that diverting runoff from parking lots and roofs will supplement current municipal supplies (Cutright, 2009).

Lakeland University has incorporated a recycling plan of runoff to create a decorative pond above an underground detention chamber next to a residential and food services building in Orillia, Ontario, Canada. The underground chamber receives runoff from the roof and also the harvested grey water from an adjacent road. The stored water is pumped up to the pond and is also used for toilet and urinal flushing. The construction began in the summer of 2009, and the building was opened over a year later. (Glist, 2010)

In urban areas, the runoff can be directed to natural depressions or reservoirs after attenuation through stormwater detention basins and ponds. It can also be stored in large underground cisterns or injected into the ground to replenish groundwater supply. The use of large underground cisterns to store surface runoff was a common practice in Persia and many other ancient civilizations. There still are many of these cisterns in villages and rural areas. Some of these underground cisterns are architecturally appealing dome-shaped buildings with open windows on the sides and towers on the roof for natural aeration. Figure 10.7 shows an old underground cistern with four aeration towers in Yazd, an ancient city in the arid central plateau of Iran. The reuse of urban runoff is yet to be exercised nationwide in the United States.

The collection of rainfall or runoff not only provides a vast supply of water for outdoor and some indoor uses but it also mitigates adverse impacts of urbanization, such as flooding, erosion, and pollution problems. The reuse of stormwater runoff in general and roof rain in particular was presented in a paper by Pazwash (1994). Subsequent papers by the author (Pazwash and Boswell, 1997, 1999, 2002) discussed the quantity of roof rain (rainwater) and stormwater runoff and presented suggestions for their collection and reuse. Information about water reclamation and reuse

FIGURE 10.7 An underground cistern with aeration towers in Yazd, Iran.

can also be found Chapter 21 in Larry Mays' Water Resources Handbook (1996). Also, a book by Amy Vickers (2001) includes ample information about indoor and outdoor conservation measures and reuse. For more information on water reuse, one is referred to recent publications by EPA including: *Rainwater Harvesting: Conservation, Credit, Codes, and Cost Literature Review and Case Studies*, dated November 16, 2015 and last updated November 29, 2023 (https://www.epa.gov/npdes/rainwater-harvesting-conservation-credit-codes-and-cost-literature-review-and-case-studies), "Green Infrastructure," issued October 4, 2024 (https://www.epa.gov/green-infrastructure), and *Green Street Handbook*, dated January 5, 2024.

Regulations for rainwater treatment and reuse are the responsibility of the states. While standards for wastewater reuse were established decades ago, many states have no regulations for the reuse of rainwater. The EPA guidelines for water reuse were first published as a research report in 1980 and were updated in 1998 and later in 2004. Recognizing the need for national guidance on water reuse planning and regulation, the EPA developed a comprehensive document on water guidelines for water reuse in September 2012 that was last updated on February 24, 2025. This document, in 643 pages, is in pdf and free of charge to be downloaded.

EPA also has prepared a report titled "Water Resue and Recycling" that was last updated July 8, 2025. This report includes water reuse, case studies, description of water reclamation of various sources, stories about sustainability in manufacturing and utilities, water reuse resources and national water reuse action plan.

10.5 RAINWATER HARVESTING

With an average annual precipitation of approximately 30 in. (760 mm), 18,700 gal of waterfalls on every 1000 ft^2 (76,000 L/100 m^2) of roof area in the United States. Collecting this rain can provide sufficient water for irrigating lawns and landscapes in many single-family homes with small and mid-size lots. Rainwater harvesting has been practiced for hundreds of years by ancient civilizations. As an organized industry, rainwater harvesting is still in its infancy, although the EPA has published the documents mentioned in the previous section. The EPA has also published "Green Infrastructure Municipal Handbook" (2017), "Soak up the rain; rain gardens" (February 2025), "Rainwater harvesting, conservation, credit, codes and cost, literature review and case studies", November 8, 2024 and Start Saving/USEPA, May 28,2025.

About two decades ago, a nonprofit organization for the rainwater industry called the American Rainwater Catchment System Association (ARCSA) was founded. Now many states and municipalities have begun adopting laws for rainwater reuse. Two states, Georgia and Texas, were pioneers in rainwater harvesting possibilities. About two decades ago, Georgia published a set of rainwater harvesting guidelines, and Texas Water Development Board published the Texas Manual on Rainwater Harvesting (2005). The following is an example of early rainwater harvesting in the United States.

Lady Bird Johnson Wildflower Center in San Antonio, Texas, appears to be the first public building where the reuse of roof rain was practiced. In this center, the rainwater is collected off roofs and terraces and is routed through a roof washer consisting of two chambers, a nonmechanical filtering system that removes leaves and pollen and then a series of surface cisterns that drain to an underground cistern. The water then goes through a filter before it is pumped up to a tank where it is stored for irrigation (O'Mally, 2007). According to the center's data, the collected rain from the 17,000 ft^2 (1580 m^2) roof amounts to 10,200 gallons (38,600 L) per inch of rainfall. Given an average annual rainfall of 30 in. (750 mm), the system can collect over 300,000 gal (1135 m^3) of rainwater annually.

Depending on its intended use, treatment requirements of rainwater vary from none at all to sophisticated systems producing drinking water. Rainwater harvesting is most suited for roof runoff, which is generally much cleaner than runoff from ground surfaces. In urban areas, and in particular cities, roof areas form a large portion of the overall impervious surfaces. If the runoff from roofs is to be stored in tanks or cisterns it should be prefiltered. Without filtration, leaves and other organic debris decompose in the rain tank (or cistern) and create ammonic conditions, which support extensive bacteria growth. The silt also builds up at the bottom and the required regular maintenance to clean the tank may become a burden.

10.5.1 Rain Barrels and Tanks

Roof rain can be stored in water tanks that can be placed above ground or underground. Water tanks come in vertical, horizontal, and cone-shaped bottoms placed on a platform. Tanks are available in plastic, HDPE, fiberglass, and stainless steel. Among these, plastic and HDPE tanks are the most economical for single-family homes. In addition to economy, these tanks have the following advantages over stainless steel tanks:

- Seamless construction, allowing easy cleaning and leak-free service
- Lightweight, less than one-half the weight of steel tanks
- Dent-free and virtually maintenance-free

Snyder, Sintex Royal, and Norwesco are among the companies that manufacture polyethylene water tanks. Texas Metal Tank is one of the companies that make galvanized and stainless-steel water tanks (www.texasmetaltanks.com). Rain tanks come in sizes ranging from 100 to over 15,000 gallons (380–56,000 L). For a mid-sized single-family home of 1350 ft^2 (125 m^2) roof area, four 750 L (200 gal) tanks retain 24 mm of rain. Snyder's 200-gallon (760 L) vertical plastic tank is a perfect choice for this amount of retainage. This tank is 40 in. (1 m) in diameter, 42 in. (105 cm) high; weighs only 33 lbs. (15 kg); comes with an 8 in (20 cm) lid, a 4 in. (10 cm) overflow, and a hose adapter; and costs roughly $400.

Some homeowners use 50 or 60 gal (190–230 L) rain barrels, which are inadequate. Specifically, two 50 gallons rain barrels can only retain 3.8 mm (0.16 in.) rain from a 100 m^2 (1075 ft^2) roof area. For a single-family dwelling with 1000 ft^2/100 m^2 of roof area, the size of a water tank to retain 1 in. (25 mm) of rain is calculated as follows:

$$1000 \times (1\text{ in.}/12) \times 7.48\text{ gal/ft}^3 = 623\text{ gal}$$
$$100 \times 25\text{ mm}/1000 = 2.5\text{ m}^3 = 2500\text{ L}$$

Of course, sizing a tank for a larger storm will result in more rainwater harvested. The size of the tank should be selected based on local rainfall pattern. This author suggests that the tanks are sized preferably based on 90th percentile rainfall. This rainfall represents the rainfall depth that is greater than or equal to than 90% of annual rainfalls. In the US states from Maine to Virginia, other than New Jersey, this rainfall ranges from 0.9 to 1.0 in. (22–25 mm) and is 1.25 in. in New Jersey. Thus, to capture most rainfalls, the rain tanks/barrels should be sized for 1.0 in. (25 mm) of rainfall in northeastern states. However, considering space limitation, ease of maintenance, and efficient use of the tanks' capacity, they maybe sized for smaller rainfalls This author suggests sizing rain barrels for 15 mm (0.6 in.) in Northeastern states. The same-sized tanks may be used in Midwestern states that have somewhat similar precipitation patterns to the Northeast. Likewise for the Pacific Northwest states that get 35 to over 80 in. of precipitation annually. The tanks in southeast/Guls states with an annual rainfall of 40 to 80 in. may be sized for 20 mm (0.8 in.). In arid southwestern states including Arizona, Nevada, California and New Mexico which receive 10 to 15 inches (250–390 mm) of flash rains annually, the tanks may be sized for 10 mm (0.4 in.) rain. Table 10.2 shows suggested rain tank sizes and estimated amount of annual harvested water from 100 m^2 and 1000 ft^2 of roof area in various regions of the US. The author also suggests that rain barrels are sizes for 10 mm (0.4 in.) rains. Thus, the barrels storage capacity would be 1.5 times smaller than the tank sizes shown on Table 10.2. Using rain tanks or rain barrels for a smaller or larger rain depth, the listed size of rain tanks on Table 10.2 can be proportionately resized. However, the volume of captured water on yearly basis will not be proportionately reduced.

The following are early examples of harvesting and reuse of roof runoff in the United States.

The School of Global Sustainability for Global Solutions of the University of South Florida (USF) installed a 30,000 gal. (113 m^3) fiberglass tank to collect the rainwater from the building roof. The water is reused for urinals and toilets in this four-story 74,788 ft^2 (6,950 m^2) building that was constructed in September 2010. This amounts to approximately 400 gal. per 1000 ft^2 of roof area. The university expects to harvest 506,000 gal (1.9×10^6 L) of rainwater, which is far more than the required 207,000 gal (780,000 L) per year to operate toilets and urinals. The rainwater, together with the condensation water collected from the building air conditioning system, is passed through a 200 μm vortex filter before it enters the cistern. After filtration, the water goes through ultraviolet (UV) treatment to eliminate bacteria. This cistern, which is sized to contain rainwater from a potential downpour, has eliminated the need for municipal water since it began operation (Cline, 2011).

A recycling rain water reuse program has been employed in Clark, New Jersey. There, rainwater from the public works building is collected in a 5,000 gallons (18,900 L) tank. With the aid of a booster pump, harvested water is used to wash the public works' cars and trucks on a concrete pad. The runoff from the concrete wash pad enters a rain garden located in the adjacent high school property to filter pollutants in runoff. Figure 10.8 depicts this rain tank.

Since roof rain is somewhat pure, it can be used for all outdoor needs without any treatment. It can also be used for toilet flushing indoors with minimal purification. The plumbing line for toilets can be simply fed by a pump placed in an underground or aboveground tank. By placing the main rain tank on high ground or installing it on a platform, the water can be supplied for all outdoor

TABLE 10.2
Suggested Size of Rain Tanks and Estimated Amount of Water Saving per Dwelling during the Growing Season per 100 m²/1000 ft² of Building Roofs

	Tank Size		Annual Water Saving	
Region	**L**	**gal**	**L**	**gal**
New Jersey	1500	375	60,000	15,000
Northeast	1500	375	60,000	15,000
Central Plains	1500	375	60,000	15,000
Pacific Northwest	1500	375	60,000	15,000
Southeast/Gulf states	2000	500	80.000	20.000
Arid Southwest	1000	250	10,000	2,500

FIGURE 10.8 Rain tank at the public works building in Clark, New Jersey.

Source: Photo by the author, 2014.

needs without a need for any pumping. In such applications, the tank may be drained at the end of the growing season to remove any sediment, and its drain valve is left open during the cold season.

The use of roof rain for toilet flushing reduces one-fourth of residential indoor water needs. The roof rain can save a lot more water in commercial and office buildings, where it can be used not only for toilet flushing but also in urinals and for washing driveways, and garage floors, if any.

In addition to its conservation benefits, harvesting rainwater from roofs will reduce the flow in municipal drainage systems and may eliminate or delay the need for upgrading stormwater management facilities such as detention basins and ponds. In a northern New Jersey town, the author suggested that the municipality consider offering rain barrels to homeowners, free of charge, to mitigate street flooding. Because of a general inadequacy of combined sewer system in that town, overflow was occurring after every heavy storm. The suggested solution would not only be less costly than upgrading the municipal sewer system but would also reduce the load on the treatment plant which in turn would save the overall operation and treatment costs.

Municipalities can take advantage of roof rain, which is a readily available and relatively pure, either directly by the collection of the roof rain from school, municipal and public buildings or indirectly by offering rebates or tax reduction incentives to homeowners. They can also require commercial and large residential developments to install a separate drainage line fed by roof rain (or surface runoff) for irrigating lawn and landscape areas and supplying other outdoor uses. In multi-family residential buildings and commercial and industrial developments where the amount of roof rain may far exceed outdoor demands, the collected water can be used for non-potable indoor demands. In such developments, stormwater management measures may also include groundwater recharge through underground retention-infiltration basins.

10.5.2 Maintenance of Rain Barrels/Tanks

Rain barrels receive pollutants from the roof during the first flush, and grit from roof shingles during heavy rainfall. Roofs containing asbestos are not suitable for harvesting drinking water, and neither are gutters containing lead solder or lead-based paint. In January 2010, the Texas Water Department Board indicated that the rainwater from various types of roofs did contain some contaminants above the EPA drinking water standards. This indicates that harvested rainwater requires treatment for potable use. However, it does not need any treatment for sprinkling outdoor and some indoor uses, such as toilet flushing.

Standing water in rain barrels creates algae and is also a food source for mosquitoes. Excessive algae growth can occur rapidly in light-color or translucent rain barrels. Although many species of algae are not harmful themselves, algae blooms can deplete oxygen in water and release toxins that can be harmful to animals.

The EPA recommendations for algae control in rain barrels are as follows:

- Keep leaves out by regularly cleaning gutters and properly filtering the water from downspouts.
- Avoid barrels with open or screen tops that allow direct light inside; likewise, avoid light-colored barrels.
- Place barrels in the shade; avoid full sun exposure.

Special consideration should be given to mosquito control in both aboveground and underground rainwater collection systems. If rainwater tanks and cisterns are improperly maintained, they can serve as breeding habitats for mosquitoes. Mosquito bites, apart from intense itching, can spread disease organisms. The most important mosquito-borne diseases in the United States are caused by viral pathogens, which include West Nile fever, St. Louis Encephalitis, eastern or western equine encephalomyelitis viruses, and Lacrosse encephalitis.

10.6 SUGGESTED ACTIONS FOR WIDESPREAD CONSERVATION AND REUSE

As individuals, we should be concerned about our limited water supplies and take measures to collect stormwater for reuse. To achieve long-term water sustainability, local and state agencies and schools need to adopt challenging actions to lead the public in promoting conservation and reuse of runoff in general, and roof rain in particular. Suggested actions include the following:

- Public education, especially at an early age
- Task force formation in schools and on campuses
- Reaching out to municipal agencies and officials
- Incentives for installing rain tanks and barrels
- Block programs to inform neighbors about water saving
- Enforcement of water savings measures in public places such as malls and universities
- Pilot projects demonstrating water harvesting measures

10.6.1 Benefits of Water Conservation and Reuse

EPA's Office of Water funded a study by the National Academy of Sciences (NAS) titled *Assessment of Water Reuse as an Approach for Meeting Future Water Supply Needs.* The study was intended mainly to revise the EPA's 2012 water reuse guidelines based on a comparison of performance, costs, energy requirements, and greenhouse gas releases for water reclamation/reuse versus desalination, long-distance transport of water supplies, or pumping from deep aquifers. The report also included case examples of using reclaimed municipal effluent as an alternative water supply by various industries.

As far as water reuse is concerned, this publication mostly follows the general trend of water reuse, which is a reclamation of wastewater. It should be noted that rainfall and runoff are far more plentiful than wastewater and require less treatment. Therefore, water reuse programs should logically pay more attention to collecting rainwater and stormwater runoff in the future.

Rooftops and driveways commonly account for 50% to 75% of total impervious area in an urban setting. Thus, retaining the rainwater from roofs alone in rain barrels, tanks, or cisterns will provide a vast source of water supply. In addition, directing runoff from driveways to lawn and landscape areas serves as an effective measure both to conserve fresh water supplies and to reduce urban stormwater runoff. These measures also result in decentralized BMPs, offering viable source control solutions to stormwater management.

Water reuse reduces both the average daily demand and the peak daily demand. Reductions in average daily demand affect the quantity of water that must be treated and the size of facilities to transport and store it. Reductions in peak daily demand impact the sizing of new water treatment plants or the expansions of existing plants and treated water storage tanks. The size of pump stations and water distribution systems is also reduced by water reuse.

Other advantages of water conservation are

- Reducing incidents of overflow from combined sewer systems
- Reducing freshwater withdrawal from lakes, rivers, and aquifers
- Lessening load to septic tanks
- Lowering energy costs
- Reducing the use of chemicals for treatment

A comprehensive discussion of the benefits of water conservation to the local community and the environment is included in AWWA Manual M52, *Water Conservation Program—A Planning Manual* (2017).

PROBLEMS

10.1 In a single-family home of four households, the old fixtures are replaced with new water-saver fixtures. The flow ratings of the old and new fixtures are

Faucets:	2.5 GPM, old;	1.28 GPM, new
Showerheads:	2.8 GPM, old;	1.8 GPM, new
Toilet tanks:	3.0 gpf, old;	1.28. gpf, new

Calculate the annual water saving in this home. The per-capita daily uses are estimated as follows:

Faucets: 5 minutes
Shower: 5 minutes
Toilet flushes: 5 times

10.2 In Problem 10.1, calculate the annual savings in water costs assuming a \$8.50 charge per CCF (100 ft^3) of municipal water.

10.3 Solve Problem 10.1 for the following data:

Old faucets: 9 LPM	Showerheads: 10 LPM	Flusher:11.0 Lpf
New faucets: 5 L	Showerheads: 7 LPM	Flusher: 4.8 Lpf

10.4 In Problem 10.3, calculate the annual savings in water costs, with the municipal water rate at \$3.0/m^3.

10.5 The sprinkler system for a single-family home consists of three zones of 12 heads in total, each rated at 2.5 GPM. The system runs three times a week for 25 minutes each zone on the average. Calculate the total amount of water used during a 6-month growing season.
If this system is replaced by a self-adjusting smart box controller that reduces the irrigation water use by 50%, what will be the amount of water saving?

10.6 Solve Problem 10.5 for sprinkler heads rated at 9 L/min. Calculate the annual cost savings for a water rate of \$3.0/m^3.

10.7 Calculate the annual amount of precipitation in ft^3 and gallons that falls on the roof of a 40,000 ft^2 commercial building in New Jersey, where the average annual precipitation is 46.6 in.

10.8 Calculate the annual amount of precipitation in m^3 and liters that falls on a 4000 m^2 roof of a commercial building in your area.

10.9 A rain tank is to collect the runoff from a 1750 ft^2 roof of a residential dwelling. Size the tank to fully retain a 0.6 in. rainfall.

10.10 An underground cistern is to be used to retain the rain from a 5000 m^2 roof of a commercial building for 20 mm rainfall. Size the cistern using:
   a. cylindrical tank
   b. cubical tank

10.11 The cistern in Problem 10.10 is installed at a location where the average annual rainfall is 1200 mm, 90% of which is equal to or less than 25 mm. Calculate the amount of rainwater harvesting by the cistern if the cistern is designed for the 90 percentile rainfall.

REFERENCES

AWWA, 2017, *Water conservation programs, a planning manual, manual 52*, 2nd ed., AWWA, 22 pp.
AWWA, July 12, 2023, Water/wastewater rate information.
AWWA, July 20, 2024, *State of water industry report.*
Brzozowski, C., March, 2009, EPA promotional partner of the year, Seattle, *Water Efficiency*, pp. 28–31.
Cline, K., July/August 2011, On campus reuse, *Stormwater Solutions*, pp. 10–11.
Cutright, E., July 2009, A first for rainwater harvesting, editorial, *Water Efficiency.*
EPA (Environmental Protection Agency), September 2012, *Guidelines for water reuse*, EPA/600/R-12/618, 643 pp., last updated February 24, 2025.
EPA (Environmental Protection Agency), January 19, 2017, *EPA water sense, Indoor water use in the United States*, https://19january2017snapshot.epa.gov.
EPA (Environmental Protection Agency), November 8, 2024, Rainwater harvesting: conservation, credit, codes, January 16 pp.; https://epa.gov > npd
EPA (Environmental Protection Agency), January 19,2017 *Outdoor water use in the United States.*
EPA (Environmental Protection Agency), June 3, 2024, *Residential toilets*, https://epa.gov>wat.
EPA (Environmental Protection Agency), July 8, 2025, *Water reuse and recycling*, https://epa.gov/waterreuse.
EPA, May 28, 2025, Start Saving.
EPA (Environmental Protection Agency), February 3, 2025, Soak up the rain; rain gardens https://www.epa.gov.
Glist, D., February 2010, Pond on a pond water harvesting system meets aesthetic functional objectives, *Urban Water Magazine*, pp. 8–9.
Hall, P.C., May/June 2005, Municipal use of stormwater runoff, *Stormwater Magazine*, pp. 74–89.
Mayer, P.W. et al., 1999, *Residential end uses of water*, AWWA Research Foundation and AWWA, Denver.
Mays, L.W., editor in chief, 1996, *Water resources handbook, chapter 21: Water reclamation and reuse*, McGraw-Hill, New York.
Obmascik, M., 1993, *A consumer guide to water conservation*, 1st ed., AWWA, Denver, CO, 63 pp.
O'Mally, P.G., March/April 2007, Stormwater harvesting: A project summary, *Water Efficiency*, pp. 51–56.
Pazwash, H., 1994, *Conservation measures in urban stormwater management*, Proceedings of the 21st Annual Conference of the ASCE Water Resources Planning and Management, Denver, CO, May 23–26, pp. 408–411.
Pazwash, H. and Boswell, S.T., April 6–9, 1997, *Management of roof runoff, conservation and reuse*, Proceedings of the ASCE 24th Annual Water Resources Planning and Management Conference, Houston, TX, pp. 784–789.
Pazwash, H. and Boswell, S.T., August 8–12, 1999, *Conservation of water, reuse of roof runoff*, Proceedings of 1999 International Water Resources Engineering Conference, Seattle, WA.
Pazwash, H. and Boswell, S.T., January 27–30, 2002, *Water conservation and reuse; an overview*, Proceedings of AWWA Water Sources Conference, Las Vegas, NV.
Poremba, S.M., July/August, 2009, Target irrigation water waste, *Water Efficiency*, p. 23.
Ramos, A.R, November/December 2006, Wild about water conservation, *Water Efficiency*, SEE pp. 46–49.

Sloan Valve Company, headquarters: Franklin Park, IL 60131, Ph. 1 (847) 671-4300, https://www.sloanvalve.com.

Solley, W.B., Pierce, R.R. and Perlman, H.A., 1998, *Estimated use of water in the United States in 1995, US Geological Survey Circular 1200*, US Department of the Interior, USGS, Reston, VA.

Texas Water Development Board, 2005, *The Texas manual on rainwater harvesting*, 3rd ed., Texas Water Development Board.

US Census Bureau Population Clock, https://www.censor.gov/popclock/.

USGS (US Geological Survey), 2018, *Estimated use of water in the United States in 2015, circular 1441*, https://pubs.usgs series no.1441.

USGS (US Geological Survey), June 18, 2018, *Trends in water use in the United States 1950–2015*, https://www.usgs.gov>science.

USGS (US Geological Survey), March 1, 2019, *Domestic water by water resources mission area*, www.usgs.gov/mission-areas/water-resources/science/domestic-water-use-overview.

Vickers, A., 2001, *Handbook of water use and conservation*, 1st ed., Water Plow Press, Amherst, MA.

Water Research Foundation, April 1, 2016, Residential End Uses of Water, Version 2, Executive Report, 16 pp., Residential End Uses of Water, Version 3, June 18, 2024.

Weinstein, G., 1999, *Xeriscape handbook*, Fulcrum Publishing, Golden, CO.

Glossary

Abstractions: Portions of the total rainfall that do not produce direct runoff. These include interception by vegetation, storage in depressions, and infiltration.
Adverse slope: Upward slope in the downstream direction in a channel or pipe.
Algae: Aquatic organisms containing chlorophyll that grow in colonies and produce mats.
Angle of repose: Angle of slope formed by granular material under the critical equilibrium condition of incipient motion.
Antecedent moisture: Water stored in the soil before the start of rainfall.
Area-elevation curve: Curve relating surface area to elevation.
Arid: Describes a climate with less than 10 in. (250 mm) of annual precipitation.
Artificial recharge: The intentional addition of water to an aquifer through injection or infiltration.
Atmospheric pressure: Force of air on the unit area of a surface.
Auger hole test: Field test for measuring hydraulic conductivity.
Backwater curve: Plot of water depth along channel.
Bacteria: Single-cell microorganisms that reproduce by fission or by spores.
Bankfull flow: Flow conditions that fill a stream to the top of its banks.
Base flow: Flow in a channel or stream due to soil moisture or groundwater.
Best management practice: A structural or nonstructural measure designed to temporarily store and treat stormwater runoff in order to avoid flooding, reduce pollution, and provide other amenities.
Biological oxygen demand: The oxygen taken by bacteria to oxidize soluble organic matter.
Bluegrass: Kentucky bluegrass, a variety of cool-season turf grass.
Buffer: Zone of vegetated ground along both sides of natural streams or wetlands to safeguard their water quality.
Bulk density: Ratio of the oven-dried mass of a sample to its original volume.
Bypass flow: Flow that bypasses an inlet on grade and is carried in the gutter to the next inlet.
Capillary rise: Height to which water rises under capillary action.
Capillary suction: Negative water pressure in soil above the water table.
Capillary water: Water in soil above the water table due to capillary forces.
Catch basin insert: Filters inserted in inlets to trap floatables, coarse and suspended solids. Some inserts also capture oil and grease.
Check dam: A small dam (berm) constructed in a gully or swale to reduce flow velocity.
Cistern: A tank or chamber to store rain or stormwater.
Clustering: A development design technique that concentrates buildings on a part of the site to allow the remaining land to be used for agriculture, recreation, and/or common open space and to preserve environmentally sensitive features.
Combination inlet: An inlet with both a curb opening and a grate.
Combined sewer: A pipe that carries both sanitary sewage and stormwater runoff.
Compaction: The closing of pore spaces among the particles of soil and rock, generally through heavy equipment during construction.
Concentration: The amount of a substance in a unit volume of solution.
Conduit: Any channel intended for conveyance of water.
Confluence: The location where two streams merge.
Constant head permeameter: Laboratory device for measuring hydraulic conductivity under a constant head.
Constructed wetland: A manmade freshwater wetland designed and constructed to serve a specific purpose.

Continuity equation: Equation based on conservation of mass; the product of cross-sectional area by flow velocity.
Contour (elevation): Line connecting points on a surface having the same elevation.
Control section: Cross-section of a channel that has a controlling structure (bridge, free outfall, or weir).
Conveyance: Measure of amount of flow carried in a channel, defined from Manning's formula, related to n, A, and P.
Correlation coefficient: An index that represents the combined effects of soil characteristics, the land cover, the hydrologic condition, and antecedent soil moisture conditions.
County Soil Map: A book prepared by the Natural Conservation Resources Service (NCRS) that presents maps and soil characteristics of a county.
Crest: Peak of a hydrograph; top of dam/spillway.
Critical depth: Depth of water at which specific energy is minimum.
Critical flow: Flow in an open channel at a minimum specific energy; Froude number = 1.0.
Critical slope: The channel slope at which uniform flow is critical.
Critical velocity: The velocity where stream flow passes from supercritical to subcritical conditions or vice versa.
Cross slope: The rate of change of roadway elevation with distance transverse to the direction of travel.
Crown: The inside top elevation of a conduit.
Culvert: A conduit used primarily to convey flow under highways and railroad embankments.
Cumulative mass curve: Graph of accumulated rainfall versus time.
Curb-opening inlet: An inlet with opening in the curb.
Curve number: An index related to soil characteristics, land cover, and antecedent soil moisture conditions in the US Soil Conservation Service (SCS) method.
Customary units (CU): Foot-pound system of units, also referred to as English units.
Dam: A barrier that impounds water for specific purposes.
Datum: Topographic reference elevation based on a benchmark (NGVD, 1929; NAVD, 1988; or arbitrary).
Dead storage: Storage in a reservoir or detention basin below the elevation of the principal outlet.
Density: Mass per unit volume of matter.
Depression storage: A depressed area that stores precipitation or runoff.
Depth of flow: Vertical distance from the bottom of a channel to the water surface.
Design discharge: Discharge associated with a selected return period.
Design storm: A hypothetical storm used in design.
Detention basin: A dry basin or other structure that detains runoff during a storm and releases it over time through an outlet structure.
Detention pond: A wet basin that temporarily stores stormwater and releases the water at a controlled rate of flow.
Detention time: The time that it takes for design flow to be discharged from a detention facility.
Deterministic: Describes a hydrologic model based on physical relations.
Development: Any use or change in the use of land due to construction of a structure or mining, excavation, landfill, or deposition, not including redevelopment.
Development density: The number of families, individuals, dwelling units, or households per unit area of land.
Dew point temperature: Temperature at which the air becomes saturated with water.
Dimensionless hydrograph: A hydrograph presented in terms of ratios of flow and time; commonly, the ordinates are the ratio of the discharge to the peak discharge, and the abscissae the ratio of time to the time to peak.
Direct runoff: The total runoff less losses.
Direct runoff hydrograph: Graph of direct runoff (rainfallÐ losses) versus time.

Directly connected impervious area: Impervious area that drains directly into a drainage system.

Disinfection: The process of removing or inactivating pathogens.

Diversion: Redirecting flow in a pipe or channel. Structure that redirects flow to another area. A channel or conduit to bypass flows above a certain elevation.

Domestic water use: The total indoor and outdoor water use on residential or industrial properties.

Drainage area: Area formed by topography that drains to a given point.

Drainage density: The concentration of streams or drainage ditches in a watershed, measured as the ratio of the total length of streams to the drainage area.

Drainage divide: Topographic boundary that directs rainfall into different basins.

Drip irrigation: An irrigation system that delivers water to plants at low pressure and in small drips through perforated plastic tubes.

Drought: An extended period of no or little rain.

Effective flow area: Portion of a cross-sectional area where water flows.

Effective precipitation: Total amount of rainfall minus evaporation and infiltration during rainfall period.

Effluent: Wastewater leaving the treatment plant.

Embankment: A manmade deposit of soil, rock, or other materials.

Emergency spillway: A structure that controls release of storm flows in excess of the design discharge from a detention basin or a reservoir.

Energy grade line: A line representing the sum of the pressure, velocity, and elevation heads.

Erosion: The detachment, wearing away, or movement of soil or rock fragments by the action of water, wind, ice, or gravity. Erosion can be sheet, rill, or gully.

Erosion control blanket (ECB): A degradable natural or synthetic material, fabricated into rolls. ECBs are used to reduce soil erosion and help growth and establishment of vegetation.

Erosive velocity: Velocity of water that is high enough to erode land surface.

Evaporation: Transformation of water from liquid to vapor. Loss of water from a surface due to vaporization.

Evapotranspiration: Water loss due to evaporation from soil and transpiration from plants.

Event: A single storm simulation.

Excavation: Any act by which soil or rock is dug, quarried, removed, or relocated.

Extended detention: Detention basins that hold and slowly release the stormwater runoff following a storm event.

Falling head permeameter: Laboratory device with a tube of dropping water for measuring hydraulic conductivity.

Faucet aerator: A screen-like device with a tube of dropping water that is enclosed in a faucet to reduce flow rate.

Field capacity: Amount of water in soil after gravitational water is drained.

Fill: A manmade deposit of soil, rock, or other materials.

Filter blanket: One or more layers of graded noncohesive material placed below riprap to prevent soil piping and allow natural seepage.

Fiter fabric: Permeable textile placed below riprap to prevent piping and to permit natural seepage

Filter media: The sand, soil, or other material in a filtration device.

Filter strip: A strip of permanent vegetation to retard the flow of runoff.

Filtration: The process of screening suspended solids from stormwater (also wastewater) through a porous medium.

Flood hazard zone: Area that will flood with a given probability. Such zones are commonly shown on FEMA or state flood maps.

Flood plain: Lands along a stream that would be inundated by a flood event.

Flow path: Common path of a fluid particle.

Force main: A pressurized conduit.

Freeboard: Vertical distance from the design water surface elevation to the top of the channel.

Frequency: Occurrence interval of a random variable.

Freshwater wetlands: An area that is inundated or saturated by surface water or groundwater.

Friction slope: Energy losses due to friction per unit of length. Equal to the total energy slope in uniform open channels.

Froude number: The ratio of inertia force to gravity force. Dimensionless parameter to characterize open channel flow regime.

Gabion: Rectangular wire mesh containers filled with stone.

Gabion mattress: A thin gabion, usually 6 to 9 in. thick, used to line channels.

Geographic information system (GIS): Computer application that displays numerous types of spatial data (such as land use, soil type, or topography) and links those data with a map.

Grading: Any stripping, cutting, filling, or any combination thereof.

Gradually varied flow: Open channel flow that changes gradually so that one-dimensional analysis can be applied in each reach.

Grate inlets: Inlets with transverse bars arranged to form an inlet structure.

Gravel: Aggregate of 1/4 in. to 3 in. stone mix.

Gravitational water: Water that will drain through the soil under the force of gravity.

Gray water/Grey water: Water from showers and sinks (including kitchen sink; British and Australian practice).

Green development: A development that integrates ecology and real estate.

Green infrastructure : A natural or semi-natural surface that reduces the runoff or improve water quality.

Green industry: Trades or stakeholders associated with landscape or irrigation or nonconventional management of runoff.

Groundwater: Water beneath the earth's surface.

Groundwater recharge: Replenishment of groundwater source, either naturally by percolation or artificially by injection.

Gully: A channel cut by concentrated runoff.

Head: The height of water above any reference plane. In hydraulics, head may be potential (pressure and elevation) and kinetic (velocity).

Hotspot: Area where land use or activities produce highly contaminated runoff.

Hydraulic conductivity: Ratio of velocity to hydraulic gradient, indicating permeability of porous media.

Hydraulic grade line: A line showing the sum of the pressure and elevation heads.

Hydraulic jump: Sudden transition from supercritical flow to subcritical flow.

Hydraulic radius: The cross-sectional area of flow divided by wetted perimeter.

Hydraulically connected impervious area: Impervious area that drains directly into a drainage system.

Hydrograph: A graph showing variation of discharge with time.

Hydrologic cycle: A representation of the physical processes that control the distribution and movement of water.

Hydrologic routing: Computation relating inflow and outflow for a detention basin/reservoir that uses the continuity equation and a storage equation.

Hydrologic soil group (HSG): An SCS soil classification.

Hydrostatic pressure: Static pressure exerted at a depth below the water surface.

Hyetograph: Graph of rainfall intensity versus time.

Hygroscopic water: Moisture that is absorbed to the surface of soil grains in the unsaturated zone.

Infiltration: Movement of water from the surface into the soil.

Infiltration basin: A detention basin that impounds stormwater and gradually exfiltrates it through the basin bed.

Infiltration capacity: Rate at which water can enter soil under inundated surface.

Infiltration rate: The rate of penetration of water through the soil surface.

Initial abstractions: The portion of rainfall that does not contribute to runoff.

Intensity: Depth of rainfall per unit of time.

Intensity–duration–frequency curve (IDF): A design chart that relates the rainfall intensity with its duration and return interval.

Invert: Bottom elevation of a channel or pipe.

Irrigation: The application of water to soil to meet water needs of plants, crops, turf, gardens, or wildlife to supplement rainfalls.

Jar irrigation: An old method of watering ground family plants, such as melons, using a water jar.

Karst: Carbonic rock, typified by the presence of limestone sinkholes and caverns.

Lag time: Time from the center of mass of rainfall to the peak of the hydrograph. In the Universal Method, developed by Pazwash, the lag time is defined as the beginning of rainfall to the onset of runoff.

Land cover/land use: The type of cover on the surface of the earth such as rooftop, pavement, grass, or tress.

Land disturbance: Any activity involving the clearing, cutting, excavation, grading, or filling of land that causes land to be exposed to erosion.

Landscape: A combination of turf and plants and mulches; may also include natural undisturbed areas.

Lining, composite: Combination of lining materials in a given cross section (e.g., riprap in low-flow channel and vegetated side slopes).

Lining, flexible: Lining material that can adjust to settlement; typically constructed of a porous material that allows infiltration and exfiltration.

Lining, rigid: Lining material that does not adjust to settlement, such as concrete or masonry.

Lining, temporary: Lining used for an interim condition (e.g., construction period).

Longitudinal slope: The rate of change of elevation with distance in the direction of travel or flow.

Low-flow faucet: A faucet that delivers no more than 1.5 GPM (5.7 LPM) under 80 $lb/in.^2$ pressure.

Low-flow shower head: A showerhead with no more than 2.2 GPM (8.3 LPM) flow.

Low-flow urinal: Urinals with no more than 0.5 gallons (1.8 L) per flush.

Low-impact development (LID): A development with minimal impacts on hydrology and water quality.

Low-volume toilet: Toilets with no more than 1.28 gallons (4.8 L) per flush.

Main channels: Large channels that carry flow from collector channels to some outlet such as a lake or stream.

Mass rainfall curve: A plot of cumulative precipitation with time.

Median: The middle value in a set of values arranged in ascending or descending order.

Micro-irrigation: An irrigation system with small sprinkler heads or emitters that delivers small amounts of water either above or below ground.

Micro pool: A smaller permanent pool incorporated into the design of larger stormwater pond.

Mild slope: Channel slope for which uniform flow is subcritical.

Mixed-use development: A tract of land with several different uses such as residential, office, manufacturing, retail, public, or entertainment.

Mulch: Bark, leaves, or straw placed around plants to reduce evaporation and weed growth.

Multifamily dwelling unit: Any building containing two or more dwelling units.

Native plants: Plants that are indigenous to an area and require little or no additional watering.

Natural drainage: Channels formed in the existing surface topography of the earth prior to changes made by man.

Natural ground surface: The existing surface of land prior to any land disturbance.

Nonpoint source pollutants: Pollutants generated from a spread-out area, rather than a localized point.

Nonstructural BMPs: Stormwater treatment techniques that use natural measures to reduce runoff volume and pollution level.

Nonuniform flow: Flow of water through a channel that changes its geometry along the flow.

Normal depth: Depth of water in a uniform channel flow.

One-hundred-year storm: A storm event that occurs on average once every 100 years.

Open channel: A natural or manmade conduit with free water surface.

Open channel flow: Flow of water through an open channel.

Open weave textile (OWT): A temporary degradable ECB made of natural or polymer yarns woven into a matrix. OWT provides erosion control and helps vegetation establishment.

Optimum channel cross section: Cross-section that requires a minimum flow area.

Orifice: An opening through which water passes under pressure.

Orifice equation: An equation that relates the discharge through an orifice to the area of the orifice and the depth of water above the center of the orifice.

Orifice flow: Flow of water into a submerged opening due to water pressure.

Outfall: The point of water discharge from a conduit, drain, or detention basin.

Overland flow: Flow of water on the land surface in a down-slope direction

Oxidation: A chemical process in which a molecule or ion loses electrons to an oxidant.

Pathogens: Disease-producing microorganisms (bacteria, fungi, viruses).

Per capita use: The amount of water used by one person during a specific time period, commonly 24 hours.

Percolation: The downward movement of water through soil.

Permeability: The capability of a soil or porous medium to transmit water; the rate of downward flow of water in soil.

Permissible shear stress: The force required to initiate movement of the channel bed or lining material.

Permissible velocity: The highest average velocity at which the water may be carried in a channel or discharged at an outfall and not cause erosion.

Pervious/porous pavement: Any pavement that allows infiltration, such as porous asphalt, pervious concrete, and pavers.

Pocket pond: A stormwater pond used for small drainage area (<5 acres, 2 hectares) that relies on groundwater to maintain a permanent pool.

Point source: The location where wastewater or any polluted water is discharged.

Pollutant: Any contaminant at concentration high enough to pose a danger to public health or endanger aquatic environments.

Pond: A vernal or perennial body of standing water, smaller than a lake.

PondPack: A comprehensive Window based hydrologic modeling for detention and infiltration basins.

Population density: The total number of residents per total area of land, excluding water bodies.

Porosity: Ratio of volume of voids to total volume of soil sample.

Porous medium: Geologic material that will allow water to flow through it.

Potable water: Water suitable for drinking.

Potential evapotranspiration: Amount of water that can be lost to evapotranspiration if water is sufficiently available.

Precipitation: Water that falls to the earth in the form of rain, snow, hail, or sleet.

Pressure flow: Flow in a closed conduit with no free water surface. The flow occurs due to pressure forces.

Pretreatment: Techniques employed to remove coarse sediment before stormwater enters a system.

Probability: Relative number of occurrences of an event after a large number of trials.

Rain sensor: A sensor that shuts off an irrigation system during rain.

Rainfall excess: Total rainfall minus the initial abstraction and losses.

Rapidly varied flow: Flow of water through a channel with rapidly changing characteristics.

Rating curve: A graph or equation that relates the stage and discharge.

Rational method: A simple, linear, rainfall-runoff relationship for estimating peak flow.

Recession curve: Portion of the hydrograph where runoff is from base flow.

Recharge: The addition of water to groundwater by precipitation or artificial infiltration and injection.

Record: A string of characters or groups of characters (fields) that are treated as a single unit in a file.

Recurrence interval: Time interval in which an event will occur once on average.

Recycled water: Reuse of water repeatedly through a closed system; also used to refer to reclaimed water.

Redevelopment: The replacement or adaptive reuse of an existing structure, or of land from which previous improvements have been removed.

Relative humidity: Ratio of water vapor pressure to the saturated vapor pressure at the same temperature.

Reservoir: Man-made storage area for flood control or water supply.

Residential water use: Water use in homes, both indoor and outdoor.

Retardance classification: Qualitative measure of the resistance to flow by various types of vegetation.

Retention: Storage reservoir or pond that retains runoff or flood water without allowing it to discharge downstream.

Retention/detention facilities: Facilities that control the quantity and rate of runoff.

Retention pond or basin: A facility designed to retain stormwater runoff on a development site.

Retrofit: To alter, adjust, or change plumbing fixtures or equipment or appliances to reduce water use.

Return period: Same as recurrence interval: time interval for which an event will occur once on average.

Reuse: The additional use of previously used water.

Riprap: Broken rock, cobbles, or boulders placed on side slopes or the bottoms of channels to prevent erosion.

Riser: A vertical pipe or structure installed on a pond/detention basin as a flow control device.

Rising limb: Portion of the hydrograph where runoff is increasing.

Roadside channel: Stabilized drainage way to collect runoff from roadways and streets.

Rolled erosion control product (RECP): A degradable or non-degradable material fabricated into rolls or mats. RECP is used to reduce soil erosion and to protect vegetation during establishment. ECBs and TRMs are collectively called RECP.

Root zone: Depth to which the vegetation draws water through its root system in soil.

Routing: The process of discharge of an inflow hydrograph through a detention system.

Runoff: Surface water from precipitation (rain or snow) and irrigation that is not absorbed by soil or retained in surface depressions.

Runoff coefficient: Ratio of runoff to precipitation.

Runoff curve number: Parameter used in the SCS method that accounts for soil type and land cover.

Saline water: Water containing salt at concentrations less than 35 g/L. (Slightly saline water contains 1–3 g/L, moderately saline water from 3–10 g/L, and highly saline water from 10–35 g/L.)

Salinity: The concentration of salts in water.

Sand filters: A structural water quality device to percolate runoff through a sand bed before discharge to drainage system.

Saturated zone: Zone of an aquifer in which the soil pores are filled with water; the soil in this zone is under pressure greater than atmospheric pressure.

Secondary treatment: With regard to wastewater and grey water, the biological process of removing suspended, colloidal, and dissolved organic matter in effluent from the primary treatment system.

Sediment: Soil and other granular materials carried by water and deposited into streams, lakes, reservoirs, and detention basins/ponds.

Sediment basin: A pond, basin, or other structure or measure that detains water to settle sediment.

Sewer: Any pipe or conduit used to collect and convey sewage or stormwater runoff from its source to a treatment plant or receiving water body.

Shallow concentrated flow: Flow that has concentrated in rills or small gullies.

Shear stress: Force exerted by flow per unit of the wetted area of the channel; stress on the channel bottom due to the hydrodynamic forces of the flowing water.
Sheet flow: A shallow movement of runoff overland.
Side slope: Slope of the sides of a channel, defined as the horizontal width per unit rise.
Site: A lot, tract, or parcel of land or a combination of contiguous lots, tracts, or parcels of land.
Slotted inlets: Roadway inlets with openings along the longitudinal axis and transverse bars spaced to form slots.
Slug test: Field test for measuring permeability by adding or removing a volume of water in a single well and observing the drop or rise of water level.
Smart growth: Urban planning that achieves environmental, community, and economic improvements.
Soil amendment: Organic and inorganic materials added to soil to improve its texture, waterholding capacity, infiltration, and nutrients.
Soil moisture storage: Volume of water held in the soil.
Soil texture: Classification of soil based on percentage of sand, silt, and clay.
Specific energy: The sum of elevation head and velocity head at a cross section from the channel bed.
Sprawl: A pattern of development characterized by inefficient access between land uses or to public facilities or services and a lack of functional open space. Sprawl is typically an automobile dependent, resource-consuming, discontinuous, low-density development pattern.
Spread: The lateral distance from the curb face to the edge of the water flowing in a gutter or on a roadway.
Sprinkler: A device for overhead delivery of water through small nozzles connected to an underground pipe or fastened to a hose. The sprinklers are commonly rotary or oscillating.
Stage–storage–discharge relationship: Variation of storage and discharge with stage for a detention basin/reservoir.
Steady flow: Flow that remains constant with respect to time.
Steep slope: Channel slope for which uniform flow is supercritical.
Storage–discharge relation: Relation between storage and outflow for a detention basin.
Storage routing: Flood routing in which discharge is uniquely related to the amount of storage.
Storm drain (storm sewer): A pipe that receives runoff from inlet(s) and conveys it downstream.
Storm drainage systems: Systems of pipes, inlets, and manholes that collect, convey, and discharge stormwater.
Stormwater management: The control and management of stormwater runoff to minimize its detrimental effects on water quantity and quality.
Stormwater wetlands: Shallow natural or manmade ponds that treat stormwater and allow for the growth of wetland-characterizing vegetation.
Streamflow: The movement of water flow at a section in streams or channels.
Streamlines: Flow lines that represent the direction of water movement in a flow.
Stripping: Any activity, such as clearing and grubbing operations, that removes or significantly disturbs vegetated or otherwise stabilized soil surface.
Sub-basins: Segments of a watershed with relatively homogeneous character.
Subcritical flow (tranquil flow): Flow of water at a velocity less than critical; Froude number less than 1.0.
Subdivision: The division of a lot, tract, or parcel of land into two or more lots or parcels.
Subsurface irrigation: The application of water below the soil surface, generally using a drip irrigation system.
Supercritical flow (rapid flow): Flow of water at a velocity greater than critical; Froude number greater than 1.0.
Superelevation: Local increase in water surface at the outer side of a channel bend.
Supplemental irrigation: The application of water to lawn or landscape area to supplement rainfall.

Surcharge: Condition in which the water level in a storm drain system rises above the grate of an inlet or manhole rim.

Sustainable development: Development that meets the needs of the present without jeopardizing the needs of future generations.

Synthetic design storm: Rainfall hyetograph obtained through statistical means.

Synthetic unit hydrograph: A unit hydrograph based on theoretical or empirical methods.

System International (SI): Also referred to as metric units; consist of meter, kilogram, or newton and second as basic dimensions.

Tailwater: Water surface level downstream of a conduit.

Technical Release No. 20 (TR-20): An SCS watershed hydrology computer model that performs runoff calculations and routing of storm through streams and ponds.

TR-55: An SCS hydrologic model that performs runoff calculations.

Time base: Total duration of direct runoff under the hydrograph.

Time of concentration: The time that it takes for runoff from all of a basin area to reach the outlet.

Time to peak: Time from the center of mass of rainfall to the peak of hydrograph.

Total dissolved solids (TDS): The quantity of dissolved material in water; usually expressed in mg/L or g/L.

Total dynamic head: The sum of static head, velocity head, and head losses.

Total energy: The sum of pressure, elevation, and velocity heads.

Total suspended solids (TSS): The quantity of suspended matter in water; expressed in mg/L or g/L.

Tractive force: Force developed due to the shear stress on the perimeter of a channel section.

Transient flow: Flow that varies with time.

Transpiration: The natural process of transferring water vapor to air by plants. Through this process the leaves and flowers keep cool and survive under the heat of the sun.

Trash rack: Grate, grill, or another device installed on an outlet structure or intake of pipe to remove large debris from entering it.

Turbidity: A suspension of fine particles in water that obscure light rays and take many days to settle.

Turf/turf grass: Hybridized grass that forms a dense growth of blades and roots.

Turf reinforcement mat (TRM): A non-degradable erosion control blanket to stabilize channel bed and banks and to promote vegetation growth

Uniform flow: Flow condition with a constant depth and velocity along the length of the channel.

Unit hydrograph: Graph of runoff versus time produced by a unit direct rainfall (1 in. or 1 cm) from a given duration storm.

Unit peak discharge: The peak discharge per unit area, with units of $m^3/s/km^2$, or cfs/mi^2.

Unsteady flow: Flow that changes with time.

Vadose zone: Zone of aeration that extends from the surface to the water table including the capillary fringe.

Vapor pressure: Partial pressure exerted by water vapor.

Varied flow: Flow where the flow rate and depth change along the channel.

Velocity head: Energy due to the velocity of water.

Velocity, mean: Discharge divided by the area of flow.

Velocity, permissible: Velocity that will not cause channel erosion.

Wastewater: The used water from residential, commercial, and industrial buildings/sites.

Water conservation: Reductions in water use, loss, or waste.

Water conservation measures: Actions, behavioral changes, devices, improved design, technology, or process implemented to reduce water use, loss, and waste.

Water-efficient landscape: A landscape that minimizes the water need. See also xeriscape.

Water harvesting: The capture and use of rainfall or runoff.

Water quality inlets: Inlets with built-in filters that remove sediment, oil and grease, and floatables from runoff.

Water reclamation: The treatment of wastewater for reuse, commonly for non-potable purposes.

Water reuse: The use of water that had previously been used for a specific purpose.

Watershed: Area of land that drains to a single outlet water course or stream and is separated from other watersheds by a topographic divide.

Watershed divide: Line that defines a watershed boundary.

Water surface profile: Plot of the depth of water along the flow in a channel.

Water table: The top of saturated soil zone in an unconfined aquifer; the surface where the soil is saturated and water pressure is exactly atmospheric.

Water use: Water that is actually used for a specific purpose or customer such as residential or agricultural users.

Water year: The continuous 12-month period that begins on October 1 and ends on September 30th of the following year and is designated by the year in which it ends.

Weed: Any unwanted or troublesome plant.

Weir flow: Gravity flow over a horizontal barrier, such as a weir, roadway, or bridge.

Well: Vertical hole dug into the soil that penetrates an aquifer.

Wet ponds: A pond with a permanent pool.

Wetted perimeter: The length of contact between the flowing water and pipe or channel.

Wilting point: Moisture content below which plants cannot extract any water and will not survive.

Wing wall: Side wall extensions of a pipe/culvert used to prevent sloughing of channel or stream banks.

Withdrawal: Water delivered or extracted from a surface or groundwater source.

Xeriscape: A trademarked term representing a landscaping that involves the selection, placement, and maintenance of low-water-use plants, turf, shrubs, and trees.

Appendix A: System International (SI)

Dimensions describe physical quantities, and units express their amounts. Units, however, vary from one system of units to another. The common systems of units are metric, English, and SI. The metric and English systems of units have been in use for a long time (for example, the metric system since 1872). In 1960, an international system of units was proposed in Europe. In this system, abbreviated as SI (System International), mass, length, time, temperature, electrical current, luminous intensity, and the amount of matter are selected as dimensions. Any physical quantity can be described in terms of these dimensions. The units of these dimensions and their symbols appear in the following table. The FHWA (Federal Highway Administration) began its conversion to the metric system in December 1990 and set a deadline for complete conversion on September 30, 1996. However, due to objections from the industry, English units are still used in the U.S. In fact, in spite of being straightforward and meaningful, there is a general lack of familiarity with metric units in this country. For simplicity and clarity it suffices to say that one kilogram is the weight of 1 liter of water, one meter equals 1000 mm and that 0 degree and100 degree Celsius, respectively represent the freezing and boiling temperatures of water at sea level.

SI Dimensions and Units

Dimensions	Units	Symbol
Length	Meter	m
Mass	Kilogram	kg
Time	Second	s
Temperature	Kelvin	K
Electrical current	Ampere	A
Luminous intensity	Candela	Cd
Amount of material	Mole	mol

In SI units, there are also two supplementary units as follows:

Radian (rad), representing planar angles
Steradian (sr), representing solid angles (used in three-dimensional space)

In hydraulic engineering, in general, and stormwater management practices, in particular, three basic dimensions are used: length, mass, and time. Every hydraulic quantity can then be expressed in terms of these three basic dimensions, abbreviated as L, M, and T, respectively. For example, the relation for acceleration

$$a = \frac{\Delta V}{\Delta t}$$

shows that acceleration, a, has the dimensions of

$$a = (L/T)/T = LT^{-2}$$

Newton's law, which relates force and mass as

$$F = ma$$

may be used to establish the dimensions of F in L, M, and T, with the following result:

$$F = MLT^{-2}$$

In hydraulic engineering, hydrology, and to some extent in stormwater management, one deals with fluid properties such as density ρ, specific weight γ, specific gravity $S_g = \gamma/\gamma_w$, viscosity μ (also called dynamic viscosity), kinematic viscosity $\nu = \mu/\rho$, and surface tension σ. Also encountered is pressure $p = F/A$ both in water at rest and in motion. In flowing water, moreover, such quantities as discharge Q, work ($W = FL$), power ($P = W/t$), angular velocity, momentum, and specific energy are involved. Dimensions of these quantities can be obtained from the laws of physics or mechanics.

As indicated, units of the aforementioned quantities depend on the system of units used. In the SI units, dimensions of quantities that are commonly used in hydraulic engineering are listed in the following table.

Quantity (Symbol)	Dimension	Units
Length (l)	L	meter (m)
Mass	M	kilogram (kg)
Time (s)	T	second (s)
Velocity (V)	LT^{-1}	(m/s)
Acceleration (a)	LT^{-2}	(m/s^2)
Density (ρ)	ML^{-3}	(kg/m^3)
Force (weight) (F)	MLT^{-2}	Newton (N) = kg weight/9.81
Pressure (p)	$ML^{-1}T^{-2} = FL^{-2}$	Pascal (Pa) (N/m^2)
Discharge (Q)	L^3T^{-1}	m^3/s
Specific weight (γ)	$ML^{-2}T^{-2}$	N/m^3
Viscosity (μ)	$ML^{-1}T^{-2}$ ($FL^{-2}T$)	kg/m·s = N·s/m²
Kinematic viscosity (ν)	L^2T^{-1}	m^2/s
Work, energy, heat (W)	$ML^2T^{-2} = (FL)$	N·m
Surface tension (σ)	$MT^{-2} = FL^{-1}$	N/m
Power (P)	$ML^2T^{-3} = (FLT^{-1})$	Watts (W), N·m/s
Momentum (mv) = M	MLT^{-1}	kg·m/s = N·s
Frequency (f)	$T^{\pm 1}$	Hertz (Hz

In this system, small or large quantities are described with the aid of a prefix, indicating the factor of 10 employed. The most common prefixes are as follows:

Prefix	Symbol	Factor
Tera	T	10^{12}
Giga	G	10^9
Mega	M	10^6
Kilo	k	10^3
Hecto	h	10^2
Deka	da	10^1
Deci	d	10^{-1}

(Continued)

Prefix	Symbol	Factor
Centi	c	10^{-2}
Milli	m	10^{-3}
Micro	μ	10^{-6}
Nano	n	10^{-9}
Pica	p	10^{-12}

On the basis of these symbols, for example, 10^5 pascals may be expressed as 100 kPa.

Conversion of some of the most commonly used units from English (also referred to as conventional units) to SI appears in the following table.

Conversion from English to SI Units

Quantity	English Units	Metric Units	Factor
Length	in.	mm	25.4 e[a]
	ft	m	0.3048 e
	yd	m	0.9144 e
	mile	km	1.609
Area	in.^{2}	cm^2	6.452
	ft^2	m^2	0.0929
	yd^2	m^2	0.8361
	acre	m^2	4047
	acre	ha	0.4047
	mi^2	km^2	2.590
Volume	in.^{3}	cm^3	16.387
	ft^3	m^3	0.0283
	yd^3	m^3	0.7646
	ounce	cm^3	29.574
	quart	liters	0.9463
	gallon	liters	3.7853
	acre-ft	m^3	1233
	CCF (100 ft^3)	m^3	2.832
Mass	slug (lb mass)	kg	0.4536
Mass density	lb/ft^3	kg/m^3	16.026
Force	lb	N	4.445
Weight	lb	kg	0.4536
	kip (1000 lb)	ton	0.4536
	ton (2000 lb)	ton	0.9072
Pressure	lb/ft^2	Pa(N/m^2)	47.88
	lb/in.^{2}	kPa	6.895
Viscosity	lb·s/ft^2	N·s/m^2	47.88
Kinematic viscosity	ft^2/s	m^2/s	0.0929
Power	ft-lb/s	watts	1.3558
	hp (horsepower)	watts	745.70
	Btu/h	watts	0.2931
Energy	1000 Btu	kWh	0.2931
Velocity	ft/s	m/s	0.3048 e
	mi/h	km/h	1.609
Rainfall intensity	in/h	mm/h	25.4 e

e reflects exact conversion factor.

Common Constants

1. Water properties
   a. At standard conditions (4°C and 760 mm Hg)
      Specific weight: 9806 N/m^3 (~1000 kg/m^3, 1 kg per liter) ≈ 62.4 lb/ft^3
      Density: 1000 kg/m^3 ≈ 1.94 slug/ft^3
      Viscosity: 1.57×10^{-6} N·s/m^2 ≈ 3.28×10^{-5} lb·s/ft^2
      Kinematic viscosity: 1.57×10^{-6} m^2/s ≈ 1.69×10^{-5} ft^2/s
   b. At normal temperature (20°C and 760 mm Hg)
      Specific weight: 9789 N/m^3 ≈ 63.2 lb/ft^3
      Density: 998.2 kg/m^3 ≈ 1.94 slug/ft^3
      Viscosity: ≈ 1.0×10^{-5} N·s/m^2 ≈ 2.09×10^{-5} lb·s/ft^2
      Kinematic viscosity: ≈ 1.0×10^{-6} m^2/s ≈ 1.08×10^{-5} ft^2/s
      Standard atmospheric pressure = 101.4 kN/m^2 ≈ 14.71 lb/in.^{2}
      Gravitational constant = 9.81 m/s^2 = 32.2 ft/s^2
   c. Freezing/boiling temperatures at sea level
      Freezing: 0°C = 32°F
      Boiling: 100°C = 212°F

REFERENCES

Pazwash, H., 2007, *Fluid Mechanics and Hydraulic Engineering*, Tehran University Press, Tehran, Iran.
Pazwash, H., 2016, *Urban Storm Water Management*, 2nd Ed, CRC Press, Boca Raton, FL, London, New York.

Appendix B: Unified Soil Classification System and Nominal Sizes of Coarse and Fine Aggregates

Soils seldom exist naturally as a single component, such as sand and gravel. Rather, they occur as a mixture; each component contributes its characteristics to the soil. In the Unified Soil Classification System (USCS), soils are classified into three major categories: coarse-grained, fine-grained, and highly organic. The USCS further divides soils into five major soil categories symbolized as

G: gravel
S: sand
M: silt
C: clay

In the ASTM (American Standards for Testing Materials) Soil Classification System D2487–11, unconsolidated soil groups are represented by a two-letter symbol in this system. The first letter is one of the preceding five symbols. The second letter can be any one of the same symbols (except for peat) or one of the following four letters describing the soil condition:

Letter	Definition
P	Poorly graded (mostly one size)
W	Well-graded (fine to coarse sizes)
H	Highly plastic
L	Low plasticity

Table B.1 lists ASTM unified soil classifications.

For a further description of USCS, one is referred to ASTM D2487-17 Standard Practices for Classification of Soils for Engineering Purposes (Unified Soil Classification System); https://www.astm.org/d2487-17.html or California Department of Transportation (CALTRANS); (https://dot.ca.gov/programs).

AASHO (American Association of State Highway Officials) classifies soils into seven groups, A-1 to A–7. These groups include gravel or sand (A-1), fine sand (A-3), and silty or clayey gravel and sand (A-2). Groups A-4, A-5, A-6, and A-7 include clay soils and silty -clay soils where more than 35% pass the No. 200 sieve.

AASHTO Soil Terminology comes from AASHTO M–145. Basic terms include:

- **Boulders & Cobbles:** Material retained on a 75-mm (3-inch) sieve.
- **Gravel:** Material passing a 75-mm (3-inch) sieve and retained on a 2.00-mm (No. 10) sieve.
- **Coarse Sand:** Material passing a 2.00-mm sieve (No.10) and retained on a 0.475-mm (No. 40) sieve.
- **Fine sand:** Material passing a 0.475-mm (No.40) sieve and retained on a 0.075-mm (No. 200) sieve.
- **Silt-Clay:** Material passing a 0.075-mm (No. 200) sieve.

TABLE B.1
ASTM Soil Groups

Symbol	Group Name
GW	Well-graded gravel
GP	Poorly graded gravel
GM	Silty gravel
GC	Clayey gravel
SW	Well-graded sand
SP	Poorly graded sand
SM	Silty sand
SC	Clayey sand
ML	Silt of low plasticity
MH	Silt of high plasticity
CL	Clay of low plasticity
CH	Clay of high plasticity
OL	Organic soil of low plasticity
OH	Organic soil of high plasticity (organic silt/clay)
PT	Peat

- **Silt Fraction:** Material passing the 0.075 mm and the larger than 0.002 mm.
- **Clay Fraction:** Material smaller than 0.002 mm.
- **Silty:** Material passing a 4.75-mm (No.4) sieve with a PI ≤ 10
- **Clayey**: Material passing a 4.75-mm (No.4) sieve with a PI ≥ 11
- **Coarse Aggregate:** Aggregate retained on the 2.00 mm sieve and consisting of hard, durable particles or fragments of stone, gravel or slag. A wear requirement (AASHTO T96) is normally required.
- **Fine Aggregate**: Aggregate passing the 2.00 mm (No. 10) sieve and consisting of natural or crushed sand, and fine material particles passing the 0.075 mm (No. 200) sieve. The fraction passing the 0.075 mm (No. 200) sieve shall not be greater than two-thirds of the fraction passing the 0.425 mm (No. 40) sieve. The portion passing the 0.425 mm (No. 40) sieve shall have a LL ≤ 25 and a PI ≤ 6. Fine aggregate shall be free from vegetable matter and lumps or balls of clay.

A coarse-grained soil is defined as that in which at least half of the material is retained on a number 200 sieve (0.075 mm). In the United States, the size designation of coarse aggregates is defined by AASHTO M43. Fine aggregates are designated based on the size of the sieve opening passing the grains. The US standard sieve (mesh) numbers are specified by the ASTM E11 and AASHTO M92 standards. According to ASTN E11, the 200 sieve differentiates coarse and fine aggregates. Note that the AASHTO Soil Terminology is slightly different from the ASTM Aggregate and Soil Terminology specified in ASTM D2487.

a. ASTM terms
    - **Coarse-Grained Soils:** More than 50% retained on a 0.075 mm (No. 200) sieve
    - **Fine-Grained Soils**: 50% or more passes a 0.075 mm (No. 200) sieve
    - **Gravel:** Material passing a 75 mm (3 in.) sieve and retained on a 4.75 mm (no. 4) sieve
    - **Coarse Gravel:** Material passing a 75 mm (3 in.) sieve and retained on a 19.0 mm (3/4 in.) sieve

- **Fine Gravel:** Material passing a 19.0 mm (3/4 in.) sieve and retained on a 4.75 mm (no. 4) sieve
- **Sand:** Material passing a 4.75 mm sieve (no. 4) and retained on a 0.075 mm (no. 200) sieve
- **Coarse Sand:** Material passing a 4.75 mm sieve (no. 4) and retained on a 2.00 mm (no. 10) sieve
- **Medium Sand:** Material passing a 2.00 mm sieve (no. 10) and retained on a 0.475 mm (no. 40) sieve
- **Fine Sand:** Material passing a 0.475 mm (no. 40) sieve and retained on a 0.075 mm (no. 200) sieve
- **Clay:** Material passing a 0.075 mm (no. 200) that exhibits plasticity and strength when dry (PI ≥ 4)
- **Silt:** Material passing a 0.075 mm (no. 200) that is non-plastic and has little strength when dry (PI < 4)
- **Peat:** Soil of vegetable matter

More information on AASHTO M43 can be found in the sections titled "703 Aggregate" and "AASHTO M-143" within the relevant construction material specifications.

b. Sieve Sizes

ASTM E11 includes Standard Specifications for women will test sieve sizes. For particle sizes 5.6 mm to 20 mm (0.02 in.), the sieve sizes are listed in Table B.2. For aggregates 6.30 mm (0.625 in.) and larger, their sizes are represented in this table as average values expressed in millimeters (inches).

TABLE B.2
ASTM Aggregate and Sieve Sizes

Standard	Altern.	Standard	Altern.	Standard	Altern.
125.00 mm	5.00	9.50 mm	3/8	425 μm	No. 40
106.00 mm	4.24	8.00 mm	5/16	355 μm	No. 45
100.00 mm	4	6.70 mm	0.265	300 μm	No. 50
90.00 mm	3 ½	6.30 mm	¼	250 μm	No. 60
75.00 mm	3	5.60 mm	No. 3 ½	212 μm	No. 70
63.00 mm	2 ½	4.75 mm	No. 4	180 μm	No. 80
53.00 mm	2.12	4.00 mm	No.5	150 μm	No. 100
50.00 mm	2	3.35 mm	No. 6	125 μm	No. 120
45.00 mm	1 ¾	2.80 mm	No. 7	106 μm	No. 140
37.50 mm	1 ½	2.36 mm	No. 8	90 μm	No. 170
31.50 mm	1 ¼	2.00 mm	No. 10	75 μm	No. 200
26.50 mm	1.06	1.70 mm	No. 12	63 μm	No. 230
25.00 mm	1	1.40 mm	No. 14	53 μm	No. 270
24.40 mm	7/8	1.18 mm	No. 16	45 μm	No. 325
19.00 mm	¾	1.00 mm	No. 18	38 μm	No. 400
16.00 mm	5/8	850 μm	No. 20	32 μm	No. 450
13.20 mm	0.530	710 μm	No. 25	25 μm	No. 500
12.50 mm	½	600 μm	No. 30	20 μm	No. 635
11.20 mm	7/16	500 μm	No. 35		

Index

Note: Page numbers followed by f, t and n indicate figures, tables and notes, respectively.

J

K

L

M

N

O

P

Q

R

U

V

W

X

Z